CHRIS NEUZIL

# EXPLORATION OF THE DEEP CONTINENTAL CRUST

Edited by

H.-J. Behr, Göttingen
C. B. Raleigh, Palisades

# The Superdeep Well of the Kola Peninsula

Edited by Ye. A. Kozlovsky

In Collaboration with
N. I. Andrianov M. I. Vorozhbitov D. M. Guberman
V. I. Kazansky E. W. Karus V. S. Lanev E. B. Nalivkina
R. A. Sumbatov

With 190 Figures

Springer-Verlag
Berlin Heidelberg New York
London Paris Tokyo

YEVGENY A. KOZLOVSKY, Editor-in-Chief
Ministry of Geology of the USSR
4/6 B. Gruzinskaya
Moscow 123812, USSR

*Translated from the Russian*
by I. P. LAVRUSHKO, G. A. BYLEVSKI, Yu. Ya. PRIZOV,
and D. E. STOLYAROV

Title of the original Russian edition:
Kol'skaja sverchglubokaja


ISBN 3-540-16416-2 Springer-Verlag Berlin Heidelberg New York
ISBN 0-387-16416-2 Springer-Verlag New York Berlin Heidelberg

Library of Congress Cataloging-in-Publication Data. The Superdeep well of the Kola Peninsula. (Exploration of the deep continental crust) Bibliography: p. 1. Geology, Stratigraphic–Pre-Cambrian. 2. Geology–Russian S.F.S.R.–Kola Peninsula. 3. Geochemistry–Russian S.F.S.R.–Kola Peninsula. 4. Borings–Russian S.F.S.R.–Kola Peninsula. I. Kozlovskiĭ, E. A. II. Andrianov, N. I. III. Series. QE653.S87 1986 551.7'1'094723 86-28015


Printed in Germany



Typesetting: Fotosatz GmbH, Beerfelden.
Offsetprinting and Bookbinding: Konrad Triltsch, Graphischer Betrieb, Würzburg
2132/3130-543210

# Preface

The present book is devoted to the study of the deep Earth's interior structure, one of the most important problems of Earth sciences today. The drilling of the Kola superdeep well inaugurated a new stage in the study of the Precambrian continental crust. The well was sunk in the northeastern part of the Baltic Shield, in an area where the Precambrian ore-bearing structures, typical of the ancient platform basements, are in juxtaposition with each other. To the present the well has been drilled to a depth of 12 km, has traversed the full thickness of the Proterozoic complex and a considerable part of the Archean stratum, and is still being worked on. This book reviews the principal results of investigations to a depth of 11,600 m; these are described in three sections: geology, geophysics, and drilling.

The book begins with a general review of the history, the present state of knowledge, and trends of further investigations in the field of study of the Earth's interior and superdeep drilling. The first section of the book considers the geology of the vicinity of the Kola superdeep well and describes its geological section based on a detailed examination both of the cores and the near-borehole area. There follows a description of new data on the petrography and geochemistry of the Precambrian complex, their zonality and conditions of metamorphism, vertical ore zonality, tectonic dislocations, gases, organic matter and subsurface waters, and finally on the evolution of the continental crust over a time span of about 3 billion years.

The second section is devoted to the geophysical investigations undertaken in the Kola superdeep well, i.e. the study of the vertical zonality of the rocks' physical properties in the section and the nature of physical boundaries; it also deals with such questions as control for borehole technological conditions, descriptions of individual units of well design and the development of technological means necessary for studying rocks of the borehole section.

The third section comprises information on the construction of the well to a depth of 11,600 m. It presents the scientific and methodological principles used in designing technological processes and drilling equipment, a description of ground installations and well-site facilities and of drilling equipment. Statistical data on drilling conditions and an interpretation of the technological and economical results of drilling are also given. The main trends and changes in the drilling process with increase in depth are analyzed, and trends in research development, as well as ways of perfecting the equipment used for deepening this unique well, are substantiated.

The book is designed for a broad range of specialists engaged in solving scientific and practical tasks in the field of investigations of the Earth's interior.

The Editor

# Contents

# Abbreviations

**of Mineral Names and Chemical Compounds Used in the Text and in Figures**

| | |
|---|---|
| Ab | albite |
| Act | actinolite |
| Al | andalusite |
| Amp | amphibolite |
| An | anorthite |
| And | andesine |
| Au | augite |
| Ax | axinite |
| Bi | biotite |
| Ca | calcite |
| Car | carbonate |
| Ch | chlorite |
| CHB | chloroform bitumen |
| Chp | chalcopyrite |
| Cl | clinozoisite |
| Cr | cordierite |
| Cu | cummingtonite |
| Dl | dolomite |
| En | enstatite |
| Ep | epidote |
| Fs | feldspar |
| Gar | garnet |
| HAM | highly aluminiferous minerals |
| Hb | hornblende |
| He | hematite |
| HHCG | heavy hydrocarbon gases |
| Hs | hastingsite |
| Hyp | hypersthene |
| Il | ilmenite |
| Kfs | K-feldspar |
| Lc | leucoxene |
| Lom | laumonite |
| Mc | microcline |
| Mont | montmorilonite |
| MP | monoclinic pyroxene |
| Mt | magnetite |
| Mu | muscovite |
| Ol | oligoclase |
| Or | orthoclase |
| Ph | phlogopite |
| Pl | plagioclase |
| Pr | prehnite |
| Pum | pumpellyite |
| Py | pyroxene |
| Pyr | pyrhotite |
| Qz | quartz |
| REE | rare earth elements |
| S | salite |
| Sd | sulphides |
| Sl | sillimanite |
| Sp | serpentine |
| Sph | sphene |
| Ss | saussurite |
| St | stilpnomelane |
| Stu | staurolite |
| Ta | talc |
| Tr | tremolite |

# Introduction

The importance of raw mineral resources in the modern world is constantly increasing. Oil, gas, and nuclear raw material, as sources of energy, are required in ever-increasing quantities; the utilization of aluminium, titanium, and alloy metals, molybdenum in particular, as well as of manganese, chromium, and nickel, is expanding (Kozlovsky 1982b). The intensification of agricultural production necessitates an increase in phosphate and potash consumption in mineral fertilizers. New branches of industry, such as radioelectronics, space technology, etc. cannot develop successfully without using rare metals. Thus the economic wealth of the countries of the world has come to depend increasingly on mineral resources, leading to a universal rise in production and consumption of raw minerals in the second half of the 20th century.

Calculations performed by Soviet and foreign specialists, as well as by UN experts and the World Energy Conference Conservation Commission, show that the consumption of mineral resources will continue to increase during the closing decades of the present century.

In the search for new reserves of mineral resources, mankind has penetrated into shelf areas and world oceans (Fig. 1.1). The depth at which exploration and exploitation of continental mineral deposits takes place is becoming ever greater; for oil and gas fields, for example, at depths of 5 – 6 km. The depth of mines is also increasing. At one gold mine, ores are extracted from a depth of more than 5 – 6 km. Even iron ores, for instance in the Krivoi Rog region, are mined at a depth of about 1 km, and their reserves have been evaluated to a depth of 2.5 km. The overwhelming majority of recently discovered mineral resources are not associated with deposits outcropping at the surface.

With this in view, a study of the deep crustal structure is becoming an increasingly urgent problem. Beyond all doubt, a clue to the mystery of origin and evolution of the Earth's crust is hidden in the unexplored depths of the Earth, which contain unknown mineral resources. Although the idea of studying the deep crustal structure has long fascinated mankind, only in the last decades has technology made its realization possible.

Three stages can be defined in the studies of the deep structure of USSR territory. The first covers the 1960's, when the general tasks were formulated, the problem was thoroughly studied from the scientific point of view, and Soviet-made technical means for ultra-deep drilling and geological-geophysical investigations in boreholes of 10 – 15 km on depth were developed. The second stage was in the 1970's, when the experimental drilling of the Kola and Saatly ultra-deep holes, as well as a number of key regional studies of deep geophysics, were undertaken. The third stage of research, begun in 1981, is a transitional stage

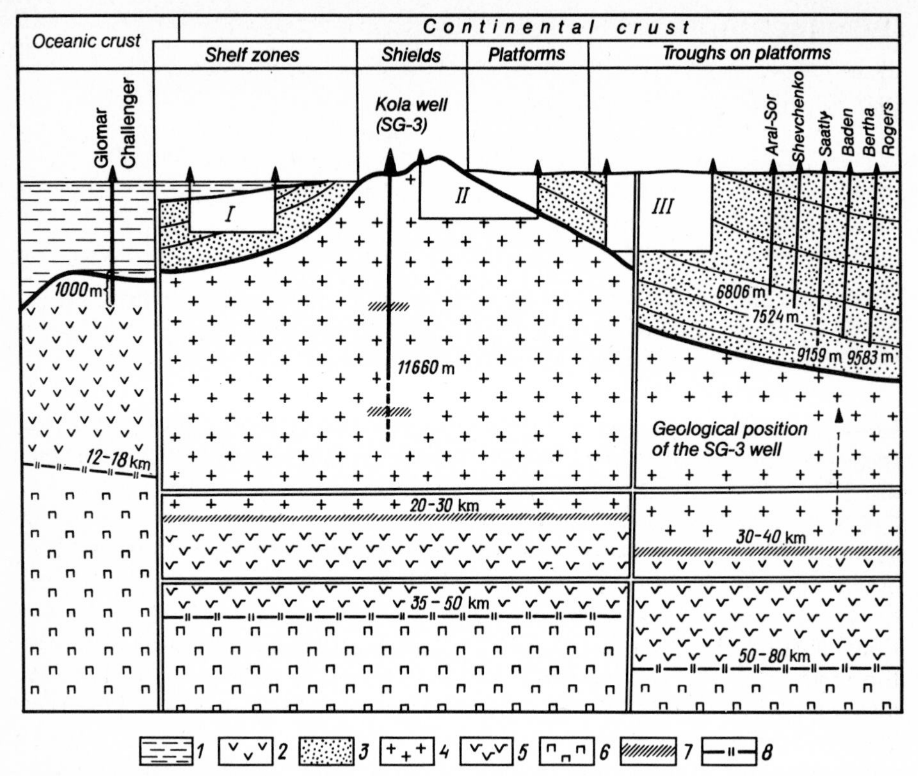


**Fig. 1.1.** Schematic section of the Earth's crust according to deep drilling data. Depths of deep exploratory drilling: *I* in the shelf zone; *II* for solid minerals; *III* oil and gas wells. *1* Earth's hydrosphere; *2* oceanic basalts; *3* Phanerozoic sedimentary and volcano-sedimentary rocks (500 Ma old); *4* Precambrian crystalline rocks of the granitic layer (1000 – 3000 Ma and older); *5* rocks of the continental basaltic layer; *6* mantle rocks; *7* high-velocity rocks – Conrad discontinuity ($V_p = 6.6 - 6.8$ m s$^{-1}$); *8* Mohorovičič discontinuity (velocity of elastic wave propagation $V_p = 8.0$ km s$^{-1}$)

planned to lead to the integrated study of the Earth's crust and upper mantle throughout the USSR.

At the beginning of the 1960's, it became possible to start work on the study of the deep interior of the Earth, as standard drilling technology made it possible to foresee the design and drilling of ultra-deep wells to a depth of 15 km within the next few years.

Several wells have been already drilled in the USSR and USA to depths of 7 – 9 km in the search for oil and gas fields. These wells were spudded in sedimentary basins, and usually penetrated through the sedimentary layers which also outcropped at the surface on the flanks of these basins. Due to this, information obtained in such wells was insufficient for fully understanding the structure and composition of the deep parts of the Earth's interior.

Special tasks to be solved by superdeep wells and possible sites for their drilling were discussed by Soviet specialists at special meetings and in the press (Belousov 1966; Belyaevsky and Fedynsky 1961; Khitarov 1961).

It was proposed that these wells could be aimed at studying:

a) the sedimentary cover of the deepest depressions on platforms;
b) the sedimentary fill of geosynclines;
c) the composition and structure of the lower granitic layer, the nature of the Conrad discontinuity, the composition of the basaltic layer;
d) the nature of the Mohorovičič discontinuity, the composition and structure of the upper mantle layers;
e) the process of the differentiation of matter in the Earth's crust;
f) sources of intrusion;
g) solutions and gases present in the Earth's crust;
h) the geothermal regime of the Earth's crust.

The following areas were proposed for drilling wells from 10 to 15 km depth: Pre-Caspian depression, the Urals, Kareliya, the Kura depression, the Caucasus, the central regions of the East European platform, the kimberlite pipes of Eastern Siberia, Tien Shan, and the Kuril Islands.

*1960–1969.* In 1960–1962 proposals for organizing investigations of the deep crustal structure were worked out.

To ensure efficient organization, coordination, and management in investigations of the deep Earth's interior, the Inter-Departmental Scientific Council for the Study of the Earth's Interior and Superdeep Drilling was set up under the auspices of the USSR State Committee for Science and Technology (GKNT). The Council consists of some 200 prominent scientists and specialists from research and industrial organizations of different ministries and departments of the USSR, including four academicians and six corresponding members of the Academy of Sciences of the Soviet Union republics, 70 doctors of science and 75 candidates of science, as well as experts from ministries and technical departments. Professor N. S. Timofeev, Doctor of Technical Science, has been Chairman of the Interdepartmental Council since 1963. He contributed much to organizing the development of equipment, technique, and technology in well drilling. Professor Ye. A. Kozlovsky, Doctor of Technical Sciences, has been the Chairman of the Council since 1975. N. I. Andrianov, chief specialist of the GKNT, is the Scientific Secretary of the Council; he is in charge of general programme coordination.

All the work on the deep study of the USSR territory has been carried out under the guidance of the Interdepartmental Scientific Council, during which time more than 80 meetings and 75 visiting sessions have been held. The Council and its departments organized and developed work on the study of the Earth's interior by geological and geophysical methods and the drilling of deep and superdeep wells.

In 1965, a comprehensive joint scientific and technical programme of study of the deep crust was planned, according to which the Kola and Saatly superdeep wells were the first to be drilled. The Kola well was planned to penetrate the oldest Archean rocks of the Baltic Shield on the Kola Peninsula in the area of the Pechenga group copper-nickel deposits, 10 km north of the town of Zapolyarny, and probably enter the basaltic layer which, according to preliminary geophysical data, was thought to occur at a depth of about 7 km. It was proposed to sink the Saatly well within the limits of the Kura depression of the Azerbaijan Soviet Socialist Republic, at a site confined to the known gravity maximum, presumed to have been caused by local uplift of the basaltic layer.

*1970–1980.* In accordance with this extended programme, a model of the Earth's crustal structure and upper mantle was planned, together with maps showing the estimated mineral resources and promising directions for further exploratory work throughout the country. Some tasks of the programme involved studies to solve those questions of applied geology closely associated with the fundamental theories of the deep structure and evolution of the Earth's crust, including further research into the processes responsible for mineral ore formation, and the laws governing their distribution pattern.

The USSR Ministry of Geology is the leading organization responsible for the study programme of the deep Earth's interior. Additionally, more than 150 research and industrial organizations are also engaged in the programme. The drilling of the Kola superdeep well was entrusted to a specialized exploration unit of the USSR Ministry of Geology – the Kola geological superdeep drilling expedition.

By the beginning of the 1970's, preparatory work for drilling the Kola superdeep well had been completed. This initiated the second stage, during which, besides drilling the Kola and Saatly wells, the Earth's interior was studied with the help of geophysical methods.

The drilling of the Kola SG-3 well (Fig. 1.2) started in May 1970. The following tasks were to be solved by this well:

1. to study the deep structure of the nickeliferous Pechenga complex and Archean crystalline basement of the Baltic Shield within the limits of the Kola Peninsula, to elucidate peculiar features of geological processes, including processes of ore formation;
2. to find out the geological nature of the seismic intersurface boundaries in the continental crust and obtain new data about the thermal regime of the Earth's interior, deep water solutions, and gases;
3. to obtain as much information as possible about the material composition of rocks and their physical state, to drill in and study a boundary zone between the granitic and basaltic layers of the Earth's crust;
4. to modernize existing equipment and technology and create new, as well as to perfect methods of geophysical investigations of both rocks and ores at great depth.

By 1980, long-term geophysical profiling, aimed at studying the crust and upper mantle, was completed in some regions. Data on industrial explosions were also used in the course of these investigations. A total of 18,000 km of deep seismic sounding profiles were conducted. On the strength of the data thus obtained, it became possible to elucidate the deep structure of the Earth's crust, i.e., trace the hypsometric position and relief of the Mohorovičič discontinuity and pre-Riphean basement in Eastern Siberia, and a series of intermediate boundaries in the consolidated crust and sedimentary cover, delineate zones of major faults

**Fig. 1.2.** Kola superdeep well. Derrick housing

and determine their vertical extent, outline with greater accuracy regional boundaries and the deep structure of positive and negative tectonic elements, which may be areas of endogenic mineral concentrations and oil and gas accumulation. New information was accumulated on the structure and physical parameters of the Earth's upper mantle within the limits of the old East European and Siberian platforms, fringing young plates and folded structures. Velocity seismograms to a depth of about 400 km were constructed, the absorbing properties of various environments were studied, and a generalized geological interpretation was given to areal zones with anomalous longitudinal seismic wave velocity propagation. Theoretical models of the crust and mantle were elaborated for some major tectonic units of USSR territory, including Western Siberia.

A combined study of different geophysical material from various geotectonic zones made it possible to establish that the previous model-based interpretation of geophysical data was too simple. For instance, the following facts were established:

a) a considerable non-uniformity in the vertical and lateral structure of the Earth's crust and lithosphere;
b) a complex relationship between the deep crustal structure and near-surface geological structures, i.e., a discrepancy between geophysical parameters (velocity, in particular) and anomalous objects caused by the presence of this discrepancy at different structural levels, which may be indicative of the bilayer disharmony of the lithospheric structure;

c) the presence of boundaries of probably differing geodynamic states of the Earth's crust, fixed at relatively shallow depth (10 – 15 km); these boundaries were encountered in addition to geological (structural-material) boundaries.

In 1980, the Kola superdeep well reached a depth of 10.7 km. As a result of the drilling, unique geological information was obtained about the deep structure of the Baltic Shield (Kozlovsky 1981; Kajiawava and Krause 1971; Kozlovsky 1981). This information substantiates previous theories. On the basis of a direct study of the mineral and geochemical composition of core samples and a series of geophysical investigations conducted in the borehole, new data were gathered on the material composition and physical state of rocks at depth; these data differ considerably from those of previous modelling data, and are important for prognosticating hidden deposits of mineral resources (iron ores, copper, nickel, mica, and rare metals) not only in the Kola Peninsula but also on other ancient massives. On the basis of these data reliable interpretation of geophysical materials became possible, which plays an important role in elaborating tectonic problems in geology.

It was established that the composition and properties of rocks change regularly with depth. For the first time the vertical zonality of rock metamorphism was revealed in one continuous section, again different from the theoretical model. The facts obtained can be used for future elaboration of the petrogenetic theory.

The experimental studies permitted an understanding of the geothermal regime of the ancient Earth's crust. The geothermal gradient proved to be higher than was anticipated; the role of mantle and radiogenic sources in the total heat balance at depth was identified. A substantial contribution was made to working out a thermal model of the formation of the Earth's crust by taking into account effective endogenic heat.

A vertical geochemical section of the Earth's crust to a depth of 11.6 km was constructed for the first time, and depth-dependent regularities of acidity and alkalinity of rocks and the behavior of ore, and rare and radioactive elements, depending on these changes, became evident. A model of the chemical composition of the primitive continental crust was built, with inclusion of data on deep seismic sounding.

New basic data on the ore-forming processes occurring in deep crustal layers were found. Commercial copper-nickel ores were discovered in an interval of 1665 – 1830 m, confined to an unknown "ore horizon"; this extends the possibilities of discovering new copper and nickel ore accumulations in the Pechenga region. Nickel ores were found to extend down to a great depth, their content of valuable elements remaining constant. At a depth of 6500 – 9500 m, zones of copper, lead, and nickel mineralization were found for the first time, showing that not only the upper layers, but also deep layers of the Earth's crust favour mineral formation. This conclusion is important for further development of the theory of mineral resources, and extends the possibility of searching for new ore deposits.

It was established for the first time that zones of highly porous, fractured rocks saturated with deep subsurface waters exist at depth in older shields; their presence had previously not been assumed. Some special features of the chemical composition of subsurface waters were found, which represent a substantial contribution to a model of the hydrophysical zonality of the Earth's crust.

The physical state and properties of rocks at a depth exceeding 10 km were identified by the new data obtained. At places a high permeability of the rocks was fixed, which is important for prognosticating underground "hollows" and assessing the possibility of their use for burying highly toxic industrial wastes.

According to data from geochemical, geophysical and nuclear-geophysical investigations conducted in the borehole, as well as from laboratory studies of core samples, a correlation was established between chemical composition, structure, and physical properties of the rocks encountered by the Kola SG-3 borehole. On the basis of this material, the real geological nature of deep seismic intersurface boundaries was revealed. These data did not confirm the geophysical model of the Pechenga region deep structure. Instead of drilling through the basaltic layer, which, on the basis of geophysical data, was expected to occur at a depth of 7000 – 7500 m, the borehole encountered compact Archean gneisses. This allows a new interpretation of data on surficial geophysical investigations, not only on the Baltic Shield, but also in other areas composed of old crystalline rocks.

The Kola section may be used as an area of reference. This will increase the reliability of geophysical survey, and particularly, the accuracy of the deep seismic sounding being carried out intensively in the USSR and ot other countries.

The unique material on the structure of the Earth's crust and the state of rocks at great depths provided by the Kola well allows a more reliable assessment of various endogenic processes in the formation of rocks and ores and consequently makes different metallogenic constructions possible on this basis. The new technical means and elaborated methods, successfully put into practice, have raised geophysical studies in wells to a higher level, which is very important for increasing the efficiency and quality of geological exploration for oil, gas and other minerals.

Another superdeep well, the Saatly well, was sunk in 1977 within the limits of the Kura depression (Azerbaijan Soviet Socialist Republic). The drilling site is confined to a gravity anomaly. This gravity maximum was assumed to have been caused by a locally uplifted top of the basaltic layer. The well penetrated loose Cenozoic and Upper Mesozoic rocks and entered a thick layer of volcanogenic rocks probably of Cretaceous and Jurassic age. These strata seem to be a cause of the gravity maximum. The well is planned to penetrate the entire thickness of the volcanogenic sequence and reach the Paleozoic basement.

In drilling the Kola and Saatly wells a number of important technological problems of superdeep drilling were solved and new equipment developed.

In other countries superdeep wells (e.g., the USA) are usually drilled with the help of units working under great strain (load capacity 6 – 8 Mega-Newton, pump pressure 50 MPa). However, superdeep wells drilled outside the USSR have not exceeded a depth of 9100 – 9600 m. An analysis of trends in constructing drilling units shows that reserves for increasing the unit load capacity further are practically exhausted. Soviet specialists worked out a technology of well drilling which made it possible to sink a well to the planned great depth (exceeding 10 km) with the help of a drilling unit of lesser load capacity (5 MN) and pump

**Fig. 1.3.** Drill string running

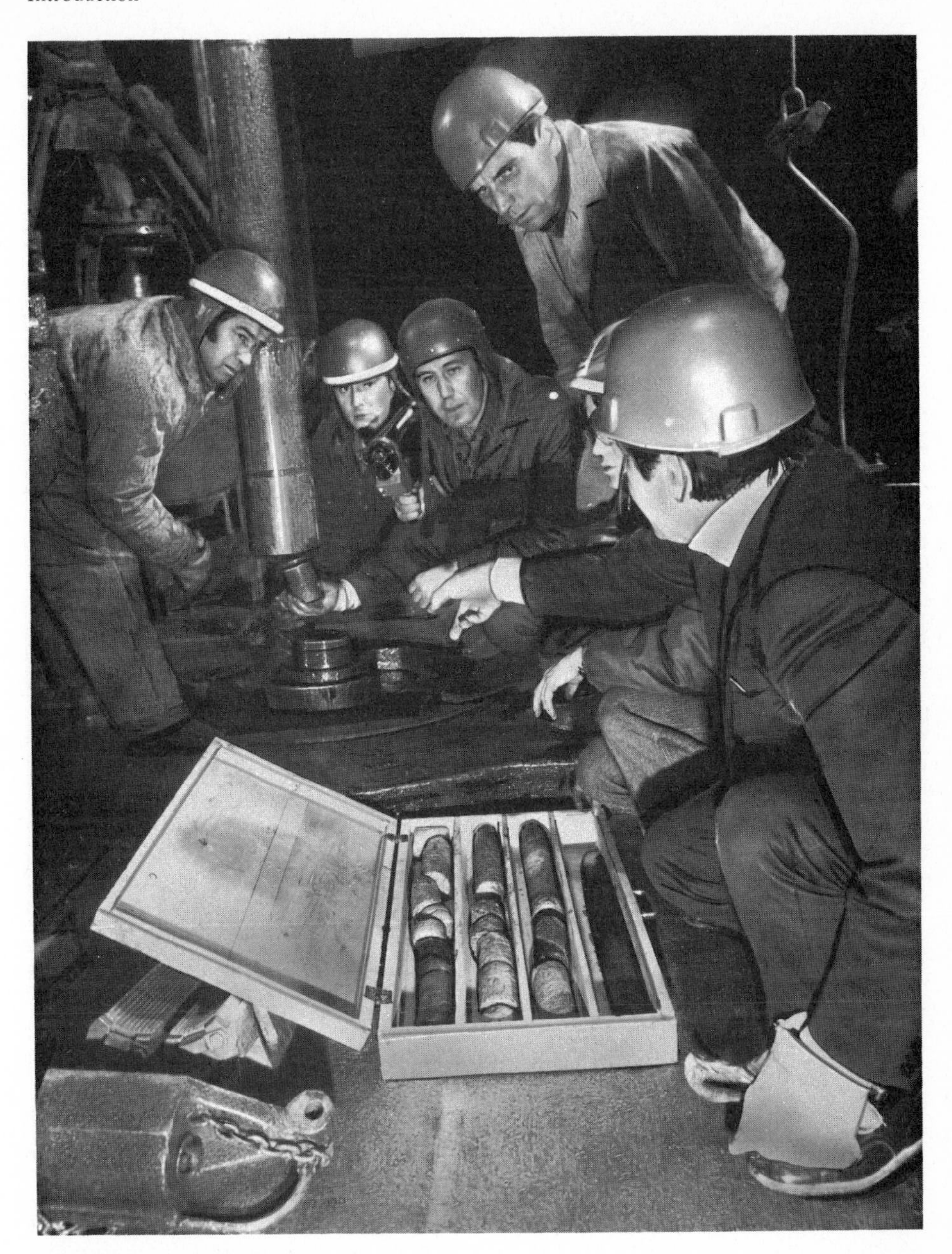

**Fig. 1.4.** Core sample lifting and extracting

pressure (40 MPa). The sinking of the Kola well has induced innovations in drilling technology, based on the principle of the advance hole. The well is run with the help of a bottom-hole turbine. The drill string is made of lightweight aluminium pipe.

A new drilling unit for driving the well to previously inaccessible depths was created by a group of scientists and workers of the Uralmashzavod (Figs. 1.3, 1.4). They have devised and put into practice rock-crushing instruments and bottom-hole engines capable of working at great depths, including bits and speed-reducing turbodrills, which work very steadily at optimum rotation speeds. Another innovation introduced in the Kola well was a system of devices which controls the turbodrill work at the well bottom. In the absence of these controlling devices, drilling of a well with the help of the down-hole motor at depths exceeding 8 – 9 km becomes impossible. Light alloy drilling pipes of high strength (yield strength reaches 500 MPa) and thermostability (up to 200 °C) have been designed and used in the well, which made it possible to drill the well at great speeds in round trip operations.

The high standard of the technical means created, their novelty and efficiency became an acknowledged fact: more than 40 patents were granted to their inventors, about 20 units of new equipment were put into serial production, their use allowing record drilling depths.

The complex scientific-technological experiment of the Kola superdeep drilling was accomplished solely by Soviet technology and technique.

The fulfillment of the drilling programme for this unique well provided important practical results: explorers have been provided with the modern technical-technological means enabling them to study and exploit the Earth's natural resources at a depth previously inaccessible.

*1981 – 1985.* Analysis of data of superdeep drilling and regional geophysical investigations undertaken in the USSR prior to 1981 shows that only a basically new approach to planning and carrying out of this work could have resulted in an increase in efficiency. However, this task could only be implemented if a regional study of the Earth's crust and upper mantle was carried out throughout the whole territory of the USSR on the basis of an integrated nation-wide system (Belousov 1982; Kozlovsky 1982).

The principal basis of the system is a network of coordinated geophysical profiles based on deep and superdeep key wells (Fig. 1.5). It has also been suggested that the prediction of geophysical testing grounds for the study of variations in geophysical fields should be included in this system. The main framework of the network is a reference basis for more detailed study within a particular region. Data from aerospace and aerogeophysical surveys for the complex interpretation of investigation results will help in constructing much-improved three-dimensional geological-geophysical models for the entire territory of the USSR.

The principle task of the studies along the extended framework profiles of long distances, so-called geotraverses, is to reveal the fundamental differences in structure and state of the lithosphere in regions of different geodynamic development throughout the country. Smaller profiles are laid within homogeneous tectonic blocks or across a system of such blocks and their confinements (junction zones, major faults, contacts). They should help in solving structural-tectonic,

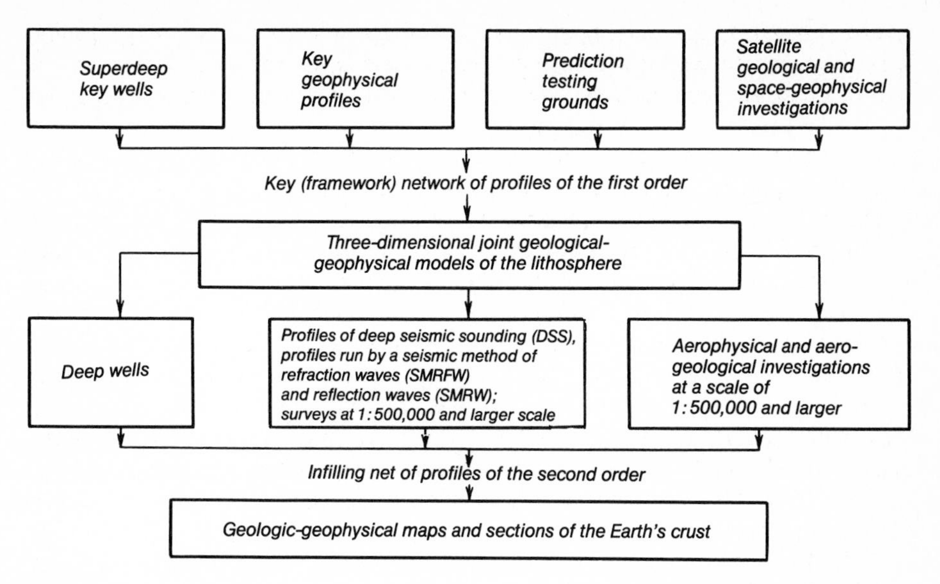


**Fig. 1.5.** Scheme of the regional study of the Earth's crust and mantle within the limits of USSR territory

lithological (material composition) and other regional problems. Surveys within the geotraverses are detailed and aimed at studying more local inhomogeneities of the upper parts of the crust, faults, and contacts. They are primarily intended to solve problems of medium- and large-scale prediction and to assist in the search for mineral deposits.

The systematic study of the crust and upper mantle is a distinctive feature of the whole programme, according to which a network of large interlinked key profiles runs over to extensive areas. This makes it possible to trace changes in physical parameters from one region to another of different geodynamic development regimes.

The integration of geological-geophysical and geochemical investigations, coinciding both spatially and temporally, allow a higher degree of reliability in the conception of the structure and dynamic state of the domain studied.

Further expansion in the drilling of superdeep and deep wells was planned. The drilling of the Kola and Saatly wells will be continued; other superdeep wells to be drilled were the Tyumen, Anastasievsko-Troitskaya and Ural wells (down to 12–15 km). At the same time, six deep wells were to be sunk in 1981–1985, three of them (Dnieper-Donetsk, Pre-Caspian, Timan-Pechora) to be drilled in oil and gas-bearing regions and the other three (Muruntau, Norilsk, Krivoi Rog) in ore-bearing regions.

According to the plan, by 1985 the Kola well was to reach a depth of 13,000 m and penetrate new intersurface boundaries in the crust, as established by seismic investigations. It was assumed that the so-called granulite-basite layer would be encountered. In deepening the borehole the further industrial testing of the *Ultramash-15000* drilling unit was planned, the technology of superdeep drilling perfected, new geophysical devices for work under conditions of high pressure and high temperature devised, and all put to practical use.

When the drilling of the Kola well is completed, the well is envisaged as becoming a unique natural laboratory for studying processes deep in the Earth's crust, for conducting permanent observations of temperature regimes, studying conditions and possibilities of burying industrial wastes at great depths, and testing and perfecting devices and methods of geological-geophysical, geochemical and hydrogeological investigations. The drilling and combined geological-geophysical investigations of the Saatly well will be continued to a depth of 11,000 m.

The drilling of the Tyumen superdeep well should offer the possibility of evaluating the oil and gas potential of the Jurassic and pre-Jurassic rocks in the northern part of Western Siberia, and elucidating the deep structure of this area and assessing its prospects. The Ural superdeep well was planned to penetrate a section of the major ore-bearing folded region, with the aim of resolving basically important problems of geology and prognosticating new deposits in this region. Results of superdeep drilling, together with geophysical investigations, should allow a higher degree of reliability in directing prospecting and exploratory work for the period 1990 – 1995 and to the year 2000.

The present state of drilling technology and techniques in the Soviet Union allows the assumption that Soviet industry is in possession of all the means necessary to implement the study programme of the country's deep structure.

In 1981 – 1985 work in research and experimental design to perfect the technology and techniques of superdeep drilling will continue. The production must be speeded up of experimental and industrial high-strength drilling pipes and tool joints; rock-crushing instruments to ensure effective drilling and core sampling at high temperatures; thermostable turbodrills with reduction gears; geophysical equipment for conducting geophysical studies at depths exceeding 10,000 m; blocks of bottom-hole devices for registering drilling regimes; chemical reagents and lubricating additives for treating thermostable drilling muds; special thermostable materials for manufacturing drilling instruments and bottom hole engines. A special model of conditions in wells drilled to a depth of 15 – 20 km, with temperature and pressure at the bottom hole at to 300° – 400 °C and 200 – 300 MPa, respectively, is now being built.

It is planned to build highly mechanized drilling units with load capacities reaching 5 Mega Newton and working pump pressures of 40 – 50 MPa for the Ural, Tyumen and Anastasievsko-Troitsk wells. The technology of further Kola drilling using the advanced-hole method to a depth below 12 km must be worked out and tried. Designs must be made for the Saatly well to secure its sinking to a depth of 13 km.

The programme plans to develop a drilling method which could effect core sampling without pipe-lifting operations.

In the programme of the study of the Earth's deep crust by deep and superdeep wells, great importance is attached to the problem of developing technical means for conducting geophysical investigations in wells under high temperature conditions (up to 350 °C) and pressures up to 30 MPa. A stationary well-logging truck hoist, needed for running and lifting operations in wells drilled to a depth of 15 km, is being put to work at present at the Kola and Saatly wells. It is also necessary to make load-bearing three-core logging cable 15,500 m in length.

There is every reason to believe that the comprehensive study of the Earth's interior within the borders of the Soviet Union will increase and deepen our knowledge of the characteristics of the crustal structure and its evolution. This will be a valuable contribution to expanding and strengthening the basis of raw minerals of the national economy of the USSR. This programme will further help to elucidate a series of general problems of the Earth's tectonosphere, structure and evolution.

This joint programme is principally aimed at studying the continental crust and upper mantle. With this in mind, it is interesting that at the time that this work was being planned and developed, American scientists proposed a project of ocean drilling, which later became an international project, in which a number of countries, including the USSR, took part. The American vessel *Glomar Challenger* drilled a total of more than 500 ocean wells; these wells were the first to give information about the composition, age, and structure of the sedimentary cover on the ocean floor and about the uppermost horizons of an underlying consolidated layer of the oceanic crust composed of basalts which had erupted on the ocean floor. The Soviet continental project, and the international oceanic project supplement each other.

There is no doubt that the oceanic project is of great scientific significance, for which reason the Soviet Union participated. It must be emphasized that the Soviet Union, as the largest continent, is particularly interested in discovering the principles of continental crustal structure and evolution. This will allow the creation of a well-grounded scientific basis for the prognostication of mineral resources. Apart from this, it should not be forgetten that mankind lives on continents and extracts the major part of raw minerals from the continental crust. In spite of the fact that in recent years sea and ocean areas are being intensively explored by man, the mineral resources beneath sea and ocean floors (preferentially resources of oil and gas), are mainly concentrated within the limits of shallow shelves, i.e. in the same continental crust covered by water. In view of the fact that the composition and structure of the continental crust are less homogeneous than those of the oceanic crust, it may be anticipated that also in the future the greater part of minerals will be discovered mainly in the continental crust.

The USSR is not the only country to recognize the importance of the study of the continental crust. American scientists, after a long period of enthusiastic oceanic research, decided to proclaim the 1981 – 1990 period as a decade of intensive study of the North American continent. When a programme for the international project *Lithosphere*, which later took over from the *Geodynamic project*, was being prepared, it was especially emphasized that continents deserve greater consideration than they had previously been given in the international projects.

The Soviet Union has always been more consistent in carrying out large-scale studies of the structure and regularities in the evolution of the continental crust than other countries. This is a deeply rooted tradition in our country, and it is still very much alive, thus promoting the further development of the basis of USSR mineral resources.

# 1 Geology

# 1.1 Geology of the North-Eastern Part of the Baltic Shield

The Baltic Shield extends over an area of 1,140,000 km$^2$. It is the largest exposed part of the crystalline basement of the Eastern European platform. The northern margin of the shield, extending along the Barents Sea coast, includes fragments of the western branch of the Timan-Kanin Baikalides constituted of Riphean rocks. In the west and north-west, the shield is overthrusted by the Scandinavian Caledonides, and in the south and east the shield surface gently plunges under the sedimentary cover of the Russian Plate.

The Kola Peninsula occupies the north-eastern part of the Baltic Shield. Here, as distinct from the southern and western areas, the most ancient geological formations predominate, thus reflecting comprehensively the main stages in the development of the Precambrian structures. Since this part of the shield was a very long and complex developmental history, it is considered to be a particular reference area for studying cyclicity in tectonic evolution, stages and tendencies in regional endogenic ore-forming processes which are here manifested very intensively. Ore deposits and ore shows, representing the metallogeny of the shield, occur within the limits of a comparatively small area of it. This accounts for the high rate of study of the area in question. This area may therefore serve as a reference for examining the main geological elements of the shield structure, its tectonic framework, magmatism and metallogeny.

## 1.1.1 Basic Features of the Geological Structure of the Kola Peninsula

The Kola Peninsula is mainly composed of Precambrian crystalline rocks (Fig. 1.6) also developed in Finland, Sweden, and Norway.

The geology of this region has been studied for many years, the results being presented in a voluminous literature. A general geological analysis was given in works by A. A. Polkanov, B. M. Kupletsky, A. E. Fersman, N. A. Eliseev, L. Y. Kharitonov, K. O. Krats, G. I. Gorbunov, A. V. Sidorenko, and many other researchers. The most complete description of the geological setting of the Murmansk region is given in *Geology of the USSR* (1958; Kharitonov, ed.).

The structural features of the Kola Peninsula are entirely determined by its geological history, which can be subdivided into two major stages: the progeosynclinal and protogeosynclinal stage and the platform stage. During the first stage, the plastic upper lithosphere was deformed, and folds, stretching mainly in a north-western direction, were formed. At the second stage, the stressed consolidated crust was broken by major fractures and faults, as a result of which the region was subdivided into blocks of differing size and configuration.

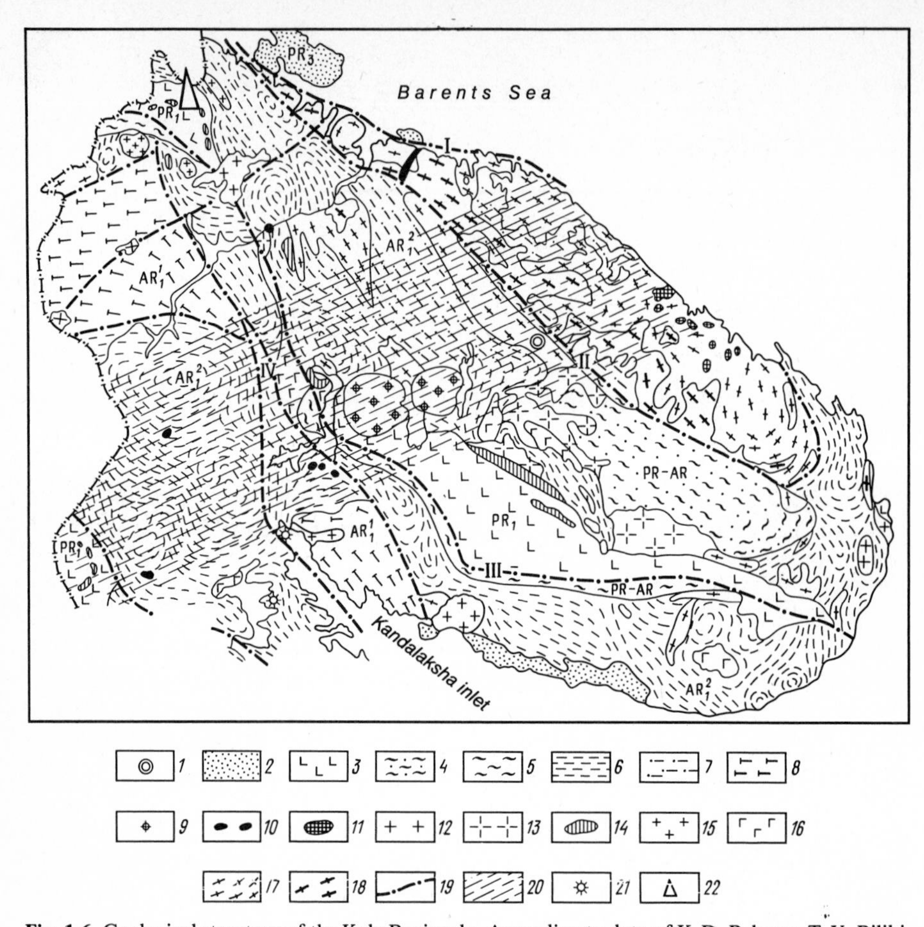


**Fig. 1.6.** Geological structure of the Kola Peninsula. According to data of K. D. Belyaev, T. V. Bilibina, V. A. Perevozchikova, L. Ya. Kharitonov. *Paleozoic: 1* Upper Devoninan and Lower Carboniferous, sedimentary-volcanogenic complex of the Kontozero structure. *Upper Proterozoic: 2* formations of the Gyperborey series: of the Rybachii Peninsula – conglomerates, sandstones, shales; of the Sredny Peninsula, Kildin and Terskaya islands – quartzites, sandstones, variegated shales, dolomites. *Lower Protozoic: 3* Pechenga, Imandra-Varzug and Kuolayarvi complexes – basic, medium-basic and ultrabasic lavas, tuffogenic formations, subordinate phyllites, siltstones, sandstones, dolomites, conglomerates metamorphosed predominantly in greenschist facies. *Lower Proterozoic – Upper Archean: 4* Keiv series – aluminiferous, carbon-bearing, micaceous and other shales, quartzites and sandstones metamorphosed predominantly in epidote-amphibolite facies; *5* Tundra series – amphibolite, micaceous, aluminiferous shales and gneisses, amphibolites, ferruginous quartzites metamorphosed predominatly in epidote-amphibolite facies. *Lower Archean, Kola series: 6* biotite-plagioclase, two-mica gneisses, biotite-plagioclase gneisses with HAM, amphibolites, predominantly amphibolite facies of metamorphism; *7* biotite-plagioclase gneisses with HAM, amphibolites, pyroxene-amphibolite-plagioclase geneisses and crystalline schists, ferruginous quartzites metamorphosed predominantly in amphibolite facies; *8* granulitic complex – cordierite garnet-biotite-plagioclase gneisses, pyroxene-amphibolite-plagioclase gneisses and crystalline schists, amphibolites and migmatized amphibolites, charnockites, enderbites metamorphosed predominantly in granulitic facies. *Intrusive and ultrametamoprhic complexes – Middle Paleozoic (Upper Devonian): 9* complex of alkaline nepheline syenites; *Lower Paleozoic: 10* complex of basic, ultrabasic and alkaline rocks; *Upper Proterozoic: 11* gabbro, diabases, granophyres; *Lower Proterozoic: 12* granites, granodiorites, *13* alkaline granites, granodiorites; *16* gabbro-anorthosites, metagabbro-norites, metagabbro; *Lower Archean: 17* granites, granite-migmatite; *18* granodiorites, gneiss-granites, plagio-granites; *19* main major faults (*I* Karpinsky; *II* Keiv-Uraguba; *III* Pechenga-Varzug; *IV* Lapland-Kolvits); *20* zone of maximum Caledonian activity (1981); *21* volcanic pipe; *22* Kola superdeep borehole

In the eastern part of the Baltic Shield, three megablocks are distinguished, the Kola, Belomor, and Karel blocks; they almost coincide with the known Kola-Norwegian, Belomor-Laplandian and Central-Karel structural zones.

The Kola megablock, occupying three-quarters of the region, is the most complex geological structure. In its northern margin, within the limits of the Barents Sea coast, the Murmansk massif can be recognized. The characteristic features of this block are widely manifested processes of granitization. In the north, this massif is bounded along the Karpinsky major fault by the Riphean sedimentary formations, which underwent subsidence during the Caledonian movements. In the south it is limited by the Keiv-Uragub major fault, which plunges steeply to the north-east. Along these tectonic zones the Murmansk block was uplifted to a considerable height and eroded. The block is composed everywhere of oligoclase plagiogranites, granodiorites, and gneisses with amphibolites. The relict schists and gneisses contain mineral relicts which clearly show that these rocks are of granulitic facies of the early evolutionary stages. The rocks commonly form isometric domes; the dominant folded structures of the north-western direction are less frequent here.

In the central part of the megablock, a wide strip (160 – 200 km) of complex folded Precambrian structures is distinguished. It belongs to the Central Kola geotectonic region which is the major synclinorium. In the north it is limited by the Keiv-Uragub major fault and in the south by the Kolvits-Lapland tectonic suture. A common feature of this region is the presence of well-preserved geological evidence that the Earth's crust of the area in question was mobile during the older stages of its development and that it was consolidated during Early Archean time. Later, in Proterozoic time, the area was involved in the processes of sedimentation, tectonic and plutonic-metamorphic activity. The latter processes did not change the structural configuration of the area, which had been formed during the Archean.

Three structural elements of the first order can be distinguished within the limits of the Central Kola region (synclinorium), the Keiv and Pechenga-Varzug synclines, and the Kola anticline, which subdivides these synclines. Due to intensive faulting and folding of the second and third order, these structures became broken into a number of segments of complex structure.

The Kola anticline, stretching throughout the peninsula, as well as the southern and northern flanks of the above-mentioned synclines, is mainly composed of Archean gneisses of the so-called Kola series (belonging to the Kola-Belomor complex). The general composition of this widely developed complex correspond to that of the lithophilic crust, integrating primitive sedimentary sandy-argillaceous rocks and derivates of andesite-dacite and basic magmas. According to Sidorenko (Sidorenko and Bilibina 1980), this complex as a whole is constituted of poorly differentiated geological formations of medium composition.

The Central Kola synclinorium is mostly made up of amphibolite-gneiss rocks composed of amphibole-biotite and biotite gneisses with numerous layers of amphibolites. The distribution of these suites in the Kola anticline is not exactly known, but the upper boundary can be set rather distinctly where these rocks are overlain by younger Archean and sometimes Proterozoic rocks.

Normally, the amphibolite-gneiss rocks are overlain by micaceous and highly aluminiferous gneisses. Associations of rocks which compose these gneiss formations differ from each other in various blocks, with highly aluminiferous rocks predominating, i.e., garnet-biotite, kyanite-garnet-biotite, cordierite- and sillimanite-bearing gneisses, alternating with biotite gneisses. Two-mica gneisses are also present, but amphibole-bearing gneisses and amphibolites occur in limited amounts. The thickness of this formation is about 1500 – 2000 m.

The jasperite-amphibolite-gneiss formation of the Kola series is distinctive among other prevailing strata known in the synclinorium. This formation is represented most fully in the central parts of the region. Its lower boundary is confined to the surface, where rocks of the formation rest on the metamorphic amphibolite gneiss. The composition of this formation is complex, and is characterized by the following rock associations:

1. garnet-biotite, sillimanite- and cordierite-bearing gneisses;
2. hypersthene-biotite amphibolites, pyroxene-amphibole gneisses and crystalline schists, and magnetite quartzites.

The thickness of this locally developed formation may be about 800 – 1000 m.

The Archean strata over extensive areas are also composed of granitoids, which form areals of migmatization and migmatite-plutonic rocks in the central parts of the domes. Massives of basic rocks are limited by folded and faulted structures. The major faults of this evolutionary stage are not well developed, which may be accounted for by the high plasticity of the gneiss resulting from granitization processes.

Different groups of rocks constitute the Archean sequence. At the same time, when studying the processes of metamorphism, it becomes obvious that in Archean time these rocks were usually subject to deformation and alteration throughout the whole are, and that these processes covered at least two stages: at the early granulitic-facies a stage rocks were subject to moderate pressures, and at the Late Archean stage they developed under conditions of the amphibolite facies. The basic structural features of this most ancient complex were formed mainly as a result of post-Archean folding.

In the south, the Central Kola synclinorium is bounded by the Lapland-Kolvits zone of major faults [the Belomor suture according to Krats (1978)], which is the regional structure separating the Kola and Belomor megablocks of the shield. This zone is characterized by distinctive zonal metamorphism and specific magmatic activity. Within this frontier zone of the Lapland and Kolvits blocks, constituting the so-called granulitic strata, rocks are represented by pyroxene-bearing crystalline schists and gneisses (garnet-diopside, garnet-bipyroxene, two-pyroxene, hypersthene), and amphibolites. The major faults in this region control the distribution pattern of the hypersthene diorite intrusions, layered anorthosites, numerous small bodies of norites, gabbronorites and ultrabasites. On the southern boundary of the Lapland block, the latter form a distinctive "serpentinite" belt. The granulitic complex has been described in detail in many publications. However, its genesis and position in the stratigraphic column still remains very uncertain.

In the south-western part of the Murmansk region, Archean formations of the Belomor megablock are developed. This block, limited in the north by the Kolvits-Lapland suture zone, appears uplifted compared with the Central Kola region, and is consequently eroded to the deeper horizons. The megablock territory is the area where only the Belomor series of the Archean age is developed. This series consists of gneissose granites, amphibolites, micaceous and highly aluminiferous gneisses, and schists intruded by the Archean and Early Paleozoic gabbro and granite bodies. The Belomor series, like the Kola series, is mostly composed of amphibolite-gneiss rocks consisting of amphibolite-biotite, biotite gneisses with layers of amphiboles, and also of overlying strata of micaceous and aluminiferous gneisses. However, as distinct from the Kola series, the ferruginous-siliceous formations are completely absent here (they might have been eroded). According to the existing theories, the lower, Verets gneissose granite-amphibolite suite and the gneiss-granodiorites of this block are attributed to the most ancient parts of the Archean geological sequence of the Karel-Kola region.

The Belomor complex also went through several stages of intensive multidirectional tectonic deformations, metamorphism and deep basic and acid magmatism in the Early and Middle Precambrian, which determined its rather complex internal structure.

Thus, in the north-eastern part of the Baltic Shield, the sialic strata of the lithosphere at the surface mainly consist of ancient Archean rocks. Their composition is similar to that of gneiss in which ultrametamorphic autochtonous sodic-calcic granitoids are widely developed. These supracrystalline rocks are of great thickness; they are deprived of clearly defined angular unconformities and characterized by only negligible facies changes. It is worthy of note that some stratification of rocks is also observed; for example, micaceous and highly aluminiferous gneisses, as well as strata with ferruginous quartzites, are usually replaced down the sequence by amphibole-gneisses. However, due to the development of complex folds, and the presence of strike-slip faults in particular, which have broken the whole Archean basement into separate mosaic blocks, it is difficult to solve the question of the stratigraphic subdivision of the Archean complex.

The absolute age of the Archean rocks is estimated to be not less than 2700 Ma. Some analytical data obtained in the course of investigations allow the age to be increased to 3300 – 3600 Ma; however, these age intervals cannot be correlated at present with concrete geological events occurring in the Early Archean.

The lower boundary at the Kola-Belomor gneiss complex is not known. It is suggested that its lowermost rocks, the so-called basement of the Archean, were completely transformed together with rocks of the described complex as a result of ultrametamorphic processes.

The Central Kola synclinorium, bounded in the north and south by major tectonic zones, differs from the adjacent Murmansk and Belomor blocks. This difference is primarily associated with a wide development of the Proterozoic sedimentary-volcanic formations. In the Central Kola depression the latter are mainly confined to the two previously described synclinal folds of the first order which began to develop in Archean time. Among these folds one can particularly distinguish the Colour Belt of the Karelides which developed on the Kola Peninsula. The Keiv synclinal zone, the second Proterozoic belt containing considerable reserves of aluminium oxide, is less in extent.

The Keiv synclinal zone is confined to the eastern part of the region. In its central part the zone is composed of crystalline highly aluminiferous schists of the Keiv series (Lower Proterozoic). The Keiv series comprises three suites: Chervurt, Vykhchur, and Pestsovo-Tundra. The first consists predominantly of silica-aluminiferous rocks, the second of rocks with a higher iron content and less aluminium oxide, and the third of silica, silica-ferruginous, and carbonate rocks. The maximum thickness of these suites, including numerous bodies of orthoamphibolites, is about 1400 m. Crystalline schists of the Keiv suite are underlain by biotite, garnet-biotite, and amphibole gneisses belonging to the so-called Tundra series (Late Archean – Early Proterozoic) and resting on the eroded crystalline basement which in the Keiv area is composed of predominantly Archean plagiogranites and granodiorites.

The Colour Belt of the Kola Peninsula extends discontinuously in diagonal direction throughout the entire peninsula in the form of a narrow strip. This belt (the Pechenga-Varzug zone) is composed of sedimentary-volcanic rocks which constitute a formation of unique character. According to the researchers's opinion, during the early stages of the Pechenga-Varzug trough development, molasse-andesite-basalt and carbonate-quartzite-trachybasalt formations accumulated here. The presence of molasse-like coarse-grained strata at the bottom and contrasting associations of volcanic rocks in two lower sedimentary-volcanic macrocycles suggests that this stage of development belongs to the orogenic tectonic type. The character of the upper sedimentary-volcanic strata (the third and fourth in the Pechenga structure) is indicative of the appearance of the picrite-basalt volcanic association similar to the intermediate tholeiites of rift zones. This means that the further development of the trough continued under conditions of lithosphere tensile stresses, which were accompanied by a related rise of the mantle matter (Duk 1977). The internal structure and character of rocks composing the Pechenga trough will be described in more detail later.

Formations of similar composition were found on the eastern extension of this structure within the limits of the Imandra-Varzug syncline. The lower parts of the sequence are represented here by the Seidorechen suite, in which andesite-basalt rocks are present in considerable amounts. In the upper parts of the sequence, metamorphites of the Umbin suite occur, with trachybasalts, tholeiitic basalts and picrites being widely developed. The uppermost part of the sequence is mainly composed of andesite-basalts of the Ilmen suite. Gabbro-peridotite intrusions are considered to be similar to picrites, the former being widely developed in the form of small bodies among sedimentary-volcanic rocks of the Pechenga graben syncline. They are represented by a group of differentiates ranging from olivinites to wehrlites and gabbro. In the process of subsequent deformation, these intrusions became foliated and underwent metamorphism. Magmatic formations, which appeared as a result of the Imandra-Varzug zone development, are major delaminated intrusions of basic and ultrabasic rocks. Massifs located in the Pansk and Fedorov tundras, as well as the Monchegorsk pluton, are assigned to these intrusions.

Another peculiar structural feature of the Colour Belt of the Karelides is also noteworthy, i.e., in the south the Proterozoic sedimentary-volcanic rocks are restricted throughout the area of their development by a major fault. South of this fault extends a narrow linear strip of gneisses and schists alternating with amphibolites which belong to the so-called Tundra series (Late Archean – Early Proterozoic).

It may thus be stated that in the epoch of the Karel diastrophism, the burst of tectonic activity in synclinal troughs of the Central Kola geotectonic zone resulted in building a new structural stage ("suprastructure") on the Archean basement. This structural stage is composed of Lower Proterozoic sedimentary-volcanic rocks; as distinct from the granite-metamorphic basement, this new structural stage is notable for its predominantly mafic rock composition.

Strike-slip faults are widely spread in the Precambrian structures of the Kola peninsula. Among the most significance are the interblock longitudinal interformational zones of deep-seated major faults stretching in western, sublongitudinal, and sublatitudinal directions and confined usually to the margins of large folded structures. They dissect the whole territory, occupied by the Precambrian crystalline strata, into massifs extending in north-westerly direction. The main longitudinal major faults of the Kola Peninsula, which border megablocks and crustal blocks of different geology and deep structure, were enumerated above. Apart from these, many faults of predominantly longitudinal direction are known inside the blocks and subdivide then blocks into separate segments. In the majority of cases, the longitudinal tectonic faults are established by the presence of tectonites of various types and are considered to be the open faults (according to A. V. Peive). The extensive faults oriented in longitudinal and latitudinal directions stretch across folded structures of the Archean and Proterozoic and only at some places do they follow the margins of crustal blocks. This complex tectonic network of major faults at some places forms a "broken saucer" structure.

Major faults developed over a long time-span, and the direction of movements along them was of an inherited character. The main systems of faults might have started to develop in the Archean and later, during a period of the maximum tectonic activity (in Proterozoic time), fault movements occurred repeatedly.

The vertical displacement of even the most major faults can only be conjectured, because distinct marker horizons in the Archean basement are lacking. For instance, according to S. I. Makievsky's data, the amplitude of the vertical displacement of major grabens and horsts appears to be 4 – 5 and even 8 – 10 km.

Gorbunov et al. (1978) are of the opinion that the geotectonic and metallogenic zoning in north-westerly direction observed throughout the Baltic Shield in the Precambrian was succeeded by tectono-magmatic activity as a result of the Caledonian and Hercynian tectogenesis manifested in the neighbouring areas. The Paleozoic tectono-magmatic activity, as distinct from that in the Precambrian, was controlled by deep-seated major faults in north-easterly direction. Detailed analysis of geological, geophysical, and geomorphological data showed that considerable vertical displacement took place along these faults, which led finally to the formation of the superimposed horst-graben structure of the shield. Of particular interest among these structural elements is the Oslo-Khibini graben, which is limited by major faults. Both in the north (Khibini, Lovozero, Kontozero) and south (the area of the town of Oslo), it comprises Paleozoic sedimentary rocks.

From this it appears that the geological structure of the Kola Peninsula and Baltic Shield as a whole, as well as the metallogeny of the latter, are generally controlled by the Precambrian folded complexes extending in north-westerly direction and, to some extent, by the mosaic-block graben-horst structure of the tectonically active region which was formed as a result of movements in the Caledonian and Hercynian geosynclines. Therefore, all deposits of mineral resources on the Baltic Shield, genetically associated with the oldest complexes of magmatic and metamorphic rocks, are generally grouped within belts extending in a north-westerly direction; these deposits are thus controlled by the presence of Precambrian geological structures.

Among these structures, the following are particularly distinguished: the Colour Belt of the Karelides with the known copper-nickel ore deposits (Kola Peninsula), the Vetryanoi Belt of the Karelia region, belts of ferruginous quartzites (Central Kola area and Kostomuksha area of the Karelia region), belts of highly aluminiferous kyanite schists (Keiv tundra), mica, ceramic pegmatites and rare-earth pegmatites on the Kola Peninsula and in the Karelia region.

In the area involved in tectonic processes in the Paleozoic (Mineral Deposits of the Soviet Union 1981) magmatic activity of the platform type was usually multi-phased in character. At early stages of this activity, massifs of the alkaline-gabbro rocks were formed (the Gremyakha-Vyrmes formation on the Kola Peninsula, the Eletozero formation in the Karelia region, the Almunge formation in Sweden). Ilmenite-titanium-magnetite ores and rare-earth mineralizations are associated with rocks of these formations.

A stage synchronous in manifestation with the Caledonian movements resulted in the formation of multi-phased intrusives of alkaline-ultrabasic rocks of the central type (Kovdor, Africanda of the Kola Peninsula, Sokli in Finland, etc.). These rocks comprise deposits of complex apatite-magnetite, titanium-magnetite-perovskite ores, rare-metal carbonatites, and sometimes concentrations of phlogopite and vermiculite. With the later stage, synchronous with the Hercynian movements, are associated massifs of nepheline-syenite formation (the Khibini and Lovozersky massifs on the Kola Peninsula and massifs in the Oslo area). Massifs of nepheline syenites contain deposits of apatite, nepheline, sphene, eudialyte and rare metals.

### 1.1.2 The Deep Structure of the Kola Peninsula

Present-day theories on the structure of the crust of the north-eastern Baltic Shield are based on comprehensive analysis data undertaken in the course of geological-geophysical investigations.

Seismo-geological sections of deep seismic sounding (DSS) and reflection seismic surveys (RSS) carried out by I. V. Litvinenko, S. A. Ankundinov, and others from 1958 – 1979 were the main sources of information about the deep structure of the Kola Peninsula. Deep seismic-sounding studies have revealed the

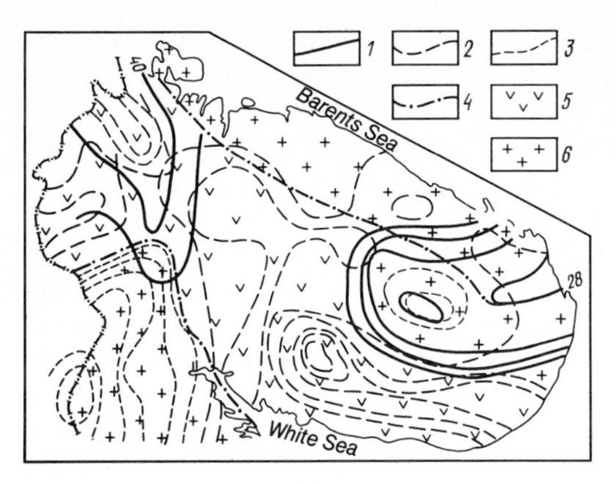


**Fig. 1.7.** Deep structure of the Kola Peninsula (Porotov et al. 1978). *1* stratigraphic contour lines of the M-discontinuity (km); *2* contour lines of the increased values of gravity field; *3* contour lines of the decreased values of gravity field; *4* main major tectonic faults; *5* crust of simatic type; *6* crust of sialic type

most distinct seismic reflector with $V_t = 7.8-8.2$ km s$^{-1}$ and $V_p = 0.3-0.4$ km s$^{-1}$ (fall velocity). This reflector corresponds to the bottom of the crust, the Mohorovičič (M) discontinuity. Gravimetric calculations show that this boundary separates the lithospheric shell with a rock density of $3.4-3.2$ gm cm$^{-3}$ from the shell with a rock density of 2.9 gm cm$^{-3}$.

The average depth to the M boundary on the Kola Peninsula, as in the whole eastern part of the Baltic Shield, is 37 km, ranging between 28 to 40 km (Fig. 1.7). Little increase in lithospheric crustal thickness is observed in the direction from east to west. For instance, in the area of the Kola superdeep borehole (SG-3) the M-discontinuity is at a depth of 40 km, while in the north-eastern part of the region a "fringe" of anomalous depth of the M-boundary is established; within the limits of this "fringe", rimming the southern flank of the Murmansk block and Keiv depression, the M-discontinuity is at a depth of about $28-30$ km. This fact has still to be explained.

There are also some less distinct seismic boundaries above the M-discontinuity, the two most important being: (1) a boundary with $V_t = 6.4-6.9$ km s$^{-1}$ which separates layers with rock density of 2.9 and 2.75 gm cm$^{-3}$; this boundary is assumed to coincide with the top of a granulite-basalt layer, Conrad (C) discontinuity; (2) a boundary with $V_t = 5.9-6.1$ km s$^{-1}$ which separates the crustal layers with a rock density of $2.60-2.75$ and $2.62-2.84$ gm cm$^{-3}$. The latter boundary is conventionally attributed to the top of a diorite layer.

According to the theories evolved in the course of deep seismic investigations and resulting from the statistical analysis of both gravimetric and seismic data, a four-layered model of the Earth's crust in the Karel-Kola region was proposed. It includes the following layers:

1. "basaltic" (granulite-basic) layer, or $\beta$-layer ($\sigma = 2.90$ g cm$^{-3}$);
2. "dioritic" layer, or $\delta$-layer ($\sigma = 2.75$ g cm$^{-3}$);
3. granite-metamorphic layer, or $\gamma$-layer ($\sigma = 2.60-2.65$ g cm$^{-3}$);
4. sedimentary-volcanic layer, or $\alpha$-layer ($\sigma = 2.62-2.87$ g cm$^{-3}$).

The isopach contours of the basaltic layer in the Karel-Kola geoblock indicate that the base level of the C-discontinuity lies at a depth of 17 km; the "relief" forms of the C-surface are developed within the HC range between 8 and 23 km.

A comparison of the pattern of structural contours of the C- and M-discontinuities shows that structures developed at different depths conform with each other. Yet some exceptions are also observed. In the eastern part of the Kola Peninsula, where the depth to the M-discontinuity does not exceed 37 km, at places 28 – 30 km, the C-discontinuity is at a level of 15 – 18 km. However, in the western part, where granulites of the Lapland block and sedimentary-volcanogenic rocks of the Pechenga graben syncline are developed, this boundary lies only at a depth of 8 – 12 km from the present surface, with the M-discontinuity established at a depth of 40 km.

Calculated gravimetric and seismic data suggest that the crust above the C-discontinuity is a layer of non-uniform density. The reason for this is that the $\delta$-, $\gamma$-, and $\alpha$-layers composing the integrated basaltic layer are discontinuous and only local. Rocks of the basaltic layers constitute magmatic, metamorphic, and sedimentary volcanic complexes. These layers have been distinguished conventionally.

The three types of block with different sets of layers on the Kola Peninsula are as follows:

1. the "dioritic" layer comprises charnockite and crystalloschist complexes. The first of them is regarded by the authors as a complex composed of the oldest rocks which could ever have been studied in geological history. The thickness of the overlying "granitic" layer within the limits of such essentially simatic blocks is very small, at some places this layer is fully eroded (Lapland-Kolvits block);
2. blocks of the Archean structural zones with the two-layered essentially sialic type of crust; the thickness of the "granitic" layer reaches 10 – 18 km (these blocks are most widely spread over the shield);
3. blocks of the karelides notable for small thickness (7 – 9 km) of the "granitic" layer. At some places the sedimentary-volcanogenic rocks directly pass into the "granulitic-basic" layer (the Colour Belt of the Kola Peninsula).

The information obtained in the Kola superdeep borehole did not confirm the previous theory of the existence of a direct transition between karelides and the granulitic-basic layer, or the unconfirmed theories which served as a basis for the very detailed subdivision of the basaltic layers in the Precambrian crystalline rocks of the crust, taking gravity and seismic data into consideration.

### 1.1.3 Geology of the SG-3 Well Site Area

The north-western part of the Murmansk region, covering the Pechenga area with the Kola superdeep borehole, is composed of old crystalline rocks characterized by various grades of metamorphism. These rocks represent formations of Precambrian complexes developed on the Baltic Shield (Fig. 1.8).

Geological study of the Pechenga area was pioneered by S. A. Konradi, A. A. Polkanov, G. Veyrinen, G. Khauzen, P. Eskola, and later continued by G. I. Gorbunov, L. Ya. Kharitonov, N. Ya. Kuryleva, G. T. Makienko, L. I. Ivanova, Yu. N. Utkin, N. A. Eliseev, V. A. Maslenikov, K. D. Belyaev, L. I. Uvadyev,

V. G. Zagorodny, S. N. Suslova, D. D. Mirskaya, I. V. Litvinenko, E. A. Polyak, M. A. Gilyarova, A. A. Predovsky, G. G. Duk, V. I. Kazansky, A. A. Kremenetsky, V. P. Petrov, O. A. Belyaev, E. D. Chalykh, and others.

Geologists have displayed a permanent interest in this area, because it was notable for its copper-nickel deposits which are confined to ultrabasic intrusions (comagmatic with volcanites) and occurred between sedimentary rocks of the Proterozoic complex. Therefore, exploration work, as well as other investigations, concentrated on the zone of sedimentary rock considered to be the most promising in the search for copper-nickel ores. The geological structure of the Archean complex, composing the folded basement of the area and fringing the Pechenga structure from all sides, has been studied to a much lesser extent.

The Pechenga area is a north-western continuation of the Central Kola geotectonic synclinorium zone. In the north and south, this zone is limited by major longitudinal faults (Kolvits-Lapland and Keiv-Uragub); towards the north-western direction it narrows, and in the Pechenga area it is of the least width in the map (see Fig. 1.6). At this one point, all the major longitudinal faults of the peninsula come together, being contiguous with each other. This remarkable fact is still in need of explanation, although some researchers note that the Pechenga area and adjacent regions of Finland, Lapland, and Norway are a distinctive center from which all the most important Precambrian structures of the eastern and central parts of the shield radiate in a fan-like pattern in south and south-eastern directions.

Among the Precambrian formations developed in the area under consideration, the following complexes of different age are defined: (a) the Archean complex of the folded basement; Archean rocks compose tectonic blocks formed of folded-dome tectonic structures; (b) the Proterozoic complex composed of volcanic-supracrystalline strata, spread over the north-western part of the Kola-Karelides zone. Each of these complexes is remarkable for particular structural features, metamorphism, magmatic activity and presence of ores.

## The Archean

The lower structural stage of the region described is composed of the oldest rocks of the Kola gneiss series. At the surface these gneisses fringe on the Pechenga graben-syncline.

To the Kola series are assigned various strata comprising suites with numerous local names. A general feature of all known stratigraphic schemes is that the lower part of the sequence is represented by suites composed predominantly of biotite-plagioclase, biotite-amphibole-plagioclase gneisses, and amphibolites, while the upper part begins with predominantly biotite-plagioclase gneisses containing highly aluminiferous minerals.

In the area of the SG-3 well site two layers are distinguished in the Kola series (from the top), namely a layer of biotite-plagioclase, biotite-amphibole-plagioclase gneisses and amphibolites and a layer of biotite-plagioclase, muscovite-biotite-plagioclase gneisses with highly aluminiferous minerals (staurolite, andalusite, sillimanite, garnet, kyanite) and rare layers of amphibolites.

The lower layer is composed of uniform biotite-plagioclase gneisses with subordinate biotite-amphibole-plagioclase gneisses, as well as of bodies of amphibolites (persistent along the strike and occurring conformably), whose thickness varies from a few meters to a few tens of meters. Amphibolites can be traced along the strike at a distance of several kilometers and can be mapped with the help of aerial photographs. At places, amphibolites are encountered, together with thin lenses of meta-ultramafic rocks. The amount of amphibolites and biotite-amphibole-plagioclase gneisses in the section varies broadly.

The presence of layers of magnetite-garnet-pyroxene-quartz-feldspathic schists and ferruginous quartzites is characteristic of the lower layer. The thickness of these rocks normally does not exceed 20 – 30 m. They are mostly confined to the upper contact zone of the layer, where it is overlain by aluminiferous gneisses. However, conformable contact is not the only mode of occurrence; in some rare instances these distinctive rocks cross-cut gneiss formations.

The upper layer of the Kola series consists of biotite-plagioclase and bimicaceous garnet-biotite-plagioclase gneisses, at places with staurolite, andalusite and kyanite. Some layers of amphibolites are also encountered. The upper layer dips gradually into the lower one, no boundary between the layers being visible. Biotite-plagioclase gneisses with aluminiferous minerals (garnet, staurolite, andalusite, kyanite) are the most widespread rocks of this layer. Cordierite is also found at those places where rocks were subject to relict granulitic metamorphism. Rocks are usually characterized by distinct banding; bands with higher and lesser content of biotite alternate with each other. In some outcrops the exposed sections are of rhythmic layering. These rhythms consist of two parts. The lower rhythms are composed of quartz-feldspathic material and the upper ones are enriched in biotite and garnet. Rock contacts within rhythms (from bottom to top) are not clearly seen, while boundaries between rhythms themselves are sharply defined. Such rhythmic stratification is preserved in the form of metamorphic banding, at places becoming more pronounced due to the presence of conformable quartz-feldspathic streaks. Banding and gneissose occurrences are conformable with bedding.

The Kola series in the area of the SG-3 well site forms a distinctive tectonic structure, i.e., a field of gneiss domes. It is developed both to the north and south of the Pechenga graben syncline. The gneiss field consists of conjugated domes. The majority of the domes are elongated in north-western direction. They are 20 – 25 km long and 10 – 15 km wide. The smaller domes are 2 – 5 km in diameter and have a notable isometric shape. The marginal parts of the domes are complicated by linear isometric, inclined and overturned folds, whose width ranges from several hundreds of meters, to a few kilometers. At places these comparatively large folds are crumpled and even form microplications. In the central parts of the gneiss domes, rocks dip at low angles, while at marginal parts the angles of dip are usually steep (60° – 70°). The central parts of the gneiss domes are composed of granites and migmatites, which have inherited the orientation of the enclosing rocks.

The gneiss layers are irregularly granitized and form fields of migmatites and gneiss-like plagiogranites. Additionally, comparatively uniform bodies of granites, crossing the Kola series rocks, are also found. Fields of aplites and pegmatites are distributed irregularly and tend to occur at those places where adjacent gneiss domes are in conjunction with each other. Aplite and pegmatite bodies are of lense- or bed-like shape. They have a thickness of several meters and sometimes reaches 30 m. These aplite and pegmatite bodies either rest conformably on other rocks or cross them, the contacts between these bodies and the enclosing rocks being sharply defined. Normally, small bodies are composed of aplites and larger ones consist of aplites and pegmatites.

In the area under consideration the Kola series rocks are metamorphosed mainly in the amphibolite facies. However, in the course of detailed surficial investigations, some zoning in facies changes was also observed with respect to the Pechenga trough. For instance, in the Western Nyassyuksky block, immediately abuting the Pechenga structure from the north, the metamorphic paragenic mineral associations correspond to the epidote-amphibolite and amphibolite facies. In the remote Eastern Nyassyuksky block, some rare relict associations of the granulitic metamorphism facies are observed among the amphibolite facies. In areas north-east of the Pechenga structure, in the Liinakhamara block, the granulite associations are represented by wider varieties; they are additionally superimposed by amphibolite and in places by greenstone metamorphic facies.

In zones of recurrent tectonic activity, manifested during different geological periods, the supracrystalline formations were either transformed into various schists or gneisses; this process was accompanied by diaphthoritic mineral changes. The lower boundary of the Kola series has not been established.

## The Proterozoic

The upper structural stage is composed of the Proterozoic sedimentary-volcanic rocks. The sedimentary-volcanic rocks of the Pechenga and the similar Varzug complexes, together with the Tundra series adjacent to the former from the south, compose a large tectonic depression which began to develop on the solid Archean basement. They are traced discontinuously from north-west to south-east diagonally throughout the Kola Peninsula.

As in the Pechenga region, in the northern part of the Imandra-Varzug regions, the Proterozoic sedimentary-volcanogenic rocks are in direct contact with the Archean gneisses of the folded basement, and they lie transgressively over conglomerates present at the base.

The southern contact zone (the so-called "southern flank") of this principal Proterozoic structure of the Kola Peninsula, notable for the long time of its development, is of a totally different structure. In the Pechenga and Imandra-Varzug regions, in an area where the comparatively low-metamorphosed sedimentary-volcanic Proterozoic rocks fringe the subplatformal rocks developed in the south, a narrow zone composed of rocks of the so-called Tundra series (Late Archean – Early Proterozoic) is defined. From geological and geophysical data it appears that linear tectonic dislocations (Pechenga Varzug major fault) which have great displacements and lie close to each other are also found in this zone.

*Tundra Series.* Rocks of the Tundra series are only locally developed in the Pechenga region, and are mainly distributed on the southern rim of the Pechenga structure. The northern boundary of this structure coincides with the zone of the Poryitash major faults (western end of the Pechenga-Varzug major fault); the configuration of the southern boundary is closely linked with tectonics of the blocked Archean basement. In addition to gneisses and amphibolites, agmatites are also widely spread within an area consisting of Archean rocks which adjoins the Proterozoic structure in the south. Agmatites consist of amphibolite fragments cemented by granitoid material. In a number of places (Kaskama Mount) it was established that rocks of the Tundra series rested with angular unconformity on Archean gneisses. Speaking generally, contacts strike in a north-western direction, although longitudinal folds intervene between blocks throughout the trend length. These blocks are shaped like narrow wedges and composed of granitoids. Tundra series rocks, about 2 – 3 km thick, occur at steep, frequently almost vertical angles, oriented mainly in a southern direction.

The Tundra series is subdivided into two suites, the Kaskam and the overlying Talyin. The Kaskam suite is represented by amphibolites, amphibole-plagioclase, biotite-amphibole-plagioclase schists in diabases, amygdaloidal diabase and andesites. These rocks occur in the lower and upper parts of the suite. Its middle part consists of biotite, epidote-biotite and biotite-amphibole gneisses, which are usually encountered together with garnet, and sometimes with magnetite and rare streaks of amphibolites. Amphibolites are notable for their high content of alumina ($Al_2O_3$ reaches 20%).

The Talyin suite occurs in small synclines which are confined to the northern and southern parts of the area composing the Tundra series. It is composed of alternating two-mica and biotite gneisses with garnet and sometimes with sillimanite and kyanite, as well as of micaceous-actinolitic schists. Gneisses, and most probably a part of the Tundra series schists are of tuffitic volcano-sedimentary origin. Among volcanic and sedimentary rocks, metamorphosed mainly in the epidote-amphibolite facies and partially granitized, there are some cross-cutting amphibolite bodies whose composition is similar to that of the aluminiferous basalts; these bodies are considered to be comagmatites of the volcanites present in this series.

In the Imandra-Varzug area, as in the Pechenga region, rocks of the Tundra series are developed (in the form of a narrow strip) only on peripheral southern flanks of the Proterozoic sedimentary-volcanic complexes. Any attempts to find rocks of the Tundra series on the northern flanks have ended unsuccessfully. The restricted occurrence of these rocks is most probably explained by the fact that rocks of the Tundra series, evidencing synclinal origin during the Late Archean – Early Proterozoic, developed only within a comparatively narrow axial part of the depression.

Origin and early growth of this structure seem to be related to the formation of the ultrabasite complex of the dunite-harzburgite formation. Such a complex, developed south and east of the Pechenga synclinorium in zones of latitudinal (Rovnin block) and north-western strike, is composed of bed-like and lens-like bodies 1000 m in length and 200 – 300 m thick. These bodies are composed of metamorphosed, boudinaged harzburgites, and appear to have been affected by granitization. It seems most probable that gabbronorites of the Mount Genralskaya type (pyroxene-gabbro formation) also appeared during this stage of regional evolution. Similar gabbronorites were also met in basal conglomerates of the Pechenga sedimentary-volcanic complex (Lower Proterozoic).

*Pechenga Complex.* The Lower Proterozoic sedimentary-volcanic rocks, developed in the central part of the region, constitute the Pechenga graben syncline. This occupies an area of 2300 km$^2$ and stretches 310° north-west at a distance of 70 km. It is asymmetrical in shape and its convex part faces north-east. Rocks dip southwards to the center of the syncline at angles of about 30° – 50°.

In the south-west the graben syncline is truncated by a zone of the deep-seated Poryitash major faults. The width of this linear zone, extending for 60 km, does not exceed 6 – 8 km. It strikes consistently towards the north-west (310° – 320°); rocks and cleavage also plunge consistently southwards with angles of depth ranging between 40° and 80°. Inside a tectonic wedge, rocks consist of the multiple-intermitted members of phyllites, siltstones, sandstones, and tuffs of basic and acid composition, metadiabases (mainly of tholeiitic composition) and numerous bedded bodies of andesite-dacite porphyrites. As a rule, rocks everywhere are strongly foliated and metamorphosed in the epidote-amphibolite facies.

The graben syncline is composed of well-stratified sedimentary-volcanic rocks of the Pechenga complex, in which four complete macrorhythms are defined, each rhythm beginning with sedimentary and ending with volcanogenic strata (Zagorodny et al. 1964). In the north, at the base of the Pechenga structure, the first sedimentary strata of polymictic coarse-grained conglomerates occur. At places they pass (up the section) into gritstones and arkose sandstones. Conglomerates transgressively overlap rocks of the Archean basement. Their thickness varies over a broad range from some tens of meters in outcrops to 200 m in an observation borehole located in the north-eastern part.

The first volcanogenic rocks continuously extend in the form of a strip along the northern periphery of the Pechenga structure. Their thickness reaches 1000 – 1300 m. They are composed of altered diabases and associated metaplagioporphyrites. The relatively low differentiation and paleotype appearance are usually characteristic features of these rocks. On the northern part of the structure, near lower contacts, the rocks at the surface are transformed into amphibolites and biotite-amphibole schists.

The second sedimentary strata rest conformably on effusive rocks; they are of comparatively small thickness, reaching 100 m. These strata extend as a continuous band along the northern flank of the structure and consist of (from bottom to top) quartzites and quartzite-sandstones, dolomites and dolomitic limestones. The process of sediment differentiation proceeded here in more perfect form that in the first sedimentary strata, and has consequently resulted in the formation of a distinctly stratified sequence.

The second volcanic rocks are composed of metadiabases, meta-andesites, metaporphyrites, and meta-amygdaloidal rocks.

This suite is of complex structure because its rocks alternate very frequently and the rock types are very varied. Increased magnetite content is the characteristic feature of volcanites. This allows identifying them very clearly at the surface with the help of magnetometers. The thickness of these rocks is not uni-

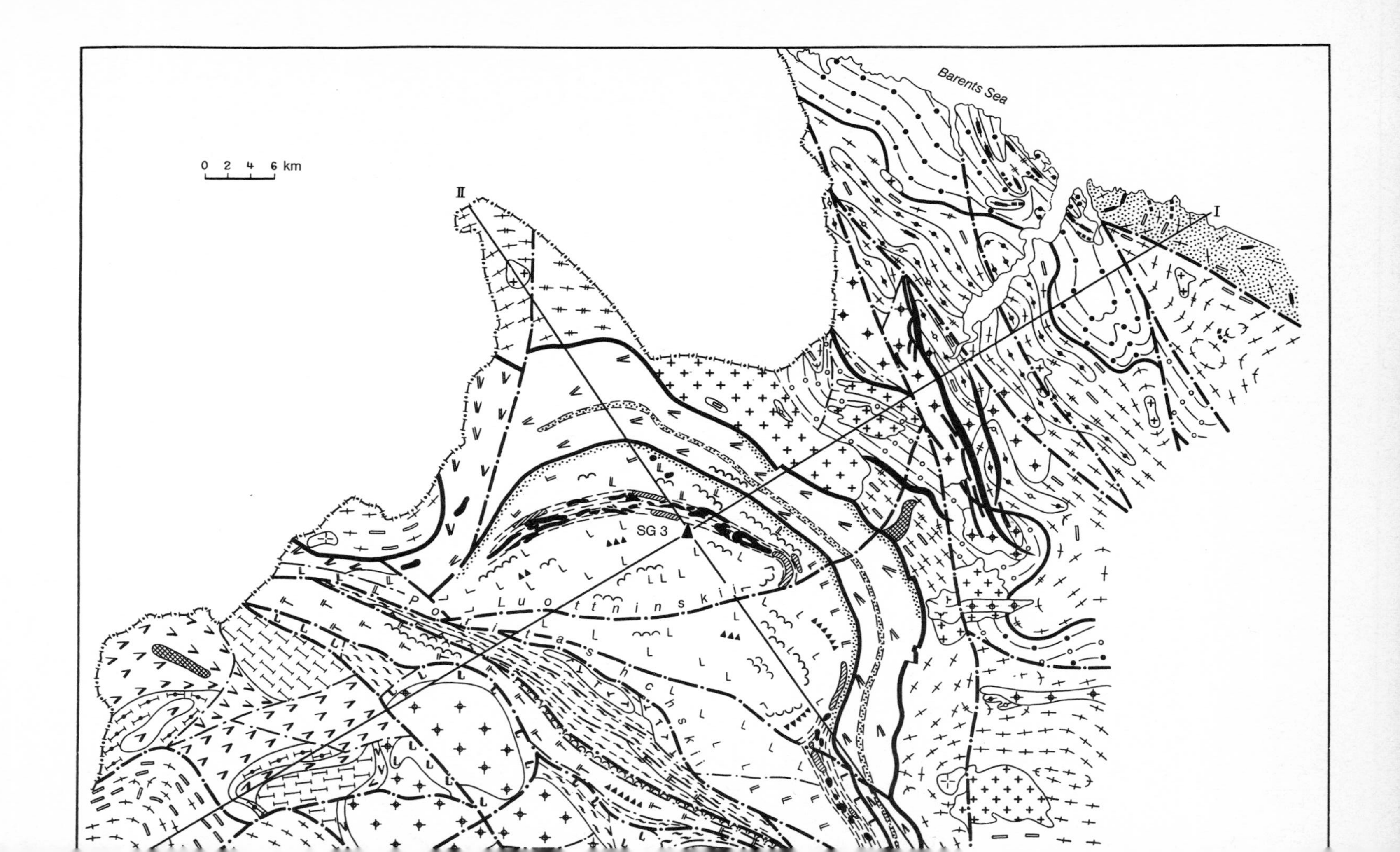
Barents Sea
0 2 4 6 km
I
II
SG 3
Luottninskii
Poritashchskii

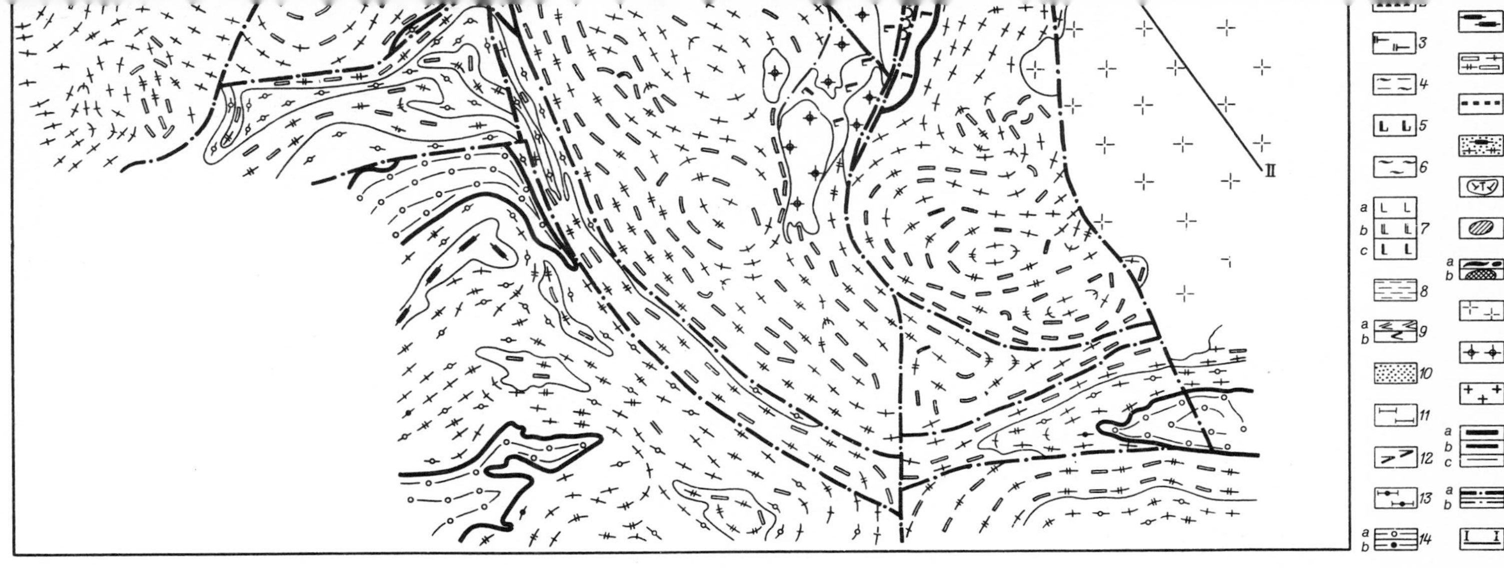

**Fig. 1.8.** Geological map of the Pechenga region. (After V. S. Lanev, E. B. Nalivkina, M. S. Rusanov, S. N. Suslova). *Pechenga complex* (*Lower Proterozoic*): *1* tuffs and sedimentary tuffogenic rocks of different composition and schists; *2* volcanites of basic-hyperbasic and hyperbasic composition and schists in them. *Southern zone, Poryitash stratum – satural tectonic block*: *3* andesite-basalt, basal metaporphyrites – schists and actinolitic amphibolites in them; *4* coaly tuffites, phyllites, siltstones, sandstones and schists in them; *Ruossel stratum – block of granitoid domes*; *5* amphibolites and amphibole-plagioclase schists in tholeiitic basalts; *6* various gneisses and schists with coaly matter in sedimentary and tuffogenic-sedimentary rocks. *Northern zone, Nickel series*: *7* metamorphosed tholeiitic basalts (*a* diabases; *b* metadiabases and greenschists; *c* amphibolites and amphibolite-plagioclase schists); *8* coaly phyllites, tuffites, siltstones, sandstones; *Luostari series*: *9* metamorphosed trachybasalts, trachyandesites, andesite-basalts (*a* metadiabases, metaleucodiabases, meta-andesites, greenschists; *b* amphibolites, amphibole-biotite-plagioclase and biotite-plagioclase schists); *10* metasandstones, metaconglomerates, quartzite-sandstones, dolomites; Tundra series (Lower Proterozoic-Upper Archean): *11 Talyin suite* – mica-chlorite-actinolite-feldspathic and mica-quartz-feldspathic schists and gneisses, frequently with garnet, *12–13 Kaskam suite*: *12* amphibolites, frequently with garnet; *13* biotite-garnet-amphibole-feldspathic schists and gneisses. *Kola-Belomor complex* (*Archean*). *Kola series, upper stratum*: *14* biotite-plagioclase gneisses with HAM (*a* without cordierite; *b* with cordierite); *middle stratum*: *15* biotite-plagioclase and amphibole-biotite-plagioclase gneisses and migmatites; *16* biotite-plagioclase gneisses with HAM (*a* without cordierite; *b* with cordierite); *17* amphibole-biotite-plagioclase gneisses with pyroxene and pyroxene-amphibole-plagioclase crystalline schists; *18* amphibolites and amphibole crystalline schists; *19* quartz-magnetite schists; lower stratum (base complex): *20* amphibolites, pyroxene-amphibole-plagioclase schists, gneisses, gneissose granites, charnocites, enderbites. *Intrusive rocks*: *21* andesite-dacite (diorite); *22* diabases and gabbro-diabases; *23* differentiated basic-hyperbasic rocks (*a* gabbro-wehrlites; *b* gabbro-norites); *24* microcline granites; *25* microcline-plagioclase granites, gneissose granites and diorites; *26* plagiogranites and migmatites; *27* geological boundaries of (*a* complexes; *b* strata, series); *28* main tectonic dislocations; *29* cross-section lines of the block diagram

form, and varies from 1000 m in the central northern part to 100 m on the flanks of the structure.

The third volcano-sedimentary sequence consists of layered sandstones (greywackes, arkoses) and siltstones in lesser amounts. The overlying dolomitized limestones in the uppermost part of the succession change into tuffites. The stratum is comparatively uniform in thickness, reaching 100 – 150 m.

The third volcanic strata are composed of actinolitic diabases, globular lava bodies, tuff breccia, and greenstone schists in diabases. Their total thickness is about 2000 m in the central northern part and 300 – 700 m on the western and eastern flanks.

The fourth tuffitic-sedimentary strata are the thickest, and are very persistent along strike. They rest on the third volcanic strata and comprise nickeliferous intrusions of basic and ultrabasic rocks; due for this reason they are called productive strata. Their true thickness is 1000 – 1500 m in the central part and 200 – 400 m on the flanks. The strata are composed of phyllites, siltstones, and sandstones with locally developed thin beds of conglomerates and limestone lenses.

The fourth volcanogenic strata are spread over the most extensive area and compose the core of the structure. In the north they overlie the third volcano-sedimentary strata and in the south, along a zone of the Poryitash regional major faults, they are in contact with various rocks of the southern flank. They are also subdivided into four structural stages (taking into account detailed data obtained at the surface). The thickness of the lower stage is about 800 – 1200 m. The alternation of globular and massive basalt lava sheets is a notable feature of the section. Picrites and pyroxene porphyrites form beds (up to 40 m thick) which are confined to the middle part of this stage. The presence of numerous layers of tuffites, tuffs of basic composition (ultrabasic in rare cases), agglomerate tuffs, and hyaloclastics is the characteristic feature of these strata. The thickness of the second, most differentiated, stage is 1200 – 1500 m. This stage is composed of globular and massive lava sheets of basic composition; thin sheets of ultrabasic lava are also present. Layers of medium-acid tuffs of about 50 – 100 m were also distinguished, together with tuffs of basic composition. The third and fourth stages are composed of similar series of rocks, i.e., massive and globular basalt-like lavas, sheets of hyaloclastics, rare tuff layers of predominantly basic composition. According to seismic data, the aggregated thickness of the whole sequence in its central part is about 3.0 – 3.5 km.

The core of the Pechenga graben syncline is of complex autonomous structure and, in the researcher's opinion, it consists of two cup-like brachysynclines of the second order separated by a narrow anticline; its crest line is complicated by a major fault. The geological and geophysical investigations carried out recently showed volcano-tectonic caldera-like depressions in the upper volcanic stages. The sizes of structures vary between 1 and 2 km. They are composed of three to five rather thick sheets of globular lavas of centroclinal type. These lava sheets fill the central cylinder-like area, which sinks along a system of concentric block faults; they also comprise a group a circular and conic dikes of gabbro-dolerites and volcanic-vent tuff agglomerates. These rocks are considered to be the first indication found among effusive rocks of the Kola Peninsula that in the geological past paleovolcanoes were represented not only by volcanoes of fissured type but also by those of conic type.

Volcanites of the strata in the graben syncline are characterized by the presence of well-preserved primary-magmatic textures and mineral associations (prehnite-pumpellyite-facies metamorphism). However, rocks become more altered on the eastern flank, i.e., in the eastern direction; outside the Lammass major fault they become metamorphosed in the greenschist facies, and outside the Kuchintundra major fault in the epidote-amphibolite facies. These very abrupt changes support the idea that vertical displacements along internal major faults of the Pechenga structure were of great throw.

All the rocks underlying volcanites of the fourth strata were metamorphosed to various facies (polyfacially metamorphosed): in the central part of the structure they are altered at the surface in the greenschist facies, while in the eastern part they are predominantly metamorphosed to the epidote-amphibolite facies.

On studying the very thick volcanic rocks of the Pechenga complex, separated from the adjacent volcanites by sedimentary rocks and frequently by a crust of weathering or by surfaces of unconformity, it becomes fairly certain that every layer is an independent formation which can be identified (from bottom to top) as follows: molasse-andesite basalt, carbonate trachy-basalt, tholeiite basalt and terrigenous tuffitic picrite basalt (Predovsky et al. 1974).

Intrusive bodies occurring in the stratified rocks of the Pechenga graben syncline are assigned to three groups.

The early intrusions are composed of gabbro-diabases; they occur in different sedimentary-volcanic rocks which compose the graben syncline. Diabases are most frequently met in the fourth volcano-sedimentary strata. All intrusions form formational bodies whose length reaches 10 km, the thickness being about 200 – 300 m.

The next intrusive phase is represented by complexes of basic and ultrabasic rocks. In the majority of cases peridotites and olivinites, usually greatly serpentinized, as well as pyroxenites, gabbro and chlorite-tremolite, and carbonate-talc schists associated with ultrabasites occur in volcano-sedimentary rocks of the productive strata. Complexes of this group are represented by the formational intrusive bodies extending for a distance from 50 to 6 km and with a thickness of between 2 and 30 m, at places 200 – 300 m. Sometimes intrusions are differentiated, with the recumbent flank of the complex normally being composed of serpentinized peridotites, and the hanging flank by gabbros. Peridotites and gabbros are separated by pyroxenites. In many instances, there are individual differences in the independent thin complexes. All the known copper-nickel deposits of the Pechenga structure are associated with ultrabasic rocks belonging to this group of intrusions.

Intrusions of late origin are represented by dioritic (andesitic) porphyrites which compose formational hypabyssal intrusions. These intrusions occur predominantly in volcano-sedimentary rocks of the so-called southern flank.

A complex of rapakivi-like granites is attributed to the post-Pechenga rock units, although all of them were formed at approximately the same time. Rocks of this complex are confined to a zone of the north-eastern major fault, which crosses the Archean rocks east of the Pechenga graben syncline. The age of these rocks, inferred from K-Ar data, was determined as being 1900–1610 Ma. The rocks are of complex structure; they are mainly composed of granites and granodiorites.

At the end of the magmatic cycle, rocks of the dike complex were formed; these are represented by diabases, fourchites, gabbro-porphyrites. Dikes of these rocks cross-cut the entire sequence of the graben syncline composed of sedimentary, volcanogenic, and intrusive rocks. At the end of the cycle hydrothermal quartz, carbonate, sulfide, and other veins were formed.

The Pechenga graben syncline is subdivided into a marginal part, which replicates its outward configuration, and an internal part, where some autonomous local structures are developed (the fourth volcanic strata). The shape of the marginal part is well defined because sedimentary and volcanic rocks alternate in the sequence. Rocks dip towards the internal part of the syncline (centroclinal dipping), the angles of dip being about 20° – 30° in the lower strata and becoming steeper (30° – 60°) in the upper sedimentary-volcanic strata.

Rocks of the Pechenga complex, particularly those of sedimentary origin, form folds of the second, third, and higher order; their width is from 3 km to 10 m.

It should be mentioned that strike-slip faults played a considerable role at all stages of the formation of the Pechenga structure. These dislocations were formed during several stages, two of which are assigned to the Proterozoic. All faults can be subdivided into the following groups: (1) marginal deep-seated major faults; (2) main intraformational major faults; (3) longitudinal faults; (4) transverse faults.

The Poryitash major fault is attributed to marginal faults. It extends for a distance of 60 km and subdivides the Proterozoic rocks of the region into two structural zones. This fault has dips between 60° and 80°. Subvolcanic intrusions of large andesite porphyrites are confined to this major fault.

The following major faults are considered to be the largest intraformational dislocations: the Western, Luotin, Kuetsyarvi, Lammas, Kuchintundra, etc. They have played a leading role in the formation of internal structures of the region, and serve as boundaries of main tectonic blocks. The displacement of these subvertical dislocations (reverse faults and normal faults), arranged radially to the Pechenga structure, has not been established, yet it seems to be considerable and to reach at least several hundred meters, taking into account the amount of horizontal displacement of rocks.

Additionally to these major faults, a number of faults of the transverse type (normal and reverse faults) are distinguished everywhere within the limits of the Pechenga structure. These dislocations were studied most thoroughly within the area of productive strata where displacements along some of these dislocations are about 30 – 50 m.

Longitudinal dislocations (overthrusts, wrench faults) were studied most thoroughly also within an area of productive strata. Here they are predominantly spread out along contacts between rocks of different physical properties. Along the strike they are conformable with rock occurrence, inherit the configuration of large folds and truncate small bends of beds.

It has been established that transverse normal and reverse faults are of later origin than folds and ores, while the longitudinal wrench faults, or at least some of them, should be considered as the pre-ore dislocations and, consequently, may serve as the most important ore-controlling structures.

### 1.1.4 Principal Results of Deep Geophysical Investigations in the Area of the SG-3 Well Site

The first studies of ancient shields of the Earth's crust by deep seismic sounding (DSS) were undertaken in the Soviet Union on the Baltic Shield, in the north-western part of the Kola Peninsula (Barents Sea – Pechenga – Lovno profile), in 1960 – 1963 (Fig. 1.9).

It was established that the thickness of the Earth's crust, identified on the Barents Sea – Pechenga – Lovno profile, which crosses the Pechenga structure along the Pechenga river (12 km east of the SG-3 well), was about 40 km. This thickness of the crust is characteristic of continents. Gentle seismic boundaries were distinguished in the crust and its block structure was suggested. Later, these results proved to be characteristic of the Baltic Shield as a whole. In former times, interpretation of the DSS data was based on the widespread idea that the Earth's crust might be conceived as a model consisting of a few thick layers, and that wave velocity abruptly changes at the boundaries between these layers and then increases with depth.

In this section of the crystalline-shield crust (south of the Rybachii Peninsula), the following seismic boundaries have been established (from top to bottom):

1. The upper boundary, distinguished with the help of both refraction and reflection waves, was detected at a comparatively shallow depth (7 – 8 km) in areas of gneiss development and beneath the Pechenga sedimentary-volcanic complex. In the southern part of the profile, in an area of the granulite-complex development, this boundary is at a depth of 3 – 5 km. Variations in velocity

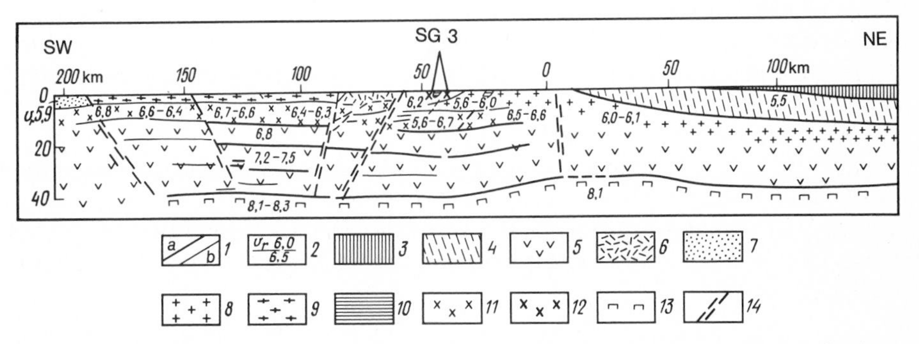


**Fig. 1.9.** Seismic cross-section of the Earth's crust along a profile Lovno-Pechenga-Barents Sea (eastern part of the Baltic Shield 1975). *1* average position of key (*a*) and other (*b*) seismic boundaries; *2* velocities (*above the line* average; *below the line* boundary, km s$^{-1}$); *3* Phanerozoic sedimentary rocks; *4* Riphean sedimentary rocks; *5* lower part of the "basaltic" layer; *6* crystalline schists and amphibolites; *7* gneiss complexes; *8* gneisses, granite-gneisses, granites; *9* granulites; *10* basic and ultrabasic intrusions; *11* middle nonuniform part of the basaltic layer; *12* Pechenga sedimentary-volcanogenic series; *13* upper mantle; *14* major faults

values observed along this interface are an important feature of the latter. It is interesting that values of $V_p$ do not depend on the depth of occurrence of this interface. It seems more probable that there is a relationship between $V_p$ and the composition of rocks outcropping in the area along the above-mentioned profile, i.e., $V_p$ increases beneath complexes of rocks with higher density values. For instance, in the southern part of the profile, beneath the granulites, $V_p = 6.4-6.7\ \mathrm{km\ s^{-1}}$; beneath the Kola gneisses and rocks of the Tundra series, $V_p$ falls to $6.0-6.4\ \mathrm{km\ s^{-1}}$; and beneath the central part of the Pechenga complex, $V_p$ again increases to $6.4-6.6\ \mathrm{km\ s^{-1}}$.

2. The seismic boundary, characterized by $V_p$ values of about $6.8\ \mathrm{km\ s^{-1}}$, is traced at somewhat higher levels in a southern direction. Its depth changes from 15 to 8–9 km.
3. The seismic velocity boundary at a depth interval of 19–25 km also rises in a southerly direction. The value of $V_p = 7.2-7.5\ \mathrm{km\ s^{-1}}$ for this boundary is only assumed; from this it seems that the fixed boundaries may be considered as being represented only by reflection but not by refraction horizons. Apart from the described subhorizontal layering of the Earth's crust, established in the course of seismic investigations conducted along this profile, deep-seated tectonic faults have been also encountered. These faults fringe in the south and north the granulitic complex additionally to the steeply dipping Pechenga-Varzug and Keiv-Uragub fault zones.

Thus, on the basis of seismic data it was previously conceived that the aggregated thickness of sedimentary-volcanic rocks in the central part of the Pechenga trough reached 7–8 km and that there was in fact no place for a typical "granitic" layer in the crustal sequence.

This last argument allowed Zhdanov (1966) to distinguish two types of the crust in the north-eastern part of the Kola Peninsula. The Pechenga structure belongs to the first type. According to Zhdanov's view, the "granulitic" layer on the Pechenga structure was displaced by thick effusive strata of basic composition and then brought down to a zone of the physicochemical stability of the "basaltic" layer. The second type of crust is represented by the Lovnin block composed of granulites; it is located about 100–200 km south of the SG-3 well, where disappearance of the "granitic" layer is related to its long-lasting rising from a great depth and to subsequent erosion.

The other 13 seismic profiles (reflection survey method) were run prior to the SG-3 well drilling with the purpose of studying the geological sequence of the Pechenga structural zone to a depth of 10–15 km. The work done allowed a more detailed interpretation of the Pechenga structure at depth, and arrived at the following conclusions:

1. The Pechenga structure, composed of sedimentary-volcanic rocks, is an asymmetrical trough. In the near-surface northern part of the trough rocks plunge in centroclinal directions at angles of 30°–60°, whereas at deep horizons rocks dip at low angles and sometimes occur almost horizontally.
2. Seismic marker horizons are represented by contact surfaces of rock layers which differ sharply from each other by velocities of the elastic waves. The fol-

lowing marker layers are distinguished: the bottom of the first volcanic layer, sedimentary rocks of the third strata, and the Pechenga complex.

3. Data on the deep structure of the Archean rocks are very contradictory and not always compatible with geological observations; this is explained, first of all, by the insufficient study of the Archean complex.

Results of gravimetric survey on a 1 : 50,000 scale carried out in 1967 – 1969 not only confirmed the basic conclusions as to the deep structure of the Pechenga graben syncline obtained during seismic investigations, but made these conclusions more detailed and introduced some corrections.

Thus, the Kola superdeep borehole was drilled with the aim of throwing light on the wide range of debatable problems in the field of geology, geophysics, and ore-prospecting of the eastern part of the Baltic Shield, and also of elucidating the same problems on a larger scale, i.e. problems associated with the Precambrian continental crust. These problems are summarized as follows:

1. The stratigraphic subdivision of the Proterozoic and, particularly, Archean complexes. According to the project, the borehole was planned to run through the entire Lower Proterozoic sedimentary-volcanic strata of the Pechenga complex. Being the deepest geological mine in rocks of the Archean basement, the well had to establish age changes in the sequence, reveal the nature of stratification and determine the thickness of strata in old complexes of the Earth's crust.
2. The internal structure of sedimentary-volcanic stages (widely spread over the Baltic Shield) "built on" the Archean basement.
3. The evaluation of prospects of deep horizons of the known ore fields of the region.
4. The time sequence of metamorphic processes in the Precambrian period and relationship between these processes and depth of rock occurrence.
5. Vertical zonation of ore mineralization and evaluation of a depth down to which the low-temperature hydrothermal mineralization is developed.
6. Decrease of rock fissuring at depth, transformation of rock structures, permeability of massives, etc.
7. The question of the emanation of gases from older rocks of the crust. This question is linked with the foregoing problem.
8. Change of paleo- and recent temperature gradients on the Baltic Shield with depth increase.
9. Lithostatic pressure (rock pressure) and its bearing on physical parameters of rock systems with depth increase.
10. Structure of deep geophysical intersurfaces in the crust and elucidation of their nature.

## 1.2 Geological Section of the Well

The integrated study of core and near-borehole space permitted characterization of the section of the Kola superdeep well. Since the Kola well was drilled with the open-hole method, and the diameter remained the same throughout the entire well depth, quality, accuracy, and correlation of geophysical data, obtained at different times and under extreme conditions were increased and greatly improved. In the course of the investigations geological, geophysical and geochemical data were collected and correlated and a series of geological, geochemical, and geophysical sections were constructed. These sections were prepared to a scale from 1:200 – 1:500 to 1:5000 – 1:25,000 (generalized scale) (Figs. 1.10, 1.11, 1.12, 1.13).

As already mentioned, the SG-3 well was planned to run through the sedimentary-volcanogenic rocks of the Pechenga complex, enter the Archean gneisses at a depth of 4700 m and encounter rocks of the high velocity granulite-basitic layer at a depth of 7 – 8 km below the surface. In fact, it turned out that the section penetrated by the well differed from the one anticipated (Fig. 1.14).

### 1.2.1 The Pechenga Complex

The Proterozoic (Pechenga) complex was encountered by the well at a depth interval between 0 and 6840 m. It is represented by rhythms of alternating sedimentary and volcanic rocks with subordinate comagmatic bodies of gabbrowehrlites and by formational intrusions of gabbrodiabase and dacite-andesite porphyrites (Fig. 1.15).

The Pechenga sedimentary-volcanic complex is subdivided into two series, the Nickel and Luostari series. The Nickel series, overlying the Luostari series over an extensive area, is the most representative rock unit developed at the surface. Its vertical thickness in the well is 4884 m. Analysis of data has shown that in the course of seismic profiling by the reflection wave method, the bottom of the Nickel series was erroneously regarded as basal rocks of the Pechenga structure. This predetermined a considerable error in anticipating the depth to the bottom of the Proterozoic sedimentary-volcanic strata. At the present time, taking these facts into account, it is possible to calculate the full vertical thickness of the Nickel series with sufficient accuracy.

It is interesting to note, that volcanogenic rocks of this series are considered to be source rocks of the nickeliferous gabbrowehrlite intrusions known in the region.

In view of the fact that the SG-3 borehole is located somewhat north of the central part of the Pechenga depression, the maximal thickness of the sedimentary-volcanic rocks of the Nickel series may be 6 – 6.5 km.

The thickness of the Luostari series in the well section is 1958 m. Judging by seismic profiling data, which allowed basal rocks of the Pechenga structure to be traced, although not very distinctly, it appears that this series does not become thicker towards the central part of the depression.

The maximum total thickness of the whole sedimentary-volcanic series of the Pechenga complex, determined by seismic investigations in south central parts of the Proterozoic structure, is therefore assessed to be 8 – 9 km.

Each new series defined in the Pechenga complex is readily subdivided into suites of sedimentary and volcanic rocks.

## The Nickel Series

The Matertin suite (9 – 1059 m) makes up the central part of the Pechenga graben syncline and is most widely developed at the surface within an area of 600 $km^2$. It is mainly represented by diabases, diabase porphyrites, and tuffs of predominantly basic composition. Ultrabasic volcanites (picritic porphyrites) and acid volcanites (quartz porphyrites and their tuffs) are encountered in negligible quantity. All of them are regarded as rocks originating from lavas erupted in the central part of the structure; they are of centroclinal occurrence and plunge southwards at angles from 20° to 70°. The borehole supplied information about the lower part of the Matertin suite.

Here numerous sheets of massive diabase bodies and occasionally globular lava bodies (more than 50) were encountered, and make up about 80% of the section. In a subordinate role are sheets of picritic porphyrites which either rest on each other or are separated by layers of pellitic, silty, at places psammitic, psephitic tuffs, tuff breccia and lava breccia of basic composition.

Diabase sheets show the same structure. Their marginal parts and the margins of individual balls are composed of thin-grained and glass-like, frequently amygdaloidal varieties, whereas the central parts are formed of comparatively coarse-grained rocks. The thickness of the sheets is 20 – 30 m. The diameter of balls in globular lava bodies usually does not exceed 20 – 30 cm.

Sheets of picritic porphyrites occur only in middle members of the suite. Their thickness reaches some tens of meters. Stratigraphically, they occupy a very strict position and are traced both in sections of other wells and at the surface, where in places massive lavas pass into globular bodies. Two types of these sheet are distinguished: differentiated and non-differentiated ones. The upper part of the differentiated sheets is composed of diabase porphyrites characterized by a distinctive dendroid texture of feldspars. These rocks gradually pass into picritic porphyrites in the lower parts of sheets. The non-differentiated sheets are wholly composed of either picritic or diabase porphyrites.

In the upper part of the section thin layers (from 0.3 to 3 m), of basic pelitic tuffs are developed. They rest on the diabase sheets with a sharply defined boundary. In places tuffs are characterized by lamination, which is seen as the

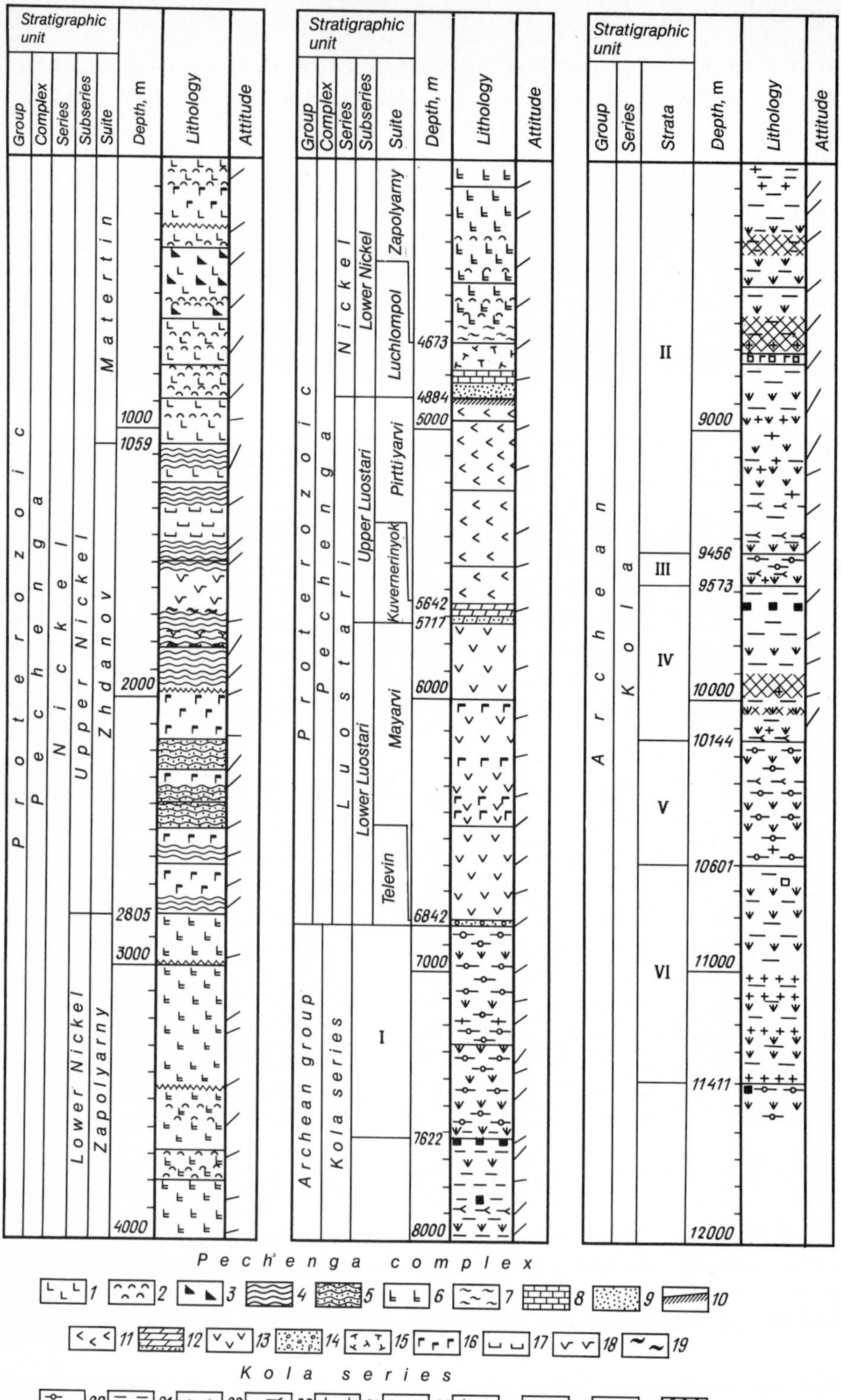

| Stratigraphic unit: Group | Complex | Series | Subseries | Suite | Depth, m | Lithology | Attitude |
|---|---|---|---|---|---|---|---|
| Proterozoic | Pechenga | Nickel | Upper Nickel | Matertin | 1000 | | |
| | | | | Zhdanov | 1059, 2000 | | |
| | | | Lower Nickel | Zapolyarny | 2805, 3000, 4000 | | |

| Stratigraphic unit: Group | Complex | Series | Subseries | Suite | Depth, m | Lithology | Attitude |
|---|---|---|---|---|---|---|---|
| Proterozoic | Pechenga | Nickel | Lower Nickel | Zapolyarny | | | |
| | | | | Luchlompol | 4673, 4884 | | |
| | | Luostari | Upper Luostari | Pirttiyarvi | 5000 | | |
| | | | | Kuvernerinyok | 5642, 5717 | | |
| | | | Lower Luostari | Mayarvi | 6000 | | |
| | | | | Televin | 6842 | | |
| Archean group | Kola series | | I | | 7000, 7622, 8000 | | |

| Stratigraphic unit: Group | Series | Strata | Depth, m | Lithology | Attitude |
|---|---|---|---|---|---|
| Archean | Kola | II | 9000 | | |
| | | III | 9456, 9573 | | |
| | | IV | 10000 | | |
| | | V | 10144 | | |
| | | VI | 10601, 11000, 11411, 12000 | | |

Pechenga complex

1, 2, 3, 4, 5, 6, 7, 8, 9, 10, 11, 12, 13, 14, 15, 16, 17, 18, 19

Kola series

20, 21, 22, 23, 24, 25, 26, 27, 28, 29

**Fig. 1.10**

alternation of thin light and dark layers. In the middle part of the Matertin suite, tuffaceous layers are thicker (from the first tens of meters to 10 m), the range of the fragment sizes also becomes greater. Layers of silty and psammitic tuffs appear in this part. At a greater depth, the section is predominantly composed of pelitic, silty, psammitic, and psephitic tuffs, and tuff and lava breccia of basic composition. Thin diabase sheets are of subordinate significance. The thickness of tuff layers changes from some tens of centimeters to several meters, but normally reaches only a few meters. Finely fragmented tuffs are distinctly layered, whereas in coarsely fragmented tuffs the layering is less obvious. Tuffs consist of large fragments of diabase, basalt, and in rare cases are represented by fragments of andesite porphyrite cemented by black pelitic material. Lava breccia is composed of altered hyalobasalts which form thin sheets or marginal parts of sheets made up of massive diabase lavas. Towards the bottom of the series, the amount of tuffaceous material decreases. Only thin layers of basic pellitic tuffs occur here, they separate thick diabase sheets.

It is to be noted that the Matertin suite is predominantly composed of rocks formed as a result of eruption of basic lavas in a shallow basin. A characteristic feature is the abundance of tuffs which compose individual horizons. The suite shows rhythmic layering. Every rhythm begins with lava eruptions of basic composition and ends with deposition of tuffogenic material. The thickness of the rhythms increases towards the middle part of the section, thus indicating a more stable volcanic regime, which was the very reason which promoted eruption of ultrabasic lavas, similar in their composition with hyperbasitic intrusions of the Pechenga ore field.

A rather variegated composition of volcanites from the Matertin suite (from diabases to quartzitic porphyres and picritic porphyrites) is indicative of the high degree of differentiation of magmas which formed this suite. Rocks of this suite are poorly altered.

---

**Fig. 1.10.** Geological section of the Kola superdeep borehole (SG-3) (After V. S. Lanev, E. B. Nalivkina, V. V. Vakhrusheva, Ye. A. Golenkina, M. S. Rusanov, Y. P. Smirnov) *Pechenga complex*: *1* augite diabases; *2* tuffs of basic composition; *3* pyroxene and picrite porphyrites; *4* phyllites of rhythmic lamination; *3* siltstones, sandstones, tuffites and tuffs; *5* sandstones of rhythmic lamination with subordinate siltstones and phyllites; *6* actinolitized diabases; *7* amphibole and biotite-chloritic schists in tuffs of ultra-basic composition; *8* dolomite; *9* arkose sandstones; *10* sericitic schists (crust of weathering); *11* metadiabases, metaandesites, metaalbitophyres, tuffs; *12* dolomites, marls and polymictic sandstones; *13* diabase porphyrites, biotite-amphibole-plagioclase schists in andesite-basalts; *14* polymictic conglomerates, gritstones, sandstones; *15*–*18* intrusive rocks (*15* andesite porphyrites; *16* gabbro-diabases; *17* essexitic gabbro; *18* wehrlites); *19* sulphide copper-nickel ores. *Kola series: 20* biotite-plagioclase gneisses, schists with HAM; *21* biotite-plagioclase and biotite-amphibole-plagioclase gneisses, schists; *22* amphibolites; *23* talc-biotite-actinolite schists (in ultra-basic rocks); *24*–*26* intrusive rocks (*24* plagiomicroclinic granites, pegmatites; *25* metagabbro; *26* plagiogranites and shadow migmatites); *27*–*29* ore manifestation (*27* magnetites; *28* titanomagnetites; *29* zones of cataclasm and mylonitization). *Strata of*: I muscovite-biotite-plagioclase gneisses with HAM (andalusite, staurolite, sillimanite, garnet); *II* biotite-plagioclase and biotite-amphibole-plagioclase gneisses and amphibolites; *III* muscovite-biotite-plagioclase gneisses with HAM (sillimanite, garnet); *IV* biotite-plagioclase and biotite-amphibole-plagioclase gneisses and amphibolites; *V* muscovite-biotite-plagioclase gneisses with HAM (kyanite, sillimanite, garnet); *VI* biotite-plagioclase gneisses and amphibolites

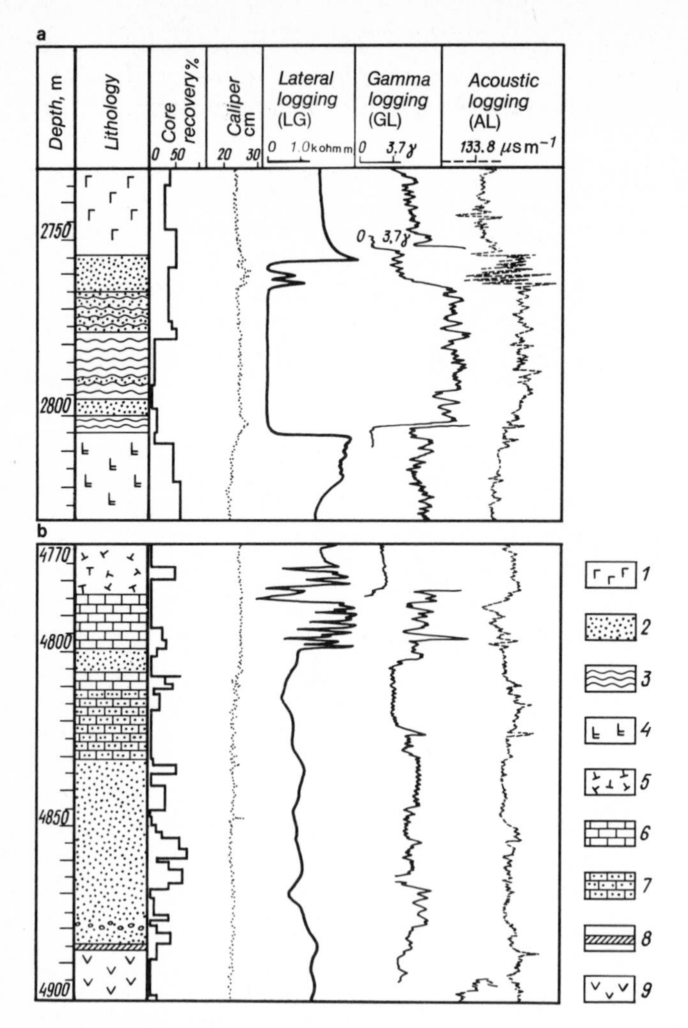


**Fig. 1.11 a, b.** Individual parts of the Kola borehole (SG-3) section in intervals 2730 – 2830 m (**a**) and 4770 – 4900 M (**b**). *1* gabbro-diabases; *2* sandstones; *3* phyllites; *4* actinolitized diabases; *5* andesitic porphyrites; *6* dolomites; *7* dolomitic sandstones; *8* sericitic schists (crust of weathering); *9* foliated diabases

The Zhdanov suite (1059 – 2805 m) rests conformably on other suites of the Pechenga complex. It forms the internal arc of the Pechenga structure, facing with its convex part to the north and north-east, and runs continuously to Kuchin tundra at a distance of 80 km away from the state boundary with Norway.

The Zhdanov suite is the most representative among other suites of the Pechenga complex composed of sedimentary rocks, and is characterized by uniform rock occurrence and great thickness. The SG-3 borehole penetrated the entire thickness of this suite in its central part for the first time. The vertical thickness of the suite together with the enclosed intrusions (gabbro, ultrabasites) is 1746 m, the volcano-sedimentary part of the sequence being 930 m thick. Contact be-

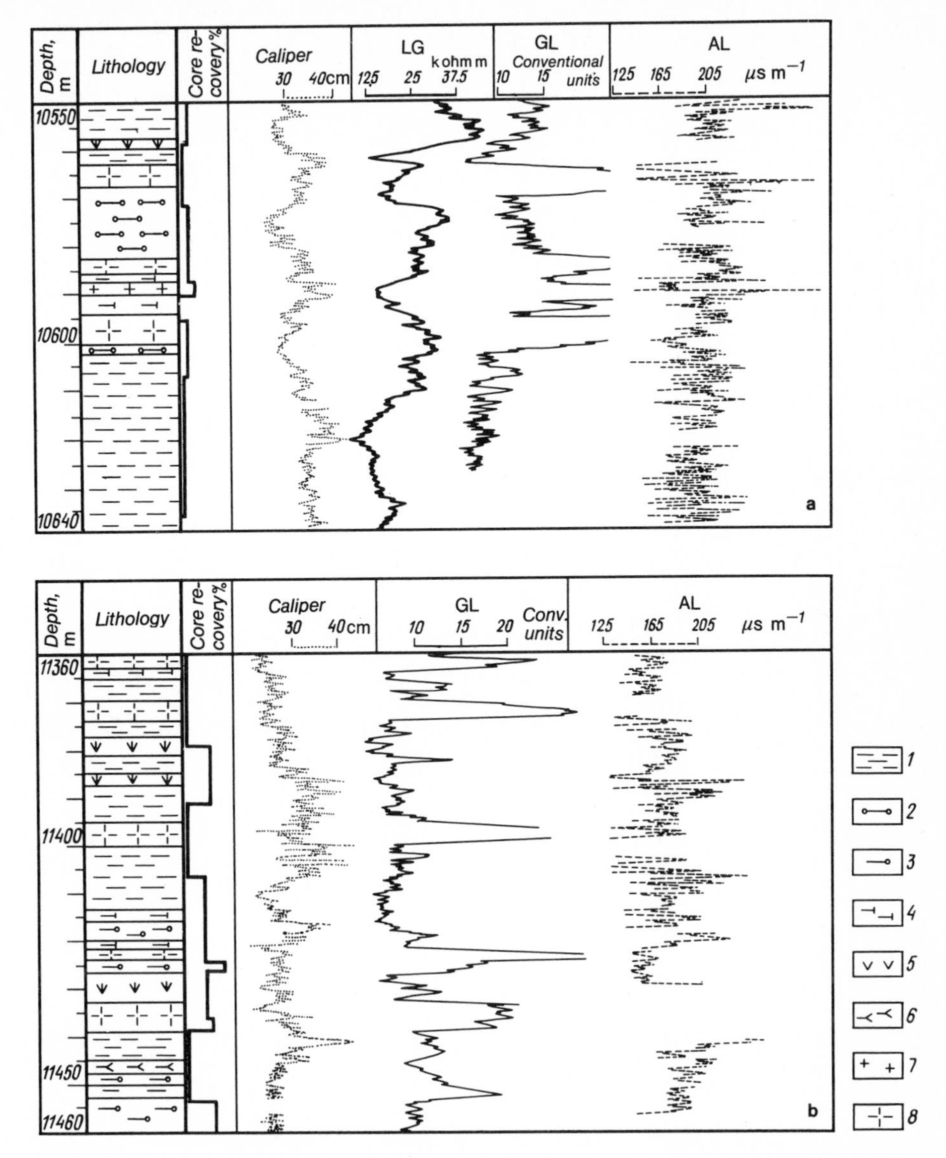


**Fig. 1.12 a, b.** Individual parts of the Kola borehole (SG-3) section in intervals of 10,550 – 10,640 m (**a**) and 11,360 – 11,460 m (**b**). *1 – 3* biotite-plagioclase gneisses (*1* with HAM; *2* with kyanite, sillimanite, garnet; *3* with garnet, sillimanite); *4* biotite-amphibole-plagioclase gneisses; *5* amphibolites; *6* talc-biotite-actinolitic schists (in ultra-basic rocks); *7* granites; *8* pegmatites

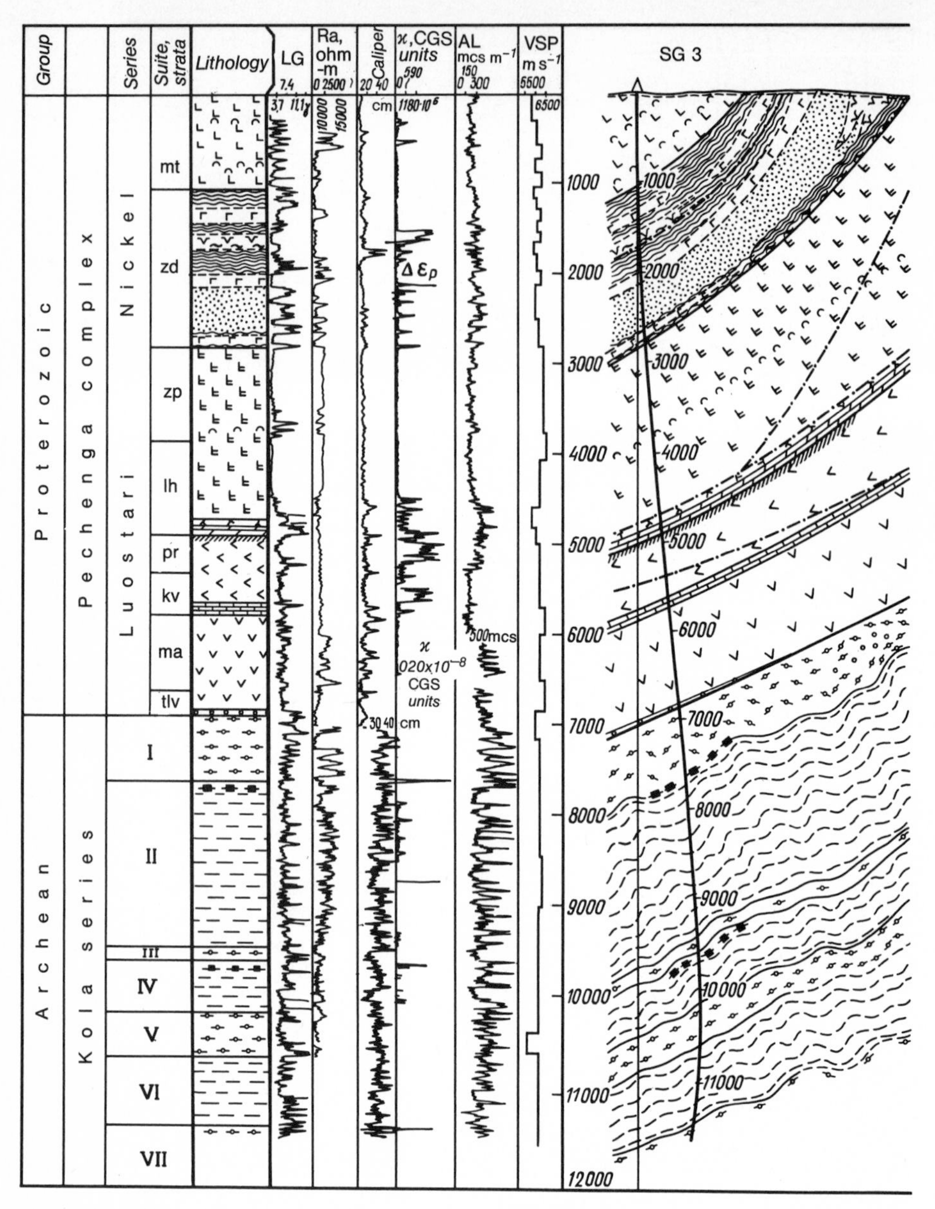


**Fig. 1.13.** Geological-geophysical section of the Kola borehole (SG-3). (After V. S. Lanev, M. S. Rusanov, Y. P. Smirnov). *1* augite diabases with layers of pyroxene and picritic porphyrites; *2* tuffs and tuffites of basic composition; *3* phyllites, siltstones with tuff layers; *4* rhythmically laminated sandstones with subordinate siltstones and phyllites; *5* actinolitized diabases; *6* dolomites, arkose sandstones; *7* sericitic schists; *8* metadiabases; *9* dolomites, polymictic sandstones; *10* diabase porphyrites and schists in them; *11* polymictic conglomerates, gritstones; *12* biotite-plagioclase gneisses with HAM; *13* magmatitized and granititized biotite-plagioclase gneisses; *14* magnetite-amphibole schists; *15–18* intrusive rocks (*15* andesite porphyrites; *16* wehrlites; *17* gabbro-diabases, crosses denote garnets; *18* tectonic dislocations). Strata *I, III, V, VII* muscovite-biotite-plagioclase gneisses with HAM (andalusite, staurolite, sillimanite, garnet) with amphibolite bodies; strata *II, IV, VI* biotite-plagioclase gneisses, biotite-amphibole-plagioclase gneisses and amphibolites

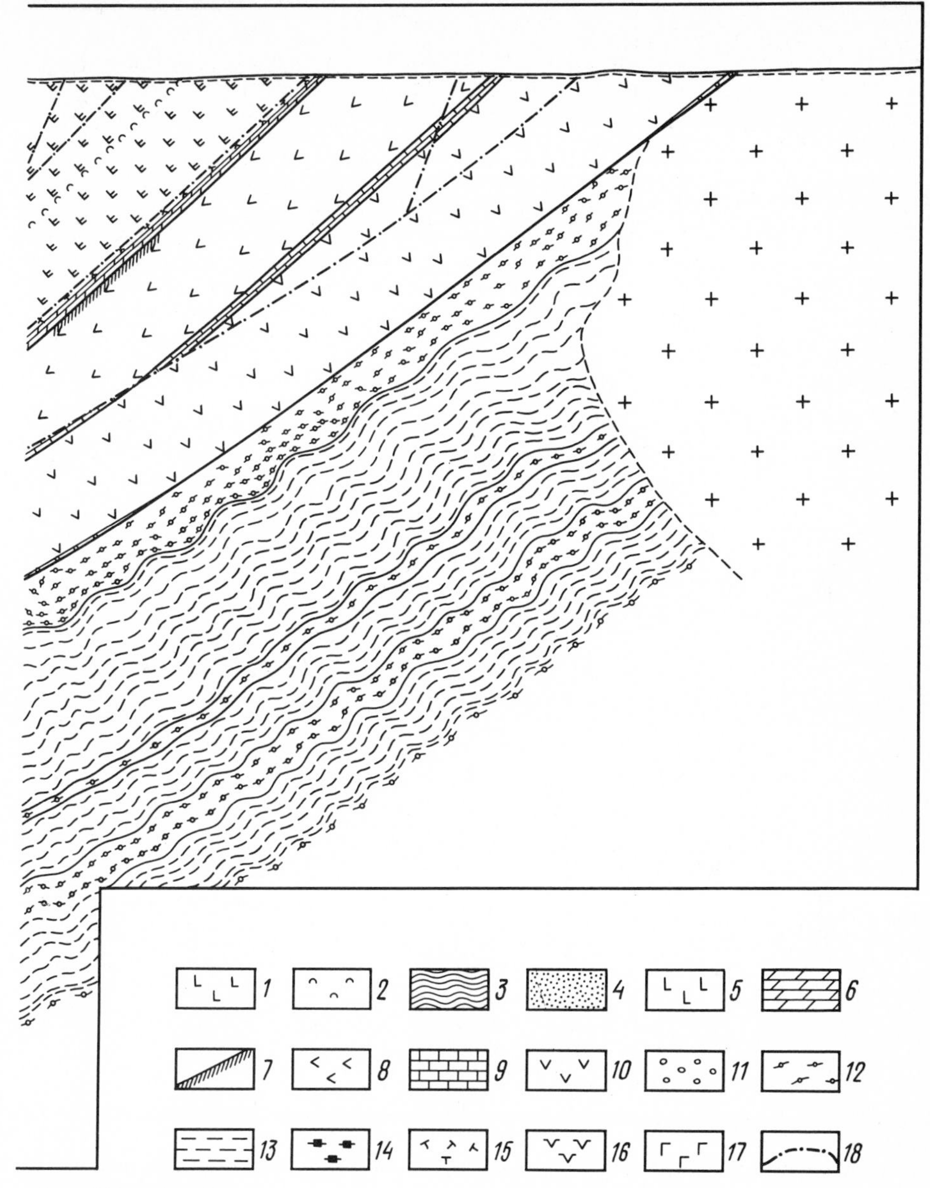


**Fig. 1.13** (continued)

tween the top of the Zhdanov suite and the overlying Matertin suite coincides with the lower boundary of the diabase sheet (with tuffs and tuffites), and is readily identified by an abrupt increase in electrical conductivity and radioactivity. Sedimentary rocks are mylonitizated in a zone of contact with diabases. The bottom of the volcano-sedimentary band rests on sheets of the actinolitized diabases of the Zapolyarny suite. At the contact, the electrical resistivity increases, whereas rock density and radioactivity decrease.

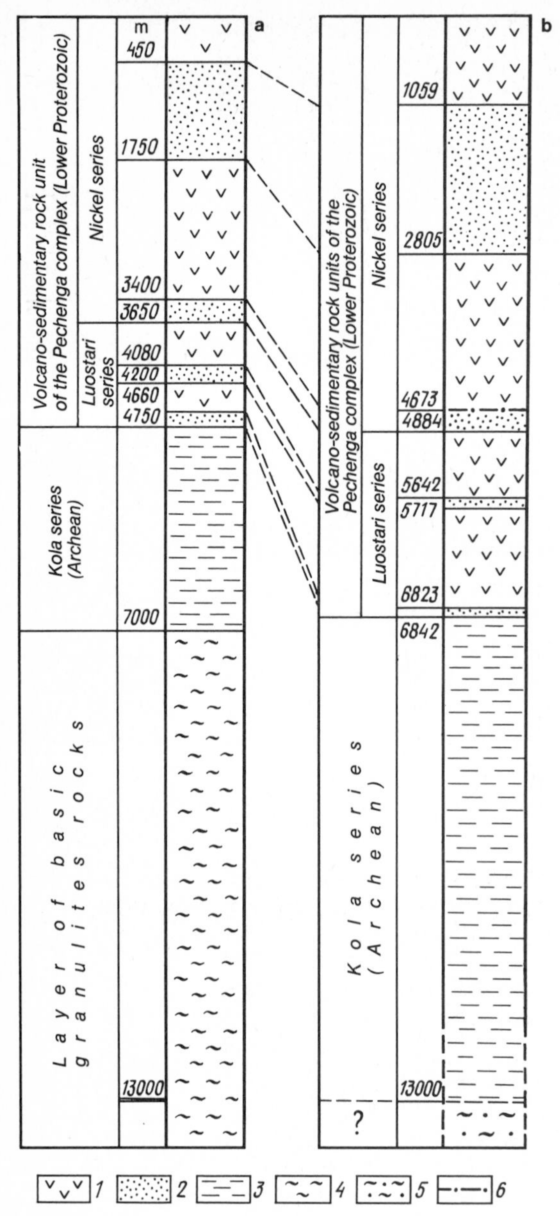


**Fig. 1.14a, b.** Correlation chart of **a** conjected SG-3 well section conceived by the data of seismic investigations, and **b** factual SG-3 well section. *1* effusive rocks of predominantly basic composition; *2* sedimentary rocks; *3* gneisses, granite-gneisses and amphibolites; *4* granulite-basalts; *5* gneisses and amphibolite of high grade of metamorphism; *6* interlayered tectonic dislocations

**Fig. 1.15.** Rocks of the Pechenga complex. *1* aphanitic diabases in contact with silty-pellitic tuff (at 27.6 m depth); *2* agglomerate tuffs (at 771 m depth); *3* phyllites and rhythmically layered sandstones (at 1266.4 m depth); *4* essexitic gabbro (at 1307.2 m depth); *5* serpentinites (at 1639.3 m depth); *6* actinolitized gabbro-diabases (at 2977.8 m depth); *7* lava breccia of actinolitized diabases (at 2977.8 m depth); *8* spheroidal partings in diabases (at 3489.5 m depth); *9* actinolitized diabases of porphyroblastic texture (at 3912.6 m depth); *10* andesite-dacitic porphyrites (at 4743.9 m depth); *11* dolomitic sandstones (at 4808.4 m depth); *12* marbled dolomites (at 4811.0 m depth); *13* metaandesites (at 5489.9 m depth); *14* tremolitized dolomites (at 6203.8 m depth); *15* chlorite-amphibole-plagioclase crystalline schists (at 6370.5 m depth); *16* metasandstones (at 6834.9 m depth)

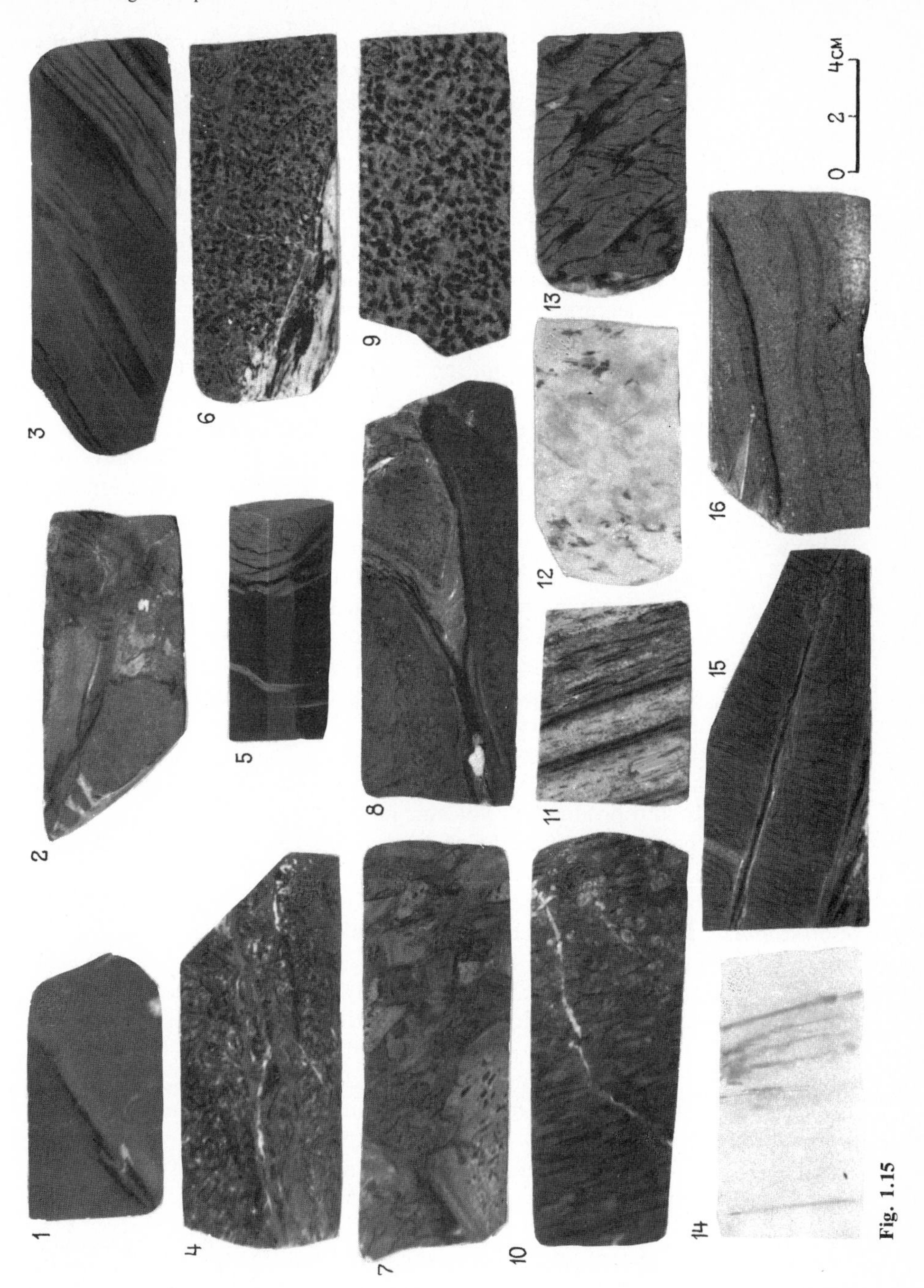


Fig. 1.15

The Zhdanov suite is composed of the following volcano-sedimentary rocks: phyllites, siltstones, sandstones, gritstones, conglomerates, sedimentary breccia, tuffs and tuffites. Sometimes carbonate rocks are encountered. Thin diabase sheets make up only a negligible part of the suite. Volcano-sedimentary rocks everywhere are subject to hydrogenic sulphide mineralization (pyrrhotite, pyrite, and chalcopyrite in lesser amounts) and contain carbon-bearing matter; this explains their low electrical resistivity.

Five members have been distinguished in the section of the SG-3 well, and each of them forms a sedimentary rhythm. The lower parts of rhythms predominantly consist of sedimentary breccia, sandstones, and gritstones, and the upper parts are composed of siltstones and phyllites. Volcano-sedimentary rocks are laminated. Thin rhythmic lamination, represented by two – three rhythms (sandstone-siltstone, siltstone-phyllite, sandstone-siltstone-phyllite) with layers from 1 mm to 5 cm thick, are the most characteristic. It is not uncommon for siltstones and sandstones to be characterized by cross-bedding, which is mainly pronounced in the upper part of the strata. In the lower subsuite cross-bedded lamination, trough-like bedding and ripple marks are present. Siltstones are notable for the wide development of underwater-slump texture. Traces of minor water erosion, roiling and mud rolls are observed at rhythm boundaries. In some layers small carbonate concretions are found. These structures show that water streams developed in shallow and, partially, in near-shore environments.

In the upper part of the suite, tuffaceous material is of considerable importance. The largest amount of this material (about 50%) is confined to the uppermost part of the suite. Phyllites and siltstones are widely developed. On the whole, the accumulation of clastics in this subsuite reflects a transgressive sedimentation interrupted at places by regressions caused by local risings. The subsuite section ends with products of volcanic activity.

In the lower part of the suite, the amount of terrigenous material increases. Generally, rocks become more coarse-grained, sandstones, gritstones and conglomerates are found in abundance. As distinct from the upper part, the character of sedimentation of the lower part of the suite is regressive and this is expressed in the fact that pelitic material decreases down the section, whereas psephitic and psammitic fractions increase sharply. The process ended by formation of conglomerates confined to the upper part of the subsuite sequence.

It should be noted that a regressive macrorhythm of considerably terrigenous sedimentation in the Zhdanov suite is interrupted by intensive volcanic activity. This leads to a change from the regressive to transgressive environment of sedimentation.

Activation of volcanism is determined by the appearance of tuffaceous rocks and thin sheets of diabases in the central part of the Zhdanov suite. As a result of detailed mapping, a marker-bed member of the Zhdanov suite has now been traced in the central part of the Pechenga ore field. Ore-bearing ultrabasite intrusions crossed in that part of the sequence which directly overlayed this rock member.

Rocks of the Zhdanov suite are metamorphosed in the prehnite-pumpellyite and greenschist facies, sedimentary rocks being less metamorphosed than volcanic rocks.

Thus, the following facts have been established:

1. volcano-sedimentary rocks of the Zhdanov suite in the SG-3 well are clearly stratified and subdivided into two subsuites, differing in composition and mode of sedimentation;
2. vertical thickness of the so-called productive strata is 1764 m; the primary idea that the thickness decreases with depth, based on seismic data, was not confirmed; it proved that in the SG-3 well section, at least to a depth of 3000 m, the thickness was constant;
3. composition of volcano-sedimentary rocks, the grade of their metamorphism and character of their occurrence at depth are analogous to those of the surficial parts of the section;
4. the Zhdanov suite, composed of volcano-sedimentary rocks, contains at depth numerous formational intrusions of gabbro, gabbro-diabases, and ore-bearing ultrabasites; the aggregated thickness of all these intrusions in a section is 800 m (45% of the productive-strata thickness);
5. the constant composition, high density of basite and ultrabasite nickeliferous intrusions are the sole evidence in favour of good prospects for the Pechenga ore field at deep horizons.

The Zapolyarny suite (2805 – 4673 m) extends as a bow-shaped band at a distance of 70 km between two sedimentary suites, the Luchlompol (lower) and Zhdanov (upper). Its width at the surface ranges from 4 km (west of the SG-3 well profile) to 500 – 800 m (in the eastern part of the Pechenga structure). Correspondingly, the actual thickness of rocks changes from 1.5 – 2 km to 300 – 400 m; they dip southwards at angles from 30° to 40°.

Both at the surface and in the SG-3 well section, this suite is represented by actinolitized diabase sheets of massive and globular lavas. The thickness of the sheets changes from a few meters to several kilometers. Metamorphosed volcano-sedimentary rocks such as tuffites and tuffaceous siltstones are of negligible development. Diabases are characterized by the same structure. Their marginal parts are composed of cryptocrystalline or fine-grained varieties, frequently of almond-shaped structure, while the central parts consist of medium-grained and coarse-grained rocks, usually with porphyroblastic texture. Globular lavas mainly contain closely spaced balls of several tens of centimeters in diameter. Cement is present in small amounts and is represented by tuffaceous or tuffaceous-carbonate material. Sometimes, globular lavas pass into tuff and lava breccias. Metamorphosed basic tuffs, tuffites, and tuffaceous siltstones in the middle part make up beds whose thickness reaches 20 m. They form rhythms which contain mainly coarse-grained material in lower parts.

On the whole, the suite also consists of rhythms. Every rhythm begins with the formation of massive lava sheets and ends with the formation of globular lava sheets; sometimes it begins with the formation of tuffs. Rocks were formed in a shallow basin.

Judging from the differences in proportion between sheets of massive and globular lavas and tuff layers, it appears that the Zapolyarny suite is subdivided into seven members. At some places its rocks are foliated; beginning from a depth of 4300 m, almost all the rocks are made up of foliated varieties. Their near-bottom parts were transformed into green schists. There are gradual transitions from the actinolitized diabases into foliated actinolitic diabases and then into carbonate-actinolite-chlorite schists, at the bottom of the suite talc-chlorite schists, formed in ultrabasitic rocks, have also been encountered.

On the whole, rocks are metamorphosed in the greenschist facies (actinolite zone). In its lowest part, in a zone of mylonitization, they are metamorphosed in the epidote-amphibolite facies.

When describing the Zapolyarny suite, it must be emphasized that its main volcanic rocks are notable for uniform composition; all the differences observed in the main part of the section are associated with structural-textural features that depend on either the massive or globular appearance of sheets, whose number in the section exceeds 100. The suite is subdivided into two parts of equal thickness by a member of sedimentary-tuffogenic rocks intercalated with sheets of the actinolitized diabases. Thin horizons of this member are remarkably uniform, both in mode of occurrence and composition. This suite extends down the dip for a distance of 6 km (from its outcrops at the surface to a depth of 3782 km); this fact, as well as the uniform composition of this suite, indicates that conditions of tectonic-volcanic activity during its formation were very stable.

The Luchlompol suite (4673 – 4884 m) extends in the form of a narrow band from the Tulpiyavr lake to the Kuetsyarvi lake. Compared with the Zhdanov suite, its thickness is small, ranging from 50 to 500 m at the surface and reaching 111 m in the vertical SG-3 well section. At the top of this suite, at the contact with the Zapolyarny suite, porphyrites occur. At the bottom, rocks of the metamorphosed weathering crust of the underlying Pirttiyarvi suite, now redeposited and reworked, are developed.

The upper part of the Luchlompol suite is composed of metamorphosed sandy dolomites and dolomites. Of the carbonates manganese carbonate is also encountered. Rocks are characterized by the well-developed rhythmic lamination of several orders. Some layers contain an admixture of carbon matter. The amount of terrigenous admixture in carbonate rocks generally increases down the section. The lower part is formed of metamorphosed arkosic sandstones with layers of gritstone. Sandstones are characterized by parallel and cross-bedded lamination, with thicknesses of layers from a few centimeters to 15 cm. It is characteristic that arkosic sandstones are enriched in hematite and magnetite; a zone confined to contact with rocks of the underlying Pirttiyarvi suite is composed of fine-grained light-coloured sericitic schists with magnetite, which contain rounded quartz fragments in some layers. These rocks are considered to be the partially redeposited metamorphosed crust of weathering of the underlying basic volcanic rocks. In this suite, in contrast to the Zhdanov suite, carbonate material is contained in a considerable amount (about 40%). Tuffaceous material and phyllites are virtually absent.

Rocks of the Luchlompol suite form a transgressive macrorhythm; they are foliated. Foliation coincides with parallel lamination and forms an angle of 60° – 70° with the core-sample axis.

Rocks are metamorphosed in the greenschist facies. Contact between the Luchlompol and Zapolyarny suites is complicated by a major intraformational fault. It is characteristic that this fault, first detected in the well and later at the surface, like the Poryitash major fault on the southern flank of the Pechenga structure, is traced at a considerable distance by the presence of hypabyssal intrusions of andesitic porphyrites. A major fault zone is clearly identified by foliated and mylonitized rocks, by an abrupt increase of borehole cavernosity, by the local increase of gas and water, as well as by a rise in temperature.

## The Luostari Series

The Luostari sedimentary-volcanic series underlies the Nickel series and outcrops at the surface in the form of a comparatively narrow band in the north-eastern part of the Pechenga structure. Its vertical thickness in the well is 1958 m. This series is also clearly subdivided into sedimentary and volcanic suites.

The Pirttiyarvi suite (4884 – 5642 m) extends at the surface in the form of a band of about 70 km in length between two suites composed of sedimentary rocks. The width of the band changes from 2.5 km in the north-western part to 100 – 200 m in the eastern part. Rocks dip southwards at angles of 20° – 30°. The suite is composed of volcanic differentiates of predominantly basitic and medium composition which alternate with each other. The amount of pyroclastic material in these strata is negligible.

The upper and lower parts of the suite are composed of apodiabase-amphibole schists, its middle part consists of apoandesite magnetite-biotite-plagioclase schists and the middle part is represented by apotrachyandesitic schists. At the top of the suite magnetite-biotite-sericitic schists of about 4 m thick, enriched with mica, are identified; these schists are the metamorphosed crust of the weathered diabases.

Despite the fact that these rocks underwent metamorphism of the epidote-amphibolite facies and became strongly foliated, they contain relicts of the primary textures and structures which allow a reconstruction of their volanic nature and make it possible to distinguish numerous sheets of intrusive eruptions. Marginal parts of sheets are represented by fine-grained varieties, at places with almond-shaped lava-brecciated and flow structures. However, most of them are composed of massive lavas. Relicts of globular lavas are found only in the lower parts of the section. Schists of all types are black in colour due to the admixture of thinly dispersed magnetite. Foliation is generally inclined at an angle of about 70° – 80° to the core-sample axis and coincides with the direction of layering in the underlying sedimentary rocks. In numerous zones of cataclasm confined to joints, rocks are of a lighter colour at a distance of 1 – 2 mm, and seem to have been subject to diaphthoresis under greenschist-facies conditions.

The Pirttiyarvi suite differs from suites of the Pechenga complex, composed of volcanic rocks, by the presence of differentiated varieties of the basitic and medium composition with somewhat alkaline character. Basitic lava marked the beginning of the eruption period, later lavas were replaced by material of medium composition and, finally, rocks of basitic composition were formed again.

The high intensity of the magnetic rock field of the Pechenga complex (up to 20 Å $m^{-1}$) is explained by their high saturation with finely dispersed magnetite.

As distinct from rocks of the massive type developed at the surface, rocks found in the well section are foliated and at places mylonitized. In many interlayers, cataclasis superimposes foliation. It seems probable that foliation, developed everywhere at depth, is in fact the cause of the abrupt changes of the rock's physical properties which were registered in the well; this particularly concerns the observed decrease in velocity propagation of longitudinal and transverse waves.

The values of thickness measured in sections at the surface compared with values calculated in wells show clearly that the thickness of this suite tends to increase down the dip towards the central part of the Pechenga structure.

The Kuvernerinyok suite (5642 – 5717 m) outcrops at the surface and extends in the form of a band from the Kuetsyarvi lake to the Gusinyi lake. Its vertical thickness in the SG-3 well is 75 m. The suite occurs between volcanic rocks of the Pirttiyarvi and Mayarvi suites. The contacts of the Kuvernerinyok suite with the overlying and underlying suites were also encountered in the drill cores, but they can be better identified by geophysical data. The suite is of uniform composition throughout the area of its development. Along a dip plane it can be traced for a distance of 9 km from an area where it outcrops at the surface. The suite is composed of metamorphosed carbonate rocks (40%) in the upper part of the section and of terrigenous rocks in the lower part. Correspondingly, two members have been distinguished, the upper one consisting of calcitic and calcite-dolomitic marls, talc-tremolite and anthophyllite marls, tremolite and anthophyllite-tremolite schists, and the lower mainly of quartz-mica-carbonate rocks and quartz-mica-schists alternating with metamorphosed quartzites and sandstones. It is characteristic that rock foliation coincides with lamination and is inclined at an angle of about 65° – 70° to the axis of the core sample.

Rocks in the section are metamorphosed in conditions of the epidote-amphibolite facies. They differ from outcropping rocks by a higher grade of metamorphism. Tremolite marbles were identified at great depth within the limits of the Pechenga structure for the first time in this part of the well section.

The Kuvernerinyok and Luchlompol suites resemble each other. Both of them were formed from identical sources and developed in near-shore marine conditions with waters of increased salinity.

The Mayarvi suite (5717 – 6823 m) is the first rock-stratigraphic unit of the volcanic part of the Pechenga complex. It lies on basal conglomerates, gritstones and sandstones. The suite forms on the map an arc-like band which is from 1 to 2 – 2.5 km and even to 5 km wide (in the north-western part of the Pechenga structure). Rocks are characterized by the centroclinal occurrence and dip southwards at angles from 20° to 30°. At the surface, the suite is composed of massive chloritized and actinolitized diabases with widely developed almond-shaped structures.

However, in the section of the SG-3 well, the Mayarvi suite consists of uniform apodiabase, amphibole-plagioclase and biotite-amphibole-plagioclase schists. In spite of the fact that rocks underwent metamorphism of the epidote-amphibolite facies and were foliated, some relics of the primary rocks structures were still preserved in them. This allowed the identification of more than 50 individual sheets of basic volcanic rocks. The thickness of the sheets changes from a few to several tens of meters. All of them are sheets of massive lavas. In addition to these rocks, the middle part of the section contains distinctive melanocratic metamorphic plagioclase-amphibole rocks. These sheets are composed of volcanic rocks of a more basic composition. They may also be composed of formational intrusions. In places rocks are poorly microactinolitized and biotitized; at one site a zone of phlogopitization of small thickness was established. In zones of cataclasis (up to 20 – 30 m thick), rocks are subject to diaphthoritic alteration which results in the formation of mineral associations of the greenschist facies.

Apodiabase amphibole-plagioclase and biotite-amphibole-plagioclase schists are characterized by distinctive foliated and lenticular-foliated metamorphic structures. Foliation is inclined at an angle of 70° – 80° to the axis of the core sample.

It should be emphasized that the thickness of the Mayarvi suite hardly changes with depth. Contrary to rocks exposed at the surface, at deep horizons rocks are foliated, metamorphosed to a rather high degree (epidote-amphibolite facies) and subject to microclinization at some places.

The Televin series (6823 – 6842 m) is represented by basal rocks of the Pechenga complex. It rests on the wavy, eroded surface of the Archean gneisses and strikes around the Pechenga graben syncline from the north. On the surface it is traced along the strike in some isolated outcrops. At depth it was encountered in two exploration wells. Its thickness is extremely variable with depth, changing from 10 to 200 m. Rocks dip southwards at 20° to 30°.

The suite is composed of conglomerates, gritstones and sandstones. It consists of sedimentary rhythms of transgressive type, comprising two to three members, and having a thickness of several meters. Rocks are characterized by parallel wavy lamination. At places unidirectional stream-type cross-lamination and ripple marks are established. The rock composition of the suite is determined by the composition of the eroding Archean rocks. Conglomerates and pebble beds are mostly represented by pebbles of gneiss-like plagiogranites, granites and gneisses. Conglomerates and pebble beds with pebbles of amphibolitized gneisses are less developed. Such rocks have been encountered in wells drilled by the Industrial Geological Organization Sevzapgeologiya in the area of Mount Generalskaya. Conglomerates and pebbles are cemented by quartz-plagioclase-mica, and sometimes mica-amphibole material. Sandstones are mostly of the arkose type, the size of the clastic material decreasing up the sequence. All these rocks are poorly metamorphosed (the grade of metamorphism does not exceed the greenschists facies).

The SG-3 borehole cut the Televin suite at a distance of more than 10 km from its outcrops (in the down-dipping direction). Due to the small amount of core recovered, it was impossible to compile the complete section of the suite. However, the top of the suite has been clearly identified by geophysical data. This suite is overlain by apodiabase amphibole-plagioclastic schists of the Mayarvi strata and underlain by the Archean gneisses with highly aluminiferous minerals which contain layers of amphibolites.

The Televin suite cored by the well is composed of metamorphosed sandstones and gritstones. Rocks are rhythmically layered, each rhythm consisting of three members; the thickness of individual layers varies from 0.5 to 3 cm, the thickness of the rhythms is 10 cm. Rhythms are of the transgressive type. Rock fragments are composed of quartz, microcline and plagioclase, cemented by biotite and muscovite. Some layers contain amphibole. Rock lamination forms an angle of 60° – 70° to the core sample axis. Lamination coincides with foliation. Rocks are metamorphosed in the epidote-amphibolite facies. However, recrystallization of rocks and formation of granolepidoblastic texture are not obstacles preventing the certain identification of the primary structures and textures of sedimentary rocks which are "visible through" the newly formed metamorphic textures.

## Intrusive Rocks

Gabbro-wehrlites, gabbro-diabases, and dacite-andesite porphyrites are found in the Pechenga complex; they occur between sedimentary and volcanic rocks. Of the rocks of the dike complex diabase and gabbro-porphyrite, dikes are most widely developed.

The main occurrence of the gabbro-wehrlite intrusions is conformable with volcano-sedimentary rocks of the Zhdanov suite. Intrusions underwent folding together with the enclosing strata. In marginal parts rocks are foliated. Intrusive bodies are usually in conformable contact with the enclosing rocks, less frequently they are cross-cutting. Tectonic contacts are observed.

Most of the units consist of individual differentiates: wehrlites, pyroxenites, and gabbro. Large intrusions are differentiated from wehrlites (at the bottom) to gabbro (at the top).

Ultrabasites are found in the SG-3 well section at a depth of 1541 – 1808 m. All of them occur in the upper part of the Zhdanov suite. The largest layer lies at the 1541 – 1680 m level. Smaller bodies are from 10 to 30 m thick. Contacts between ultrabasites and the enclosing rocks are sharp. In zones of exocontacts leucoxene-chloritic rocks are present. Rocks are foliated both in exo- and endocontact zones. Bodies of ultrabasites are asymmetrically zoned, caused both by their primary non-uniform composition and by secondary alteration of a different degree. Rich copper-nickel ores of the impregnated type are confined to the lower endocontact of the largest ultrabasite body. Its central part, about 90 m thick, is represented by serpentine peridotites which pass into serpentinites 10 m thick towards the upper and lower contacts. In contact zones 13 – 16 m thick, serpentinites are talcose, chloritized, amphibolitized, and form zones of serpentine-chlorite-talc, chlorite-tremolite and chlorite-talc-tremolite rocks.

Gabbro and gabbro-diabase bodies have been encountered in the well section. Most of them occur in rocks of the Zhdanov suite. A few small bodies are present in apodiabase and amphibole-plagioclase schists of the Mayarvi suite. These are formational bodies which are notable for similar composition and structure. Their thickness reaches 200 m. Rocks of all the bodies are altered, the type of alteration depending on their position in the SG-3 well section: in the upper part of the section gabbro and gabbro-diabases are epidotized, chloritized and scapolitized, in the middle part rocks are intensively chloritized, actinolitized, and in the lower part they are transformed into plagioclase-amphibolite rocks.

Dacite-andesite porphyrites compose formational intrusions, the largest of which are confined to zones of the Poryitash and Luchlompol major faults. A body occurring in a zone of the Luchlompol major fault was encountered at a depth of 4673 m. Its vertical thickness is 111 m. The body is confined to a boundary between the Zapolyarny and Luchlompol suites. Dacite-andesite rocks, like enclosing rocks, are intensively foliated, particularly in marginal parts of bodies.

### 1.2.2 The Archean Complex

In the SG-3 well section, the Archean complex is represented by the Kola series, which is composed of muscovite-biotite-plagioclase and biotite-plagioclase gneisses with highly aluminiferous minerals and, less frequently, by biotite-amphibole-plagioclase gneisses (Fig. 1.16). Numerous beds of amphibolites and rare thin layers of meta-ultramafites lie conformably on the latter rocks and are in very close association with them. Rocks of this series were multiply metamorphosed and underwent regional granitization of varying degree which led to the formation of plagiogranites, granites, pegmatites, aplite pegmatites and conjugated basificates, i.e., biotites and phlogopites.

As was previously assumed, a layer of biotite-plagioclase gneisses with highly aluminiferous minerals is distinguished in the upper parts of the series (6842 – 7622 m); it passes down the section into a thick layer of biotite-plagioclase gneisses. The same type of intercalation of aluminiferous gneisses with biotite-plagioclase gneisses continues further downwards; the only difference is that one layer is more frequently replaced by another and, consequently, their thickness becomes less as compared with that of the two upper layers.

In the light of the data obtained it is fairly certain that the previous supposition that the folded structures close in the borehole area at a depth of 7 – 8 km was not confirmed. The rock dipping in the well was found to be nonpersistent: angles of dip, like those at the surface, change from subhorizontal to subvertical. Nevertheless, for the whole series, medium angles of dip (40° – 60°) predominate. On the strength of three-vector magnetic measurements, which allow determination of the spatial position of the highly magnetic beds, it appears that gneisses of the Kola series strike predominantly in a north-western direction and dip southwards even at such a great depth.

Amphibolites present in gneisses belong in composition to different types; they constitute beds of varying thickness. There is no strict regularity in the distribution of different amphibolite types in individual layers. The degree of saturation of the gneiss by amphibolites and the degree of rock granitization in the well section are not high enough to ensure detailed subdivision of the series. However, it is worthy of note that only the upper part of the gneiss section (6840 – 7270 m) contains a small quantity of amphibolites, not more than 6 – 8% of the total thickness. With increase in depth, the proportion of the various rocks abruptly changes, and the amount of basic rocks rises from 25% to 40% (Fig. 1.17), a fact that appears to be an important characteristic of the SG-3 well section.

The presence or complete absence in gneisses of staurolite and garnet, indicators of minerals, or such minerals as orthite and epidote, is the leading criterion used for subdividing the Kola series into individual units.

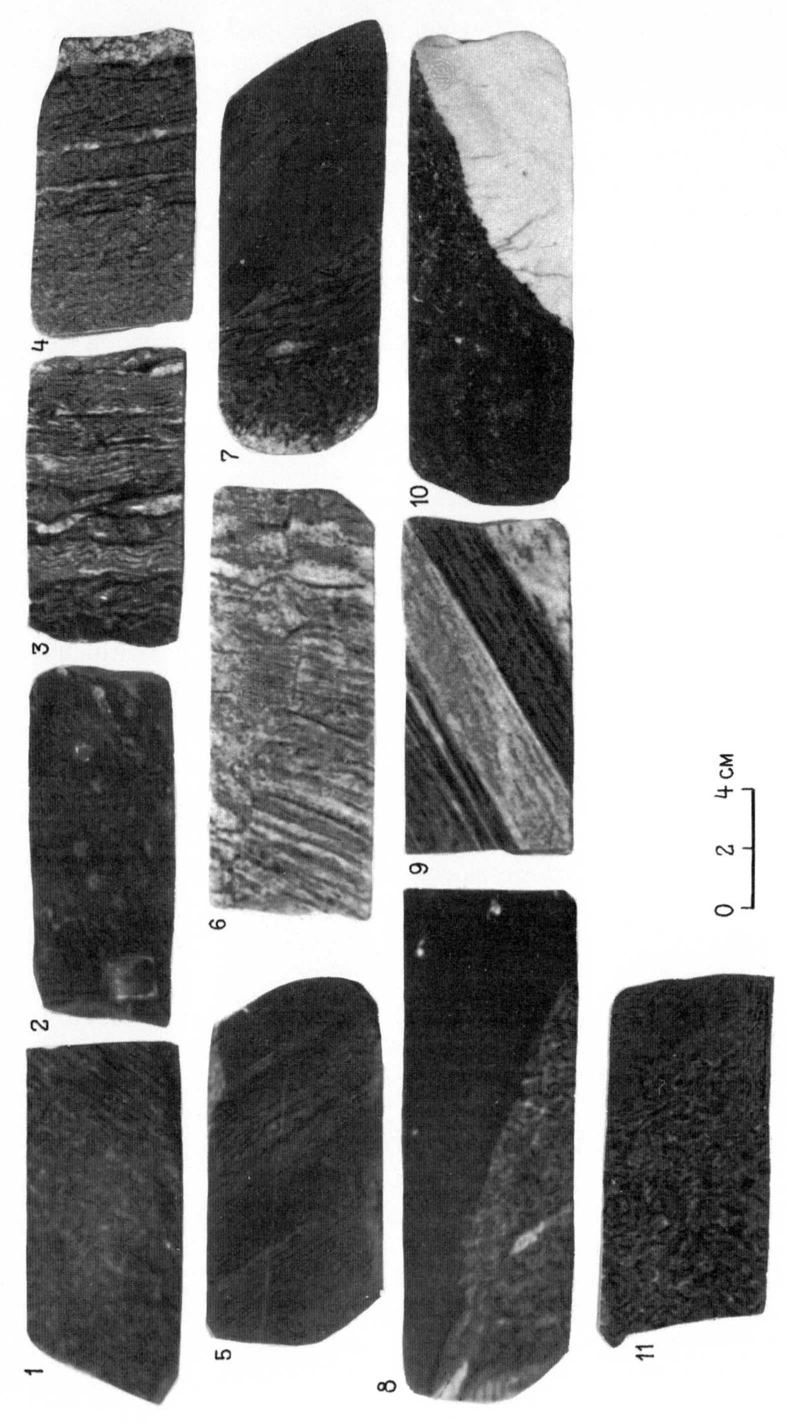


**Fig. 1.16.** Rocks of the Kola series. *1* amphibolites (depth 6848.5 m); *2* garnet-biotite-amphibole-plagioclase schists (depth 7389.3 m); *3* bimicaceous garnet-sillimanite gneisses (depth 7446.4 m); *4* aluminiferous biotite-plagioclase gneisses (depth 7469.8 m); *5* clinopyroxene amphibolites (depth 7600.1 m); *6* migmatized biotite-plagioclase gneisses (depth 7770.5 m); *7* phlogopite-tremolite schists (depth 8021.4 m); *8* amphibolites in contact with biotite-amphibole-plagioclase schists (depth 8868.9 m); *9* migmatized epidote-biotite-amphibole-plagioclase gneisses (depth 8772.1 m); *10* aplite in contact with amphibolite (depth 9061.3 m); *11* magnetite-epidote-biotite-plagioclase schists (depth 9651.3 m)

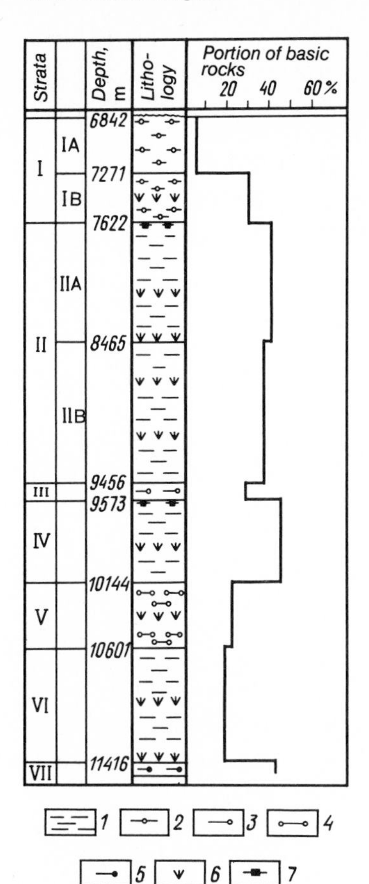


**Fig. 1.17.** Degree of abundance of amphibolites in the Kola series of the SG-3 well section. *1* biotite-plagioclase gneisses; *2* with highly aluminiferous minerals (andalusite, staurolite, sillimanite, garnet); *3* with sillimanite and garnet; *4* with kyanite, sillimanite and garnet; *5* granite-biotite-plagioclase gneisses with sillimanite; *6* amphibolites; *7* magnetite-amphibole schists

Seismic investigations carried out at the surface detected seismic boundaries at a depth of 12.5 – 13.5 km and led to the assumption that these boundaries correspond to lower boundaries of the typical gneiss developed in the Kola series, and that the borehole has crossed the upper and middle parts of this series. By assuming this, it appears that the total vertical thickness of the series in the SG-3 well section beneath the Pechenga graben syncline is not less than 6 km. It is certain that at those places which are not overlain by sedimentary-volcanic rocks of the Pechenga complex, the vertical thickness may be twice as much.

In the well section, the Kola series is subdivided into the following seven units (from top to bottom):

1. muscovite-biotite-plagioclase gneisses with highly aluminiferous minerals (andalusite, staurolite, sillimanite, garnet);
2. biotite-plagioclase gneisses and amphibolites;
3. muscovite-biotite-plagioclase gneisses with highly aluminiferous minerals (sillimanite, garnet);
4. biotite-plagioclase, biotite-amphibole-plagioclase gneisses and amphibolites;
5. muscovite-biotite-plagioclase gneisses with highly aluminiferous minerals (kyanite, sillimanite, garnet);

6. biotite-plagioclase gneisses and amphibolites;
7. biotite-plagioclase gneisses with highly aluminiferous minerals (sillimanite, staurolite, garnet).

Two upper layers, the most representative ones, are correlated with the Kola series outcropping in the north-western part of the Kola peninsula. Other layers, penetrated by the SG-3 borehole, add to the total thickness of the series.

It seems probable that stratigraphic analogues of these strata are not exposed at the surface within the limits of the region.

The muscovite-biotite-plagioclase gneisses with highly aluminiferous minerals (andalusite, staurolite, sillimanite, garnet) directly underlie the Pechenga complex in the well section (6842 – 7622 m). The contact of gneisses with the overlying rocks of the Pechenga complex was penetrated by the well; it was established by an abrupt increase of radioactivity above the background values characteristic of the gneiss-complex rocks. It should be noted that in the contact zone lamination in gritty sandstones of the Televin suite of the Proterozoic complex, as well as lamination and gneissic character of rocks in the underlying Archean gneisses, are of one and the same orientation with respect to the core sample axis, i.e., there are no angular unconformities between these most important Precambrian complexes in the SG-3 well section.

Gneisses are commonly represented by alternating leucocratic layers rich in quartz and plagioclase and melanocratic layers enriched by biotite. Their thickness varies from millimeters to 2 – 7 cm. Layers form binary rhythms which are also grouped in thicker rhythms and form beds of melanocratic, leucocratic and mesocratic gneisses. Highly aluminiferous minerals are confined mainly to those parts of the rhythms which are enriched by mica. Rocks are of the gneissic type. Their gneissic character is caused by the oriented disposition pattern of biotite scales, quartz and plagioclase grains. Foliation coincides with lamination. Banding also coincides with lamination and forms angle of 40° – 60° with the core sample axis. The thickness of conformable amphibolite bodies encountered among gneisses, is from a few meters to 25 m. They are foliated and usually biotitized, in places characterized by mottled structure. Rare thin layers (up to several meters) of biotite-amphibole-plagioclase gneisses are met among muscovite-biotite-plagioclase gneisses with highly aluminiferous minerals. They tend to occur at the near contact parts of amphibolites. Gneisses with highly aluminiferous minerals are muscovitized. Due to the specific distribution pattern of amphibolites the layer is somewhat nonuniform: in the upper part occasional amphibolite layers are met, their total amount does not exceed 8%; in the lower part the total amount of amphibolites is about 30%. There are amphibolites which by their chemical composition mainly correspond to basalts of a common series. Two insignificant bodies of meta-ultramafites composed of phlogopite-actinolitized schists were also found.

Down the well section, the layer of biotite-plagioclase gneisses and amphibolites passes conformably into the underlying unit. It was encountered in an interval of 7622 – 9456 m and it proved to be the thickest. A bed of quartz-magnetite-amphibole rocks was drilled through in the upper part of this layer; according to data of the magnetic survey its thickness is about 15 m.

The layer is composed of biotite-plagioclase gneisses alternating with bodies of amphibolites whose content reaches 30%. Gneisses and amphibolites lie conformably on each other and contacts between them are sharp. The thickness of the amphibolites changes from some meters to tens of meters. These main types of rock are of multi-alternating character and consequently form a lamination structure of the section. Beds of biotite-amphibole-plagioclase gneisses and meta-ultramafites are present in a subordinate amount.

Biotite-plagioclase gneisses are usually poorly banded. Well-developed foliation and rare banding are in conformable orientation with contacts and commonly form an angle of 40° – 60° with the core sample axis. Gneisses are muscovitized and biotitized, which causes their spotted greyish-white colour. They are irregularly granitized and form banded migmatites which at places pass into gneiss-like plagiogranites.

These gneisses are considered to be a very characteristic type of rocks widely developed in the Archean formations on continents. Their mineral composition in the SG-3 well section is also comparatively simple; it is represented by biotite, plagioclase, quartz and accessory orthite.

The amphibolites present in this layer are considerably differentiated. Three types can be distinguished according to mineral composition, proportion of leucocratic and melanocratic components and chemical composition.

Hornblende amphibolites, or amphibolites proper, correspond by their composition to basalts of the normal series. They are widely extended and encountered throughout the layer at all depths. In large bodies, the observed differentiation of amphiboles results in the subdivision of layers into leucocratic and melanocratic parts.

Amphibolites with cummingtonite are characterized by the increased content of aluminium and magnesium and their chemical composition is analogous with that inherent to aluminiferous-magnesium basalts (AM-amphibolites). They are developed in the upper part of the section; this means that they occupy a fixed stratigraphic position.

Ferrous amphibolites (Fe-amphibolites) are rocks rich in brownish hornblende. They are distinguished by their black colour and by the irregular distribution pattern of rock-forming minerals within the limits of one body. In places amphibolites gradually pass into hornblendites, which are the rocks almost fully composed of hornblende. By their chemical composition these amphibolites correspond to ferruginous basalts. They occur in the lower part of the layer.

All types of amphibolite are irregularly epidotized, biotitized and in places pierced by plagioclase streaks. In marginal parts, biotite rims of several centimeters thick are observed rather frequently. Magnetite segregations are also observed, with their amount ranging from several percent to 20%. As a rule, these segregations are confined to the marginal parts of the amphibolite bodies where magnetite and quartz form pseudomorphs in hornblende.

Actinolitic and talc-actinolitic rocks (meta-ultramafites) are met in the form of thin bodies (a few meters), which are in conformable occurrence with gneisses and amphibolites. Larger bodies are of zonal structure: their marginal parts are composed of actinolite rocks, while the central parts consist of talc-actinolites. Phlogopite rims several tens of centimeters thick are confined to the marginal parts.

An irregular distribution pattern of main rock types in the layer of biotite-plagioclase gneisses and amphibolites makes it possible to subdivide it into two parts. In the upper part magnesium-aluminiferous amphibolites are widely spread. In the lower part biotite-plagioclase gneisses alternate with ferruginous amphibolites.

Down to a depth of 9456 m, the Archean complex in the Kola borehole section is composed of the two thickest strata, which by their mineral associations and rock composition are the most representative ones. They are confined to the upper gneissic part of the Kola series and rather widely developed at the surface, where they compose the gneiss basement fringing the Proterozoic Pechenga structure. Judging by seismic data, the borehole has crossed the middle part of the Kola series gneisses at the interval of 9456 – 11662 m. It is also represented by layers of biotite-plagioclase, and biotite-amphibole-plagioclase (less frequently) gneisses with highly aluminiferous minerals. The thickness of the middle part is considerably less than that of the two upper layers and, consequently, the frequency of one layer shifting into another becomes higher here, so that the impression may be gained that the middle part of the Kola series is more differentiated.

The composition of gneiss rocks in this part of the section is mostly similar to that of gneisses present in the upper strata; this is particularly true for the biotite-plagioclase and biotite-amphibole-plagioclase gneisses. For the muscovite-biotite-plagioclase gneisses with highly aluminiferous minerals, it can be said that individual layers are characterized by different associations of mineral indicators, i.e., in the third gneiss layer only sillimanite and garnet are found, in the fifth layer sillimanite, kyanite and garnet are present, and in the seventh layer sillimanite, staurolite and garnet are observed.

The saturation of gneiss with amphibolites is still not high here. About 30 – 35% of the total thickness of individual layers are made up of mafic rock layers. Amphibolites correspond without exception to basalts of the normal composition in all layer. Individual occurrences of amphibolites of considerably ferruginous and magnesium-aluminiferous composition, as well as layers of talc-actinolitic rocks (meta-ultramafites), are met at different hypsometric levels, irrespective of gneiss composition. Contrary to gneisses, which are very irregularly migmatized throughout the entire section, bodies of amphibolites were only partially involved in the migmatization process; in zones of contact with gneisses, amphibolite bodies exhibit a peculiar banded structure (alternation of quartz-feldspar and considerable amphibolite streaks). More frequently, in zones of contact with gneisses, amphibolites form a rim of a higher biotite content. At contacts between meta-ultramafites and gneisses, as in the upper layers, predominantly phlogopite is developed.

Amphibolites of all varieties are mainly in conformable occurrence with gneisses. Gneissosity, banding, and contact surfaces are commonly parallel to each other and normally form an angle of about 40° – 60° to the core sample axis; this means that in gneiss layers, medium angles of dip with respect to a horizon dominate. The only exception is the fifth gneiss layer with highly aluminiferous minerals (at a depth of 10,144 – 10,601 m), where parallel rock structures form an angle of 70° – 80° to the core sample axis, which is indicative of its predominantly subhorizontal attitude. The reason for such angular reorientation is not clear; one may only suppose that it is related to plicate tectonics which are revealed in the gneiss layer in a complex manner.

Gneissose plagiogranites and migmatites are widely developed in gneisses of the Kola series. Transition zones between gneisses and migmatites are gradual. In places, lighter-coloured rocks are observed among gneisses in the form of spots; these rocks merge with each other, whereas darker rocks are seen only as shadows. Gneissose plagiogranites with nebulitic structures appear here. At places, shadow structures fully disappear and only comparatively uniform gneissose plagiogranites are seen. Simultaneously, the sizes of mineral grains increase. Gneissosity in plagiogranites is preserved without any change in orientation, which remains that of gneisses. Such transformations are of multi-recurrent character throughout all layers.

Nevertheless, it seems evident that layers are mainly composed of granitized gneisses, while gneissose plagiogranites can be distinguished only conventionally.

Small bodies of granites and numerous types of aplite pegmatites have also been encountered in the Kola series. Granites are represented by biotite-bearing, compositionally uniform rocks. They form several isolated bodies of 10 – 15 m thick which occur at a depth below 9600 m. These bodies occur conformably within gneiss layers, and contacts between them and enclosing rocks are usually sharp. Characterized by high values of radioactivity, these granites are similar to porphyry-like granites of the Proterozoic complex outcropping east of the Pechenga structure. Aplite pegmatites are in both conformable and non-conformable occurrence with the gneissosity of the enclosing rocks. The following types of transition from gneisses to aplite pegmatite are found: in gneisses one can see aplite pegmatites which are in conformable occurrence with gneisses; these aplite pegmatites are characterized by irregular distribution of bands which are replaced by zones of gneisses with non-uniform, spotted distribution of aplite-pegmatite areas surrounded by gneisses with porphyroblastic plagioclases. Such zones are conjugated with aplite pegmatites in which gneisses have survived only in the form of relics. A parallel-oriented gneiss structure is traced in aplite-pegmatoid rock in the form of shadows without any change in orientation. Aplite-pegmatoid rock is composed of quartz, plagioclase, muscovite, and accessory minerals (garnet, sillimanite). Biotite relics are preserved in small amounts. The sizes of mineral grains are not uniform and grain distribution is irregular in pattern.

Amphibolites of veined bodies are met in gneiss layers at different depths, for instance, at 7470, 9900, and 10,050 m. These rare bodies of 10 – 15 m thick (they cross-cut gneisses) are of approximately the same age as rocks of the Proterozoic complex. Textures of the largest bodies of amphibolites differ from one zone to another: in marginal parts they are represented by fine-grained dark green amphibolites, in central parts by medium-grained “speckled” varieties with well-defined linearity oriented in subvertical direction.

### 1.2.3 The Physical Character of Rock Foliation

This problem will be especially dealt within the second section of the book. Here we shall briefly dwell on the distinctive geological features of the section which are responsible for the physicochemical properties of rocks.

The upper part of the Pechenga complex, 70% composed of igneous, poorly altered rocks of massive structure, is distinguished by the highest density (average values range from 2.88 to 3.01 gm $cm^{-3}$), low porosity and permeability. Due to the presence of very irregularly distributed sulphides and magnetites and the abundance of graphite, this part of the section is strongly differentiated by electromagnetic data. For instance, the specific electrical resistivity of rocks changes over an unusually broad range, i.e., from $1 \times 10^{-6}$ to $1 \times 10^{-9}$ Ohm $m^{-1}$. Magnetic properties also change over a broad range, particularly at places where gabbro-wehrlites with ore-bearing rocks are developed. According to data of gamma logging (GL), natural radioactivity of igneous rocks in this interval is expressed by its lowest value; it changes between 0.3 and 1.2 n · C $kg^{-1}$ $h^{-1}$ and increases in rocks of the Zhdanov series. The section is mainly composed of elastic-plastic rocks of high strength and abrasive resistance, which has predetermined the normal elliptical shape of the borehole cross section, its diameter being approximately equal to the size of the drill bit.

These parts of the section are characterized by the highest velocities of elastic oscillations in rocks, reaching 7 km $s^{-1}$ in some beds. According to data from seismic-acoustic studies, the average velocity of elastic-wave propagation is 6.4 – 6.6 km $s^{-1}$. Such high velocity values in rocks of the upper part of the crustal sequence, usually inherent in the basaltic layer, can be logically explained, i.e., they are caused by poorly altered massive tholeiite-basalt rocks which erupted at final stages of the Pechenga graben-syncline formation.

The lower part of the Pechenga complex is composed of trachyandesite basalts (80%). It is characteristic that they are metamorphosed in the epidote-amphibolite facies and mainly composed of foliated varieties. Correspondingly, these rocks are of much lower density than the overlying rocks. The average density here is 2.78 – 2.89 g $cm^{-3}$, the open porosity and permeability rise slightly. It appears from the measured electrical parameters that the section is poorly differentiated and is characterized by high values of specific resistivity. Magnetic properties of rocks indicate that a layer of metavolcanites of the subalkaline petrochemical type, containing abundant hematite and magnetite impregnations, can be readily distinguished here; this has resulted in a high magnetic susceptibility of the rocks. GL data also support the conclusion that the radioactivity values here are more differentiated; in spite of the fact that these values are not high (up to 2.2 n · C $kg^{-1}$ $h^{-1}$), they are twice as great as radioactivity values in volcanites of the upper layer. It is notable that increase in intensity of the prograde metamorphism in sedimentary strata is accompanied by abrupt decrease in radioactivity. For instance, according to GL data, radioactivity of rocks in the second sedimentary stratum, composed of quartz-sandstones and marbled dolomites, ranges only from 0.3 to 0.9 n · C $kg^{-1}$ $h^{-1}$.

The lower, Archean stage, is distinguished by a high degree of rock differentiation. The rocks are characterized by the widest range of rock density changes,

2.69 g $cm^{-3}$ for gneisses and 2.93 g $cm^{-3}$ for amphibolites. The porosity of the rocks somewhat increases (up to 1% on average), permeability also rises. The section is characterized by high values of specific resistivity and is in fact not differentiated by electrical properties. Magnetic susceptibility of the Archean complex on the whole is decreased with respect to the Early Proterozoic complex. With increase in depth, when a grade of metamorphism increases and magnetic minerals are destroyed, the difference between average values of the magnetic susceptibility of igneous and sedimentary rocks becomes less pronounced. However, it should be noted that strata of low magnetic rocks contain beds of ferruginous-silicate rocks which are characterized by maximal values of the magnetic susceptibility ($50-84/4\pi$).

These strata are notable for their increased radioactive background, from average values of $2\,n \cdot C\,kg^{-1} \cdot h^{-1}$ in amphibolites to $7.8\,n \cdot C\,kg^{-1} \cdot h^{-1}$ and higher in pegmatites and microclinic granites. Since the Archean rocks can be differentiated rather clearly by their content of radioactive elements, this means that radioactive logging methods should be considered the primary tools to be used for subdividing the Archean strata into various lithological units.

Due to the fact that the Kola well section is composed of gneisses and, since the strength properties of rocks are not uniform, the borehole cross-section increased by 20–30% over its nominal diameter, and the elliptical shape of the borehole became more elongated (at many intervals the ratio of small to large diameter is 1:2, and even 1:3).

The three-layered structure of the section is established more clearly by the elastic properties of rocks than by other physical parameters. Against all expectations, velocity values of wave propagation in the lower part of the well section, composed of the so-called granite-metamorphic layer of the Earth's crust, were approximately of the same order as those established for the overlying, considerably more uniform, andesite-basalt layer. The average velocity in the gneissic Archean rocks is rather high, and corresponds to the conventionally defined lower velocity boundary of the basaltic layer of the crust; this can be accounted for, to some extent, by a wide development of amphibolites (up to 35%) in the well section. It seems most probable that when calculating velocity values for rocks at the depth reached, the effect of temperature and rock lithostatic pressure should be taken into account.

### 1.2.4 The Stratigraphic Subdivision of the Section

The study of sedimentary-volcanic strata of the Pechenga graben syncline in the SG-3 well section makes the stratigraphic scheme of the Pechenga complex more accurate.

On the whole, this complex is subdivided by its facial-lithological composition, magmatism, and tectonic development into two megarhythms, each of which corresponds to a certain stage of the Pechenga trough development. The first megarhythm is associated with the formation of polymictic (basaltic) conglomerates, quartzites, carbonate rocks and rocks of the trachyandesite-basalt series. The second megarhythm is formed by arkose, greywacke sandstones, siltstones, phyllites with a considerable admixture of tuffaceous material, and volcanic rocks of the tholeiite-basalt series. Every megarhythm started its development under conditions of a more active tectonic regime and the formation of coarsely fragmented conglomerate-sandstone sedimentary strata; later, each of them was succeeded by a quiet tectonic environment where silty-clayey and carbonate rocks were developed. A lengthy hiatus separates these two megarhythms, evidenced by the presence of a weathering crust, local angular unconformities, and the major interscale tectonic zone which was identified both in the SG-3 well and at the surface (Luchlompol major fault).

Two series are distinguished in the Pechenga complex, i.e., the Nickel and Luostari series. Each of them corresponds to a particular stage (megarhythm) of the development of the Pechenga structure. These series are clearly subdivided into suites and composed of either sedimentary or volcanic rocks.

**Table 1.1.** Stratigraphic subdivision of the Pechenga complex (Lower Proterozoic) in the SG-3 well section

| Series | Suite | Interval drilled, m | Description of rocks | Vertical thickness, m |
|---|---|---|---|---|
| Nickel | Matertin (mt) | 9 – 1059 | Diabases of globular and massive lavas with subordinate layers of tuffs, picrite and pyroxene porphyrites | 1050 |
| | Zhdanov (zd) | 1059 – 2805 | Thin rhythms of siltstones and phyllites with layers of tuffs and sandstones. Oligomictic and polymictic sandstones with layers of conglomerates and siltstones | 1746 |
| | Zapolyarny (zp) | 2805 – 4673 | Actinolitized diabases of globular and massive lavas with subordinate layers of tuffaceous rocks | 1868 |
| | Luchlompol (lh) | 4673 – 4884 | Dolomites, dolomitic sandstones, arkose sandstones with layers of siltstones | 211 |
| Luostari | Pirtiyarvi (pr) | 4884 – 5642 | Metadiabases, meta-andesites, schists in diabases and andesites (banded lavas included) | 758 |
| | Kuvernerinyok (kv) | 5642 – 5717 | Tremolite-carbonate schists, marbled limestones, quartz-sandstones | 75 |
| | Mayarvi (ma) | 5717 – 6823 | Amphibole-plagioclase and biotite-amphibole-plagioclase schists (in diabases) | 1106 |
| | Televin (tlv) | 6823 – 6842 | Metasandstones and metagritstones | 19 |

Variations in rhythmic character, lithology, grain sizes, etc., allow subdivision of the suites into rock members which differ from contiguous members; however, it is difficult to trace them down the dip and along the strike, due to the scarcity of barren rocks and complex fault tectonics.

By taking into consideration the stratigraphic position of the Pechenga complex rocks and their bulk, the conclusion may be drawn that they correspond to rocks of the Early Proterozoic complex of the Karel epoch of diastrophism. The basal conglomerates are similar to rocks of the Sarioly series of the Karelia, the lower part of the section resembles rocks of the Segozero and Onega series of the Karelia and the upper part is correlatable with the Suiisari (sandstone-clay-pycrite-diabase) series.

Regarding the thicknesses of individual suites, series, and the whole Pechenga complex, data obtained in the SG-3 well have made it possible to give concrete figures on thickness. Previously, thickness values could only be assumed on the basis of a free interpretation of seismic data (Table 1.1). Since the borehole was sunk in the central part of the Pechenga structure, it seems probable that the calculated true thicknesses of suites and series in this borehole correspond to an average thickness of these stratigraphic units developed in the central part of the Pechenga trough. The uppermost Matertin suite is the only exception, the borehole being located in its subbottom part. Data from the SG-3 well indicate the remarkably persistent composition and thickness of sedimentary strata. Volcanic strata of the complex are less persistent. Their horizontal thickness along the strike and vertical thickness down the dip change over a broad range. There are good reasons to assume that the second volcanic strata wedge out in eastern and southern directions.

In the section of the Kola superdeep borehole, as distinct from the sequence of the Lower Proterozoic sedimentary-volcanic rocks, stratification of gneisses is not distinct. In fact, the section is devoid of the characteristic marker horizons or clearly seen contact zones between individual strata. Such criteria as migmatization-granitization of individual strata, and their saturation by amphibole layers of different types are not differentiated, being the same for all strata.

Nevertheless, in the SG-3 well section it became possible to distinguish layers composed of either biotite-plagioclase gneisses with stauroite, sillimanite, kyanite, andalusite, garnet, or biotite-plagioclase and biotite-amphibole-plagioclase gneisses, where orthite and epidote are observed instead of aluminiferous minerals (Table 1.2). In the upper part of the Kola series (6842 – 7622 m), the thickest layer was found; it is characterized by the greatest variety of highly aluminiferous minerals. Similar strata are also present in intervals of 9456 – 9573 m, 10,144 – 10,601 m and below 11,416 m. These facts refute the existing view that layers of gneisses with highly aluminiferous minerals are confined to the upper parts of the Kola series.

### 1.2.5 The Age of the Pechenga Complex and Kola Series

Micropaleontological determinations and geochronological studies of rocks and processes of metamorphism in the Kola borehole section were performed under the active support and scientific guidance of K. O. Krats and E. K. Gerling.

**Table 1.2.** Scheme of stratigraphic subdivisions of the Kola series (Archean) in the SG-3 well section

| Number of layer | Interval drilled, m | Description of rocks | Vertical thickness, m |
|---|---|---|---|
| I | 6842 – 7622 | Muscovite-biotite-plagioclase gneisses with HAM (andalusite, staurolite, sillimanite, garnet) | 780 |
| | 6842 – 7271 | Upper part of the layer: muscovite-biotite-plagioclase gneisses with HAM | 430 |
| | 7271 – 7622 | Lower part of the layer: muscovite-biotite-plagioclase gneisses with HAM and bodies of amphibolites | 350 |
| II | 7622 – 9456 | Biotite-plagioclase and biotite-amphibole-plagioclase gneisses and amphibolites | 1830 |
| | 7622 – 8465 | Upper part of the layer: biotite-plagioclase gneisses and amphibolites mainly with cummingtonite | 840 |
| | 8465 – 9456 | Lower part of the layer: biotite-plagioclase, biotite-amphibole-plagioclase gneisses and amphibolites | 990 |
| III | 9456 – 9573 | Muscovite-biotite-plagioclase gneisses with HAM (sillimanite, garnet) | 120 |
| IV | 9573 – 10,144 | Biotite-plagioclase gneisses and amphibolites | 570 |
| V | 10,144 – 10,301 | Muscovite-biotite-plagioclase gneisses with HAM (kyanite, sillimanite, garnet) | 460 |
| VI | 10,601 – 11,416 | Biotite-plagioclase gneisses | 820 |
| VII | 11,416 – 11,662 | Biotite-plagioclase gneisses with HAM (garnet, sillimanite) and amphibolites with garnet) | 280 |

To assemble a micropaleontological description of the Pechenga complex Timofeev (1979) has studied more than 90 rock samples taken mainly from metasedimentary rocks. As a result of this study, the evolution of genera and species of microfossils, sphaeromorphidas in particular, was established along the well section. The deepest horizons (Kuvernerinyok suite) were studied over the interval of 5710 – 5687 m (6 samples). Here, sphaeromorphidas, *Protosphaeridium rigidulum* Tim., *Trematosphaeridium* sp., *Symplassosphaeridium* sp. were found.

Over an interval of 4884 – 4793 m (Luchlompol suite), 16 samples were studied. They contain the sphaeromorphidas *Protosphaeridium tuberculiferum* Tim., *Protosphaeridium* sp., *Orygmatosphaeridium* sp. and *Gloeocapsomorpha* sp. This complex of microfossils can be correlated with a similar one found in the lower and middle parts of the Krivoy Rog and Kursk series.

Over the interval of 3882 – 3759 m (middle section of the Zapolarny series) ten samples were studied. They contain the sphaeromorphidas *Protosphaeridium laccatum* Tim., *P. tuberculiferum* Tim., *P. densum* Tim., *P. flexuosum* Tim.,

*P. accis* Tim., *P. rigidulum* Tim., *Orygmatosphaeridium* sp., *Gloeocapsomorpha priscata* Tim., *Gloeocapsomorpha* sp., *Symplassosphaeridium* sp., *Synsphaeridium* sp.

At a depth of 3755 m, one species in a state of division, *Protosphaeridium aciss* Tim., was found. This is the first finding of a eukaryot (species *Nucellosphaeridium minutum* Tim.); earlier in 1966, it was found in the Besovetsky suite in the area of Petrozavodsk town and described by B. V. Timofeev.

Over an interval of 2805 – 2172 m (Zhdanov suite), 28 samples were studied. The following sphaeromorphidas were found in this part of the section: *Protosphaeridium rigidulum* Tim., *P. aciss* Tim., *P. tuberculiferum* Tim., *P. densum* Tim., *P. flexuosum* Tim., *P. planum* Tim., *Protosphaeridium* sp., *Ocridosphaeridium* sp., *Orygmatosphaeridium distributum* Tim., *Trematosphaeridium* sp., *Trachysphaeridium laminaritum* Tim., *Gloeocapsomorpha* sp., *Trachysphaeridium laminaritum* Tim., *Gloeocapsomorpha* sp., *Synsphaeridium* sp. Thirty samples were extracted from the uppermost part of the Zhdanov suite (1981 – 1000 m), which is considered to be an area where sedimentary-metamorphic rocks (metasiltstones, meta-aleuropelites and phyllitic schists) were developed. Plant remnants of the greatest variety were met here. The following sphaeromorphidas were distinguished in 13 samples: *Protosphaeridium flexuosum* Tim., *P. tuberculiferum* Tim., *P. rigidulum* Tim., *P. laccatum* Tim., *P. patelliforme* Tim., *P. asaphum* Tim., *P. planum* Tim., *P. aciss* Tim., *Protosphaeridium* sp., *Stichosphaeridium implexum* Tim., *S. sinapticuleferum* Tim., *Orygmatosphaeridium* sp., *Trematosphaeridium* sp., *Gloeocapsomorpha* sp., *Pterospermopsimorpha* sp., *Trachysphaeridium laminaritum* Tim., *Zonosphaeridium* sp., *Ethmosphaeridium* sp.

An analysis of the evolutionary development of sphaeromorphidas (from top to bottom) permits the following conclusions:

1. at a depth exceeding 4700 m (over an interval of 5800 – 4700 m microfossils from the second and third sedimentary strata were studied) rocks of the pre-Yatuly series (Lower Proterozoic according to K. O. Krats) are developed; this allows correlation of the above-mentioned and lower parts of the Pechenga series with the Tungud series of southern Karelia, the Bolsheozersk series of western and southern Karelia, the Khirivinavolsk series of Northern Karelia, as well as with the Krivoy Rog series of the Kursk magnetic anomaly (KMA) according to their age;
2. rocks of the Yatuly and Suissar series (Middle Proterozoic according to K. O. Krats) are developed from the surface to a depth of 4700 m; rocks of the Yatuly series (sedimentary-volcanic section of the Segozero type) compose the lowermost parts of this section (4670 – 2800 m).

Of particular importance is the discovery of the ancient eukaryot (sp. *Nucellosphaeridium minutum* Tim.) extracted from an interval of 3882 – 3759 m; an analogous specimen was found earlier in the Besovetsk suite (area of the town of Petrozavodsk) and described by B. V. Timofeev, who assigned it to the Iotnic age (Upper Proterozoic according to K. O. Krats). The presence of such types of sphaeromorphidas in volcano-sedimentary strata of the third (from bottom to top) sheet of the Pechenga complex (i.e. among rocks of the Yatuly series) allows the assumption that at least the upper parts of the Pechenga graben-syncline section are of thrust nature (allochthone?). With this knowledge, it became possible to understand the nature of the great thickness (about 4000 m) of the Upper Karelides (Yatuly, Suiisary series) in the Pechenga zone which is two to three times greater than that of the analogous rocks of the Karelia where their stratotypes have been distinguished.

The results of the micropaleophytological studies in the Kola borehole support the previous conclusion (Duk 1977) that rocks of different age and type are present in the Pechenga complex: the lower part of the section is composed of the pre-Yatuly rocks which were deposited and affected by volcanism 2.3 – 2.2 Ga ago, while the younger rocks are 2.2 – 2.8 Ga old.

Indeed, according to K. O. Krats, O. A. Levchenkov, and others, a lower boundary of the Yatuly series can be defined by determining the age of the quartz porphyrites which underlie this series in the Lekht area of Karelia. The bottommost part of the series is represented by a crust of weathering. It was established that the age of the quartz porphyrites is 2455 Ma. It should also be noted that the age of the gabbro-diabase intrusions in the Yatuly sedimentary complex of the Karelia region is $2180 \pm 60$ Ma, as determined by the U-Ph isochron method. According to earlier determinations made by Finnish geologists, the age of the Yatuly series is 2050 – 2100 Ma. According to data published by Tugarinov et al. (1980), the age of the basal conglomerates of the Kursk series is 2180 – 2200 Ma. Of particular interest is the upper age limit of the Iotnic rock unit (not more than 1700 Ma) determined from the age of diabase dikes (1670 Ma; Tugarinov and Bibikova 1980).

In view of the information presented above, it appears that the boundary "4700 – 4800 m" in the Kola borehole section acquires a particular meaning. It is no accident that disaggregated rocks were encountered below this level, that the high-temperature greenschist facies passes into epidote-amphibolite facies at exactly this boundary, that the andesite-basalt magmatism is succeeded here by picrite-basalt magmatism, and that a major thrust is also confined to this level. It had earlier been established by seismic data that the bottom of the Pechenga complex corresponds to this deep level. Finally, it appears obvious that all anomalous age data responsible for the incorrect determination of the geological age were obtained over this same interval of 4700 – 5700 m (Gerling et al. 1983). In this report, all this information has served as a basis for distinguishing two major megarhythms of volcanic activity (the Nickel and Luostari series in the Kola well section).

The isotope age of the sedimentary-volcanic rocks of the Pechenga complex in the Kola SG-3 well section was determined mainly by the K-Ar technique, which is the most sensitive indicator of the superimposed processes. It is thus no accident that the complex and lengthy structural-metamorphic evolution of rocks is reflected in a wide range of K-Ar age values determined from the study of amphiboles, biotites, plagioclases, muscovites, etc. and rocks as a whole.

To determine the age of minerals and rocks by the K-Ar technique, the argon content was mainly measured by the volumetric method; however, from a depth of 7768 m, the isotope dilution method with $Ar^{38}$ as a tracer was applied. For calculations of age, the decay constant of $^{40}K$ was used: $\lambda k = 0.581 \times 10^{-10}$ years$^{-1}$;

$\lambda_\beta = 4.962 \times 10^{-10}$ years$^{-1}$; the amount of $^{40}K$ was calculated by using a formula: $^{40}K = K_{total} \cdot 1.167 \times 10^{-4}$ g year$^{-1}$; mean square root error ($\sigma$) in age determination aggregates the corresponding errors of K and Ar datings, while for the isotope-dilution method it is 3 – 5%. For the age between 1400 and 200 Ma, the mean square root error is 40 – 50 Ma (biotites, muscovites) and 70 – 100 Ma (amphiboles, plagioclases) if the volumetric method is used, and 30 – 40 and 70 – 80 Ma with the isotope dilution method. For the age 600 – 1000 Ma the error was no more than 20 – 30 Ma for amphiboles, plagioclases and for a large sample.

The age determined by the K-Ar technique was corrected by applying data obtained by other methods. As a result of the study of core samples from the SG-3 borehole, it was established that the K-Ar radiometric age of minerals and rocks changes over a wide range, varying from 600 to 2100 Ma. Age values determined by K-Ar hardly change with depth.

An analysis of statistical age maxima determined by the K-Ar technique, together with Rb-Sr isochron data (Gorokhov et al. 1982), and Pb isotope data for sulphides [$^{207}Pb/^{206}Pb$; $^{208}Pb/^{206}Pb$; (Vinogradova et al. 1959)] allows dividing the age dates obtained for the Pechenga complex into the three following groups:

I – from 2100 to 1500 Ma;
II (rejuvenated) – from 1100 to 600 Ma;
III (anomalously ancient) – from 2300 Ma to 13 Ga, representing incorrect (unreal) age due to local surplus of argon.

Ages established by the K-Ar technique for a depth interval of 0 – 6842 m in the Kola well (groups I and II) correspond fairly well with the known geochronological dates for the entire eastern part of the Baltic Shield (Pushkarev et al. 1978; Tugarinov and Bibikova 1980), and mainly represent a time when the Pechenga rocks were undergoing metamorphic transformations.

Over a depth interval of 4700 – 5700 m, the K-Ar method gave an exaggerated age, reaching anomalous values (13 Ga) and even exceeding the age of the solar system by $5 \times 10^9$ years. Anomalous age values were obtained for the strongly foliated metabasites, meta-andesites, metasandstones and marbles which compose a large zone of major faults permeable to fluids confined to a zone between the base of the Zapolyarny suite and the upper part of the Mayarvi suite. Age values determined by the U-He method for chlorite, carbonate, tourmaline and tremolite either correspond to those established by the K-Ar technique or exceed them due to surplus of argon and helium in the minerals examined. The origin of the surplus argon and helium in the minerals studied is of common interest; the Pechenga complex has served as an example in examining this problem (Gerling et al. 1983).

To determine the age of the Kola series over an interval of 6842 – 10,315 m, K-Ar and Pb-Pb techniques were applied (Tables 1.3 and 1.16). The K-Ar technique, which fixed the time of the Svecofennian (Karelian) structural-metamorphic reconstruction of the Kola series rocks, cannot be used for determining the absolute age of the basement rock. In view of this fact, it became necessary to apply other radiochronological methods. For this purpose we have taken seven series of zircons out of the biotite-muscovite-epidote migmatized gneisses over an

**Table 1.3.** Radiometric age of the granulitic metamorphism stage of the Kola series

| SG-3 well section | | | Kola-Norwegian and Belomorsk-Lapland zones (surface) | | |
|---|---|---|---|---|---|
| Depth, m | Method (mineral) | Age, Ma | Method (mineral) | Age, Ma | Reference |
| Biotite plagiogneisses and biotite plagiogneisses with highly aluminiferous minerals | | | Gneisses and amphibolites of the Kola series | | |
| 7150 – 7200 | $^{207}Pb/^{206}Pb$ (zircon) | 2704 ± 27 | $^{207}Pb/^{206}Pb$, isochron | 3150 ± 50 | Sobotovich et al. (1963) |
| 7150 – 7200 | $^{207}Pb/^{206}Pb$ (zircon) | 2725 ± 25 | $^{207}Pb/^{206}Pb$ (zircon) | 2700 | Tugarinov and Bibikova (1980) |
| 7600 – 7656 | $^{207}Pb/^{206}Pb$ (zircon) | 2785 ± 21 | Rb-Sr (gross sample) | 2700 | Gerling and Matveeva (1964) |
| 7600 – 7656 | $^{207}Pb/^{206}Pb$ (zircon) | 2800 ± 12 | | | |
| 9745 – 9755 | $^{207}Pb/^{206}Pb$ (zircon) | 2640 ± 15 | Rb-Sr, isochron | 2660 ± 40 | |
| 10,301 – 10,315 | $^{207}Pb/^{206}Pb$ (zircon) | 2650 ± 20 | K-Ar (hornblende, plagioclase) | 2430 – 2460 | |
| Amphibolites and amphibole schists | | | Granitoids of the lower part of the Pechenga river | | |
| 7973 – 7980 | $^{207}Pb/^{206}Pb$ (zircon) | 2620 ± 20 | Rb-Sr, model | 2830 ± 100 | Gorokhov et al. (1976) |
| 7967 | K-Ar (hornblende) | 2340 ± 80 | **Gabbro-anorthosites, enterbites** | | |
| 8409 | K-Ar (hornblende) | 2280 ± 70 | $^{207}Pb/^{206}Pb$ (zircon) | 2750 | Tugarinov and Bibikova (1980) |
| | Gneiss-plagiogranites | | U-Pb, isochron | 2803 ± 15 | Puskkarev et al. (1978) |
| 6900 | K-Ar (plagioclase) | 2770 ± 70 | | | |

*Note:* Determination of the radiometric age of zircons by the method of thermo-ionic emission was performed by V. G. Zaslavsky; the age of hornblende and plagioclase by the K-Ar method was determined by T. V. Koltsova.

interval of 7150 – 10,315 m (Table 1.3). The first two series were collected in intervals of 7150 – 7200 and 7600 – 7656 m. Zircons are small in size (0.1 mm), transparent, pink in colour, without inclusions, well faced. These zircons are characteristic of the Kola gneisses as a whole. The age determined proved to be 2700 – 2800 ± 20 Ma.

Three other series of zircons, collected over intervals of 7973 – 7980, 9745 – 9754 and 10,301 – 10,315 m, gave 2620 – 2650 ± 20 Ma. It should be noted that zircons with 2640 ± 15 Ma are similar in their morphology to the zircons previously described for the first two series; zircons of the other two series are poorly faced, sometimes very well rounded, contain inclusions, and are strongly dimmed.

Geological and petrological data show that the rocks examined are represented by diaphthorites (of the Kola age) in rocks of the granulitic facies. From this it appears that the determined dates should correspond only to the granulitic stage of metamorphism; an age of 2.7 – 2.8 Ga is considered to be the most reliable.

Granulitic metamorphism of the Kola gneisses in the Kola-Norwegian zone was manifested at approximately the same time (see Table 1.3); this was determined with the help of the Pb-Sr isochron technique (Gorokhov et al. 1976, 1982). According to Gorokhov et al. (1981) the Pb-Sr model isochron age of granitoids, which were formed synchronously with the manifestation of granulitic metamorphism, is 2830 ± 100 Ma (in the vicinity of the Kitsa railway station and the lower part of the Pechenga river). Data on these series of zircons are in good accordance with the age of zircons extracted from rocks of the Kola series (2830 ± 15 Ma); this age was fixed with respect to a ratio of $^{207}Pb/^{206}Pb$ by Bibikova et al. (1973), and was also determined for enderbites by using the U-Pb isochron technique (Pushkarev et al. 1978).

Thus the material examined demonstrates that the Baltic Shield craton was formed 2.7 – 2.8 Ga ago and later existed as the stable block which had undergone unidirectional rising and active denudation. During later stages of platformal magmatism, manifested in zones of deep-seated major faults, layered intrusions of basic composition and intrusions of charnockites formed; they penetrated into the shield body about 2.4 Ga ago (Tugarinov and Bibikova 1980).

## 1.3 Rocks and Rock-Forming Minerals

The study of rocks and minerals permits the identification in the SG-3 well section of volcanic, sedimentary and intrusive rocks which underwent metamorphism of various grades, as well as a description of metamorphic and ultrametamorphic rocks.[1] In the upper part of the section these are formed of the Proterozoic metamorphosed sedimentary-volcanic complex and in the lower part of the section they are represented by the Archean polymetamorphic complex.

### 1.3.1 Metamorphic and Intrusive Rocks of the Pechenga Complex

The proportion of volcanic rocks in the Pechenga complex exceeds 70%, the rest consists of volcano-sedimentary rocks. Volcanic rocks are mostly represented by basic rocks which are present throughout the entire section of the Pechenga complex. Volcanic rocks of medium composition in the lower part of the section are less developed. Even less developed are volcanic rocks of ultrabasic composition, being mainly confined to the upper part of the Pechenga complex. Volcano-sedimentary rocks contain tuffs, tuffbreccia of basic composition, conglomerates, gritstones, sandstones, siltstones, pelites and carbonates. Rocks are metamorphosed throughout the region, their grade of metamorphism increasing down the SG-3 well section from the prehnite-pumpellyite to the epidote-amphibolite facies; at places amphibolite facies of metamorphism was also observed.

#### Volcanic Rocks

The Matertin, Zapolyarny and Mayarvi suites are almost completely composed of volcanic rocks of basic composition. These are widely developed in the Pirttiyarvi suite and form thin individual sheets in the Zhdanov suite. In the upper part of the sequence they are represented by diabases, in which basalt relicts (Matertin suite) are preserved at some places, partially in marginal parts of individual sheets. Further down the section, diabases are intensively saussuritized and chloritized (Zhdanov suite). Even furtherdown, they are represented by actinolitized diabases (Zapolyarny suite). Over the interval of 4884 – 5642 m volcanites are formed of apodiabase magnetite-amphibole-plagioclase schists (Pirttiyarvi suite). Finally, in the lower part of the section they are represented by apodiabase

[1] The following researchers also participated in investigations carried out: M. I. Khotina, B. K. Kasatov, K. P. Sokolova, K. S. Mischenko, V. S. Kozlov, L. G. Kuznetsova, and V. N. Kurganova.

amphibole-plagioclase schists (Mayarvi suite). All these rocks differ from each other both in structural-textural features and in mineral composition. Fine-grained diabases and apodiabase rocks with grains of 0.1 mm and smaller occupy different positions in geological bodies. They form either marginal parts of large sheets or thin sheets. Medium-grained rocks with grains of 0.1 – 1 mm form the major part of sheets. Coarse-grained diabases and apodiabase rocks with grains of more than 1 mm are usually confined to the central parts of the largest sheets. Marginal parts of sheets and diabase balls are characterized by an amygdaloidal structure. These regularities in distribution of rocks by grain sizes within the limits of geological bodies are observed in apodiabase amphibole-plagioclase schists of the Mayarvin suite; this allows identification of individual sheets even among strongly metamorphosed rocks.

Volcanic rocks of the upper part of the section are characterized by the presence of magmatic textures and minerals. In the middle parts of the section rocks are characterized by features of primary magmatic and newly formed metamorphic textures and minerals together. In the lower part of the section, rocks are mostly represented by newly formed metamorphic minerals, textures and structures, while features of the primary rocks have survived only in the form of relicts.

Basalts are represented by fine-grained or glass-like varieties. They consist of monoclinic pyroxene-augite as well as of poorly and irregularly crystallized volcanic glass. Sometimes very thin laths of albitized plagioclase are met in the basalt mass.

Diabases are formed of monoclinic pyroxene-augite, salite (50 – 40%), plagioclase (30 – 20%); occasional grains of brown hornblende and quartz are also found. The proportion of chlorite, saussurite, epidote, prehnite, pumpellyite, calcite, scapolite, and axinite is insignificant. The amount of accessory minerals such as sphene and leucoxene ranges from 1 to 6%. Rare grains of apatite, zircon, olivine and fluorite are observed in samples of crushed rocks.

Ore minerals, normally in an amount of 1 – 5%, are represented by pyrrhotite with small ingrowths of chalcopyrite. Sphalerite and pyrite are also found. Diabases are mostly characterized by diabase, poikilophyric and porphyric textures.

Basalts are characterized by microlite-glassy, variolite-glassy, radiated and spinifex textures. The structure of the rocks is massive or amygdaloidal, the amygdales being filled with chlorite, quartz, calcite, and prehnite.

Actinolitized diabases consist of actinolite (50 – 70%), plagioclase, saussurite, epidote, and chlorite; sometimes calcite is present. Relicts of salite are also encountered. Beginning from a depth of 3700 m, rocks contain bluish-green hornblende which replaces actinolite. Sphene and leucoxene represent accessory minerals. Ore minerals, to the amount of 1 – 5%, are represented by pyrrhotite, chalcopyrite, pyrite, sphalerite, and pentlandite, with pyrrhotite predominating. Rocks are predominantly of relict, diabase, and poikilophytic texture in which idiomorphic laths of plagioclase (or pseudomorphs of saussurite in pyroxene) are combined with pseudomorphs of actinolite in pyroxene or with epidote-saussurite-chlorite rock mass. Actinolitized diabases are characterized by massive or amygdaloidal structure, but in the lower part of the Zapolyarny suite foliated and granonematoblastic textures appear. The amygdaloidal structure of the primary magmatic rocks is deformed and amygdales are lens-like in shape.

Apodiabase magnetite-plagioclase-amphibole schists are composed of plagioclase (30 – 50%), bluish-green hornblende (50 – 60%). The proportion of magnetite and ilmenite is from 5 to 15%. Hematite, occasional grains of chalcopyrite, bornite, and chalcocite are also found. Quartz, epidote, carbonate, biotite, and chlorite are observed in small amounts. The rocks are notable for their combination of magmatic and metamorphic textures. The rocks are granonematoblastic in texture, at places with relicts of diabases. Locally, rocks are characterized by a porphyroblastic texture, while the bulky rock mass is represented by a granonematoblastic texture. Porphyroblasts are formed of bluish-green hornblende and biotite of foliated, at places relict amygdaloidal structure.

Apodiabase amphibole-plagioclase schists are of simple mineral composition, containing about 50% hornblende and approximately 50% plagioclase. At places quartz, epidote and biotite are observed. Locally, rocks are irregularly biotitized and microclinitized. In cataclasmic zones apodiabase amphibole-plagioclase schists are chloritized, with carbonate, quartz and epidote being widely spread. Sphene is the characteristic accessory mineral. Ore minerals amount to 1 – 5%, and are represented by ilmenite, magnetite, hematite, pyrrhotite, chalcopyrite, and pyrite. The rocks are characterized by foliated structure and a granonematoblastic texture.

Volcanic rocks are developed in the middle part of the Pirttiyarvi suite, represented by magnetite-biotite-plagioclase apoandesite schists. The rock composition is as follows: plagioclase ca. 50%, biotite 20 – 30%, quartz – several percent, magnetite and ilmenite up to 10 – 20%. Sometimes rocks contain grains of potash feldspar, calcite, epidote and zoisite. Of the accessory minerals, sphene, zircon, apatite and leucoxene are present, the size of the mineral grains reaching only hundredths of a millimeter. Impregnated particles attain a size of tenths of a millimeter. Rocks are characterized by a combination of magmatic and metamorphic structures and textures, i.e., massive, amygdaloidal, fluidal structure, and porphyritic, trachytic, felsitic texture; foliated, cataclastic structure and granolepidoblastic, at places porphyroblastic, texture. Magmatic structures and textures are preserved in apoandesite magnetite-biotite-plagioclase schists in the form of relicts.

Ultrabasic volcanic rocks are represented by picrite porphyrites and occur only in the upper part of the section in the Matertin suite. They are composed of idiomorphic grains of monoclinic pyroxene, augite, and salite which are contained in considerably talc-serpentinite-chlorite fibrous fine-grained rock matrix. This thin-fibrous mass seems to have been formed at the expense of volcanic glass. Brown hornblende is met in rocks in small amounts. Accessory minerals are represented by occasional small grains of pyrite, sphene, ilmenite and apatite. Abundant magnetite is present in the form of magnetite "dust" in the talcose and serpentinitic rocks. The rocks are characterized by their porphyritic texture, with the main rock mass being represented by a microfibrous texture.

Down the SG-3 well section rock-forming minerals change. Examples of their chemical composition are shown in Table 1.4.

**Table 1.4.** Chemical composition of minerals from metabasites of the Pechenga complex

| Oxides | Diabase | | Actinolitized diabase | | | Apodiabase-amphibole plagioclase rock | | | Apodiabase amphibole-plagioclase schists | | |
|---|---|---|---|---|---|---|---|---|---|---|---|
| | Prehnite-pumpellyite facies of metamorphism | | Greenschist facies of metamorphism | | | Epidote-amphibolite facies of metamorphism | | | Amphibolite facies of metamorphism | | |
| | Sample 226 Rock | Sample 2305 Augite-salite | Sample 9585 Rock | Sample 9508 Actinolite | Sample 9508 Plagioclase | Sample 22,115 Rock | Sample 19,663 Hornblende | Sample 19,663 Plagioclase | Sample 21,704 Rock | Sample 21,704 Hornblende | Sample 21,194 Plagioclase |
| $SiO_2$ | 47.70 | 48.90 | 47.42 | 54.98 | 64.11 | 49.90 | 49.25 | 62.56 | 54.04 | 47.70 | 58.61 |
| $TiO_2$ | 1.36 | 0.92 | 0.83 | 0.07 | 0 | 0.79 | 0.48 | not ds. | 0.86 | 0.48 | not ds. |
| $Al_2O_3$ | 13.67 | 4.37 | 15.72 | 1.68 | 21.99 | 12.54 | 6.52 | 24.06 | 14.01 | 8.52 | 25.59 |
| $Fe_2O_3$ | 2.01 | 3.05 | 3.28 | 1.41 | – | 6.66 | 3.16 | – | 1.15 | 4.09 | not ds. |
| FeO | 12.01 | 10.50 | 9.70 | 11.27 | | 8.91 | 14.59 | – | 9.05 | 13.02 | not ds. |
| MnO | 0.20 | 0.25 | 0.17 | 0.18 | not ds. | 0.17 | 0.35 | – | 0.15 | 0.30 | not ds. |
| MgO | 7.00 | 14.66 | 7.00 | 15.97 | not ds. | 6.62 | 10.66 | – | 5.72 | 11.03 | not ds. |
| CaO | 6.30 | 12.60 | 9.50 | 12.84 | 13.84 | 9.65 | 11.11 | 5.32 | 9.23 | 11.88 | 7.49 |
| $Na_2O$ | 1.66 | 0.40 | 1.08 | not ds. | 9.86 | 2.73 | 0.77 | 8.69 | 2.52 | 0.81 | 7.21 |
| $K_2O$ | 1.66 | 0.07 | 0.47 | 0.08 | 0.10 | 0.97 | 0.47 | 0.16 | 1.27 | 0.60 | 0.19 |
| $H_2O^-$ | 0.22 | not ds. | not ds. | not ds. | not ds. | not ds. | – | not dt. | – | – | not dt. |
| $H_2O^+$ | not dt. | 0.82 | not dt. | 2.13 | not ds. | not dt. | 2.02 | not ds. | – | – | not dt. |
| $P_2O_3$ | not ds. | not dt. | 0.10 | not dt. | not ds. | 0.76 | not dt. | not ds. | 0.12 | not dt. | not dt. |
| l.o.c.[a] | 5.22 | not dt. | 3.47 | not dt. | not ds. | 1.16 | – | not ds. | 2.10 | 1.90 | not dt. |
| Total: | 99.01 | 100.14 | 98.69 | 100.61 | 99.19 | 100.45 | 99.45 | 100.77 | 100.22 | 100.33 | 99.09 |

*Note.* Analyzed by V. S. Belova and A. B. German.
[a] Loss on calcination.
not ds. = not discovered.
not dt. = not detected.

Monoclinic pyroxene is developed in diabases and basalts of the upper part of the section. From a depth of 3.5 km onwards it disappears and is completely replaced by amphibole.

Monoclinic pyroxene is represented by augite in basalts, by augite and salite in diabases, and by salite only in chloritized and saussuritized diabases. Gradually, with the increase of the grade of metamorphism, augite is replaced by salite and salite by amphibole. Augite and salite, when found together, both constitute the same grains of spotty zonation. Augite is brownish in colour and salite is greenish. Augite is characterized by: $Ng = 1.736 - 1.719$; $Np = 1.706 - 1.690$; $cNg = 48° - 42°$. Salite is distinguished by the lesser values of refraction indices: $Ng = 1.715 - 1.707$; $Np = 1.687 - 1.680$; $cNg = 40° - 38°$. Silicate chemical analysis data established that monoclinic pyroxene is a mixture of augite and salite (see Table 1.4). These subcalcium augite-salites are characterized by a considerable admixture of chromium, nickel, and cobalt. Microprobe analysis has shown a difference between augite and salite, by establishing that when augite is transformed into salite, the titanium and aluminum content in the mineral composition decreases. The complex composition of monoclinic pyroxenes was also established by X-ray analysis; these pyroxenes proved to be subcalcium augite-salites. All of them lack calcium and have a similar iron to magnesium ratio. Monoclinic pyroxene extracted from picritic porphyrites is also represented by subcalcium augite-salite, but is distinguished by a higher magnesium content.

Amphibole in diabases and in saussurized diabases is predominantly represented by actinolite. It appears only in the form of occasional grains. Actinolite is widely developed in the actinolitized diabases. It replaces salite and, as a result of this, pseudomorphs are formed. Actinolite is also developed in the form of spicular grains in saussurite-chlorite rock matrix or it forms large poikiloporphyroblasts. In the lower part of the section, actinolite is replaced by bluish-grey hornblende and almost fully disappears; it appears again in zones of cataclasm and diaphthoresis during a retrograde metamorphism. With the use of a microprobe, it was established that when salite is replaced by actinolite, the amount of titanium and calcium decreases (Fig. 1.18). In the transitional zone these elements form small grains of sphene-leucoxene (Fig. 1.19). The iron and aluminum content rises. Actinolite is characterized by slight changes in the refraction indices: $Ng = 1.650 - 1.660$, $Np = 1.625 - 1.640$ in rocks occurring in the upper part of the section; down the section these indices increase slightly – $Ng = 1.660 - 1.670$, $Np = 1.638 - 1.662$. The chemical composition of actinolite is characterized by medium values of iron content. Thermal analysis of actinolite has shown that at a temperature of about 1000 °C, an endothermal effect is observed. The appearance of this effect is shifted to a higher temperature as depth of actinolite occurrence increases down the well section. It was established by infraspectroscopy data that all actinolites studied are similar in composition. Some increase of ferruginous material in a mineral is observed down the section, and this effect is manifested in the increase of the relative intensity of the absorption band at $687 - 694\ cm^{-1}$.

Bluish-green hornblende appears in the actinolitized diabases from a depth of 3700 m. It is developed in actinolite during the prograde metamorphism. Bluish-green hornblende is mostly developed in apodiabase magnetite-amphibole-pla-

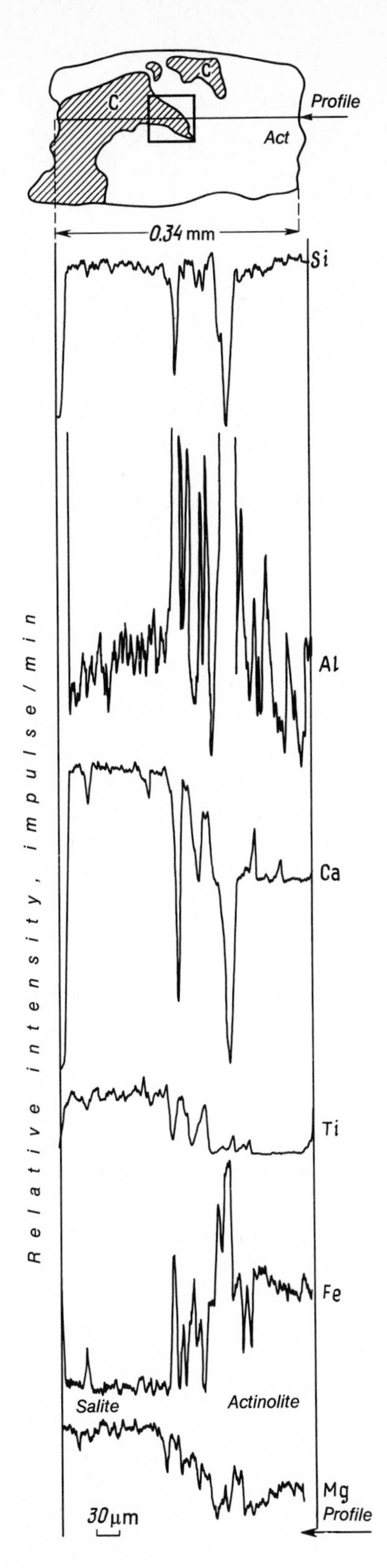


**Fig. 1.18.** Distribution of elements ($Ka_{1+2}$) along the profile, with salite (*C*) being replaced by actinolite (*Act*), sample No. 7729

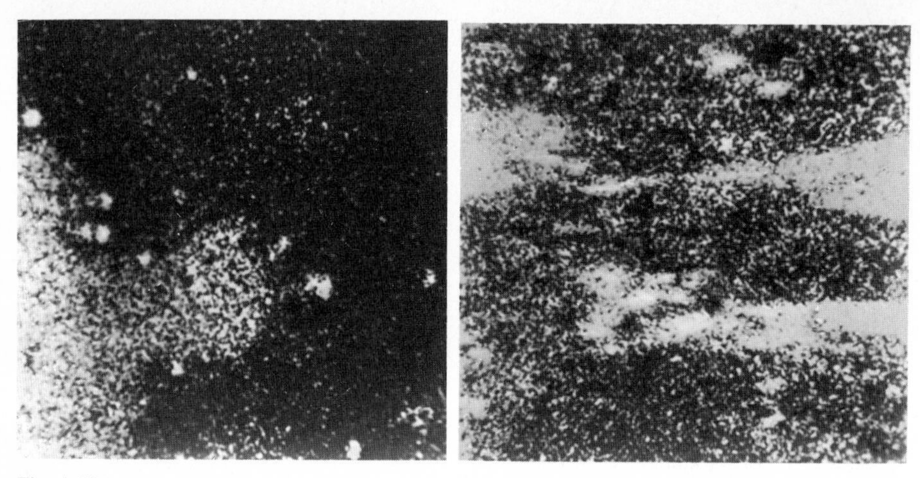

Fig. 1.19 Fig. 1.20

**Fig. 1.19.** Distribution of lines of the characteristic X-ray radiation ($Ka_{1+2}$) Ti in salite (*light colour*) and in actinolite (*dark colour*)

**Fig. 1.20.** Distribution of ($Ka_{1+2}$) Al in hornblende (*light colour*) and actinolite (*dark colour*)

gioclase schists (Pirrtiyarvi suite), where actinolite is preserved only in the form of relicts. Hornblende is also widely developed in the lower part of the section in apodiabase amphibole-plagioclase schists of the Mayarvi suite, where it replaces green hornblende during the retrograde metamorphism. Replacement of actinolite by bluish-green hornblende is accompanied by decrease in iron and titanium content; simultaneously, the content of magnesium and silicon increases (Fig. 1.20). The released ferrum is precipitated in the form of magnetite "dust". Refraction indices of the bluish-green hornblende change as follows: $Ng = 1.690 - 1.677$; $Np = 1.665 - 1.655$; $Ng\text{-}Np = 0.025 - 0.021$; $cNg = 16° - 18°$ (up to 22°).

The chemical composition of hornblende differs from that of actinolite by its higher of aluminum, sodium, and ferric iron content. X-ray examination shows that hornblende is characterized by parameter values of an elementary cell higher than those of actinolite. Study of bluish-green hornblende by infrared spectrometry demonstrates that infrared spectra are of intermediate character, i.e. they are between the infrared spectrum of actinolite and the brownish spectrum of green hornblende. This spectrum offers the possibility of establishing a gradual transition between these amphiboles. When heated, bluish-green hornblende is destroyed; the temperature for destruction is higher than that for actinolite, exceeding 1000 °C, fixed by the manifestation of an endothermal effect. Green hornblende with a brownish tinge is found in apodiabase amphibole-plagioclase schists in the lower part of the Pechenga complex. It is irregularly replaced by bluish-green hornblende, with no sharp boundaries between the two types. Green hornblende with a brownish tinge is unevenly coloured and characterized by refraction indices changing in the following way: $Ng = 1.686 - 1.676$;

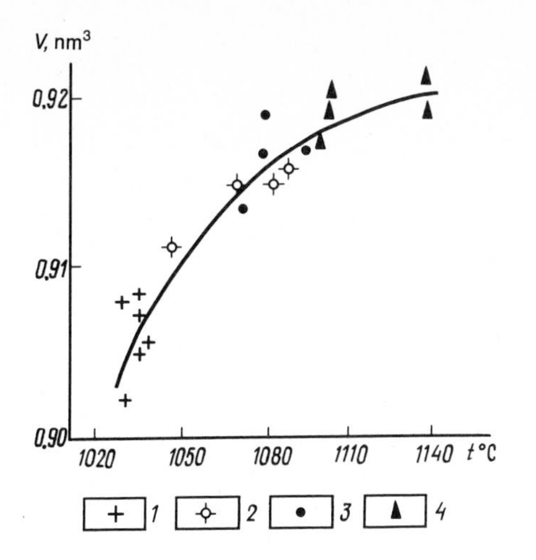


**Fig. 1.21.** Volume of an elementary cell versus temperature of the maximum of an endothermal effect of amphiboles belonging to a series of actinolite-hastingsite. *Facies*: *1* greenschist; *2* epidote-amphibolite; *3* amphibolite; *4* granulitic

$Nm = 1.670 - 1.665$, $Np = 1.660 - 1.648$; $Ng\text{-}Np = 0.024 - 0.028$; $cNg = 17° - 22°$. The chemical composition of hornblende (see Table 1.4) is distinguished by its high iron and aluminum content. A considerable amount of aluminum is contained in the sixfold coordinate of the mineral. X-ray data show that this hornblende is characterized by non-uniform parameters of an elementary cell. This means that the ratio of magnesium, iron, silicon and aluminum in the fourthfold coordinate varies. However, it must be noted that in general, the volume of an elementary cell increases in amphiboles from 0.906 to 0.912 $m^3$ in the direction from actinolite to hornblende, i.e., with increase in depth in the SG-3 well section (Fig. 1.21).

It must be emphasized that in metamorphosed basic volcanic rocks of the upper and middle parts of the SG-3 well section amphiboles are mainly represented by actinolite (Matertin, Zhdanov and Zapolyarny suites). In the lower part of the section bluish-green hornblende (Pirttiyarvi suite) and green hornblende with a brownish tinge (Mayarvi suite) are the main constituents of amphiboles (Fig. 1.22). Additionally, in areas of green hornblende with a brownish tinge, bluish-green hornblende and actinolite are also present in the form of diaphthorites; cummingtonite is very rarely found. In areas filled with bluish-green hornblende, actinolite serves as a diaphthoritic mineral.

Plagioclase in diabases is found in the form of chloritized, saussuritized and actinolitized idiomorphic laths. Sometimes these secondary minerals form pseudomorphs in plagioclase. In apodiabase magnetite-amphibole-plagioclase and magnetite-amphibole-plagioclase schists plagioclase is usually present in the form of rounded, lens-like xenomorphic grains. At places they are characterized by relicts of plagioclase present in the form of laths. In the upper part of the section plagioclase is represented by albite with $Ng = 1.538$, $Nm = 1.533$, and $Np = 1.530$. Plagioclase of more basic composition is preserved here only occasionally, usually it is represented either by oligoclase No. 30 with $Ng = 1.550$, $Nm = 1.546$, $Np = 1.543$ or by andesite No. 32 with $Ng = 1.552$, $Nm = 1.548$, $Np = 1.544$. Plagioclase extracted from apodiabase magnetite-amphibole-plagioclase schists is non-uniform in composition and changes from albite to oligoclase

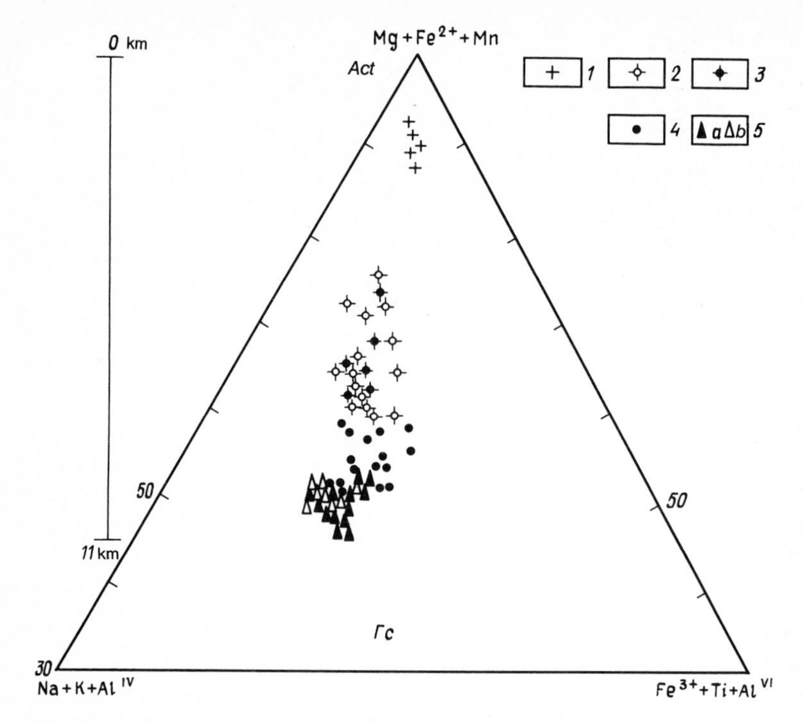


**Fig. 1.22.** Alteration trend in amphibole composition; amphiboles belong to a series of actinolite-gastinsite and actinolite-metabasite (vertical SG-3 well section). *Proterozoic complex*: *facies 1* greenschist; *2* epidote-amphibolite. *Archean complex*: *facies 3* epidote-amphibolite; *4* amphibolite; *5* granulitic (*a* from the SG-3 well; *b* from the area in the vicinity of the SG-3 well)

within individual parts of grains. For albite $Ng = 1.536$, $Np = 1.530$; for oligoclase $Ng = 1.542$, $Np = 1.535$. Plagioclase of the apoandesite magnetite-biotite plagioclase schists has the same composition. Plagioclase of the apodiabase amphibole-plagioclase schists also has a variable composition which changes irregularly from andesine to albite within one grain. For andesine $Ng = 1.554$, $Nm = 1.551$, $Np = 1.546$; for albite $Ng = 1.536$, $Nm = 1.530$, $Np = 1.528$. However, it must be noted that in the Proterozoic complex albite is replaced by andesine further down the section (see Table 1.4, Fig. 1.23).

X-ray data established that with the increase in depth, a parameter $\Delta 2\theta_{131-131}$ in plagioclase increases from 1.10 to 1.8, which indicates that the distribution of aluminum on octahedrons $T_2O$, $T_1M$, $T_2O$, $T_2M$ changes with depth. The variation in aluminum content in plagioclases with depth from 0.85 (Fig. 1.24) indicates that the degree of order in plagioclase decreases.

Analysis of X-ray data on the studied plagioclases, together with information about plagioclase composition, allows the conclusion that albites of the upper part of the section are characterized by a higher order of crystallinity structure and that they were formed at low temperatures. The degree of their order is equal to 100, their temperature of formation is about 350 °C. Belonging to the lowest order are the more basic plagioclases present in apodiabase-amphibole-plagio-

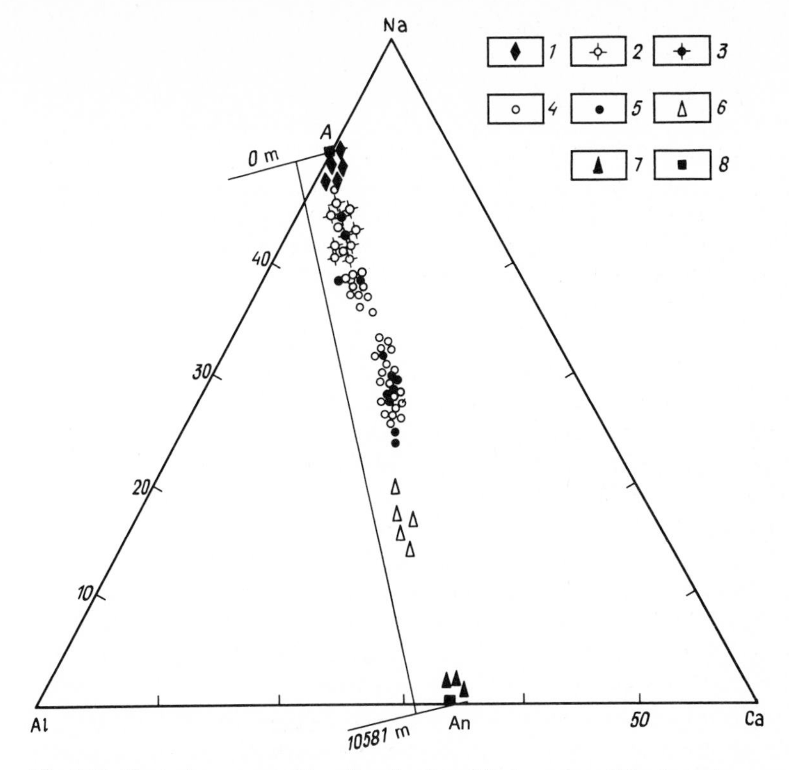


**Fig. 1.23.** Plagioclase composition alteration trend (in formula units); plagioclases are extracted from metabasites (vertical SG-3 section). *Proterozoic complex*: *1* diabases (Matertin suite); *2* actinolititized diabases (Zapolyarny suite); *3* apodiabase magnetite-amphibole-plagioclase schists (Pirttiyarvi suite); *4* apodiabase amphibole-plagioclase schists (Mayarvi suite). *Archean complex*: *5* epidotized amphibolites; *6* amphibolites; *7* garnet-amphibolites; *8* nominal compositions of albite and anorthite

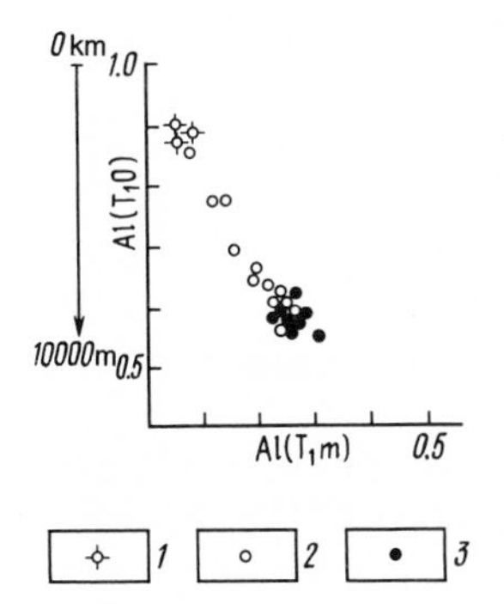


**Fig. 1.24.** Distribution alteration trend of Al in the structure of plagioclase from metabasites (vertical SG-3 well section). *Proterozoic complex*: *1* diabases (Matertin suite) and actinolititized diabases (Zapolyarny suite); *2* apodiabase magmetite-amphibole-plagioclase schists (Pirttiyarvi suite), apodiabase amphibole-plagioclase schists (Mayarvi suite); *3* amphibolites. Amount of Al present in crystalline lattice of plagioclase in tetrahedron $T_1O$ and $T_1m$

clase schists of the lower part of the section. The temperature of their formation is in the range of 500° – 550 °C.

Chlorite, widely developed in diabases, is found in the form of small scales in the main rock matrix. Sometimes it is developed in plagioclase together with saussurite. In actinolitized diabases it is present in small amounts. In these rocks chlorite is represented by a monotonous magnesium-iron type with

$Ng = Nm = 1.630 - 1.640$ and is characterized by the bluish-gray colour of the anomalous interference.

According to heat curves obtained by thermal analysis, it appears that in all cases a thermal effect is established at a temperature of 580° – 650°; this effect is caused by a process of mineral disintegration.

In apodiabase magnetite-amphibole-plagioclase and amphibole plagioclase schists, chlorite is met in zones of cataclasm and is developed at the expense of amphiboles and biotites, being regarded as a diaphthoritic mineral. It is distinguished by lesser refraction indices ($Ng = Nm = 1.612 - 1.626$), indicative of its poorer iron content.

Saussurite is developed in plagioclase in the form of micro-grained aggregates, it forms pseudomorphs and is also present in the main mass of diabases. Epidote and clinozoisite are tightly combined with saussurite and form idiomorphic grains. Epidote is represented by a variety rich in iron with $Ng = 1.770 - 1.778$, $Np = 1.722 - 1.729$; an endothermal effect at the heat curves is observed at a temperature of about 980 °C. Minerals of the epidote group in apodiabase magnetic-amphibole-plagioclase schists are predominantly associated with zones of rock cataclasm.

Volcanic rocks of the SG-3 well section are attributed to two megarhythms of volcanic activity: the lower trachytoandesite-basalt rhythm, which includes rocks of the Mayarvi and Pirttiyarvi suites, and the upper picrite-basalt rhythm, which combines rocks of the Zapolyarny and Matertin suites. The rocks are irregularly metamorphosed. They are characterized by a combination of signs inherent in magmatic and metamorphic rocks; the composition of minerals is usually nonuniform. Further down the well section, massive rocks are replaced by foliated ones.

Data on mineral paragenesis and mineral composition indicate that further-down the SG-3 well section the grade of prograde regional metamorphism of rocks increases and passes from the prehnite-pumpellyite to epidote-amphibolite and amphibolite facies.

## Sedimentary Rocks

Tuffs, tuffbreccia, phyllites, siltstones, and sandstones dominate among the volcano-sedimentary rocks, whereas gritstones, conglomerates, sedimentary breccia, carbonates, and lava breccia are less developed.

Tuffs, tuffites, and tuffbreccia are widely developed in the Matertin suite, whereas in the Zhdanov, Zapolyarny and Pirttiyarvi suites they are less frequent. Tuffs of basic composition predominate. In the Pirttiyarvi suite metamorphosed tuffs of andesite composition are developed. The following types of tuff are distinguished by the composition of their fragments: lithoclastic, crystalloclastic, crystallovitroclastic, and vitroclastic. According to the size of fragments, the tuffs are subdivided into the following types: tuffbreccia (size of fragments 1 – 10 cm), gritty-psammitic (0.1 – 1 cm), psammitic (0.01 – 0.1 cm), silty-pelitic and pelitic (less than 0.01 cm). Fragments formed of volcanic glass, diabase, diabase porphyrite, and quartzic diabase amount to 70 – 80%; tuffs of medium composition are represented by andesite porphyrite, disintegrated volcanic glass, fragments of plagioclase, and sometimes by quartz. The main part of the rock fabric is composed of disintegrated ashy material which is frequently chloritized and actinolitized. The grading of tuff material is poor, fragments are not well rounded. Fragments are represented by albite, diabase, crystallized glass, plagioclase, quartz in chess-board pattern; leucoxene and sulphides are also present. Cement of the basal type is formed of pyroclastic and clay material and for this reason the majority of tuffites resemble chloritic phyllites and ashy tuffs.

Phyllites and phyllite-like schists are widely developed in the Zhdanov suite. Macroscopic study shows that they are represented by fine-grained, frequently rhythmically layered dark grey or almost black rocks. Phyllites usually contain sulphides (pyrite-pyrrhotite, etc.) which are either seen as impregnations or as streaks, pockets and nodules. The presence of carbonaceous matter (0.2 – 3.5%) is a notable feature. The size of fragments serves as a criterion for distinguishing between pelitic and silty-pelitic varieties of phyllites. By mineral composition they are represented by chloritic (most widely developed), sericite-chlorite, and sericitic varieties.

Siltstones and sandstones are distinguished from each other by the size of the terrigenous particles. Such intermediate varieties as sandy siltstones and silty sandstones are also present. Fragments are mostly formed of quartz, plagioclase, microcline, and occasionally of amphibole, pyroxene and garnet. The Zhdanov suite contains fragments of volcanic glass, ashy material and angular lath-like fragments of plagioclase; all these fragments are considered to have been formed as a result of pyroclastic outbursts. Cement present in siltstones and sandstones is usually of basal type, occasionally of contiguous or pore type. The composition of the cement is similar to that of phyllites; in places cement can be attributed to the quartz-sericite, carbonate, chlorite-carbonate or siliceous types. Sandstones and siltstones of different suites are distinguished by their fragment composition. The upper part of the Zhdanov suite is formed mainly of tuffaceous and quartz-plagioclase varieties, whereas its lower part is represented by oligomictic rocks. Arkosic sandstones and siltstones are widely developed in the Luchlompol suite; quartzites and metamorphosed quartz-feldspar sandstones and siltstones are abundant in the Kuvernerinyok suite. Rocks are also distinguished by their ore mineral composition. In the Zhdanov suite ore minerals, which amount at places to 10 – 15%, are represented by sulphides (pyrite, pyrrhotite, etc.); sphene and leucoxene are also present. Arkosic sandstones of the Luchlompol suite are enriched with magnetite and hematite. Some layers of the Kuvernerinyok mica-quartz schists which were developed in siltstones contain apatite (up to 3 – 5%) and epidote.

Sedimentary breccias were identified for the first time on the basis of data obtained in the SG-3 well, and are developed in the Zhdanov suite. Macroscopically, they are layered grey-coloured silty-psammitic rocks with lens-like fragments of pelites. The main rock fabric is composed of tuffaceous siltstone, tuffaceous sandstone or their tuffaceous varieties. The whole complex of the investigations indicates that these rocks were formed in a period of sedimentogenesis. Shallow, strong bottom currents are considered to be the most favourable environments for their formation. Limestones, dolomites, limy schists, sandy dolomites and carbonate sandstones are not widely developed in the Zhdanov suite; in the Luchlompol and Kuvernerinyok suites they form individual members. In the Luchlompol suite pure limestones and dolomites are developed in some places only, most of them containing various amounts of clastic material which is mainly represented by quartz, plagioclase and also, although in lesser amounts, by microcline. Rocks of the Kuvernerinyok suite are predominantly composed of calcitic and dolomitic marbles which frequently contain tremolite, phlogopite, sometimes scapolite and talc. Tremolitic and tremolite-carbonate schists are also present.

Gritstones and conglomerates in the Zhdanov suite are encountered in the form of layers in sandstones which occur at the bottom of the section. They form the lower parts of rhythms. In the Luchlompol suite they are confined to a member of arkosic sandstones.

Gritstones are characterized by their irregularly grained structure. The sizes of terrigenous material usually range from 0.2 to 2 – 5 mm. In the Zhdanov suite gravel fragments, up to 5 mm in size, are well-rounded and mainly represented by quartz, although fragments of phyllites and diabases are also present in lesser amounts; the Luchlompol suite contains gravel fragments of microcline. Conglomerates of the Televin suite have been studied in most detail in outcrops. The presence of different sizes of fragments allows their subdivision into pebble-, boulder- and block-like types. They are polymictic in composition. Pebbles are formed by the basement rocks: granitoids, amphibolites, gneisses; quartz and microcline pebbles are also found. Pebbles are either well- or medium-rounded. Cement is represented by sandstones.

Volcano-sedimentary rocks of various suites are characterized by common features of mineral composition; however, some differences are also observed. Rocks of the Zhdanov suite are composed of the following main rock-forming minerals: quartz, plagioclase, sericite, chlorite, carbonates (mainly calcite); microcline is also met and the presence of carbonaceous matter is a rather notable feature. Minerals of heavy fractions of the crushed rocks (from individual values to 3%) are represented by actinolite, hornblende, hastingsite, epidote, monoclinic pyroxene (diopside, salite, ferrosalite, hedenbergite, chromium diopside, aegirine diopside, aegirine), rhombic pyroxene (bronzite, hypersthene), almandine, pyrope, grossular almandine, staurolite, prehnite, pumpellyite and olivine. Apatite, zircon, sphene, fluorite, tourmaline, barite, corundum and martite are also found. A characteristic feature of these rocks is the permanent presence of pyrrhotite and, in lesser amounts, of pyrite; chalcopyrite, ilmenite, sphalerite, galenite, arsenopyrite, hematite and limonite were also discovered.

The mineral composition of the rocks of the Luchlompol suite differs from one part of the section to another because the amount of different minerals in the rocks varies. Dolomite is the most characteristic mineral for the upper member, whereas quartz, plagioclase, microcline and calcite dominate in the lower member. Heavy fractions of the crushed rocks contain hornblende, hypersthene, augite, ferrosalite, aegerine diopside, aegerine, almandine, staurolite, epidote, anhydrite, zircon, rutile, orthite (up to 15.3% of the heavy fraction), brookite, fluorite, tourmaline, sphene, apatite and anatase in sandy dolomites. The Luchlompol rocks are enriched by hematite and magnetite; martite, leucoxene and ilmenite are also present. Carbonate rocks are distinguished from sandstones and siltstones by the composition of the heavy fraction. They are enriched by such accessory minerals as tourmaline, rutile, anatase and apatite.

The data on mineralogical composition show that the rocks of the Luchlompol suite differ from those of the Zhdanov suite by a greater development of dolomite, the presence of an accessory mineral such as anatase, the almost complete absence of ore minerals such as sulphides and by the great amount of hematite and magnetite. Their heavy fraction contains a lesser amount of epidote, rutile, ilmenite, actinolite, tourmaline; prehnite, pyrope and chromite are absent.

The mineral composition of rocks from the Kuvernerinyok suite is clearly distinguished from that of rocks from the Zhdanov and Luchlompol suites: (1) carbonate rocks are characterized by a more variable composition of carbonates, represented by calcite, dolomite, occasionally by magnetite and ferrodolomite; tremolite, anthophyllite and phlogopite are also present; (2) clastic material is represented predominantly by quartz; (3) the heavy fraction is characterized by less diversity of minerals. Rocks of this suite contain varieties enriched in orthite (in individual cases amounting to 65% of the heavy fraction); ore minerals are met in negligible amounts, magnetite being the most widely spread mineral, and hematite less frequent; pyrite, chalcopyrite, limonite and marcasite are also found.

The principal minerals of the Televin suite are plagioclase, quartz, microcline, biotite and muscovite; apart from these, chlorite and epidote are present. The following minerals have been discovered in the heavy fraction: monoclinic pyroxene, staurolite, almandine, zircon, apatite, rutile, tourmaline; pyrrhotite, pyrite and ilmenite (predominates) are the representatives of ore minerals; leucoxene and magnetite are found as occasional grains. The poorest composition of the heavy fraction is characteristic of the Televin suite.

Quartz is one of the most widespread minerals, mainly of clastic origin. In diagenesis quartz grains underwent regeneration and were recrystallized during the processes of metamorphism.

Plagioclase is met in amounts approximately equal to those of quartz. In rocks of the Zhdanov suite it is present in the form of lath fragments or poorly rounded grains. In rocks of the Kuvernerinyok and Televin suites plagioclase grains are frequently round in shape. Plagioclase is sericitized and saussuritized.

In rocks of the Zhdanov suite plagioclase is represented by albite and oligoclase (Nos. 3 – 27). Andesine (No. 35) is met occasionally. In rocks of the Luchlompol suite the composition of plagioclase is more uniform (oligoclase No. 18 predominates); in rocks of the Kuvernerinyok and Televin suites oligoclase and oligoclase andesine (Nos. 17 – 30) are developed; in rocks of the Televin suite andesine (No. 40) is also present.

Microcline is found in small amounts in individual varieties of sandstones and siltstones of the Zhdanov suite. In greater quantity it is present in sandstones of the Luchlompol and Televin suites. Refraction indices of microcline ($Ng = 1.527 - 1.529$, $Np = 1.519 - 1.521$) from rocks of the Zhdanov suite are higher than those of microclines from the underlying suites ($Ng = 1.522 - 1.526$, $Np = 1.515 - 1.518$). This is caused by the presence of greater amount of sodium components (23 – 28%) in microcline of the Zhdanov rocks. Microclines from rocks of other volcano-sedimentary suites contain a lesser amount of sodium components, and sometimes these are completely absent. On the basis of X-ray studies, it was established that microcline deprived of sodium admixture is characterized by $\Delta p = 1.05$, which corresponds to 100% molar content of the potassium phase. These data also indicate that this microcline is of a higher order; this allows its attribution to the maximal microclines.

Chlorite is the main constituent of the cement in rocks of the Zhdanov and Luchlompol suites. In the Kuvernerinyok and Televin suites the main part of the cement is represented by mica, which is developed in feldspars, dark-coloured minerals, pelitic matter, fragments of volcanic rocks and biotite; intergrowths between chlorite and mica are also found. In rocks of the Zhdanov suite, two types of chlorite are established: iron-magnesian and magnesian chlorites. The first ($Ng = Nm = 1.620 - 1.642$, with high iron content ranging from 48 to 75%) is a commonly developed type, the second ($Ng = Nm = 1.560 - 1.590$, with iron content ranging from 0 to 14%) is also met. In rocks of the Luchlompol, Kuvernerinyok and Televin suites, iron-magnesian chlorites with iron content between 47 and 71% ($Ng = Nm = 1.625 - 1.645$) are present.

Sericite is widely developed in the cement of siltstones, sandstones and phyllites of the Zhdanov suite, where it is associated with chlorite. High values of refraction indices ($Ng = 1.614 - 1.616$) are a characteristic feature of mica of this type. Endothermal effects of sericite are within a temperature range of 700° – 750 °C, which allows it to be attributed to phengite. Additionally, in rocks of the Zhdanov suite large flakes of colourless mica are met, confined to cleavage fissures. They are distinguished by smaller refraction indices ($Ng = 1.606$, $Nm = 1.599$, $Np = 1.558$) and $T_{b.e} = 900\,°C$.[2]

Sericite from sandstones of the Luchlompol suite is characterized by an intense endothermal effect at a temperature of about 850 °C, which is typical for muscovite. Colourless mica from rocks of the Kuvernerinyok and Televin suites is notable for low refraction index values ($Ng = Nm = 1.587 - 1.603$) and in the Kuvernerinyok suite it is found in association with mica of the phlogopite-biotite core. Thus, down the sequence of volcano-sedimentary rocks, mica of the hydromuscovite and phengite type passes into muscovite; this is explained by the increased grade of metamorphism.

Biotite is present in small amounts in volcano-sedimentary rocks in the lower part of the Zhdanov suite. In the underlying volcano-sedimentary suites it is developed in considerable quantity. In rocks of similar composition, which belong to various suites (for example, quartz-feldspar sandstones), biotite is characterized by somewhat different properties. In rocks of the Zhdanov suite dark brown biotite has high refraction indices ($Ng = Nm = 1.645 - 1.671$) and is characterized by high total iron content (from 56 to 71%). In rocks of the Luchlompol suite biotite is of lighter colour (light brown) and its refraction indices decrease ($Ng = 1.631 - 1.645$); the total iron content is 49 – 56%. In rocks of the Kuvernerinyok suite light-coloured mica of the phlogopite-biotite series is developed; it is represented by two varieties: one, found in sandstones and siltstone, is characterized by higher iron content ($Ng = 1.600 - 1.622$, total ferrum content is

[2] $T_{b.e}$ = temperature of the beginning of an endothermal effect.

31.42%) and the second, met in marbled dolomitic limestones, is represented by light-coloured phlogopite ($Ng = 1.587 - 1.597$, total iron content is 24 – 28%).

In sandstones of the Televin suite, as compared with biotites from sandstones of the Kuvernerinyok suite, iron content in biotites (48 – 52%) slightly increases, but is still less than that of biotites from sandstones of the Luchlompol suite. A correlation of iron content of biotites from analogous rocks of various suites shows that it decreases down the section; this appears to reflect the increase in temperature of metamorphism with depth.

Carbonaceous-bearing matter is present in phyllites and siltstones of the Zhdanov suite and in tuffaceous rocks of the Matertin suite. A small amount of carbonaceous matter is contained in sedimentary rocks of the Luchlompol and Kuvernerinyok suites, dispersed in rocks in the form of dust or small scales of 0.01 – 0.03 mm. According to thermal analysis data, the starting temperature of final combustion of the carbonaceous matter rises down the section as the grade of metamorphism increases. In rocks of the Matertin suite and in the upper part of the Zhdanov suite (to a depth of about 1700 m), $C_{org}$ is represented by its least metamorphosed variety with $T_{b.e} = 480° - 520\,°C$, with mineral temperature (480 °C) being characteristic of $C_{org}$ of the uppermost part of the section. At a deeper level, rocks of this suite predominantly contain cryptocrystalline graphite ($T_{b.e} = 650\,°C$). Thus, the present information shows that down the section degree of order in carbonaceous matter increases. Carbonates are included in the sandstone and siltstone cement. They form calcareous schists, limestones, dolomites and marbles. Calcite is the most widespread mineral (No. = 1.658 – 1.662), it can be determined on the base of thermal analysis data by the presence of an endothermal effect at a temperature of about 820 °C. Calcite is followed by dolomite (in abundance) and is characterized by No. = 1.687 – 1.689. Its endothermal effects appear at temperatures of 820° and 920 °C. Magnesium carbonate (No. = 1.689) and ferrodolomite were met in marblized limestones of the Kuvernerinyok suite in small amounts.

The study of the mineral composition of volcano-sedimentary rocks made it possible to arrive at a conclusion as to the source of sediments. It was established that terrigenous material was formed as a result of the erosion of the granite-gneiss basement of the Pechenga complex and of the underlying strata mainly composed of magmatic rocks of basic and subalkaline composition. The basement rocks were a source of the following minerals: microcline, rounded fragments of disintegrated plagioclase, quartz, hornblende, hypersthene, diopside, almandine, staurolite, zircon, orthite, tourmaline. Rocks of the Pechenga complex (basic and ultrabasic magmatic rocks) were a source of plagioclase (angular, semi-angular fragments), titanium-augite, salite, ferrosalite, hedenbergite, chrome diopside, leucoxene, and of minerals formed as a result of alteration of feldspars or volcanic glass (pumpellyite, prehnite, and partially chlorite). A part of the material supplied (angular fragments of volcanic glass, fragments of plagioclase laths, ashy material) was a result of pyroclastic outbursts. Subalkaline and alkaline rocks produced aegirine, aegirine-diopside, and hastingsite. Basement rocks contributed material to two lower layers; however, a part was brought there at the expense of pyroclastic outbursts. In the two upper layers terrigenous material is more variable; it was derived from the eroded basement rocks as well as from rocks of the underlying strata of the Pechenga complex and from subalkaline rocks. The amount of material formed as a result of the erosion of the Pechenga complex increases up the section and predominates in the upper part of the Zhdanov suite. The proportion of material brought to this part of the Zhdanov suite at the expense of pyroclastic outbursts also increases.

In volcano-sedimentary rocks of similar composition (in sandstones and siltstones), the iron content decreases in a biotite group as the grade of metamorphism rises; hydromica is replaced by muscovite (phengite was an intermediate variety which succeeded hydromica). As the grade of metamorphism rises, the low-grade metamorphic varieties of $C_{org}$ are replaced by finely scaled graphite, with cryptocrystalline graphite having been an intermediate rock in the transformation process.

## Intrusive Rocks

In the SG-3 well section, intrusive rocks are represented by hyperbasites, gabbro-diabases and dacite-andesite porphyrites. In the SG-3 well, hyperbasite bodies are represented by wehrlites composed of fully serpentinized olivine (50 – 70%) and monoclinic pyroxene (30 – 40%). Wehrlites in the SG-3 well are more or less serpentinized, talcized, chloritized and amphibolitized. The least altered hyperbasites in the SG-3 well are represented by serpentinized peridotites (wehrlites), which consist of the following minerals: monoclinic pyroxene, serpentinite, hornblende, chlorite, in places with tremolite and biotite, sulphides (which are always present); pyrrhotite, pentlandite, chalcopyrite, pyrite and accessory minerals (magnetite, ilmenite, apatite, garnet, chromite). Rocks are characterized by a relict poikilophyric texture. Sulphide copper-nickel mineralization has been established in hyperbasite bodies (see Chap. 1.6).

Serpentinites are represented by altered wehrlites, in which primary olivines and pyroxenes are fully serpentinized. The rock consists of serpentine, chlorite (about 60‰), talc (up to 20%) and tremolite (up to 10%). Of the primary minerals, relicts of pyroxenes and hornblende are met occasionally; accessory minerals and sulphides are the same as in the previous group. They are characterized by a reticulate texture; in places sulphides fill a space between pseudomorphs in olivine and monoclinic pyroxene. Strongly altered, talcized, chloritized and amphibolitized peridotites, in which primary minerals and textures are scarcely preserved, have received their names from the presence and prevalence of one or other secondary mineral. In the SG-3 well section they are represented by serpentine-chlorite-talc, chlorite-tremolite-talc, serpentine- or chlorite-carbonate-talc, chlorite-tremolite and hornblende-tremolite rocks.

The textures of rocks are distinguished by their mineral composition; if talc predominates, then the texture is of the microlepidoblastic type; at those places where newly formed tremolite and actinolite are present, rocks are characterized by a nematoblastic texture.

There are two varieties of monoclinic pyroxene which differ from each other by colour and refraction indices. Pinkish brown pyroxene ($Ng = 1.725 - 1.728$,

$Np = 1.697 - 1.701$) is the most widespread mineral; brownish-green pyroxene ($Ng = 1.717 - 1.722$, $Np = 1.688 - 1.629$) is present in much lesser amount. Monoclinic pyroxenes of hyperbasites from the SG-3 well are very similar in their optical and morphological properties to the pyroxenes of hyperbasites known in the Zhdanov massif. Monoclinic pyroxenes belong to a series of augites. It is observed that the amount of iron, titanium, aluminum and sodium regularly increases from the bottom to the top of the massif, whereas the amount of magnesium, nickel, cobalt and chromium decreases.

Olivine in wehrlites of the SG-3 well is fully serpentinized. In wehrlites of the Zhdanov massif it is represented by chrysolite – $Fa_{20-25}$ ($Ng = 1.714$, $Nm = 1.696$, $Np = 1.674$) and sometimes by $Fa_{10}$.

Brown hornblende is met in the form of individual grains in serpentinized wehrlites and serpentinites of the SG-3 well. The iron content in hornblende ranges from 25 to 40% ($Ng = 1.696 - 1.679$, $Np = 1.688 - 1.655$, $cNg = 12° - 18°$). Serpentine is the most widespread secondary mineral of wehrlites which form pseudomorphs in olivine and pyroxene. It is represented by light green foliated, occasionally finely scaled antigorite ($2V = 30°$), $Ng = 1.570 - 1.580$, $Np = 1.561 - 1.570$. An endothermal peak is seen at heat curves at a temperature interval of 780° – 800 °C.

Gabbro-diabases in the SG-3 well section are met in an interval of 93.3 – 2754 m and they form six bodies whose thickness varies from 50 to 150 m; these bodies occur between diabases of the Matertin suite and volcano-sedimentary rocks of the Zhdanov suite. Apodiabases of the Mayarvi suite contain a few small bodies of the altered gabbro-diabases represented by melanocratic amphibole-plagioclase rocks. These bodies have clear, sharp contacts; sometimes in exocontacts, siltstones and phyllites were transformed into nodular schists, of similar structure and similar composition. Their marginal parts are composed of fine-grained rocks and the central parts of medium- and coarse-grained rocks. The trachytoid texture is clearly seen. Large bodies consist of different rocks: the upper parts are made of leucocratic quartzic gabbro-diabases which are replaced downwards by non-quartzic gabbro-diabases; the near-bottom parts of these bodies are made of melanocratic pyroxene rocks. Sometimes, essexitic gabbro-diabases are met in the upper parts of bodies. The character of the secondary alteration of gabbro-diabases depends on their position in the section: in the upper parts of the section they are epidotized, chloritized and scapolitized; in the middle part they are intensively chloritized; in the lower part they are transformed into plagioclase-amphibole rocks.

Leucocratic rocks are predominantly composed of plagioclase (about 60%); melanocratic rocks consist of monoclinic pyroxene (about 60%), chlorite, saussurite, carbonate (30%); quartz, potassium, feldspar, actinolite, hornblende are also met in small amounts; accessory and ore minerals are represented by apatite, leucoxenized sphene and ilmenite, sulphides; pyrrhotite, pyrite and chalcopyrite are present in individual grains. Plagioclase is almost fully saussuritized and represented by albite – 8 – 10% $An$ ($Ng = 1.539$, $Np = 1.526$).

Pyroxene augite is brown or light green in colour ($Ng = 1.733 - 1.727$, $Np = 1.699 - 1.698$), and resembles the pyroxene of diabases. Hornblende is represented by two types: one greenish brown in colour ($Ng = 1.684$, $Np = 1.672$) and the other bluish green ($Ng = 1.705 - 1.700$, $Np = 1.675 - 1.673$); the second type replaces the first type.

Actinolite is developed in all primary minerals: $Ng = 1.645$. All gabbro-diabase bodies are characterized by the presence of the stable ore association represented by impregnations of sulphides, magnetite and ilmenite. The mineral composition is as follows: pyrrhotite 40–60%, chalcopyrite 35–60%, pyrite 5–10%, magnetite 10–15%, ilmenite 5–25%. On the whole, the contents of ore elements in gabbro-diabases are close to those in volcanic rocks and are characterized by the prevalence of copper (0–2%), which exceeds nickel (0.01%), as well as by the increased amount of titanium (0.7%) and by the low quantity of chromium (0.2%) and vanadium (0.03%).

Andesite-dacite porphyrites were met in an interval of 4673–4784 m among magnetite-chlorite schists and limestones of the Luchlompol suite.

The rock contains impregnations of plagioclase and quartz (20–30%) and a fine-grained main mass which consists of quartz, albite-oligoclase, sericite, chlorite, biotite and calcite. The rock is mainly of porphyric texture; micro-felsitic and lepidoblastic textures are characteristic of the main rock mass.

Rocks of the Pechenga complex are metamorphosed. In the SG-3 well metamorphism of rocks increases with depth from the prehnite-pumpellyite to amphibolite facies, which results in the formation of the zonal metamorphic complex at medium pressures. The grade of metamorphism of sedimentary rocks is less than that of magmatic rocks, and does not exceed the epidote-amphibolite facies of the bottom of the section. One part of this work is devoted to metamorphism (see Chap. 1.5).

### 1.3.2 Polymetamorphic Rocks of the Archean Complex

In the Archean complex penetrated by the SG-3 well, beginning from a depth of 6842 m, the following formations have been encountered: the Kola series formed of muscovite-biotite-plagioclase gneisses with highly aluminiferous minerals, biotite-plagioclase gneisses, biotite-amphibole-plagioclase gneisses, amphibolites, meta-ultramafites, granitized gneisses and amphibolites; in granitized gneisses, bodies of gneiss-like plagiogranites, granites, aplitic pegmatites and pegmatites are present; small intrusions of granites were crossed at a depth of 9600 m; in zones where meta-ultramafites and amphibolites contact with gneisses, phlogopite rocks are developed.

#### Gneisses and Granitized Gneisses

Gneisses form the major part of the Archean section of the SG-3. They contain muscovite-biotite-plagioclase gneisses with highly aluminiferous material, biotite-plagioclase gneisses and biotite-amphibole-plagioclase gneisses.

Muscovite-biotite-plagioclase gneisses with highly aluminiferous minerals are developed at four stratigraphic levels of the section. They consist of biotite (20–30%), partially replaced by muscovite, quartz (30–40%), plagioclase (about 30%). Muscovite is present everywhere. Highly aluminiferous minerals such as andalusite, staurolite, sillimanite and garnet are met in the upper part of the section. Their total share in the mineral composition of rocks does not exceed 3–5%. Gneisses of the middle part contain only sillimanite (fibrolite) and garnet; gneisses of the lower part contain sillimanite, kyanite and garnet. Accessory minerals are represented by grains of apatite, magnetite, ilmenite and zircon. Due to the rhythmically layered rock structure, rock-forming minerals are irregularly distributed. Banding, being the result of rock enrichment with melanocratic or leucocratic minerals, coincides with layering. At places banding is complicated by lens-like mineral forms. Rocks are not only characterized by banding, they are also gneissose. Gneissosity also coincides with layering and banding. Mineral grains are of different sizes, ranging from tenths of a millimeter to several millimeters. Rocks are characterized by granoblastic, heteroblastic, lepidoblastic and porphyroblastic textures.

Clearly banded muscovite-biotite-plagioclase gneisses with highly aluminiferous minerals predominate in the upper part of the SG-3 well section. In the middle and lower parts these gneisses are characterized by a uniform distribution of minerals. By taking into account these structural peculiarities of gneisses, this type of rock can be subdivided into three groups: melanocratic, leucocratic and mesocratic muscovite-biotite-plagioclase gneisses with highly aluminiferous minerals. Muscovitized biotite-plagioclase gneisses with highly aluminiferous minerals gradually pass into biotite-plagioclase gneisses with highly aluminiferous minerals, and this is expressed by the colour of the rocks becoming lighter, and the size of the grains increasing. They contain 15% muscovite, 2–3% epidote and several percent microcline. The lightest in colour and recrystallized rocks are represented by plagiogranites, which at places pass into granites; the content of potassium feldspar in granites is 20–30%.

When the rocks described above are muscovitized, highly aluminiferous minerals disappear. Rocks are characterized by a gneissose structure and lepidoheteroblastic texture. Many minerals exist in the mutual "state of reaction" (muscovite and biotite, muscovite and plagioclase, muscovite and staurolite) which leads to the formation of symplectites. In places in these gneisses aplite-pegmatoid rocks appear. They also gradually pass into gneisses. Simultaneously, the gneiss colour becomes lighter, the size of plagioclase and quartz grains increases. Microcline and muscovite appear, replacing biotite. Staurolite and andalusite disappear; first, they are replaced by muscovite and then by quartz, plagioclase and microcline. Biotite is met in small amounts. The size of all the minerals enumerated changes from tenths of a millimeter to a few centimeters. The rocks have heteroblastic texture.

Biotite-plagioclase gneisses with highly aluminiferous minerals, their muscovitized varieties, gneissose plagiogranites, granites and aplite-pegmatoid rocks are linked with each other spatially, by mineral composition and genetically; therefore, minerals of these rocks are described together.

Biotite in biotite-plagioclase gneisses with highly aluminiferous minerals is represented by brown with reddish shade (the first generation) passing into brown and light brown (on *Ng*). This is accompanied by a change of refraction indices of biotite from 1.637 to 1.627. Correspondingly, as a result of biotite phlogopit-

**Table 1.5.** Chemical composition of minerals from main types of rocks of the Archean complex

| Oxides | Granulitic facies | | | | | | | | Amphibolite facies | | | | | |
|---|---|---|---|---|---|---|---|---|---|---|---|---|---|---|
| | Biotite-plagioclase gneiss with HAM | | Biotite-plagioclase gneiss | | Pyroxene-bearing amphibolite (bipyroxene-amphibole-plagioclase crystalline schist) | | | | Biotite-plagioclase gneiss with HAM | | | Biotite-plagioclase gneiss | | |
| | Biotite 38,063 [a] | Plagioclase 38,063 | Biotite 36,337 | Plagioclase 36,337 | Hypersthene 36,045 | Diopside 36,045 | Hornblende 37,309 | Plagioclase 36,045 | Garnet 23,809 | Biotite 23,825 | Plagioclase 22,514 | Garnet 24,630/1 | Biotite 24,630 | |
| $SiO_2$ | 34.40 | 62.04 | 37.93 | 61.65 | 53.26 | 52.28 | 40.50 | 55.47 | 37.63 | 35.50 | 60.41 | 37.37 | 35.94 | 63.91 |
| $TiO_2$ | 3.05 | not ds. | 2.36 | not ds. | 0.82 | 0.20 | 2.10 | not ds. | not ds. | 1.88 | not ds. | not ds. | 1.94 | not ds. |
| $Al_2O_3$ | 18.71 | 23.69 | 15.50 | 24.06 | 0.30 | 1.87 | 11.50 | 28.33 | 20.67 | 19.82 | 24.87 | 20.54 | 18.07 | 23.15 |
| $Fe_2O_3$ | | | | not ds. | not ds. | not ds. | 6.84 | not ds. | 0.83 | 1.14 | not ds. | 0.99 | 4.68 | not ds. |
| FeO | | | | not ds. | not dt. | not dt. | 16.62 | not ds. | 26.76 | 17.25 | not ds. | 29.76 | 16.70 | not ds. |
| $\sum FeO$ | 21.26 | not ds. | 19.56 | not ds. | 19.77 | 6.79 | | | | | | | | |
| MnO | 0.43 | not ds. | 0.28 | not ds. | 0.66 | 0.14 | 0.38 | not ds. | 1.17 | 0.14 | not ds. | 1.26 | 0.22 | not ds. |
| MgO | 8.64 | not ds. | 11.13 | not ds. | 24.76 | 15.11 | 6.50 | not ds. | 1.43 | 11.43 | not ds. | 1.27 | 8.74 | not ds. |
| CaO | 0.18 | 5.05 | 0.17 | 5.42 | 0.44 | 22.72 | 10.51 | 10.20 | 11.27 | 0.05 | 6.61 | 8.79 | 0.12 | 4.28 |
| $Na_2O$ | 0.27 | 8.73 | not ds. | 8.71 | not ds. | not ds. | 1.29 | 5.59 | not ds. | 0.30 | 7.86 | not ds. | 0.24 | 9.45 |
| $K_2O$ | 9.25 | 0.13 | 9.77 | 0.12 | not ds. | 0.08 | 1.37 | 0.19 | not ds. | 8.40 | 0.05 | not dt. | 8.89 | not dt. |
| $H_2O^-$ | not dt. | not dt. | not dt. | not dt. | not dt. | not dt. | not dt. | not dt. | not dt. | 0.80 | not dt. | not dt. | 1.35 | not dt. |
| $H_2O^+$ | 2.88 | not dt. | 2.93 | not dt. | not dt. | not dt. | 1.76 | not dt. | not dt. | 3.39 | not dt. | not dt. | 2.35 | not dt. |
| l.o.c. [b] | not dt. | not dt. | not dt. | not dt. | not dt. | not dt. | not dt. | not dt. | not dt. | | not dt. | not dt. | not dt. | not dt. |
| $P_2O_5$ | not dt. | not dt. | not dt. | not dt. | not dt. | not dt. | not dt. | not dt. | not dt. | 0.20 | not dt. | not dt. | 0.18 | not dt. |
| F | not dt. | not dt. | not dt. | not dt. | not dt. | not dt. | 0.13 | not dt. | | | | | | |
| Total: | | | | | | | 99.50 | | | 100.30 | not dt. | not dt. | 99.42 | not dt. |
| $-O=2F$ | | | | | | | 0.05 | | | 0.08 | not dt. | not dt. | 0.07 | not dt. |
| Total: | 99.77 | 99.64 | 99.63 | 99.96 | 100.01 | 99.19 | 99.45 | 99.88 | 99.76 | 100.22 | 99.81 | 99.98 | 99.35 | 100.83 |

**Table 1.5** (continued)

| Oxides | Amphibolite | | | | Epidote-amphibolite facies | | | | | | | | |
|---|---|---|---|---|---|---|---|---|---|---|---|---|---|
| | Amphibolite facies | | | | Biotite-plagioclase gneiss with HAM | | | Biotite-plagioclase gneiss | | | Amphibolite | | |
| | Garnet | Horn-blende | Biotite | Plagio-clase | Garnet | Biotite | Plagio-clase | Garnet | Biotite | Plagio-clase | Horn-blende | Biotite | Plagio-clase |
| | 24,945 | 24,945 | 24,945 | 24,945 | 23,900 | 23,900 | 23,900 | 24,630 | 26,467 | 26,399 | 37,849 | 37,849 | 37,309 |
| $SiO_2$ | 38.21 | 44.70 | 34.74 | 59.85 | 37.00 | 37.58 | 62.23 | 36.92 | 38.86 | 66.03 | 48.51 | 37.84 | 60.36 |
| $TiO_2$ | 0.17 | 0.83 | 2.45 | not ds. | not ds. | 1.68 | not ds. | not ds. | 0.75 | not ds. | 0.55 | 2.49 | not ds. |
| $Al_2O_3$ | 21.03 | 10.79 | 19.50 | 24.81 | 20.99 | 16.18 | 23.21 | 20.90 | 17.02 | 21.44 | 7.79 | 17.53 | 24.58 |
| $Fe_2O_3$ | 0.85 | 4.60 | not ds. | not dt. | not dt. | not dt. | not dt. | not ds. | 1.90 | not dt. | 4.14 | not dt. | not dt. |
| FeO | 23.95 | 13.98 | not dt. | not dt. | 32.33 | not dt. | not dt. | 30.90 | 10.96 | not dt. | 9.70 | not dt. | not dt. |
| $\sum FeO$ | | | 19.72 | | | 12.32 | | | | | | 13.51 | not dt. |
| MnO | 0.58 | 0.27 | 0.04 | not ds. | 5.80 | 0.31 | not ds. | 8.23 | 0.09 | not ds. | 0.19 | not ds. | not ds. |
| MgO | 3.96 | 9.74 | 10.31 | not ds. | 2.04 | 17.99 | not ds. | 1.75 | 16.15 | not ds. | 13.71 | 15.07 | not ds. |
| CaO | 10.98 | 11.59 | 0.36 | 7.06 | 1.85 | 0.23 | 4.54 | 1.37 | 0.09 | 2.67 | 12.66 | 0.50 | 6.32 |
| $Na_2O$ | not ds. | 1.33 | 0.20 | 7.72 | not ds. | 0.91 | 8.98 | not ds. | 0.19 | 9.80 | 0.40 | 0.17 | 7.40 |
| $K_2O$ | not ds. | 0.48 | 9.19 | not ds. | not ds. | 9.07 | 0.08 | not ds. | 10.00 | 0.04 | 0.26 | 9.80 | 0.93 |
| $H_2O^-$ | not dt. | not dt. | not dt. | not dt. | not dt. | not dt. | not dt. | not dt. | 0.64 | not dt. | not dt. | not dt. | not dt. |
| $H_2O^+$ | not dt. | 2.01 | 3.23 | not dt. | not dt. | 3.85 | not dt. | not dt. | 2.80 | not dt. | 2.07 | 2.57 | not dt. |
| l.o.c.[b] | not dt. | 2.01 | 3.23 | not dt. | not dt. | not dt. | not dt. | not dt. | not dt. | not dt. | not dt. | not dt. | not dt. |
| $P_2O_5$ | not dt. | not dt. | 3.23 | not dt. | not dt. | not dt. | not dt. | not dt. | not dt. | not dt. | not dt. | not dt. | not dt. |
| F | not dt. | 0.05 | 3.23 | not dt. | not dt. | not dt. | not dt. | not dt. | 0.17 | | not dt. | not dt. | not dt. |
| Total | not dt. | 100.37 | 3.23 | not dt. | | | | | 99.72 | | | | |
| $-O=2F$ | | 0.02 | | | | | | | 0.07 | | | | |
| Total | 99.73 | 100.35 | 99.74 | 99.44 | 100.01 | 100.12 | 99.04 | 100.07 | 99.65 | 99.98 | 99.98 | 99.48 | 99.59 |

*Note.* Analysts: A. B. German, K. K. Gumbar, G. F. Petrova.
[a] Sample nos.
[b] l.o.c. = loss on calcination.
not ds. = not discovered.
not dt. = not detected.

ization, which proceeded prior to the processes of biotite muscovitization, its iron content decreases from 40 to 35%. Thermal investigations reveal the presence of more iron biotite, with a temperature of the maximum of an endothermal effect of about 1160 °C. According to data from chemical analysis (Table 1.5), all biotites from the rocks examined are characterized by an increased amount of $Al_2O_3$. A small amount of $Cr_2O_5$, SrO and BaO is also observed. The Fe content is usually constant. The following three generations of biotite are established in these gneisses: primary reddish biotite characterized by a high content of titanium; brown, greenish brown, less iron-and less titanium-bearing biotite of the second generation, which replaces the biotite of the first generation; and magnesian biotite of the third generation. The first biotite is considered to have been formed under granulite-facies metamorphism. It was found in gneisses at a depth of 10,220 m. The second biotite appears in amphibolite facies conditions and is in fact met in these rocks at all depths. The third biotite was formed in conditions of the epidote-amphibolite facies and is the most widespread biotite, particularly in the upper part of the section (Fig. 1.25).

Biotite is replaced by muscovite, which frequently forms pseudomorphs in biotite. When this happens, small dust-like grains of magnetite, titanium-magnetite and ilmenite are fixed. At some places sillimanite is developed in biotite. Sometimes biotite is replaced by chlorite. Muscovite also replaces staurolite and andalusite. It is also characteristic that the interaction between muscovite and quartz results in the formation of symplectites which are seen as reaction rims. Muscovite is also present as individual porphyroblasts, scales or accumulations of scales. Refraction indices of muscovite do not change within the limits of one grain. Some muscovites have $Ng$ reaching approximately 1.606 – 1.608 and others $Ng = 1.600$; this demonstrates that their composition is not uniform. Two types of muscovite are established by thermal analysis data: for the first type an endothermal effect appears at heat curves at a temperature of about 870 °C and for the second type at about 1000 °C.

X-ray data show that muscovite is the double-layered modification of $2\,M_1$. The presence of the intermediate phase indicates that it is closely associated with biotite. Chlorite is also formed at the expense of biotite and after muscovite; however, it is met in rocks only occasionally and its amount is very small. It is represented by a light green variety, characterized by a greenish brown anomalous interference colour. Chlorite is notable for $Ng = Nm = 1.605$ and 1.607 which makes it possible to consider it as an iron-poor variety. Plagioclase forms rounded grains, somewhat elongated in the direction of gneissosity. It comprises about 30 – 40% of the rock mass. Polysynthetic twins in plagioclase are poorly developed. Triads are absent. Refraction indices of plagioclase change: $Ng = 1.545 - 1.554$, $Np = 1.538 - 1.547$. Correspondingly, the plagioclase composition varies from No. 35 to No. 17 (see Table 1.5). These changes take place within the limits of one and the same grain in which an abrupt boundary between plagioclases of different composition is absent. Sometimes, it is established that the composition of plagioclase changes within the limits of one and the same grain many times. This gives plagioclase a spotty appearance. According to X-ray data, plagioclase of more basic composition is characterized by a not very high index of structural order and by a higher temperature of formation (about

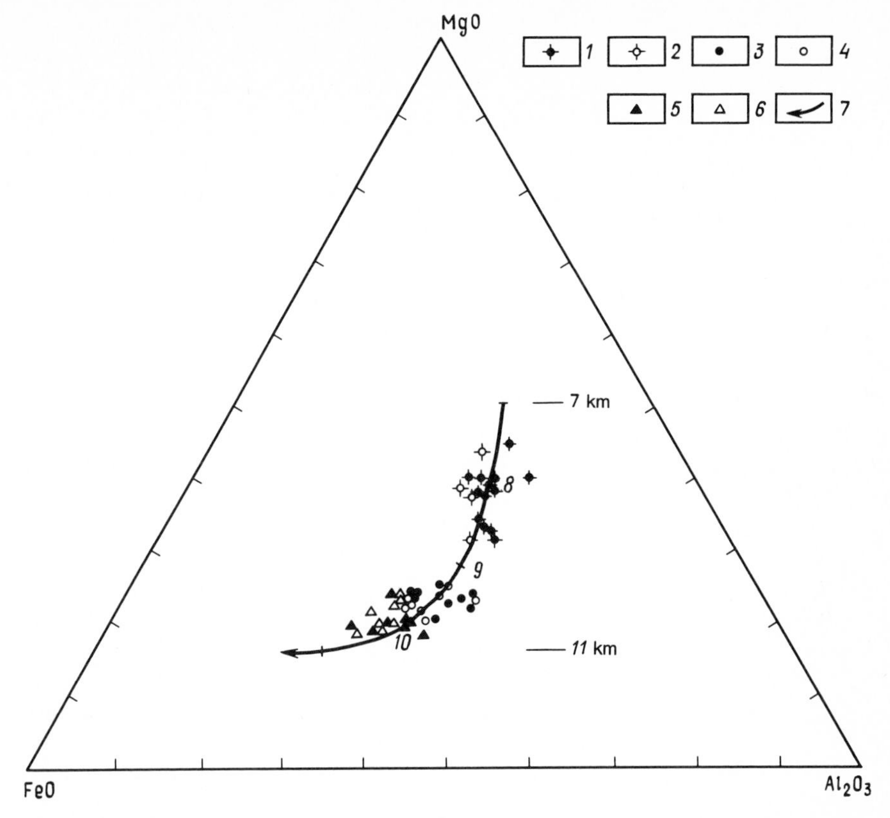


**Fig. 1.25.** Diagram of mica composition for a series of phlogopite-biotite from metamorphic rocks (in percent). *1* phlogopite from phlogopite, phlogopite-actinolitic rocks, epidote-amphibolite facies (SG-3 well); *2* biotite from biotite rocks, epidote-amphibolite facies (SG-3 well); *3* biotite from biotite-bearing amphibolites, biotite-plagioclase gneisses, biotite-plagioclase gneisses with HAM, amphibolite facies (SG-3 well); *4* biotite from biotite-bearing amphibolites, biotite-plagioclase gneisses, amphibolite facies (area of the Titovka river); *5* biotite from biotite-bearing amphibolites, biotite-plagioclase gneisses, biotite-plagioclase gneisses with HAM, granulitic facies (SG-3 well); *6* biotite from biotite-bearing amphibolites, biotite-plagioclase gneisses, biotite-plagioclase gneisses with HAM, granulitic facies (area of the Not river); *7* trend of mica composition alteration with depth (mica belongs to a series of phlogopite-biotite) in rocks of the SG-3 well section

550 °C), whereas its more acid variety has a higher index of structural order and a lower temperature of formation (450 °C).

Quartz makes up about 30% of the rock mass. It is present in the form of rounded grains, which are elongated in the direction of grain gneissosity; the size of grains varies from tenths of a millimeter to some millimeters. In places, quartz grains from lens-like accumulations. Quartz thermoluminescence of these rocks begins at 200 °C and reaches its maximum at a temperature of 350° – 380 °C. The intensity of thermoluminescence is 25 – 35 units. At higher values of the infrared spectroscopic index (a ratio of optical densities of the absorption bands with frequency of 800 and 1100 $cm^{-1}$), reaching 2.2 and 2.0, the quartz from these and other gneisses of the Archean complex differs from the quartz of the Pechenga

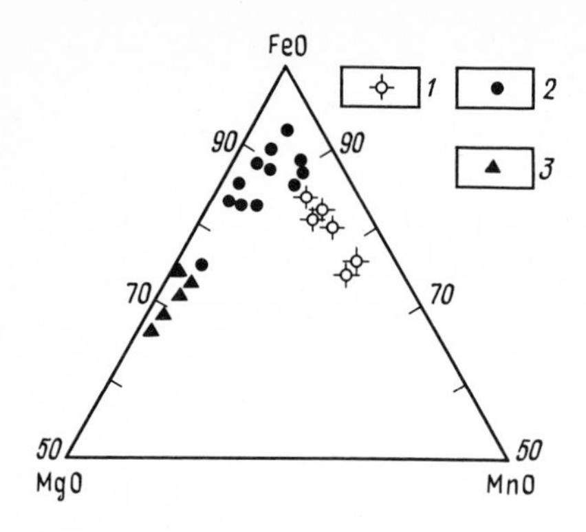


**Fig. 1.26.** Diagram of composition of garnets from metamorphic rocks of the SG-3 well section and of the area in the vicinity of the well (in percent). *1* biotite-plagioclase gneisses, biotite-plagioclase gneisses with HAM, epidote-amphibolite facies; *2* amphibolites, biotite-plagioclase gneisses, biotite-plagioclase gneisses with HAM, granite-biotite-amphibole-plagioclase gneisses, amphibolite facies; *3* garnet-biotite-plagioclase rocks, granulitic facies (area of the Not river)

complex rocks. Staurolite is seen in rocks only occasionally in the upper layer. It forms large poikiloblasts, up to several millimeters. Grain margins are uneven. Staurolite is replaced by muscovite. Poikilitic ingrowths in staurolite are represented by quartz. Staurolite has a constant composition with $Ng = 1.780$, $Nm = 1.744$, $Np = 1.738$. Its pleochroism is yellow in colour on *Ng*, and almost colourless on *Np*. Andalusite is met in rocks even more rarely than staurolite, and is also present only in the upper part. It is found in the form of large poikiloporphyroblasts (up to several millimeters) with coroded grain margins, which are replaced by muscovite and fibrolite. Andalusite is colourless with $Ng = 1.638$, $Nm = 1.633$ and $Np = 1.628$. Two generations of garnet are met in the rocks: the first generation is represented by rare poikiloporphyroblasts dispersed in the rocks, and the second generation is observed as individual grains in small quartz-feldspathic lenses or as rims surrounding garnet grains of the first generation. Garnet of the first generation is represented by almandine, which contains pyrope (11%) and grossular (30%) in a rather significant amount. We are of the opinion that it was formed during the amphibolite stage of metamorphism. Garnet of the second generation contains a lesser amount of pyrope (8 – 9%), but is enriched by spessartite (5 – 6%). Its appearance is the result of gneiss diaphthoresis in conditions of the epidote-amphibolite facies (see Table 1.5 and Fig. 1.26).

Sillimanite is met in rocks mainly in the form of spicular accumulations which develop in plagioclase and biotite. In the upper layer it is present in paragenetic association with andalusite, staurolite and garnet, in the middle stratum with garnet, and in the lower stratum with kyanite.

In the SG-3 well section biotite-plagioclase gneisses are widespread in strata which alternate with strata of muscovite-biotite-plagioclase gneisses with highly aluminiferous minerals. They are comparatively uniform, grey, medium-grained rocks, notable for their gneissose structure and lepidogranoblastic texture. Rocks with lenticular-banded and banded structure are of subordinate significance. The thickness of the bands reaches several millimeters; mineral grains have a size some parts of a millimeter. The pattern of their distribution in rocks is comparatively uniform, and only at places do they form monomineral lens-like accumulations about 1 cm thick, which are elongated in the direction of gneissosity. The mineral composition of rocks is rather simple: plagioclase-andesine and oligoclase about 40%; quartz 30 – 40%; biotite 20%; epidote. Of the accessory minerals, orthite and sphene are the most characteristic. Apatite, zircon, ilmenite, magnetite and sulphides such as pyrrhotite and pyrite are also found. However, plagioclase and biotite are characterized by a changeable chemical composition. Biotite-plagioclase gneisses are regarded as a peculiar petrochemical rock type of andesite-dacite or tonalite composition. Granitized biotite-plagioclase gneisses are closely linked with biotite-plagioclase gneisses by gradual transition, and are developed in the section together with biotite-plagioclase gneisses, as has been already mentioned. They differ from biotite-plagioclase gneisses by the presence of muscovite, the predominance of oligoclase and their considerable amount of epidote (up to 10%). Mineral grains are characterized by their differing sizes, due to which rocks possess heteroblastic textures. Aggregates of mineral symplectites, formed as a result of the reaction between different minerals, are also met. These aggregates can be exemplified by the combination of epidote with quartz or muscovite with quartz. A non-uniform distribution of minerals is also typical. The gneissose structure of these rocks is considered to be of inherited origin. At places granitized biotite-plagioclase gneisses gradually pass into gneissose plagiogranites, and if microcline appears they also pass into gneissose granites. Both of them, as distinct from granitized biotite-plagioclase gneisses, are more coarse-grained rocks, but they are considered to be more equi-grained and uniform as compared with granitized biotite-plagioclase gneisses.

Biotite-plagioclase gneisses and granitized gneisses are chloritized at places. In zones of rock cataclasm, plagioclase is pellitized along very thin fissures. In large fissures calcite, prehnite, zeolites, chlorite and epidote are encountered.

Plagioclase is present in biotite-plagioclase gneisses and in their granitized varieties in the form of small grains of tenths of a millimeter in size and in the form of large porphyroblastic grains of several millimeters. Grain boundaries are uneven. The composition of plagioclase is not uniform, and changes within the limits of one and the same grain from andesine to oligoclase, occasionally to albite, plagioclase with $Ng = 1.547 - 1.542$ predominates. As a rule, it forms the marginal parts of grains and porphyroblasts.

Plagioclase of more basic composition with $Ng = 1.553$ and $Np = 1.545$ is of subordinate significance. It fills small grains in the main rock mass or forms central parts of large grains, and is characteristic of non-granitized biotite-plagioclase gneisses. Twins are rarely met in plagioclases of these rocks, triads are absent. Plagioclase-antiperthite is present in granitized gneisses, and contains small ingrowths of potassium feldspar. Grains of microcline, which replaced plagioclase, are usually found in rocks in association with plagioclase-antiperthite. Muscovite and epidote are also developed in plagioclase. In some parts of the SG-3 well section plagioclase is pellitized. The chemical composition of plagioclase is given in Table 1.5.

Potassium feldspar-microcline is present in granitized gneisses in small amounts. It forms xenomorphic grains, corrodes plagioclase, "penetrates" into it along cleavage planes and forms antiperthite ingrowths inside the plagioclase. For microcline $Ng = 1.524$ and $Np = 1.518$. The chemical composition of microcline is characterized by the presence of an albite admixture. According to X-ray data, the granitized biotite-plagioclase gneisses contain two textural modifications of potassium feldspars. The first is a monoclinic modification; by its textural state it corresponds to the high orthoclase and is characterized by a non-ordered distribution of Al in the crystalline lattice ($\mathrm{Al}\,T_1 \simeq \mathrm{Al}\,T_2$; $\Delta p = 0-0.2$). It comprises not more than 25% of the potassium-feldspathic phase mass and is present as relicts of grains in which microcline is developed. The second triclinic generation of potassium feldspar is represented by low microcline; according to its textural state it corresponds to the most ordered modification ($\mathrm{Al}\,T_1 = 0.97-1.00$; $\Delta p = 0.95-0.99$). Since a temperature regime is the main factor exerting influence on an "order-disorder" state in textures of feldspars present in metamorphic rocks, it may be concluded that the formation of potassium feldspars of the rocks examined proceeded in two phases. During a high-temperature stage a non-ordered plagioclase formed under granulitic-facies conditions. During the later low-temperature stage the ordered plagioclase, distributed in rocks in greater amounts, was formed. Thus, it appears that in granitized biotite-plagioclase gneisses, relicts of the granulitic facies can be identified by taking into account the character of potassium feldspar.

Biotite is present in the form of individual subparallel scales or their lens-like accumulations. In the middle parts of the section it is represented by a reddish brown variety with $Ng = Nm = 1.645$ (second generation). In gneisses of the lower layer, beginning from a depth of 9600 m, dark-coloured, reddish brown biotite with $Ng = Nm = 1.654$ (first generation) is present. Biotite is phlogopitized everywhere and its colour on $Ng$ becomes brown; simultaneously, refraction indices $Ng = Nm$ change and reach 1.630 (third generation). Data on the thermal study of biotite demonstrate that this mineral is disintegrated at temperatures of about 1170 °C (first generation) and about 1160 °C (second and third generations). On the basis of chemical composition data, the following biotites can be distinguished in some places: biotites with a high content of titanium and iron (first generation) and biotites with high content of magnesium (second and third generations) (see Fig. 1.5). According to the characteristics described, biotite from biotite-plagioclase gneisses is similar to that from biotite-plagioclase gneisses with highly aluminiferous minerals. Biotite of the first generation corresponds to biotite of the granulitic facies (see Fig. 1.25), whereas biotite of the second generation corresponds to biotite of the epidote-amphibolite facies. From this it appears that in biotite-plagioclase gneisses biotites are the relicts of the granulitic phase. Biotite of the second and third generations is replaced by muscovite. Muscovite is present in rocks in the form of "biotite replacements", scales developed in plagioclase, and in the form of large porphyroblasts with reaction rims at the boundaries with plagioclase, where symplectites of the "lace" type are formed. The muscovite content is 1 – 5%, decreasing somewhat down the SG-3 well section. Muscovite, closely associated with biotite, is characterized by the following refraction indices: $Ng = Nm = 1.606-1.608$; $Np = 1.568$. Refraction indices of muscovite, which develops at the expense of plagioclase, are lower: $Ng = 1.600$; $Np = 1.564$.

On heat curves obtained for rocks enriched in muscovite and for pure muscovites, two types of muscovite are established according to the endothermal effects. One endothermal effect was produced at a temperature of 880 °C, and the second one corresponds to a temperature of 1000 °C or about 1170 °C. In muscovites of the first type associated with biotite, $\Delta t = t_e 2 - t_e 1 = 230°$; in muscovites of the second type $\Delta t = 130°$ ($t_e2$ is the temperature of the maximum of a high-temperature endothermal effect for muscovite; $t_e1$, the temperature of the maximum of a low-temperature endothermal effect for muscovite). In other words, it appears that a muscovite variety transitional from biotite to muscovite is characterized by the second endothermal effect at heat curves at a temperature similar to that for biotite.

Quartz is present in rocks in the form either of xenomorphic grains of different sizes or their lens-like accumulations due to which the rock acquires a heterograined texture.

Epidote is a characteristic mineral of biotite-plagioclase and granitized biotite-plagioclase gneisses. It is present in granitized gneisses in an appreciably greater amount. It forms individual grains and porphyroblasts of complex texture and is represented by two generations. Epidote of the first generation occupies the central parts of grains and is distinguished by higher refraction indices: $Ng = 1.750$, $Np = 1.724$. Epidote of the second generation is "covered by" symplectites, and epidote of first generation forms a "lace" rim of symplectite intergrowths of epidote and albite. This reaction rim is confined to a boundary between epidote of the first generation and plagioclase of the andesine and oligoclase composition. Epidote of the second generation is characterized by lower refraction index values: $Ng = 1.735$, $Np = 1.722$.

Rather frequently epidote crystals of complex structure contain grains of orthite with pleochroism and green colour on *Ng*. Its refraction indices are as follows: $Ng = 1.761$, $Np = 1.750$. The chemical composition of orthite is notable for its high lanthanum (4%) and cerium (13.5%) content.

Sphene is present in rocks in the form of rare isolated grains. There are two generations of this mineral. The generation of a darker colour is inside grains with $Ng > 2$ and $Np = 1.890$. A dark-coloured generation is covered by a rim of lighter colour with smaller refraction index values. In places, sphene is rimmed by rare grains of titanium magnetite which are present in the rocks. In its chemical composition, sphene of the first generation is enriched in titanium.

## Amphibolites

Three types of amphibolite have been identified in the Archean complex of the SG-3 well section: a hornblende type, whose chemical composition corresponds to that of basalts of the alkaline-earth series; a cummingtonite-hornblende (aluminiferous amphibolite) type, whose chemical composition corresponds to that of aluminiferous basalts enriched by magnetite, and a type of amphibolite which passes into hornblendite, whose chemical composition corresponds to that of iron basalts (iron amphibolites). These types of amphibolite differ in mineral composition, accessory mineralization, texture and chemical composition.

Hornblende amphibolites are widely spread throughout the Archean strata at different depths in the SG-3 well section. They are dark green in colour and have a simple mineral composition: hornblende about 70%, plagioclase with a small amount of biotite 30%. In some cases grains of monoclinic pyroxene-salite and hypersthene (in crushed rocks) have been encountered. Grains of quartz are seen occasionally; in places grains of garnet were found in small amounts. Of the accessory minerals the following are typical: sphene amounting to 5%, apatite, magnetite, ilmenite, and sulphides such as pyrite and chalcopyrite. Pyrrhotite always contains rare ingrowths of pentlandite. The ore mineral content does not exceed 1%. Of the secondary minerals, epidote is present. At some places chlorite and quartz are met in small amounts. In cataclasm zones, along fissures, amphibolites contain prehnite, calcite and zeolites; plagioclase is pellitized. These amphibolites are notable for their linear structure. At places, a banded structure is seen, a result of the alternation of mainly plagioclase and amphibole bands whose thickness reaches 1 cm. A lenticular banding structure with lenses of 1 – 2 mm thick is also observed. The texture of the rocks is of a granonematoblastic type.

Cummingtonite-hornblende amphibolites are confined mainly to a layer of biotite-plagioclase gneisses and aluminiferous amphibolites, although they are also met in other part of the section in the form of rare bodies. These rocks have a greyish-green colour, and differ from hornblende amphibolites by their more leucocratic character. Cummingtonite-hornblende amphibolites contain about 50% plagioclase. Hornblende and cummingtonite, which replaces the former, also amount to 50%. Biotite is present in rocks in a small quantity. The accessory mineral content is less than 1%. They are represented by sulphides such as pyrrhotite with pentlandite ingrowths, pyrite and chalcopyrite. Ilmenite is met in small amounts. Cummingtonite-hornblende amphibolites are characterized by their nematoblastic texture.

Amphibolites and hornblendites (ferruginous amphibolites) appear at a depth of 8455 m, and differ by their black colour from the two types of amphibolite described. The amount of hornblende in them varies from 70 to 90%, the other part being represented mainly by plagioclase. Biotite and quartz are contained in small amounts. Accessory minerals such as sphene, magnetite and ilmenite (up to 3%), and sulphides (pyrrhotite with ingrowths of pentlandite, pyrite and chalcopyrite) are present. The rocks are characterized by a granonematoblastic texture. At a depth of more than 10,000 m, garnet appeared in bodies of iron amphibolites. These bodies also contain quartz, biotite, magnetite and ilmenite; in places this type of rock passes into biotite-garnet-quartz-amphibole-plagioclase rock with ilmenite. Amphibole is replaced here by biotite and garnet, and is preserved only in the form of relicts. These rocks acquire a porphyroblastic texture, with the main rock fabric being characterized by a granoblastic texture.

Amphibole of amphibolites is represented mainly by hornblende, present in the form of idiomorphic rod-like parallel grains. Hornblende, present in the epidotized, chloritized and silicified rocks, acquires corroded configurations and becomes lighter-coloured. Hornblende is green in colour on *Ng*, in places it is brownish-green or bluish-green. Quite often its colour is not uniform and changes within the limits of one and the same grain from darker to lighter shades of irregular spotty appearance; this happens when dark-coloured amphibole is replaced by a lighter one during diaphthoresis. Correspondingly, refraction indices of hornblende also vary. The highest values of refraction indices ($Ng = 1.689$, $Nm = 1.675$, $Np = 1.663$) correspond to brownish green (on *Ng*) hornblende. Green hornblende is characterized by some lower (on *Ng*) refraction indices: $Ng = 1.680$, $Nm = 1.670$, $Np = 1.658$. They become even lower for bluish green and light green hornblende. At places hornblende is replaced by pale green actinolite with $Ng = 1.660$; $Np = 1.637$. From this it appears that the composition of amphibole is not uniform and that it changes as colour intensity, refraction indices and iron content decrease. However, an intensively green-coloured hornblende predominates in rocks, characterized by a rather high iron and aluminum content (see Table 1.5). The chemical composition of the lighter-coloured hornblendes, replacing the intensively coloured ones, is distinguished by lesser iron and higher silicon content.

Brown hornblende is distinguished by a small amount of silicon and high content of titanium, and corresponds in its composition to magnesian hastingsite hornblende [according to Liak's (1978) classification]. Its role in rocks of the SG-3 well section increases with depth. The composition of this hornblende is indicative of its generation under granulitic facies conditions. On the composition diagram (see Fig. 1.22) it is confined to a field of hornblendes which have been chosen, for comparison, from metabasites of the granulitic phases developed in the vicinity of the SG-3 well area.

By X-ray analysis data, this hastingsite hornblende is characterized by parameters inherent to an elementary cell typical of hornblendes of the granulitic facies (see Fig. 1.23). Its formation under granulitic facies conditions follows from thermal analysis data. The endothermal reaction shows that it is attributed to a group of hornblendes formed in conditions of granulitic facies in the vicinity of the SG-3 well area. According to thermal analysis data, green hornblende is destroyed at a temperature about 1080 °C. A lighter hornblende, replacing a green one, is distinguished by a lesser temperature (1060 °C).

Apart from hornblende, cummingtonitized amphibolites contain cummingtonite, which replaces hornblende, and forms idiomorphic colourless grains with polysynthetic twins. Its content in rocks is 5–10%. For cummingtonite $Ng = 1.673$, $Np = 1.664$. The chemical composition of cummingtonite is characterized by medium iron content values. In heat curves and endothermal effect is established at a temperature of 1000 °C which is typical for cummingtonite.

Plagioclase forms rounded or elongated xenomorphic grains usually deprived of twins. It has a non-uniform composition expressed in a poorly zonal structure. The central parts of the grains are normally characterized by a more basic composition whereas their marginal parts are more acid in composition. In places plagioclase is of spotty zonal fabric in which mineral parts of more acid and more basic composition alternate with each other. Its composition changes from andesine with $Ng = 1.553$, $Np = 1.545$ to oligoclase with $Ng = 1.549$, $Np = 1.542$. Sometimes labradorite with $Ng = 1.568$, $Np = 1.561$ and newly formed albite with $Ng = 1.540$, $Np = 1.530$ are encountered. Andesine predominates in the amphibolites considered, coexisting with hornblende. Oligoclase is developed in andesine contemporaneously with the formation of bluish green hornblende, which replaces green hornblende. As a rule, plagioclase is not cloudy, but of fresh appearance. However, in some spots of the rock, usually in zones of cataclysm, it is strongly pellitized along very thin fissures; this indicates that processes such as hypergenesis occurred here. Chemical analysis data (see Table 1.5) show that the rocks contain andesine and labradorite; in rare cases anorthite is also met (see Fig. 1.23).

Taking into account the data of X-ray analysis, it appears that labradorite can be attributed to plagioclase with a not very high index of structural order, whereas andesine is characterized by a higher order. Thus, in the process of deanorthization, a plagioclase structural order becomes more perfect (see Fig. 1.24).

Biotite is found everywhere in amphibolites in small amounts. It forms scales replacing amphibole. The colour of biotite on *Ng* changes from reddish brown to brown and light brown. These fluctuations are sometimes seen within the limits of one and the same grain. Correspondingly, refraction indices also change: $Ng = Nm$ from 1.640 to 1.630. In the lightest-coloured mica they fall to 1.623 and sometimes to 1.612. These changes are caused by a decrease in iron content during the process of phlogopitization. By chemical composition data (see Table 1.5) it has also been established that the composition of biotite varies. Here a distinction is made between biotites with high and medium values of iron content, and biotites rich in magnesium, i.e. biotites which in their composition approach phlogopites. Biotite from amphibolites is characterized by an endothermal effect at a temperature of about 1150° – 1170 °C, and this biotite is similar to biotite from biotite-plagioclase gneisses. Phlogopitization of biotite in amphibolites proceeded together with actinolitization of hornblende. In both cases, the iron content of these femic minerals decreases.

Monoclinic pyroxene is met in amphibolites in rare cases. It is represented by salite. Grains are xenoform in nature and are replaced by amphiboles. In rare cases, at a depth of about 9600 m, amphibolites contain hypersthene in crushed rocks. Hypersthene and salite from amphibolites of the SG-3 well are analogous in chemical composition to the corresponding minerals from bipyroxene-amphibole-plagioclase crystalline schists of the granulitic facies which are developed at the surface north-east of the SG-3 well in the area of Linnakhamari. Sphene is a typical accessory mineral of amphibolites. It forms lens-like clustered grain accumulations elongated in the direction of rock lineation. In places, sphene contains grains of magnetite which cover it from all sides, and together with them, sphene forms aggregates of a symplectite type. In rocks where lighter hornblende is developed at the expense of the intensively coloured hornblende, sphene grains are surrounded by rims of magnetite and ilmenite.

Ilmenite is present in amphibolites in two generations: in the form of small grains covered on all sides by sphene, and in the form of dust-like grains dispersed in rocks, or in the form of rims around sphene grains in those rocks in which comparatively ferruginous hornblende is replaced by less ferruginous amphibole.

### Meta-Ultramafites

Meta-ultramafites are represented by actinolite and talc-actinolite rocks.

All small bodies are composed of actinolite rocks. Talc-actinolite rocks are placed inside larger bodies. Actinolite rocks contain 80 – 85% actinolite and 15 – 20% phlogopite. Grains are 0.1 – 0.2 mm in size. The texture of the rocks is of a felty type. Talc-actinolite rocks are distinguished by the presence of talc (up to 30%). Accessory minerals are represented by ilmenite, pyrrhotite with ingrowths of pentlandite, chalcopyrite, and pyrite.

Actinolite forms spicular idiomorphic grains with poorly expressed pale green pleochroism. Its refraction indices on $Ng$ changes from 1.658 to 1.645. Actinolite with $Ng = 1.645 - 1.648$ is the most widespread variety. The iron content in a mineral is 22 – 23%. Actinolite from talc-actinolite rocks is even more ferruginous.

Phlogopite forms individual scales or their accumulations in rocks. It replaces actinolite developed on grain margins or penetrates into actinolite along cleavage planes. Phlogopite from actinolite rocks is characterized by $Ng = Nm$, which changes from 1.620 to 1.607. A fluctuation of refraction indices takes place within the limits of one and the same grain. Correspondingly, the iron content decreases from 25 to 15%. Phlogopite from talc-actinolite rocks is characterized by a smaller amount of iron.

Talc is present in the form of small scales and their accumulations with $Ng =$ 1.596, $Np = 1.598$. In the heat curves for talc, an endothermal effect is established at a temperature of 970 °C. Talc-actinolite and actinolite rocks are non-metamorphosed, and ultrabasic rocks are represented by picrites, which most probably originated as a result of volcanic-glass metamorphism; this may be surmised from the presence of the typical felt and flaser textures.

Phlogopite rocks are confined to contacts between amphibolites and meta-ultramafites. This determines their minor variations in mineral composition. They contain 80 – 100% phlogopite; actinolite is also present. In rocks associated with amphibolites, plagioclase is found. Sometimes, phlogopite rocks contain muscovite which replaces phlogopite. Magnetite, rutile, orthite, apatite, graphite and occasional chromite are the accessory minerals discovered. The size of mineral grains varies from tenths of a millimeter to some millimeters. The rocks are characterized by an imbricated structure and a lepidoblastic texture. Phlogopite is represented by comparatively idiomorphic scales. At places they contain dust-like inclusions of magnetite and orthite; phlogopite is sometimes replaced by actinolite.

Phlogopite is characterized by $Ng = Nm$, changing from 1.608 to 1.620; this indicates that the rocks contain both low-iron and ordinary iron phlogopite. Phlogopites with increased iron content are characteristic of the endocontact zones in amphibolites. According to the chemical analysis data (see Fig. 1.25), two types of phlogopite have been identified: low-iron and iron. Iron phlogopites contain small admixture of titanium.

## Alteration of Rocks in Polymetamorphic Complex

The study of the Archean rocks drilled by the SG-3 well has revealed that they are subject to the following main processes: (1) metamorphism of the granulitic facies; (2) regional metamorphism proceeding contemporaneously with the folding of the amphibolite facies; (3) regional retrograde metamorphism of the epidote-amphibolite facies; (4) regional granitization in conditions of the epidote-amphibolite facies; (5) diaphthoresis of the greenschist facies manifested locally along major fault zones; (6) low-temperature metamorphism manifested locally along catyclasm zones.

In biotite-plagioclase gneisses with highly aluminiferous minerals relict minerals of the granulitic facies are represented by highly iron- and highly titanium-bearing reddish brown biotite (see Table 1.5). These rocks contain $Pl_{35} + Q_3 + Bi_f = 45 - 60 + Gar$ mineral association, which is considered to be the earliest in origin of other widespread mineral associations. This mineral association corresponds to the amphibole facies of metamorphism. Later, at a stage of plagioclase deanorthization and biotite phlogopitization, staurolite and andalusite appear in the rocks, and the mineral association becomes $Pl_{21} + Q_3 + Bi_f = 35 + Gar + Stu \pm And$, corresponding to the beginning of the epidote-amphibolite facies. All these alterations were followed by the muscovitization of the rocks. Muscovite develops in biotite, and this process is accompanied by the segregation of small grains of magnetite; muscovite develops also in staurolite and andalusite. Deanorthization of plagioclase continues, microcline appears, and rocks are crystallized, which results in grain size increase; finally, the following mineral association is formed: $Pl_{17} + Q_3 + Bi_f = 35 - 22 + Gar + Mu + Mt + Mc$.

As a result of this, andalusite and staurolite fully disappear, whereas muscovite, plagioclase and microcline develop in the rocks very intensively. Gneissose plagiogranites and granites are formed instead of gneisses. We consider this process as one of regional granitization when the rocks become enriched in Na, K, Si, while the content of Mg, Fe, and Ca decreases. In view of the fact that some minerals replace others and form pseudomorphs, and that the newly originating parallel structures inherit previously existing structures, it may be concluded that the process of rock granitization proceeding in a metasomatic way. A characteristic feature of the rock alteration process is muscovitization, which is regarded as the beginning of granitization. The fact is that mineral association of all three stages of rock alteration are frequently met in one and the same rock, with associations of earlier origin being present among associations of later origin in the form of relicts.

In biotite-plagioclase gneisses the granulitic facies in rocks is represented by relicts of highly ferruginous and titanium-bearing biotite (see Table 1.5). The mineral association $Pl_{35-40} + Q_3 + Bi_f = 42 - 60 \pm Gar$ is the earliest by origin association among these which are widely spread in biotite-plagioclase gneisses. It is assigned to the amphibolite facies of metamorphism. Later, a mineral association $Pl_{21-25} + Q_3 + Bi = 25 - 20 + Ep_1 + Mt_1 + Sph_1$ was formed.

The appearance of this mineral association corresponds to the beginning of the epidote-amphibolite facies of the regional retrograde metamorphism. At this stage of metamorphism, biotite is phlogopitized, plagioclase is deanorthitized, epidote, magnetite and sphene of the first generation appear. This mineral association is followed by the next: $Pl_{17-20} + Q_3 + Bi_f + Mu + Ep_2 + Mt_2 + Sph_2 \pm Mc$.

A process of mineral transformation during this stage begins with their muscovitization, which is accompanied by continuing deanorthization of plagioclase. Epidote, sphene and magnetite of the second generation are formed. Microcline appears. The stage of rock alteration is a process of granitization under conditions of the epidote-amphibolite facies. Muscovitization is the beginning of granitization. Gradual changes in composition of rock-forming minerals, the presence of pseudomorphs of newly formed minerals and reaction rims, the enrichment of rocks with Na, K, Si and the decrease of Mg, Ca and Fe makes it possible to consider the process of rock alteration at this stage as a process of metasomatic granitization.

In amphibolites, the following relict minerals of the granulitic facies (beginning from a depth of 9600 m) were found: hastingsite hornblende, diopside, hypersthene, highly ferruginous and titanium-bearing biotite, anorthite (see Table 1.5). The next mineral association of the amphibolite facies is as follows: $Pl_{52-43} + Hb_f = 40-35 + Bi_f = 40-60$. The metamorphism of the amphibolite facies was followed by the metamorphism of the epidote-amphibolite facies, which led to the formation of the following mineral association: $Pl_{28-33} + Hb_f = 38-30 + Bi_f = 38-28 + Ep_1 + Mt_1 + Sph_1$.

The mineral composition and texture of amphibolites became more complex. In addition to green hornblende, bluish green hornblende appeared, and replaced the green hornblende. The iron content in amphiboles and biotite decreased, and at places it fell to 18%; this means that biotite was in fact transformed into phlogopite. At the stage of retrograde metamorphism of the epidote-amphibolite facies a process of phlogopitization took place.

Metamorphism of the epidote-amphibolite facies of granitization was followed by granitization under conditions of the epidote-amphibolite facies. At this time amphibolites were biotitized and muscovitized, plagioclases were deanorthitized, at places the plagioclase and quartz content increased. In some cases microcline was discovered in the rocks. Rock granitization was irregular in character, and this favoured the process of matter differentiation. In some cases, subcontact parts of amphibolites became enriched with biotite and phlogopite, in other cases they acquired a lighter colour and amphibole-plagioclase gneisses formed. During the process of amphibolite granitization two types of mineral associations were formed: (1) $Pl + Hb$; (2) $Pl_{17-20} + Hb + Bi + Q_3 + Ep_2 + Mt_2 + Sph_2$.

Hence, it may be concluded that the process of amphibolite phlogopitization began at a stage of retrograde metamorphism under conditions of the epidote-amphibolite facies; however, it only became the process of ore generation when the granitization of rocks led to the formation of potassium, which was brought into the rocks.

From this it appears that one and the same process of metamorphism and metasomatism were proceeding in the main rock types of different chemical composition which constitute the Archean structure in the SG-3 well section. These processes are as follows: metamorphism of the granulitic facies, succeeding metamorphism of the amphibolite facies, metamorphism of the epidote-amphibolite facies, granitization in conditions of the epidote-amphibolite facies (Table 1.6). Metamorphic processes, proceeding without any essential change in the chemical composition of rocks, were later transformed into metasomatic processes; at this stage of rock granitization, material was supplied and evacuated. This has resulted in the formation of the complex structure of the Archean polymetamorphic rock-stratigraphic unit.

**Table 1.6.** Paleotemperatures of metamorphism determined by mineralogical geothermometers

| Rocks | Sample nos. | Paragenetic mineral association | Processes of rock transformation | Temperature °C |
|---|---|---|---|---|
| Biotite-plagioclase gneisses with aluminiferous minerals | 23,908 | Garnet-biotite $K_{Mg}^{Gar} = 0.15$; $K_{Mg}^{Bi} = 0.49$ | Regional metamorphism. Epidote-amphibolite facies | 600 |
| | | Garnet-biotite $K_{Mg}^{Gar} = 0.15$; $K_{Mg}^{Bi} = 0.49$ | Regional metamorphism. Epidote-amphibolite facies | 600 |
| | 24,271 | Garnet biotite $K_{Mg}^{Gar} = 0.10$; $K_{Mg}^{Bi} = 0.50$ | Retrograde regional metamorphism. Epidote-amphibolite facies | 550 |
| | 24,287 | Garnet biotite $K_{Mg}^{Gar} = 0.08$; $K_{Mg}^{Bi} = 0.49$ | Retrograde regional metamorphism. Epidote-amphibolite facies | 550 |
| | 22,730 | Muscovite-biotite $K_{Na}^{Mn} = 19$; $K_{Na}^{Bi} = 5$ | Granitization | 650 |
| Biotite-plagioclase gneisses | 26,431 | Garnet-biotite $K_{Mg}^{Gar} = 0.35$; $K_{Mg}^{Bi} = 0.62$ | Regional metamorphism; amphibolite facies | 800 |
| | | Garnet-biotite $K_{Mg}^{Gar} = 0.35$; $K_{Mg}^{Bi} = 0.62$ | Regional metamorphism; amphibolite facies | 700 |
| | 26,339 | Garnet-biotite $K_{Mg}^{Gar} = 0.08$; $K_{Mg}^{Bi^2} = 0.56$ | Retrograde metamorphism; epidote-amphibolite facies | 500 |
| | 24,301 | Muscovite-biotite $K_{Na}^{Mu} = 22$; $K_{Na}^{Bi} = 9$ | Granitization | 650 |
| Garnet-biotite-amphibole-plagioclase gneisses | 23,873 | Garnet-biotite $K_{Mg}^{Gar} = 0.25$; $K_{Mg}^{Bi} = 0.54$ | Regional metamorphism; amphibolite facies | 750 |
| Pyroxene-bearing amphibolites (relicts of bipyroxene-amphibole-plagioclase crystalline schists) | 36,045 | Salite-hornblende $K_{Mg}^{Hb} = 0.54$; $K_{Mg}^{S} = 0.80$ Hypersthene-hornblende $K_{Mg}^{Hyp} = 0.70$; $K_{Mg}^{Hb} = 0.54$ | Regional metamorphism; granulitic facies | 1200 |
| Cummingtonite-bearing amphibolites | 38,166 | Anorthite-hornblende $K_{Ca}^{Pl} = 0.97$; $K_{Ca}^{Hb} = 0.82$ | Regional metamorphism; granulitic facies | 1000 |
| Pyroxene-bearing amphibolites | 37,615 | Salite-hornblende $K_{Mg}^{S} = 0.066$; $K_{Mg}^{Hb} = 0.43$ | Regional metamorphism; granulitic facies | 900 |
| Amphibolites | 24,945 | Garnet-biotite $K_{Mg}^{Gar} = 0.22$; $K_{Mg}^{Bi} = 0.43$ | Regional metamorphism; amphibolite facies | 800 |
| | 26,776 | Hornblende-plagioclase $K_{Mg}^{Hb} = 0.67$; $K_{Ca}^{Pl} = 0.57$ | Regional metamorphism; amphibolite facies | 800 |
| | 38,133 | Hornblende-plagioclase | Retrograde metamorphism; amphib- | 500 |

## On the Temperature of Metamorphism

In polymetamorphic rocks of the SG-3 well section, minerals of early origin are preserved everywhere in the form of relicts among mineral parageneses of later origin. This polyfacial character of paragenetic minerals within the limits of one and the same rock indicates that processes of metamorphic transformation did not reach their final stages. With this in view, a very careful approach is needed to assess the temperatures of rock metamorphism in order to distinguish mineral parageneses in equilibrium. To determine temperatures of metamorphism of rocks in the SG-3 well, the following mineral pairs were used: amphibole-plagioclase, amphibole-biotite, garnet-biotite, muscovite-biotite, plagioclase-potassic feldspar (see Table 1.6), as well as Glebovitsky's (1977) and Perchik's (1970) geothermometers. However, it must be said that the temperature values obtained are considered relative.

The temperature of metamorphism was assessed for mineral associations of biotite-plagioclase gneisses with highly aluminiferous minerals, biotite-plagioclase gneisses and amphibolites. Archean metabasalts of the SG-3 well section are the most convenient rocks to used for the evaluation of temperatures of metamorphism, being predominantly composed of two minerals of alternating composition, i.e. amphibole and plagioclase, which are sensitive to changes in temperatures of metamorphism. The process of metamorphism, proceeding in the interval between the greenschists and granulitic facies, leads to a successive replacement of actinolite (in equilibrium with acid plagioclase) by actinolitic (normal) hornblende with andesite labradorite, hastingsite hornblende with labradorite-anorthite (Nalivkina and Vinogradova 1980).

The highest-temperature assemblage is represented by hypersthene, diopside, hastingsite hornblende, labradorite-anorthite; it was formed at a temperature of 1000° – 1200 °C (diopside-hornblende and hypersthene-hornblende geothermometers (Perchuk 1970). Sometimes almandine and biotite are present. The temperatures of this mineral assemblage formation were determined on the basis of a study of the potassium distribution between plagioclase and hornblende. The results obtained have shown that the regional metamorphism of rocks of primary basalt composition proceeded under conditions of the amphibolite facies at a temperature of 700° – 750 °C [amphibole-plagioclase geothermometer; Perchuk (1970)]. The garnet-biotite association was formed at a temperature of 750 °C [garnet-biotite geothermometer; Glebovitsky (1977)]. The mineral association, bluish-green hornblende and oligoclase were formed at a temperature of about 650 °C. The process of biotite phlogopitization took place at the same temperature. The temperature of actinolite formation (of later origin) in these rocks was 500° – 550 °C.

The application of the amphibole geothermometer method, based on the assessment of the changing properties of amphiboles (in a series of actinolite-hastingsite) from metabasites of various metamorphic facies (Nalivkina and Vinogradova 1981), allows the conlusions that the temperature of the formation of hastingsite hornblende of the metabasites studies in the SG-3 well section is about 1000°C.

An analysis of the temperature data on the formation of the polymetamorphic Archean complex using different mineralogical geothermometers allowed the conclusion that this complex was formed over a broad temperature interval. The temperature regime is assessed as follows: about 1000° – 1200 °C granulite facies; 800° – 650 °C amphibolite facies; 650° – 500 °C epidote-amphibolite facies; 650° – 600 °C granitization in conditions of the epidote-amphibolite facies.

### 1.3.3 Rock-Forming Minerals in Vertical Deep Section

Rock-forming minerals of the metamorphic rocks of the Archean in the SG-3 well section are represented by amphiboles, plagioclases, biotites and granites. They represent genetic series of rocks composed of minerals whose composition and properties were changing in time. The studied section of the crust penetrated by the SG-3 well from 0 to 11,662 m is characterized by the presence of rocks whose age ranges from 1.6 to 2.8 Ga. It was established that the grade of metamorphism of the Proterozoic rocks increases with depth from the prehnite-pumpellyite to the amphibolite facies, and a zonal metamorphic complex of moderate pressures is formed. The metamorphism of the underlying Archean complex is of a polymetamorphic character. Mineral associations of the granulitic amphibolite and epidote-amphibolite facies have been discovered in this complex.[3] It was established that down the section the role of mineral associations of the epidote-amphibolite facies decreases, whereas mineral relicts of the granulitic facies of metamorphism acquire more importance. From this it appears that on the whole the vertical section of the Precambrian crust in the SG-3 well is composed of metamorphic rocks, which change with depth from the prehnite-pumpellyite to granulitic facies. Moreover, it is worthy of note, that the greater the age of rocks and the deeper their occurrence, the more importance is attached to a higher grade of metamorphism. Let us now examine the changes of main rock-forming minerals in metabasites of isochemical composition which are known at all depths of the SG-3 well section.

Ca-amphiboles belong to a mineral series of an alternate composition: from older to younger metabasites and from deep to more shallow horizons, hastingsite hornblende is replaced by actinolitic hornblende and, finally, by actinolite (see Fig. 1.22). It is established that an evolutionary process is characterized by the following distinctive features in the chemical and structural changes of amphiboles. In amphiboles of this series (from hastingsite hornblende to actinolitic hornblende), the content of $SiO_2$ and MgO increases and the content of $Al_2O_3$ (predominantly $Al^{IV}$), $TiO_2$, F and alkali decreases; the degree of iron oxidation also decreases from 1/3 to 1/5.

The X-ray structural characteristic of amphiboles of this series (from hastingsite hornblende to actinolite) displays a decrease of an elementary cell (medium values, $nm^3$): from 0.92580 (for granulitic facies) to 0.91540 (for amphibolite facies), to 0.91190 (for epidote-amphibolite facies), and to 0.90560 (for greenschists facies). These values are directly proportional to the temperatures of metamorphism and to the maximum of an endothermal effect in heat curves of

[3] At places, in zones of cataclasm, mineral associations of low stages of metamorphism are developed.

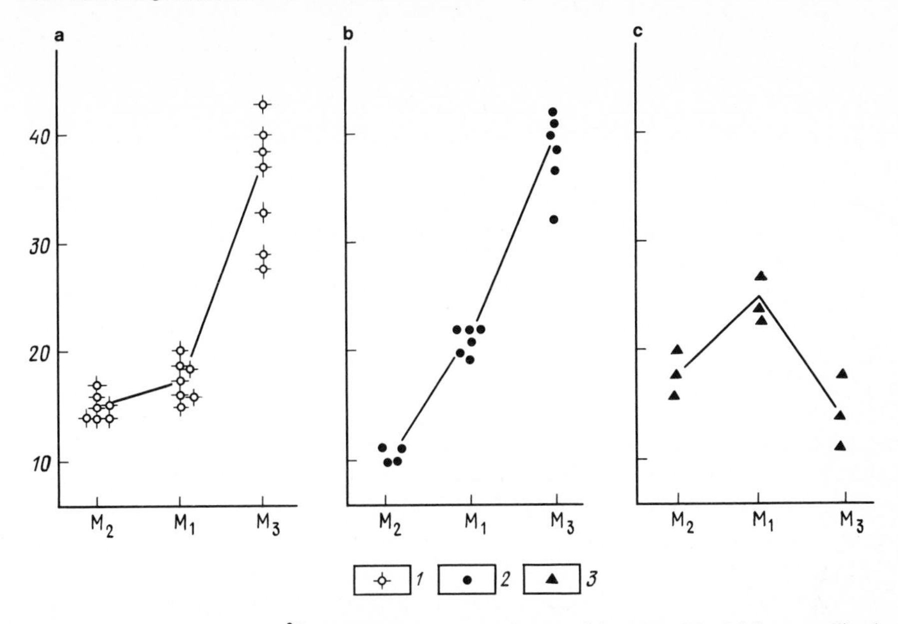


**Fig. 1.27a–c.** Scheme of $Fe^{2+}$ distribution in octahedron positions $M_1$, $M_2$, $M_3$ in crystalline lattice of amphiboles of the actinolite-hastingsite series. *Facies*: **a** epidote-amphibolite; **b** amphibolite; **c** granulit; the ratio of $Fe^{2+}$ in each of $M_1$, $M_2$ and $M_3$ positions to total content of $Fe^{2+}$ (in percent)

amphiboles (see Fig. 1.21), and they mainly reflect the thermal regime of metamorphism. In the series of amphiboles considered, i.e. from hastingsite hornblende to actinolitic hornblende, a distribution pattern of iron in the octahedron positions $M_1$, $M_2$ and $M_3$ acquires more perfect order (Fig. 1.27).

Amphiboles of the genetic series of Ca-amphiboles, which underwent changes in geological time, have acquired new features which reflected the thermodynamic conditions of their formation. This series is of complex structure.

Plagioclases from metabasites of the SG-3 well section form a genetic series of anorthite-albite which have developing under thermodyamical conditions of the granulitic to prehnite-pumpellyite facies. It is characteristic that the content of CaO and $Al_2O_3$ decreases and the content of $N_2O$ increases in plagioclases as their age and depth of occurrence decreases (see Fig. 1.23). Regular changes in plagioclase structure and in the distribution of aluminum in octahedrans between $T_1O$, $T_1M$, $T_2O$, and $T_2M$ are established. As anorthite passes into albite, the content of Al in a position of $T_1O$ increases from 0.61 to 0.90, whereas its content in other tetrahedrons decreases from 0.30 to 0.05 (see Fig. 1.24). Simultaneously, $\Delta$Al changes in plagioclase from 0.33–0.38 to 0.85 up the section; this indicates that the degree of order in the plagioclase structure increases up the section as the grade of metamorphism decreases.

Biotites are developed mainly in the Archean part of the section at a depth of 6842–11,662 m. Biotites from metabasites and gneisses are identical in composition and differ from each other only by their position in the vertical section of the crust, and depending on the grade of metamorphism to which they were subject.

Biotite-phlogopitobiotite forms a genetic series in which the iron and titanium content decreases, and the magnesium content increases as the age of the rocks and the depth of their occurrence decrease (see Fig. 1.25).

Garnets are present mainly in gneisses of the Archean part of the section. They form a genetic series of almandine-pyrope-bearing almandine in which the pyrope content decreases and spessartine increases as the degree of metamorphism and the depth of rock occurrence decreases (see Fig. 1.26).

From this it follows that rock-forming minerals present in metamorphic rocks either throughout the SG-3 well section (amphiboles, plagioclases) or within a considerable part of this section are represented by minerals of changing composition, and form the following series: actinolite-hastingsite; biotite-phlogopitobiotite; albite-anorthite; almandine-pyrope-bearing almandine.

Each of these mineral series makes a certain genetic sense when considered in connection with the studied Precambrian section formed in a time interval between 2.8 and 1.6 Ga. It should be noted that their composition and structure change along the SG-3 well section as depth increases. The redistribution of rock-forming minerals also takes place. For instances, in some minerals, the content of various elements systematically falls, whereas in other minerals it increases, but nevertheless, all these changes observed in all the mineral series examined lead to an increase of density. It seems probably that the primary members of these series were characterized by maximal densities.

The composition and properties of the main types of rock-forming minerals from the Precambrian metamorphic rocks depend on their position in the deep vertical section of the crust, and they allow an assessment of pressure-temperature conditions under which metamorphic rocks were formed at different stages of crustal evolution.

According to mineralogical and petrographical data, the main rock types of the Proterozoic complex in the SG-3 well section are represented by the following rocks: magmatic rocks – metabasalts (diabases, actinolitized diabases, apodiabase magnetite-amphibole-plagioclase schists, apodiabase amphibole-plagioclase schists), metapicritites, metagabbro, metawehrlites; metamorphosed sedimentary rocks – pelites, siltstones, sandstones, conglomerates and carbonate rocks. The intensity of metamorphism of the rocks of the Pechenga complex increases with depth from the prehnite-pumpellyite to the amphibolite facies.

In the Archean complex the main types of rock are represented by biotite-plagioclase gneisses with highly aluminiferous minerals, biotite-plagioclase gneisses – grey gneisses, amphibolites, meta-ultramafites and their granitized varieties, plagiogranites and granites. They were subject to recurrent metamorphism, which resulted in the formation of a polymetamorphic complex characterized by a spotty zonal structure. The following sequence of mineral associations is distinguished in this complex: associations of the granulitic, amphibolite and epidote-amphibolite facies as well as associations resulting from granitization under epidote-amphibolite-facies conditions.

Rock-forming minerals of the Archean and Proterozoic rocks (amphiboles, plagioclases, biotites, garnets) form the following genetic series of changing composition: actinolite-hastingsite, albite-anorthite, biotite-phlogopite and almandine-pyrope.

## 1.4 Geochemistry and Conditions of Formation of the Precambrian Complexes

A detailed description of the Pechenga complex rocks was given for the first time by Zagorodny et al. (1964), who adduced evidence in support of a theory of the geological integrity and synchronous formation of the northern and southern zones of the Pechenga structure. It was suggested that this structure originated as a result of the successive formation of four major rhythms which started their development with sedimentation and ended with volcanism. The chemical study of volcanites revealed some peculiar features, on the basis of which they can be regarded as products of the cyclic, multi-staged and complex process of basalt magma evolution in the direction from the spilite-keratophyric series enriched with alkaline elements to low-alkaline tholeiites. Basite-ultrabasite intrusions, including nickeliferous and andesitic extrusions, are considered to be differentiates of the same basalt lava, the only exception being that this lava intruded the rocks at the final stages of magmatic activity, i.e. during the inversion stage of the Pechenga structure formation.

Later, on the basis of geological and geochemical data, Predovsky and his colleagues (1974) subdivided the Pechenga complex into the following five sedimentary-volcanic formations (from bottom to top): the molasse-andesite-basalt, carbonate-quartzite-trachybasalt, carbonate-ferruginous-silica-tholeiite-basalt and terrigenous-tuffaceous-picrite-basalt formations developed in the northern zone, and the tuffaceous-picrite-andesite formation known in the southern zone of the Pechenga structure. The geochemical reconstruction of the primary nature of metamorphosed rocks has also been accomplished. This revealed some indications that the basal horizons of metasedimentary rocks underwent weathering, and that the process of sedimentation was affected by volcanic activity.

The geochemistry of subcrystalline rocks of the older age developed on the south-western and north-western rims of the Pechenga structure (Tundra and Kola series) has not been fully studied; some details of the chemical composition and genesis of these rocks were brought to light in works by Bondarenko and Dagelaisky (1968), Kremenetsky (1979) and Suslova (1976).

A systematic petrological and geochemical study of these complexes in the Kola super-deep well section and an investigation of the adjoining area were carried out by layer-by-layer synthesis and analysis of those rocks which had been distinguished by logging, and examined in core samples and at outcrops; during this study different types of rock were described and tested. The elaborate study procedure involves subdivision of each layer into unaltered or slightly altered metamorphic rocks, into their migmatized or metasomatic altered varieties and, finally, into all kinds of veins according to their thickness and position in the section. Subsequently, analogous rocks were combined in appropriate groups and studied in detail with the purpose of finding out their petrographical and geochemical characteristics and revealing the primary nature and subsequent changes in rock composition, dependent on depth of occurrence and intensity of the superimposed processes.

By this method, the SG-3 well section was subdivided into stratigraphic and formational units and then correlated with the surrounding rocks. In the course of the investigations, regularities in distribution of elements during stages of magmatic and sedimentary differentiation and zonal metamorphism were studied, and the relationship between geochemical and elastic-density characteristics of rocks established. This was accomplished to elucidate the nature of crustal non-uniformity.

The initial description of core samples and preparation of necessary documentation was performed by the authors of this section of the book together with Yu. P. Smirnov; geochemical studies were conducted by the authors and by M. Z. Abdrakhimov, V. I. Afanasova, D. D. Budyansky, T. V. Kopeikina, V. D. Nartikoyev, A. G. Nekhorosheva, T. V. Ryabova, L. K. Samodurova, S. F. Sobolev, I. Y. Shirokova, V. Y. Elevich. Chemical analyses of rocks were performed by means of a standard method (I. S. Razina, G. I. Fadeev et al.); quantitative determinations of elemental admixtures were made by methods of the flame spectrometry (L. P. Gulba), by a special rapid spectrographic method (A. A. Gusselnikov), by a neutron-activation method (A. V. Gurevich, S. M. Lyapunov) and by a chemical spectrographic method (E. T. Kataeva, L. I. Serdobova). Thermobarochemical investigations involved application of such methods as homogenization, cryometry, gas chromatography and microspectrographic analysis (Y. V. Vasiliev, Y. G. Chekhovskikh). Study of elastic-density properties of rocks were fulfilled by Y. I. Kuznetsov and A. P. Trofimov. Statistical processing of geochemical data was accomplished by V. S. Voronov and E. N. Pozdnyakov with the help of a computer.

### 1.4.1 The Pechenga Complex

Drilling of the Kola SG-3 well throughout the entire section of the Paleozoic complex (0 – 6842 m) for the first time permitted a reconstruction and study of its volumetric geological and geochemical model and, on the basis of data obtained, what was already known of its structure became more accurate. It became possible to interpret the evolution and tendencies of the Precambrian sedimentation, intrusive and effusive magmatic processes from new aspects and to determine the geochemical regime of the zonal metamorphism and the accompanying metasomatic processes.

The Pechenga complex section begins with meta-effusive rocks of the Matertin suite (0 – 1059 m)[4], whose stratigraphic position corresponds to the fourth volcanic layer exposed at the surface. These effusive rocks are mainly represented by tholeiitic basalts (87%), including olivine metabasalts, normal metabasalts and low-alkaline metabasalts (Table 1.7), which compose sheets of massive and

---

[4] According to the principle accepted in constructing and describing the SG-3 well section, here and further sequences of suites and rocks are given from top to bottom.

**Table 1.7.** Mean compositions of metasedimentary-volcanic rocks of the Pechenga complex from an interval pf 0 – 6842 m of the SG-3 well section

| Components | Metasedimentary rocks | | | | | | Meta-effusive rocks | | | | | |
|---|---|---|---|---|---|---|---|---|---|---|---|---|
| | 1 | | 2 | 3 | | 4 | 5 | 6 | 7 | 8 | 9 | |
| | L | N | L | L | N | L | N | N | N | N | L | N |
| $SiO_2$ | 71.62 | 74.46 | 74.76 | 52.64 | 49.36 | 15.41 | 57.71 | 53.92 | 44.97 | 43.93 | 44.44 | 43.48 |
| $TiO_2$ | 0.22 | 0.40 | 0.72 | 0.25 | 1.05 | 0.07 | 1.18 | 1.50 | 1.50 | 1.40 | 3.12 | 2.22 |
| $Al_2O_3$ | 13.52 | 6.66 | 8.72 | 5.71 | 7.36 | 1.35 | 14.91 | 16.77 | 13.08 | 5.66 | 8.12 | 6.39 |
| $Fe_2O_3$ | 0.72 | 0.89 | 4.20 | 0.79 | 0.85 | 0.08 | 1.93 | 1.77 | 2.42 | 2.90 | 5.80 | 1.77 |
| FeO | 3.13 | 7.05 | 1.79 | 1.60 | 7.20 | 0.43 | 10.19 | 10.52 | 15.83 | 11.99 | 10.42 | 10.78 |
| MnO | 0.03 | 0.06 | 0.05 | 0.05 | 0.49 | 0.08 | 0.07 | 0.05 | 0.16 | 0.14 | 0.21 | 0.22 |
| MgO | 1.56 | 2.40 | 1.19 | 10.04 | 2.82 | 18.22 | 2.89 | 1.15 | 8.26 | 20.43 | 12.23 | 15.78 |
| CaO | 1.32 | 2.86 | 1.54 | 13.07 | 15.30 | 28.22 | 1.72 | 2.91 | 3.30 | 6.04 | 9.08 | 14.32 |
| $Na_2O$ | 3.63 | 0.64 | 0.54 | 0.59 | 1.00 | 0.15 | 1.99 | 1.89 | 1.34 | 0.42 | 0.20 | 0.22 |
| $K_2O$ | 2.85 | 0.62 | 4.69 | 2.67 | 0.70 | 0.19 | 2.48 | 3.49 | 0.08 | 0.06 | 0.01 | 0.06 |
| $P_2O_5$ | 0.08 | 1.40 | 0.12 | 0.27 | 0.08 | 0.10 | 0.16 | 0.09 | 0.15 | 0.38 | 0.26 | 0.25 |
| $CO_2$ | 0.16 | 0.82 | 0.61 | 10.06 | 11.00 | 34.16 | 0.55 | 0.24 | 1.40 | 0.10 | 0.20 | 0.02 |
| $H_2O^-$ | 0.10 | 0.20 | 0.04 | 0.09 | 0.14 | 0.11 | 0.12 | 0.18 | 0.38 | 0.34 | 0.24 | 0.24 |
| $H_2O^+$ + l.o.c.[a] | 0.66 | 1.39 | 0.59 | 1.18 | 1.20 | 0.95 | 3.76 | 5.30 | 5.18 | 6.32 | 4.75 | 4.05 |
| $S_{total}$ | 0.12 | – | 0.08 | 0.10 | 1.30 | 0.08 | 2.14 | – | 1.29 | 0.02 | – | 0.02 |
| Li | 34.0 | 29.3 | 8.6 | 26.8 | 21.0 | 16.8 | 39.9 | 43.8 | 32.5 | 14.5 | 19.8 | 15.6 |
| Rb | 71.0 | 25.6 | 93.8 | 73.6 | 25.8 | 38.0 | 72.2 | 97.9 | 26.2 | 12.7 | 20.0 | 8.7 |
| Sr | 80.0 | 59.7 | 26.4 | 49.4 | 162.0 | 73.8 | 46.4 | 15.4 | 26.3 | 13.4 | 25.1 | 15.7 |
| Ba | 575.0 | 106.2 | 705.5 | 333.3 | 113.1 | 320.0 | 165.3 | 283.3 | 61.8 | 10.0 | 201.2 | 17.2 |
| B | 9.0 | 38.6 | 16.4 | 32.0 | 29.3 | 83.8 | 33.6 | 40.0 | 10.5 | 16.0 | 7.0 | 12.9 |
| Sc | 50.0 | 12.2 | 21.8 | 21.8 | 12.5 | 11.6 | 11.9 | 17.9 | 17.0 | 4.1 | 7.6 | 6.9 |
| Ga | 27.5 | 13.6 | 15.7 | 20.7 | 12.7 | 7.0 | 19.2 | 24.5 | 19.6 | 10.6 | 20.0 | 12.8 |
| La | 22.9 | 21.9 | 33.4 | 17.4 | – | 3.8 | 23.4 | – | 53.0 | – | – | 20.5 |
| Ce | 64.1 | 37.6 | 65.6 | 36.5 | – | 11.3 | 44.3 | – | 66.7 | – | – | 35.5 |
| Sm | 3.7 | 5.4 | 5.7 | 2.0 | – | 0.70 | 7.1 | – | 9.8 | – | – | 7.2 |
| Eu | 1.0 | 1.3 | 1.8 | 0.70 | – | 0.30 | 1.5 | – | 1.4 | – | – | 1.2 |

[a] l.o.c. = Loss of calcination.

**Table 1.7** (continued)

| Components | Metasedimentary rocks | | | | | | Meta-effusive rocks | | | | | |
|---|---|---|---|---|---|---|---|---|---|---|---|---|
| | 1 | | 2 | 3 | | 4 | 5 | 6 | 7 | 8 | 9 | |
| | L | N | L | L | N | L | N | N | N | N | L | N |
| Yb | 1.7 | 2.8 | 2.4 | 1.5 | – | 0.30 | 3.0 | – | 3.0 | – | – | 1.9 |
| Lu | 0.10 | 0.50 | 0.40 | 0.20 | – | 0.10 | 0.70 | – | 0.60 | – | – | 0.40 |
| Th | 12.9 | 2.9 | 6.1 | 8.0 | – | 1.2 | 4.2 | – | 4.8 | – | – | 5.5 |
| U | 3.7 | 2.0 | 1.7 | 0.20 | – | 0.40 | 2.5 | – | 3.3 | – | – | – |
| Ge | 0.50 | 2.4 | 0.46 | 1.1 | 1.9 | 0.33 | 2.6 | 2.0 | 2.3 | 1.9 | 0.79 | 2.0 |
| Zr | 85.0 | 64.0 | 71.0 | 98.9 | 70.2 | 34.4 | 70.3 | 94.0 | 64.1 | 48.9 | 50.0 | 71.3 |
| Sn | 3.1 | 5.2 | 3.4 | 3.5 | 4.2 | 3.0 | 32.7 | 5.2 | 5.8 | 4.1 | 3.0 | 5.7 |
| Hf | 6.3 | 3.0 | 4.5 | 10.7 | – | 1.8 | 3.2 | – | 4.7 | – | – | 2.5 |
| Nb | 5.2 | 7.0 | 4.0 | 4.4 | 5.3 | 2.0 | 5.7 | 7.7 | 4.9 | 3.9 | 2.4 | 7.9 |
| F | 290.0 | 586.0 | 504.1 | 819.4 | 526.7 | 1176.7 | 1071.1 | 720.0 | 817.5 | – | – | – |
| Cu | 10.0 | 119.0 | 6.5 | 48.7 | 104.8 | 12.8 | 162.3 | 130.0 | 222.9 | 307.8 | 242.5 | 329.5 |
| Zn | 26.9 | 191.5 | 51.4 | 78.9 | 137.2 | 24.9 | 199.8 | 230.8 | 213.5 | 253.9 | 30.0 | 209.1 |
| Ag | 0.09 | 0.05 | 0.02 | 0.05 | 0.06 | 0.05 | 0.08 | 0.07 | 0.10 | 0.01 | 0.10 | 0.19 |
| Pb | 20.8 | 16.3 | 21.3 | 15.6 | 17.0 | 16.4 | 20.6 | 22.2 | 11.7 | 4.4 | 11.0 | 6.6 |
| Mo | 1.8 | 4.2 | 0.69 | 1.2 | 3.3 | 0.88 | 7.0 | 5.0 | 10.1 | 1.9 | 1.2 | 3.0 |
| V | 69.5 | 190.4 | 190.2 | 108.9 | 161.0 | 49.7 | 233.6 | 255.4 | 247.5 | 136.3 | 250.0 | 241.5 |
| Cr | 37.2 | 150.6 | 240.0 | 138.9 | 195.8 | 79.8 | 209.7 | 202.8 | 367.6 | 958.0 | 220.0 | 254.7 |
| Co | 12.8 | 32.4 | 18.8 | 15.9 | 29.8 | 11.4 | 36.0 | 31.1 | 49.9 | 137.5 | 32.2 | 77.3 |
| Ni | 19.8 | 88.7 | 38.6 | 34.7 | 72.6 | 23.5 | 112.1 | 81.8 | 226.7 | 1080.4 | 59.0 | 181.7 |
| Number of samples | 4 | 128 | 73 | 9 | 32 | 8 | 259 | 24 | 17 | 28 | 4 | 39 |

**Table 1.7** (continued)

| Components | Metaeffusive rocks | | | | | | | | | | |
|---|---|---|---|---|---|---|---|---|---|---|---|
| | 10 | | 11 | | 12 | 13 | 14 | 15 | | 16 | 17 |
| | L | N | L | N | N | L | L | L | N | L | L |
| $SiO_2$ | 50.07 | 47.72 | 50.86 | 47.73 | 47.66 | 49.66 | 48.27 | 54.43 | 55.27 | 54.21 | 58.08 |
| $TiO_2$ | 1.41 | 1.16 | 2.00 | 1.60 | 1.33 | 1.73 | 2.27 | 0.98 | 1.38 | 1.67 | 1.24 |
| $Al_2O_3$ | 11.98 | 13.93 | 12.83 | 13.29 | 13.31 | 13.98 | 14.46 | 13.55 | 11.40 | 14.79 | 13.42 |
| $Fe_2O_3$ | 5.74 | 3.04 | 4.34 | 2.81 | 2.80 | 6.25 | 5.94 | 1.84 | 1.85 | 4.26 | 2.08 |
| FeO | 7.29 | 9.97 | 9.89 | 11.71 | 11.30 | 8.81 | 9.85 | 8.94 | 9.64 | 7.57 | 9.20 |
| MnO | 0.20 | 0.19 | 0.16 | 0.20 | 0.16 | 0.25 | 0.23 | 0.16 | 0.12 | 0.13 | 0.14 |
| MgO | 7.28 | 7.96 | 6.58 | 6.62 | 6.12 | 5.44 | 5.97 | 5.98 | 4.11 | 4.33 | 3.34 |
| CaO | 7.83 | 9.98 | 6.20 | 9.23 | 10.35 | 5.84 | 4.83 | 7.92 | 6.40 | 4.43 | 6.35 |
| $Na_2O$ | 3.56 | 2.20 | 3.03 | 2.46 | 0.60 | 5.26 | 4.09 | 3.14 | 2.65 | 4.13 | 3.19 |
| $K_2O$ | 0.37 | 0.29 | 1.17 | 0.30 | 0.39 | 0.47 | 1.63 | 1.24 | 0.76 | 2.41 | 1.44 |
| $P_2O_5$ | 0.11 | 0.09 | 0.18 | 0.16 | 0.22 | 0.28 | 0.22 | 0.16 | 0.13 | 0.20 | 0.17 |
| $CO_2$ | 2.33 | 0.21 | 0.23 | 0.50 | 0.32 | 0.43 | 0.18 | 0.30 | 2.39 | 0.48 | 0.05 |
| $H_2O^-$ | 0.08 | 0.13 | 0.14 | 0.18 | 0.20 | 0.09 | 0.12 | 0.13 | 0.06 | 0.22 | 0.09 |
| $H_2O^+$ + l.o.c. | 3.54 | 3.24 | 2.37 | 3.74 | 5.26 | 2.02 | 2.23 | 1.63 | 4.22 | 1.84 | 1.42 |
| $S_{total}$ | 0.07 | 0.09 | 0.07 | 0.11 | 0.18 | 0.09 | 0.05 | 0.03 | 0.37 | 0.07 | 0.03 |
| Li | 12.5 | 12.4 | 14.6 | 11.9 | 8.8 | 9.1 | 12.5 | 13.0 | 18.5 | 15.1 | 11.5 |
| Rb | 15.0 | 6.9 | 31.8 | 7.6 | 12.3 | 14.2 | 37.5 | 40.2 | 8.8 | 57.5 | 41.7 |
| Sr | 75.0 | 38.1 | 28.4 | 52.9 | 62.0 | 36.8 | 28.1 | 42.5 | 103.8 | 29.1 | 34.0 |
| Ba | 300.0 | 85.7 | 75.0 | 67.0 | 54.0 | 223.5 | 195.9 | 255.0 | 87.5 | 443.1 | 190.0 |
| B | 7.5 | 35.6 | 3.5 | 17.6 | 9.1 | 8.0 | 100.6 | 6.3 | 12.0 | 7.5 | 5.6 |
| Sc | 13.0 | 13.0 | 6.3 | 14.4 | 13.0 | 11.2 | 9.2 | 4.3 | 6.8 | 5.2 | 5.8 |
| Ga | 20.0 | 14.2 | 16.3 | 17.5 | 15.5 | 16.2 | 16.4 | 18.4 | 22.5 | 17.0 | 19.5 |
| La | – | 9.4 | – | 12.4 | 6.2 | 17.7 | 30.5 | 19.9 | 40.0 | 34.9 | 30.5 |
| Ce | – | 24.3 | – | 25.9 | 21.0 | 50.4 | 68.1 | 52.2 | 45.0 | 80.2 | 68.1 |
| Sm | – | 3.8 | – | 5.3 | 4.2 | 5.8 | 5.5 | 3.9 | 5.9 | 5.4 | 5.5 |
| Eu | – | 1.4 | – | 1.7 | 2.1 | 2.0 | 1.6 | 1.3 | 2.5 | 2.2 | 1.6 |

**Table 1.7** (continued)

| Components | Metaeffusive rocks | | | | | | | | | | |
|---|---|---|---|---|---|---|---|---|---|---|---|
| | 10 | | 11 | | 12 | 13 | 14 | 15 | | 16 | 17 |
| | L | N | L | N | L | L | L | L | N | L | L |
| Yb | – | 2.8 | – | 3.2 | 2.7 | 3.4 | 2.9 | 1.9 | 2.8 | 2.3 | 2.9 |
| Lu | – | 0.60 | – | 0.60 | 0.60 | 0.10 | 0.60 | 0.40 | 0.40 | 0.40 | 0.60 |
| Th | – | 0.80 | – | 1.2 | 1.2 | 2.0 | 5.7 | 3.4 | 5.1 | 4.8 | 5.7 |
| U | – | 0.20 | – | 0.20 | – | 1.1 | 1.1 | 1.5 | – | 0.60 | 1.1 |
| Ge | 0.05 | 0.90 | 0.79 | 0.93 | 0.84 | 0.08 | 0.74 | 0.70 | 2.2 | 0.12 | 0.44 |
| Zr | 70.0 | 41.4 | 51.6 | 51.6 | 49.0 | 53.8 | 48.8 | 54.6 | 45.0 | 64.0 | 58.4 |
| Sn | 3.0 | 5.7 | 2.7 | 4.6 | 3.2 | 3.0 | 3.2 | 2.8 | 2.9 | 2.8 | 2.9 |
| Hf | – | 2.1 | – | 2.3 | 2.0 | 3.5 | 4.9 | 3.8 | 2.9 | 5.5 | 4.9 |
| Nb | 0.50 | 3.2 | 2.4 | 3.0 | 1.8 | 0.62 | 1.0 | 1.9 | 3.5 | 1.4 | 2.1 |
| F | – | 82.0 | 480.0 | 120.2 | 158.0 | 46.9 | 205.9 | 73.6 | 640.0 | 55.0 | – |
| Cu | 150.0 | 153.7 | 266.6 | 202.7 | 202.5 | 76.2 | 164.0 | 148.3 | 575.0 | 112.8 | 137.4 |
| Zn | 30.0 | 118.5 | 88.4 | 131.3 | 122.0 | 40.0 | 62.4 | 77.4 | 130.0 | 47.7 | 81.4 |
| Ag | 0.01 | 0.12 | 0.05 | 0.05 | 0.02 | 0.02 | 0.03 | 0.03 | 0.08 | 0.02 | 0.05 |
| Pb | 13.0 | 16.8 | 18.0 | 11.0 | 8.9 | 14.5 | 21.0 | 23.5 | 8.2 | 29.3 | 24.1 |
| Mo | 1.0 | 0.72 | 0.70 | 1.08 | 0.95 | 0.98 | 1.1 | 0.87 | 1.7 | 0.99 | 1.7 |
| V | 95.0 | 260.8 | 221.7 | 301.9 | 280.5 | 182.3 | 520.6 | 221.6 | 292.5 | 198.9 | 223.0 |
| Cr | 620.0 | 155.1 | 238.4 | 83.9 | 97.2 | 207.3 | 116.1 | 216.7 | 120.5 | 67.8 | 39.6 |
| Co | 50.0 | 51.3 | 63.2 | 47.0 | 44.3 | 43.4 | 67.1 | 40.7 | 45.5 | 36.6 | 32.9 |
| Ni | 150.0 | 116.0 | 94.2 | 65.2 | 66.0 | 83.5 | 64.6 | 69.3 | 67.0 | 35.0 | 19.9 |
| Number of samples | 2 | 117 | 6 | 386 | 20 | 13 | 17 | 105 | 4 | 35 | 18 |

[a] l.o.c. = Loss on calcination.

1 feldspar-quartz metasandstones; 2 – arkose metasandstones; 3 – metasandstones and meta-silty sandstones with carbonate cement; 4 – sandy dolomites; 5 – metasiltstones, including those with carbonate cement and sulphides; 6 – metapelites, including those with sulphides; 7 – metatuffites, including those with carbonate cement and sulphides; 8 – metapicrites; 9 – picritic metaporphyrites (picritobasalts); 10–13 – metadiabases, plagioclase-amphibole schists and amphibolites (10 – olivine basalts, 11 – basalts, including ferrobasalts, 12 – low-alkali basalts, 13 – spilitic basalts); 14–16 – biotite-amphibole plagioclases, including those with magnetite (14 – trachybasalts, 15 – andesite basalts, including those of a pyroxene-olivine type; 16 – trachyandesite basalts); 17 – quartz-amphibole-biotite plagioclase schists, including those with magnetite (andesites); L – rocks of the Luostari series; N – rocks of the Nickel series. Content of petrogenic elements, in percent; elemental admixtures, in $g\,t^{-1}$; "–" content was not determined.

globular lavas. Metapicritobasalts and metapicrites are of subordinate significance; they form a continuous differentiated series associated with metaperidotites and metapyroxenites. Because of this, and also due to the presence of spinifex textures and the similarity in chemical composition of these rocks with the Archean komatiites of South Africa, they have been assigned to the komatiites (Suslova 1976). However, as distinct from the reference rocks, the Pechenga komatiites are characterized by a higher Ti and Fe content. If, moreover, a ratio of MgO/CaO is taken into account, then they can be subdivided into peridotite ($>2$), pyroxenite ($1-2$) and basalt ($<1$) varieties. The distribution pattern of petrogenic and accompanying elemental admixtures in this series displays a regular decrease in content $SiO_2$, $Al_2O_3$, Ti, V, and Nb against the general background of increased concentration of Co, Cr, and Ni. The secondary members of the suite are the lavas of dacite and dacite-liparite composition and their tuffs which are developed only at the surface. In view of the fact that these rocks and picrites occur in combination with each other, acid lavas may be assigned to the marginal differentiates of the picritic series.

On the petrochemical diagram of $SiO_2$ versus $Na_2O+K_2O$ (Fig. 1.28), volcanites of the Matertin suite are grouped only partially in a field of picrites and picritobasalts, but mainly in a field of tholeiitic basalts, including alkaline basalts. A study of the geochemistry of these rocks, by taking into account their position in the SG-3 well section, allows the conclusion that alkalinity and magnesium content slightly increase in the lower part of the suite and that the iron content increases in the upper parts of the section; this increase is seen as a recurrent trend. The distribution pattern of elemental admixtures also depends on the stratigraphic position of volcanites, i.e. in tholeiites associated with picrites the content of Ti is high and the content of Cr, Ni and Nb is low; tholeiites associated with acid volcanites are enriched with Pb.

Volcano-sedimentary rocks of the Matertin suite are represented by silty-pelitic, occasionally lithoclastic tuffs and also by subordinate tuffites, which are genetically related with alternating volcanites. Tuffs, as distinct from diabases, are characterized by a higher content of $SiO_2$ (52.02%) and a lesser content of $TiO_2$ (1.24%), MgO (3.38%) and CaO (3.30%). Tuffs are also characterized by a very low degree of iron oxidization ($Fe_2O_3/FeO = 0.02$ as compared with 0.2 in diabases), and by higher alkalinity, which results in higher concentrations of Li ($33\ g\ t^{-1}$), Sr ($300\ g\ t^{-1}$) and Ba ($77\ g\ t^{-1}$). Tuffs associated with ultrabasic rocks differ from the above-described pyroclastic rocks by a higher content of $TiO_2$ (1.7%), Cr ($400\ g\ t^{-1}$), Ni ($240\ g\ t^{-1}$) and Co ($85\ g\ t^{-1}$).

In bulk, structure and chemical composition of rocks, the Zhdanov suite fully corresponds to the fourth sedimentary strata developed at the surface. Its lithogeochemical features allow us to subdivide it into four rhythmically layered members (gritty sandstone – 2619–2805 m, sandy silty pelitic – 2155–2619 m, volcano-silty pelitic – 1263–2155 m and pelitic-tuffaceous – 1059–1203 m); these members also form two heterogeneous rhythms: the lower terrigenous rhythm (composed mainly of sandstones) and the upper volcanogenic-terrigenous rhythm (composed of finely rhythmic siltstones and pelites with considerable admixture of volcanic material). Some general features of metapelites resemble those of silty pelitic tuffs of the Zhdanov and Matertin suites, i.e. similar

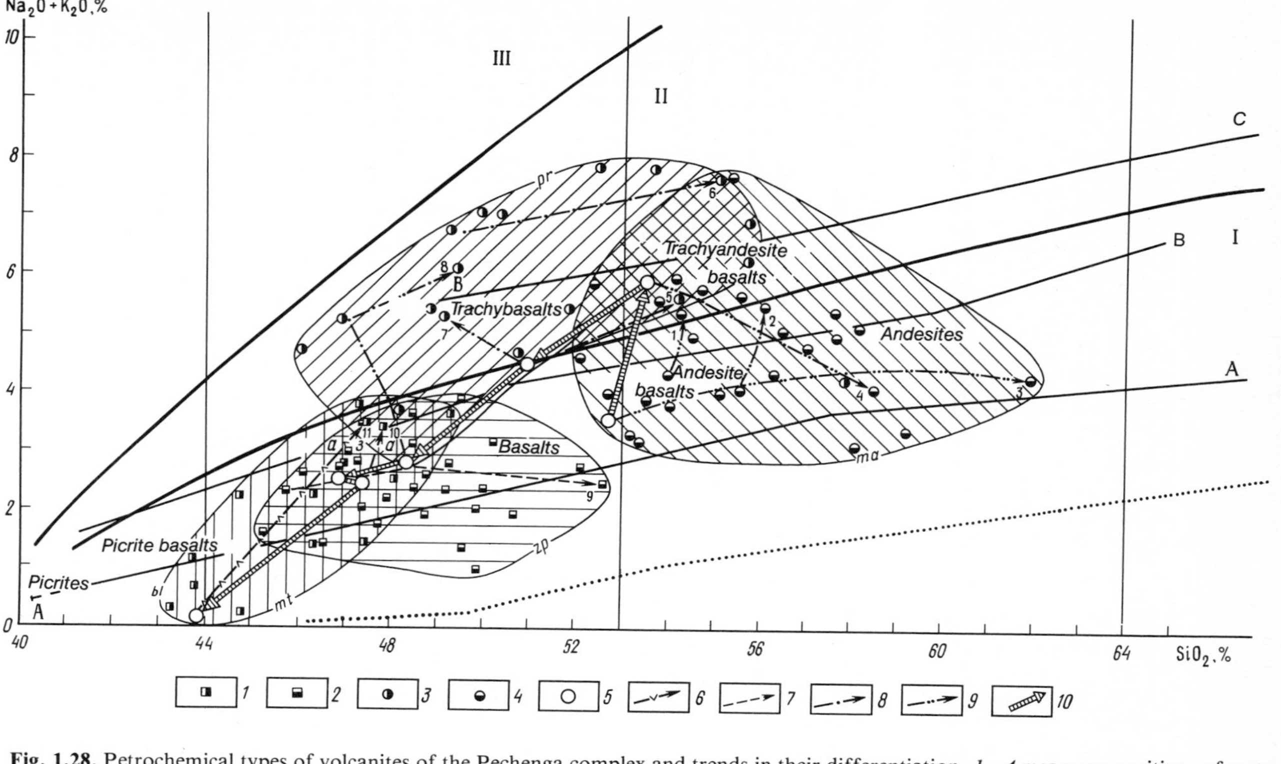


**Fig. 1.28.** Petrochemical types of volcanites of the Pechenga complex and trends in their differentiation. *1–4* mean compositions of meta-effusive flows in suites: *1* Matertin; *2* Zapolyarny; *3* Pirttiyarvi; *4* Mayarvi; *5* compositions of the first portions of melts of effusive series, *6–9* trends of the meta-effusive rocks differentiation in suites: *6* Matertin; *7* Zapolyarny; *8* Pirttiyarvi (trend number denotes position of effusive series in the section from bottom to top); *10* evolution of initial melt portions of the effusive series during formation of the Pechenga complex. *I* area of rocks with normal alkalinity; *II* area of subalkali rocks; *III* area of alkali rocks. Differentiation trends of the effusive series of island areas: *A* tholeiitic; *B* calcareous-alkali; *C* alkali (Luts 1980)

basicity, increased content of aluminiferous minerals (about 16% $Al_2O_3$) and the same $TiO_2$ content and other elemental admixtures of the iron group. Metasilty pelites are characterized by a lower content of aluminiferous minerals (11.9%), higher content of $SiO_2$ (65.19%) and a similar content of alkaline elements. Metasandstones are notable for their high $SiO_2$ content (up to 74%), low content of $Al_2O_3$ (7.08%), their total alkalinity $Na_2O + K_2O = 1.46\%$ and content of $TiO_2 = 0.48\%$. Nevertheless, concentration levels of elemental admixtures in a series of sedimentary differentiation (metasandstones-metasiltstones-metapelites) are similar to each other (see Table 1.7). A characteristic feature of metasedimentary rocks of the Zhdanov suite is the presence of carbonaceous matter (at places even individual layers of sulphide-carbonaceous schists are formed).

Tuffs and tuffites are characterized by their high content of total Fe (19.8%), MgO (7.04%), Cr (up to 0.08%) and Ni (up to 0.07%), which indicates that these tuffs resemble tuffs of the Matertin suite and that this period of sedimentation was characterized by the manifestation of active volcanic processes. As a result of this, all metasedimentary rocks of this suite, metapelites in particular, as compared with analogous rocks of the underlying suites, became enriched with sulphides and, consequently, by chalco- and siderophile elements, i.e. Cu ($130-165\ g\,t^{-1}$), Zn ($160-230\ g\,t^{-1}$), Pb ($17-25\ g\,t^{-1}$), Mo ($5-7\ g\,t^{-1}$), V ($190-225\ g\,t^{-1}$). Sulphide minerals are present in the form of impregnations and are subdivided into two genetic groups: syngenetic and epigenetic ones. Sulphur of the syngenetic pyrrhotite and pyrite contains predominantly heavy isotope $^{34}S$; average values of a $\delta^{34}S$ are equal to +16.6 and 17.24‰ respectively; since the parent rocks are enriched in organic matter, it can be concluded that these sulphides were formed in the reducing environment synchronously with sedimentation. The sulphur isotope composition of the epigenetic sulphides is characterized by greater homogeneity and they are considerably enriched in a light isotope $^{32}S$. Average values of $\delta^{32}S$ of pyrrhotite and pyrite are equal to +4.03 and +4.88‰ respectively; this indicates that the source of ore matter for this group of sulphides and copper-nickel ores of the Pechenga area was of deep-seated origin ($\delta^{34}S = 1.4$ and +5.2‰).

Effusive rocks (apparent thickness about 90 m), represented by tholeiitic basalts, ferrobasalts and, occasionally, by andesite-basalts, are the least developed in the Zhdanov suite. With respect to chemical composition they are analogous to volcanites of the Matertin and Zapolyarny suites.

Basic meta-effusives of the Zapolyarny suite (2805 – 4673 m) present in the Pechenga complex have the maximum apparent thickness (1.9 km) and are characterized by a uniform composition. Ninety five percent of them are formed of tholeiite basalts, including olivine basalts, normal basalts and low-alkaline basalts (see Table 1.7); 5% of them are composed of volcano-sedimentary metarocks and basic tuffs which differ from the associated effusive rocks notable for the increased content of $SiO_2$, $Al_2O_3$ and $K_2O$ and decreased content of MnO, MgO and CaO. Meta-effusive rocks are represented by massive and globular lavas, occasionally by lava breccia. Some sheets are clearly differentiated; normal basalts are present in the central parts and olivine basalts are at endocontacts. On the petrochemical diagram (see Fig. 1.28) they form a condensed swarm inside a field of basalts of normal alkalinity. In the SG-3 well section, geochemical trends of tholeiites, reflecting the process of magma differentiation (up the section), are characterized, firstly, by successive increase of the magnesium content and, secondly, by increase in iron content and alkalinity. Sedimentary-volcanogenic rocks are not widely developed, and are known only in the central part of the suite. In chemical composition they are similar to volcanic rocks of the Zhdanov suite but differ from them by higher content of $TiO_2$ (1.1%) and higher values of K/Na (1.0%).

The Luchlompol (4784 – 4884 m) and Kuvernerinyok (5642 – 5717 m) suites have a similar thickness, analogous structure (at the bottom sandstones, at the top carbonates) and the same composition of sedimentary rocks (arkosic meta-sandstones, including dolomitic ones, and dolomites, including sandy ones) which allows these suites to be considered together. A general characteristic feature of the chemical composition of metaterrigenous rocks (see Table 1.7) is the considerable predominance of potassium over sodium (K/Na = 2.3 – 11.8) and their enrichment in Ca (up to 21%) and MgO (up to 19%). Consequently, these rocks contain an increased amount of Rb (up to 98 $g\,t^{-1}$), Ba (up to 795 $g\,t^{-1}$) and Sc (up to 34 $g\,t^{-1}$); this determines the specific character of the geochemistry of the suites of the Pechenga complex under consideration.

The Pirttiyarvi suite (4884 – 5642 m) is made of meta-effusive rocks represented by trachybasalts (32%), alkaline basalts (25%) and trachyandesite basalts (7%). Of subordinate importance are pyroclastic rocks of mixed composition (1%).

On the classification diagram (see Fig. 1.28) symbols denoting effusive rocks are concentrated mainly in the area of subalkaline rocks, and are spread over this area, according to their degree of basicity, from normal basalts to trachytes. As compared with volcanites of the overlying suites, meta-effusive rocks of the Pirttiyarvi suite, its lower part in particular, are shifted towards the alkaline side, and correspond to a sodium type; the exceptions are represented by trachytes ($Na_2O$ – 5.5%, $K_2O$ – 4.2%) which have so far been found only at the surface. A few rhythms of the magmatic differentiation (basalts→alkaline basalts→trachybasalts), distinguished by a successive increase in iron content and total alkalinity, are distinctly seen in the section (from bottom to top).

The Mayarvi suite (5717 – 6823 m) is composed of strongly metamorphosed effusive rocks with much less metamorphosed (less than 1%) metatuffs. Normal andesites predominate among meta-effusive rocks, including pyroxene-olivine andesite basalts and trachyandesite basalts, whose content does not exceed 2%; the content of andesites is much less (3%).

On the petrochemical diagram (see Fig. 1.28), symbols denoting these rocks are located at a boundary between areas of normal and increased alkalinity. Andesites, as distinct from andesite basalts, contain a high amount of $SiO_2$, $K_2O$, and $Fe_2O_3$ and a correspondingly small amount of FeO, MgO, and CaO (see Table 1.7). Leucocratic varieties of andesite basalts occupy an intermediate position between these types and are distinguished by higher content of $TiO_2$ (1.57%). In the SG-3 well section (from bottom to top) effusive rocks of the Mayarvi suite are characterized by geochemical trends similar in general to those of the Pirttiyarvi suite, i.e. their iron content increases successively (including Sc, V and Co), and this is accompanied by a slight increase in alkalinity in the lower part of the suite and its decrease up the section.

The Televin suite (6823 – 6842 m) is represented by a thin member of metamorphosed gritty plagiosandstones which are distinguished by the increased content of $Na_2O$ (3.63%, K/Na – 0.73 – 1.25%), $Al_2O_3$ (13.5%), and $SiO_2$ (71.6%) observed against the general background of very low values of $\Sigma Fe$, CaO, and MgO (see Table 1.7).

On the basis of the geological, petrographical, mineralogical, and geochemical criteria, we subdivide intrusive rocks of the Pechenga complex into four units: the Prematertin gabbro-diabase unit, two Synmatertin units (gabbro-wehrlite and gabbro-diabase) and the Postmatertin unit of granodiorites (andesite-dacite).

The Prematertin gabbro-diabase unit was defined for the first time. It is represented by individual bodies of apogabbro-diabase amphibolites developed only among metavolcanites of the Mayarvi suite; rocks of this unit (Prematertin amphibolites) were metamorphosed at the time of the Mayarvi rock metamorphism under conditions of the epidote-amphibolite facies. This fact, as well as the absence of similar rocks in overlying metavolcanites of the Pirttiyarvi suite, substantiates the theory of an existing genetical relation between the gabbro-diabases considered and the final stages of the andesite-basalt volcanic activity. For instance, from an analysis of the petrochemical diagram (Fig. 1.29) it appears very clearly that there is a similarity between a trend of the Pirttiyarvi effusive rocks and composition of the Prematertin gabbro-diabases; sharp differences exist between the former trend and composition of the Synmatertin gabbro-diabases, whose geochemical trend is oriented in the opposite direction. Pre- and Synmatertin gabbro-diabase units can also be distinguished from each other by

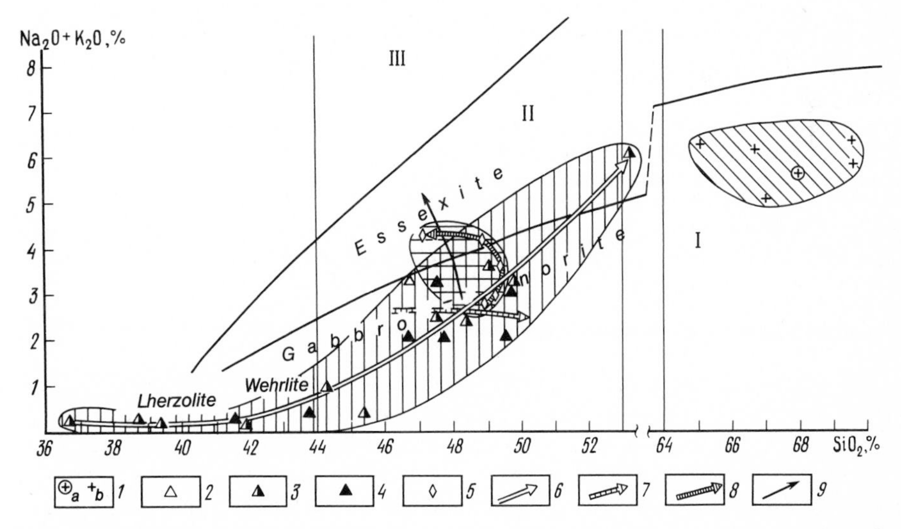


**Fig. 1.29.** Petrochemical trends of the metaintrusive rocks of the Pechenga complex and trends of their differentiation. *1 – 5* mean compositions of metaintrusive rocks present in suites: *1* Luchlompol (*a* mean; *b* individual composition of the Post-Matertin complex); *2* Matertin; *3* Zhdanov; *4* Zapolyarny; *5* Mayarvi; *6 – 8* differentiation trends of the metaintrusive rocks; complexes: *6* Syn-Matertin gabbro-wehrlites; *7* Syn-Matertin gabbro-diabases; *8* Pre-Matertin gabbro-diabases; *9* differentiation trend of metavolcanites of the Pirttiyarvi suite

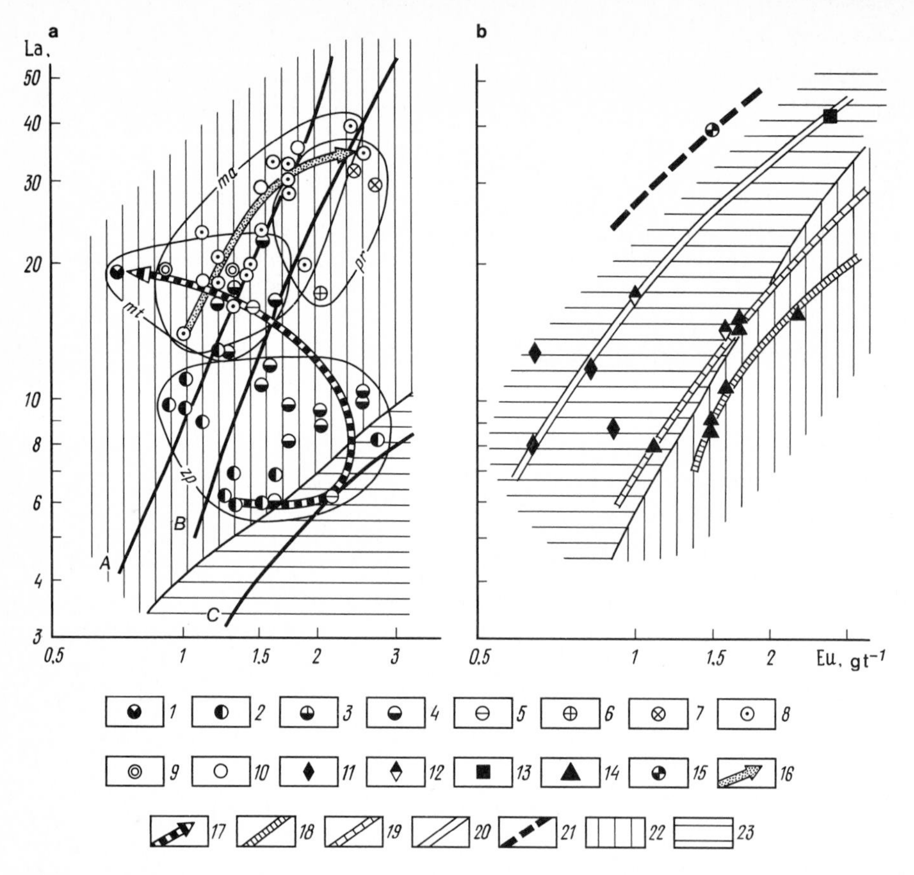


**Fig. 1.30 a, b.** Types of distribution of rare earth elements in meta-effusive (**a**) and metaintrusive (**b**) rocks of the Pechenga complex. *1–10* mean compositions of meta-effusive rocks; *1* metapicrite-basalts; *2* olivine metabasalts; *3* differentiated metabasalts; *4* metabasalts; *5* low-alkali metabasalts; *6* spilites; *7* metatrachybasalts; *8* meta-andesitobasalts; *9* metatrachyandesitobasalts; *10* meta-andesites; *11–15* mean compositions of meta-intrusive rocks: *11* metalherzolites; *12* metalherzolito-wehrlites; *13* essexinitic metagabbro; *14* metagabbronorites; *15* dacitic metaporphyrites; *16–17* compositions of rare earth elements in meta-intrusive rocks of the Pre-Matertin gabbro-diabase (*18*), gabbro diabase (*19*) and Syn-Matertin gabbro-wehrlite (*20*) and Post-Matertin dacite-metaporphyrite (*21*) complexes; *22* areas of occurrence of continental and island-arc basalts (for **a**) and differentiated trapp intrusions (for **b**); *23* areas of occurrence of oceanic basalts (for **a**) and ultrabasites (for **b**); *A–C* trends of volcanites [after Luts (1980)] of: *A* island arcs and active continental margins; *B* oceanic islands; *C* mid-ocean ridges

their typomorphic accessory mineral content (apatite, ilmenite, orthite, and pyrite) and elemental admixtures (Sc, B, F, Ge, Cr, Pb, Zn, and Zr), as well as by the content and distribution of rare-earth elements (Table 1.8). On the diagram, La versus Eu (Fig. 1.30b), a trend of Prematertin gabbro-diabases, as distinct from Synmatertin ones, is shifted into the area of continental basites.

The Synmatertin unit of gabbro-diabase fully corresponds to the tholeiitic basalts of the third and fourth volcanic layers by grade of metamorphism, mineral composition and geochemical features (see Tables 1.7, 1.8).

**Table 1.8.** Mean composition of intrusive rocks of the Pechenga complex from the SG-3 well section

| Components | 1 | 2 | 3 | | 4 | 5 | 6 |
|---|---|---|---|---|---|---|---|
| | Sm | Sm | Pm | Sm | Sm | Sm | Ptm |
| $SiO_2$ | 37.85 | 42.99 | 48.77 | 47.60 | 46.93 | 52.30 | 67.81 |
| $TiO_2$ | 1.01 | 2.28 | 1.54 | 1.69 | 3.80 | 1.96 | 0.65 |
| $Al_2O_3$ | 3.67 | 7.07 | 13.39 | 13.22 | 10.41 | 14.81 | 14.74 |
| $Fe_2O_3$ | 7.51 | 4.44 | 2.26 | 3.02 | 2.47 | 2.17 | 1.73 |
| FeO | 10.38 | 12.25 | 12.83 | 12.08 | 14.74 | 11.05 | 3.19 |
| MnO | 0.20 | 0.19 | 0.24 | 0.22 | 0.21 | 0.19 | 0.07 |
| MgO | 26.17 | 18.12 | 6.74 | 6.24 | 7.82 | 2.64 | 2.10 |
| CaO | 3.20 | 4.58 | 8.22 | 8.88 | 5.44 | 4.38 | 2.45 |
| $Na_2O$ | 0.17 | 0.37 | 2.72 | 2.68 | 0.13 | 3.86 | 2.77 |
| $K_2O$ | 0.11 | 0.14 | 0.89 | 0.28 | 0.03 | 2.09 | 2.77 |
| $P_2O_5$ | 0.15 | 0.30 | 0.14 | 0.15 | 0.33 | 0.56 | 0.21 |
| $CO_2$ | 0.47 | 0.20 | 0.14 | 0.56 | 0.04 | 0.26 | 0.41 |
| $H_2O^-$ | 0.41 | 0.11 | 0.15 | 0.24 | 0.28 | 0.41 | 0.21 |
| $H_2O^+$ + l.o.c.[a] | 8.77 | 7.13 | 2.04 | 3.67 | 7.55 | 3.39 | 1.43 |
| $S_{total}$ | 0.28 | 0.44 | 0.05 | 0.11 | 0.09 | 0.27 | 0.15 |
| Li | 8.9 | 12.5 | 10.2 | 19.2 | 36.5 | 15.1 | 26.2 |
| Rb | 13.3 | 8.6 | 27.1 | 9.7 | 5.0 | 42.7 | 115.5 |
| Sr | 5.0 | 13.0 | 17.7 | 33.9 | 12.5 | 67.6 | 47.2 |
| Ba | 13.5 | 23.5 | 108.0 | 75.3 | 10.0 | 256.7 | 425.9 |
| B | 67.7 | 12.8 | 6.9 | 14.6 | 9.2 | 18.6 | 24.0 |
| Sc | 4.7 | 6.5 | 4.5 | 13.0 | 9.0 | 2.0 | 5.0 |
| Ga | 5.7 | 7.5 | 18.6 | 13.9 | 11.5 | 28.1 | 22.7 |
| La | 12.0 | 17.3 | 10.4 | 12.5 | – | 41.8 | 39.4 |
| Ce | 21.0 | 32.0 | 32.4 | 15.4 | – | 51.3 | 96.9 |
| Sm | 2.4 | 4.3 | 4.1 | 5.6 | – | 12.6 | 7.6 |
| Eu | 0.60 | 1.1 | 1.6 | 1.4 | – | 2.8 | 1.5 |
| Yb | 2.0 | 1.8 | 1.9 | 2.9 | – | 3.5 | 5.0 |
| Lu | 0.40 | 0.30 | 0.70 | 0.50 | – | 0.60 | 0.50 |
| Th | 1.2 | 1.2 | 1.0 | 0.60 | – | 4.9 | 15.6 |
| U | 0.20 | 0.50 | 0.70 | – | – | 0.60 | 6.4 |
| Ge | 1.2 | 2.1 | 0.58 | 1.4 | 2.8 | 1.6 | 0.54 |
| Zr | 30.6 | 55.0 | 26.7 | 48.0 | 70.0 | 139.0 | 186.7 |
| Sn | 6.0 | 4.3 | 2.8 | 4.7 | 7.0 | 6.3 | 6.3 |
| Hf | 1.0 | 2.3 | 3.0 | 1.9 | – | 4.9 | 6.9 |
| Nb | 3.4 | 7.1 | 9.9 | 3.2 | 10.0 | 17.6 | 7.5 |
| F | – | – | 141.3 | 55.1 | – | – | 472.2 |
| Cu | 1022.3 | 237.9 | 192.7 | 22.7 | 400.0 | 54.8 | 17.9 |
| Zn | 190.7 | 160.3 | 94.7 | 144.6 | 215.0 | 396.2 | 35.4 |
| Ag | 0.27 | 0.03 | 0.05 | 0.03 | 0.03 | 0.01 | 0.14 |
| Pb | 15.6 | 5.0 | 17.6 | 10.0 | 4.5 | 8.5 | 35.4 |
| Mo | 4.9 | 2.1 | 0.93 | 1.36 | 1.5 | 5.6 | 1.0 |
| V | 183.9 | 184.1 | 180.7 | 362.0 | 160.0 | 48.0 | 74.7 |
| Cr | 1603.5 | 977.0 | 254.7 | 83.8 | 155.0 | 20.1 | 29.4 |
| Co | 207.7 | 90.3 | 58.1 | 50.1 | 40.5 | 20.5 | 17.8 |
| Ni | 2519.2 | 1121.6 | 147.3 | 103.0 | 115.0 | 17.7 | 12.5 |
| Number of samples | 26 | 34 | 16 | 128 | 3 | 39 | 27 |

[a] l.o.c. = Loss on calcination.
1–2 – metaperidotites, chlorite-serpentinite-talc and chlorite-tremolite schists (1 – lherzolites, 2 – lherzolitowherlites); 3–4 – metagabbro-diabases, melanocratic plagioclase-amphibole schists and amphibolites (3 – gabbronorites, 4 – norites); 5 – essexinitic metagabbro-diabases; 6 – andesite-dacitic metaporphyrites; Pm, Sm, Ptm – intrusive complexes (Pm – Pre-Matertin, Sm – Syn-Matertin, Ptm – Post-Matertin). Content of petrogenic elements is given in percent; content of elemental admixtures is given in $g\,t^{-1}$, "–" content was not determined.

The Synmatertin unit of gabbro-wehrlites is similar to the ultrabasic effusive rocks of the Matertin suite in chemical and mineral composition and also in differentiation (see Tables 1.8 and Fig. 1.29). The existing difference is pronounced only in types of contact (intrusive and effusive), structures (massive, globular and amygdaloidal) and textures (high degree of crystallization, almost complete absence of devitrified glass and spinifex textures characteristic of effusive komatiites). Individual intrusions of the gabbro-wehrlite complex are normally represented by one of its differentiates (peridotite, pyroxenite, gabbro-or essexitic gabbro), and occasionally by the reduced or complete series of differentiates. A complete series is inherent to large bodies, which are confined to the "productive" Zhdanov suite, and found at the surface, where, going from bottom to top, peridotites are replaced, firstly by pyroxenites, then by gabbro, and finally by essexitic gabbro (see Fig. 1.29). On the basis of the two types of volcanite differentiation in the Matertin suite revealed by us (see Chap. 1.10), which exhibits slight rhythmic character in its lower part and consists of well-defined rhythms in the upper part it is possible to arrive at a conclusion that comagmatic intrusive units were formed in an analogous sequence: gabbro-diabases→gabbro-wehrlites. Such an order of formation is also in good correspondence with the geological relationships between these units which have been established at the surface.

The Postmatertin unit of andesite-dacite porphyrites in the SG-3 well section is represented by a single subvolcanic body (apparent thickness 111 m), which serves as a boundary between the Luchlompol and Zapolyarny suites. In mineral and chemical composition these rocks most resemble the andesite-dacite plagioporphyrites developed at the surface in a zone of the Poryitash major fault. As distinct from the latter rocks, porphyrites encountered by the Kola well are characterized by a higher $SiO_2$ content (see Table 1.8); from this it follows that they form one single whole series which begins with andesite-dacitic and ends with dacitic porphyrites.

## Geochemical Conditions of Formation

Geochemical data obtained in studying the Pechenga complex in the SG-3 well section make it possible to examine the general evolutionary trend of Proterozoic sedimentation and magmatism and, on the basis of it, to distinguish two stages reflecting an abrupt change in the tectonic regime of the development of the Pechenga structure. A comparative geological and geochemical analysis of the Proterozoic rocks was performed with the help of original genetic diagrams.

In order to assess the abundance of the initial analogues of metasedimentary rocks in the SG-3 well section, a litho-petrochemical diagram was constructed (see Fig. 1.83); distribution of petrogenic elements ($SiO_2$, CaO and "free"[5] $Al_2O_3$), reflecting various types of sedimentary differentiation, is shown at the triangles. It is seen on this diagram that the Televin suite is characterized by a trend of only terrigenous differentiation; the Kuvernerinyok and Luchlompol

[5] "Free" $Al_2O_3$ is determined with the help of a coefficient $10[Al_2O_3\text{-}(K_2O+Na_2O+CaO)]$ which does not take into account a portion of alumina fixed in feldspars.

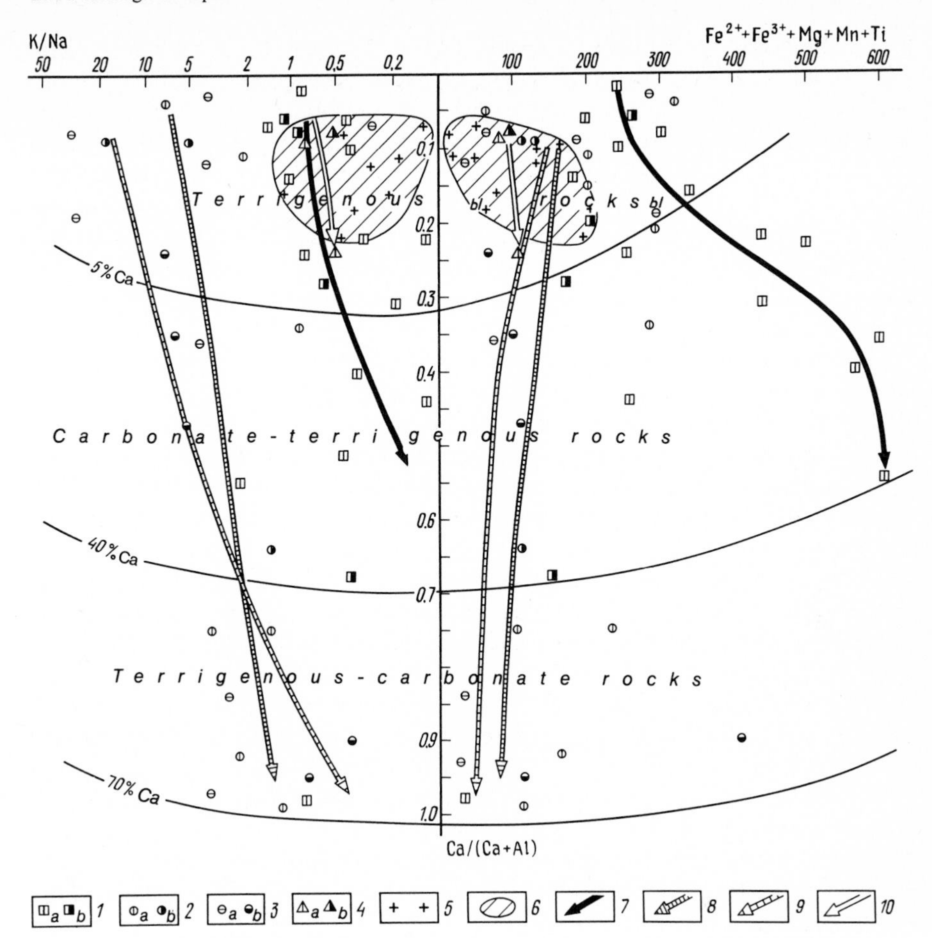


**Fig. 1.31.** Petrochemical differentiation trends of sedimentary rocks of the Pechenga complex. *1–4* compositions of rocks from the SG-3 well section (*a*) and surrounding surface (*b*): *1* Zhdanov suite; *2* Luchlompol suite; *3* Kuvernerinyok suite; *4* Televin suite; *5–6* plagiogneisses of the Kola complex (*5*) and area of occurrence of their main types (*6*); *7–10* differentiation trends of sedimentary rock suites: *7* Zhdanov; *8* Televin; *9* Kuvernerinyok; *10* Luchlompol

suites are notable for identical trends of terrigenous-chemogenic differentiation, and finally, the Zhdanov suite is characterized by a trend of terrigenous, and partially chemical-terrigenous differentiation.

To reveal more detailed features of similarity or difference in chemical composition of the metasedimentary rocks, a petrochemical diagram was used (Fig. 1.31); this discloses and makes a quantitative evaluation of such important signs of the rocks studied (from the genetical point of view) as the presence of volcanic material ($\Sigma Fe^{3+}+Fe^{2+}+Mn+Mg+Ti$) or calcium ($Ca/Ca+Al$) contained in an amount exceeding that typical for feldspars, as well as to assess the proportion of arkose or considerably plagioclase material (K/Na) present in psammites. The distribution of the metasedimentary rocks on this diagram generally corresponds to the established trends of sedimentary differentiation for each suite. It is clearly seen that metaterrigenous rocks of the Televin suite inherit the mineral and chemical compositions of underlying plagiogneisses of the basement (hatched field) and are characterized by a low degree of differentiation, so that they may be considered as a crust of weathering (which was probably not displaced) formed in low areas of the basement paleorelief. This indicates that sediments of the Televin suite fix the Pre-Pechenga stage of stabilization of the basement tectonic regime. All these facts, together with the observed fact that at the surface meta-effusive rocks of the Mayarvi suite directly overlap the Archean basement, cast doubts on the theory that the Televin suite can be assigned to the lower sedimentary-volcanic strata of the Pechenga complex.

The petrochemical trends of the Kuvernerinyok and Luchlompol suites are identical in composition of terrigenous and, correspondingly, terrigenous-carbonate deposits. The clearcut difference between compositions of terrigenous rocks of these suites and rocks of the underlying Archean basement in the SG-3 well section indicates that these rocks were formed by sediments supplied by different sources located at a comparatively great distance from each other. It seems probable that the Archean granite-gneiss blocks, developed north of the Pechenga structure, might have served as sources of sediment. The small thickness of terrigenous sediments, their persistent character down the dip and along the strike, as well as the presence among them of carbonate facies, undoubtedly demonstrates that both suites were formed in conditions of shallow continental lagoons.

Other conditions of sedimentation characterize a trend of the metasedimentary rocks of the Zhdanov suite. The terrigenous part of this suite, on the one hand, is similar in composition to that of the Archean plagioclase ($K/Na<1$) which underlie the Pechenga complex, but on the other hand, it differs sharply from all underlying rocks by its higher content of Fe, Mg, Mn, and Ti, which is indicative of their enrichment by volcanic material. Tuffogenic rocks and associated effusive rocks of the Zhdanov suite are characterized by an analogous chemical composition, with volcanites developed in the underlying and overlying suites; this indicates that they have been formed during one single tectono-magmatic cycle. The great thickness and flysch-like character of the sediments of the Zhdanov suite, as well as the presence of organic matter and salts dissolved in its underground waters (see Chap. 1.9), indicates that sediments were deposited in marine conditions; the presence of volcanogenic material and syngenetic sulphides illustrates that sedimentation was greatly affected by volcanism.

The distinctive litho-petrochemical features of sedimentary rocks considered above are in compliance with data showing that the degree of concentration of elemental admixtures in rocks and the character of their distribution differ. The litho-geochemical diagram may serve as an example (Fig. 1.32), with the help of this diagram a group of metasandstones, widely spread in all sedimentary strata of the Pechenga complex, was studied. The diagram clearly shows three isolated fields distinguished by the proportion of clastic minerals (quartz, plagioclase, microcline), by the composition of the main rock mass (mica, chlorite, carbonates) and by the concentration levels of Ba, Li, B, Zn, and Ni. The first field is occupied by sandstones of the Televin suite, the second by sandstones of the Kuvernerinyok and Luchlompol suites, and the third by sandstones of the Zhdanov suite; this fully confirms the validity of distinguishing three different sources of sediment for this class of sedimentary rock, and substantiates the conclusion that they have been deposited in domains characterized by a different tectonic regime. The same geochemical character of the sedimentation stages is displayed in the distribution of rare-earth and radioactive elements (Fig. 1.33).

From this it appears that a litho-geochemical analysis of the metasedimentary rocks of the Proterozoic complex identifies three formational series in the SG-3 well section: (1) a greywacke series (Televin suite) formed in the pre-Pechenga stage of Archean basement stabilization; (2) carbonate-terrigenous series (Kuvernerinyok and Luchlompol suites), which marks the initial origin of the Pechenga structure; (3) volcanogenic-terrigenous series (Zhdanov suite), which characterize a stage of the subsequent protoactivation.

These tectonic stages are also reflected in the chemical composition of volcanic formations. On the petrochemical diagram $SiO_2 = Na_2O + Ka_2O$ (see Fig. 1.28) points denoting metavolcanites form four fields where the andesite-basalt formation (Mayarvi suite) is partially overlapped by the trachybasalt (Pirttiyarvi) tholeiite-basalt (Zaplyarny) and picrite-tholeiite (Matertin) formations. In addition to this, pairs of the spatially conjugated volcanic formations exhibit a regular change in types of differentiation, i.e. silicic type in the lower pairs and alkali-silicic type in the upper pair. Trends of magmatic differentiation of the andesite-basalt and trachybasalt formations distinctly fix a growth in iron content in conditions of increased alkalinity values, whereas trends of the tholeiite-basalt and picrite-tholeiite formations reflect rhythmic changes in the increase in iron content (in lower horizons of the formation) and magnesium content (in upper horizons) when alkalinity is comparatively low. These distinctive features in the evolution of volcanism show that it was developed during two successive stages, i.e. first the andesite-basalt stage, which includes the andesite-basalt and trachybasalt formations, and second the picrite-basalt stage, which includes the tholeiite-basalt and picrite (komatiite)-basalt formations.

The character of distribution of elemental admixtures in a series of crystal differentiation was studied by taking into account values of general rock acidity. It was established that the differentiated series of the andesite-basalt stage of volcanism are characterized by higher contents of Rb, Ba, and Cr, and that the picrite-basalt stage is notable for higher contents of Sc, Zr and V. Additionally, higher concentrations of Hf, Pb, Mo, and lower concentrations of Sr, B, Zr, Co, and Ni in rocks of the first stage (Fig. 1.34) bring them closer to the volcanic series of continental rifts, whereas rocks of the second stage are similar to volcanites of oceanic rifts.

The difference between two stages is also in agreement with the different levels of content of rare-earth elements and with different fractiation trends of these elements. For instance, on the diagram La-Eu (see Fig. 1.30a), symbols denoting effusive rocks of the andesite-basalt stage of volcanism successively shift direction from continental to island-arc series, and the picrite-basalt stage is characterized by another tendency in development, i.e. it shifts from the island-arc compositions to the oceanic ones (third layer) and again to the continental compositions in the upper parts of the section (fourth layer).

The formation of the vertical metamorphic zonation in the SG-3 Kola well section can be linked with these two stages of volcanic activity.

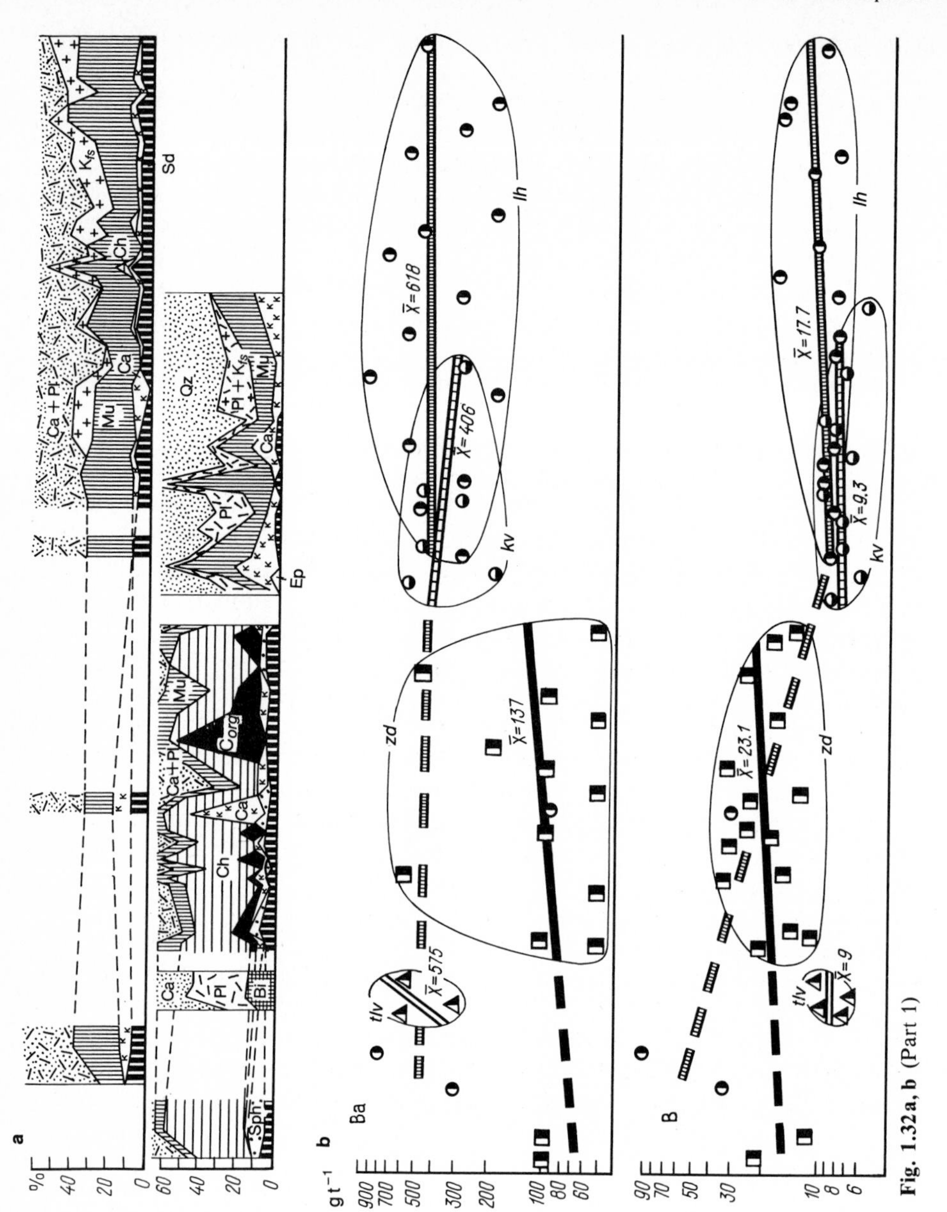


Fig. 1.32 a, b (Part 1)

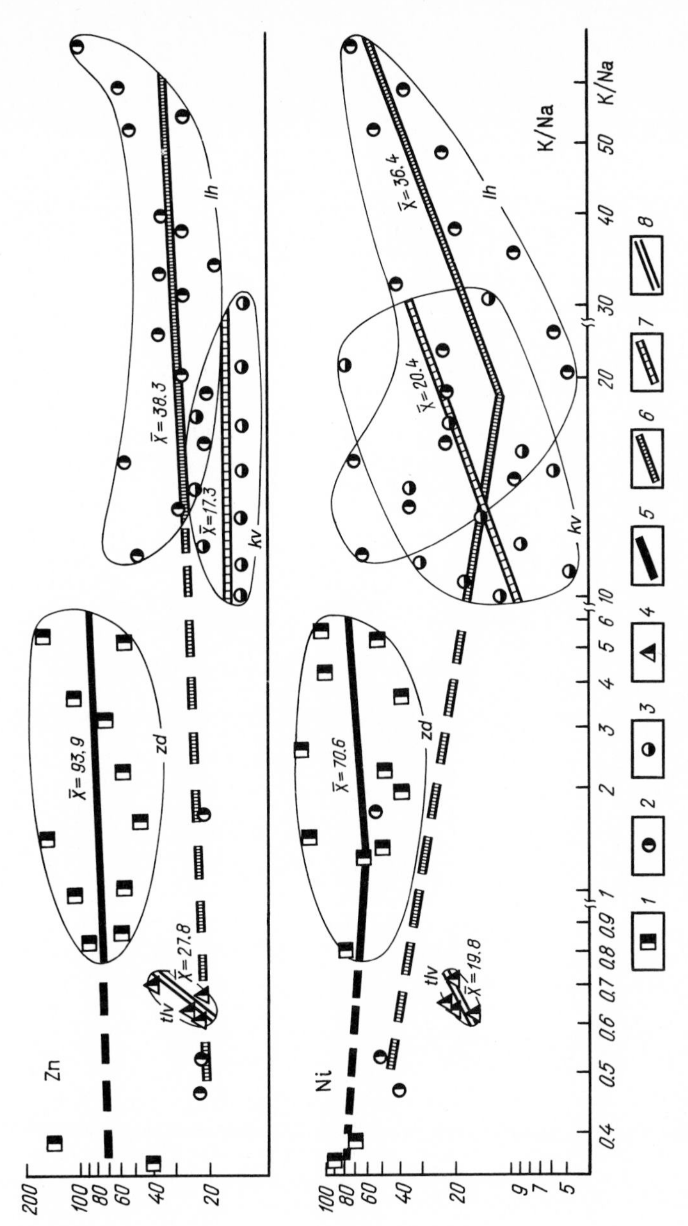


**Fig. 1.32 a, b.** Lithological composition (**a** in %) and geochemical differentiation trends (**b**) of sedimentary rocks of the Pechenga complex. *Suites: 1* Zhdanov; *2* Luchlompol; *3* Kuvernerinyok; *4* Televin; *5 – 8* geochemical trends

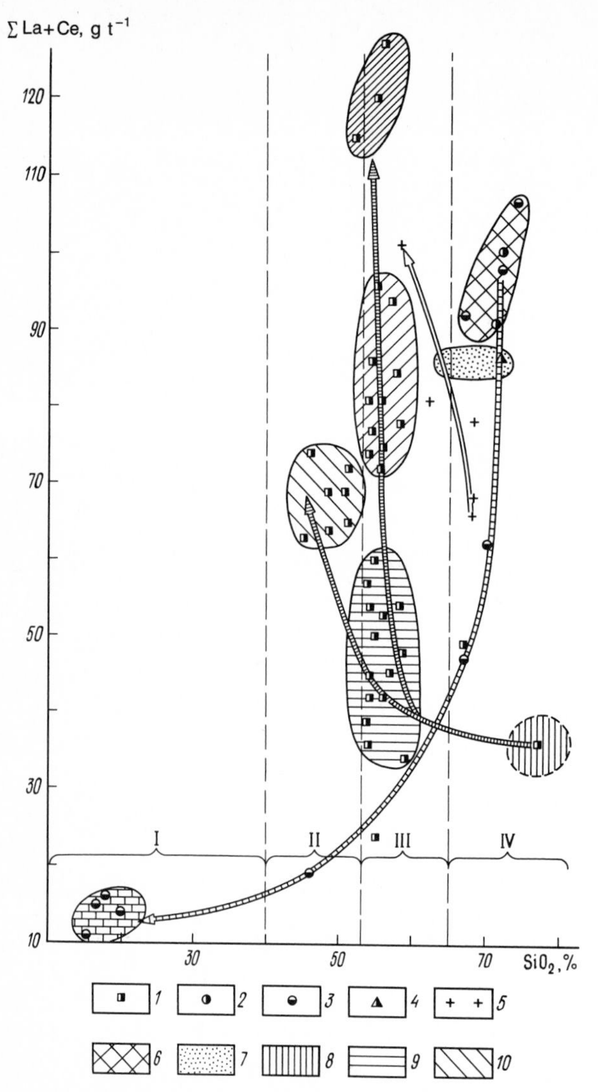


**Fig. 1.33.** Distribution types of rare earth elements in metasedimentary rocks of the Pechenga complex. *Suites: 1* Zhdanov; *2* Luchlompol; *3* Kuvernerinyok; *4* Televin; *5* gneisses and plagiogneisses of the Kola complex; *6–13* fields of metasedimentary rocks of the Pechenga complex (*6* arkose metasandstones; *7* quartz-plagioclase metasandstones; *8* plagioclase-quartz metasandstones; *9* metasiltstones; *10* metapellites; *11* metasiltstones with admixture of tuff material; *12* metatuffites; *13* sandy dolomites and limestones); *14–16* trends of rare elements in rocks of the Kola complex (*14*), Luchlompol and Kuvernerinyok suites (*15*), Zhdanov suite (*16*). *I* sandy dolomites; *II* aleuropellites; *III* aleurosandstones; *IV* sandstones

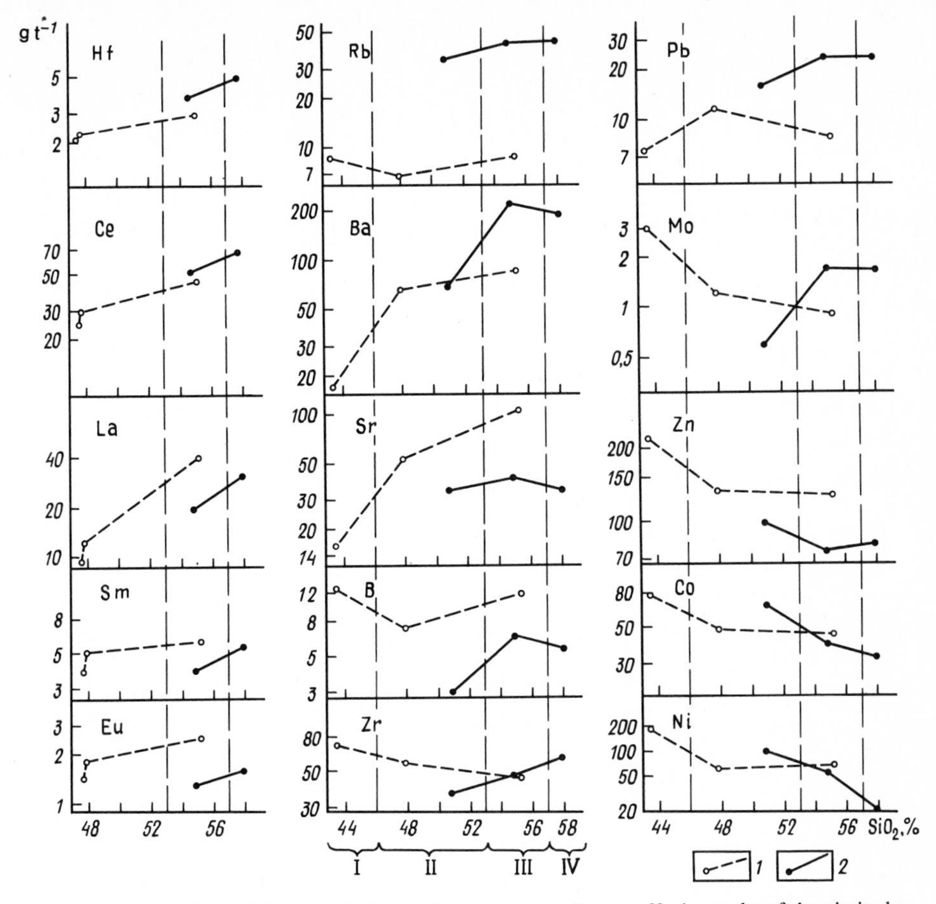


**Fig. 1.34.** Distribution of element-admixtures in main types of meta-effusive rocks of the picrite-basalt (*1*) and andesite-basalt (*2*) stages of volcanism of the Pechenga complex. *I* metapicrite-basalts; *II* metabasalts; *III* meta-andesite-basalts; *IV* meta-andesites

Stage of andesite-basalt volcanism ended with low-temperature autometasomatic and metamorphic alteration of the rocks of the Luostari series (4563 – 6842 m). An interval of ceasing volcanic activity was followed by the second picrite-basalt stage of volcanism which, in its turn, ended with a zonal greenschist metamorphism of the Nickel series rocks (0 – 4563 m); the degree of metamorphism becomes less moving from the periphery of the structure towards its centre and up the section of the SG-3 well. This stage was also associated with the second stage of metamorphism which affected the underlying rocks of the Luostari series, and was a result of local temperature rise in a layer of the Luostari series overlain by thick (about 2 km) sheets of tholeiites, and their simultaneous sinking into the Archean basement. A study of temperature and pressure conditions (see Chap. 1.5) has showed that conditions of the second stage of metamorphism of the Luostari series correspond to the epidote-amphibolite and amphibolite facies ($T = 500\,°C$; $P = 0.2 – 0.4$ GPa). The increase of temperature in this closed system has led to the increase of pressure according to the autoclave effect principle; as a result of this, a considerable volume (6 – 7%) of synmetamorphic fluids enriched with Cl, I, Ba, Sr, and K was released. Due to the high fluidal pressure existing at insufficient geostatic pressure, microhydrofractures have been formed in metamorphosed rocks and rock-forming minerals were partially dissolved in the newly formed synmetamorphic fluid (see Chap. 1.9). A part of the fluid released during the metamorphic process might also have been pushed out of the area of development of the Luostari rocks both up and down the sequence, for instance, along fissure zones in the Archean basement, which was initially metamorphosed under granulite-facies conditions. When rocks of the Archean basement were affected by this fluid, the retrograde changes in rocks began to develop down to a considerable depth; these retrograde changes are considered as isofacial with respect to the second stage of metamorphism of the Luostari series.

## Geochemical Zonality of Metamorphism

Details characteristic of the vertical metamorphic zonality of the Pechenga complex are given below (see Chap. 1.5). The results of our observations are generally compatible with these data. Tendencies in changing mineral and chemical compositions of metabasites, developed in different zones of the prograde metamorphism in the SG-3 well section, are considered below. To achieve this goal, we have chosen a narrow series of the initial rocks (tholeiite basalts and their intrusive counterparts) which contain 46 – 51% SiO and 60 – 73% Fe. The metabasite petrography of each zone had been studied previously; a scheme displaying their quantitative mineral composition is shown in Fig. 1.35. It was established that there were two parageneses in the prehnite-pumpellyite and sometimes in the epidote-chlorite zones, i.e. relict and metamorphic parageneses; a portion of the first paragenesis successively decreases and is fully destroyed at a depth of about 2 km. The relict paragenesis is represented by the initially magmatic minerals: subcalcium augite, andesine (Nos. 41 – 46) and ilmenite. Two groups of minerals are distinguished in the composition of metamorphogenic paragenesis. These minerals are the replacement product of relict minerals such as pyroxene (actinolite and chlorite), plagioclase (oligoclase, albite, prehnite, clinozoisite) and ilmenite (sphene). Actinolite occurs in both zones together with hornblende, the amount of which is much less than actinolite. In the lower part of the epidote-chlorite zone metabasites contain considerable amounts of quartz and calcite which are released as a result of the full deanorthization of plagioclase.

A biotite-amphibole zone is characterized by the decomposition of relict minerals and the appearance of a newly formed biotite. The chlorite content is somewhat reduced; this is accompanied by a decrease in iron content and a sharp increase in actinolite (up to 40 – 50%). Actinolite coexists with hornblende within the limits of the described zone as well as in the overlying epidote-chlorite zone. This may be explained by the fact that they mix with their isomorphic series under different conditions, rather than by involving temperature changes as a cause. The same explanation also stands for an association of albite with oligo-

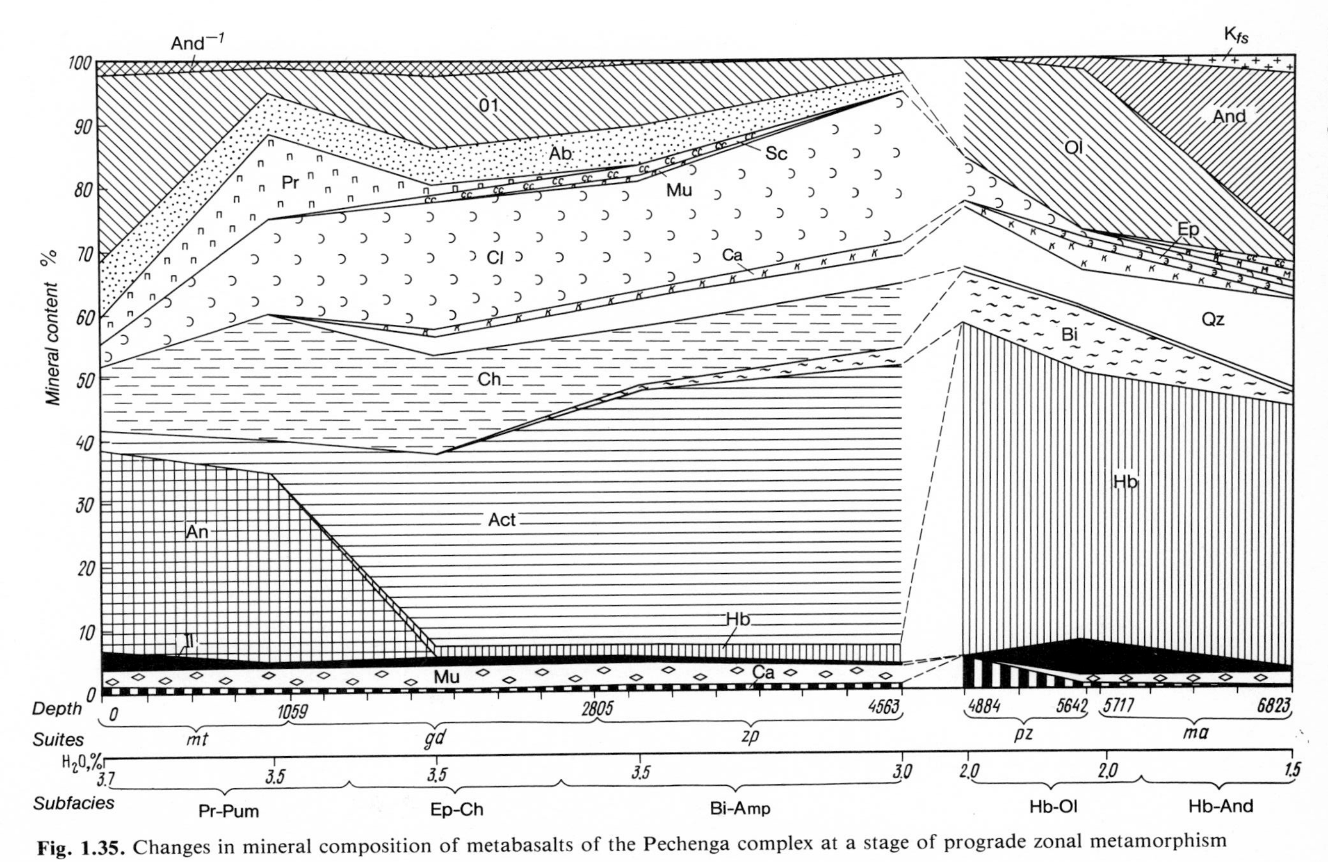


**Fig. 1.35.** Changes in mineral composition of metabasalts of the Pechenga complex at a stage of prograde zonal metamorphism

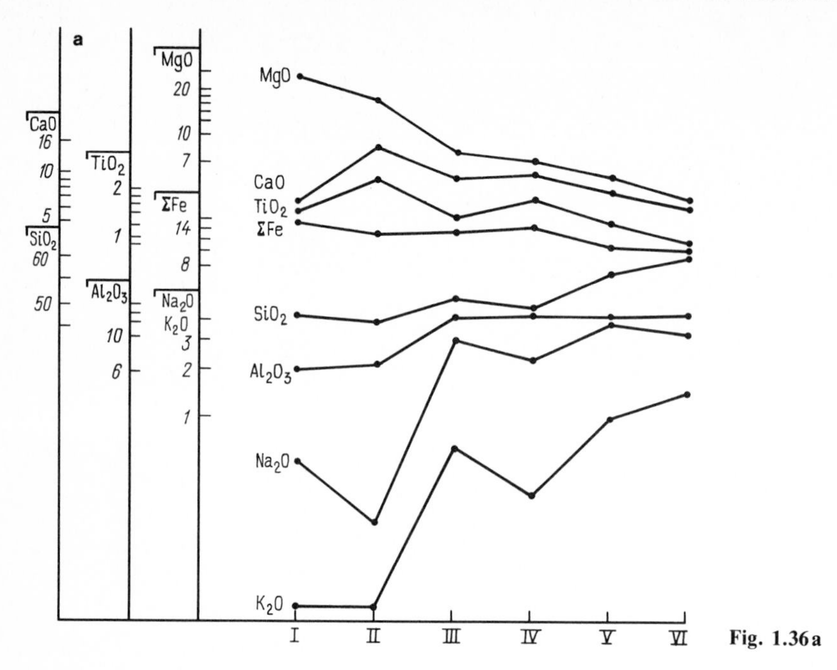


**Fig. 1.36 a**

clase. This fact, as well as the absence of sharp leaps in composition of the coexisting minerals and their quantitative ratio in the described part of the section, allow the conclusion that metabasalts of the picrite-basalt stage of volcanism (Nickel series) have been metamorphosed only under greenschist-facies conditions within a full temperature range inherent to this stage.

Sharply distinct compositions of the mineral parageneses are observed in the basic rocks of the Luostari series (4563 – 6835 m). For instance, the hornblende-oligoclase zone consists predominantly of amphibolites composed of hornblende, oligoclase (Nos. 20 – 27), epidote, quartz and biotite. It should be noted that initial equivalents of amphibolites of this zones had a somewhat more silicic and subalkali composition. In the hornblende-andesite zone, amphibolites, corresponding by composition to normal basalts, contain biotite, which is paragenetic with ferruginous hornblende and ferruginous epidote. This zone differs from the previous one by its wide development of newly formed andesite (Nos. 35 – 40). These parageneses indicate that prograde metamorphism of the andesite-basalt stage of volcanism (Luostari series) was manifested in conditions of medium- and high-temperature zones of the epidote-amphibolite facies. In the lower part of the section metasomatic potassium feldspar appears. The presence of a small amount of chlorite and carbonate in individual zones is associated with the retrograde metamorphism.

In view of the fact that metabasites, uniform in their initial composition and similar to tholeiite basalts, are widely spread in the section, it becomes possible to use this group of rocks to reveal laws in the behaviour of petrogenic and minor elements in prograde metamorphism and, consequently, to solve one of the

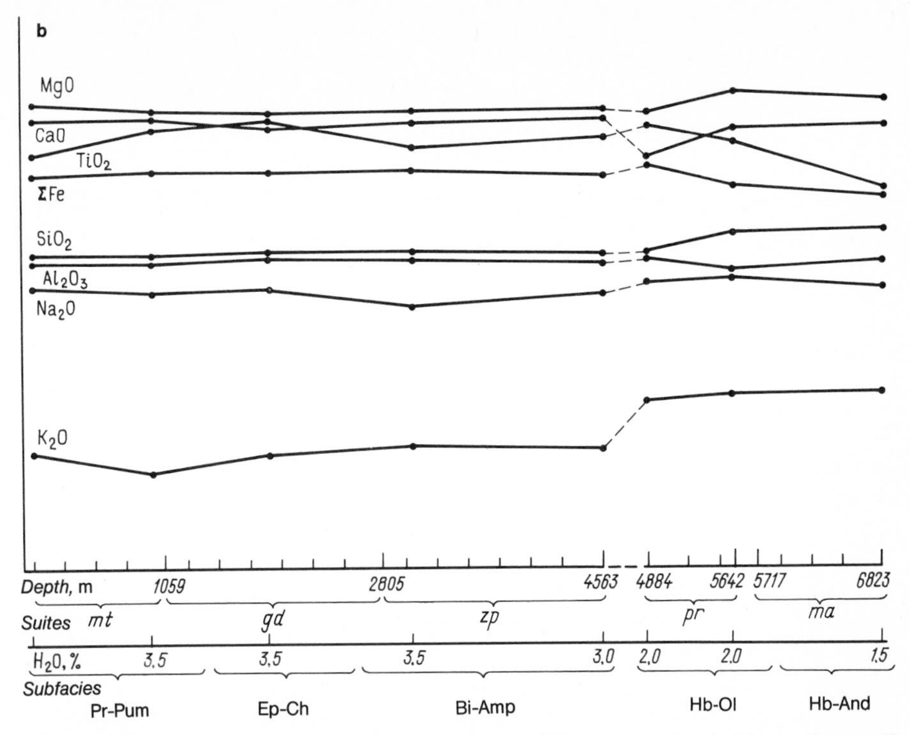


**Fig. 1.36 a, b.** Distribution types of petrogenic elements in different petrochemical types of meta-effusive rocks **(a)** and metabasalts of a normal series **(b)** during prograde metamorphism of the Pechenga complex. *I* metapicrites; *II* metapicrite-basalts; *III* olivine metabasalts; *IV* metabasalts; *V* meta-andesite-basalts; *VI* meta-andesites

cardinal problems of present-day petrology, the nature of the geochemical regime (isochemical or allochemical) of regional metamorphism.

To assess the influence which fluctuation of the initial rock composition exerts on the distribution of chemical elements during metamorphism (metabasites are closely associated with metapicrite basalts in the upper parts of the section and with apoandesite basalt amphibolites in the lower part), distribution trends of petrogenic elements and elemental admixtures in a series of compositionally different initial rocks (from picrites to andesites) were established in advance. Later, tholeiite metabasites, a narrow petrochemical class, were analyzed to elucidate the behaviour of the same elements, depending on their grade of metamorphism (Fig. 1.36).

These trends and the statistical appraisal of their significance are the sole evidence in support of an isochemical regime in the behaviour of petrogenic elements, including K and Na. As was expected, metamorphism exerts substantial influence only on water content, which increases sharply (from 2.4 to 6.8%) in conditions of the prehnite-pumpellyite facies when initially magmatic pyroxene and plagioclase are replaced by water-bearing minerals, and then gradually decreases to 2.5% in conditions of the greenschist facies; a sharp leap in water content (from 2.5 to 1.5%) is established at the boundary between the greenschist and epidote-amphibolite facies (Fig. 1.37).

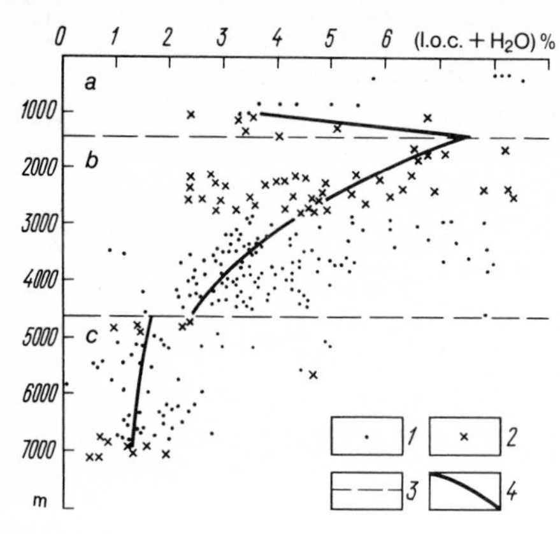


**Fig. 1.37.** Change of $H_2S$-content with depth in conditions of prograde metamorphism in rocks of the Pechenga complex. *1* metavolcanic rocks; *2* metatuffogenic-sedimentary rocks; *3* boundaries of facies occurrence (*a* prehnite-pumpellyite; *b* greenschist; *c* epidote-amphibolite); *4* trend of the changing mean $H_2S$ content

The pattern of distribution of elemental admixtures is of a heterogeneous character. By analyzing their trends from top to bottom of the SG-3 well section, it becomes obvious that the majority of elemental admixtures fall into three areas of change in concentration (Fig. 1.38).

The first area (1060 – 1800 m) coincides with a zone of the full decomposition of the initially magmatic minerals during which the content of Cr, Ni, Yb, B, Sn, Ga, and Pb basically decreases (evacuation) against the general background of Ce and Ge accumulation.

The second area (4563 – 4884 m) is fixed at the boundary between the greenschist and epidote-amphibolite facies; it coincides with a zone of contact between volcanites of the andesite-basalt and picrite-basalt stages of volcanism and is characterized by sharp changes in concentration of all elemental admixtures in corresponding metabasalts. Analogous tendencies are also observed for other petrochemical types of volcanite of the Pechenga complex (e.g. meta-andesite-basalts, see Fig. 1.34); from this it appears that a distribution pattern of chemical elements in this part of the SG-3 well section was mainly affected by a geochemical specialization of initial magmatic rocks but not by metamorphism.

The third area (5642 – 6323 m) of redistribution of initial concentrations of elements is marked at the boundary between the hornblende-oligoclase and hornblende-andesine zones of the epidote-amphibolite facies and its presence supports the predominant influence of metamorphism (Kremenetsky and Dmitrenko 1982). It should also be noted that high and anomalously high concentrations of Pb, Ba, and B, established in metabasalts of the lower part of the Mayarvi suite, are associated with the later manifested metasomatosis.

Some additional data on conditions of metamorphism of the Pechenga complex have been obtained as a result of thermal-pressure geochemical study of isofacial hydrothermal veins and streaks (Table 1.9).

In the Nickel series (0 – 4568 m) syngenetic inclusions of the prograde stage of metamorphism are known throughout the entire section; their phase composition gradually changes (down the section) as follows: $L > G (0-1500\ \mathrm{m}) \rightarrow L > G + K (L \geqslant G \geqslant K_{L+G} (1500-4000\ \mathrm{m}) \rightarrow G \geqslant L + K (G \geqslant K \geqslant G)$ (4000 – 4563 m). The liquid phase of inclusions ($L$) is represented by a saline solution with 30% concentration; the gase phase ($G$) contains methane; carbon dioxide is found only in the presence of carbonate rocks; solid phases ($K$) are represented by one-three iso- and anisotropic crystals whose temperature of dissolution exceeds that of the homogenization of liquid and gaseous phases. In the upper part of the section, syngenetic inclusions are homogenized into the liquid phase and in the lower part they are homogenized into the gase phase, with concentration of the saline solution and the homogenization of inclusions being increased down the section. For instance, the prehnite-pumpellyite subfacies formed by salinity >30% and $T = 190° - 340\,°C$[6]. The temperature gradient of the subfacies formation exceeded 100 °C per km.

The epidote-chlorite subfacies (1400 – 2600 m) also formed in the presence of water-saline solutions with salt concentration ⩾30% in conditions of a temperature increase down the section from 340° to 455 °C. A temperature gradient of this subfacies changed down the section within the range of 100° – 80 °C per km. The biotite-muscovite subfacies (2600 – 4563 m) formed near the upper boundary of the subfacies under the action of solutions of the non-uniform aggregate state at a temperature of 455° – 510°C when the salt concentration did not exceed 30%, and it also formed near the lower boundary of the subfacies under the action of gaseous solutions at a temperature $T \geqslant 500° - 560\,°C$ when the salt concentration was much higher than 30%. A temperature gradient in the upper part of this interval was 70° per km, whereas it was 60° per km in its lower part.

In addition to this, temperature-pressure data show that the temperature-dependent character of regional metamorphism near intrusive bodies of gabbro-diabases and ultrabasites was not disturbed; this indicates that these bodies had been intruded much earlier than the main phase of metamorphism made clear. At a fluidal stage of metamorphism, pressure in an interval of 3200 – 330 m did not exceed a lithostatic pressure (0.1 – 0.12 GPa).

In the section of the Luostari suite (4563 – 6842 m), streaks synchronous with a stage of prograde metamorphism (amphibole-quartz streaks in meta-effusive rocks and quartz-sericite streaks in metasedimentary rocks) are of local occurrence and small in size. Inclusions of solutions in these rocks are represented by interstitial micro-voids; normal syngenetic inclusions were not established, and hence pressure-temperature ($P - T$) parameters of metamorphism for this part of the section cannot yet be experimentally assessed; however, by extrapolating a trend of paleotemperatures in the overlying Nickel series, it became possible to determine the temperature of the hornblende-oligoclase paragenesis (610° – 650 °C) and that of the hornblende-andesine paragenesis (610° – 685 °C).

In vein minerals of retrograde metamorphism, inclusions are characterized by the following phase composition: $L \geqslant K \geqslant G$, in more rare cases $K \geqslant L \geqslant G$; the latter phase composition is established when these vein minerals are overfilled with minerals as "prisoners" (halite, carbonate) and minerals as "associates" (chlorite?); inclusions are homogenized at $T = 260° - 460\,°C$. Values of the paleotemperature gradient are 50° – 60 °C per km. The process proceeded in the presence of liquid solutions with a very high degree of salt concentration at $T = 260° - 460\,°C$. The gaseous phase contains a small amount of methane.

---

[6] Temperature intervals of facies were calculated on the basis of concrete values of homogenization temperatures and by taking into account the temperature gradient of the main phase of regional metamorphism.

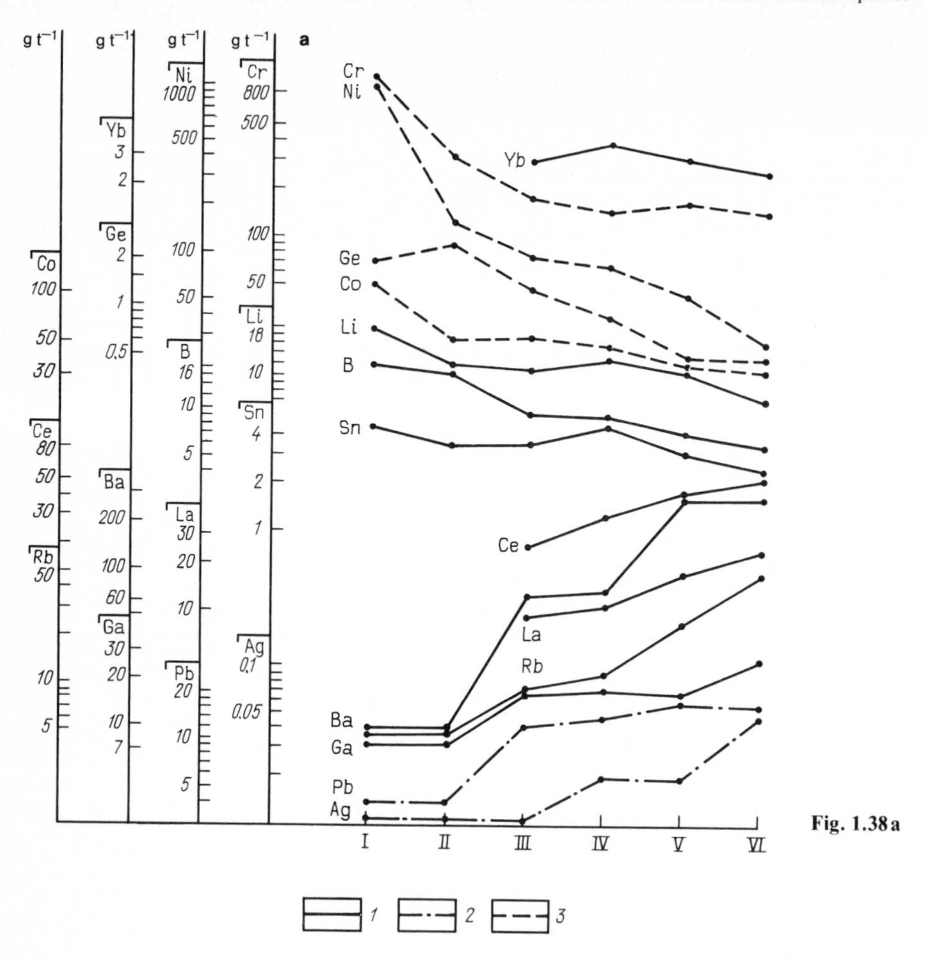


**Fig. 1.38 a**

Biotitization, silicification and microclinization of rocks from the lower part of the Pechenga complex, as well as formation of microcline-quartz streaks and zones of vein granulation (isofacies with quartz) are associated with metasomatic solutions which differ sharply in composition (substantially $CO_2$-bearing) and density (high density) from those of the main phase of regional metamorphism.

Pressure of the liquid phase ($P_{CO_2}$) in zones of granulation reached 0.37 – 0.55 GPa (depth 4150 – 6050 m). Additionally, the presence of one- and two-phased inclusions of compacted $CO_2$-bearing and $CO_2 - H_2S$-bearing solutions of different age in these zones indicates that a state of increased pressure has existed for a long time.

Chloritization and carbonatization of rocks and coexistent streaks were linked with diaphthoresis and proceeded under the action of substantially sodium chloride solutions, whose concentration also increased down the section. In was assessed that epidote-quartz streaks were formed at temperatures of 340° – 370 °C (6800 m); epidote-carbonate and carbonate streaks of later origin were formed at temperatures of 200° – 315 °C (1500 – 7200 m).

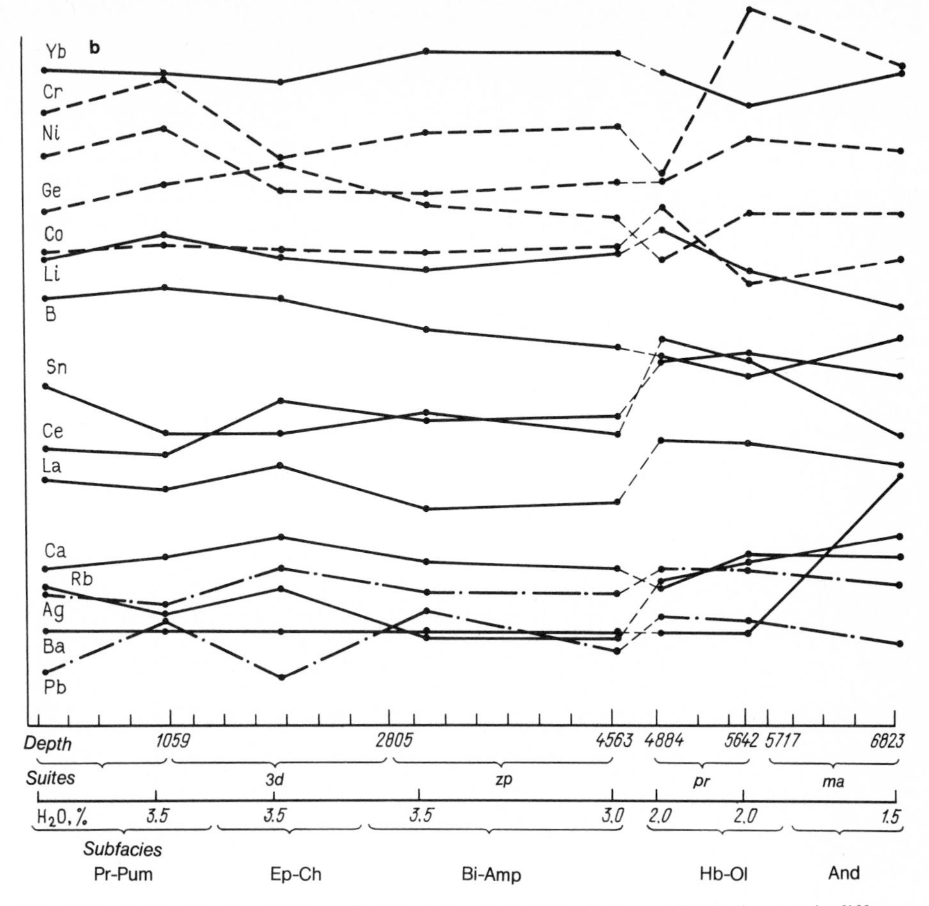


**Fig. 1.38 a, b.** Distribution of lithocyclic (*1*), chalcophylic (*2*) and siderophile (*3*) elements in different petrochemical types of meta-effusive rocks **(a)** and in metabasalts of a normal series **(b)** during prograde metamorphism of the Pechenga deposits. Symbols as in Fig. 1.36

All vein inclusions of the Pechenga complex were affected by bitumen- and hydrogen-hydrocarbon-bearing solutions ($CH_4$, $C_2H_6$, $C_2H_4$, $C_3H_6$, $C_4H_{10}$) of the substantially methane base (a ratio of $CH_4$ to a sum of other hydrocarbons $= 1:0.2-1:0.02$) during a period when the latest faults were being formed.

### 1.4.2 The Archean Complex

The main part of the Archean rocks is composed of plagiogneisses and migmatites within them (66%), amphibolites and amphibole schists (30%) and granitoids (4%).

Plagiogneisses and migmatites. The petrographic subdivision of rocks was accomplished with the help of original diagrams (Fig. 1.39) reflecting the quantitative relationships between quartz, feldspars and mica (for rocks lacking CaO) or

**Table 1.9.** Physical conditions and compositions of metamorphic solutions from various stages of the

| Prograde line, main stage | | | | Local diaphthoresis, early stage |
|---|---|---|---|---|
| Depth, m | Subfacies | Metamorphic rocks, isofacies, simultaneously folded stringers (I) | | simultaneously folded diaphthoritic stringers |
| | | autogenic | | autogenic |
| 0 | | gaseous, liquid; aqueous-saline; salt concentration ← ≫30% →← >30% →← ≃30% → | $T_h$ 190° | liquid aqueous-saline of very high concentration |
| 1000 | prehnite-pumpellyite ~340–190°; 1400 | | Temp. gradient ~120°/km; $T_h$ 300° | $T_h$ 260 |
| 2000 | epidote-chlorite ~450–340°; 2600 | | Temp. gradient ~100°/km; $T_h$ 415° | |
| 3000 | biotite-actinolite ~560–455° | | Temp. gradient ~70°/km; $T_h$ 510° | $T_h$ 395° |
| 4000 | 4563 | | Temp. gradient ~60°/km; $T_h$ 555° | $T_h$ 460° |
| 5000 | hornblende-oligoclase hornblende-andesine ~685–560° | | | |
| 6000 | 6835 | | | |
| 7000 | | | | |

ration of the rocks of the Pechenga complex

| | Retrograde line | Local diaphthoresis, late stage | | Zone of cataclasis | Inclusion from solutions in the late fracture fillings |
|---|---|---|---|---|---|
| | Stage of high temperature/ high pressure processes | | | | |
| | Granulated zones: quartz (III), microcline-quartz veins (III′) | late-diaphthoritic stringers (IV, V′) | | | |
| | allogenic | allogenic | | allogenic | |
| | $T_h$ 400° | | | | $T_h$ 205 – 30° |
| | | | | | $T_h$ 135 – 100° |
| | | | ← ⩽30% → | | $T_h$ 175 – 125° |
| | | | $T_h$ 280° | | |
| | | | aqueous-saline ← ⩾30% → | bitumen-hydrogen-hydrocarbon and essentially methane | $T_h$ 215 – 100° aqueous-saline |
| | | | $T_h$ 280° | | |
| $CO_2$ ~3.7 · $10^8$ Pa | $T_h$ 310° | | salt concentration (NaCl: ← >30% →) | | $T_h$ 215 – 120° |
| | $P_{CO_2+H_2O}$ ~ ~2.2 · $10^8$ Pa | | | | |
| $CO_2$ ~5,2 · $10^8$ Pa | $T_h$ 390° | aqueous-saline | | | |
| | $P_{CO_2+H_2O}$ ~ ~2.9 · $10^8$ Pa | | | | $T_h$ 180 – 105° |
| $CO_2$ ~5.5 · $10^8$ Pa | | $T_h$ 340° | | | |
| | | $T_h$ 370° | | | |
| | | | $T_h$ 310° | | |

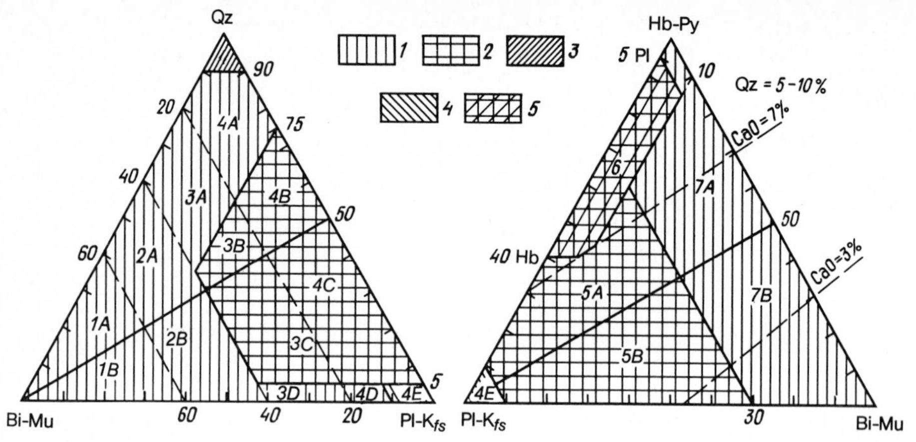


**Fig. 1.39.** Mineralogic classification of metamorphic rocks of the Kola complex. *Rock groups: 1* schists; *2* gneisses and plagioclases; *3* quartzites; *4* feldspathic rocks; *5* amphibolites. *Rock types: 1A, 1B* anchimicaceous schists (*1A* Fs-Qz; *1B* Qz-Fs); *2A, 2B* melanocratic micaceous schists (*2A* Fs-Qz; *2B* Qz-Fs); *3A, 3D* mesocratic micaceous schists (*3A* Fs-Qz; *3D* Fs); *3B, 3C* mesocratic gneisses and plagiogneisses (*3B* Fs-Qz; *3C* Qz-Fs); *4A, 4D* leucocratic schists (*4A* Fs-Qz; *4D* Fs); *4B, 4C* leucocratic gneisses and plagiogneisses (*4B* Fs-Qz; *4C* Qz-Fs); *4E* feldspathic rocks (plagioclasites); *5A* micaceous-amphibole (pyroxene) gneisses and plagiogneisses; *5B* amphibole (pyroxene)-micaceous gneisses and plagiogneisses; *6* amphibolites; *7A, 7B* amphibole schists, hornblendes, pyroxenites (*7A* micaceous-amphibole (pyroxene) schists; *7B* amphibole (pyroxene)-micaceous schists

quartz, feldspars and minerals of amphibole and pyroxene groups (for rock rich in CaO). Both triangles are subdivided into fields which correspond to the main groups of metamorphic rocks (1 – 5). Plagiogneisses and gneisses are subdivided on the basis of quantitative mineral ratios into types whose names are given in Table 1.10.

A characteristic feature of meso- and melanocratic plagiogneisses is the presence of andalusite, staurolite, sillimanite, garnet and kyanite; their total amount rarely exceeds 8%. These rocks contain an almost similar assemblage of accessory minerals; here it is worth noting that in passing from the melanocratic to mesocratic plagiogneisses, the content of some minerals decreases, i.e. pyrite ($g\,t^{-1}$) from 55 to 49, pyrrhotite from 1410 to 1078, chalcopyrite from 24 to 7, zircon from 18 to 17 and monazite from 50 to 20, while the content of other minerals increases, i.e. ilmenite + sphene from 155 to 284 and orthite from 4 to 8; this is indicative of their genetical affinity. This conclusion is also supported by the similarity in isotope composition of sulphur in sulphides. In mesocratic plagiogneisses the amount of $\delta^{34}S$ is +4.88‰, that of pyrrhotite +3.41‰. Similar isotope values are also characteristic of leucocratic biotite plagiogneisses (+3.38‰ for pyrite and +4.78‰ for pyrrhotite). It is observed that sulphur becomes somewhat isotopically "lighter" in leucocratic biotite gneisses ($\delta^{34}S$ of pyrite = +1.69‰, $\delta^{34}S$ of pyrrhotite = 0.52‰).

Data from the morphological analysis of accessory zircon from leucocratic biotite plagiogneisses and from mesocratic plagiogneisses with highly aluminiferous minerals show that the amount of fragmental grains (41 – 83%) considerably exceeds that of idiomorphic crystals. They are represented by uniformly small-

**Table 1.10.** Nomenclature of metamorphic and granitoid rocks of the Archean complex of the SG-3 well

| Group | Subgroup | Type | Field Nos./Symbols of rocks are shown in Fig. 1.40 |
|---|---|---|---|
| Plagiogneisses and gneisses | Bimicaceous schists and plagiogneisses with HAM: garnet, staurolite, andalusite, sillimanite (HAM) | Bimicaceous schists and melanocratic plagiogneisses with HAM | 1A, B+2, A, B |
| | | Mesocratic plagiogneisses with Ham | 3B, C |
| | | Leucocratic plagiogneisses with HAM | 4C, C |
| | Biotite plagiogneisses and gneisses | Leucocratic biotite plagiogneisses ($Q_3<10\%$ and $Q_3>10\%$) | 4C |
| | | Leucocratic biotite gneisses | 4C |
| | Epidote-biotite plagiogneisses | Leucocratic epidote-biotite plagiogneisses | 4C |
| | | Mesocratic epidote-biotite plagiogneisses | 3C |
| | Hornblende-biotite schists and plagiogneisses | Hornblende-biotite schists | 7B |
| | | Hornblende-biotite plagiogneisses | 5B |
| Amphibolite and amphibole schists | Amphibolites | Fe-Mg-amphibolites<br>Fe-amphibolites<br>Si-amphibolites | 6-biotitized varieties |
| | Gabbro-amphibolites | Porphyroblastic Fe-amphibolites | 6 |
| | | Actinolitic schists<br>Talc-actinolitic schists | 7A, B-biotitized varieties |
| Granitoids | Plagiogranites | Plagiogranites | – |
| | | Plagiopegmatoids | – |
| | Granites | Granites | – |
| | | Granite pegmatoids | – |
| | | Granodiorites | – |
| | Subalkali granites | Porphyry granites | – |

sized grains (0.09×0.03 mm) of a hyacinth type exhibiting an almost equal ratio of transparent to non-transparent varieties (Fig. 1.40). This fact, as well as the distinctly visible unimodal distribution pattern of zircon crystal lengthening, clearly shows that effusive material is contained in initial plagiogneisses, whereas the presence of regenerated fragmental grains indicates that these effusives (at least part of them) had undergone weathering and partial redeposition at the pre-metamorphic stage.

Unaltered varieties of plagiogneisses (paleosome) rarely occur in the SG-3 well section; usually they are recrystallized (neosome) to varying extent and even transformed into characteristic migmatic structures. The latter are represented by

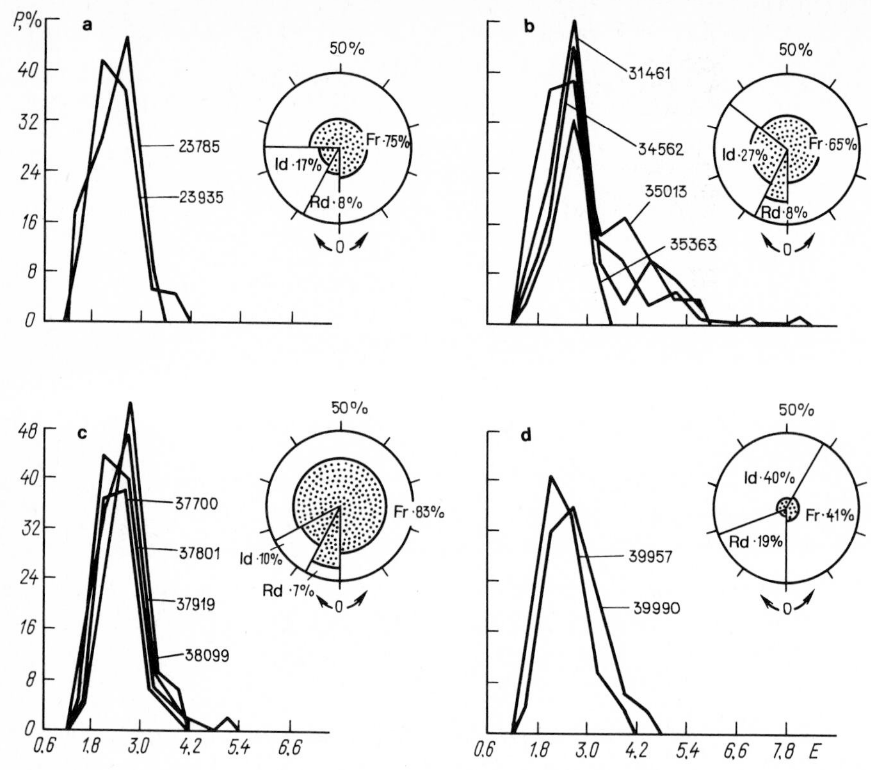


**Fig. 1.40 a – d.** Elongation histogram (*E*) and cyclogram of distribution of idiomorphic (*id*), rounded (*rd*) and fragmented (*fr*) grains of accessory zircon in plagiogneisses of the Archean complex. Bimicaceous schists and plagiogneisses with HAM: **a** I stratum; **b** III stratum. Leucocratic biotite gneisses: **c** II stratum; **d** IV stratum. Areas with *dots* denote portions of non-transparent zircons; *P, %* frequency of occurrence

regularly alternating melanosomes and leucosomes in the ratios 1:1 in melanocratic plagiogneisses, 1:2.3 in mesocratic plagiogneisses, and 1:4.5 in leucocratic biotite plagiogneisses. The mineral composition of melanosomes corresponds to that of the associated paleosomes; the only difference is in the high content of mica (60 – 80%) in meso- and melanocratic varieties). The thickness of leucosomes varies from a few mm to 10 cm. They consist of plagioclase, quartz, biotite and microcline (up to 10%), which form plagioclase migmatites. Leucosomes of leucocratic biotite plagiogneisses and gneisses in particular contain microcline, whose amount ranges from 10 to 30%, respectively; these are microcline-plagioclase migmatites.

The usually conformable occurrence of plagioclase migmatites, as well as the observed equality between the contents of biotite in a melanosome with garnet, staurolite and other minerals, typical also of a paleosome paragenesis, are indicative of the synmetamorphogenic formation of plagioclase migmatites, which is a result of differentiation of the initial material in leucosome and melanosome.

On the albite-quartz-orthoclase diagram (Fig. 1.41) leucosomes of plagioclase migmatites are overlapped by fields of natural plagiogranites whose melting tem-

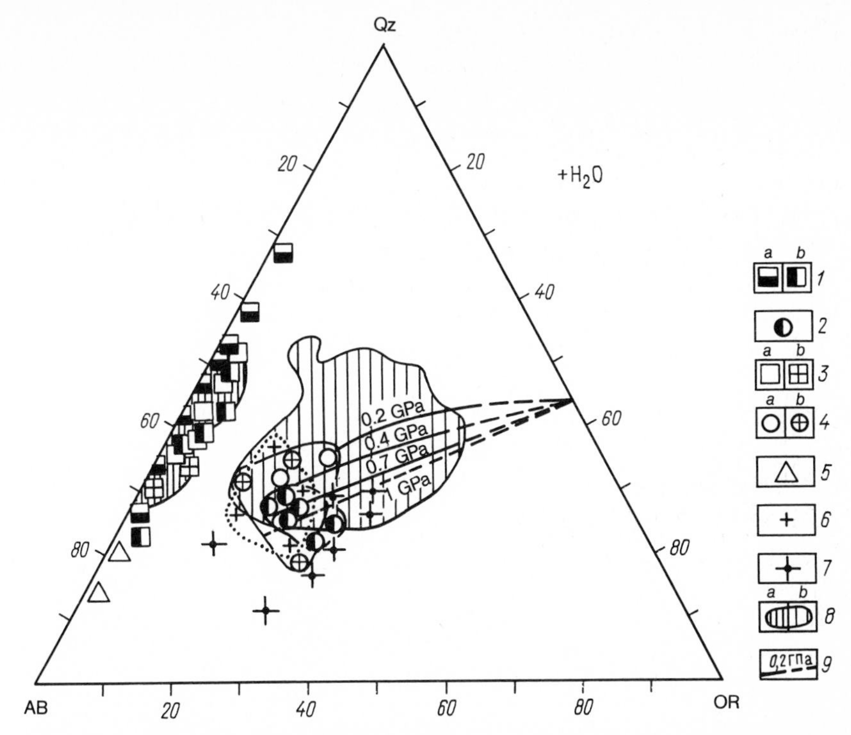


**Fig. 1.41.** Modal compositions of migmatites and granitoids of the Archean complex on a diagram albite-quartz-orthoclase (Winkler 1979). *1* leucosomes of plagioclase migmatites from bimicaceous plagioclases and schists with HAM (*a*) and leucocratic biotite and epidote-biotite plagiogneisses (*b*); *2* leucosomes of microcline-plagioclase migmatites; *3–5* Archean complex of granitoids: *3a* plagiogranites; *3b* plagiopegmatoids; *4a* granites; *4b* granite-pegmatoids; *5* granodiorites; *6* late-folding granites; *7* porphyry granites and pegmatites; *8* a field of compositions of plagiogranites, granodiorites (*a*) and granites (*b*); *9* projections of cotectic lines in a system of quartz-albite-arthoclase-water of $P_{H_2O}$ = 0.2, 0.4, 0.7 HPa (Winkler 1979)

perature is not less than 690 °C at $P$ $H_2O$ = 0.5 GPa (quartz-albite eutectic). However, the melanosome parageneses of these migmatites and associated paleosomes (Pl + Q + Bi + And + Stu ± Mu ± Gar ± Sl) correspond to a temperature not exceeding 550° – 650 °C, which demonstrates that migmatites cannot be formed by melting. Leucosomes of microcline-plagioclase migmatites form a field near cotectic lines of the area of magmatic granitoid rocks. The similarity between conditions of metamorphism of this complex and the $P–T$ parameters inherent at the beginning of an anatexis (Winkler 1979) shows that the formation of these leucosomes might have been a result of partial biotite gneisses.

The mean chemical compositions of unaltered gneisses of the Archean complex are given in Table 1.11. They show that in passing from leucocratic biotite plagiogneisses to leucocratic, mesocratic and melanocratic plagiogneisses with HAM, the content of oxides gradually increases, i.e. oxides of $\Sigma$Fe increases from 2.27 to 6.94%, $K_2O$ from 1.58 to 3.67%, and MgO from 0.58 to 3.66%, whereas the content of other oxides decreases, i.e. $Na_2O$ decreases from 7.76 to

**Table 1.11.** Mean compositions of Archean rocks from SG-3 well section (6842 – 11,662 m)

| Components | Plagiogneisses and gneisses | | | | | | | | | |
|---|---|---|---|---|---|---|---|---|---|---|
| | 1 | 2 | 3 | 4 | 5 | 6 | 7 | 8 | 9 | 10 |
| $SiO_2$ | 62.01 | 63.99 | 69.02 | 64.22 | 70.48 | 70.96 | 65.05 | 62.28 | 64.36 | 51.25 |
| $TiO_2$ | 0.68 | 0.68 | 0.46 | 0.26 | 0.34 | 0.27 | 0.50 | 0.59 | 0.49 | 0.37 |
| $Al_2O_3$ | 16.40 | 16.33 | 15.23 | 20.17 | 15.68 | 15.35 | 16.22 | 17.56 | 16.91 | 18.09 |
| $Fe_2O_3$ | 1.23 | 0.87 | 0.47 | 0.48 | 0.51 | 0.55 | 1.14 | 0.57 | 1.02 | 1.82 |
| FeO | 5.71 | 5.55 | 4.14 | 1.79 | 2.18 | 1.59 | 3.19 | 4.54 | 3.47 | 6.90 |
| MnO | 0.07 | 0.07 | 0.05 | 0.02 | 0.03 | 0.04 | 0.06 | 0.05 | 0.05 | 0.11 |
| MgO | 3.47 | 2.88 | 1.79 | 0.58 | 0.88 | 0.65 | 1.85 | 1.94 | 1.84 | 8.04 |
| CaO | 1.74 | 1.99 | 2.07 | 2.42 | 2.09 | 1.68 | 4.22 | 4.00 | 4.57 | 4.23 |
| $Na_2O$ | 3.08 | 3.31 | 3.58 | 7.76 | 5.54 | 5.32 | 4.00 | 4.96 | 4.61 | 3.47 |
| $K_2O$ | 3.67 | 2.76 | 1.64 | 1.48 | 1.48 | 2.80 | 1.62 | 2.14 | 1.35 | 4.48 |
| $P_2O_5$ | 0.07 | 0.09 | 0.08 | 0.08 | 0.11 | 0.07 | 0.13 | 0.13 | 0.12 | 0.08 |
| $CO_2$ | 0.20 | 0.15 | 0.16 | 0.16 | 0.16 | 0.39 | 0.17 | 0.16 | 0.16 | 0.17 |
| $H_2O^-$ | 0.13 | 0.12 | 0.07 | 0.05 | 0.11 | 0.17 | 0.13 | 0.17 | 0.13 | 0.16 |
| $H_2O^+$ | 0.90 | 0.67 | – | 0.07 | 0.26 | 0.30 | 0.28 | 0.19 | 0.16 | 0.48 |
| l.o.c.[a] | 1.48 | 1.06 | 1.13 | 0.65 | 0.56 | 0.32 | 0.90 | 0.77 | 0.74 | 1.57 |
| S | 0.24 | 0.11 | 0.29 | 0.09 | 0.13 | 0.75 | 0.07 | 0.08 | 0.07 | 0.07 |
| Li | 51.1 | 44.6 | 21.8 | 9.8 | 15.7 | 17.9 | 57.0 | 36.2 | 29.0 | 30.1 |
| Rb | 185.7 | 131.6 | 82.2 | 34.0 | 54.4 | 95.2 | 126.5 | 123.4 | 77.6 | 122.5 |
| Cs | 6.9 | 7.1 | 2.6 | 1.1 | 1.8 | 1.0 | 1.9 | 3.9 | – | 4.5 |
| Sr | 85.0 | 110.4 | 158.9 | 186.0 | 218.2 | 212.0 | 208.5 | 258.8 | 136.5 | 249.4 |
| Ba | 534.3 | 296.3 | 230.0 | 400.7 | 406.9 | 558.9 | 232.0 | 421.0 | 395.3 | 303.1 |
| B | 6.0 | 6.4 | 7.7 | 8.2 | 8.1 | 8.3 | 6.2 | 6.3 | 6.9 | 5.6 |
| Sc | 16.8 | 14.1 | 8.8 | 2.2 | 2.6 | 3.2 | 13.9 | 11.3 | 11.6 | 15.8 |
| Ga | 32.1 | 23.8 | 20.6 | 29.3 | 29.0 | 29.9 | 30.2 | 33.9 | 27.1 | 20.6 |
| La | 26.0 | 24.0 | 24.0 | 25.0 | 26.0 | 25.0 | 14.0 | 17.0 | – | 25.0 |
| Ce | 54.0 | 47.0 | 51.0 | 44.0 | 46.0 | 44.0 | 26.0 | 28.0 | – | 25.0 |
| Nd | 13.0 | 17.0 | – | 14.0 | 19.0 | 13.0 | 13.0 | 13.0 | – | 17.0 |
| Sm | 4.0 | 3.5 | 3.4 | 2.1 | 2.1 | 2.3 | 2.8 | 2.8 | – | 3.4 |
| Eu | 1.2 | 1.2 | 1.0 | 0.81 | 0.78 | 0.72 | 1.1 | 1.2 | – | 1.3 |
| Tb | 0.67 | 0.59 | 0.62 | 0.22 | 0.26 | 0.19 | 0.49 | 0.39 | – | 0.90 |
| Yb | 1.8 | 1.8 | 1.9 | 0.38 | 0.40 | 0.21 | 1.7 | 1.1 | – | 2.7 |
| Lu | 0.25 | 0.27 | 0.29 | 0.06 | 0.06 | 0.05 | 0.30 | 0.22 | – | 0.34 |
| Th | 7.1 | 5.8 | 5.6 | 8.4 | 9.1 | 8.4 | 0.70 | 1.2 | – | 1.8 |
| U | 6.0 | 5.0 | 8.0 | 3.7 | 4.7 | 2.1 | 1.5 | 2.0 | – | 2.6 |
| Ge | 0.50 | 0.51 | 0.50 | 0.50 | 0.50 | 0.50 | 0.52 | 0.53 | 0.50 | 0.69 |
| Zr | 71.4 | 74.7 | 72.2 | 98.7 | 104.5 | 69.1 | 57.0 | 60.0 | 42.9 | 39.4 |
| Sn | 3.8 | 3.0 | 3.0 | 2.6 | 2.9 | 3.2 | 4.2 | 3.8 | 3.5 | 3.3 |
| Hf | 4.2 | 4.1 | 4.1 | 3.7 | 3.6 | 3.1 | 4.6 | 4.3 | – | 4.1 |
| Nb | 3.2 | 7.6 | 5.2 | 2.1 | 2.9 | 2.5 | 4.7 | 2.5 | 2.4 | 3.2 |
| Ta | 0.66 | 0.54 | 0.52 | 0.19 | 0.24 | 0.12 | 0.33 | 0.39 | – | 0.45 |
| F | 1152.8 | 872.5 | 244.4 | 197.3 | 172.9 | 169.2 | 683.5 | 509.4 | 312.1 | 931.9 |
| Cu | 40.4 | 89.4 | 50.9 | 24.3 | 15.7 | 17.3 | 83.4 | 26.7 | 29.2 | 48.3 |
| Zn | 145.7 | 89.6 | 52.2 | 35.7 | 43.1 | 44.5 | 71.1 | 115.3 | 95.0 | 90.3 |
| Ag | 0.06 | 0.08 | 0.05 | 0.03 | 0.05 | 0.06 | 0.11 | 0.05 | 0.04 | 0.07 |
| Pb | 21.8 | 31.6 | 17.9 | 20.7 | 21.7 | 31.0 | 20.9 | 20.2 | 25.1 | 21.4 |
| Mo | 1.1 | 1.6 | 1.3 | 0.77 | 0.81 | 0.80 | 1.5 | 0.98 | 0.80 | 1.0 |
| V | 112.8 | 115.5 | 76.9 | 31.0 | 32.8 | 33.7 | 139.2 | 130.4 | 88.9 | 156.4 |
| Cr | 216.7 | 150.7 | 99.3 | 37.7 | 31.6 | 43.1 | 95.4 | 44.0 | 97.1 | 160.9 |
| Co | 27.7 | 20.8 | 14.0 | 3.8 | 4.6 | 5.3 | 19.6 | 15.1 | 17.6 | 26.4 |
| Ni | 112.3 | 67.1 | 44.2 | 15.1 | 9.5 | 9.9 | 42.8 | 29.6 | 72.3 | 107.2 |
| No. of samples | 23 | 60 | 11 | 15 | 130 | 91 | 20 | 32 | 17 | 16 |

[a] l.o.c. = Loss on calcination.

1 – 3 – two-mica plagiogneisses with HAM (1 – melanocratic, 2 – mesocratic, 3 – leucocratic); 4 – 5 – leucocratic biotite-plagioclase plagiogneisses (4 – quartz content is less than 10%, 5 – more than 10%); 6 – leucocratic biotite gneisses; 7 – 8 – epidote-biotite plagiogneisses (7 – leucocratic, 8 – mesocratic); 9 – 10 – hornblende-biotite plagiogneisses (9 – mesocratic, 10 – melanocratic);

| Amphibolites and amphibole schists | | | | | | | Granitoids | | | |
|---|---|---|---|---|---|---|---|---|---|---|
| 11 | 12 | 13 | 14 | 15 | 16 | 17 | 18 | 19 | 20 | 21 |
| 48.15 | 50.01 | 49.60 | 54.48 | 51.72 | 52.73 | 58.01 | 72.11 | 72.32 | 72.83 | 72.19 |
| 2.23 | 1.40 | 1.64 | 0.80 | 0.41 | 0.19 | 0.07 | 0.20 | 0.11 | 0.15 | 0.27 |
| 10.86 | 14.20 | 13.64 | 14.28 | 17.57 | 7.18 | 2.12 | 14.78 | 15.31 | 14.62 | 14.24 |
| 3.50 | 3.10 | 3.70 | 2.37 | 2.20 | 2.90 | 1.60 | 0.79 | 0.28 | 0.15 | 0.73 |
| 11.33 | 10.31 | 10.09 | 7.83 | 6.92 | 6.78 | 5.56 | 2.32 | 1.62 | 2.14 | 1.59 |
| 0.22 | 0.22 | 0.22 | 0.16 | 0.15 | 0.23 | 0.16 | 0.04 | 0.09 | 0.03 | 0.12 |
| 9.60 | 6.00 | 6.15 | 5.89 | 7.95 | 12.21 | 24.85 | 0.63 | 0.33 | 0.33 | 0.50 |
| 9.85 | 9.95 | 10.04 | 8.76 | 9.31 | 14.26 | 3.49 | 2.79 | 1.25 | 1.02 | 1.07 |
| 1.89 | 2.31 | 2.36 | 2.50 | 2.02 | 0.54 | 0.25 | 5.28 | 4.29 | 4.63 | 3.30 |
| 0.68 | 0.63 | 0.64 | 0.89 | 0.51 | 0.43 | 0.08 | 1.30 | 3.50 | 3.63 | 5.40 |
| 0.19 | 0.12 | 0.14 | 0.11 | 0.06 | – | 0.06 | 0.04 | 0.05 | 0.09 | 0.07 |
| 0.14 | 0.20 | 0.17 | 0.26 | 0.15 | – | 0.16 | 0.16 | – | 0.26 | 0.22 |
| – | 0.08 | 0.09 | 0.12 | 0.04 | – | 0.06 | – | – | 0.14 | 0.16 |
| – | 0.17 | 0.28 | 0.35 | 0.78 | 0.98 | 2.08 | – | – | 0.46 | 0.30 |
| 1.53 | 1.32 | 1.14 | 1.47 | 1.20 | 1.52 | 1.58 | 0.07 | 0.03 | 0.61 | 0.29 |
| 0.22 | 0.11 | 0.11 | 0.08 | 0.06 | – | 0.07 | – | – | 0.05 | 0.10 |
| 17.7 | 13.8 | 14.4 | 15.3 | 12.6 | 5.8 | 3.5 | 8.4 | 5.0 | 21.8 | 15.6 |
| 26.3 | 17.4 | 17.6 | 43.4 | 10.4 | 5.0 | 16.7 | 22.0 | 73.3 | 99.4 | 229.1 |
| 2.6 | 2.0 | 2.0 | 2.0 | 2.0 | – | 1.0 | 0.32 | 0.56 | 1.1 | 2.5 |
| 186.9 | 157.3 | 180.0 | 199.3 | 227.1 | 200.0 | 22.5 | 172.0 | 268.1 | 111.8 | 81.8 |
| 154.6 | 119.2 | 138.2 | 201.3 | 167.1 | 35.0 | 33.8 | 382.0 | 1085.6 | 310.0 | 342.6 |
| 6.8 | 6.3 | 5.9 | 7.0 | 4.7 | 7.0 | 4.5 | 7.7 | 11.1 | 8.8 | 8.3 |
| 44.0 | 37.3 | 36.9 | 31.8 | 32.9 | 9.0 | 5.2 | 3.3 | 1.7 | 6.1 | 2.6 |
| 21.3 | 20.9 | 23.5 | 19.6 | 22.1 | 20.6 | 8.2 | 21.0 | 26.7 | 25.6 | 31.0 |
| 29.0 | 12.0 | 13.0 | 14.0 | 7.0 | – | 0.97 | 33.0 | 6.0 | 4.0 | 62.0 |
| 58.0 | 27.0 | 27.0 | 28.0 | 12.0 | – | 5.0 | 64.0 | 9.0 | 8.0 | 138.0 |
| – | 12.0 | 14.0 | 15.0 | 8.0 | – | 10.0 | 22.0 | 7.0 | 5.0 | 54.0 |
| 8.3 | 3.8 | 4.1 | 3.1 | 1.2 | – | 0.55 | 3.9 | 0.66 | 1.0 | 8.6 |
| 2.7 | 1.4 | 1.5 | 1.1 | 0.58 | – | 018 | 0.87 | 0.64 | 0.39 | 0.62 |
| 0.99 | 0.73 | 0.80 | 0.59 | 0.35 | – | 0.10 | 0.51 | 0.14 | 0.54 | 0.97 |
| 2.0 | 2.5 | 2.5 | 2.0 | 1.7 | – | 0.39 | 0.44 | 0.20 | 0.48 | 1.1 |
| 0.25 | 0.36 | 0.45 | 0.34 | 0.31 | – | 0.05 | 0.15 | 0.04 | – | 0.15 |
| 2.2 | 1.9 | 1.5 | 2.0 | 1.5 | – | 1.0 | 18.0 | 1.7 | 3.2 | 41.2 |
| 2.4 | 2.0 | 2.0 | 2.2 | 2.0 | – | 2.0 | 2.0 | 1.0 | 9.1 | 5.6 |
| 1.0 | 0.75 | 0.77 | 0.57 | 0.50 | 1.0 | 1.8 | 0.50 | 0.50 | 0.50 | 0.50 |
| 73.1 | 68.7 | 58.7 | 55.8 | 27.5 | 5.0 | 7.5 | 114.0 | 43.3 | 61.2 | 81.5 |
| 5.9 | 2.9 | 5.6 | 3.4 | 2.3 | 6.0 | 2.2 | 2.7 | 3.8 | 3.2 | 4.1 |
| 5.5 | 2.6 | 2.6 | 2.9 | 1.1 | – | 0.30 | 4.1 | 1.6 | 2.7 | 6.0 |
| 5.5 | 3.2 | 3.5 | 4.6 | – | 3.0 | 1.5 | 8.3 | 2.4 | 7.8 | 4.8 |
| 1.8 | 0.53 | 0.51 | 0.34 | 0.27 | – | 0.10 | 0.16 | 0.06 | 1.2 | 0.40 |
| 590.8 | 428.5 | 273.1 | 566.8 | 212.1 | 1600.0 | 750.0 | 67.5 | 61.1 | 417.5 | 213.5 |
| 192.8 | 129.2 | 114.6 | 72.9 | 35.6 | 5.5 | 4.4 | 39.8 | 12.7 | 83.0 | 6.8 |
| 150.4 | 124.8 | 126.0 | 97.0 | 160.0 | 180.0 | 115.0 | 34.0 | 23.3 | 37.3 | 40.2 |
| 0.03 | 0.06 | 0.06 | 0.06 | 0.04 | 0.05 | 0.02 | 0.06 | 0.04 | 0.13 | 0.06 |
| 103.1 | 13.0 | 12.3 | 18.7 | 17.8 | 15.0 | 4.1 | 27.2 | 32.2 | 22.9 | 53.4 |
| 1.6 | 1.0 | 1.6 | 0.89 | 0.78 | 1.0 | 0.81 | 0.78 | 0.75 | 1.3 | 3.4 |
| 188.3 | 239.1 | 361.6 | 182.2 | 77.8 | 83.5 | 14.2 | 29.8 | 11.9 | 43.3 | 21.3 |
| 414.6 | 92.1 | 80.7 | 29.6 | 106.8 | 540.0 | 980.0 | 24.6 | 40.2 | 48.4 | 35.8 |
| 61.0 | 41.1 | 50.8 | 38.7 | 47.4 | 36.0 | 95.0 | 5.1 | 3.3 | 11.9 | 3.1 |
| 169.1 | 70.1 | 64.0 | 76.1 | 160.6 | 790.0 | 2175.0 | 7.2 | 8.1 | 26.1 | 12.1 |
| 26 | 68 | 101 | 60 | 15 | 4 | 4 | 5 | 15 | 16 | 34 |

11 – Fe-Mg-amphibolites; 12 – Fe-amphibolites; 13 – porphyroblastic Fe and Fe-Mg-amphibolites; 14 – Si-amphibolites; 15 – Al-Mg-amphibolites; 16 – actinolitic schists; 17 – talc-actinolitic schists; 18 – plagiogranites; 19 – granites; 20 – granite pegmatoids; 21 – porphyry granites. Content of petrogenic elements is given in percent, content of elemental admixtures – in $g\,t^{-1}$; "–" – content was determined.

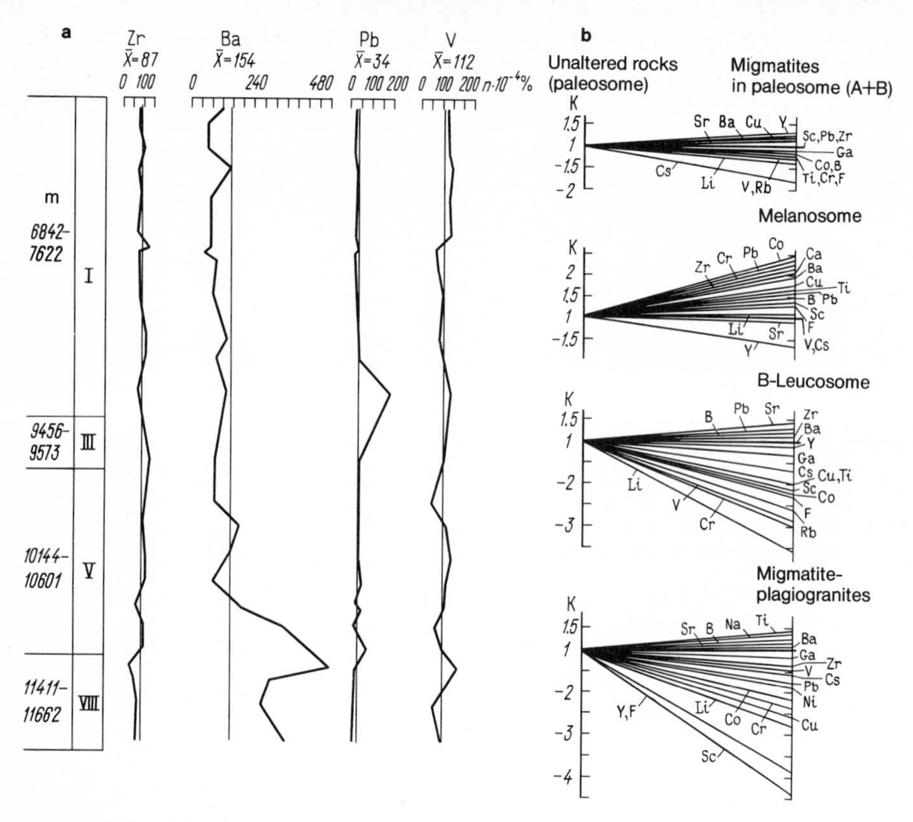


**Fig. 1.42 a, b.** Distribution of element-admixtures in paleosomes of mesocratic plagiogneisses with HAM in the Kola complex section **(a)** and in associated migmatites **(b)**

3.08%, CaO from 2.42 to 1.74%, and $H_2O_3$ from 20.17 to 16.40%. Analogous regularities are characteristic of elemental admixtures (see Table 1.11) which are chemically similar to the lithophile elements: the amount of Rb, Li, Cr, V, Co, Sc, and Zn increases by 3.5 – 7 times, whereas the content of Sr, B, and partially, Ba decreases by 1.4 – 2.6 times. These tendencies are accounted for by the increased content of biotite against the background of the decreased content of plagioclase in rocks of the series examined.

Thus, it appears that the presence of one single geochemical tendency, which brings rocks of leucocratic biotite plagiogneisses into one and the same series with alternating mesomelanocratic plagiogneisses containing HAM, is indicative of the genetical affinity of the initial rocks.

When tracing meso- and melanocratic plagiogneisses to deep horizons, it becomes evident (see Fig. 1.42) that the distribution pattern of Zr, Pb, V, and other elements in these rocks is characterized by minimal variations in average values for each layer compared with the average value of the whole section. It should be noted that a number of elements exhibit a wide range of fluctuations in average value, due mainly to changes in plagioclase content.

As was already mentioned earlier, leucocratic biotite gneisses, leuco- and mesocratic epidote-biotite and hornblende-biotite plagiogneisses form isolated layers in strata of leucocratic biotite plagiogneisses. Their geochemical features are also determined by taking into consideration variations in mineral composition. For instance, despite general similarity between chemical composition of leucocratic biotite gneisses (microcline content does not exceed 5%) and that of the enclosing biotite plagioclases, the leucocratic biotite gneisses are distinguished by a high average content of $K_2O$ (2.8%) and, correspondingly, of Rb (95 g $t^{-1}$), Ba (559 g $t^{-1}$) and Pb (31 g $t^{-1}$). Considerable differences in chemical composition are established for epidote-biotite plagiogneisses, particularly for hornblende-biotite plagiogneisses and schists which, as distinct from leucocratic biotite plagiogneisses, are characterized by a higher mean content of CaO (4.0 – 4.5% respectively), MgO (1.8 – 8.0%), $Fe_2O_3$ (4.3 – 8.7%), K (1.6 – 4.5%), and of Li, Rb, F, V, Zn, Cr, and Ni (see Table 1.11) geochemically associated with them. These associations of elements are characteristic of basic rocks and they may indicate that an admixture of tuffaceous material is present in the initial rocks.

According to the distribution of rare-earth elements (REE), leucocratic biotite plagiogneisses and gneisses are similar to the Early Archean tonalite series, and associated meso- and melanocratic plagiogneisses with HAM are successively shifted into the area of sedimentary rock composition, i.e. psammites-pelites (see Fig. 1.81). This fact, as well as higher values of the initial ratio of $^{87}Sr/^{86}Sr$ (0.7073 in leucocratic biotite plagiogneisses and 0.7428 in leucocratic plagiogneisses with HAM) (see Table 1.17) makes it possible to consider leucocratic biotite plagiogneisses and gneisses constituting the main part of the SG-3 well section as products of tonalitic magma crystallization, while their remaining part and all meso- and melanocratic plagiogneisses with HAM are regarded as products of disintegration and redeposition of volcanic-plutonic rocks of an older basement.

The geochemical study of migmatites was conducted according to the following scheme: paleosome→neosome→migmatite (leucosome + melanosome). Normally, the mean compositions of paleosomes and associated neosomes and migmatites are characterized by close values of all petrogenic elements, which is indicative, in general, of the isochemical regime of migmatite formation (Fig. 1.42b). By comparing the mean compositions of paleosomes with the final products of their metamorphic differentiation (leucosomes and melanosomes), it becomes possible to obtain more accurate ratio values of lithophile elements during the process of migmatization. For instance, inspection of the petrochemical diagram (Fig. 1.43) shows that the mean paleosome composition of different rocks from the plagioclase series forms one single trend of successively increasing content of $SiO_2$ and Na/K value in the direction from melanocratic plagiogneisses with HAM to leucocratic biotite plagiogneisses. Symbols denoting leucosomes and melanosomes are placed on either side of paleosomes which are parallel to this trend; this supports the conclusion as to a general complementary character of their compositions with respect to the initial paleosome. However, calculation of migmatite composition on the basis of real leuco- and melanosomes, taken in natural proportions (46 : 54 for a, 67 : 33 for b, see Fig. 1.43), and comparison of

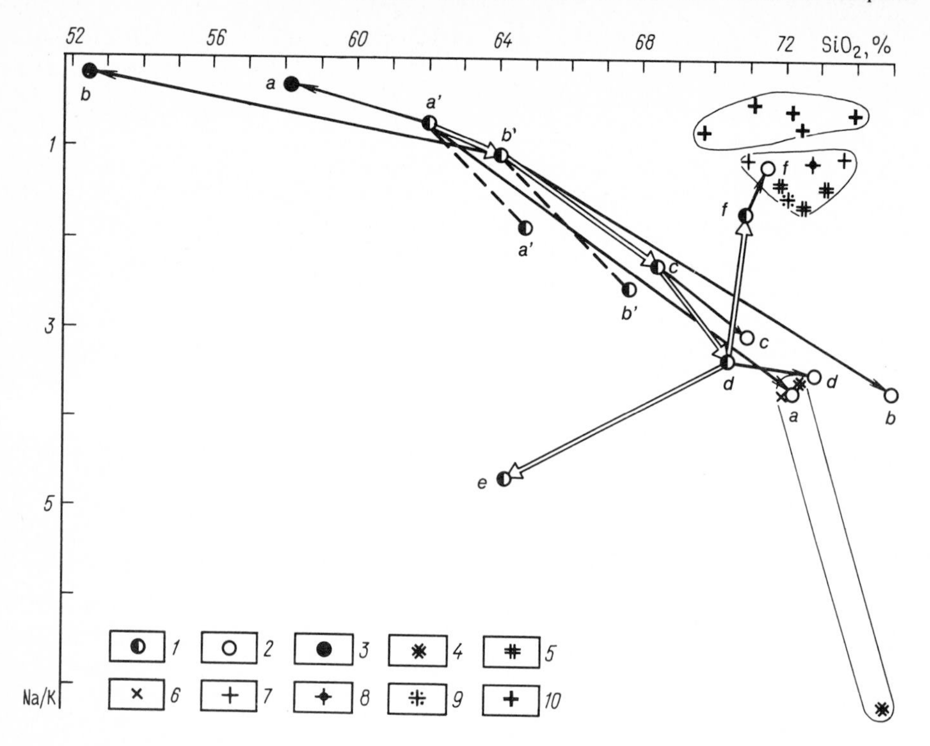


**Fig. 1.43.** Chemical composition of migmatites and granitoids of the Archean complex. *1–3* migmatites (*1* paleosome; *2* leucosome; *3* melanosome) from bimicaceous schists with HAM (*a*), mesocratic plagiogneisses with HAM (*b*), leucocratic plagiogneisses with HAM (*c*), leucocratic biotite plagiogneisses with Qz>10% (*d*) and Qz>10% (*e*) and leucocratic biotite gneisses (*f*); *4–8* granitoids (*4* plagiogranites, thickness<1 m; *5* granites, thickness<1 m; *6* plagiogranites, thickness>2 m; *7* granites, thickness>2 m; *8* pegmatoids); *9* late-folded granites; *10* porphyry granites; calculated compositions of migmatites from bimicaceous schists with HAM (*a*) and mesocratic plagiogneisses with HAM (*b*)

this composition with the initial paleosome, reveal an insignificant deficiency of $K_2O$ and a corresponding surplus of $N_2O$. From this it appears that the process of migmatization of plagioclases led to a partial withdrawal of potassium out of plagioclases and their enrichment in sodium and silicon.

This law is particularly emphasized by the characteristic distribution of elemental admixtures typomorphic for mica. For instance, when comparing migmatites with their initial paleosomes, it becomes clear that Li, Rb, V, F, and, partially, Cr are being withdrawn. This tendency is most pronounced in the migmatization of leucocratic biotite plagiogneisses and mesocratic plagiogneisses with HAM, which exhibit maximal decrease (4 times) in concentration of the elements mentioned accompanied by negligible increase (1.4 times) in B and Pb content. A similar tendency to withdrawal is characteristic also of the whole REM groups (Fig. 1.44c, d).

The distribution of elemental admixtures between leuco- and melanosomes is complementary in character (see Fig. 144b) and corresponds to that of the litho-

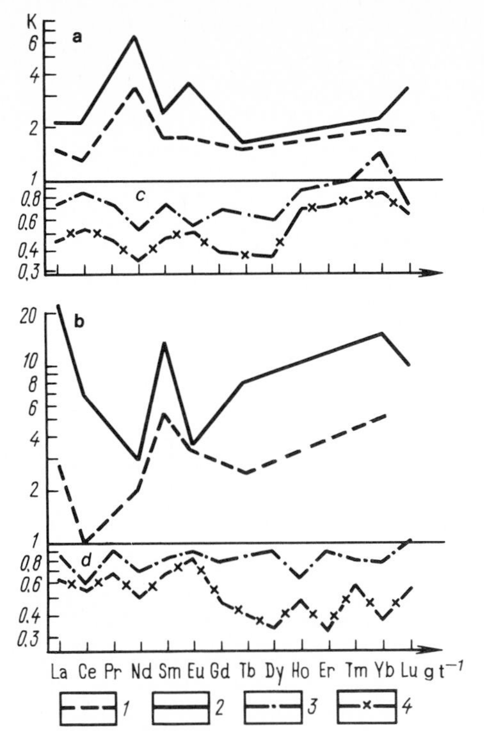


**Fig. 1.44a, b.** Distribution of rare earth elements during biotitization of amphibolites (**a**) and amphibole schists (**b**), as well as during migmatization of bimicaceous plagiogneisses with HAM (**c**) and leucocratic biotite plagiogneisses (**d**). $K = C_{Bi}(M)/Cp$, where $C_{Bi}(M)$ = content of REE in biotitized (*Bi*) or migmatized (*M*) varieties; *Cp* content of REE in a paleosome; degree of biotitization: *1*, 5–25%; *2*, 25–50%; *3* migmatites; *4* leucosome

phile elements. Thus, synmetamorphic migmatization of rocks of the plagiogneiss group is considered on the whole as the isochemical process which leads only to partial withdrawal of K, Li, Rb, F, and all REM.

All types of plagiogneisses, including migmatites, are characterized by superimposed muscovitization with distinct replacement of biotite and plagioclase, and also by partial quartz recrystallization, which can result in the formation of a banded structure. In melanosome, synmetamorphic muscovite is formed together with muscovite of later origin which is cross-oriented with respect to gneissosity. Rock muscovitization and silicification are seen most distinctly in leucocratic biotite plagiogneisses and gneisses; through mineral replacements (Bi→Mu, Pl→Mu+Qz) they have maximal affinity with regional greisenization of low intensity. Geochemically, this process is distinctly expressed in the decrease in content of $\Sigma$Fe (2.6 times) and MgO (2.2 times), as well as in $TiO_2$, MnO, V, Co, Ni, Sc, La, and accompanied by enrichment of the initial rocks with sodium (1.5 times) and partially with $Al_2O_3$, $SiO_2$, Pb, Ba, and B. Chloritization, sericitization and poorly manifested carbonatization are the usual subsequent processes in zones of cataclasis where, in addition to these processes, muscovite and quartz underwent recurrent recrystallization.

Amphiboles and amphibole schists constitute respectively 24 and 6% of the Archean rock bulk and are represented in the SG-3 well section by individual bodies of small thickness. According to the $SiO_2$ content, they form two subgroups. The first subgroup ($SiO_2$<53%) includes the following types of rocks:

Fe-Mg amphibolites ($Fe_2O_3$ + FeO > 10%, MgO > 7%), apoolivine basalt); Fe amphibolites ($Fe_2O_3$ + FeO > 10%, MgO < 7%, apobasalt): Al-Mg amphibolites ($Al_2O_3$ > 16%, MgO > 7%, apohypersthene basalt) and actinolite schists (CaO > 10%, MgO < 10%, apopyroxenite). The second subgroup ($SiO_2$ > 53%) comprises Si amphibolites (MgO < 7%, apoandesitobasalt) and talc-actinolite schists (MgO > 20%, apoolivine pyroxenite).

In strata of leucocratic biotite plagiogneisses, which are the most voluminous, all the above-mentioned types of amphibolite are present in the following proportion: Fe-Mg and Fe amphibolites 66 – 69%, Al-Mg amphibolites 8 – 13%, Si amphibolites 3 – 12%, amphibole schists 6 – 22%. In strata of meso-, melanocratic plagiogneisses with HAM Al-Mg amphibolites and talc-actinolite schists are completely absent; other types are present in the following proportion: Fe-Mg and Fe amphibolites 62%, Si amphibolites 30%, and actinolite schists 8%.

The central parts of Fe-Mg amphibolite bodies are frequently composed of coarse-grained porphyroblastic varieties, the latter additionally forming isolated bodies. A characteristic feature of porphyroblastic amphibolites is a blastogabbraic texture, indicating their initially intrusive nature; Al-Mg amphibolites contain typomorphic minerals such as cummingtonite in association with low ferruginous hornblende. Porphyroblastic amphibolites are characterized by their massive structure, blastodioritic texture and the presence of relicts of initially magmatic plagioclase (core sample No. 60, rim No. 38); these features, as well as their specific chemical composition, permit their classification as subvolcanic rocks. The above-mentioned amphibolites are biotitized to different degrees. Si amphibolites are characterized by a maximal content of quartz (up to 15%) and by an equal amount of biotite and hornblende (3 – 10%). Amphibole schists are represented by anchimonoactinolites and talc-actinolites (actinolite content up to 90%, talc content up to 20%); cummingtonite and anthophyllite are met much more rarely. Phlogopite is of a distinctly superimposed nature and its content varies over a broad range; sometimes, mainly in endocontact zones and in cross-cracks, even phlogopite rocks are formed.

The comparative analysis of accessory minerals shows that the following variations in mean mineral content are observed in a series of Fe-Mg→Fe→Si amphibolites: a mean content of sphene with ilmenite decreases from 5010 to 1700 ($g\,t^{-1}$) apatite from 318 to 4, zircon from 9 to a few units, magnetite from 1.5 to a few units, chalcopyrite from 78 to 27, pyrrhotite from 542 to 340; pyrite content increases from 34 to 54 $g\,t^{-1}$. Talc-actinolite schists differ from amphibolites by the presence of initially magmatic chromite, the higher content of sulphides and the complete absence of accessory silicate minerals.

Mean values of $\delta^{34}S$ of pyrite in a series of Fe-Mg amphibolites→Fe amphibolites→amphibole schists vary from +2.51 to 3.50‰, and values of pyrrhotite range from +3.39 to +3.60‰; these values are of the same order as those of sulphides of plagiogneisses ($\delta^{34}S$ of pyrite is +3.38 – +4.88‰, $\delta^{34}S$ of pyrrhotite is 3.41 – 4.78‰). A somewhat different sulphur-isotope composition was established for pyrrhotite from Si amphibolites ($\delta^{34}S$ = +1.46‰).

The mean chemical compositions of amphibolites from the SG-3 well are given in Table 1.11. Their comparative analysis shows that in a tholeiite-basalt series of Fe-Mg→Fe-Si amphibolites $SiO_2$ and $Al_2O_3$ content gradually increases, and this increase is accompanied by a decrease in $\Sigma Fe$, MgO, CaO, and $TiO_2$ content. Elemental admixtures exhibit analogous tendencies, i.e. slight increase of Ba and Rb content and distinct (14 times) decrease in content of Cr, Co, Cu, Zn, Sc, Zr, and Mo. The chemical composition of porphyroblastic amphibolites is similar to that of Fe-Mg and Fe amphibolites. This fact, as well as the analogous chemical composition and the observed close geological relationship between them in the SG-3 well section (Fe-Mg and Fe amphibolites constitute endocontacts, and porphyroblastic amphibolites compose the central parts of bodies), speak in favour of the genetic affinity of these rocks; their structural-textural differences were the result of different conditions of the initial melt recrystallization in marginal and central zones of magmatic bodies.

Fe amphibolites and Si amphibolites, mostly developed in the SG-3 well section, were also statistically compared with amphibolites developed at the surface. It was established that at least some of them are associated with Proterozoic magmatism.

Al-Mg amphibolites are distinguished by their higher content of $Al_2O_3$, MgO, Sr, Zn, and low content of $\Sigma Fe$ and $Na_2O$, Rb, Zr, F, Cu, and V (see Table 1.10). Their chemical composition clearly resembles that of amphibole schists, which are notable for the presence of cummingtonite, magnesian amphibole and biotite. This fact, as well as similar initial conditions of their occurrence (closely spaced or conjugating layers), allow the combination of Al-Mg amphibolites and amphibole schists in one single independent series.

Thus, by taking into account all the geological petrographical and mineralogical geochemical features, the rocks of this group here described can be subdivided into the two following series: the first is the Fe-Mg amphibolites→Fe amphibolites (including porphyroblastic amphibolites)→Si amphibolites series and the second the Al-Mg amphibolites→actinolite schists→talc-actinolite schists series. This first series of amphibolites corresponds in its chemical composition to the volcano-plutonic association of tholeiite type, comprising respectively olivine basalts→normal basalts (including gabbro)→andesite basalts; moreover, a part of the rocks of this series (Fe amphibolites) might be a product of the same magmatic process which led to the formation of the Pechenga complex. The second series of amphibolites is represented by a subvolcanic association, i.e., pyroxene basalts→pyroxenites→olivine pyroxenites, and is fully assigned to rocks synchronous with those of the Kola complex. The primary mantle genesis of rocks, considered as the initial analogies of the Archean amphibolites, is confirmed by the low primary ratio of $^{87}Sr/^{86}Sr = 0.7040$. In their composition and content, REE amphibolites are correlated with volcano-plutonic associations of basic rocks of the Early Archean greenstone belts (see Fig. 1.82).

It was noted earlier that all types of amphibolite are characterized, although to differing extents, by superimposed biotitization. In a series of Fe-Mg amphibolites, a process of hornblende replacement by biotite was caused by the supply of $K_2O$ accompanied by the accumulation of $SiO_2$, $Al_2O_3$, and, partially, of MgO. It was established that the inflow of elemental admixtures such as Li, Pb, Ba and B was the most stable in character; Zr, Nm, Cu, and, partially, V have evacuated from rocks.

On the contrary, in a series of Al-Mg amphibolites→actinolite schists→talc-actinolite schists, the process of biotitization and muscovitization is accompanied by the successive withdrawal of $SiO_2$ and enrichment in $\Sigma Fe$. When this process operates in Al-Mg amphibolites and actinolite schists, CaO is evacuated from the rocks and MgO is accumulated (calcium amphiboles are replaced by biotites), but in talc-actinolite schists the process is reversed (such predominantly calcium-free silicates as anthophyllite and talc are replaced by phlogopite). The redistribution of elemental admixtures is fixed in the form of the inflow of Li, Rb, Ba, V, Cu, Nb, Zr, and evacuation of Zn.

Biotitization and muscovitization are also accompanied by the inflow of REE into amphibolites and amphibole schists, in particular (see Fig. 1.43 a, b). The inflow of the main part of elemental admixtures into amphibolites compensates for the above-mentioned withdrawal of these elements (see Fig. 1.43 c, d) during migmatization of the enclosing plagiogneisses; this means that these processes can be considered as complementary, and, consequently, also as autometasomatic, leading to the redistribution of one and the same elements between plagiogneisses (withdrawal) and amphibolites (inflow) of the Archean complex.

Granitoids make up only 4% of the entire bulk of the Archean complex; they constitute a leucosome of these migmatites and individual bodies. These latter are also composed of plagiogranites, granites, pegmatoid and porphyritic granites.

On the quartz-albite-orthoclase diagram, symbols denoting plagiogranites and plagiomigmatoids form one separate field with leucosomes of plagioclase migmatites which extends along a triangle side "quartz-albite" (see Fig. 1.41). This fact, as well as the absence of inclusions of melts, shows that plagiogranites and plagiopegmatoids might have been formed as a result of metamorphic differentiation. Granites and granitic pegmatoids (as well as leucosomes of microcline-plagioclase migmatites) are located near cotectic lines in the area of magmatic melts, thus indicating that they have passed through a stage of partial melting; the presence of melted inclusions in them supports this conclusion. Porphyritic granites also possess a composition resembling that of the corresponding area of the magmatic melts, however, as distinct from the granites described above, porphyritic granites are shifted towards cotectic lines which correspond to the increased fluid pressure ($\geqslant 0.7$ GPa). According to temperature-pressure data, porphyritic granites contain melted inclusions (quartz + feldspar) which have been homogenized at a temperature of 840° – 845 °C and under a fluid pressure of 0.65 GPa.

The features considered are indicative of the distinct magmatic nature of porphyritic granites and of peculiar P-T conditions of formation which were not linked with the processes of ultrametamorphism of the Archean rocks.

The chemical composition of granitoids is given in Table 1.11. Granites differ from plagiogranites in increased content of $K_2O$ (3.1 times), Ba (7.7 times), partially, Rb, and in the somewhat decreased content of $Al_2O_3$, $\Sigma Fe$, CaO, Zr (2.7 times), Nb (3.4 times), Cu (3.1 times), and they display a close petrochemical relationship with granitic pegmatoids. On the whole, judging by petrochemical features, the Archean granitoids are assigned to rocks of a normal series.

Porphyritic granites are characterized by a pronounced predominance of potassium over sodium at the alkali sum exceeding 8.0%, and hence they belong to rocks of a subalkali series. Apart from this, they clearly differ from the Archean granitoids by their higher Mo content (4.5 times), Zn (2.2 times), Rb (6 times), Zr (1.9 times). By their lesser Cu content (4.8 times), as well as by their composition and chemical characteristics, including the distribution of REE (see Fig. 1.83), they are similar to finely porphyritic granites of the Litsk-Aragub complex.

As a result of geochemical temperature-pressure studies of the Archean rocks, interstitial and contour inclusions have been discovered in them. On applying methods of cryometry and differential gas chromatography, it was established that inclusions are filled with overcompacted water-saline solutions and sometimes contain a small amount of primary inclusions of liquified methane. Primary inclusions of solidified magmatic melt were found in quartz and plagioclase of migmatites. In plagioclase they are finely recrystallized and characterized by a true negative shape and frequently contain two – three gas bubbles. It proved that the process of homogenization of the inclusions could not be brought to its final stage even in conditions of very slow heating, because inclusions became unsealed due to the formation of cracks. Inclusions in quartz are an isometric in shape, and are partially recrystallized; they are represented in four – five anisotropic crystals and by a small gas bubble. Melting of the crystal phases takes place at a temperature of over 650 °C. Full homogenization of the primary inclusions in quartz occurs at a temperature of 830° – 820 °C. Quartz also contains melt inclusions of the silicate-saline melts and saline melt brines. They are confined to the sealed cracks and characterize the later stages of migmatite formation. Analogous inclusions were established in plagiogranites.

Syngenetic inclusions in plagiopegmatites are rarely encountered. In the marginal parts of veins they are represented mainly by melt varieties and, in rarer cases, by recrystallized varieties. The latter are filled with anisotropic and low-anisotropic crystals (75 – 90% of the total inclusion volume) and contain two – three gas bubbles (10% of the volume). On heating, the first crystal is dissolved at a temperature below 500 °C, the others at a temperature between 650° and 800 °C. Full homogenization occurs at a temperature of about 810 °C. Most of the melt inclusions contain four to six anisotropic and isotropic crystals (85% of the volume), liquid phase (less than 10% of the volume) and a small gas bubble. Full homogenization occurs at temperatures of 780°, 775°, and 765 °C. In quartz cores of plagiopegmatites, subsyngenetic melt inclusions of near-negative shape are present. Solid phases are represented by large isotropic and small anisotropic crystals and occupy 75 – 65% of the volume; the liquid and gas phases occupy 10 – 25% and 15 – 20% of the volume, respectively. A gas bubble is homogenized at a temperature interval of 300° – 340 °C. Full homogenization comes between 705° and 675 °C. Melt inclusions in plagiopegmatites are usually met, together with almost synchronous conservation by one-phase inclusions of the liquid carbon dioxide whose specific volume ranges from 0.98 to 1.01 $cm^3 g^{-1}$.

In porphyritic granites the primary solidified inclusions of the magmatic melt and associated carbon dioxide inclusions were established in grains of magmatic quartz. They are recrystallized and consist of quartz and feldspar. Complete homogenization of inclusions occurs at temperatures of 845° – 840 °C. The specific volume of the liquid carbon dioxide of the one-phase inclusions is 1.02 – 1.00 $cm^3 g^{-1}$.

In some pegmatite veins, zones of granulated quartz are developed. Quartz contains the one-phase inclusions of liquid carbon dioxide; sometimes these inclusions contain the second phase represented by a water-saline solution. Judging by the temperature of ice-melting (from $-56°$ to $-58\,°C$), the one-phase inclusions are filled with carbon dioxide. Temperatures of carbon dioxide homogenization reach $-33\,°C$, thus being indicative of its high density. Specific volumes of the predominantly carbon dioxide fluid in the one-phase inclusions change from 0.91 to 0.95 $cm^3\,g^{-1}$. Pressure of the fluid phase ($P_{CO_2}$) in formation of the granulation zones (determined by taking into account the homogenization temperatures of the crystallized inclusions) was 0.7 – 0.8 GPa.

Inclusions in quartz from zones of diaphthoresis (chloritization, carbonatization) are the multi-phase in type; solid phases are represented by crystals of halite. The temperature of complete homogenization changes within the range of 380° – 250 °C. It is not uncommon for minerals from the cataclasis zones to contain bitumen-hydrocarbon inclusions. The morphological features of these inclusions are similar to those inherent to inclusions from the Pechenga complex.

## Geochemical Conditions of Formation

On the strength of the geological-geochemical and geochemical temperature-pressure studies of metamorphic and ultrametamorphic rocks, two major stages in evolution can be distinguished: (1) a stage of sedimentation and volcanism, and (2) a stage of metamorphism and ultrametamorphism.

The first stage (more than 2.8 Ga ago) was characterized by the accumulation of terrigenous deposits and acid terrigenous-volcanic rocks (70% of the thickness of the Archean complex in the SG-3 well section) as well as by the simultaneous manifestation of tholeiite volcanism (23%) and intrusion of comagmatic basite-ultramafic bodies (7%).

The second stage was notable for the evident process of regional metamorphism occurring under granulite-facies conditions (identified in the SG-3 well section by the presence of individual relicts of the corresponding mineral associations; see Chap. 1.3) and processes of the retrograde metamorphism and ultrametamorphism of the andalusite-sillimanite type. Metamorphic reactions, resulting in the formation of mineral parageneses of the epidote-amphibolite and amphibolite facies, proceeded in the presence of the water-saline fluid with an admixture of methane. Judging by the absence of the recrystallized melt and carbon dioxide inclusions in leucosomes and plagiogranites, it appears most probable that rock migmatization, which accompanied the metamorphism, had been proceeding during the process of the metamorphic differentiation of the primary rocks, which were not melted. It was proved that a process of synmetamorphic migmatization of plagiogneisses was isochemical in character and was accompanied by a partial withdrawal of K, Li, Rb, F, and all REE. On the contrary, the biotitization of amphibolites, proceeded synchronously with migmatization, was characterized by inflow of the above-cited elements and by redistribution of the primary components ($SiO_2$, $\Sigma Fe$, CaO); this makes it possible to consider biotitization as an autometasomatic process, complementary to migmatization. Within the entire Archean complex it was initiated by the exchange of the most movable major and minor elements (metamorphogenic-metasomatic associations) between plagiogneisses and amphibolites.

The vein granites, formed synchronously with the migmatization, were crystallized out of the anatectic melt at an initial temperature of 820° – 830 °C, and the associated paleopegmatites were crystallized out of the silicate-saline melts at a temperature of 810° – 675 °C and under the fluid-phase pressure ($P_{CO_2}$) ranging from 0.66 at the beginning of crystallization to 0.56 GPa at the end.

The formation of porphyritic granites of the Lutsk-Araguba type is associated with the Pechenga complex. Porphyritic granites were crystallized out of the silicate magmatic melts at an initial temperature exceeding 845 °C and under the pressure of 0.65 GPa, exerted by the accompanying carbon dioxide fluid ($P_{CO_2}$). Granite veins of the later phases were crystallized at a temperature of 825° – 820 °C and under the fluid-phase pressure ($P_{CO_2}$) of 0.8 GPa. Carbon dioxide solutions of high density, accompanying the formation of these granites, initiated muscovitization and silification of metamorphic rocks of the Archean complex (regional greisenization). Geochemically, this process has led to significant enrichment of plagiogneisses in $Na_2O$ and partially in $Al_2O_3$, $SiO_2$, Pb, Ba and B against the background of a distinct decrease in content of $\Sigma$Fe, MgO, $TiO_2$, MnO, V, C, Ni, Sc, and La.

Diaphthoresis processes (chloritization and carbonatization) occurring in rocks of the Archean and Proterozoic complexes were induced by the action of the substantially chlorine sodium solutions with a homogenization temperature ranging from 380° to 280 °C. In the cataclasis zones, linked with the late faults, rocks were subject to the action of bitumen-hydrogen-hydrocarbon solutions (mainly by methane solutions). The simultaneous manifestation of this process in rocks of the Archean and Proterozoic complexes is an indication of the presence of the common deep source of hydrocarbon solutions.

### 1.4.3 Geochemical Section of the Earth's Crust

*Metasedimentary Rocks.* The Proterozoic complex is subdivided into the following two lithological-geochemical formations, which is in general accordance with an established distribution pattern of $SiO_2$, $\Sigma$Fe, $K_2O$, Li, and Sr in the rocks: (1) volcanic-terrigenous (Zhdanov suite) and (2) carbonate-terrigenous (Luchlompol and Kuvernerinyok suites). The first formation is characterized by a low $SiO_2$ and $K_2O$ content, and a high $\Sigma$Fe, Li, Sr and Gi content; the amount of $SiO_2$ and Sr gradually rises and the amount of $\Sigma$Fe and Cr decreases down the section, caused by the transition of metapelites and metatuffites dominating in the upper part of the section into metasandstones, mostly developed in the lower part. The second formation is characterized by a high $SiO_2$ and $K_2O$ content, decreased $\Sigma$Fe content and minimal Li and Sr content. The Televin suite is characterized by a high $SiO_2$ content, increased Li and Sr content, and minimal $\Sigma$Fe content.

In passing to the Archean complex, the most abrupt changes are established for $SiO_2$, $\Sigma$Fe, Sr, and Cr. On the whole, metasedimentary rocks of the Archean complex are characterized by a reversed coupling between $SiO_2$ and $\Sigma Fe$, and by direct coupling between $\Sigma Fe$ and $K_2O$, Li, and Cr.

The Archean metaterrigenous strata are also distinguished by the monotonous pattern of distribution of the elements studied throughout almost the entire depth of the section. A considerable difference is seen only in the fifth layer (decreased amount of $\Sigma Fe$ and increased amount of Sr and Cr), which is accounted for by the appearance of amphibole- and epidote-bearing varieties in its composition (admixture of the primary material in the initial sediments).

*Meta-Effusive Rocks.* The distribution trend of $SiO_2$, $\Sigma Fe$, $K_2O$, Li, Cr, and Zr in sections of the Proterozoic effusive suites fully supports a geological-geochemical subdivision of these rocks into two formational series corresponding (from top to bottom) to the picrite-basalt and andesite-basalt stages of volcanism. By correlating these series, one can see that, with depth increase, the content of $SiO_2$, $K_2O$, Li continually rises and the content of $\Pi Fe$ falls. The observed irregular distribution of Cr and Zr in some suites (Matertin and Pirttiyarvi) is caused by transition of some types of magmatic differentiation into others (see Fig. 1.89).

At the contact between the Proterozoic and Archean complexes, the content of $SiO_2$, $K_2O$ and Zr increases and the $\Sigma Fe$ content decreases; this is a result of an abrupt transition from Proterozoic basic to Archean acid metavolcanites. The Archean meta-effusive strata are characterized by a poorly expressed distribution pattern of the elements studied throughout the entire section. Down the section, $SiO_2$ and Li content slightly increased and Zr content decreases; fluctuations in $K_2O$ content indicate that plagiogneisses (potassium feldspar ($K_{fs}<5\%$) and gneisses ($K_{fs}>5\%$) alternate in the section.

Analysis of the distribution of the total acidity ($Ac$)[7] in the SG-3 well section shows that the Proterozoic complex is characterized by a rhythmic alternation of the basic meta-effusive rocks ($Ac<1$) with metasedimentary rocks ($Ac>1$). It should be noted that the constant acidity level is inherent to the homogeneous andesite basalt and tholeiite basalt formations, whereas in the trachybasalt and picrite basalt formations the acidity firstly decreases down the section and then rises (for instance, in the Matertin suite), or vice versa (Pirttiyarvi suite). These non-uniformities in distribution of $Ac$ in meta-effusive suites (as in the metasedimentary rocks) are associated with the different character of the magmatic or sedimentary differentiation respectively.

On the whole, the total acidity of the Archean rocks is caused by the presence of basic effusive rocks which dominate in the section (70%); these effusives are characterized by two distinct levels of $Ac$: the low level (0.50 – 0.85 in the Nickel series) and the higher level (0.80 – 1.00 in the Luostari series). Rocks of the underlying Archean complex are characterized by an even higher level of acidity. Distribution of $Ac$ in these rocks exhibits a rhythmic character caused by the alternation of the primarily metaterrigenous and meta-effusive strata.

---

[7] Parameter $Ac$ is the quantitative evaluation of the total acidity of the igneous rocks; when it is calculated, all components determined with the help of the chemical analysis, with the exception of volatile components, are used (Borodin, 1981).

Thus, a step-like increase of the total acidity is established down the SG-3 well section: 0.5 – 0.8→0.8 – 1.0→1.1→1.6. This tendency is accompanied by a corresponding decrease in rock density ($\rho$gm cm$^{-3}$): 2.95 – 3.1→2.75 – 2.90→2.65 – 2.80. The distribution pattern of the longitudinal wave velocities ($Vp$) measured in corresponding rock samples is as follows: in the upper part of the section their values are in direct relationship with the rock density, and are 6.0 – 6.7 km s$^{-1}$; further down the section, the observed interrelation between changing values of $Ac$ and $\rho$ is not accompanied by regular changes of $Vp$, which is explained by the increase in porosity and fissuring of rocks of the Pechenga and Archean complexes.

The distribution of Na/K ratio values versus $Ac$ in the well section is of differing types: in the Proterozoic complex there is an indirect relationship between these indices, whereas in the Archean complex the relationship is direct. The different types of relationship reflect differences in the mineral composition of these two complexes. The total alkalinity of the rocks ($Na_2O + K_2O$) increases step-wise with depth: 1.1 – 3.6% (0 – 4.5 km)→3.9 – 7.5% (4.5 – 6.8 km)→5.0 – 9.5% (6.8 – 11.6 km).

The behavior of Th in the section is closely related to rock acidity, particularly in the Proterozoic complex, exhibiting a rhythmic character synchronous with $Ac$: in meta-effusives of the Nickel series 0.5 – 2.5 g t$^{-1}$, in metasedimentary rocks 1.0 – 6.0 g t$^{-1}$, in the Luostari series 1.5 – 6.0 and 5.5 – 8.0 g t$^{-1}$, respectively. In rocks of the Archean complex the content is generally higher (2.5 – 9.0 g t$^{-1}$), particularly in its upper part (up to 17 g t$^{-1}$).

As distinct from the radioactive Th, $\Sigma$TR trends in the SG-3 well section are less expressed and exhibit a close indirect relationship only with the ratio Na/K in the Proterozoic complex. Of particular importance are the data obtained for the first time on the distribution of the bound water in metamorphic rocks of the Pechenga complex and Kola series. The behaviour of $H_2O^+$ in the upper part of the Pechenga complex (Nickel series) differs sharply from that in the lower part of the complex (Luostari series) and in the underlying Archean complex. In the Nickel series the $H_2O^+$ content decreases successively in a rhythmic manner down the section in each of the suites: 5.0 – 3.2% in the Matertin suite, 5.5 – 4.5% in the Zhdanov suite, 4.8 – 3.0% in the Zapolyarny suite. In the Luostari series and underlying the Archean complex, the $H_2O^+$ content is constant, although much lowered: 2% in the Luostari series and ≤1% in the Archean complex; the latter also exhibits a distinct rhythmic character in the distribution of $H_2O^+$ in metaterrigenous (1 – 5%) and meta-effusive (0.5%) strata. This tendency to decrease of $H_2O^+$ with depth reflects the distinctive nature of the metamorphism of the Pechenga structure at the time of its origin and during its further development on the Archean basement. It has already been stated that a prograde metamorphism of the Proterozoic rocks began at $T = 400\,°C$ and continued at progressively increasing temperatures up to 500° – 550 °C in conditions of the hydrodynamically closed system which was formed when the metamorphosing andesite basalts of the Luostari series were sinking into the Archean basement and then covered by thick (1.9 km) tholeiite sheets of the Nickel series. At that time, lithostatic pressure did not exceed 0.08 – 0.12 GPa. An increase of temperature by 100° – 150 °C under quasi-isochore conditions should lead to the dehydration of highly hydrous minerals and the release of a great volume of free water; this should be accompanied by a simultaneous increase in $P_{fl}$ (pressure of fluids) up to 0.15 – 0.30 GPa. The developing $P_{fl}$ exceeds not only $P_{sd}$ (pressure of solid rocks) but also the limit of the tensile strength of the metamorphosed rocks (0.04 – 0.05 GPa); in the closed system this should inevitably lead to an abrupt increase in porosity as a result of microfracturing (Fyfe et al. 1981), and, consequently, to a decrease in *Vp*. A petrophysical examination of rocks from the SG-3 well section supports this assumption. For instance, the measured values of the closed porosity, which do not depend on technogenic factors, increase abruptly down the section in passing from the greenschists facies (0.5 – 1%) to the epidote-amphibolite facies (2 – 3%), and this tendency is observed throughout the entire thickness of the Luostari series and most of the Kola complex. From this it appears that a considerable part of the metamorphic fluid released became preserved in situ inside the disaggregated strata, and its remaining part, together with the disseminated metals extracted from the rocks, was displaced into more favourable parts of the basement, where higher concentrations of metals have been formed.

From this it appears that the disaggregation of rocks during the process of their dehydratation under closed system conditions is one of the reasons responsible for the observed seismic non-uniformity in the Earth's crust.

These litho-geochemical features of the metasedimentary rocks of the Pechenga complex permit the identification of three formations: (1) greywacke (crust of weathering) formation (Televin suite); (2) carbonate-terrigenous formation (Kuvernerinyok and Luchlompol suites); (3) volcanic-terrigenous formation (Zhdanov suite).

In the metasedimentary rocks of the ore-bearing Zhdanov suite, two groups of sulphide mineralization are recognized, i.e. the syngenetic and epigenetic groups. Data on the isotope abundance of sulphide suggest that sulphides of the epigenetic group and ores of the copper-nickel deposits of the Pechenga region are derived from one and the same source.

The meta-effusive rocks of the Pechenga complex are subdivided by their litho-geochemical features into four formations, i.e. andesite-basalt (Mayarvi suite), trachybasalt (Pirttiyarvi suite), tholeiite-basalt (Zapolyarny suite) and picrite (komatiite)-basalt (Matertin suite) formations.

Different geochemical trends in matter differentiation and their recurrence throughout geological time have been established for each volcanic formation, viz. a silicic trend in the lower member of pairs (in the andesite-basalt and tholeiite-basalt formations) and an alkali-silicic trend in the upper members of pairs [in the trachybasalt and picrite (komatiite) formations]. This recurrent character of matter differentiation and the information available on the radiological age of rocks indicate that two stages of volcanism, separated from each other by a time interval, can be distinguished in the Pechenga structure, i.e. (a) the andesite-basalt stage (Luchlompol series) and (b) the picrite-basalt stage (Nickel series).

The amount of silicic matter and the value of total alkalinity in the first portions of melts of each formation tend to decrease in the course of time; this petrochemical tendency of volcanites from the Pechenga complex is indicative of the antidromal character of volcanism within the limits of the Pechenga structure.

The following new scheme of the stratigraphical and formational subdivision of the Pechenga complex has been suggested (from bottom to top); the Luostari series comprises two sedimentary-volcanic rhythms (Mayarvi + Kuvernerinyok and Pirttiyarvi + Luchlompol suites) and is assigned to the carbonate-terrigenous trachybasalt-andesite-basalt formational series; the Nickel strata also consist of sedimentary-volcanic rhythms (Zapolyarny + Zhdanov and Matertin + Kolloyaur suites) and form the volcanic-terrigenous-picrite (komatiite)-basalt formational series.

The Proterozoic meta-intrusive formations are subdivided into three complexes, i.e. Pre-, Syn-, and Post-Matertin complexes. The Pre-Matertin complex, recognized for the first time, is represented by apogabbro-diabase amphibolites which are genetically related to a stage of andesite-basalt volcanism. The Syn-Matertin complex of gabbro-diabases and gabbro-wehrlites is comagmatic by geochemical parameters with metavolcanites of the Matertin suite; the insignificant differentiation of the material in the lower part of the suite and the rhythmic character of differentiation (from picrites to tholeiites) in its upper part show that the corresponding intrusive rocks have been formed in the same sequence: ore-free gabbro-diabases→nickeliferous gabbro-wehrlites. The Post-Matertin complex in the SG-3 well section is represented by a subvolcanic body of andesite-dacite porphyrites genetically related to andesite-dacite plagioporphyrites, which mark a zone of the Poryitash major fault at the surface.

The following two stages of Proterozoic metamorphic zonality have been established and characterized in detail: the first stage (low-temperature autometasomatic and metamorphic changes in the Luostari series) is related to the final stage of andesite-basalt volcanism; the second stage (epidote-amphibolite metamorphism of the Luostari series and zonal greenschist metamorphism of the Nickel series) has appeared in association with a stage of picrite-basalt volcanism. The physicochemical conditions of Proterozoic metamorphism and metasomatism have been reconstructed and the following stages characterized on the basis of temperature-pressure data: (a) prograde metamorphism of the epidote-amphibolite facies (560° – 680 °C); (b) zonal metamorphism of the greenschist facies occurring in the presence of the water-saline and gaseous solutions; salt concentration during this stage was about 30%; temperature in the well section varies as follows: from 190° to 340 °C in an interval of 0 – 1400 m (prehnite-pumpellyite subfacies); from 340° to 455 °C in an interval of 1400 – 2600 m (epidote-chlorite subfacies), and from 455° to 560 °C in an interval of 2600 – 4563 m (biotite-actinolite subfacies); (c) early diaphthoresis (liquid solutions with very high salt concentration and a temperature exceeding 460 °C); (d) potassium metasomatism of the lower part of the Proterozoic complex (carbon dioxide and highly dense solutions – $P_{CO_2} = 0.37 - 0.55$ GPa); (e) late diaphthoresis (chloride-sodium solutions at a temperature of 340° – 370 °C); (f) the latest cataclasis (bitumen-hydrogen-hydrocarbon, substantially methane solutions).

The results of the investigations made it possible to trace the evolution of the mineral composition of the Proterozoic basalts during a process of prograde metamorphism, and reveal the following principles in the behaviour of the main elements and elemental admixtures: (a) the water content increases sharply in the prehnite-pumpellyite zone (from 2.4 to 6.8%), decreases in the zone of greenschist facies to 2.5% and falls sharply (from 2.5 to 1.5%) at the boundary between the greenschist and epidote-amphibolite facies; (b) main elements, including potassium and sodium, in all zones of metamorphism display a character similar to the isochemical character; (c) elemental admixtures are notable for the presence of two zones where their initial concentrations are redistributed: the upper zone (a zone of full decomposition of the primary magmatic minerals), characterized by a withdrawal of chromium, nickel, ytterbium, boron, tin and lead from the system, and the lower zone (at the boundary between the low- and high-temperature zones of the epidote-amphibolite facies), which shows that the mobility of the elements increases during metamorphism.

The classification and nomenclature of the metamorphic rocks of the Archean complex has been compiled, based on the quantitative ratios between main rock-forming minerals. The main types of metamorphic rocks of this complex (plagiogneisses and migmatites, amphibolites and amphibole schists, Archean granitoids and Proterozoic porphyritic granites) have been geochemically identified.

It was probed that rocks of the Archean complex are characterized by the generally isochemical synmetamorphic migmatization which leads to partial withdrawal of K, Rb, F and other components from plagiogneisses; contrary to this, biotitization of amphibolites, proceeding synchronously with migmatization, is characterized by the inflow of the same elements, as also by the redistribution of the initial components (Si, Al, Fe, Ca). This means that biotitization can be considered as an autosomatic process, complementary to migmatization.

On the basis of the studies performed of mineral inclusions, a model of the Archean complex formation has been made and the P-T parameters evaluated.

The geochemical section of the Earth's crust thus constructed allowed the identification of geochemical trends in the metasedimentary and meta-effusive rocks at different depths, assessment of changes in total acidity (Ac) and alkalinity ($Na_2O + K_2O$) of rocks with depth increase and an evaluation of the behaviour of rocks depending on the presence of radioactive and rare-earth elements; this resulted in establishing the following principles:

a) a step-wise increase of total rock acidity down the section has been established; it reflects a peculiar feature of the primary composition of the Pechenga (Nickel and Luostari series) and Archean complexes: 0.5 – 0.8 (down to 4.5 km) →0.8 – 1.0 (4.5 – 6.8 km)→1.1 – 1.6 (6.8 – 11.6 km); there is also a synchronous step-wise increase of (1) total alkalinity ($Na_2O + K_2O$ = 1.1 – 3.6→3.9 – 7.5→5.0 – 9.5%), thorium content (0.5 – 6.0→1.5 – 8.0→2.5 – 17.0 $g\ t^{-1}$), (2) content of rare-earth and other elemental admixtures.

b) the different distribution pattern at the bound ($H_2O^+$) and free ($H_2O^-$) water has been established: the amount of $H_2O^+$ decreases successively down the section in the Nickel series (5.0 – 3.5→5.5 – 4.5→4.8 – 3.0%); the amount of $H_2O^+$ decreases abruptly in transition to rocks of the Luostari series (2%) and to the Archean complex (<1%); at the same time, the content of $H_2O^-$ remains practically constant throughout the section. All this has determined the peculiar features of the manifestation of the Proterozoic zonal metamorphism.

When studying the relationship between the chemical composition of rocks and their elastic-density properties, it was established that there are both direct and indirect relationships between the total acidity and density of rocks throughout the section; a direct relationship between the velocity of the elastic longitudinal waves and the density of rocks (and respectively, an indirect relationship between the velocity and density) is observed only to a certain depth. At lower levels this relationship is disturbed, i.e. an insignificant increase of density (by $0.1-0.2\ g\ cm^{-3}$) corresponds to an anomalously large increase of velocity values (by $1.5-2.4\ km\ s^{-1}$).

# 1.5 Zonality and Age of Metamorphism

The first detailed description of metamorphic rocks of the Pechenga complex was made by Zagorodny et al. (1964). Based on geological mapping data, metamorphic rocks were subdivided into the greenschist facies (comprising two subfacies), amphibolite-epidote and amphibolite facies. It was established that isogrades of regional metamorphism cross-cut the lithological boundaries and that temperatures of metamorphism increase from upper stratigraphic horizons to lower ones, i.e. from the greenschist to amphibolite facies. Narrow zones of diaphthorites of the greenschist facies were established against the general background of the regional metamorphism.

Duk (1977) recognizes two stages of metamorphism of the Pechenga rocks; the first is characterized by high-grade metamorphic zonality of the andalusite-sillimanite type, and the second by low-grade zonality of the kyanite-sillimanite type. Seven facies and subfacies are identified for the first stage of metamorphism: from the zeolite (assumed) and prehnite-pumpellyite to the epidote-amphibolite facies. Six facies were identified for the second stage, i.e. from the prehnite-pumpellyite facies to the facies of garnet amphibolites. Boundaries of the metamorphic zones of the first stage correspond to the stratigraphic boundaries whereas isogrades of the second stage cross both the stratigraphic boundaries and isogrades of the first stage.

According to Belyayev and Petrov (1980) and Belyayev et al. (1977), the zonality of metamorphism in the Pechenga structure is expressed in the form of isogrades of biotite, hornblende, garnet, staurolite, kyanite and monoclinic pyroxene (these latter three are observed only on the south-eastern end of the northern flank and on the southern flank of the Pechenga structure). Isogrades cross stratigraphical boundaries, although in general they follow the present-day configuration of the Pechenga structure; metamorphism is assigned to the kyanite-sillimanite type.

The Kola series was metamorphosed in Late Archean and Early Proterozoic time under conditions of the amphibolite facies, in places, under conditions of the granulitic facies. In the immediate vicinity of the Pechenga structure there is a zone of low-temperature amphibolite facies. This facies is replaced by the high-temperature amphibolite zone and then, in the Pechenga bay area and within the limits of the Lapland block, it is replaced by the granulitic facies.

This zonality of Late Archean (or Early Proterozoic) metamorphism is superimposed by the diaphthoresis which is synchronous with the middle Proterozoic prograde zonal metamorphism of the Pechenga complex. Diaphthoresis processes were mostly developed near contacts with the Pechenga structure and also in individual tectonic zones at some distance from this structure. The grade of diaphthoresis corresponds to the greenschist, epidote-amphibolite and to the low-temperature amphibolite facies with moderate pressures.

The metamorphism of rocks in the SG-3 well section was studied by researchers from different organizations. The compelling results of these investigations have demonstrated that the grade of metamorphism superimposed on the Pechenga complex increases down the section. Further manifestations of this metamorphism in rocks of different composition at great depth were examined, and signs of low-temperature retrograde processes were revealed. However, some differences of opinion arose as to the facies of metamorphism, age of metamorphism and conditions under which it appeared.

The vertical zonality and physicochemical conditions of metamorphism were identified on the basis of investigations carried out by A. A. Glagolev, V. I. Kazansky, K. V. Prokhorov, and V. L. Rusinov. The age of zonal metamorphism was identified by using results of studies conducted by V. A. Maslennikov, S. N. Voronovsky, and L. N. Ovchinnikov.

### 1.5.1 Vertical Metamorphic Zonality

The problem of metamorphism zonality in the Kola superdeep well was investigated by means of documentation of prepared special core samples, comparative study of metamorphic minerals by micro-X-ray spectral analysis, and paragenetic diagrams constructed for different facies and subfacies. The results of these studies have been summarized for the SG-3 well section (Fig. 1.45). At the same time, the relationships between metamorphism and deformation of rocks were studied.

Since the Pechenga complex is predominantly composed of effusive, intrusive and tuffaceous rocks of basic composition and of rocks of medium composition similar in mineral association to the latter, the zonality of metamorphism can be established by taking metabasites as reference rocks. The underlying Archean complex also contains a sufficient amount of metabasites, which can also be taken as characteristic of the vertical zonal metamorphism throughout the drilled interval. The metamorphism of the sedimentary rock of the Pechenga complex was examined in less detail. When studying the Kola series, considerable emphasis was placed on examining the mineral associations of "highly aluminiferous" gneisses and granitization phenomena.

In all petrographical examinations mineral compositions were determined by the X-ray microanalyzer *Cameca MS-46* (analyzed by N. B. Troneva and B. A. Boronikhin). Minerals such as chlorite, epidote, biotite, amphibole, mica, etc., occurring throughout the section, were systematically studied. More than 200 microprobe analyses of minerals were made. The results of these analyses are presented on the composition-paragenesis diagrams, partially given in Tables 1.12 and 1.13, and were used to recognize and describe facies and subfacies.

A microanalyzer determines only the total iron content which is given in the tables in the form of FeO or, less frequently, $F_2O_3$. The following coefficients are presented in the tables and used in the text: $f = Fe + Mn/Fe + Mn + Mg$ for minerals in which Fe is contained mainly in the form of FeO; $F = Fe/Fe\text{-}Al$ for min-

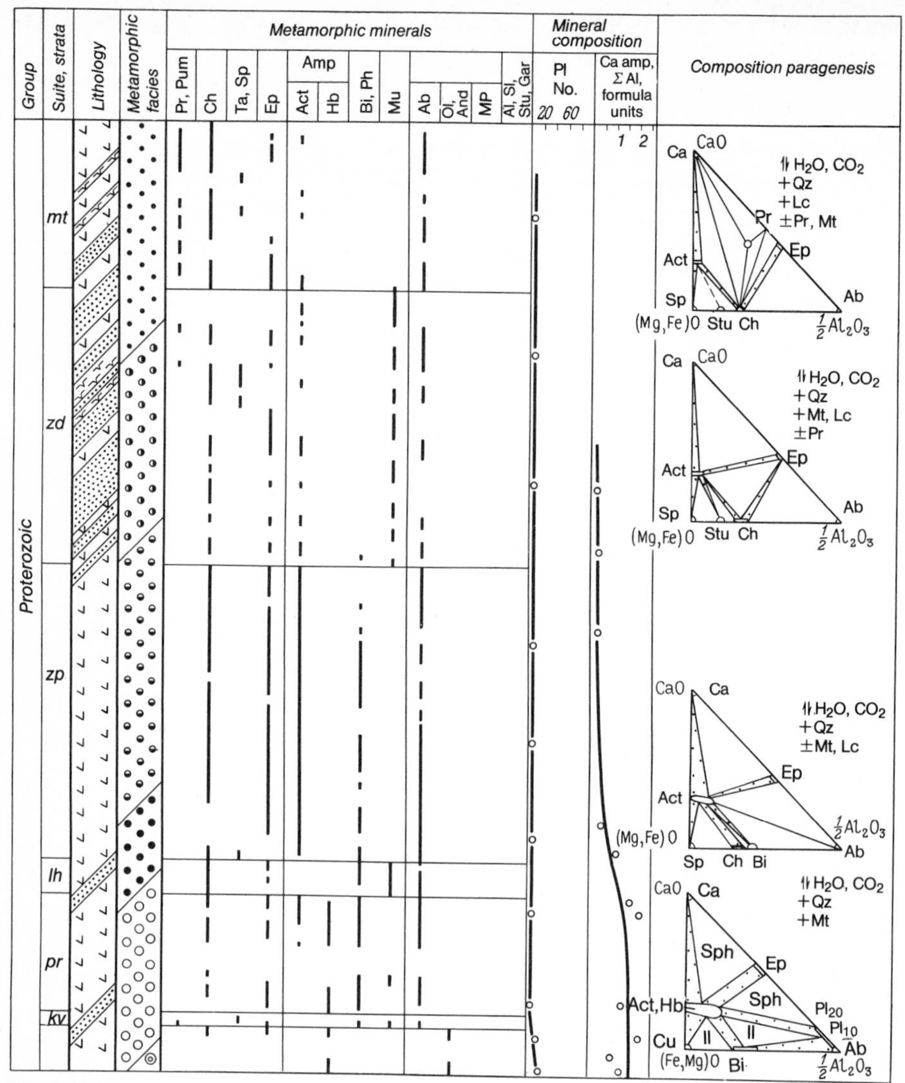


**Fig. 1.45 a**

erals where $Fe_2O_3$ predominates, and $Al = A_2O_3/Fe\text{-}Al$ for minerals where $Fe_2O_3$ predominates, and $Al = A_2O_3/A_2O_3 + SiO_2$ for chlorites.

As a result of the investigations, four facies of metamorphism have been identified within an interval from 0 to 11,500 m: I prehnite-pumpellyite facies (0–1400 m); II greenschists facies (1400–4900 m); III epidote-amphibolite facies (4900–6000 m); IV amphibolite facies (below 6000 m).

The greenschist facies is also subdivided into three subfacies: epidote-chlorite (1400–3200 m), biotite-actinolite (3200–4340 m) and biotite-amphibole (4340–4900 m). Additionally, in the lower part of the Pechenga complex and in the Kola series, manifestations of low-temperature retrograde metamorphism

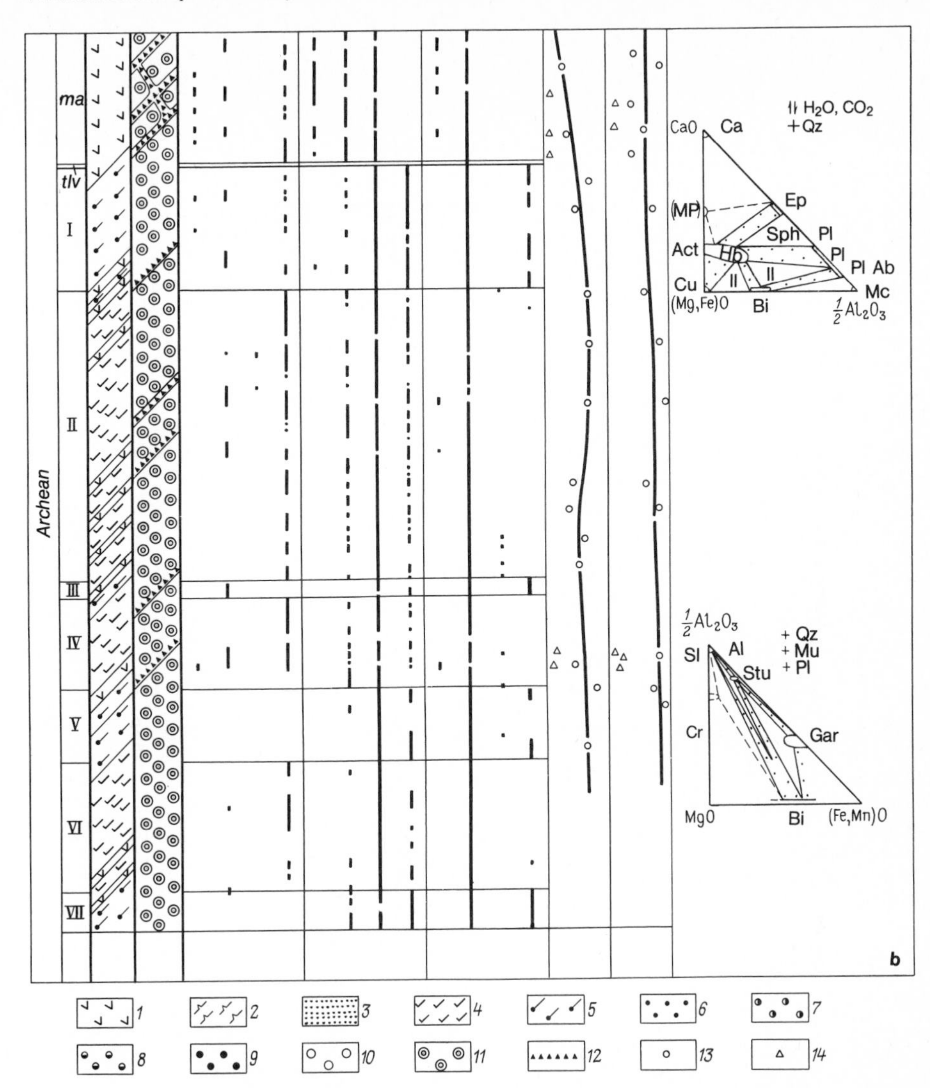


**Fig. 1.45 a, b.** Vertical zonality of metamorphism in SG-3 well section. (After A. A. Glagolev, V. I. Kazansky, V. L. Rusinov). *1–5* metamorphic rocks (*1* metabasites; *2* meta-ultrabasites; *3* metasedimentary rocks; *4* biotite plagiogneisses; *5* sillimanite gneisses); *6–11* metamorphic facies of the prograde metamorphism; *6* prehnite-pumpellyite; *7–9* greenschist (subfacies: *7* epidote-chlorite; *8* biotite-actinolite; *9* biotite-amphibolite); *10* epidote-amphibolite; *11* amphibolite; *12* regrade greenschist facies; compositions of metamorphic minerals; *13* of prograde stage; *14* of regrade stage. On diagrams "composition – paragenesis" are shown: mineral associations of metabasites for the prehnite-pumpellyite, greenschist and epidote-amphibolite facies; mineral associations of metabasites (*top*) and gneisses with HAM (*bottom*) for the amphibolite facies

**Table 1.12.** Chemical composition of some minerals of prograde metamorphism

| Components | Prehnite | Pumpellyite | Stilpnomelane | | Epidote | Clinopyroxene | | Cummingtonite | | Muscovite | | Staurolite | | Garnet | |
|---|---|---|---|---|---|---|---|---|---|---|---|---|---|---|---|
| | 1 | 2 | 3 | 4 | 5 | 6 | 7 | 8 | 9 | 10 | 11 | 12 | 13 | 14 | 15 |
| $SiO_2$ | 46.55 | 38.68 | 46.26 | 46.30 | 40.47 | 52.61 | 52.04 | 54.23 | 56.17 | 45.03 | 47.66 | 24.73 | 27.61 | 37.21 | 38.60 |
| $TiO_2$ | – | – | – | – | 0.03 | 0.06 | 0.04 | – | 0.11 | 0.25 | – | 0.55 | 0.57 | – | – |
| $Al_2O_3$ | 24.83 | 18.01 | 7.11 | 6.88 | 26.91 | 1.65 | 1.13 | 0.24 | 0.59 | 38.56 | 36.65 | 54.70 | 52.50 | 20.32 | 21.56 |
| $Fe_2O_3$ | 0.74 | 14.76 | – | – | 8.87 | – | – | – | – | – | – | – | – | – | – |
| FeO | – | – | 30.89 | 30.45 | – | 11.61 | 15.09 | 24.10 | 27.18 | 0.71 | 0.65 | 17.47 | 14.27 | 32.62 | 31.08 |
| MnO | – | – | 0.79 | 0.99 | 0.12 | 0.42 | 0.46 | 0.75 | 0.37 | – | 0.45 | 0.37 | 0.33 | 0.65 | 2.07 |
| MgO | – | 1.26 | 4.49 | 4.78 | 0.03 | 10.95 | 8.53 | 18.95 | 14.92 | 0.28 | 0.45 | 1.45 | 1.56 | 2.56 | 2.67 |
| CaO | 29.22 | 22.18 | 0.04 | – | 23.53 | 23.48 | 22.41 | 0.65 | 0.60 | 0.01 | – | 0.03 | – | 3.18 | 5.29 |
| $Na_2O$ | – | – | 0.28 | 0.22 | 0.25 | 0.56 | 0.47 | 0.13 | 0.07 | 0.79 | 0.80 | 0.07 | – | – | – |
| $K_2O$ | – | – | 1.40 | 2.17 | 0.05 | – | – | – | – | 9.54 | 10.13 | – | – | – | – |
| Total | 101.34 | 94.89 | 91.26 | 91.79 | 100.26 | 100.34 | 100.17 | 99.05 | 100.01 | 95.17 | 96.79 | 99.37 | 96.84 | 101.53 | 101.27 |

1 – large-foliated variety from the prehnite-quartz zone in metabasite; associated minerals – quartz, chlorite, albite, relicts of clinopyroxene (sample No. 1849, depth 757 m); 2 – fine-grained masses in plagioclase of metabasite; associates with prehnite, chlorite, quartz, albite (sample 117, depth 70,5 m); 3 – light green with brown shade; aggregates of thin nonoriented plates in association with chlorite, albite, epidote and quartz in essexite gabbro-diabase (sample K4480, depth 1367 m); 4 – radiated or sheaf-like aggregate of thin plates in metabasites: associates with biotite, actinolite, chlorite, albite (sample 10320, depth 2699 m); 5 – from actinolitized essexite gabbro-diabase; associates with stilpnomelane, ferruginous chlorite, albite, microcline and leucoxene (sample 4480, depth 1367 m); 6 – light grey, from epidote amphibolite; associates with hornblende, epidote, plagioclase (labradorite, bytownite), quartz and sphene (sample 24824, depth 7652.6 m); 7 – colourless, from poorly altered amphibolite; associates with hornblende, actinolite, epidote, plagioclase (albite) and chlorite (sample 36795, depth 9936.3 m); 8 – small prisms in chlorite-amphibole-quartz schist; present together with chlorite, biotite, light-coloured hornblende and plagioclase (sample 19391, depth 5874 m); 9 – from garnet-cummingtonite amphibolite; associates with hornblende, basic plagioclase, garnet and biotite (sample 38167, depth 10253.8 m); 10 – large-foliated, from highly aluminiferous gneisses; associates with garnet, staurolite, andalusite, plagioclase and quartz (sample 22799, depth 6983.1 m); 11 – from garnet-biotite gneiss; present in rock with garnet, biotite, plagioclase and quartz (sample 38569, depth 10480.7 m); 12 – from highly aluminiferous gneiss; associates with garnet, biotite, andalusite, plagioclase, muscovite and quartz (sample 22799, depth 6983.1 m); 13 – from highly aluminiferous gneiss; associates with garnet, biotite, plagioclase, muscovite, sillimanite and quartz (sample 24056, depth 7421.2 m); 14 – from highly aluminiferous gneiss; associates with biotite, staurolite, quartz, plagioclase and sillimanite (sample 24051, depth 7419.4 m); 15 – marginal part of a grain depleted in Mn and slightly enriched in Mg as compared with its central part (from cummingtonite-garnet amphibolite); associates with hornblende, plagioclase (bytownite No. 87), (cummingtonite and biotite).

**Table 1.13.** Chemical composition of some minerals from zones of retrograde metamorphism

| Components | Prehnite | | Pumpellyite | | Epidote | | Actinolite | | Muscovite | | Chlorite | | Plagioclase | |
|---|---|---|---|---|---|---|---|---|---|---|---|---|---|---|
| | 1 | 2 | 3 | 4 | 5 | 6 | 7 | 8 | 9 | 10 | 11 | 12 | 13 | 14 |
| $SiO_2$ | 40.15 | 44.50 | 39.29 | 37.51 | 37.21 | 39.85 | 52.99 | 51.18 | 47.67 | 46.14 | 26.11 | 27.08 | 69.45 | 67.57 |
| $TiO_2$ | 0.15 | – | – | – | – | 0.04 | – | 0.41 | 0.65 | 0.88 | – | – | – | – |
| $Al_2O_3$ | 20.48 | 24.88 | 23.64 | 23.45 | 23.75 | 22.38 | 3.01 | 1.95 | 30.31 | 30.77 | 20.51 | 20.98 | 18.28 | 20.81 |
| $Fe_2O_3$ | 1.68 | 0.35 | 12.19 | 7.24 | 15.16 | 14.18 | – | – | – | – | – | – | – | – |
| FeO | – | – | – | – | – | – | 18.16 | 21.53 | 4.53 | 3.06 | 29.44 | 24.50 | – | – |
| MnO | 0.03 | – | 0.10 | 0.14 | – | 0.17 | 0.31 | 0.38 | 0.04 | – | 0.40 | 0.27 | – | – |
| MgO | 0.27 | – | 0.30 | 2.10 | – | – | 12.05 | 9.79 | 2.33 | 0.99 | 12.84 | 13.98 | – | – |
| CaO | 26.63 | 27.82 | 22.83 | 23.06 | 23.35 | 23.22 | 11.74 | 11.87 | 0.09 | 0.06 | – | 0.06 | 0.89 | 1.97 |
| $Na_2O$ | – | 0.57 | 0.35 | – | 0.03 | 0.12 | 0.27 | 0.96 | 0.28 | 0.19 | – | 0.09 | 10.78 | 10.99 |
| $K_2O$ | – | 0.18 | 0.02 | – | – | – | 0.24 | 0.77 | 11.01 | 10.68 | – | 0.10 | 0.04 | 0.13 |
| Total | 89.39 | 98.30 | 98.72 | 93.50 | 99.50 | 99.96 | 98.77 | 98.84 | 96.91 | 92.77 | 89.30 | 87.06 | 99.44 | 101.47 |

1 – from altered amphibolite; associates with chlorite, epidote, actinolite, albite, sericite and low-temperature potassium feldspar (sample 37184, depth 9999.5 m); 2 – from altered granite-gneiss; associates with albite, quartz, epidote, chlorite, potassium feldspar and relict biotite (sample 37186, depth 10,000 m); 3 – greenish-grey, from micaceous polymict sandstone; associates with quartz, microcline, carbonate, epidote, phlogopite and muscovite (sample 19352, depth 5713 m); 4 – light greenish-blue, from altered schistose amphibolite; associates with albite, actinolite and epidote (sample 20679, depth 6396.6 m); 5 – pale yellow, from altered foliated amphibolite; present together with actinolite hornblende, albite and pumpellyite (sample 20679, depth 6396.6 m); 6 – colourless, from altered amphibolite; associates with actinolite, chlorite, albite and secondary potassium feldspar (sample 37250, depth 10,020 m); 7 – light green, from chloritized and albitized foliated amphibolite, associates with epidote and secondary potassium feldspar (sample 19652, depth 6094 m); 8 – pale-green, from altered amphibolite; associates with chlorite, albite, epidote and potassium feldspar (sample 37250, depth 10,020.2 m); 9 – from greisenized biotite gneiss; associates with quartz, albite, biotite and carbonate (sample 31170, depth 8725.3 m); 10 – from poorly altered gneiss; associates with chlorite, microcline, biotite and quartz (sample 35697, depth 9552.4 m); 11 – from altered gneiss; associates with muscovite, epidote, plagioclase and quartz (sample 30492, depth 8644.4 m); 12 – from poorly altered amphibolite; associates with hornblende, epidote, albite, secondary potassium feldspar and quartz (sample 37818, depth 10,166.1 m); 13 – from altered foliated amphibolite; associates with actinolite, chlorite, potassium feldspar and epidote (sample 19652, depth 6094 m); 14 – from altered amphibolite; associates with hornblende, epidote, chlorite, potassium feldspar (sample 37818, depth 10,166.1 m).

were established; it is clear that this metamorphism was superimposed on the high-temperature mineral associations of the amphibolite facies.

The zonal metamorphism is considered to be prograde in character in the Pechenga complex. According to age determinations it is 2100 – 1500 Ma old. In rocks of the Kola series it is superimposed on the earlier metamorphism of the granulitic facies. A description of the relict mineral associations of the granulitic facies has already been given (see Chap. 1.3). The question of stages of metamorphism is considered in a later chapter (see Chap. 1.10), taking into account studies performed both in the well and at the surface.

*The Prehnite-Pumpellyite Facies.* The temperature boundaries marking the limits of this occurrence were not clearly defined and many assumptions have been made in geological literature about this problem. Coombs (1961), who recognized this facies for the first time, was of the opinion that its lower temperature limits is determined by the equilibrium state between laumontite and prehnite (when temperature increases, laumonite is decomposed):

$$3.3 \text{ lau} = \text{pr} + 3.3 \text{ Q} + 2 \text{ mont} + 5.1 \text{ H}_2\text{O} .$$

According to the experimental data of L. P. Gurevich, this equilibrium corresponds to a temperature of about 260 °C. A high-temperature limit of the facies is usually assigned to the boundary of pumpellyite occurrence or to a transitional zone where the pumpellyite-chlorite paragenesis is replaced by the epidote-actinolite paragenesis at a temperature of about 360 °C. The well has encountered only the middle and lower parts of the facies (i.e., temperature conditions of the facies formation), but we failed to find a zone transitional to the "zeolite part" of the facies. Zeolites were not found in the core samples at all. It is characteristic of the rocks studied that they contain only a small amount of pumpellyite, which is irregularly distributed in them. This causes difficulties in determining the lower boundary of the facies with proper accuracy. The greatest depth at which pumpellyite in metabasites is met is 1315 m where it occurs in paragenesis with stilpnomelane and epidote, and corresponds to the pumpellyite-chlorite-quartz or pumpellyite-stilpnomelane subfacies (Dobretsov et al. 1972). At a higher level, at a depth of 992 m, pumpellyite is associated with prehnite and chlorite. Prehnite is traced only to a depth of 1000 m. A transitional zone from the prehnite-pumpellyite to the greenschist facies was assigned to a depth of 1400 m, that is, to the bottom of the essexite-gabbro sill from which a sample with pumpellyite has been extracted and then examined (depth 1315.5 m).

Comparison of data obtained in studying core samples from the SG-3 borehole with the material provided by satellite wells 1885 and 1886 has showed that a lower boundary of the prehnite-pumpellyite facies goes up to the surface in a north-eastern direction, together with contacts of rocks of different lithology. In well 1886 it corresponds to a depth of about 800 m, and in well 1885 to a depth of 360 m (Fig. 1.46). Lines of approximately the same degree of the monoclinic pyroxene replacement by actinolite follow almost parallel to the lower limits of the facies; the configuration of these lines is similar to that of isotherms of metamorphism. The degree of actinolitization increases down the section. Unaltered

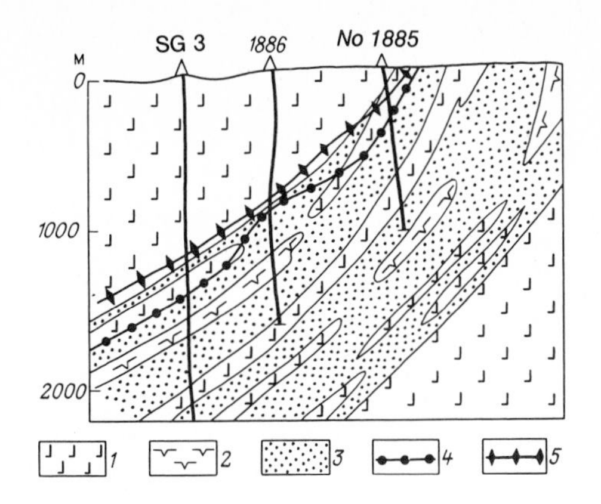


**Fig. 1.46.** Lithological section of the SG-3, 1886 and 1885 wells. *1* metabasites; *2* meta-ultrabasites; *3* metasedimentary rocks; *4* boundary between prehnite-pumpellyite (*above*) and greenschist (*below*) facies; *5* boundary at which clinopyroxene is replaced by actinolite

pyroxene is traced to the following depths: 1100 m in the SG-3 well, 800 m in well 1886 and 200 m in well 1885; a high degree of replacement is fixed in these wells from a depth of 1700, 1600 – 1650, and 900 m, respectively.

In metabasites of the prehnite-pumpellyite facies, textures and some minerals of the primary magmatic rock are well preserved. Secondary minerals constitute from 10 to 70% of the rock bulk. Plagioclase is considered to be the least preserved of the magmatic minerals in conditions of metamorphism; plagioclase impregnations are pseudomorphosed and fully replaced by albite, sometimes other minerals are also present. Monoclinic pyroxene, however, was not usually subject to alteration. The main metamorphic minerals are as follows (in order of abundance): chlorite, albite, prehnite, stilpnomelane, pumpellyite, epidote, leucoxene, quartz (small amount), at places calcite and actinolite. Pyrrhotite is a widespread mineral. The diagnosis of some minerals was confirmed by X-ray structural analysis data. The degree of rock alteration is non-uniform. Metamorphism has affected the rocks most intensively in zones of increased cracking, distinguished by the presence of many quartz-carbonate veins and traces of intense prehnitization; it should be noted that both main rock mass and impregnations of pyroxene and plagioclase were subject to replacement.

Parageneses of minerals in the prehnite-pumpellyite facies are characterized by the stable paragenesis of chlorite with prehnite in upper intervals and of pumpellyite with epidote and with stilpnomelane near the lower boundary of the facies. The paragenesis of pumpellyite with stilpnomelane usually appears in highly ferruginous rocks and is accompanied by ferruginous chlorite.

Pumpellyite is represented by fine-grained short-prismatic crystals, sometimes by spherulites. It is dark green in a core sample and bright green in thin sections. Pleochroism is usually highly developed (from bright green to colourless) but sometimes it is indistinctly seen due to the fine-grained fabric of an aggregate. An optical diagnosis was confirmed by X-ray structural analysis. According to results by electronic microanalyzer (see Tables 1.12 and 1.13), the total content of iron (F) in this mineral proved to be 0.34. In the majority of the published analyses the total content of $Fe_2O_3$ and $Al_2O_3$ varies from 24 to 28%, whereas in pumpellyite the total content is 32.77%. From this it may be assumed that this mineral contains at least 4 – 4.5% FeO.

The iron content in stilpnomelane, which frequently associates with pumpellyite, is about 0.79 (see Table 1.12). Its refraction indices, measured in larger lamella [$Ng = 1.592$, $Np = 1.545$, $Ng = Np$ (calculated) $= 0.047$] correspond most probably to ferruginous stilpnomelane with

$$\frac{Fe^{3+} + Al}{Fe^{2+} + Mn + Mg + Fe^{3+} + Al} = 0.25 .$$

Epidote is represented by non-zonal crystals of elongated-prismatic habit or, in rarer cases, by small rose-like growths of crystals. Normally, it is colourless in thin sections but possesses non-uniform optical properties inherent to ferruginous varieties: $2V$ from $-68°$ to $-78°$ $Ng' = 1.732 - 1.740$. The content of iron in epidote is equal to 7.88 – 9.18% which corresponds to $F = 0.16 - 0.18$. The mineral was seen in the form of aggregates with chlorite, prehnite, stilpnomelane, less frequently with calcite, and sometimes with pumpellyite. Epidote in the epidote-calcite veins is characterized by a higher iron content ($F = 0.20 - 0.24$).

Chlorite is developed only in the main mass and represented by delessite-diabantite. It is pale green to green in colour with slight pleochroism; elongation is positive, the optical sign is negative. The iron content varies within the range of 0.35 – 0.73, aluminum content (Al) changes from 0.48 to 0.50. The highest iron content is characteristic of chlorites in paragenesis with stilpnomelane and pumpellyite. According to electronic microprobe data and as a result of applying a method of thin section colouring by complex colour, it was established that carbonates are represented by calcite and iron-poor calcite. The late carbonate veinlets are composed of pure calcite, whereas metasomatic carbonate excretions, disseminated in rocks, are constituted of both varieties.

Within the prehnite-pumpellyite facies sedimentary rocks are usually represented by phyllites. They are characterized by the following paragenesis: chlorite + sericite + carbonate + quartz + pyrrhotite. Organic matter is frequently present in great amounts. Prehnite was discovered in tuff layers which occurred among phyllites. The intensity of alteration of the metamorphic sedimentary rocks is much less than that of the basic volcanites. Small bodies of ultrabasites in an interval of 350 – 602 m are transformed into serpentinous, chlorite-serpentinous, thremolite-serpentinous, talc-thremolite rocks containing relict clinopyroxene. It may be assumed that metamorphism of the ultrabasites was also developing under conditions of the prehnite-pumpellyite facies. However, any attempts to separate it from other possible transformations of the ultrabasic rocks on the basis of core examination have failed.

The greenschist facies is developed in an interval of 1400 – 4900 m and subdivided into three subfacies.

The epidote-chlorite subfacies is characterized by an active replacement of pyroxenes by actinolite. Beginning from a depth of about 2000 m, pyroxene is practically completely actinolitized. This replacement results in the formation of pseudomorphs, although the primary porphyry texture of magmatic rocks is left unchanged. Epidote, chlorite, quartz, in places stilpnomelane and carbonate are developed in the main rock mass and in pyroxene grains; actinolite is devel-

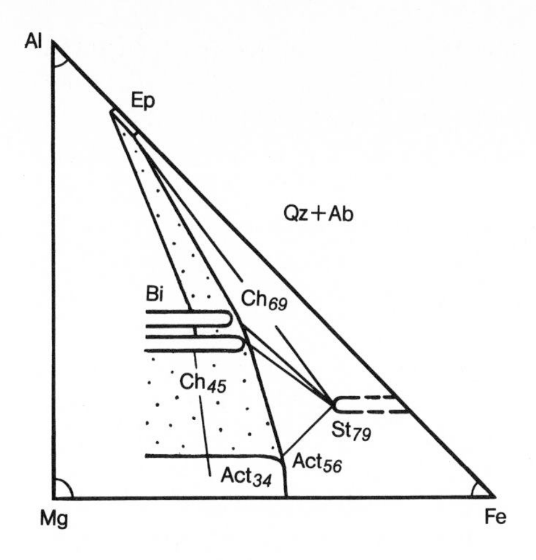


**Fig. 1.47.** Diagram "composition – paragenesis" of metabasites from the greenschist facies. *Symbols* denote ferrum content of minerals

oped in pyroxene grains. Instead of a paragenesis of pumpellyite + stilpnomelane, one can observe such parageneses as epidote + actinolite and epidote + stilpnomelane which are very stable here (Fig. 1.47). Stilpnomelane does not differ in composition from analogous mineral characteristic of the prehnite-pumpellyite facies. In this association epidote possesses a higher iron content (0.20 – 0.25) and, respectively, a higher reflection index (*Ng* is up to 1.744) and small angle $2V = 75°$. These minerals associated with highly ferruginous chlorite ($F = 0.69 - 0.73$); actinolite contains a small amount of aluminum, whereas its iron content is about 0.56. In associations deprived of stilpnomelane, the iron content in these minerals is much less. At places, actinolite and chlorite exhibit an irregular distribution of magnesium and iron in individual grains; in addition to actinolite, brown and green amphiboles are also present. For instance, in actinolite pseudomorphs in pyroxene there are small areas of irregular shape filled with brown titanium-bearing and dark green highly aluminiferous amphiboles. Differences in composition of these areas have been established with the microprobe. In shape they resemble relicts, and hence one may assume that actinolite, being a product of metamorphism, superimposes amphibole of a kaersutite type.

The biotite-actinolite subfacies is characterized by the intensive rock recrystallization which resulted in the disappearance of primary textures. A tendency is observed for comparatively uniformly grained textures to replace porphyry textures. This is considered to be an indication of the beginning of the granulation process. The iron content in biotite in paragenesis with actinolite and epidote is within the range of 0.37 – 0.58, whereas that of actinolite and epidote is 0.4 – 0.5 and about 0.25, respectively. Actinolite contains rather a high amount of aluminum. In some intervals (2100 – 2700, 3098 – 3104, 3310 – 3316 m) it displays chemical and optical non-uniformity: rims of brighter green colour appear around actinolite. These rims are composed of amphibole, which contains higher amounts of aluminum and iron than are in the core part of actinolite. These green zones with sharp boundaries are frequently seen along cracks inside actinolite grains. A distribution pattern of components (Fe, Al) in sections across these grains demonstrates that two amphiboles of different composition are in contact with each other. Aluminum-bearing amphiboles may originate as a result of the following reaction: $ep_{15} + bi + act = hb$. It was also observed that in the lower parts of this subfacies (3400 – 3800 m) ferruginous epidote was covered by a rim of clinozoisite. In other areas of a grain, epidote is usually of uniform fabric.

The study of mineral compositions has revealed a distinct correlation between values of iron content in the co-existing minerals. On a diagram Al-Fe-Mg (see Fig. 1.47) two groups of paragenesises are distinguished: (1) in ferruginous metabasites with stilpnomelane and (2) in metabasites with medium iron content. The first group is represented by an association of stilpnomelane with epidote, chlorite, actinolite and biotite, which contain the greatest amount of iron. The least iron content in chlorites in paragenesis with stilpnomelane and epidote is 0.69. Chlorites containing greater amounts of magnesium, associate either with actinolite or with biotite deprived of stilpnomelane. In iron-poor aluminiferous rocks chlorite ($F = 0.20$) occurs in stable state with muscovite and biotite ($F = 0.24$).

The biotite-amphibole subfacies is considered to be a transitional zone between the greenschist and epidote-amphibolite facies. It is composed of actinolite, actinolite hornblende, albite, epidote and biotite. Chlorite, quartz, in places calcite, are present in the form of admixtures (see Fig. 1.45). A new tendency is displayed in the evolution of the rock texture, i.e. collective recrystallization of actinolite takes place and large porphyroblasts are formed. The green rims described above are seen around actinolite, although the intensity of their development here is much higher. In places actinolite was fully transformed into green aluminiferous amphibole-hornblende containing up to 12% $Al_2O_3$. Along with the increase of aluminum in the marginal rim around amphibole, the content of iron and sodium (to a less extent) also rises. From this it appears that, in addition to actinolite, the described interval of the well section contains actinolite hornblende and normal hornblende. It may be that the appearance of hornblende is caused by the ferruginous nature of the rocks in which the iron content is somewhat greater. Epidote with a higher iron content, and ferruginous chlorite ($F = 0.50$) occur in association with hornblende rather than in association with actinolite in contiguous intervals.

As in the prehnite-pumpellyite facies, so in the biotite-amphibole subfacies, carbonates are represented mainly by calcite and iron-poor calcite. The iron content does not exceed a few percent. In some coloured thin sections grains of ferruginous calcite, cemented by pure calcite, were found. In other places they occur separately. Ferruginous calcite tends to develop near grains of hornblende, whereas pure calcite is met in plagioclase crystals, together with quartz grains, etc. However, this tendency was observed only in some thin sections, and therefore needs to be confirmed by studying a great amount of data. In some quartz pockets small crystals of dolomite were discovered. In a core sample extracted from a depth of 2866 m, larger segregations of dolomite-calcite aggregates were found.

In sedimentary rocks of the Zhdanov suite, metamorphic minerals of the greenschist facies are represented by sericite, muscovite, chlorite, and carbonate. Finely squamose sericite, characterized by double refraction of about 0.015, is met down to a depth of 2460 m. Below this depth rather large flakes of muscovite, characterized by high values of double refraction (0.025 – 0.030), are present. Metasedimentary rocks of the Luchlompol suite are assigned to biotite-muscovite schists. They have undergone recrystallization, which resulted in the formation of larger grains and in the redistribution of the carbonaceous matter. In arkose sandstones, in addition to mica, quartz and carbonate, potassium feldspar is also present. Feldspar is represented by microcline or orthoclase and constitutes fragments. Thin veinlets of potassium feldspar, which cross-cut albite grains, are also met. Everywhere albite is replaced by muscovite, and survives only in the form of a relict mineral, which is not stable in these conditions. Judging by its optical properties and X-ray microanalyses, light-coloured mica in an interval of 4809 – 4840 m is represented by phlogopite. Magnetite from the same interval is characterized by a rather high content of titanium (up to 6.87%).

Relict magmatic textures (poikilophytic or porphyric) and massive structures of meta-ultrabasites, encountered in an interval of 1540 – 1805 m, are partially preserved. However, in subcontact zones these textures disappear; simultaneously, foliated structures appear. Brown kaersutite and green actinolitic hornblende, formed at the expense of clinopyroxene during epimagmatic or early autometamorphic stage, are the earliest in origin among minerals of meta-ultrabasites. Kaersutite contains 0.24 – 0.57 formular units of Ti, the iron content is 0.22 – 0.40. Serpentine and chlorite are the most widespread minerals of meta-ultrabasites. Serpentine was not analyzed in detail; the iron content in it varies from 0.10 to 0.14. Chlorite is represented by a few varieties. According to Hey's diagram (Hey 1954) they are assigned to diabantite, jenkinsite, delessite and talc-chlorite with iron contents from 0.14 to 0.32. Chlorites with higher content of iron (0.40 – 0.44) are found less frequently. These minerals are of at least partially autometamorphic origin. There is no doubt that talc and tremolite belong to the metamorphic minerals of meta-ultrabasites. The composition of talc corresponds to its theoretical composition, with the iron content varying from 0.07 to 0.20. The iron content in tremolite ranges from 0.09 to 0.20, with the exception of individual samples, where the iron content in colourless amphibole is as high as 0.38. In one of the meta-ultrabasite samples (No. 6522 m, depth 1802 m), stilpnomelane was met in association with talc, chlorite, and calcite. Talc and chlorite are distinguished by the increased content of iron (0.20 – 0.42). However, iron content in stilpnomelane (0.55 – 0.58) is much less than that in metabasites (0.78 – 0.79). Metamorphism has affected not only ultrabasites, but also the associated copper-nickel mineralization. This topic is considered in more detail in a later chapter (see Chap. 1.6).

It should be emphasized that a part of the transitional section from the greenschists to epidote-amphibolite facies coincides with a thick zone of foliation which embraces the lower parts of the Zapolyarny suite, the entire sequence of the Luchlompol suite and the upper parts of the Mayarvi suite. In this zone a sharp anisotropy of newly formed structures and textures appears in metamorphic rocks (see Chap. 1.7).

*The Epidote-Amphibolite Facies.* The upper boundary of this facies coincides approximately with the top of Pirrtiyarvi suite (about 4900 m), the lower boundary is placed inside the Mayarvi suite (about 6000 m). Rocks are represented by uniform foliated amphibolites and, in much rarer cases, by massive varieties.

Apart from amphibole and plagioclase, also biotites and magnetite constitute both rock types; sphene, in places quartz, epidote, tourmaline and carbonate are present in the form of admixtures. Plagioclase is usually represented by albite, but in some areas the basicity of plagioclase increases (Nos. 10 – 15). Sometimes amphibolites of this facies contain cummingtonite (depth 5874 and 5968 m), showing that these rocks are undersaturated with CaO. The diagram composition-paragenesis is shown in Fig. 1.45. Despite the general uniform character of metamorphism inherent to this facies, there are some intervals where either properties of minerals change or paragenesises replace each other. For instance, the upper interval is characterized by the development of an association such as hornblende + albite, alternating with an association of albite + magnetite + biotite.

Hornblende is represented by idiomorphic crystals with bright pleochroism changing from bluish green on *Ng* to light yellowish green on *Np*. Below a depth of 5100 m, the rock contains bright green hornblende ($Ng = 26° - 29°$). In places its crystals are dissected by streaks of calcite with quartz. Near these streaks hornblende becomes colourless. Epidote and sometimes chlorite are met in individual intervals. In the presence of epidote hornblende has a paler colour. Near the boundary between amphibolites and underlying biotite-muscovite schists (at about 5200 m), hornblende exhibits signs of chloritization. This sample contains quartz, an admixture of carbonate, hornblende, chlorite, and albite.

Depending on the rock composition, the content of iron in hornblende varies ($F = 0.26 - 0.75$). The iron content in biotite also changes over a broad range ($F = 0.29 - 0.75$); cummingtonite is usually less ($F = 0.33 - 0.45$) than that in the associated hornblende.

Meta-andesites and sedimentary rocks constitute a considerable part of the epidote-amphibolite facies in the SG-3 well section. Meta-andesites are represented by two varieties: one variety is characterized by the same parageneses as those inherent to amphibolites, and the second possesses the composition of the biotite plagioclase fine-grained schists, sometimes with magnetite. The second variety contains biotite and acid plagioclase. Calcareous metasandstones are composed not only of quartz, plagioclase, mica and microcline but also of calcite, epidote and magnetite. It is common for minerals to be represented by rather magnesian varieties, i.e. by phlogopite with iron content 0.04 – 0.11 and associated phengite which contains 6.28% FeO and 2.23% MgO (0.33 and 0.36 formula units). Carbonate calcareous-magnesian rocks, poor in iron, are regarded as metamorphosed dolomitic siliceous limestones. At present they contain associations made of calcite, dolomite, tremolite, talc, phlogopite and chlorite. The compositions of these minerals are characterized by extremely low values of iron content. Chlorite is the most ferruginous mineral ($F = 0.08$). There is no doubt that these rocks belong to the epidote-amphibolite facies of metamorphism. The absence of diopside indicates that temperature conditions of the amphibolite facies were not attained. However, quartz, being associated at some places with calcite, was not found in any thin section together with dolomite, which could have been possible in conditions of the greenschist facies. Paragenesises of carbonate-silicate rocks are shown in Fig. 1.48.

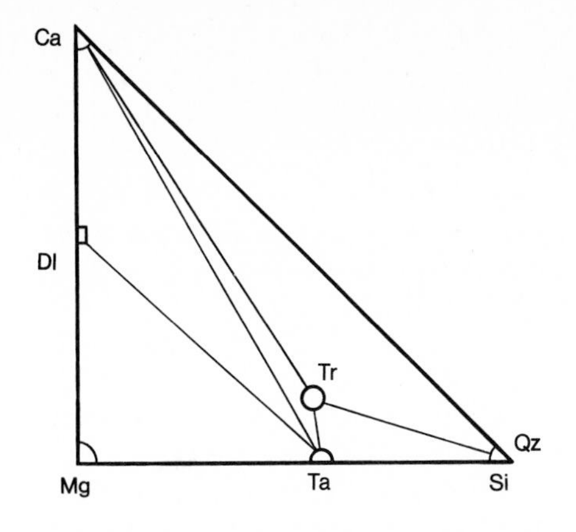


**Fig. 1.48.** Diagram "composition – paragenesis" of the metamorphosed silicic-carbonate rocks

*The Amphibole Facies* is represented in the SG-3 well section by rocks formed at the lowest temperatures inherent to this facies; these rocks possess almost the same character as rocks of the epidote-amphibolite facies. This facies is characterized by the systematic presence of highly aluminiferous hornblende and by more acid plagioclase than plagioclase No. 15. The bulky parts of the Mayarvi suite and the granite gneiss complex are included in this facies. Amphibolites of this facies possess the following textures: (1) thin-foliated textures, both fine- and coarse-grained; (2) poorly foliated textures of subordinate significance.

In mineral (and chemical) composition, amphibolites of the Mayarvi suite are subdivided into two groups: (1) amphibolites containing calcite-bearing minerals such as epidote or sphene which are saturated with calcite; these are the most widespread amphibolites; (2) amphibolites which contain not only hornblende of plagioclase and biotite but also cummingtonite, that is an amphibole deprived of calcium. An accessory titanium-bearing mineral is represented here not by sphene but by ilmenite (titanomagnetite). These rocks can be considered as amphibolites undersaturated with calcium.

Hornblendes of amphibolites from the Mayarvi suite are represented by highly aluminiferous varieties (1.5 – 2.5 Al in formula units). The iron content in hornblendes varies over a broad range (0.26 – 0.66). The iron content in biotite changes from 0.29 to 0.59, these values corresponding approximately to those of hornblende. The iron contents in these minerals may vary. In amphibolites, particularly in those characterized by a coarse-grained structure, one may see a zonal character of hornblende which is characteristic of prograde metamorphism, i.e. marginal parts of grains are intensively coloured and enriched in aluminum. The plagioclase composition of amphibolites (unaffected by prograde metamorphism) from the Pechenga complex ranges between compositions inherent to No. 19 and No. 37.

The major part of the amphibolites of the granite gneiss complex penetrated by the well is composed of hornblende and plagioclase in almost equal proportions. Biotite and quartz are the permanent constituents of amphibolites, although they are met in various, usually small, quantities. Accessory minerals are represented by apatite, sphene and ilmenite. Among other types of amphibolite one can recognize cummingtonite and epidote amphibolites as well as clinopyroxene and garnet amphibolites, which were not encountered in the previous rock units. The Kola complex also contains ore-bearing magnetite-ilmenite amphibolites (see Chap. 1.6).

Differences in the chemical composition of amphibolites are distinctly pronounced in a ratio of Ca/(Ca + Al + Mg + Fe). In the clinopyroxene amphibolites this ratio exceeds 0.30, in ordinary varieties it changes between 0.21 and 0.24; in the cummingtonite amphibolites it is less than 0.20.

Amphibolites contain hornblende which is characterized by normal content of alumina (1.6 – 2.36 formula units) and moderate alkalinity (Na + K = 0.33 – 0.68 formula units). The iron content changes in the range of 0.28 – 0.66%. Taking a ratio of $Al^{IV}$ to alkali, it appears that some hornblendes resemble a series of hastingsite-tschermakite (oligoclase). Plagioclases are assigned to andesine or labradorite, and in rarer cases to more basic varieties. The content of anorthite changes between 36 and 67%, as was established by microanalyzer data, and it is between 30 and 80% when angles of extinction in the $\perp$ (a) zone were measured. The plagioclase of amphibolites is characterized by a zonal character in the distribution of its different types: more acid plagioclase is located in the central part, whereas more basic plagioclase is confined to the peripheral part. Sometimes this zonality is of more complex nature: the rim composed of more basic plagioclase is again followed by a rim made up of more acid plagioclase. The last case is considered to be a result of migmatization or subsequent prograde processes.

The biotite or plagioclases hardly differs from that of gneisses. Its iron content changes from 0.32 to 0.67 and depends on the total content of iron in rocks in association with minerals.

Dark-coloured minerals present in cummingtonite amphibolites are characterized by a higher content of magnesium as compared with ordinary amphibolites, whereas plagioclase is notable for its more basic composition. The iron content in hornblende of this association is 0.26 – 0.39, and the amount of biotite is 0.32 – 0.34.

The content of magnesium in cummingtonite is rather considerable (Mg = 0.31 – 0.40). Plagioclase contains 50 – 70% An. This type of amphibolite never contains either epidote or sphene. It is clearly seen that the cummingtonite amphibolites are lacking in calcium as compared with other types of amphibolites: a ratio of Ca/(Ca + Mg + Fe + Al) is less than 0.20.

The epidote amphibolites are the most widespread amphibolites. Individual grains of epidote are also present in ordinary amphibolites. The primary epidote occurs in association with hornblende and basic plagioclase. If secondary epidote is formed in conditions of prograde metamorphism, then plagioclase is albitized, and hornblende is actinolitized and chloritized. At places this is clearly seen on examining the mutual relationships between minerals in thin sections, but normally it is difficult to distinguish one epidote from another.

Epidote in amphibolites is colourless or yellowish, and its iron content is about 0.22, which is also characteristic of the secondary epidote.

The clinopyroxene epidotes are a rather rare variety of amphibolite present in the SG-3 well section. In the metamorphic strata of the Pechenga series, clinopyroxene was not met even in the lowermost parts where the amphibolite facies begins. In the granite gneiss complex clinopyroxene was encountered in five or six places together with abundant amounts of epidote. The comparatively rare occurrence of clinopyroxene in the thin sections examined is probably accounted for by the small amount of calcium present in them. This explanation is supported by comparing data of the chemical analyses of clinopyroxene amphibolite and other types of amphibolite. For clinopyroxene amphibolite a ratio of Ca/(Ca + Al + Mg + Fe) exceeds 0.30, whereas it changes from 0.17 to 0.24 for other varieties.

The clinopyroxene amphibolites are composed of hornblende, plagioclase, epidote, biotite, clinopyroxene, actinolite and sphene. Hornblende and biotite possess no particular features on the basis of which one could distinguish them from the same minerals contained in ordinary amphibolites. The iron content in hornblende and actinolite is 0.47 – 0.48 and 0.26 – 0.36, respectively. According to the chemical composition, clinopyroxene can be regarded as salite characterized by an iron content equal to 0.32 – 0.56. Normally, the basicity of plagioclase is not higher than that of ordinary amphibolite (35 – 60% An), although in some plagioclases the amount of An reaches 92%. Actinolite, like some epidotes, was formed as a result of the superimposed retrograde process.

Garnet amphibolites are also met only in the granite gneiss complex, with their relative role becoming more significant as depth increases. In sample No. 38167 (depth 10,253 m) garnet amphibolite is composed of hornblende (F = 0.52), plagioclase-bytownite (No. 87), biotite (F = 0.48), cummingtonite (F = 0.51) and garnet (F = 0.82 – 0.86). Cummingtonite is characterized here by the highest values of iron content, and garnet is notable for the highest values of grossular content (up to 19.8% in the central part of a grain).

Parageneses of minerals in metabasites of the amphibolite facies are illustrated in Fig. 1.45.

A detailed petrographic description of gneisses and migmatites of the Kola series has been presented earlier. Therefore, we shall given here only a brief description of their mineral assemblages.

In addition to biotite, muscovite, quartz, and plagioclase, highly aluminiferous gneisses contain andalusite, garnet, staurolite and sillimanite. Sillimanite is represented by fibrolite, which sometimes passes into thinly prismatic spicular sillimanite. Andalusite forms large grains of irregular shape surrounded by muscovite flakes. Garnet is represented by irregular, isometric grains, and sometimes by isolated grain fragments; it is hardly ever met in the form of rectilinear polyhedrons. Garnet consists of almandine (66 – 70%) with admixture of pyrope (7 – 16%), spessartine (8 – 15%), grossular (up to 7%), and andradite (up to 6%) in rarer cases. Staurolite is met almost as frequently as garnet; it is found in the form of irregular grains characterized by a poikiloblastic texture with inclusions of quartz.

As a rule, highly aluminiferous gneisses contain biotite and muscovite. The content of iron in biotite (in the presence of garnet or staurolite) is 0.51 – 0.56. The total content of Mg and Fe in muscovite of the highly aluminiferous gneisses does not exceed 0.18 formula units; this means that the content of phengite is not high. Some muscovite is certainly of secondary origin and is developed in andalusite and sillimanite. Not infrequently, highly aluminiferous gneisses contain large laminae of chlorite which does not form pseudomorphs in biotite. The iron content in chlorite is almost similar to that of biotite (0.48 – 0.58). The paucity of potassium in gneisses may be adduced as the reason for the co-occurrence of chlorite and biotite in these rocks. By taking basicity values into account plagioclase may be attributed to oligoclase, which contains 20 – 30% anorthite. Cordierite was not found in highly aluminiferous gneisses, and this fact may be accounted for by the absence of gneisses with a considerable amount of magnesium. It is worth noting that although andalusite and sillimanite are frequently present in rocks, kyanite was met in only one sample. It may be assumed that the highly aluminiferous gneisses have been formed during the metamorphism of sedimentary rocks (sandstones with clay cement and layers of clay) (see Fig. 1.45).

Biotite gneisses are usually of bimicaceous nature, i.e. they contain muscovite, although in small amounts. The compositions of minerals are the same as those contained in gneisses. Plagioclases can be characterized by a higher basicity, i.e. they can contain less than 20% An. These rocks easily become migmatized and usually a certain part of them is represented by the uniformly grained leucocratic aplite-like quartz-feldspar rocks which may be regarded as the remelted substrata. The amphibole-biotite gneisses are considered to be transitional rocks between the biotite gneisses to amphibolites. In addition to biotite, plagioclase, quartz and muscovite, they contain hornblende, and frequently epidote. The composition of plagioclase in these rocks may vary from that inherent to plagioclase, as in the highly aluminiferous and biotite gneisses, to andesine (with 30 – 40% An), as in amphibolites. Some of the biotite-amphibole gneisses-migmatites were probably formed as a result of the plagiomigmatization of amphibolites. Of very rare occurrence are the biotite-amphibole gneisses which contain, in addition to hornblende, cummingtonite. These rocks are the leucocractic varieties of the cummingtonite amphibolites.

According to chemical analysis data, the content of $Al_2O_3$ in the highly aluminiferous gneisses is not much higher than it is in the ordinary biotite gneisses (18.4% versus 17.30%); the molar ratio of Al to a sum of Ca + Na + K in them is much higher (1.78 – 2.26 versus 0.86 – 1.21). The paucity of alkali and calcium in the primary sediments, with their inert state in the course of the metamorphic processes being taken into consideration, is the reason for the formation of highly aluminiferous minerals.

Meta-ultrabasites of the Kola series are usually represented by the tremolite-talc-phlogopite, tremolite-phlogopite-cummingtonite and tremolite-talc-chlorite schists. Magnetite and ilmenite are found in them in small amounts. Cummingtonite is the most ferruginous mineral (F = 0.25 – 0.27) of meta-ultrabasites, although it contains higher amounts of magnesium than cummingtonite of amphiboles (0.31 – 0.40). Calcic amphiboles contains a small amount of aluminum (less than 0.29 formular units) and its iron content is 0.12 – 0.17, which allows it to be assigned to tremolite. The iron content in talc does not exceed 0.06. Chlorite is represented by its highly aluminiferous variety (up to 2.1 formula units Al), with iron content being equal to 0.16 – 0.19. The iron content of phlogopite is 0.22 – 0.23. In rather rare instances, meta-ultrabasites contain anthophyllite.

Thus is appears that the mineral composition of meta-ultrabasites corresponds to the composition of the same amphibolite facies.

Migmatites, highly characteristic of the Kola series, consist of plagioclase, potassium-sodium feldspar, quartz and biotite. Epidote, clinozoisite and muscovite are present in variable quantities. In contacts with the granitizated rocks, migmatites are of transitional composition. Among biotite gneisses, migmatites are characterized by a composition inherent to tonalite-granodiorite. When migmatites occur near highly aluminiferous gneisses, they contain relict crystals of staurolite and garnet. At the contact with amphibolites, migmatites are characterized by diorite composition. Large relict grains of hornblende are surrounded by biotite flakes, and they are gradually replaced by an aggregate made up of feldspars and quartz. In the continuous amphibolites, a process of migmatization has led to the formation of quartz-feldspar veinlets along foliation planes and to the biotitization of hornblende. A leucosome of nebulitic migmatites of granodiorite composition is represented by granite enriched with potassium-sodium feldspar as compared with the majority of migmatites.

Plagioclases in migmatites exhibit a zonal character in their compositional nature, with the core being composed of more basic material. In migmatites of diorite composition andesine (Nos. 33 – 35) occupies the marginal zones of plagioclase (Nos. 28 – 30). In migmatites of granite composition a core of plagioclase (Nos. 12 – 18) is surrounded by an envelope of albite (Nos. 8 – 10). Potassium-sodium feldspar is represented by the latticed nonperthite microcline. A high amount of potassium oxide (14.46 – 15.60%) and low amount of the plagioclase minal (0.86 – 11.1%) are indicative of the relatively low temperatures of crystallization. In its content of aluminum (1.73 formula units), alkali (0.59) and iron (0.46), hornblende of migmatites corresponds to hornblendes of gneisses. The iron content in biotite is 0.45 – 0.66, i.e. the same in metamorphic rocks. The iron content in epidote varies between 0.09 and 0.21, and is less than that in epidote of metamorphic rocks. Muscovite is characterized by a low content of the phengite minal (Mg + Fe = 0.31 formula units), which is the same as in gneisses, and much less than in zones of prograde metamorphism. In their chemical composition, migmatites correspond to an alkali-earth series of the plagiogranite-diorite-granodiorite-granites (Nockolds 1954). A correlation of the chemical compositions of migmatite and initial rocks (gneisses, amphibolites) shows that migmatites, as compared with the enclosing rocks, contain more $SiO_2$, $Al_2O_3$ and alkali ($Na_2O$ in particular) but less $TiO_2$, iron, MgO and CaO.

### 1.5.2 Low-Temperature Retrograde Metamorphism

Beginning at a depth of 6000 m, manifestations of retrograde metamorphism are distinctly fixed in amphibolites and granite gneisses; these manifestations are confined to zones of gently and steeply inclined faults. The areas of the most intensive retrograde metamorphism coincide with zones of vigorous crushing and cataclasm of rocks. Such zones of rock alteration were observed in the following intervals: 5977 – 5984, 6065 – 6120, 6250 – 6268, 6336 – 6410, 6615 – 6677, 7652 – 7656, 8117 – 8118, 9370 – 9375, 9921 – 9940 and 9994 – 9999 m. Investiga-

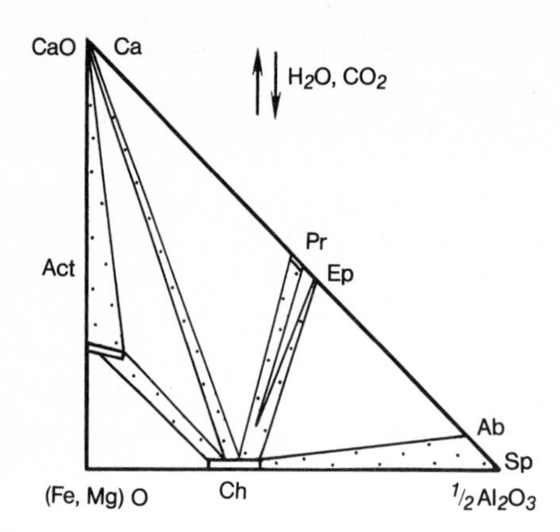


**Fig. 1.49.** Diagram "composition – paragenesis" of amphibolite metamorphism

tion of the altered amphibolites makes it possible to recognize a relatively high-temperature step in the retrograde metamorphism which corresponds to the biotite subfacies of the greenschist facies and to a low-temperature step corresponding to the biotite-free subfacies of the greenschist facies. Signs of even more low-temperature alterations, similar to the alterations usually fixed in conditions of the prehnite-pumpellyite facies, have also been established. The intensive retrograde-metamorphic processes occurred at the level of the biotite-free subfacies of the greenschist facies (Fig. 1.49).

Relatively high-temperature alteration is expressed in some decrease of the plagioclase basicity and hornblende aluminum content. Epidote appears (or its quantity increases) in amphibolites rich in calcium, and cummingtonite appears in amphibolites lacking calcium. Some biotite is still preserved but in places it is partially replaced by chlorite. Usually, the intensity of this process is not strong, but in places (particularly in the granite gneiss complex) accumulations of epidote bands and zones of continuous epidotization and silicification are found. Epidotization is frequently not accompanied by plagioclase albitization, and only hornblende is partially replaced by actinolite. During this undoubtedly metasomatic process, CaO flowed into sediments, while other components replaced each other.

In zones of more low-temperature alteration, biotite has not survived, plagioclases are albitized and hornblende is replaced by actinolite. In some narrow zones, amphibolites are transformed into epidote-chlorite rock; this becomes possible if some amount of aluminum and calcium are displaced. It is necessary to indicate differences in composition of epidote observed in the upper parts of the Pechenga complex and in zones of retrograde alteration confined to the lower parts of the complex. The iron content of epidote in the upper parts of the complex varies over the range of 0.16 – 0.18, whereas in the retrograde zones it is usually much higher (0.25 – 0.43).

The alteration of rocks at the level of the prehnite-pumpellyite facies is fixed by the appearance of prehnite and pumpellyite in the form of isometric accumulations and thin streaks. As distinct from the bright green pumpellyite present in the upper part of the section, pumpellyite of the retrograde stage has a distinct blue colour of varying intensity; it comprises individual grains or radiated aggregates.

The alteration of the Ca-rich amphibolites at a level of the biotite-free step of the greenschist facies of metamorphism occurred according to the following reactions: $Hb + Pl_{20-30} + Bi + Act + Ch + Ab + Ep + Kfs \rightarrow Ch + Ab + Kfs$.

Plagioclases of amphibolites, containing 20 – 40% An (and even greater amounts), are usually replaced by non-twinned albite or, more frequently, by very fine-grained brownish feldspathic mass. According to X-ray microanalyzer data, this mass consists of a thin admixture of acid plagioclase and potassium feldspar. These rocks also contain pure albite and potassium feldspar. Measurements made with the help of the Fedorov table showed that these inclusions are represented by non-latticed microcline. Chlorite from Ca-rich amphibolites, which were subject to alteration during retrograde metamorphism, contains 0.40 – 0.71 Fe and 2.30 – 2.47 Al (in formula units).

In the process of intensive alteration, amphibolites are transformed in some places into the albite-k-feldspar chlorite rocks characterized by textures like those inherent to foliated amphibolite. Signs of actinolite replacement by chlorite are frequently seen. Sometimes calcite appears in rocks. In zones of strong tectonic stresses, many lens-like inclusions and veinlets of calcite are met. From this it appears that calcium could be withdrawn from rocks and be later deposited at some areas of restricted extension.

A process of retrograde metamorphism has affected amphibolites, which were lacking in calcium, and resulted in the chloritization of biotite and albitization of plagioclase. The manifestation of higher-temperature retrograde metamorphism has led here to the additional cummingtonitization and partial actinolitization of hornblende, whereas in the calcium amphibolites it has resulted in the epidotization of calcium amphibolites. At a depth of 6617 m (sample No. 21560), in an interval composed of cummingtonite amphiboles, the following mineral assemblage has been formed in the process of retrograde metamorphism: hornblende, actinolite hornblende, cummingtonite, biotite, plagioclase, chlorite and actinolite. The coexistence of hornblende and actinolite, biotite and chlorite can be easily explained as a result of the action of retrograde processes which have led to alteration of the rocks. It is most astonishing that cummingtonite coexists with epidote because, as a rule, they do not occur together. It appears probable that they were formed as a result of the relatively high-temperature retrograde alteration of hornblende and basic plagioclase and, therefore characterize the extreme irregularity of this transformation. Chlorite, being formed as a result of transformation of calcium-poor amphibolites during a process of retrograde metamorphism, contains more magnesium than is present in chlorite from Ca-rich amphibolites.

It is interesting to note that the age of small veinlets present in brecciated zones and the age of the observed successive processes of retrograde metamorphism occurring in amphibolites correspond to each other. In zones of intensive alteration, rock is subject to complete transformation, but outside these zones it becomes altered only on contact with veinlets. A characteristic of the rock alteration is usually corresponds to veinlets filling. Veinlets originate in the following order: quartz-epidote, chlorite-k-feldspar, carbonate and prehnite-pumpellyite veinlets.

In gneisses and migmatites of the granite gneiss complex, the manifestation of retrograde metamorphism at a relatively high-temperature step has led to muscovitization, accompanied by silicification. In these rocks fibrolite appears rather frequently. All these facts indicate that the process possesses features of feeble greisenization. The composition of secondary muscovite differs from that of regional-metamorphic muscovite by its considerable content of a phengite component (Fe + Mg = 0.47 versus 0.07 – 0.18 formula units). The most intensive greisenization has affected the following intervals: 6830 – 6860, 7520 – 7540, 7858 – 7876, 8635 – 8680, 8980 – 9030, 9775 – 9850 m, etc.

The alteration of gneisses at a lower temperature stage is expressed in the chloritization of biotite, epidotization and serpentinization of potassium feldspar. In individual zones and veinlets prehnite is found. The amount of plagioclase in potassium feldspar is less than that in the microcline of migmatites. Prehnite has its usual composition and the iron content in it varies from 0.01 to 0.03. Zones of cataclasm and retrograde changes are frequently characterized by the presence of the sulphide mineralization.

### 1.5.3 Physico-Chemical Conditions of Metamorphism

The study of core samples from the Kola superdeep borehole made it possible for the first time to trace a transition from the prehnite-pumpellyite to amphibolite facies in one and the same section. The genetic relationship between these facies is proved not only by the regular replacement of some mineral assemblages by others, but also by a regular alteration of metamorphic minerals such as plagioclases and amphiboles (see Fig. 1.45).

However, there are great difficulties to face in determining the temperatures and pressures that existed at the stage when this metamorphic zonality was formed. These difficulties have two causes: (1) the mineralogical geothermometers and geobarometers used for zones of low-temperature metamorphism are imperfect; (2) the mineral assemblages in these zones are not fully balanced (Glagolev et al. 1983).

The study of metabasites of the prehnite-pumpellyite facies shows that they contain unbalanced minerals. It should be noted that glassy mass, laths of plagioclase and impregnations of pyroxene become irregularly altered during the process of metamorphism. The main glassy mass is the first to be transformed. In most cases, it is replaced by chlorite; in the course of this process surplus silica is released in the form of quartz grains which are frequently characterized by the radiate extinction. The iron content of chlorite usually exceeds that of the rocks by not more than 2 – 5%. From this it may be assumed that the residual liquid of magmatic rocks of basic composition might have been enriched in iron as compared with the rock on the whole.

Plagioclases, present in the form of thin and relatively long laths with extensive contact surface, were previously represented by labradorite-bytownite. In the poorly altered rocks they are everywhere replaced by albite, prehnite, pumpellyite, epidote, etc. In the prehnite pumpellyite facies plagioclase is replaced by prehnite and albite or, much less frequently, by pumpellyite. It may be supposed that pumpellyite could be formed providing plagioclase had interacted with the metamorphosed glassy mass enriched in Mg and Fe.

In conditions of the greenschist facies, epidote, which partially replaced the previously albitized plagioclase, becomes widely spread in plagioclase laths and around them. When epidote develops in albite, calcite is a product of clinopyroxene which is replaced by actinolite, and iron is a product of chlorite. In conditions of the prehnite-pumpellyite facies, impregnations of the monoclinic pyroxene are hardly ever subject to alteration, whereas in conditions of the upper stages of the greenschists facies they are replaced by actinolite and disappear at a depth of 2000 m.

Different textures correspond to various facies, i.e. the relict porphyritic texture corresponds to the prehnite-pumpellyite facies, the texture transitional between porphyritic, nemato- and porphyroblastic textures correspond to the greenschist facies, the idiogranoblastic texture corresponds to the beginning of the epidote-amphibolite facies. Thus, it is evident that the porphyritic texture has survived only at lower depths due to the low velocity of the components' movements in the rocks. When pyroxene is replaced by actinolite, a ratio of $MgO/(MgO+FeO)$ in an impregnation changes abruptly: the iron content exceeds 50%, the actinolite content is less. However, alumina and silica display no signs of displacement. These two components form a conservative base of porphyritic texture. With increasing metamorphic grade, $Al_2O_3$ and $SiO_2$ are partially redistributed between actinolite impregnations and chlorite of the main mass. Beginning from a depth of 2100 m, rocks exhibit signs of the porphyritic texture "creeping out". When this happens, actinolite needles become longer and extend beyond the limits of pyroxene impregnations; at the same time a felted actinolite aggregate appears in the main mass. During this process iron is withdrawn from impregnations, and $Al_2O_3$, MgO and $H_2O$ flow into them. The shape of albite impregnation is left unchanged here. Below a depth of 3180 m, not only pyroxene but also plagioclase impregnations disappear. In places a collective recrystallization of actinolite takes place and its porphyroblasts are formed. This zone, coinciding with the biotite-actinolite subfacies, is characterized by the presence of stable metamorphic idiogranoblastic or lepidogranoblastic textures.

Correspondingly, three zones have been distinguished in the section; they characterize various degrees at which an equilibrium state was reached during metamorphism. These zones are as follows: (1) the zone of chemical and textural equilibrium (0 – 2100 m); (2) the zone with relict primary textures and bimetasomatism of microscale manifestation (2100 – 3200 m); (3) the zone of full recrystallization of rocks under conditions similar to the equilibrium state (below 3200 m).

The non-homogeneity of metamorphic minerals is also indicative of incomplete equilibrium. For instance, in the upper parts of the greenschist facies, colourless crystals of actinolite in metabasites become covered by a hornblende rim of green colour. The boundary between this rim and the core is sharp; the rim is moreover characterized by a higher content of $Al_2O_3$, Fe, $Na_2O$, and a lower content of $SiO_2$, MgO than the core. Hornblende is developed not only in the rim but also along actinolite cleavage planes. Microprobe data show that amphibolites with 1.2% and 5.2% alumina content have direct contact with each other here; correspondingly, the iron content also differs.

For this reason, the evaluation of temperatures and pressures for the prehnite-pumpellyite and greenschist facies is based on results obtained in studying gas-liquid inclusions in metamorphic veins and veinlets (see Chap. 1.4). It is worth the reminder that, according to data of I. V. Banshchikova, inclusions present in minerals of the isofacial veins change their phase composition down the section from a type $L>G$(0 – 1500 m) to a type $L>G+K$ or $L\geqslant G>K_{1+s}$ (1400 – 4000 m) and then to a type $G\geqslant L+K$ or $G\geqslant K>L$ (below 4000 m). The homogenization of the early groups of syngenetic inclusions in an interval of 170 – 3100 m is directed towards the liquid phase at temperatures of 190° – 460 °C, and in an interval of 3100 – 4100 m at a temperature range of 480° – 512 °C; below a depth of 4100 m homogenization is directed towards the gaseous phase at a temperature of 510 °C and higher.

Beginning from a depth of 4900 m and down to the borehole bottom, that is, within a range of occurrence of the epidote-amphibolite and amphibolite facies, possibilities for determination of temperatures and pressures in conditions of metamorphism become better. Likhoidov (1981) has prepared a geothermobarometer which measures temperature and pressure by taking into account compositions of plagioclase and hornblende which occur in equilibrium with epidote. A geothermobarometer operates by using experimental data. Such a mineral association is typical of the majority of metabasites in the SG-3 well, beginning approximately from a depth of 5000 m. At higher levels, plagioclase of metabolites is represented by albite, and the mineral association characterizes temperatures less than 500 °C, which are regarded as the lowermost limits of geothermobarometer sensitivity. When processed, data from micro-X-ray-spectral analyses show that values of different determinations change within a considerable range, although there is a general rise in temperature of metamorphism as depth increases, and only a slight growth of pressure within a range of 0.2 – 0.4 GPa (Table 1.14). The scattered character of the determined values is probably caused by an incomplete equilibrium state of mineral assemblages (this assumption is supported by the observed changes in mineral composition in one and the same sample) or by the influence of the later superimposed processes.

In determining the temperatures of metamorphism of rocks from the granite gneiss complex by bimineral geothermometers (Perchuk and Ryabchikov 1976), it becomes evident that the measured values are considerably scattered (Table 1.15). Nevertheless, the majority of figures are within a range of 540° – 620 °C; these are the values of the probable maximal temperatures of the regional metamorphism. A distribution pattern of Mg and Fe in three pairs of hornblende and clinopyroxene (sample nos. 24824, 34405 and 36795; depth 7652.6, 9200.0 and 9936.3 m, respectively) is indicative of too high temperature values (690°, 640° and 725 °C).

Thus, the majority of figures obtained for rocks of the lower parts of the Pechenga and granite gneiss complex indicate that rocks of the amphibolite facies were formed in conditions of the epidote-amphibolite and lower temperature stages. Pressure values (0.2 – 0.4 GPa) have confirmed that rocks were subject to an andalusite-sillimanite type of metamorphism established earlier by studying mineral assemblages.

The development of andalusite and fibrous sillimanite-fibrolite in the absence of kyanite shows that pressure did not reach a triple point of equilibrium of an-

**Table 1.14.** Evolution of temperature and pressure of the prograde metamorphism in amphibolites of the SG-3 well by geothermobarometer (Likhoidov et al. 1981)

| Sample nos. | Depth, m | Plagioclase No. | Al in hornblende, formular units | Temperature, °C | P, GPa |
|---|---|---|---|---|---|
| 18483 | 4887.6 | 2 | 1.88 – 2.37 | <500 | – |
| 19383 | 5765.0 | 11 – 17 | 1.56 – 2.21 | 500 – 515 | 0.4 – 0.7 |
| 19487 | 6004.0 | 19 | 1.65 – 2.37 | 520 | 0.4 – 0.8 |
| 19828 | 6166.0 | 37 | 1.21 – 1.65 | 560 | 0.2 – 0.3 |
| 19943 | 6216.6 | 30 | 1.60 – 2.36 | 540 | 0.3 – 0.7 |
| 20771 | 6457.2 | 27 | 1.11 | 535 | 0.2 |
| 22325 | 6783.9 | 25 | 1.06 – 1.44 | 530 | 0.2 – 0.3 |
| 22706 | 6946.0 | 32 – 67 | 1.98 – 2.03 | 545 – 7650 | 0.2 – 0.5 |
| | 8337.4 | 64 | 2.54 | 650 | 0.4 |
| 36811 | 9940.3 | 33 | 2.06 | 550 | 0.5 |
| 37687 | 10,141.4 | 79 | 1.95 | 650 | 0.2 |
| 38620 | 10,504.5 | 61 | 1.95 | 650 | 0.2 |

dalusite-sillimanite-kyanite, i.e. it did not reach a value of 0.3 – 0.4 GPa. The presence of staurolite, in both the upper and lower parts of the section, shows that temperatures of metamorphism in these intervals were almost of the same order. The coexistence of garnet ($f = 0.84 - 0.90$) with staurolite ($f = 0.82 - 0.87$), biotite ($f = 0.52 - 0.62$) and andalusite is indicative of a manifestation of the andalusite-staurolite subfacies of the staurolite facies, which corresponds to upper parts of the epidote-amphibolite and lower parts of the amphibolite facies within the limits of the generally adopted system.

It has already been mentioned that the Kola series differs from the Pechenga complex by intensive rock granitization resulting in the formation of banded and nebulitic migmatites. The newly formed minerals are represented by microcline, plagioclase and quartz. Migmatites enriched in these minerals are notable for a texture transitional to granitic texture. Plagioclases are characterized by normal zonality and some of their crystals (in the form of tables) are found, together with idiomorphic biotite flakes, inside microcline porphyroblasts; this is usually characteristic of magmatic rocks. From all these facts it appears that a neosome is probably of magmatic nature. In their chemical composition migmatites are placed within an ordinary series of rare-earth rocks. The composition of an outer zone of plagioclase in the microcline-bearing migmatites is more or less stable. All this shows that in the process of granitization a general state of the physico-chemical equilibrium was usually reached.

Figure 1.50 displays mineral parageneses of migmatites. Conodes are drawn in agreement with parageneses of a neosome. It is obvious that these parageneses are typical of the normal alkaline-earth magma. A cotectic line of crystallization of the normal alkaline-earth magma is drawn on the same diagram. Symbols denoting the migmatites studied are located near this line. If the composition of a paleosome is extracted from the aggregated composition of granitic migmatite, then a symbol denoting a neosome almost fully coincides with the eutectic. From

**Table 1.15.** Determination of temperatures of metamorphism by bimineral geothermometers (Perchuk and Ryabchikov 1976)

| Sample nos. | Depth, m | Ca(Ca + Na + K) | | Temperature, °C | Mg(Mg + Fe + Mn) | | | | Temperature, °C |
|---|---|---|---|---|---|---|---|---|---|
| | | Hb | Pl | | Car | Stu | Hb | Bi | |
| 22799 | 6983.1 | – | – | – | 0.08 – 0.14 | 0.14 | – | 0.38 | 530 – 650 |
| 23153 | 7119.9 | 0.80 – 0.84 | 0.49 | 540 – 590 | – | – | – | – | – |
| 23742 | 7367.0 | – | – | – | 0.10 | – | – | 0.48 – 0.49 | 490 – 500 |
| 24051 | 7419.4 | – | – | – | 0.11 | – | – | 0.44 | 550 |
| 24824 | 7652.6 | 0.84 – 0.86 | 0.39 | 500 – 520 | – | – | – | – | – |
| 27044 | 7964.0 | 0.82 – 0.84 | 0.60 – 0.67 | 580 – 610 | – | – | – | – | – |
| 31100 | 8716.4 | 0.72 – 0.73 | 0.22 – 0.24 | 505 – 520 | – | – | – | – | – |
| 31596 | 8858.0 | 0.80 – 0.82 | 0.36 – 0.39 | 530 – 560 | – | – | 0.71 – 0.72 | 0.56 | 750 |
| 31596 | 8858.0 | 0.75 – 0.76 | 0.36 – 0.40 | 570 – 610 | – | – | 0.46 – 0.48 | 0.50 – 0.54 | 550 – 575 |
| 36811 | 9940.3 | 0.70 | 0.33 | 600 | – | – | – | – | – |
| 37148 | 9990.4 | 0.78 – 0.81 | 0.45 – 0.53 | 595 – 625 | – | – | – | – | – |
| 37480 | 10,093.5 | 0.79 | 0.19 | 460 | – | – | 0.70 | 0.68 | 660 |
| 37687 | 10,141.4 | 0.77 | 0.79 | 700 | – | 0.58 | 0.58 | – | 660 |
| 38167 | 10,253.0 | 0.80 | 0.87 | 750 | 0.14 – 0.18 | 0.48 | 0.52 | – | 550 – 610 |
| 38569 | 10,480.7 | – | – | – | 0.10 | – | 0.48 | – | 520 |
| 38620 | 10,504.5 | 0.80 | 0.61 | 640 | | 0.54 | 0.54 | – | 590 |

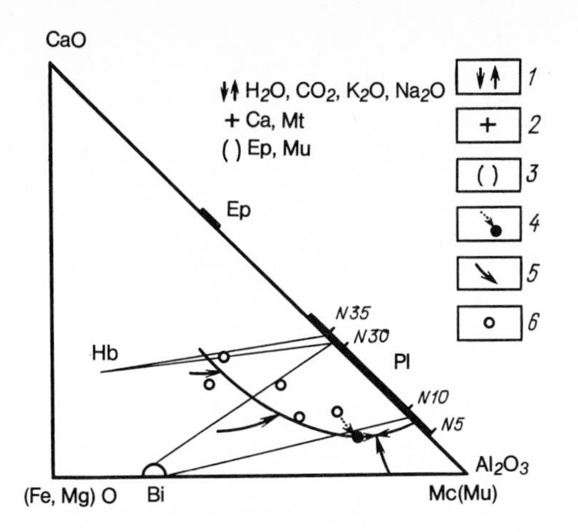


**Fig. 1.50.** Parageneses and modes of migmatite crystallization. *Components: 1* movable; *2* surplus; *3* relict; *4* symbols denoting composition of a neosome of the granite migmatite; *5* modes of crystallization of normal earth-alkaline granitoids; *6* points indicating compositions of the studied migmatites

this it appears that a neosome of migmatites is characterized by an anchieutectic composition.

The material presented indicates the generally active behaviour of alkalines during the process of granitization. However, the presence of microcline in association with albite No. 8 – 10, and the composition of albite show that the normal activity of potassium decreased at a stage when the migmatites were forming. A local increase in potassium activity takes place near amphibolites at those places where microcline is associated with andesine No. 30 – 33, yet this activity does not exceed the normal earth alkaline level.

Touching upon the question of the temperatures reached during granitization, one should bear in mind that there is a relationship between the content of titanium and aluminum in the sixfold coordination in biotite and the temperature of their formation. The ratio of these components is normal for biotites from metamorphic rocks of the epidote-amphibolite facies and from those of the amphibolite facies of low-metamorphic grades. The content of aluminum in hornblendes of migmatites is as follows: $Al^{VI}$ 0.35, $Al^{IV}$ 1.38 formula units, which is characteristic of this mineral occurring in the high-temperature parts of the epidote-amphibolite facies (Dobretsov et al. 1972). All these data show that the process of granitization proceeded under the same thermodynamical conditions as those inherent to the metamorphism of amphibolites and gneisses of the Kola series.

It is an established fact that the contact between the Kola series and Pechenga complex is a boundary which abruptly limits the development of migmatites; however, the reasons for this are not fully clear. It might be that the lithology of metamorphic rocks played a certain role, as is observed in the Kola series, in which metabasites are much less granitized than gneisses. However, it cannot be ruled out that the Kola series was subject to migmatization prior to and after the formation of the Pechenga complex, the granitization conjugated with the prograde metamorphism of the Pechenga complex having been less intensive.

### 1.5.4 Age of the Zonal Metamorphism

The results of the K-Ar and Rb-Sr age determinations of metamorphic rocks from the Kola superdeep well section are presented in Tables 1.16 – 1.18.

**Table 1.16.** K-Ar age of metamorphic rocks of the Pechenga complex from the SG-3 well

| Series | Suite | Sample, nos. | Depth, m | Rock | Mineral | K, % | $^{40}Ar \times 10^{-3}$ $cm^3$ $year^{-1}$ | Age, Ma |
|---|---|---|---|---|---|---|---|---|
| 1 | 2 | 3 | 4 | 5 | 6 | 7 | 8 | 9 |
| Nickel | zd | 4318[a] | 1325 | Essexite metagabbro | Plagioclase | 0.268 | 0.036 | 1960 ± 60 |
| | | 9524[a] | 2535 | Gabbro-diabase | Actinolite | 0.077 | 0.0084 | 1710 ± 80 |
| Luostari | | 12760[a] | 3318 | Metadiabase | Plagioclase | 0.280 | 0.028 | 1620 ± 60 |
| | zp | 14662[a] | 3901 | Metadiabase | Actinolite | 0.075 | 0.0068 | 1520 ± 70 |
| | | 14668[a] | 3901 | Metadiabase | Hornblende | 0.100 | 0.013 | 1930 ± 70 |
| | | 17860 | 4715 | Dacite | Biotite | 6.860 | 0.814 | 1813 ± 35 |
| | | | | Metaporphyrite | Biotite | 7.740 | 0.773 | 1620 ± 40 |
| | | 18167[a] | 4800 | Dolomitic metasandstone | Muscovite | 7.250 | 0.919 | 1890 ± 35 |
| | lh | 18276[a] | 4841 | Arkose metasandstone | Mineral aggregate | 4.400 | 0.248 | 1070 ± 20 |
| | | | | | Sericite from cement | 5.600 | 0.267 | 950 ± 15 |
| | | 18405[a] | 4865 | Arkose metasandstone | Mineral aggregate | 4.410 | 0.283 | 1190 ± 13 |
| | | | | | Sericite from cement | 5.200 | 0.264 | 1000 ± 10 |
| | | 18628[a] | 5006 | Amphibolite-biotite plagio-clase schist | Mineral aggregate | 0.770 | 0.109 | 2020 ± 100 |
| | | | | | Amphibole | 0.835 | 0.153 | 2035 ± 80 |
| | pr | 19041[a] | 5495 | Biotite-magnetite plagioclase schist | Biotite | 7.270 | 0.812 | 1750 ± 30 |
| | | 19043[a] | 5507 | Epidote-magnetite plagio-clase schist | Biotite | 7.100 | 0.678 | 1570 ± 32 |
| | | 20776 | 6465 | Metagabbrodiabase | Biotite | 5.580 | 0.615 | 1730 ± 50 |
| | | 20779 | 6460 | Metagabbrodiabase | Hornblende | 0.390 | 0.0400 | 1630 ± 45 |
| | | 22233 | 6761 | Amphibolite | Hornblende | 0.380 | 0.0470 | 1876 ± 70 |
| | ma | 22273[a] | 6767 | Amphibole-biotite plagio-clase schist | Biotite | 7.600 | 1.179 | 2130 ± 40 |
| | | | | | Hornblende | 0.350 | 0.0196 | 1860 ± 70 |
| | | | | | Plagioclase | 0.100 | 0.0151 | 2100 ± 70 |
| | | 22374[a] | 6793 | Amphibolite-biotite plagio-clase schist | Biotite | 7.370 | 1.142 | 2130 ± 40 |
| | | | | | Hornblende | 0.370 | 0.0532 | 2040 ± 70 |
| | | | | | Plagioclase | 0.110 | 0.0297 | 2890 ± 80 |
| | | 22387 | 6791 | Amphibolite | Biotite | 4.200 | 0.560 | 1949 ± 70 |
| | | 22441 | 6800 | Amphibolite | Biotite | 5.700 | 0.873 | 2118 ± 45 |

*Note:* Analyses were made by T. V. Koltsova[a], S. I. Voronovsky, and L. V. Ovchinnikova.

**Table 1.17.** K-Ar age of metamorphic rocks of the Kola complex from the SG-3 well section

| Stratum | Sample, nos. | Depth, m | Rock | Mineral | K, % | $^{40}Ar \times 10^{-3}$ $cm^3$ $year^{-1}$ | |
|---|---|---|---|---|---|---|---|
| 1 | 2 | 3 | 4 | 5 | 6 | 7 | 8 |
| I | 22544[a] | 6900 | Plagiopegmatoid | Muscovite | 8.100 | 0.913 | 1750 ± 30 |
| | | | | Plagioclase | 0.650 | 0.162 | 2770 ± 70 |
| | 22660 | 6918 | Mesocratic plagiogneiss with HAM | Biotite | 7.000 | 0.665 | 1569 ± 65 |
| | 22709 | 6952 | Fe-amphibolite | Amphibole | 0.530 | 0.072 | 1970 ± 60 |
| | 22748a | 6967 | Mesocratic plagiogneiss with HAM | Biotite | 7.300 | 0.711 | 1600 ± 30 |
| | 22778 | 6979 | Biotite-hornblende schist | Biotite | 6.100 | 0.762 | 1872 ± 70 |
| | 22947 | 7028 | Migmatite-pegmatoid | Biotite | 6.990 | 0.648 | 1544 ± 48 |
| | 23290 | 7185 | Leucocratic biotite gneiss | Biotite | 6.750 | 0.606 | 1511 ± 50 |
| | 23308 | 7186 | Granite | Biotite | 7.340 | 0.592 | 1404 ± 43 |
| | 23470[a] | 7281 | Fe-amphibolite | Amphibole | 0.460 | 0.066 | 2020 ± 70 |
| | 23550 | 7352 | Biotitized actinolite schist | Biotite | 6.580 | 0.894 | 1972 ± 50 |
| | 23617 | 7360 | Mesocratic plagiogneiss with HAM | Biotite | 7.140 | 0.712 | 1621 ± 38 |
| | | | | Muscovite | 5.900 | 0.599 | 1640 ± 40 |
| | 23869 | 7388 | Si-amphibolite | Biotite | 6.900 | 0.813 | 1805 ± 50 |
| | 24067[a] | 7431 | Mesocratic plagiogneiss with HAM | Biotite | 7.520 | 0.779 | 1660 ± 30 |
| | | | | Muscovite | 7.520 | 0.710 | 1560 ± 30 |
| | 24216 | 7468 | Fe-Mg-amphibolite | Biotite | 6.060 | 0.697 | 1778 ± 50 |
| | 24305 | 7485 | Melanocratic plagiogneiss with HAM | Muscovite | 7.400 | 0.762 | 1648 ± 50 |
| | 24461[a] | 7508 | Si-amphibolite | Biotite | 8.070 | 0.875 | 1710 ± 30 |
| | | | | Amphibole | 0.656 | 0.074 | 1750 ± 55 |
| | 24618 | 7535 | Mesocratic plagiogneiss with HAM | Biotite | 7.550 | 0.789 | 1684 ± 40 |
| | | | | Muscovite | 8.030 | 0.812 | 1635 ± 40 |
| II | 24784 | 7658 | Glimmerite (in talc-actinolite schist) | Biotite | 7.850 | 1.030 | 1930 ± 45 |
| | 24823 | 7660 | Fe-amphibolite | Hornblende | 0.260 | 0.0378 | 2054 ± 88 |
| | 24915 | 7678 | Biotite-hornblende plagiogneiss | Biotite | 6.810 | 0.924 | 1970 ± 45 |
| | 24919 | 7679 | Al-Mg-amphibolite | Biotite | 7.220 | 0.782 | 1710 ± 35 |
| | | | | Hornblende | 0.480 | 0.0568 | 1810 ± 50 |
| | 24964 | 7691 | Fe-amphibolite | Hornblende | 0.314 | 0.0501 | 2169 ± 50 |

**Table 1.17** (continued)

| Stratum | Sample, nos. | Depth, m | Rock | Mineral | K, % | $^{40}Ar \times 10^{-3}$ $cm^3$ $year^{-1}$ | |
|---|---|---|---|---|---|---|---|
| 1 | 2 | 3 | 4 | 5 | 6 | 7 | 8 |
| II | 25214[a] | 7786 | Leucocratic biotite plagiogneiss | Plagioclase | 0.439 | 0.057 | 1910 ± 60 |
| | | | | Biotite | 7.970 | 0.663 | 1430 ± 30 |
| | 25464 | 7792 | Leucocratic biotite gneiss | Muscovite | 8.250 | 0.782 | 1567 ± 50 |
| | 26509-D | 7895 | Glimmerite (in actinolitic schist) | Biotite | 7.940 | 0.978 | 1856 ± 35 |
| | 26558 | 7905 | Leucocratic biotite plagiogneiss | Biotite | 7.480 | 0.811 | 1712 ± 40 |
| | 27029 | 7967 | Al-Mg-amphibolite | Hornblende | 0.247 | 0.045 | 2340 ± 80 |
| | | | | Cummingtonite | 0.123 | 0.018 | 2030 ± 70 |
| | 26873 | 7980 | Leucocratic biotite plagiogneiss | Biotite | 7.090 | 0.542 | 1352 ± 27 |
| | | | | Muscovite | 8.330 | 0.756 | 1523 ± 50 |
| | 26918 | 7990 | Epidote-biotite | Biotite | 8.020 | 0.936 | 1790 ± 40 |
| | 27248 | 8020 | Leucocratic biotite plagiogneiss | Muscovite | 7.810 | 0.677 | 1475 ± 38 |
| | | | | Biotite | 6.390 | 0.620 | 1591 ± 40 |
| | 27741 | 8120 | Leucocratic biotite gneiss | Muscovite | 8.350 | 0.801 | 1579 ± 50 |
| | 27875 | 8128 | Leucocratic biotite plagiogneiss | Biotite | 7.700 | 1.010 | 1930 ± 42 |
| | 28004 | 8189 | Pegmatite granite | Muscovite | 6.870 | 0.848 | 1858 ± 50 |
| | 28107 | 8229 | Leucocratic biotite gneiss | Biotite | 6.340 | 0.683 | 1704 ± 50 |
| | 28353 | 8265 | Fe-amphibolite | Biotite | 7.260 | 1.360 | 2377 ± 60 |
| | 28825[a] | 8376 | Fe-Mg-amphibolite | Amphibole | 0.740 | 0.091 | 1850 ± 60 |
| | 28918 | 8406 | Granite | Biotite | 4.780 | 0.442 | 1541 ± 45 |
| | | | | Muscovite | 7.140 | 0.800 | 1748 ± 40 |
| | 28933[a] | 8410 | Fe-amphibolite | Amphibole | 0.860 | 0.150 | 2280 ± 70 |
| | | | | Biotite | 7.640 | 0.756 | 1610 ± 35 |
| | 29067 | 8418 | Al-Fe-amphibolite | Hornblende | 0.452 | 0.0691 | 2116 ± 60 |
| | | | | Biotite | 5.100 | 0.595 | 1794 ± 40 |
| | 30654 | 8678 | Leucocratic epidote-biotite plagiogneiss | Biotite | 6.910 | 0.640 | 1543 ± 45 |
| | 31318 | 8775 | Hornblende-biotite plagiogneiss | Biotite | 7.420 | 0.714 | 1583 ± 50 |
| | 32000 | 8967 | Si-amphibolite | Biotite | 6.900 | 0.802 | 1785 ± 40 |
| | 33400 | 9030 | Granite | Muscovite | 7.860 | 0.710 | 1517 ± 50 |
| | 33727[a] | 9049 | Fe-amphibolite | Amphibole | 0.930 | 0.109 | 1800 ± 50 |

**Table 1.17** (continued)

| Stratum | Sample, nos. | Depth, m | Rock | Mineral | K, % | $^{40}Ar \times 10^{-3}$ cm$^3$ year$^{-1}$ | |
|---|---|---|---|---|---|---|---|
| 1 | 2 | 3 | 4 | 5 | 6 | 7 | 8 |
| II | 33830 | 9074 | Leucocratic biotite gneiss | Biotite | 7.150 | 0.986 | 1990 ± 50 |
| | | | | Muscovite | 6.080 | 0.553 | 1524 ± 50 |
| | 34855 | 9267 | Biotitized actinolitic schist | Biotite | 7.080 | 0.849 | 1825 ± 50 |
| | 35403 | 9449 | Porphyroblastic Fe-amphibolite | Biotite | 7.250 | 1.010 | 2002 ± 55 |
| | 35413 | 9451 | Porphyroblastic Fe-amphibolite | Amphibole | 0.680 | 0.105 | 2120 ± 60 |
| III | 35515[a] | 9481 | Porphyroblastic granite | Biotite | 8.030 | 0.688 | 1460 ± 40 |
| | 35552 | 9499 | Porphyroblastic granite | Muscovite | 8.730 | 0.791 | 1520 ± 40 |
| IV | 35891 | 9672 | Glimmerite (in talc-actinolitic schist) | Phlogopite | 5.460 | 0.623 | 1769 ± 40 |
| | 36158 | 9748 | Fe-amphibolite | Hornblende | 0.320 | 0.0486 | 2108 ± 80 |
| | 36580 | 9906 | Porphyroblastic Fe-amphibolite | Hornblende | 0.812 | 0.0980 | 1832 ± 24 |
| | 36923 | 9860 | Porphyroblastic Fe-amphibolite | | 0.768 | 0.980 | 1897 ± 21 |
| V | 38378[a] | 10315 | Hornblende-biotite plagiogneiss | Amphibole | 1.200 | 0.124 | 1660 ± 50 |
| | | | | | 8.100 | 0.840 | 1660 ± 40 |

*Note:* I, III, V – layers of the meso-, melanocratic plagiogneisses with HAM; II, IV – layers of the leucocratic biotite plagiogneisses and gneisses. Analyses were made by T. V. Koltsova[a], S. N. Voronovsky, L. V. Ovchinnikova.

A wide range of age determinations performed with the help of the K-Ar technique is characteristic of rocks from the Pechenga complex. Rocks of the upper part of the Pechenga complex (Nickel suite) are 1.9 – 1.5 Ga old, while more intensively metamorphosed rocks of the lower part (Luostari series) are older than 1.9 Ga (Table 1.16).

The age of rocks from the Kola series (Table 1.17), determined by the K-Ar technique, generally corresponds to that of those from the Pechenga complex; it also varies over a broad range, and only in amphibolites does it approach a mean value (1830 ± 110 Ma). It is also worth noting that the values determined for the muscovite age differ only slightly from each other. These differences have only a negligible bearing on the age determination of the associated biotite. As already mentioned above, the muscovite present in the leucocratic biotite plagiogneisses is of a superimposed nature, and in the meso- and melanocratic plagiogneisses with highly aluminiferous minerals, it is in a state of equilibrium with biotite. Therefore, the data obtained, having been processed, were grouped together, accounting for a factor of superimposed muscovitization. A process of regional greisenization of rocks from the Kola series, which occurred 1532 ± 135 Ma ago, has resulted in lowering the mean age values for biotite from the muscovitized rocks (1572 ± 108 Ma). Thus the latest age maximum appears to be statistically determined as 1.5 Ga.

Biotites from rocks are confined to an older maximum of the lower part of the Pechenga complex (1827 ± 178 Ma), mica (1830 ± 110 Ma) and amphiboles (1015 ± 134 Ma). It should be noted that the K-Ar technique has allowed determination of the age of the prograde metamorphism of the Pechenga complex. However, due to the manifestation of this prograde metamorphism and subsequent regional greisenization, the initial age of the Kola series rocks, determined by this technique, was halted.

Additional age determinations for the metamorphic processes in the Kola series have been obtained by the Rb-Sr isochron method (Table 1.18).

Data received for the leucocratic biotite gneisses correspond to an isochron of 1663 Ma and an initial ratio of $^{87}Sr/^{86}Sr = 0.7073$. The isochron equation is as follows: $^{87}Sr/^{86}Sr = 0.7073 + 2.39 \times 10^{-2}\, ^{87}Pb/^{86}Sr$. Isotope data for amphibolites are also confined to an isochron; however, this isochron lies below an isochron for gneiss; the isochron equation is as follows: $^{87}Sr/^{86}Sr = 0.745 + 2.32 \times 10^{-2}\, ^{87}Pb/^{86}Sr$. The age of 1617 Ma and the initial strontium ratio of 0.7045 correspond to this equation. An isochron built for the leucocratic bimicaceous plagioclases with HAM lies at an even lower level; it corresponds to a minimal age (1360 Ma) and to a maximal value of the initial strontium ratio (0.7428). Isotope data for the biotitized actinolite schist (sample Nos. 34852 – 34855) do not fit any of the isochrons built and, consequently, the age of this schist was calculated separately (22,252 Ma); a value of 0.7073 is taken as the initial strontium ratio. Data on three glimmerite samples from the endocontact zones of amphibolite schists also do not form an isochron; hence, their age was calculated by taking individual values into consideration, with the initial ratio being equal to 0.7073. As a result, the age of glimmerite from the talc-actinolite schist (sample No. 24784) was determined as 2182 Ma, and age of glimmerite from the biotite-amphibolite schist (sample No. 27209) proved to be similar to an

**Table 1.18.** Isochron age and initial ratio $^{87}Sr/^{86}Sr$ in metamorphic rocks of the Kola complex from the SG-3 well section

| Nos. | Sample (depth interval, m) | $^{86}Sr$, g t$^{-1}$ | $^{87}Rb$, g t$^{-1}$ | $^{87}Sr/^{86}Sr$ | $^{87}Sr/^{86}Sr$ | Age, Ma |
|---|---|---|---|---|---|---|
| 1 | 2 | 3 | 4 | 5 | 6 | 7 |
| Isochron No. 1 (leucocratic two-mica plagiogneisses with HAM) | | | | | | |
| 1 | 35570 | 21.30 | 46.64 | 0.7674 ± 5 | 2.190 | |
| 2 | 35571 | 22.52 | 36.14 | 0.7557 ± 5 | 1.604 | |
| 3 | 35574 (9500 – 9517) | 23.19 | 40.38 | 0.7593 ± 3 | 1.741 | |
| 4 | 35575 | 34.13 | 26.43 | 0.7239 ± 1.5 | 0.774 | |
| Isochron age at $(^{87}Sr/^{86}Sr)_0 = 0.7248$ | | | | | | 1360 ± 100 |
| Isochron No. 2 (leucocratic biotite plagiogneisses) | | | | | | |
| 5 | 26927 (7979 – 7985) | 29.54 | 7.55 | 0.7131 ± 8 | 0.255 | |
| 6 | 26950 (7984 – 7991) | 18.18 | 27.36 | 0.7427 ± 2 | 1.504 | |
| 7 | 27465 (8064 – 8067) | 24.00 | 18.47 | 0.7286 ± 2 | 0.769 | |
| 8 | 28107 (8214 – 8222) | 44.69 | 22.35 | 0.7180 ± 5 | 0.500 | |
| Isochron age at $(^{87}Sr/^{86}Sr)_0 = 0.7073$ | | | | | | 1663 ± 50 |
| Isochron No. 3 (amphibolites, including biotitized amphibolites Nos. 9, 10) | | | | | | |
| 9 | 24948 (7670 – 7678) | 24.74 | 2.13 | 0.7069 ± 2 | 0.0861 | |
| 10 | 28353 (8257 – 8261) | 26.81 | 4.78 | 0.7093 ± 2 | 0.230 | |
| 11 | 29671 (8532 – 8551) | 18.44 | 1.72 | 0.7063 ± 5 | 0.0932 | |
| 12 | 31125 (8714 – 8734) | 16.91 | 9.87 | 0.7181 ± 5 | 0.584 | |
| 13 | 31536 (8836 – 8865) | 13.8 | 1.82 | 0.7072 ± 1 | 0.132 | |
| 14 | 31998 + 32000 (8952 – 8959) | 26.67 | 3.31 | 0.7079 ± 8 | 0.124 | |
| Isochron age at $(^{87}Sr/^{86}Sr) = 0.7045$ (6.7044)[a] | | | | | | 1617 ± 40 (1648 ± 46)[a] |
| Dating of individual samples | | | | | | |
| 15 | 24784 (7643 – 7648 glimmerite-biotite gneiss) | 0.15 | 91.00 | 19.8007 ± 10 | 606.6 | 2182 ± 75 |
| 16 | 27209 (7998 – 8006) glimmerite-biotite gneiss | 24.29 | 79.92 | 0.7866 | 3.29 | 1677 ± 65 |
| 17 | 27562 (8096 – 8100) glimmerite-biotite gneiss | 0.45 | 44.16 | 2.7332 ± 5 | 98.1 | 1439 ± 55 |
| 18 | 34852 (9066 – 9075) biotitized actinolitic schist (Bi = 15%) | 5.26 | 17.02 | 0.8111 ± 7 | 3.23 | 2251 ± 70 |

[a] In brackets are given values of: $^{87}Sr/^{86}Sr$ and age of amphibolites deprived of the biotitization signs (Nos. 9 – 10). Analyses were made by S. N. Voronovsky and E. K. Khokhlov.

isochron for plagiogneisses (1677 Ma). The age of glimmerite from the actinolite schist (sample No. 27563) was established as 1439 Ma (at $^{87}Sr/^{86}Sr = 0.7073$).

The data presented above suggest that the period of the chemical closing of the "rubidium-strontium" system corresponds to the age of 1650 Ma, and the period of closing of the "potassium-argon" system corresponds to the age of 1530 Ma.

Additional information on the structural-petrological evolution of the metamorphic complexes may be obtained by examining the internal structure of fault systems and inclusions present in rock-forming minerals, as well as by correlating all geological, geochemical and geochronological data received in the well and the adjacent area.

It should be said that, on the one hand, the study of metamorphism on the basis of data obtained in the well has provided results which in general confirm the investigations carried out at the surface, but on the other hand, they allow some new conclusions.

By examining the data of the SG-3 well, it became possible for the first time to study in detail the metamorphic zonality in one continuous section within a range from the prehnite-pumpellyite to amphibolite facies, to characterize mineral parageneses of different facies and subfacies, and the physicochemical conditions of their formation.

It was established that conditions of metamorphism do not change in passing from metabasites of the Pechenga complex to granite-gneisses and amphibolites of the Kola series, despite the presence of a distinct stratigraphical break. In the majority of cases, the association of the metamorphic minerals of the Kola series also corresponds to that of the amphibolite facies. Hence, the prograde metamorphism superimposed on the Pechenga complex was intensively manifested also in older rocks.

The granitization of the Kola series was completed under the same thermodynamical amphibolite-facies conditions by way of metasomatic replacement and partial melting of rocks at normal and decreased potassium activity.

The primary lithological composition of rocks exerted a considerable influence on rock metamorphic transformation and granitization. In one and the same suite of the Pechenga complex, metamorphism of greater intensity was manifested in more basic magmatic rocks than in sedimentary rocks; in the Kola series the grade of metamorphism of gneisses is much higher than that of amphibolites.

Down the section both the temperatures of the prograde metamorphism and the degree of the mineral association equilibrium increase. Correspondingly, three zones are distinguished in the SG-3 well section: (1) a zone of chemical and structural non-equilibrium (0 – 2100 m); (2) a zone with relict primary textures and manifestation of biometasomatism (2100 – 3200 m), and (3) a zone of full recrystallization of rocks in conditions approaching equilibrium (below 3200 m).

Zones of retrograde metamorphism, coinciding with zones of faulting, were discovered in the well below a depth of 6000 m by examining core samples. It was established that the retrograde metamorphism proceeded under conditions of the greenschist and prehnite-pumpellyite facies.

On the basis of K-Ar and Rb-Sr dating, it became possible to determine the age of the prograde metamorphism of rocks of the underlying Kola series (2.1 – 1.7 Ga) as well as the age of the later regional greisenization (1.6 – 1.5 Ga).

# 1.6 Ore Mineralization

The sulphide copper-nickel deposits of the Pechenga ore field are confined to a profile running across the north-western part of the Kola peninsula (Atlas 1973; Gorbunov 1968; Yeliseev 1961). Most of the deposits are localized within the limits of the Pechenga ore field, being confined to the complex of the same name, and a lesser part was discovered in the Kola series. Ferruginous quartzites, muscovite-bearing pegmatites and hydrothermal lead-zinc veins are also known in the area under consideration.

Gorbunov (1968) established that sulphide copper-nickel ores of the Pechenga ore field are spatially and genetically associated with the altered basic-ultrabasic intrusions; these ores are of primary-magmatic origin, but were subject to later metamorphism and hydrothermal influence which facilitated the formation of rich epigenetic ores. The localization of the copper-nickel mineralization was a result of a complex combination of lithological, stratigraphical, tectonic and magmatic factors. The majority of the nickeliferous intrusions is concentrated in the productive Zhdanov suite and confined to its lower and upper parts; these deposits form two linear ore concentration areas, i.e. the Western and Eastern.

Transverse folds, made up of volcano-sedimentary rocks, as well as interstratal tectonic zones, play an important role in the distribution of nickeliferous intrusions. These latter are usually confined to the bottom parts of basic-ultrabasic intrusions and determine the position of the epigenetic breccia-like ore deposits (Structures of copper-nickel are fields 1978).

One of the main objectives in drilling the Kola super-deep borehole in the central part of the Pechenga ore field was to evaluate the copper-nickel potential of its deep horizons. This task was successively solved. The well encountered a previously unknown zone of ultrabasites with copper-nickel ores in the middle part of the Zhdanov suite. Further drilling has proved the presence of ore mineralization not only in the Pechenga complex but also in the underlying Kola series, and has revealed great varieties in genetic types.

Ore minerals occur throughout the entire well section penetrated by the Kola superdeep borehole. The principal are as follows: pyrrhotite, chalcopyrite, pyrite and pentlandite of the sulphides, and magnetite, ilmenite and leucoxene of the oxides. The content of the sulphides is not uniform, and changes up to 50% in individual parts of single grains. Volcano-sedimentary rocks of the Zhdanov suite are characterized by the highest sulphide concentration; here pyrrhotite mineralization predominates. Masses of ultrabasic rocks with copper-nickel mineralization are also confined to this suite. Below a depth of 3000 m, rocks of the Pechenga complex usually contain less than 1% sulphides, and only in an interval of 6000 – 6835 m does this rise to 4%; here chalcopyrite predominates. Morphologically, sulphides occur in the form of the bylayer impregnations (from hundredths of a millimeter to 2 mm in size), streaks, bands and lens-like segregations, sometimes associated with calcite and quartz. The amount and size of sulphide bodies become noticeably greater in zones with cataclastic structures and rock foliation.

Sulphides are also found irregularly distributed in rocks of the Kola series. Their relative content changes as follows: in gneisses from single grains to 0.5%; in individual samples 1 – 2%; in amphibolites 0.1 – 0.3%; in areas of sulphide accumulation 3 – 5%; in meta-ultrabasites 0.1 – 0.5%. In foliated or poorly foliated rocks sulphides form thin uniformly disseminated impregnations, rare large impregnations and streaks. In gneisses and amphibolite impregnations, lens-like and veinlet-like segregations are more frequently confined to foliated planes. Veinlets and cross-cutting foliation are also met.

The distribution of oxide mineralization, also developed throughout the well section, is more uniform than sulphide mineralization. The content of ore oxides varies in single grains by 5 – 10%. They dominate over sulphides throughout the section, the only exception being the Zhdanov suite.

Of the oxide minerals, magnetite, titanomagnetite, chrome-spinels, ilmenite, rutile, hematite and leucoxene are established. The highest content of oxide minerals, principally magnetite, in the Proterozoic complex was established in metadiabases of the Pirttiyarvi suite as well as in metaperidotites; the latter also contain ilmenite, titanomagnetite and chrome-spinels. Leucoxene predominates among oxides in the volcano-sedimentary rocks.

In rocks of the Archean complex oxide mineralization is represented by ilmenite, magnetite and rutile. Their distribution in rocks is not uniform. In gneisses and amphibolites only ilmenite is usually present, in schists and titanium-bearing orthoamphibolite both minerals are met; in iron quartzites only magnetite occurs. Regarding the quantity of minerals it may be said that ilmenite and magnetite are considered to be either accessory or sometimes main ore minerals.

In more rare instances, ore minerals are represented by compounds of lead (galenite), zinc (sphalerite), tellurium (altaite), cobalt (cobalt-pentlandite, siegenite, linnaeite, glaucodot, cobaltite), nickel (millerite, violarite, gersdorffite), molybdenum (molybdenite), silver (argentopentlandite, silver sulphide and telluride), chromium (chrome-spinel). The distribution of these minerals in the SG-3 well section is shown in Table 1.19.

On the basis of the geological, petrographical and mineralogical study of core samples from the Kola well, several types of ore mineralization have been distinguished.

1. Sulphide copper-nickel mineralization associated with basic-ultrabasic intrusions in the Zhdanov suite, and copper-nickel mineralization in metabasites and meta-ultrabasites of the Kola series.
2. Iron-titanium mineralization in metabasites of the Kola series.
3. Ferruginous quartzites among granite gneisses of the Kola series.
4. Hydrothermal sulphide mineralization in zones of retrograde dislocation metamorphism in metabasites of the Pechenga complex and in rocks of the Kola series.

**Table 1.19.** Distribution of ore minerals in the SG-3 well section

| Rocks | Minerals | |
|---|---|---|
| | Main | Accessory and rare |
| *Pechenga complex* | | |
| Metabasites | Pyrrhotite, leucoxene | Chalcopyrite, pyrite, sphalerite, pentlandite, cobalt-pentlandite, argento-pentlandite, marcasite, linnaeite (?), cobaltite, siegenite, mackinawite, bornite, molybdenites, galenite, sulphide silver (?), ilmenite, hematite, magnetite, graphite |
| Metaperidotites | Pyrrhotite, pentlandite, magnetite, chalcopyrite, ilmenite | Sphalerite, cobaltite, mackinawite, silver telluride (?), chrome-spinellids, titanomagnetite, leucoxene |
| Metamorphosed volcano-sedimentary rocks | Pyrrhotite, pyrite, leucoxene | Chalcopyrite, sphalerite, pentlandite, cobaltite, arsenopyrite, molybdenite, galenite, glaucodot, altaite, graphite, hematite, ilmenite |
| *Archean complex* | | |
| Gneisses | Pyrrhotite, pyrite, chalcopyrite | Pentlandite, sphalerite, siegenite, molybdenite, galenite, gersdorffite, millerite, intermetallic compound of Cu, Zn, Ni, Co; native copper, graphite; ilmenite, magnetite, rutile, hematite, sphene, ilvaite |
| Amphibolites | Pyrrhotite, pyrite, chalcopyrite, magnetite, ilmenite, rutile, sphene | Pentlandite, siegenite, sphalerite, molybdenite, marcasite, argentopentlandite, bornite, galenite, violarite, leucoxene |
| Meta-ultrabasites | Pyrrhotite, pyrite, chalcopyrite, pentlandite, ilmenite | Siegenite, galenite, cobalt-pyrite, magnetite, rutile, sphene |

The investigation of ore mineralization of the first type is of interest from genetical and practical aspects. Studies of other types of ore mineralization are important in tackling the general problem of the distribution of ore concentrations in vertical sequence in the Earth's crust.

### 1.6.1 Sulphide Copper-Nickel Mineralization

#### Conditions of Localization

Sulphide copper-nickel mineralization in association with metaperidotites was encountered by the Kola superdeep well in the middle part of the Zhdanov suite, which is characterized by thin rhythmic alternation of phyllites, siltstones, sandstones and the presence of abundant volcanic material.

Three massifs of metaperidotites have been established in the Zhdanov volcano-sedimentary suite by the well in intervals of 1541 – 1677, 1756 – 1788 and 1802 – 1808 m.

The largest intrusion (1541 – 1647 m) of about 100 m true thickness is composed of considerably serpentinized peridotites. The marginal parts of this massif have undergone the strongest alteration (talcose, albitized).

Peridotites have been transformed into serpentinites, chlorite-serpentinite, tremolite-serpentinite, talc-tremolite and other analogous rocks. Clinopyroxene is the only magmatic rock-forming mineral met in the form of relicts: however, it disappears at a depth of 1563 m. In places porphyritic textures are preserved, characterized by the presence of short prismatic brains of clinopyroxene occurring among serpentine. Relict porphyritic textures are notable for the presence of serpentine pseudomorphs in olivine in the chlorite-tremolite main mass. Reticulate or cellular structures are common in serpentinites. Titanium-bearing brown amphibole-kaersutite, epimagmatic or early autometamorphic by origin, is considered to be one of the earliest newly formed minerals. Intensively coloured actinolite hornblende may have been formed together with amphibole-kaersutite. Other minerals are of late-autometamorphic and regional metamorphic origin. The most interesting bodies of copper-nickel ores encountered by the SG-3 well are associated with this same massif of ultrabasic rocks.

The second massif of ultrabasic rocks, encountered in an interval of 1756 – 1788 m, is also composed of fully altered (serpentinized, tremolitized and talcose) peridotites containing only a poorly mineralized body of the impregnation type in the middle part of the section. The third body of ultrabasic rocks established in an interval of 1802 – 1808 m is represented by foliated peridotites transformed into serpentinite-talc and tremolite-talc schists. A zone of foliation is filled with veinlets and sulphide selvages, and contains brecciated ore in its upper part.

Sulphide copper-nickel mineralization, associated with the first ultrabasic massif, is also developed in the enclosing phyllites met in the upper and lower boundaries of massif. A study of core samples from the SG-3 well has made it possible to recognize several zones in this massif, which differ from each other in peridotite metamorphism and in copper-nickel mineralization (Fig. 1.51).

The upper part of the massif, 13 m thick (1541 – 1554 m), is composed of metaperidotites, which contain not only serpentine and chlorite but also talc and tremolite; kaersutite and green actinolite hornblende, developed in clinopyroxene, are also present in small quantities. Clinopyroxene is fully replaced by serpentine (bastite) and amphiboles. Ore mineralization is represented by poor sulphide impregnation. As a rule, rocks clearly exhibit relict magmatic textures, i.e. poikilitic and panidiomorphic. They are seen as pseudomorphs of bastite serpentine and magnetite in clinopyroxene, and as pseudomorphs of reticulate serpentine in olivine.

The main part of the intrusion, about 100 m thick (interval 1554 – 1654 m), is represented by serpentinized and chloritized peridotite, in which not only relict magmatic textures but also relict clinopyroxene have survived. Relict clinopyroxene disappears in narrow zones of intensive serpentinization and carbonitization. Kaersutite and green actinolitic hornblende are also occasionally met. Talc is absent. Tremolite occurs in rare cases. Small veinlets or chains of magnetite grains, together with serpentinite and chlorite, constitute pseudomorphs in olivine and clinopyroxene. The lower part of the intrusion was subject to the strongest alteration and deformation; here a zone of talc-bearing metaperidotites is separated from the talc-tremolite foliated metaperidotites by chlorite schists and blastomylonites.

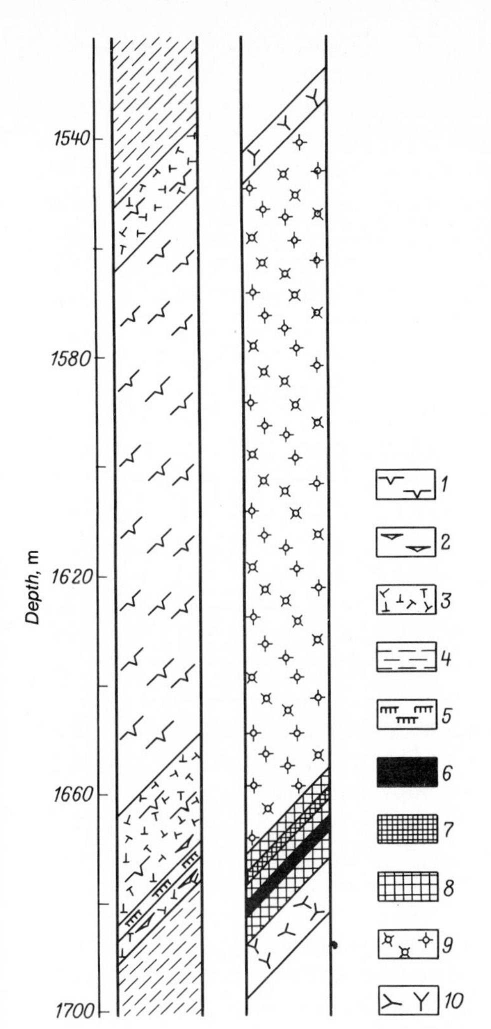


**Fig. 1.51.** Structure of nickeliferous massif of ultrabasites in the SG-3 well section: Zonality of metaperidotites and zonality of the sulphide copper-nickel mineralization: *1*–*3* metaperidotites (*1* massive; *2* foliated; *3* talcose); *4*–*5* phyllites and siltstones (*4* layered; *5* foliated); *6*–*8* ores (*6* brecciated; *7* densely impregnated; *8* disseminated-impregnated); *9*–*10* mineralizations (*9* poorly impregnated in metaperidotites; *10* veinlet-impregnated in phyllites and siltstones)

The character of the copper-nickel mineralization also changes down the section. In the upper part of the metaperidotite massif, to a depth of 1665 m, poor mineralization of impregnation type is developed; the content of sulphides here is 1–3% of rock bulk, that is almost the same as in the middle part of the intrusion. Then, metaperidotites become enriched in sulphides amounting to 10% in an interval of 1665–1668 m, and 15–40% in an interval of 1668–1671 m, the average content being 20%. In the lower part of this zone (1671–1674 m), represented by the strongly foliated talcose and tremolitized peridotites, the amount of sulphides again falls to 5–7%. A peculiar feature of this zone is the presence of sulphides tightly intergrown with chaotic prismatic crystals of tremolite; these intergrowths are widely spread. It is also characteristic of the upper part of the zone that magnetite here forms widely spread intergrowths with tremolite and sulphides; magnetite also replaces tremolite and sulphides. Pyrrhotite, the predominant mineral among sulphides, is represented only in its hexagonal modification.

In an interval of 1674 – 1675.5 m, metasedimentary rocks were encountered. They constitute either a xenolith or a tectonic fragment, which became held among ultrabasites while moving along a zone of blastomylonites. Rocks are represented by chloritic schists and sericitic siltstones. The former are composed of unidirectionally oriented lamellae and flakes of chlorite among which small lens-like accumulations of leucoxene occur. Siltstones consist of flaky sericite containing small fragments of quartz and feldspar. Rocks are full of the quartz-sulphide lenses and veinlets extending along planes of foliation. In places the content of sulphides reaches 10 – 15%; they cement rock fragments and form characteristic brecciated structures.

Blastomylonites in meta-ultrabasites are characterized by their tremolite-talc-carbonate composition and contain a subordinate amount of serpentine and chlorite. Relict magmatic textures of these rocks, notable for the presence of foliated and brecciated structures, did not survive here at all. Sulphides form elongated or veinlet-like segregations, oriented along foliation planes, or constitute the breccia cement. The breccia-like ores with fragments of foliated metaperidotites adjoin breccia in phyllites. The presence of grains of nickel and cobalt sulpho-arsenides, as well as sphalerite, which occur here in addition to pyrrhotite, pentlandite and chalcopyrite, is a peculiar feature of the sulphide mineralization in this zone.

A typical pentlandite-pyrrhotite association in the enclosing phyllites is traced at a distance of 6 m in the upper boundary and at about 11 m in the low boundary of the massif, where it is then replaced by sedimentary pyrrhotite mineralization.

Several small bodies of metabasites and meta-ultrabasites with a typical copper-nickel mineralization (in intervals of 7340.0 – 7352, 7920.0 – 7941.5 m, etc.) have been recognized in the Kola series. Meta-ultrabasites of the granite gneiss complex are represented by a few varieties. Phlogopite-tremolite, talc-tremolite, phlogopite-talc and cummingtonite tremolite (with phlogopite) schists are the most characteristic varieties. All other intermediate varieties are also met. Chlorite and magnetite are present in all meta-ultrabasites. The subcontact parts of bodies and thin layers of meta-ultrabasites in places are turned into the phlogopite (biotite) schists or hornblende-biotite (phlogopite), sometimes quartz-bearing rocks. The sulphide segregations in zones of copper-nickel mineralization amount to 0.1 – 1.5%, but more frequently are less than 0.3%. Sulphides are usually irregularly distributed and form segregations of a lens-like or elongated streak-like shape, with the long part of these bodies being extended along the plances of rock foliation.

## Types of Ores and Their Mineral Composition

Three types of copper-nickel ore have been established in the Proterozoic complex: a disseminated-impregnated and densely impregnated type in the altered peridotites; a breccia type in the tectonic zones; and a streaky-impregnated type in phyllites. Only in the Archean complex has the disseminated-impregnated type of mineralization in substantially amphibole-rich rocks (metabasites and meta-ultrabasites) been recognized.

*Impregnated Ores.* Poor impregnated mineralization in peridotites is characterized by the following mineral content: sulphide minerals less than 2 – 3%, oxidized minerals about 5 – 7%. The altered peridotites with sulphide content reaching 5 – 10% are assigned to the disseminated impregnated ores, and those with sulphide content exceeding 15% to the densely impregnated ores (Fig. 1.52). Pyrrhotite, mainly of hexagonal type, is the main ore mineral; it is the basis of the impregnated mineralization, and occurs in the form of tight intergrowth with pentlandite and chalcopyrite. The latter is irregularly distributed in the form of small isolated segregations. In places the chalcopyrite content rises considerably. In serpentinites sulphides are always associated with magnetite, which forms segregated bodies of an irregular shape, streaks and rims at boundaries between sulphides and silicates. Sulphides are often replaced by magnetite. The amphibolitized peridotites are notable for the presence of the tightly intergrown sulphides and lamellar crystals of tremolite. Sulphide pseudomorphs in olivine are met very rarely. Grains of ilmenite, titanomagnetite and chrome-spinel are the primary minerals which are met in peridotites, the latter frequently becoming covered by rims of magnetite.

*Breccia Ores.* A typical breccia ore is shown in Fig. 1.53. Sulphides serve as cementing matter for phyllite fragments of different sizes (from hundredths of a millimeter to 12 cm) and shape, and also for ultrabasic rocks. The sulphide content varies from 20 to 90%. Pyrrhotite is the main ore mineral, represented by intergrowths of its monoclinic and hexagonal modifications; it constitutes irregularly grained allotriomorphic aggregates of 0.1 – 0.5 mm in size. Pentlandite is met in the form of lamellar inclusions present in pyrrhotite, rims on the peripheral parts of its grains, and isometric grains. Chalcopyrite, forming small grains and bands, is irregularly distributed; it occurs in the form of single grains or may form bodies in which its content reaches 50%. In places chalcopyrite contains grains of sphalerite.

Veinlet-impregnated mineralization in phyllites and siltstones is represented by thin uniformly disseminated impregnations, mottled accumulations, lenses and veinleits of sulphides, which occur in conformity with rock layering and foliation. The shape and size of the impregnations vary, depending on the grain sizes of the enclosing rocks. The size of sulphide grains in the clayey-micaceous phyllites does not exceed 0.01 mm. In addition to impregnations, thin sulphide veinleits, both parallel to and cross-cutting the rock layering, are also met here; they often contain quartz and carbonates. The size of the segregations present in the coarse-grained sandstone layers reaches 2 mm. Sulphides are located between grains of quartz and other rock-forming minerals. The sulphide content in individual layers is from 2 to 60%. Pyrrhotite predominates here (90 – 95% of the bulk sulphide volume); the content of pentlandite does not exceed 3%, and that of chalcopyrite is 3.5%. Sphalerite, individual grains of cobaltite, and other rare minerals are met in the form of occasional segregations; these minerals are tightly intergrown with chalcopyrite.

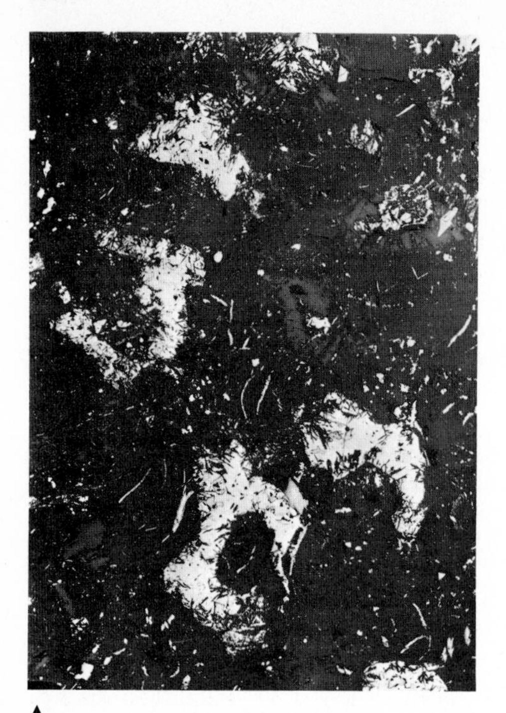

▲
**Fig. 1.52.** Disseminated-impregnated copper-nickel ore in metaperidotite. Polished thin section, ×20; sample No. 5565, depth 1643.5 m

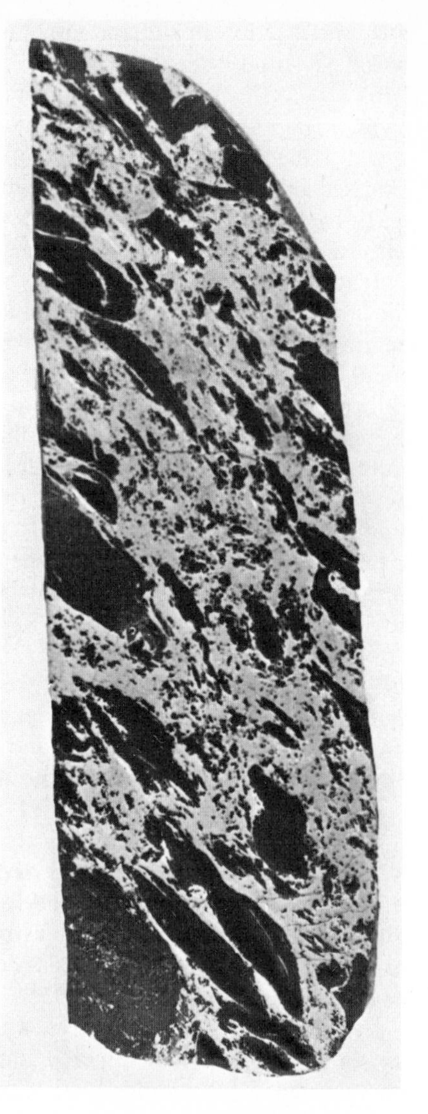

**Fig. 1.53.** Brecciated copper-nickel ore. Phyllite ► fragments (*black*) in sulphide cement (*light grey*). Polished lump of ore, 3/4 of the natural size, sample No. 6096, depth 1734 m

Pyrrhotite is a mineral which predominates in ores of all types. Its relative amount, taken with respect to the total bulk of sulphides, ranges from 60 to 90%. In the strongly altered peridotites it is met in the form of the xenomorphic segregations, and is tightly intergrown with spicular tremolite; in very rare cases it is met in the form of small rounded segregated bodies. In the breccia ores pyrrhotite serves as a cement, and in the veinlet-impregnated ores, occurring among phyllites, it constitutes accumulations, lenses and veinlets of different shape. Pyrrhotite aggregates are characterized by their fine-grained, allotriomorphic texture. Grain sizes vary from 0.01 to 0.2 mm. The presence of lamellar pentlandite inclusions is characteristic feature of the internal structure of pyrrhotite grains. These inclusions are almost always present in the impregnated ores, but rare in the breccia ores. Pyrrhotite is represented by its two

**Table 1.20.** Composition of pyrrhotite and pentlandite (X-ray spectral analysis, %)

| Sample, nos. | Depth, m | Type of ore | Mineral | Fe | Ni | Co | S | Total |
|---|---|---|---|---|---|---|---|---|
| 5110 | 1543.9 | Impregnated | Hexagonal pyrrhotite | 60.10 | 0.65 | – | 39.1 | 99.85 |
| | | | Pentlandite | 29.0 | 35.0 | 2.0 | 35.2 | 101.2 |
| 5565 | 1643.5 | Impregnated | Monoclinic pyrrhotite | 59.7 | 0.68 | 0.77 | 40.6 | 101.75 |
| | | | Hexagonal pyrrhotite | 60.0 | 0.68 | – | 39.3 | 99.98 |
| | | | Pentlandite | 29.0 | 35.0 | 1.10 | 35.0 | 100.10 |
| 5719 | 1666.9 | Impregnated | Pentlandite | 29.4 | 35.7 | 1.3 | 33.2 | 99.6 |
| 5273 | 1669.7 | Impregnated | Hexagonal pyrrhotite | 60.0 | 0.70 | – | 39.0 | 99.70 |
| 5762 | 1675.4 | Brecciated | Pentlandite | 29.7 | 36.5 | 1.3 | 33.1 | 100.6 |
| | | | Pentlandite | 30.3 | 36.2 | 0.76 | 33.8 | 101.0 |
| | | | Hexagonal pyrrhotite | 59.9 | 0.6 | 0.04 | 39.4 | 99.9 |
| | | | Monoclinic pyrrhotite | 59.4 | 0.35 | 0.03 | 40.7 | 100.5 |
| 5800 | 1682.9 | Veinlet-impregnated in phyllites | Pentlandite | 35.3 | 35.1 | 1.2 | 33.5 | 100.1 |
| | | | Pentlandite | 32.7 | 32.9 | 0.81 | 33.5 | 99.9 |
| | | | Hexagonal pyrrhotite | 60.0 | 0.55 | 0.04 | 39.6 | 100.2 |
| 5856 | 1692.5 | Veinlet-impregnated in phyllites | Hexagonal pyrrhotite | 60.4 | 0.05 | 0.05 | 39.0 | 99.5 |
| 5856 | 1693.6 | Veinlet-impregnated in phyllites | Hexagonal pyrrhotite | 60.2 | 0.08 | – | 39.0 | 99.28 |

Analysts: I. P. Laputina, G. N. Muravitskaya, Ya. A. Pakhomovsky.

modifications, i.e. hexagonal and monoclinic, which are met either separately or in the form of tight intergrowth. In all types of copper-nickel ores hexagonal pyrrhotite predominates. The relative content of monoclinic pyrrhotite in aggregates composed of ingrowths does not exceed 3 – 5%. Only in individual impregnations of larger size, met among the altered perioditites and at places in the cement of the breccia ores, does the content of monoclinic pyrrhotite reach 20 – 25%. Monoclinic pyrrhotite is usually seen in the form of rims in the peripheral parts of grains of hexagonal pyrrhotite, in fissures and around pentlandite inclusions in pyrrhotite.

The composition of pyrrhotite was studied with the help of an electronic probe *Cameca MS-46*. Results of the investigations are presented in Table 1.20. The composition of hexagonal pyrrhotite from ores of different types varies considerably in the amount of its main components. The iron content is 58.9 – 60.2%. The minimal content of iron is characteristic of hexagonal pyrrhotite from the breccia ores; this pyrrhotite is intergrown with monoclinic pyrrhotite. The iron content in monoclinic pyrrhotite is 59.4 – 59.7%. The results of compositional studies of hexagonal pyrrhotites, taking into account values of interplanar space, performed by P. Arnold's method, confirms the results of the analyses, which showed that the composition of hexagonal pyrrhotites, intergrown with monoclinic pyrrhotite, is characterized by a higher sulphur content. The atomic composition of hexagonal pyrrhotite changes from 48.1 to 47.9% Me; the Me content in hexagonal pyrrhotite, present in the form of intergrowths with monoclinic pyrrhotite, does not exceed 47.5%. These data are in agreement with a phase equilibrium diagram for the low-temperature area (below 320 °C) in the system Fe-S (Taylor 1970).

The relatively high nickel content in pyrrhotite (up to 0.70%) and low cobalt content are characteristic features. The nickel content in coexisting hexagonal and monoclinic pyrrhotites of peridotites is the same; in monoclinic pyrrhotite of breccia ores this content is half that in hexagonal pyrrhotite, which is intergrown with the former; this regularity and the mode of occurrence of monoclinic pyroxene (rims around the peripheral parts of the hexagonal pyrrhotite grains), if considered together, confirm the supposition that monoclinic pyrrhotite has been formed as a result of the replacement of hexagonal pyrrhotite under conditions, which facilitated the withdrawal of not only iron but also nickel (Genkin et al. 1965).

Pentlandite, tightly intergrown with pyrrhotite, is met in ores of all types. It forms narrow rims (0.05 mm width) along boundaries of pyrrhotite grains, isometric grains of 0.2 mm across in pyrrhotite or in serpentinous peridotite, and very small wedge-shaped flame-like lamellae in pyrrhotite grains. The relative content of pentlandite in various ores changes over a broad range. By chemical composition and by X-ray microanalysis data, pentlandites from ores of different types are shown to be similar to each other (see Table 1.20). The content of nickel extracted from pentlandite, which occurs in the impregnated ores in the altered peridotites, varies from 35,0 to 36,5%, the cobalt content is between 0.10 and 2.0%. It is characteristic that the non-uniform composition was established not only in different grains in one and the same thin section, but also in individual points of one and the same grain. This non-uniformity is particularly pronounced in pentlandites of veinlet-impregnated mineralization in phyllites, where the nickel content in three grains, measured at a great number of points, changes from 26.5 to 35.9%; the iron content is within the range of 21.3 – 37.5%, the cobalt content 0.35 – 1.0% and the sulphur content 32.0 – 33.0%. This compositional non-uniformity seems to indicate that pentlandite was formed under non-equilibrium conditions.

Chalcopyrite is met in ores of all types in quantities varying from individual grains (from 0.01 to 0.3 mm in size) to an amount of 3%. It forms very small inclusions and xenomorphic segregations in pyrrhotite grain, and rims in the peripheral parts of segregations at contacts with silicates. Chalcopyrite, like pyrrhotite, is met in the altered peridotites in the form of veinlet-like segregations tightly intergrown with tremolite; if often contains very small inclusions of sphalerite. In the mineralized phyllites, on contact with the impregnated ores in the intrusion, chalcopyrite forms thin, small lenses, isolated from pyrrhotite, which occur conformably with the phyllite layering. By microprobe analysis data, the composition of chalcopyrite is close to the theoretical composition. In addition to the main elements, admixtures of nickel (up to 0.01%) and cobalt (0 – 0.1%) where also established.

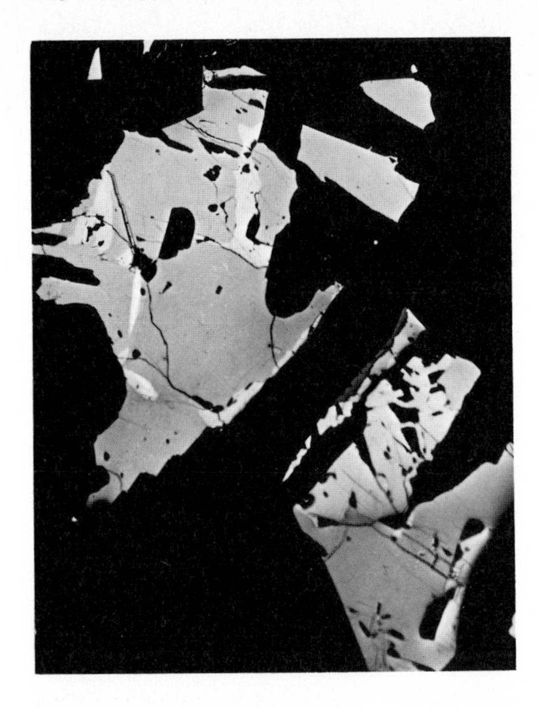

**Fig. 1.54.** Lamellar segregations of pentlandite (*white*) in pyrrhotite (*light grey*), in meta-ultrabasite. Polished thin section, ×20; sample No. 36023, depth 9674 m

Sphalerite belongs to those minerals which do not form large segregations, although they are widely spread in ores of all types. In xenomorphic segregations of small size (not greater than 0.2 mm) sphalerite is usually met in the form of tight intergrowth with chalcopyrite and pyrrhotite, sometimes with magnetite.

Individual isometric or elongated grains of cobaltite (up to 0.1 mm) were discovered among pyrrhotite and chalcopyrite in impregnated ores of altered peridotites. In the reflected light cobaltite is white in colour and isotropic. The character of the relationship between cobaltite and tremolite indicates that the former is later in origin. In addition to cobalt, all analyzed grains of cobaltite contain nickel (up to 8.1%) and iron (up to 5.0%). The nickel content is expressed by a maximal value, that of iron is minimal. Individual grains possess zonal internal structure. External zones are enriched in iron.

Chrome spinel, similar to ferrichromite, is present everywhere in the impregnated ores of the altered peridotites in the form of isometric grains (0.1 mm in size). Its relative content does not exceed 0.4%. Not infrequently chrome spinel grains are surrounded by rims of magnetite.

Magnetite is represented by thin dust-like impregnations, veinlets, small isometric grains intergrown with silicates and sulphides; it is also met in the form of narrow rims around chrome-spinel grains.

Pyrrhotite, represented mainly by its monoclinic modification, is the predominant mineral of the pentlandite-bearing association known in meta-ultrabasites and enclosing rocks. Pentlandite is developed in two main forms, i.e. thin lamellar segregations and porphyry formations (Fig. 1.54).

Lamellar and lens-like inclusions of pentlandite in pyrrhotite are the most widespread form of its segregation in copper-nickel mineralization. The size of

**Table 1.21.** Composition of pentlandite and coexisting pyrrhotite (X-ray spectral analysis, %)

| Sample, nos. | Depth, m | Mineral | Ni | Co | Fe | S | Total | $\frac{Ni+Co}{Fe}$ |
|---|---|---|---|---|---|---|---|---|
| 23482[a] | 7291.0 | Pentlandite | 35.19 | 2.12 | 30.27 | 33.70 | 101.28 | 1.23 |
| | | Pyrrhotite | 0.95 | 0.11 | 59.37 | 40.06 | 100.49 | |
| 24209 | 7460.5 | Pentlandite | 31.82 | 7.42 | 27.45 | 34.02 | 100.71 | 1.43 |
| | | Pyrrhotite | 0.99 | 0.20 | 59.35 | 40.00 | 100.54 | |
| 26639 | 7928.1 | Pentlandite | 36.75 | 0.67 | 29.61 | 33.91 | 100.94 | 1.26 |
| | | Pyrrhotite | 2.8 | 0.07 | 57.49 | 39.53 | 99.89 | |
| 28210 | 8238.6 | Pentlandite | 37.90 | 0.70 | 28.06 | 33.25 | 99.91 | 1.38 |
| | | Pyrrhotite | 4.00 | 0.12 | 56.36 | 40.09 | 100.57 | |
| 28420[a] | 8293.0 | Pentlandite | 32.67 | 1.19 | 31.71 | 33.34 | 98.91 | 1.06 |
| | | Pyrrhotite | 1.77 | 0.04 | 58.44 | 39.53 | 99.78 | |
| 30142 | 8385.8 | Pentlandite | 32.48 | 4.47 | 28.90 | 32.25 | 98.10 | 1.27 |
| | | Pyrrhotite | 1.50 | 0.18 | 59.59 | 38.04 | 99.31 | |

[a] Copper-nickel mineralization in gneisses near meta-ultrabasites. Analysts: Z. P. Laputina, G. N. Muravitskaya.

lamellae and lenses of pentlandite, forming rare isolated or individual segregations in pyrrhotite, is $0.01-0.03 \times 0.7-0.15$ mm. They are most typical of amphibolites but are also met in meta-ultrabasites and conjugating gneisses. Many sulphide segregations are completely deprived of pentlandite and composed of ingrowths of pyrrhotite and chalcopyrite; sometimes they are of oval drop-like shape. Porphyritic segregations of pentlandite in pyrrhotite and chalcopyrite were met only in meta-ultrabasites (sample nos. 26639 and 34856). The size of pentlandite segregations reaches $0.1 \times 0.5$ mm. In shape and mode of occurrence (they occur among pyrrhotite and chalcopyrite) they are similar to analogous segregations met in typical copper-nickel ores.

Compositional studies of the coexisting pentlandite and pyrrhotite from meta-ultrabasites and metabasites of the Archean complex, and a comparison of their composition with that of pentlandite and pyrrhotite from the copper-nickel mineralization of the eastern ore field of the Pechenga area are presented in Table 1.21.

According to the compositional analysis of pentlandite, the cobalt content in this mineral varies over a great range, i.e. from 0.67 to 7.42%. The least cobalt concentrations were established in pentlandite, and the greatest in amphibolites. Similar values of a (Ni + Co)/Fe ratio in analyses Nos. 1, 4 and 6, despite different nickel and cobalt content in pentlandite, indicate Ni $\rightleftarrows$ Co isomorphism. In analysis No. 1, cobalt seems also to replace iron. The pentlandite from gneisses is the most ferruginous.

Pyrrhotite from copper-nickel mineralization is also characterized by varying nickel content (from 0.95 to 4%) and cobalt (from 0.04 to 0.020%) (Table 1.21). The highest content of nickel, 2.8 and 4% (analyses Nos. 3 and 4), was established in pyrrhotite from meta-ultrabasites.

## Genetic Peculiarities

Intrusions of ultrabasic rocks were obviously the source of copper-nickel mineralization in the Proterozoic complex; this is supported by the geological position of mineralization, the distribution pattern of the productive pentlandite-bearing associations and data on the isotope content of sulphide sulphur. An association of pyrrhotite with pentlandite and the increased (more than 0.5%) content of nickel in pyrrhotite, which occurs beyond the limits of ore-bearing serpentinite peridotites, are seen in the underlying phyllites only at a distance of 10.5 m.

It should be mentioned again that the enclosing volcano-sedimentary strata contain pyrite-pyrrhotite mineralization. Textures of this mineralization and modes of sulphide segregation in tuffites, phyllites and siltstones are rather different, i.e. streaky, impregnated and breccia-like. When met in streaks, sulphides are usually associated with quartz and carbonate. It is characteristic that sulphides of the streaky and impregnated types are represented by subordinate bodies with respect to phyllite layers, which are irregularly enriched with sulphides. Peculiar structures arise in zones of phyllite foliation, characterized by the presence of thinly dispersed sulphide impregnations in combination with streaky and lens-like segregations oriented along foliation planes.

Pyrrhotite is the main ore mineral. It is represented by its hexagonal modification or ingrowths of hexagonal and monoclinic pyrrhotite in which the content of the latter does not exceed 10 – 15%. The composition of pyrrhotite, as distinct from copper-nickel mineralization, is characterized by a comparatively low nickel content (less than 0.1%) and low content or absence of cobalt. A study of the sulphur isotope composition of pyrrhotite[8] from the volcano-sedimentary rocks has showed the presence of rather heavy sulphur ($\delta\,^{34}S$) in pyrrhotite; its content reaches +20.97‰ in tuffaceous sandstones and +19.46‰ in phyllites.

The distinctive features observed in the distribution of pyrrhotite and other massive sulphide mineralization in volcano-sedimentary rocks of the productive strata of the Pechenga series, the low content of nickel in pyrrhotite and the similarity between sulphur isotope abundance ($\delta\,^{34}S$) of pyrrhotite and seawater sulphates indicate that massive sulphide mineralization in sedimentary rocks originated together with the enclosing rocks as a result of diagenetic, epigenetic and subsequent metamorphic processes at the level of the greenschist facies.

Changes in the sulphur isotope abundance of sulphides in a zone of contact between metaperidotite and metasedimentary rocks indicates that massive sulphide mineralization has affected the formation of the copper-nickel mineralization (Fig. 1.55). The values of $\delta\,^{34}S$ of sulphides in impregnated ores and peridotites, as well as in conjugated brecciated and veinlet ores, vary from +0.65 to +2.92‰, whereas in phyllites, which occur under contact with peridotites, these values change from +5.29 to +8.54%. From this it appears that, as distinct from sulphur which is similar to meteoric sulphur present in ore-bearing metaperidotites, sulphur of the mineralized phyllites, containing massive sulphide mineralization, becomes distinctly heavier as the distance from contact with

---

[8] Determination of the sulphur isotope abundance in sulphides was made by L. P. Nosik with the help of MI-mass-spectrometer.

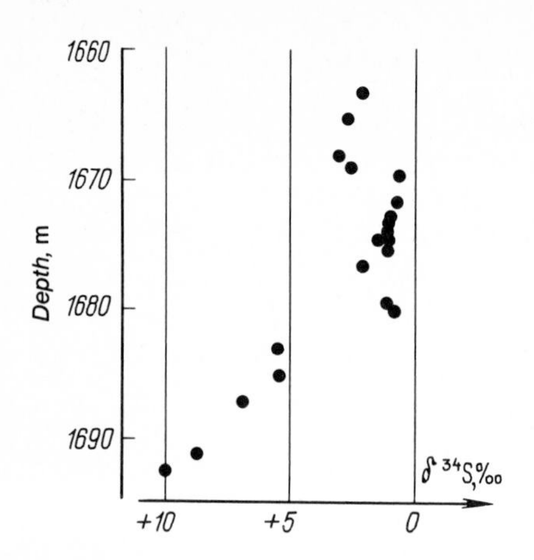


**Fig. 1.55.** Variations of sulphur isotope abundance in sulphide copper-nickel ores encountered by the SG-3 well

peridotites increases. The heaviest sulphur (+8.54%) was established in mineralized phyllites at the boundary of copper-nickel mineralization. Below the boundary, at a distance of 11 m from the contact with peridotites, the value of $\delta\ ^{34}S$ is already +9.76‰. This may be a result of the mixing of sulphide sulphur (similar to meteoric sulphur), penetrated from the overlying peridotites, with the sedimentary sulphur of massive sulphide mineralization of phyllites. The disappearance of pentlandite and the abrupt decrease of nickel in pyrrhotite at a distance of 11 m from the contact (1682.5 m) is also in agreement with this supposition.

Thus, copper-nickel ores of the Proterozoic complex encountered by the SG-3 well are similar to ores of the Pechenga deposits in geological position and mineral composition. This similarity is expressed, first of all, in that the known mineralization is localized in subbottom parts of the ultrabasic intrusions and in their nearest exocontact, with the highest concentrations being confined to these zones, and, secondly, in the character of the metamorphic alterations of the nickel-bearing peridotites. Analogous types of ore and distinctive modes of their occurrence (preservation of the primary-magmatic textures of the sulphide-silicate ingrowths) are met only in the poorly impregnated ores in peridotites, in the densely impregnated ores in peridotites which underwent intensive redeposition and recrystallization at the time of their formation, and in the breccia and streaky-impregnated ores in phyllites. The position of these richer ores is undoubtedly controlled by zones of dislocation metamorphism, along which recurrent tectonic displacements took place.

The copper-nickel mineralization in the Archean complex is undoubtedly similar to ores of the Allarechin deposit, located north of the Pechenga structure, in view of the relationship of this mineralization with ultrabasites occurring among granite gneisses (Yakovler and Yakovleva 1974). Ultrabasites and copper-nickel ores are also subject to metamorphism here under conditions of the amphibolite facies. It is known that metamorphic processes, before they reach the amphibolite stage, do not disturb the sulphur isotope ratio in sulphides (Ohmoto and Rye 1979).

In meta-ultrabasites of the Kola series, sulphur composition is similar to its meteoric standard. For instance, at a depth of 7928 m (sample No. 26639) $\delta\ ^{34}S$ proved to be +0.85‰ in pyrrhotite and +0.7‰ in chalcopyrite.

**Fig. 1.56.** Foliated structure of amphibolite with iron-titanium mineralization (*light grey*). Polished lump of ore, 9/10 of the natural size, sample No. 31078, depth 8711.5 m

## 1.6.2 Iron-Titanium Mineralization in Metabasites

Oxide mineralization, represented by magnetite, ilmenite and sometimes by rutile, is met throughout the SG-3 well section. High concentrations of these minerals in the iron-titanium ores were discovered in the Kola series at a depth of 8711 m in the foliated biotite amphibolites (Fig. 1.56). The chemical composition of the non-feldspar amphibolites is shown in Table 1.22. The high content of iron ($FeO + Fe_2O_3 = 28.53\%$) and $TiO_2$ (5.17%) deserves special notice. On the whole, the chemical composition corresponds to that of alkaline gabbroids (low content of $SiO_2$, high content of $TiO_2$, moderate content of $Al_2O_3$), although the alkali content is not higher than in ordinary amphibolites present in the section studied. It may be supposed that prior to metamorphism, these non-feldspar amphibolites were represented by their melanocratic ore-bearing gabbro varieties.

The iron-titanium mineralization[9] is represented by an association of magnetite and ilmenite. They are intergrown with pyrrhotite and chalcopyrite. The amount of oxide minerals (ilmenite, magnetite) at some places reaches 40–50%, the amount of sulphides is 5–7%.

The composition of iron-magnesium minerals of the ore-bearing amphibolites differs from that of minerals of ordinary amphibolites by its considerably higher content of iron and by the increased amount of $TiO_2$. Hornblende is characterized by the highest content of iron (0.78), as compared with all other hornblendes contained in other rocks. The $TiO_2$ content in hornblendes varies from 0.83 to 1.16%. Biotites differ from ferruginous quartzites by their reddish colour, and are characterized by a high content of iron (0.64–0.75) and by the highest content of $TiO_2$ of biotites of other rocks (2.98–3.47% versus 1.87–2.22% in ordinary amphibolites). Plagioclase of the titanium-magnetite metagabbro is rather acid in nature (22–24% An).

The quantitative ratio of ilmenite to magnetite in ore aggregates is about 60:40. Magnetite grains contain small inclusions of silicates, and this distinguishes them from grains of ilmenite deprived of such inclusions.

[9] Study of the iron-titanium mineralization was performed by O. V. Karpova.

**Table 1.22.** Chemical composition of the titanomagnetite metagabbro and ferruginous quartzite (%)

| Oxides | Ore-bearing amphibolite | Ferruginous quartzite |
|---|---|---|
| $SiO_2$ | 38.60 | 38.60 |
| $TiO_2$ | 5.17 | 0.22 |
| $Al_2O_3$ | 9.60 | 6.55 |
| $Fe_2O_3$ | 6.33 | 17.75 |
| FeO | 22.20 | 18.27 |
| MnO | 0.31 | 0.13 |
| MgO | 3.54 | 2.41 |
| CaO | 8.04 | 0.36 |
| $Na_2O$ | 1.37 | 0.85 |
| $K_2O$ | 1.88 | 2.25 |
| $H_2O^-$ | – | – |
| $H_2O^+$ | 1.19 | 1.50 |
| $P_2O_5$ | 0.32 | 0.18 |
| $CO_2$ | 0.28 | 0.32 |
| l.o.c.[a] | 1.37 | – |
| Total | 100.20 | 99.39 |

[a] l.o.c. = Loss on calcination.

Ilmenite is more widespread in amphibolites than magnetite, and it is met not only in ore-bearing but also in other amphibolites. Ilmenite grains are usually arranged in the form of isolated groups oriented along the enclosing rock banding. Their sizes vary from 0.02 to 0.2 mm. Rims of the secondary non-ore minerals are seen in the marginal parts of ilmenite grains; sometimes small grains of ilmenite are fully replaced.

Ilmenite is met in amphibolite (sample No. 31100, depth 8716.4 m) in the form of individual isolated grains, which form lens-like segregations in rocks. Within the limits of individual accumulations ilmenite grains differ in size and in mode of segregation. Large lamellar grains (0.1 – 0.2 mm), sometimes of sinuous configuration, and small elongated and isometric grains of ilmenite (0.01 – 0.03 mm) are the most typical. Ilmenite segregations, analogous in shape and size, were seen in titanomagnetite of the metamorphosed ores in amphibolites of the Kusinsky deposit in the Northern Urals (Myasnikov 1977).

Ilmenite in the ore-bearing amphibolite (sample Nos. 31077, 31078, 8711, 5) is represented by an aggregate of grains which are tightly intergrown with magnetite, thus forming the closely spaced (in many cases they are linked with each other) lenticular-banded accumulations in rocks (Fig. 1.57). Small isometric grains (0.01 – 0.03 mm) disseminated in silicates are met less frequently. The amount of ilmenite in amphibolites changes from individual grains to 10%, or to 30% in the ore-bearing amphibolite.

Ilmenites from different depths are characterized by similar compositions (Table 1.23). The only difference is in lower content of magnesium and manganese in ilmenite than in ore-bearing amphibolite (samples nos. 31077, 31078).

**Fig. 1.57.** Banded inclusions of magnetite, ilmenite and pyrrhotite in foliated amphibolite. Polished thin section, ×60; sample No. 30077, depth 8711.0 m

As a rule, magnetite is met only in the ore-bearing amphibolites containing dense impregnations of ore minerals (up to 60%). Of these minerals, magnetite is contained in amounts reaching 40% (grain size 0.1 – 0.2 mm).

The composition of magnetite (see Table 1.23), samples nos. 31077, 31078) is characterized by low (although higher than ferruginous quartzites) content of titanium dioxide, elemental admixtures (Mg and Mn) and by its high vanadium content. It is of interest that a similar composition is inherent in magnetite from the massive metamorphic ores in amphibolites of the Kussin massif of the Southern Urals; this magnetite corresponds to iron-vanadium ores. The relatively low content of titanium and vanadium in magnetite, occurring in association with ilmenite, is apparently a result of rock metamorphism.

An examination of the textures of the oxide mineral aggregates and the internal structure of their grains in the iron-titanium ores of the Kussin deposit has shown that during the process of rock metamorphism titanomagnetite is recrystallized, and titanium is released; moreover, collective recrystallization results in the formation of ilmenite (Myasnikov 1977). In the formation of titanium-magnetite mineralization, encountered by the Kola superdeep well, processes of metamorphism also promoted the transformation of less valuable titanomagnetite ores into high grade ilmenite-magnetite ores.

### 1.6.3 Ferruginous Quartzites

Ferruginous quartzites (samples Nos. 24733 and 24734, depth 7635 m) are characterized by banded structures caused by the presence of biotite and quartz

**Table 1.23.** Composition of ilmenite and magnetite (X-ray spectral analysis, %)

| Sample, nos. | Depth, m | Mineral | $TiO_2$ | FeO | MgO | MnO | CaO | ZnO | $Al_2O_3$ | $Cr_2O_3$ | $V_2O_5$ | $Nb_2O_5$ | $SiO_2$ | Total |
|---|---|---|---|---|---|---|---|---|---|---|---|---|---|---|
| 24733 | 7635.0 | Magnetite | 0.03 | 99.23 | – | 0.06 | – | 0.01 | 0.14 | 0.02 | 0.10 | – | 0.37 | 99.96 |
| | | Magnetite | – | 98.58 | – | 0.06 | – | – | 0.9 | 0.01 | 0.12 | – | 2.31 | 101.17 |
| 29646 | 8524.6 | Ilmenite | 50.06 | 47.79 | 0.14 | 1.99 | – | 0.03 | – | 0.02 | 0.15 | 0.04 | 1.17 | 99.39 |
| | | Ilmenite | 50.48 | 45.78 | 0.23 | 2.05 | – | 0.03 | – | 0.02 | 0.11 | 0.04 | 1.00 | 99.74 |
| 33981 | 9083.2 | Ilmenite | 52.10 | 45.88 | 0.18 | 1.68 | – | – | – | 0.02 | 0.13 | – | 0.67 | 100.66 |
| | | Ilmenite | 51.19 | 45.38 | 0.18 | 1.72 | – | – | – | 0.02 | 0.13 | 0.04 | 0.99 | 99.65 |
| 31100 | 8716.4 | Ilmenite | 51.39 | 45.63 | 0.18 | 1.81 | – | – | – | 0.01 | 0.11 | 0.04 | 1.10 | 100.24 |
| 31077 | 8711.0 | Ilmenite | 51.15 | 47.36 | 0.09 | 1.10 | – | 0.02 | – | 0.26 | – | – | – | 99.746 |
| | | Magnetite | 0.19 | 99.21 | 0.06 | 0.19 | 0.06 | 0.03 | 0.03 | 0.19 | 1.95 | – | – | 102.21 |
| 31078 | 8711.0 | Ilmenite | 49.99 | 47.21 | – | 0.85 | 0.08 | – | 0.12 | 0.03 | 0.28 | 0.04 | 1.35 | 102.56 |
| | | Magnetite | 0.20 | 98.85 | – | – | 0.10 | – | 0.24 | 0.22 | 1.36 | – | 0.89 | 101.86 |

Analyzed by A. I. Tsepin.

layers which alternate with streaks enriched in magnetite. In addition to biotite, quartz and mgnetite, sample No. 24733 contains accessory minerals such as apatite, zircon, orthite, which are not minerals characteristic of the ferruginous-siliceous formation. In addition to quartz, magnetite and biotite, sample No. 24734 contains an insignificant amount of plagioclase, epidote and garnet.

Chemical analysis of ferruginous quartzite (sample No. 24734) has revealed high iron content ($FeO + Fe_2O_3 > 35\%$). The relatively high (for the ferruginous-siliceous formation) content of $Al_2O_3$ is accounted for by the presence of plagioclase, biotite, epidote and garnet. It is important to emphasize that the amount of $TiO_2$ (0.22%) is insignificant (see Table 1.22).

Biotite in both samples is grassy-green in colour with sharp pleochroism, and is characterized (sample No. 24734) by a high iron content ($f = 0.63$) and relatively small titanium content (0.05 formular units). Plagioclase is represented by andesine (33% An). The content of iron in epidote is high ($f = 0.27$), higher than in any other rocks studied.

Magnetite in samples of ferruginous quartzite is contained in amounts of 20 – 30% in the form of squeezed grains of 0.1 – 0.3 mm in size. Grains form intergrowths not larger than 0.5 – 1 mm. Accumulations of small (0.03 – 0.07 mm) crystals, extending along foliation planes, are also met. Magnetite of the ferruginous quartzites is almost deprived of admixtures, which is very characteristic of magnetite of the ferruginous-siliceous formation (see Table 1.23, sample No. 24733).

### 1.6.4 Hydrothermal Sulphide Mineralization in Zones of the Retrograde Dislocation Metamorphism

Zones of retrograde dislocation metamorphism, associated with hydrothermal sulphide mineralization, were established at a depth of 6 – 10 km in metabasites of the Pechenga complex and in amphibolites and granite gneisses of the Kola series (see Chap. 15 and 1.7). Sulphide mineralization occurs in the form of thin, uniformly disseminated impregnations; in rare cases it is represented by larger lens-like and streaky impregnations, streaks and veinlets, both parallel to and cross-cutting the rock foliation. It is characteristic that sulphide streaks occur in combination with the associated thin sulphide impregnation.

The mineral composition of sulphide mineralization is represented predominantly by pyrrhotite, pyrite and chalcopyrite, galenite, bornite, molybdenite, argentopentlandite and siegenite. The same paragenetic association also contains vein minerals such as quartz, carbonate, prehnite, epidote, chlorite, actinolite, albite and potassium feldspar, which are confined to zones of cataclastic structure crushing and cracking of the metamorphic rocks.

The sulphide mineralization in foliated metabasites of the Pechenga complex is shown in Fig. 1.58. In the carbonate-quartz veinlets 2 – 3 cm thick, which contain chlorite and epidote, pyrrhotite and pyrite form rather large segregations. Argentopentlandite found in these sulphide-carbonate-quartz veinlets is of particular interest. This rare mineral forms small segregations among pyrrhotite and chalcopyrite in association with pentlandite, sometimes with sphalerite.

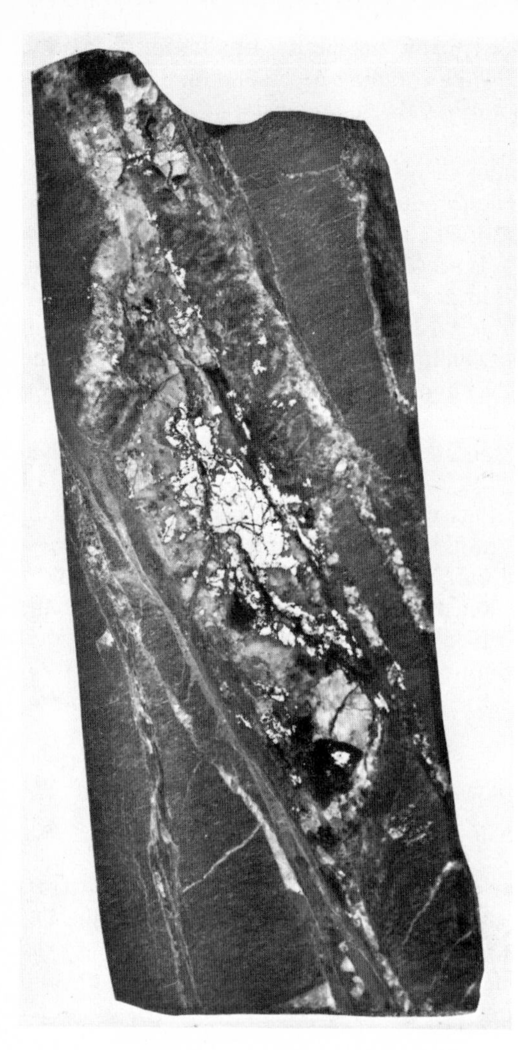

▲

**Fig. 1.59.** Hydrothermally altered migmatite with chalcopyrite impregnation. Polished lump ore, 9/10 of natural size, sample No. 31876, depth 10,000 m

◄

**Fig. 1.58.** Carbonate-quartz veinlet with sulphides (*light*) in foliated amphibolite (*black*). Polished lump of ore, 9/10 of natural size, sample No. 20568, depth 6380 m

It is important to emphasize that argentopentlandite, having originated at great depth, is always met as a primary-hypogenic mineral.

Hydrothermal mineraliatzion in migmatites of the Kola series from a depth of 10 km is shown in Fig. 1.59. Here, near the mylonites, extending along the contact between amphibolites and migmatites, the latter underwent cataclasis , and they are crossed by thin fissures filled with prehnite, secondary quartz, carbonate, chlorite and sericite. The altered migmatites contain chalcopyrite impregnation and are remarkably similar in structural features to the low-temperature metasomatites of the copper-porphyry and other stockwork hydrothermal deposits which were always considered to be the near-surface formations.

The data presented unreservedly demonstrate the epigenetic character of sulphide mineralization and indicate that the latter is closely associated with zones of low-temperature retrograde metamorphism.

**Table 1.24.** Content of cobalt and nickel in pyrites (g $t^{-1}$)

| Rock | Depth, m | Co | Ni |
|---|---|---|---|
| SG-3 well in rocks of the Kola series[a] | | | |
| Gneiss | 7921.0 | 200.0 | 3000 |
| Gneiss | 7367.2 | 400.0 | 2000 |
| Gneiss | 7377.6 | 5000.0 | 2500 |
| Amphibolite | 7443.8 | 2000 | 2000 |
| Amphibolite | 7489.8 | 3000 | 400 |
| Amphibolite | 7544.9 | 12000 | 5000 |
| Gneiss | 8486.4 | 13000 | 200 |
| Amphibolite | 9526.5 | 6000 | 2000 |
| Gneiss | 9968.0 | 2000 | 3000 |
| Metasedimentary rocks of the Pechenga complex[b] | | | |
| Subarkose metaphyllites | | 50 | 70 |
| Subarkose metaphyllites | | 60 | 50 |
| Meta-aleuropsammites | | 120 | 150 |
| Meta-aleuropsammites | | 90 | 120 |
| Meta-aleurolites | | 340 | 180 |
| Carbonaceous-sulphide schists | | 180 | 156 |
| Carbonaceous-sulphide schists | | 145 | 180 |

[a] Data from laser microspectral analysis.
[b] Data from qualitative spectral analysis (Akhmedov and Ozhegova 1974).

**Table 1.25.** Isotope abundance of sulphide sulphur

| Sample, nos. | Depth, m | Mineral | $^{34}S$, ‰ | Sample, nos. | Depth, m | Mineral | $^{34}S$, ‰ |
|---|---|---|---|---|---|---|---|
| 23742 | 7367.0 | Pyrrhotite | −0.36 | 31890 | 8931.8 | Pyrrhotite | −0.30 |
| 24361 | 7489.8 | Pyrite | +0.66 | 33374 | 9029.0 | Pyrrhotite | −0.13 |
| 24710 | 7552.1 | Pyrrhotite | −0.55 | 34125 | 9123.6 | Pyrrhotite | 0.27 |
| 26639 | 7928.1 | Pyrrhotite | +0.85 | 34856 | 9264.8 | Pyrrhotite | 0.32 |
| | | Chalcopyrite | +0.71 | 35667 | 9526.5 | Pyrrhotite | 0.19 |
| 27148 | 7991.4 | Pyrite | +2.09 | 36795 | 9936.3 | Pyrrhotite | +1.76 |
| | | Pyrrhotite | +1.06 | 37004 | 9968.0 | Pyrrhotite | +0.87 |
| 31077 | 8711.0 | Pyrrhotite | −0.82 | | | | |

To elucidate the origin of this mineralization, in addition to the investigations undertaken to establish its composition and geological structural position, the following questions have been examined: the distribution of cobalt and nickel in pyrite, the isotope abundance of sulphur in sulphides and the evaluation of the temperatures of mineral formation.

According to the published data, it appears that the distribution of cobalt and nickel in pyrite and the value of the Co/Ni ratio are of genetic nature (Akhmedov and Ozhegova 1974; Bratia et al. 1979). Data on the content of cobalt and nickel in pyrite from gneisses, amphibolites and ultrabasites in the SG-3 well, as well as their correlation with data on pyrite from the metasedimentary rocks of the Pechenga complex and on pyrites of sedimentary origin, are presented in Table 1.24. The significant difference between pyrite of rocks from the SG-3 well in an interval of 7–10 km and pyrite of the metasedimentary rocks of the Pechenga complex is mainly expressed in a considerably higher content of cobalt and nickel in pyrite from the well, although values of the Co/Ni ratio in both cases are similar. This is indicative of the endogenic origin of the pyrite examined.

A study of the sulphur isotope abundance of sulphides in an interval of 6–10 km was carried out in all the rocks. It became possible to recognize 15 monomineral samples throughout the entire section, i.e. ten samples of pyrrhotite, four samples of pyrite and two samples of chalcopyrite (both from veinlets in rocks and from the relatively uniformly distributed sulphide impregnations).

The sulphur isotope compositions determined are given in Table 1.25. The most important characteristic of the values obtained is their uniformity and similarity with the meteoric standard. Despite the varying composition of the rocks, out of which sulphides have been extracted, and the differences between the sulphides themselves, their $\delta^{34}S$ values change from +2.09 to −1.42‰, less than 1‰ for most of sulphides. Only for two samples (6400–6500 m) did $\delta^{34}S$ of pyrrhotite and chalcopyrite reach 5‰. These values differ considerably from the $\delta^{34}S$ values in sulphides of the Pechenga sedimentary-metamorphic complex in the SG-3 well section, where they change over a broad range, i.e. from values similar to those of the meteoric standard in sulphides of the peridotite intrusions with copper-nickel mineralization to +21‰, typical of sulphides of sulphide mineralization of sedimentary origin. The observed close correspondence of $\delta^{34}S$ values of sulphides for all rocks studied in the Kola series to values of the meteoric standard is indicative of the presence of one and the same sulphur source.

The formation temperatures of sulphide mineralization were assessed by cobalt geothermometer (Bezmen et al. 1975) by establishing the character of the cobalt distribution in the coexisting pyrrhotite and pyrite (Table 1.26). The cobalt content in these sulphides was determined by E. A. Korina on the basis of data of laser microspectral analysis in the spectral laboratory, and by V. A. Boronikhin by electronic probe in the X-ray spectral laboratory of the IGEM of the USSR Academy of Sciences.

In almost all samples, the cobalt content in pyrite exceeded that in pyrrhotite, therefore a coefficient of the cobalt distribution is as follows

$$K^{Co} = \frac{\text{Co in pyrrhotite (\%)}}{\text{Co in pyrite (\%)}} < 1 \, .$$

Temperatures of formation of the coexisting pyrrhotite and pyrite, calculated on the basis of this coefficient, are considered approximate, because homogenic distribution of cobalt was not fixed in all pyrite grains. However, it is characteristic that the temperature of formation, calculated for pairs of these sulphides, did not exceed 490 °C in all the rocks studied; this means that it is considerably lower than temperatures inherent to the amphibolite facies of metamorphism which affected all the enclosing rocks. Most of the calculated temperatures change between 184° and 323 °C. The lowest temperature (184 °C) was determined for the pyrite-pyrrhotite veinlet-like segregations found in a quartz vein in gneisses, which are confined to a zone of retrograde dislocation metamorphism.

**Table 1.26.** Temperatures of formation of coexisting pyrrhotite and pyrite measured by cobalt geothermometer (Bezmen et al. 1975)

| Sample, nos. | Depth, m | Co % | $K_D^{Co}$ | $\lg K_D^{Co}$ | t °C |
|---|---|---|---|---|---|
| 23482 | 7291.0 | 0.15[a] / 0.20 | 0.75 | −0.12 | 268 |
| 23742 | 7367 | 1.19 / 1.80 | 0.65 | −0.18 | 279 |
| 23750 | 7367.2 | 0.15 / 0.40 | 0.37 | −0.43 | 323 |
| 23811 | 7377.6 | 0.20 / 0.50 | 0.40 | −0.40 | 318 |
| 24152a | 7443.8 | 0.10 / 0.20 | 0.50 | −0.30 | 300 |
| 24359p | 7498.8 | 1.00 / 0.30 | 3.33 | 0.52 | 184 |
| 24782 | 7644.9 | 0.10 / 1.20 | 0.08 | −0.10 | 485 |
| 27148 | 7991.4 | 0.80 / 0.10 | 0.08 | −0.10 | 266 |
| 29552 | 8486.4 | 0.10 / 1.30 | 0.077 | −1.11 | 490 |
| 35667 | 9526.5 | 0.10 / 0.60 | 0.17 | −0.77 | 397 |

[a] Co content: numerator – in pyrrhotite; denominator – in pyrite.

On the basis of data on sulphur isotope abundance of the coexisting pyrrhotite and pyrite, extracted from a sample represented by the hydrothermally altered gneiss from a depth of 7991.4 m, a formation temperature was roughly assessed by applying the sulphur-isotope geothermometer (Kajiawara and Krause 1971). The correlation of temperature (260 °C), determined by this method, with temperature established for the same pair of minerals by applying the cobalt pyrite-pyrrhotite geothermometer (266 °C), confirms the reliability of the results.

It was mentioned earlier that sulphide mineralization, represented by both disseminated sulphide segregations and their veinlet-like formations, was confined to areas of superimposed cataclasis and hydrothermal alterations, that temperatures of pyrrhotite and pyrite formations, determined by the cobalt pyrite-pyrrhotite geothermometer, were relatively low, and that values of $\delta^{34}S$ of sulphides were low. All these facts confirm the existence of a relationship between sulphide mineralization and the retrograde events manifested in the metamorphic rocks of the Pechenga complex and the Kola series. Sulphur of solutions which formed sulphides was of juvenile, probably of mantle origin.

The results of the Kola well drilling have confirmed the prognosis of the presence of sulphide copper-nickel ores in the lower horizons of the Pechenga ore field. It was established that the geological position of these copper-nickel ores is the same as in other Pechenga deposits; they are confined to the contact zone between metaperidotites and phyllites, which is complicated by a zone of foliation occurring in conformity with this contact. According to their structural-textural features, mineral composition and genesis they are similar to ores outcropping at the surface; this demonstrates that copper-nickel mineralization is traced continuously down the dip at least to a distance of 2.5 km. It is also important that the Kola superdeep well has encountered a new, previously unknown zone of ultrabasites and copper-nickel ores confined to the middle part of the volcano-sedimentary Zhdanov suite.

These results change the existing theories of the vertical zonality of endogenic ore formation in the Earth's crust and offer an explanation of the distribution pattern of different types of mineralization at depth by treating it as a result of the successive action of the superimposed magmatic, metamorphic and hydrothermal processes which occurred at different times and under various conditions.

## 1.7 Shear Zones and Mineralized Fissures

All researchers recognize the important implication of faulting for the geological setting of the Pechenga region. These were studied in most detail in the Pechenga ore field. According to Gorbunov et al. (1978), the geological structure of the Pechenga ore field was formed in five stages. During the first stage basic effusive rocks were erupted and gabbro-diabased sills intruded. The second stage is notable for the formation of an asymmetric synclinorium, which originated as a result of the Karel folding in conditions of compressive stresses acting in the direction from east to west; during this stage poorly mineralized basic and ultrabasic intrusions were injected. The third stage is characterized by changes in direction of the compression stresses, the appearance of transverse folds and associated ore-bearing intrusions, and copper-nickel deposits. During the fourth stage enclosing rocks and ores were subject to metamorphism under conditions of brittle deformation. During the fifth stage ore bodies were displaced along faults and formed gouges. The so-called Main Tectonic Zone, which occurs inside the Zhdanov suite and extends both parallel and at an acute angle to the layering of the volcano-sedimentary rocks, played an important role in the localization of the nickeliferous basic-ultrabasic intrusions. It is assumed that ore-bearing intrusions have penetrated into this zone along the Portyitash and Luostari major faults. Recurrent tectonic movements along this zone were accompanied by the displacement of sulphide masses which were brought into underlying rocks, by the formation of late hydrothermal mineral assemblages and by the intensive effect of metamorphic processes.

The present study of cores from the Kola superdeep well did not trace the entire history of the formation of the shear zones, or identify and characterize all their types recognized at the surface. Nevertheless, this study provides a rare opportunity to explore the internal structure of faults, to reveal the relationship between metamorphism and rock dislocations, to establish the vertical zonality in the distribution of the mineralized fissures and to give a quantitative description of the latter.

The internal fabric of shear zones and the distribution pattern of the mineralized fissures were studied systematically down to a depth of 7 km (Fig. 1.60). The study was possible for the following two reasons: (1) the monoclinal occurrence of the Pechenga complex and the reliable correlation between the sedimentary-volcanic strata which compose the well section and outcrop at the surface; (2) the use of a straight core barrel for core sampling, which guarantees the correct orientation and positioning of core samples during a round trip. At lower depths information is sporadic, but is nevertheless, also of interest.

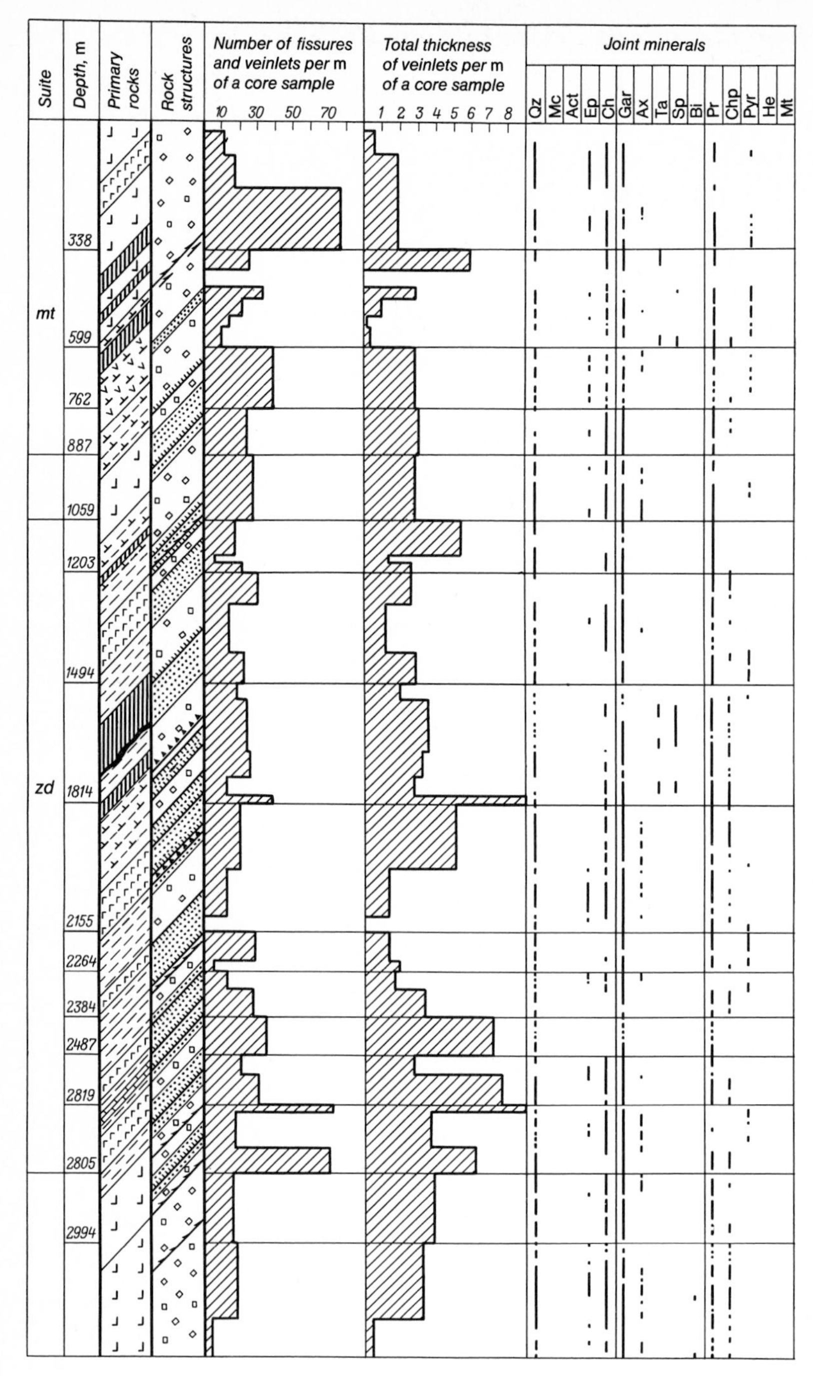
Suite
Depth, m
Primary rocks
Rock structures
Number of fissures and veinlets per m of a core sample
10 30 50 70
Total thickness of veinlets per m of a core sample
1 2 3 4 5 6 7 8
Joint minerals
Qz
Mc
Act
Ep
Ch
Gar
Ax
Ta
Sp
Bi
Pr
Chp
Pyr
He
Mt
mt
zd
338
599
762
887
1059
1203
1494
1814
2155
2264
2384
2487
2819
2805
2994

Fig. 1.60a

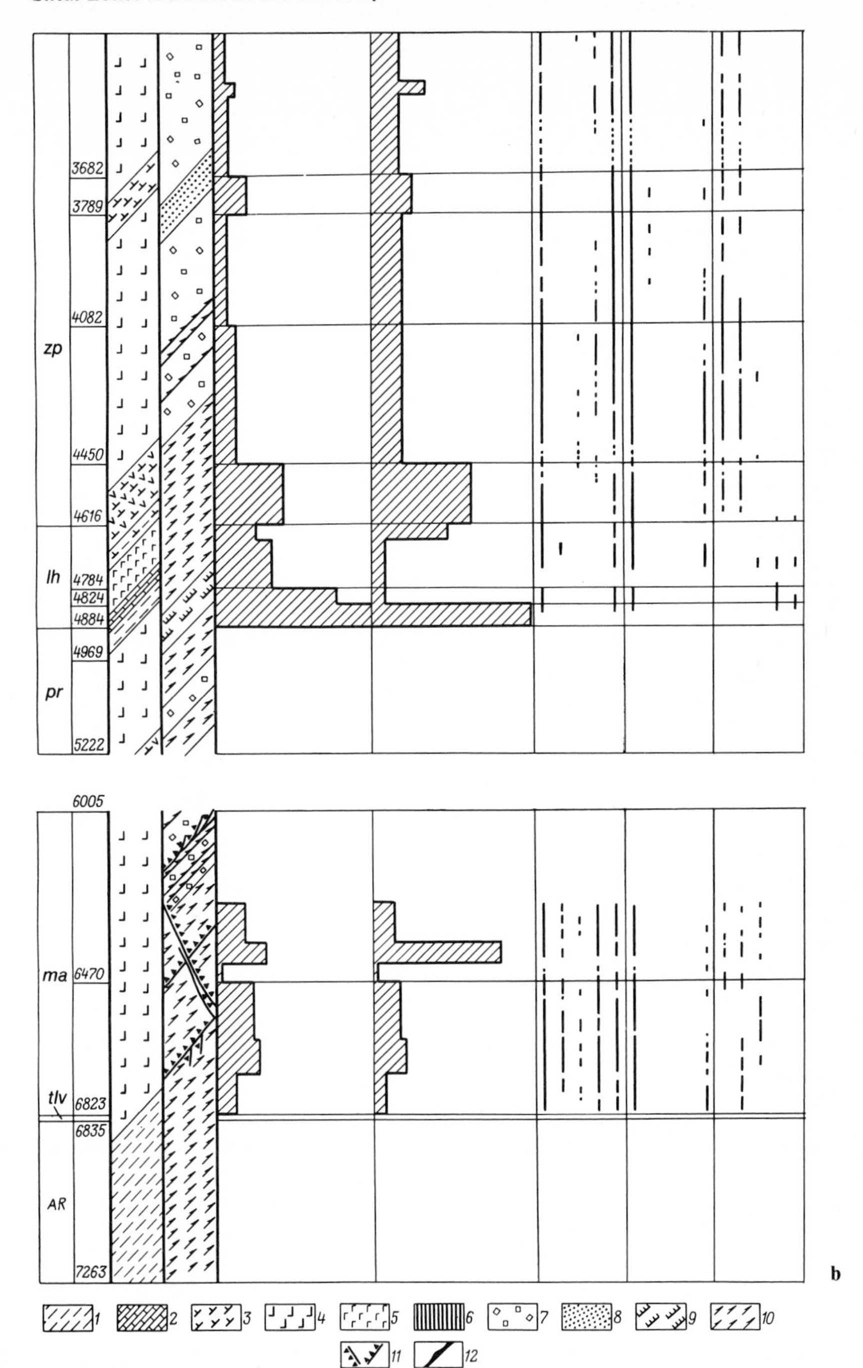


**Fig. 1.60 a, b.** Distribution of shear zones and mineralized fissures in the SG-3 well section. (After V. I. Kazansky and Y. P. Smirnov) *Primary rocks: 1* sandstone; *2* carbonate rocks; *3* basic tuffs; *4* basic lavas; *5* basic intrusive rocks; *6* ultrabasic intrusive rocks. *Rock structures: 7* relict massive and globular; *8* relict layered; *9* layered-foliated; *10* crystalline-foliated; *11* cataclastic; *12* ore bodies

### 1.7.1 Shear Zones and Dislocation Metamorphism

A reconstruction of the evolution of shear zones and the conditions of their formation in the Precambrian crystalline rocks is largely based on studying the processes of dislocation metamorphism. Generally speaking, dislocation metamorphism can be identified as a process expressed in changes in the structure, texture and mineral composition of the rocks in those parts of major faults which occur below zones of diagenesis and weathering. These changes take place without any significant rock melting and proceed under conditions differing from those in which the rocks have been formed.

The study of dislocation metamorphism in the Kola superdeep well was based on the documented core data, the correlation of structures and textures of metamorphic rocks, together with their mineral assemblages and the examination of the spatial orientation of metamorphic minerals (quartz, carbonate, chlorite, talc, mica, amphiboles). Microstructural studies were performed by L. P. Gordienko and A. V. Kuznetsov, who examined those core samples for which a position "upper part – lower part" was known.

Correspondingly, a "rock structure" column was introduced into the SG-3 well section chart, and these structures were divided into five groups: (1) relict massive and globular; (2) relict layered; (3) layered-foliated; (4) crystalline-foliated; (5) cataclastic.

The three latter groups are associated with dislocation metamorphism. The layered-foliated structures are characteristic of sandstones, siltstones and fine-grained tuffs, which underwent foliation and partial recrystallization but preserved features of the layered clastic rocks. The crystalline-foliated textures are characteristic of the intensively deformed and metamorphosed sedimentary and pyroclastic rocks, as well as of intrusive rocks (sheets of basic lavas or intrusions of basic and ultrabasic composition) affected by fault metamorphism. This group includes blastomylonites, blastocataclastics and foliated amphibolites and gneisses whose tectonic nature was proved by micro-textural analysis. Finally, the group of cataclastic structures combines breccia, micro-breccia, cataclastics and mylonites with distinct features of brittle deformations.

The most comprehensive data on changes in conditions of fault metamorphism with depth were obtained for those faults which occurred conformably with the Pechenga complex. It was established that intensity of deformation and of rock recrystallization increases with depth, thus being in agreement with the general temperature zonality of prograde metamorphism (Fig. 1.61). To a depth of 4340 m, manifestations of fault metamorphism are restricted by the presence of comparatively narrow tectonited zones, whose internal fabric mainly depends on the composition of the primary rocks. Below this depth the influence of lithology is much less pronounced, and all rocks are subject to foliation, which results in the appearance of deformation textures, the regular orientation of metamorphic minerals and rock anisotropy.

Tectonic deformations in volcanic and sedimentary rocks of the Matertin suite, metamorphosed under prehnite-pumpellyite facies conditions, are poorly developed, and tectonic zones up to 10 cm thick have been identified only in meta-ultrabasites. Intermediate ultrabasites, despite their intense serpentiniza-

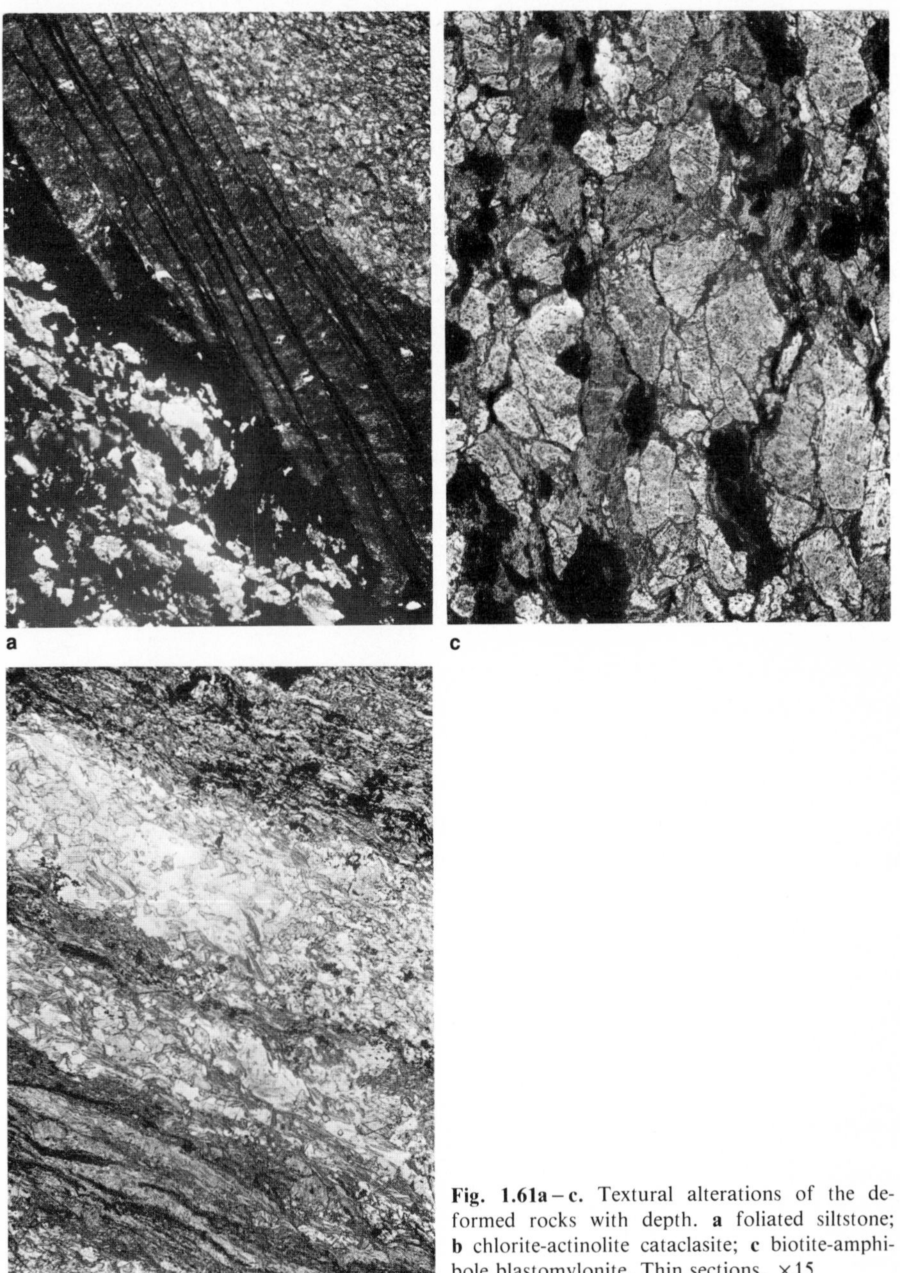

**Fig. 1.61a – c.** Textural alterations of the deformed rocks with depth. **a** foliated siltstone; **b** chlorite-actinolite cataclasite; **c** biotite-amphibole blastomylonite. Thin sections, ×15

tion, preserve their initial structure; under the microscope the latter are represented by distinct relatively large (0.7 – 1 mm) segregations of porphyric monoclinic pyroxene and smaller (0.2 – 0.5 mm) pseudomorphs of serpentine, tremolite and talc in olivine. Tectonized zones exhibit sharp foliation, and are composed of calcite, talc, tremolite, serpentine and non-transparent ore minerals. Foliation is caused by the alternation of lens-like, gently curved, considerably carbonaceous bands 1 – 8 mm thick and considerably talc-bearing bands with gliding planes which separate the former bands from each other. The microtextural study of thin sections has revealed a varying character of calcite and talc orientation in different aggregates. For instance, in chaotically grained talc-carbonate ingrowths, neither mineral displays any regular orientation. In lepido-granoblastic aggregates, corresponding to planes of predominant gliding, optical axes of talc flakes form a distinct maximum perpendicular to foliation, whereas optical axes of calcite grains are arranged chaotically. Finally, it appears that in large calcite lenses optical axes of carbonate grains tend to arrange in the form of a belt so that the microtextural diagram is represented by a type transitional from *R*-tectonite to *S*-tectonite. The experimental studies undertaken by Rozanov (1962) showed that *S*-tectonite is produced from *R*-tectonite when the unidirectional stresses increase. This allows the assumption that recrystallization and enlargement of calcite grains took place under the influence of intensive compressive stresses.

In the Zhdanov suite the metamorphism of faulting was stronger than in the Matertin suite, with various rocks being affected by it to different degrees. Metabasites are the least deformed rocks, while sedimentary rocks and the conjugated meta-ultrabasites are the most deformed.

Thin tectonized zones (0.5 – 10 cm), represented by cataclasites and blastocataclasites with mineral assemblages characteristic of the biotite-amphibole grade of the greenschist facies, are developed in metabasites. They contain fully destroyed dark-coloured primary minerals and plagioclase tables which are albitized, crushed, destroyed, cut by hairline cracks and filled with epidote, chlorite and calcite. The fragments are of an isometric angular shape, and their cross-sectional sizes are some few millimeters. In places the deformed elongated lens-like albite grains exhibit a parallel arrangement and even form foliated chlorite-actinolite cataclasites in which the chlorite-actinolite-carbonate fabric cements porphyroclasts of the altered plagioclase. The carbonate mass contains lens-like and oval ingrowths of quartz grains characterized by sinuous-serrated boundaries and wavy extinction, as well as by the presence of chlorite accumulations. In the fine-grained fabric, optical axes of calcite grains display the distinct belt orientation inherent to the *R*-tectonite type. An analogous pattern of the microtextural diagram is also characteristic of quartz. Large lens-like calcite grains possess more distinct and perfect orientation, i.e. their optical axes are grouped in the form of one maximum which is perpendicular to the plane of a tectonized zone. Consequently, the growth of calcite grains is caused by the action of intensive compressive stresses. The same phenomenon was observed in tectonized zones present in meta-ultrabasites.

Thin selective foliation is the most characteristic manifestation of the shear zones in sedimentary rocks of the Zhdanov suite. It is most intensively pronounced in phyllites. However, if faults affect two or three rock units composed of fragments of different sizes, then foliation is mainly developed in the finest-grained rocks.

Foliated phyllites and phyllites proper cannot be studied microtexturally. Foliated sandstones, siltstones and conformably occurring quartz-calcite-sulphide veinlets, transformed into blastocataclasites, are considered to be more favourable objects of investigation. Thin sections of foliated rocks exhibit echelon-like gliding planes spaced 0.5 – 0.2 mm apart and composed of non-transparent coaly matter. The deformed veinlets at gliding planes were recrystallized into fine-grained (0.01 – 0.05 mm) quartz-calcite aggregates which also contain chlorite and iron sulphides, mainly pyrrhotite. On microtextural diagrams drawn for zones of foliation, the optical axes of quartz and carbonate display patterns of the *R*-tectonite type with a horizontal axis, thus being indicative of the differential tectonic movements which took place along foliation planes.

The content of sulphide minerals in the conformably occurring calcite-chlorite blastocataclastics is less than in the surrounding siltstones and claystones. Some cataclasite zones are enriched in pyrite and not in pyrrhotite. This is considered to be additional proof in support of the diagenetic nature of sulphide mineralization and its redistribution as a result of prograde metamorphism in the Pechenga complex. Foliation zones, confined to contacts between nickeliferous ultrabasites and other rocks, also originated in the process of prograde metamorphism, and at present they occupy a definite position in tectonic structures and in the metamorphic zonality of the Pechenga complex (see Chap. 1.6).

As a whole, foliation zones in rocks of the Zhdanov suite can be assigned to the category of dispersed conformable faults. There are two intervals with the highest concentration of these zones. The first interval is confined to a depth of 1500 – 2000 m, where the Kola superdeep well has encountered nickeliferous ultrabasites. The second interval is between 2250 and 2800 m, and corresponds to the lower part of the Zhdanov suite; at the surface the Main Tectonic Zone is confined to this part of the suite.

Contrary to sedimentary-volcanic rocks, the underlying metabasites of the Zapolyarny suite were hardly subject to the action of fault metamorphism in an interval of 2800 – 4000 m. A basically different situation is observed in the zone of the Luchlompol major fault, which passes through the lower part of the Zapolyarny suite, the entire sequence of the Luchlompol suite and underlying volcanites (at a depth of 4340 – 5100 m). As the distance to the zone of the Luchlompol major fault decreases, rocks become affected, and some massive and globular lavas, which underwent alteration of the greenschist grade of metamorphism, become slightly foliated; any indications of the primary structures and textures in these altered lavas have not survived. In places they pass into chlorite-actinolitic blastocataclasites composed of actinolite, epidote, chlorite, biotite, albite, calcite, and then they are again replaced by non-deformed rocks, etc. A depth of 4340 m is the limit below which blastocataclasites and blastomylonites obviously predominate. The foliation of blastomylonites results from the alternation of bands of different grain size and mineral composition, and it becomes even more pronounced due to the presence of wavy-carbonate marks which amount to 25 percent of the bands's bulk. Under the microscope, these blastomylonites are

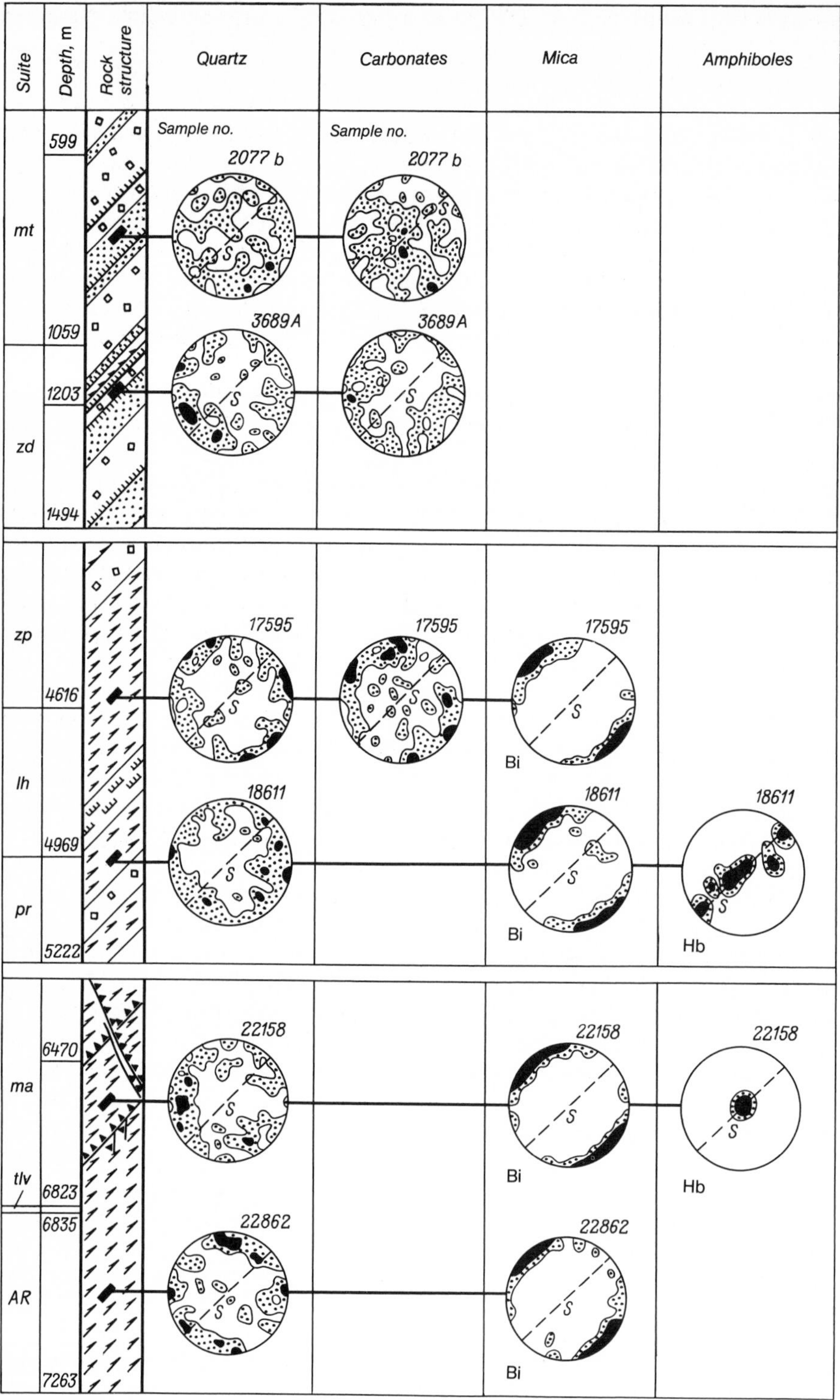
Suite
Depth, m
Rock structure
Quartz
Carbonates
Mica
Amphiboles
Sample no.
Sample no.
2077 b
2077 b
3689A
3689A
mt
zd
599
1059
1203
1494
zp
lh
pr
4616
4969
5222
17595
17595
17595
18611
18611
18611
Bi
Bi
Hb
ma
tlv
AR
6470
6823
6835
7263
22158
22158
22158
22862
22862
Bi
Hb
Bi
S

Fig. 1.62

seen as rather thin-grained (0.05 – 0.3 mm) rocks with a lepidogranoblastic main fabric.

Quantitative interrelations between minerals of the main fabric are not constant, and they change sharply, particularly across foliation planes. For instance, epidote-chlorite bands conjugate with biotite bands, and they are rather sharply separated from quartz-feldspar bands containing only individual chlorite lamellae; actinolite is usually intergrown with chlorite, etc. The porphyroblast segregations larger in size, mainly represented by highly aluminiferous actinolite, epidote, biotite or calcite, are also met in the fine-grained main mass.

The microtextural study of blastomylonite samples shows that they are characterized by a distinct and uniform orientation of minerals, both by shape and by internal structure. For instance, optical axes of quartz and calcite grains are arranged according to the *R*-tectonic type with a horizontal axis of rotation, and flakes of chlorite, biotite and muscovite are located along planes of foliation. An analogous arrangement is seen in the foliated plagioporphyrites and metasandstones of the Luchlompol suite (Fig. 1.62).

The outward appearance of the Luchlompol rocks resembles that of ordinary fine-grained sandstones and siltstones characterized by hardly perceptible foliation. However, when studied under the microscope, they display everywhere signs of recrystallization which results in the appearance of layered schistose structures. In the least deformed rocks, the primary layered distribution of coal, sharp differences between cement and clastic material and angular shapes of the non-rounded grains are preserved, although they acquire an unusual lens-like configuration. The main sandstone mass was transformed into augen blastocataclasites which contain larger (0.25 – 0.75 mm) oval and lens-like segregations of microcline, acid plagioclase and quartz covered by a fine-grained (0.05 – 0.2 mm) lepidogranoblastic texture which is constituted of the same minerals, light-coloured mica and carbonate.

At the boundary between the Luchlompol suite and the underlying volcanic rocks of the Mayarvi suite, the greenschist facies is replaced by the epidote-amphibolite facies. Metabasites are strongly foliated here and represented by biotite-amphibole blastocataclasites and blastomylonites. They consist of oligoclase, highly aluminiferous hornblende, biotite, magnetite and quartz. Foliation is less sharply pronounced than in chlorite-amphibolite blastomylonites; nevertheless, it is distinctly seen in the form of multi-alternating bands and striae enriched in either feldspars and quartz, or hornblende, biotite and magnetite, and also in the form of a distinct parallel arrangement of biotite scales. The hornblende occurrence has another character. This mineral forms aggregates of small, comparatively well-shaped grains, but is sometimes found in the form of large crystals of about 3 mm length, generally conformable with the fine-grained (0.01 – 0.05 mm) lepidogranoblastic main fabric; however, these large crystals are deprived of visible traces of tectonic action. Such crystals of hornblende might have originated somewhat later, after the formation of the main fabric. Quartz in blasto-

**Fig. 1.62.** Orientation of minerals in a zone of the Luchlompol major fault. *Black rectangles* denote samples in which orientation of quartz and carbonates optical axes, mica cleavage and crystallographic *c* axes of amphiboles were measured. *S* foliation. Density 1 – 4%. Symbols as in Fig. 1.60

clasites and blastomylonites is oriented according to the *R*-tectonite and *S*-tectonite "flattened" types; flakes of biotite are oriented according to the *S*-tectonite type with one maximum perpendicular to the foliation planes.

Regular spatial orientation is also characteristic of grains of highly aluminiferous hornblende. On the stereographic projection *c* axes are grouped in a foliation plane in the form of several maxima. The main maximum is located in the foliation strike line, and coincides with an axis of the *R*-tectonite in quartz.

From this it appears that the structural-textural features of the deformed rocks from a zone of the Luchlompol major fault, as well as the distinct and regular orientation of quartz, carbonates, biotite, muscovite, chlorite and hornblende show beyond doubt that fault movements were taking place during the process of metamorphism, and that the line of movement was perpendicular to the line of the fault strike.

Sharp anisotropy, caused by the foliation and recrystallization of the deformed rocks, as well as by the appearance of the regular orientation of their minerals, is the main characteristic feature of rocks in the zone of the Luchlompol major fault. Transition from massive to schistose texture is followed by the anisotropy of longitudinal and transverse waves. The appearance of anisotropy in the deformed rocks, rather than the different character of the Pechenga complex, more pronounced in upper horizons, as well as an abrupt increase of metamorphism intensity, most readily explain the geological nature of the first seismic boundary.

Below the Luchlompol major fault the degree of foliation of volcanic rocks decreases, and metabasites with relict massive and globular structures are the main rocks here as compared with crystalline-foliated rocks. The latter compose tectonic sutures and zones in the first 10 meters, which occur conformably with the rocks of the Pechenga complex. Massive and crystalline-foliated metabasites may be assigned to the epidote-amphibolite facies on the basis of the presence of mineral parageneses. The intensive foliation and recrystallization (resulting even in the disappearance of primary textures) are characteristic also of the sandstones and carbonate rocks which separate the Mayarvi and Pirttiyarvi suites.

Beginning from a depth of 6230 m and down to the bottom of the Mayarvi suite, almost all volcanic rocks were transformed into thinly foliated amphibolites. They consist of highly aluminiferous hornblende, andesine, biotite, small amounts of quartz, sphene ilmenite, and possess lepidogranoblastic textures. Hornblende crystals are often characterized by linear orientation. This is seen on microtextural diagrams where the *c* axes of hornblende grains form a sharp maximum coinciding with the line of foliation strike, and the *Nm* axes are grouped in the form of a belt. Inside this belt there is an isolated maximum which is located in the foliation plane. The tectonic nature of the foliated amphibolites is confirmed by a regular arrangement of the optical axes of quartz grains according to the *R*-tectonite type; this was also observed in the zone of the Luchlompol major fault. Microstructural diagrams for biotite also show the same character.

The correlation between the faulting metamorphism in the lower parts of the Pechenga complex and that of the underlying granite gneisses is not fully clear. However, two facts deserve attention. It was established that in passing to granite gneisses, the type of rock deformation and grade of metamorphism do not change radically. The tilt angles of foliation with respect to the core sample axis on both sides of the contact proved to have the same value. Conglomerates are absent in core samples extracted from the well. At a depth of 113 – 118 and 142 – 146 m and below, contact with the Pechenga series, thin-foliated amphibolites, analogous to foliated metabasites of the Mayarvi suite, occurs. Moreover, granite gneisses exhibit the same patterns in microtextural diagrams for quartz (*R*-tectonite with a horizontal *B* axis) and for mica (*S*-tectonites with one maximum perpendicular to foliation) as those characteristic of products of the fault metamorphism of the lower part of the Pechenga complex. From this it follows that the foliation of granite gneisses is of tectonic origin, although this assertion does not rule out the possibility that foliation could have inherited the layering of sedimentary rocks.

The results of microtextural analysis of the Kola series rock samples extracted at a depth of 8 – 10 km by straight core barrel have shown that rock-forming minerals in gneisses and migmatites possess a regular orientation with respect to foliation. In a vertical section, parallel to the core sample axis and perpendicular to foliation, optical axes of quartz grains form a belt at the peripheral part of the large circle of projections. Biotite flakes lie mainly in a foliation plane, whereas prisms of hornblende are elongated in horizontal direction. This allows the supposition that the deformation pattern of the Kola series rocks was one and the same for different tectonic levels at a considerable distance away from contact with the Pechenga series.

In some instances, signs of deformation of different ages, corresponding to various tectonic levels, are also identified against the general background described above. For example, linear textures of metamorphic minerals and axes of small folds in foliated metabasites of the Pechenga series and in amphibolites of the Kola series are usually parallel to the foliation strike. However, in some intervals at a shallow depth, both of them are arranged along the dip line or in diagonal direction. Similar anomalies are identified in the course of microtextural analysis of metamorphic rocks. As a whole, studies of the fault metamorphism in cores of the SG-3 well confirm Väyrynen's (1938) theories on the imbrication thrust structure of the Pechenga complex. Based on data on the composition of the coexisting metamorphic minerals (see Chap. 1.5), it may be assumed that the thick foliation zone of the Luchlompol major fault has served as a screen for deep heat flows; this idea is fully compatible with the theory of the synchronous character of prograde metamorphism which occurred during tectonic displacements of the Pechenga complex.

For the question of correlation between data obtained in the well and observed at the surface, it may be said that a section of the SG-3 well is readily correlated with the Main Tectonic Zone in the lower part of the Zhdanov suite, and also with the Luchlompol major fault which runs along the boundary between the Nickel and Luostari series. The structure of the Main Tectonic Zone is the same in the well and at the surface. In the Luchlompol major fault, the intensity of sedimentary and effusive rock foliation is higher at depth than at the surface. The reason for this is not clear. Two explanations may be advanced, either a decrease in the grade of metamorphism up the fault tilting plane, or a cross-cutting pattern of metamorphism isogrades with respect to the imbrication thrust structures.

### 1.7.2 Mineralized Fissures and Their Distribution at Different Depth Levels

The presence of fissures filled with different minerals throughout the entire depth penetrated is considered to be one of the most unexpected results of the Kola superdeep drilling. These fissures were the object of systematic observations. It is beyond doubt that non-mineralized fissures also are developed in the SG-3 well section. However, they cannot be identified in the area adjacent to the well by known logging methods. Only non-mineralized fissures of technogenic origin can be established with certainty in core samples.

On the basis of core study it was established that mineralized fissures belong to two classes, lithogenetic and tectonic, which are subdivided into groups and types.

Class A – lithogenetic fissures. These include primary "premetamorphic fissures" which are subdivided into two groups: diagenetic and contraction.

Group A-1 – diagenetic fissures. These are developed only among poorly metamorphosed sedimentary rocks and are mainly present in the Zhdanov suite. They are subdivided into three types: injection (neptunian dikes), subsidence and gravitational.

Group A-2 – contraction fissures. These are represented by two types: fissures of globular partings and fissures of the top part of the diabase sheets.

Class B – tectonic fissures. Fissures of this class are the most widespread and diverse. They formed during prograde and retrograde stages of metamorphism. As distinct from lithogenetic fissures, they stretch over a considerable distance, cross-cut different rocks and are filled with varied associations of vein and ore minerals.

Group B-1 – fissures associated with prograde metamorphism. The main characteristic feature of fissures of this group is the resemblance between the mineral composition of vein filling and that of enclosing metamorphic rocks; any hydrothermal changes outside the veins are absent. Two types of fissure are distinguished in this group by taking into account the mechanism of their formation, i.e. shear and tear fissures.

Group B-2 – fissures associated with retrograde metamorphism. As distinct from fissures associated with prograde metamorphism, fissures of this type were formed during the later period, and are composed of mineral associations which differ sharply in mineral composition from that of the enclosing metamorphic rocks. Near these fissures textures and minerals of metamorphic rocks are destroyed. Fissures of the group B-2 are less abundant than those of the group B-1, and they are also subdivided by mode of formation into shear and tear fissures.

The mineralized tectonic fissures are largely responsible for the development of the technogenic fissures which originated firstly, as a result of the dynamic action of drilling mud on rocks at the bottom of the well, and secondly as a result of the removal of lithostatic pressure when core samples are lifted to the surface. Very frequently, tectogenic fissures inherit veinlets and thin foliation zones filled with talc, serpentine, chlorite and calcite.

Sampling of the partially oriented core in drilling through the Pechenga complex has made it possible to subdivide the mineralized fissures according to their spatial position into four series:

I series – fissures occurring conformably with contacts and with the primary structural elements of the enclosing rocks (formational fissures which correspond to major zones of the interstrata and intrastrata gliding);

II series – steep fissures oriented almost perpendicular to layering or contacts, and corresponding to transverse shear zones;

III series – steeply dipping, usually vertical fissures, stretching at angles not less than 30° – 60° with respect to layering and contacts, and corresponding to diagonal dislocations;

IV series – gently dipping fissures, cross-cutting layering and contacts; in their attitude and character of occurrence they correspond to gentle reverse thrust faults.

Fissures were studied in intervals of similar character, whereby the following main indices were used: number of fissures per m of a core ($\Sigma_n$); total thickness (in cm) of veins and streaks ($\Sigma_m$), aggregated thickness and also thickness of each series. The results of individual measurements applying the weighted mean of the thickness were used to calculate integrated data for a layer, member and intrusive body. These data have been combined for 600 uniform intervals, correlated with physical properties of rocks and processed by electronic computer.

Systematic observations for the mineralized fissures at a depth from 0 to 7 km have demonstrated that their number, spatial orientation and mineral association depend on many factors which are as follows: the initial lithology of rocks, the grade of their metamorphism, the presence of major faults and finally the depth of occurrence.

By means of mass measurements it was established that the number of mineralized fissures regularly decreased in passing from metamorphosed rocks to ultrabasites, tuffs of basic composition and basic intrusive rocks (Fig. 1.63). The thickness of the filled veins in fissures decreases in almost the same manner; the only exceptions are the tuffs of basic rocks in which the aggregated thickness of veins per meter of core is less than that in the metasedimentary rocks.

Against this general background, the following descending series in the productive volcano-sedimentary strata is recognized: sandstone-phyllites-tuffs. The main lavas of the Matertin and Zapolyarny suites are characterized by less intensive fissuring than metasedimentary rocks; however, the intensity of fissuring in lavas is higher than in tuffs; the thickness of vein rocks of the above-mentioned suites is considerably less than that of volcano-sedimentary rocks. Intrusive rocks are even less fissured; they are also characterized by the following series of decreasing fissuring: peridotites-gabbro-diabases-plagioporphyrites.

The lithological composition of rocks affects not only the intensity of fissuring but also the spatial orientation of fissures. In volcano-sedimentary rocks, formational fissures obviously predominate, steeply dipping faults of the I and III series are met less frequently, fissures of the IV series are practically absent. In tuffs, conformable fissures (I series) are the most widespread, fissures of the IV series occur less frequently and shear zones of the II series, cross-cutting the layering, are met very rarely. In effusive rocks formational fissures (I series) and faults of the II series, perpendicular to the former, predominate; steeply dipping fissures of the III series are not met so frequently; and gently dipping fissures of the IV series are met very rarely. In intrusive rocks, the orientation of fissures is

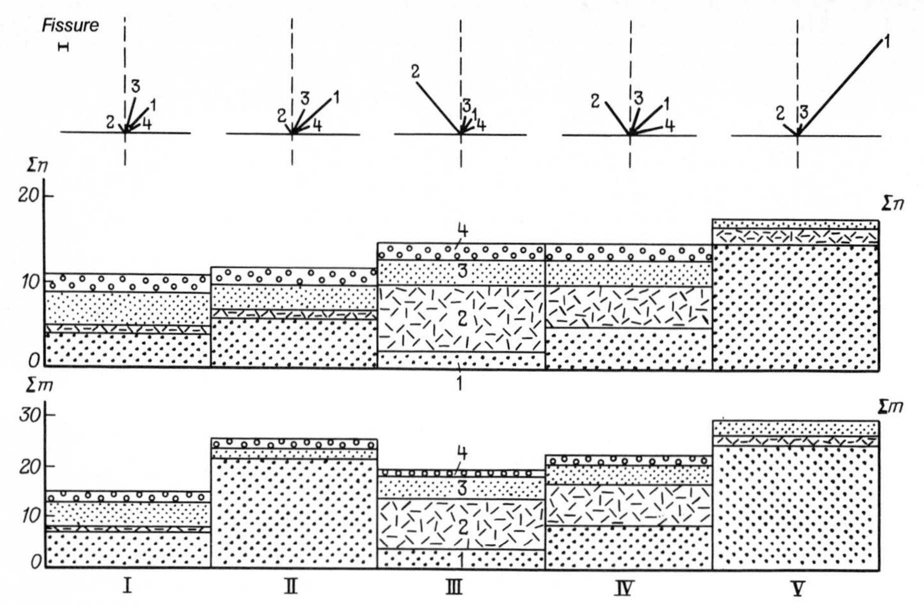


**Fig. 1.63.** Effect of rock lithological composition on density of the mineralized fissures ($\Sigma_n$) and on thickness of veinlets ($\Sigma_m$) for fissure series from 1 to 4. *Rocks: I* gabbro-diabase; *II* tuffs of basic rocks; *III* peridotites; *IV* basic lavas; *V* sedimentary

different, and is as follows: in gabbro-diabases, formational and steeply dipping fissures of the III series predominate, while fissures of the IV and II series are met very rarely. In metabasites, fissures of the II and III series predominate, whereas gently dipping fault zones of the I and IV series are of subordinate significance.

To establish a relationship between the depth of rock occurrence and the development of the mineralized fissures, a number of fissuring parameters were selected; these parameters were taken for intervals composed of metabasites in various units of the Pechenga complex. A total of 15 intervals were correlated (Table 1.27, Fig. 1.64). This made it clear that the number of fissures is still sufficiently great even at a considerable distance from the present-day surface (4 – 6.8 km), and that in some instances they are present in amounts of many tens per meter of core sample. The zones of intensive fissuring in the metamorphosed basic lavas at depths of 5900 – 6800 m may serve as the most striking example. In these intervals, the foliated metabasites are cut by tectonic dislocations associated with numerous sinous tear fissures and by microbrecciated zones consisting of epidote, chlorite, quartz, microcline, calcite and, also, quite frequently, of pyrrhotite and chalcopyrite (Fig. 1.65). The presence of microbrecciated, stockwork and reticulate-streaky structures of veins indicates that the minerals were deposited in open hollows.

While quantitative indices of fissuring change insignificantly with depth, the distribution of fissures in the series is different. Irrespective of the fact that basic lavas contain mainly formational shear zones and faults perpendicular to the former ones (of the I and II series), in the upper part of the section, in an interval

**Table 1.27.** Distribution of mineralized fissures in metabasites at different depth intervals

| Representative intervals, m | Number of uniform intervals | Parameters of fissures | | | | | | | | | | Percentage of shear (SH) and tear (TR) fissures, % | |
|---|---|---|---|---|---|---|---|---|---|---|---|---|---|
| | | In the aggregate | | I series | | II series | | III series | | IV series | | SH | TR |
| 93.0 – 178.1 | 4 | 11 | 1.3 | 7 | 1.0 | – | – | 3 | 0.2 | 1 | 0.1 | 20 | 80 |
| 233.0 – 334.0 | 11 | 51 | 1.3 | 37 | 1.0 | 5 | 0.1 | 8 | 0.2 | 1 | – | 16 | 84 |
| 605 – 762.4 | 9 | 25 | 1.7 | 9 | 0.7 | 2 | 0.2 | 11 | 0.3 | 3 | 0.5 | 14 | 86 |
| 889.0 – 1017.6 | 11 | 20 | 2.2 | 7 | 1.1 | 3 | 0.1 | 7 | 0.8 | 3 | 0.2 | 9 | 91 |
| 1282.0 – 1417.0 | 15 | 10 | 0.9 | 4 | 0.5 | – | – | 5 | 0.3 | 1 | 0.1 | 35 | 65 |
| 1987.0 – 2133.0 | 22 | 9 | 1.0 | 2 | 0.3 | 1 | 0.1 | 2 | 0.2 | 4 | 0.4 | 35 | 65 |
| 2487.0 – 2558.0 | 7 | 14 | 1.9 | 4 | 0.6 | 2 | 0.3 | 8 | 1.0 | – | – | 25 | 75 |
| 2641.2 – 2754.5 | 14 | 12 | 2.5 | 6 | 1.3 | 2 | 0.4 | 4 | 0.8 | – | – | 44 | 56 |
| 3005.0 – 3196.7 | 27 | 12 | 2.2 | 1 | 0.3 | 6 | 1.3 | 3 | 0.1 | 2 | 0.5 | 35 | 65 |
| 3531.4 – 3682.0 | 15 | 5 | 0.7 | 2 | 0.6 | 1 | 0.1 | 2 | – | – | – | 52 | 48 |
| 3851.0 – 4014.0 | 30 | 7 | 1.4 | 1 | 0.3 | 3 | 0.7 | 2 | 0.2 | 1 | 0.2 | 42 | 58 |
| 4389.4 – 4498.0 | 8 | 7 | 1.0 | 1 | 0.2 | 3 | 0.5 | 1 | 0.1 | 2 | 0.2 | 60 | 40 |
| 6357.5 – 6398.0 | 5 | 18 | 2.3 | 1 | 0.1 | 10 | 1.5 | 4 | 0.5 | 3 | 0.2 | 73 | 27 |
| 6533.0 – 6616.0 | 8 | 13 | 0.7 | 2 | 0.3 | 8 | 0.2 | 2 | 0.1 | 1 | 0.1 | 70 | 30 |
| 6734.0 – 6823.0 | 6 | 8 | 0.4 | 3 | 0.2 | 3 | – | 1 | 0.1 | 1 | 0.1 | 62 | 38 |

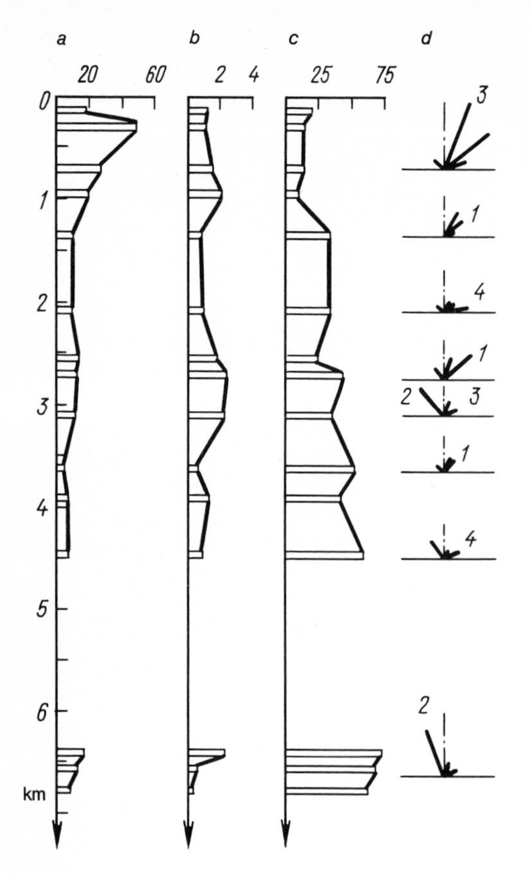


**Fig. 1.64a–d.** Effect of depth on distribution of the mineralized fissures in metabasites. **a** number of fissures per 1 m ($\Sigma_n$); **b** total thickness (in cm) of veinlets per 1 m ($\Sigma_m$); **c** relative amount of shear fissures, %; **d** distribution of fissures in series 1–4

from 0 to 3000 m, conformable fissures dominate over cross-cutting fissures, while in an interval from 3000 to 5000 m steeply dipping fissures of the II series are predominantly developed. The corresponding regularity is inherent to fissures belonging to the IV and III series. This peculiar feature can be explained by the action of lithostatic rock pressure, which increases with depth.

It was also established that the ratio of shear to tear fissures changes with depth. In an interval of 0–3200 m, the number of shear fissures ranges from 10 to 35%. At deeper horizons (3500–6700 m) it increases to 90%. Below a depth of 6260 m, the relative number of shear fissures is 76%. Shear ruptures at such depths are concentrated in groups of faults which are "feathered" by short, thin tear fissures. The thickness of the groups does not exceed several meters, the thickness of the zones is 25–40 m.

G. I. Gorbunov has drawn attention to the close relationship existing between mineral associations of veins and the composition of the enclosing rocks of the Pechenga complex at the surface. The study of core samples from the Kola superdeep well supports this conclusion. Serpentine, talc, chrysotile-asbestos, axinite, epidote, actinolite, microcline, hematite and magnetite are the vein minerals whose formation was strongly dependent on the composition of the enclosing rock. Serpentine with talc and chrysotile-asbestos are met only in ultrabasites.

a

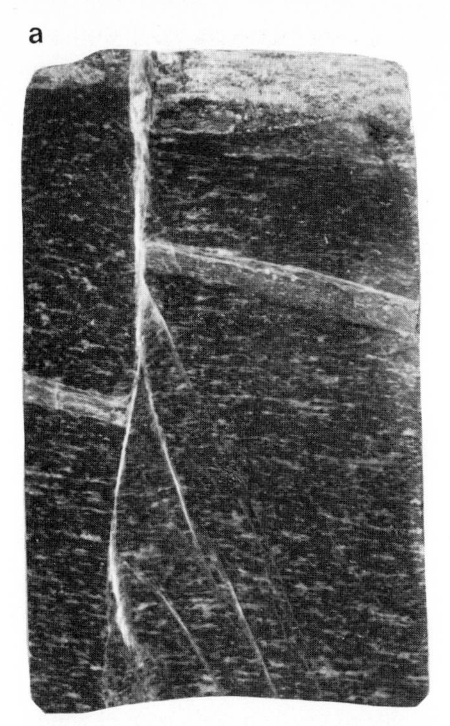

b

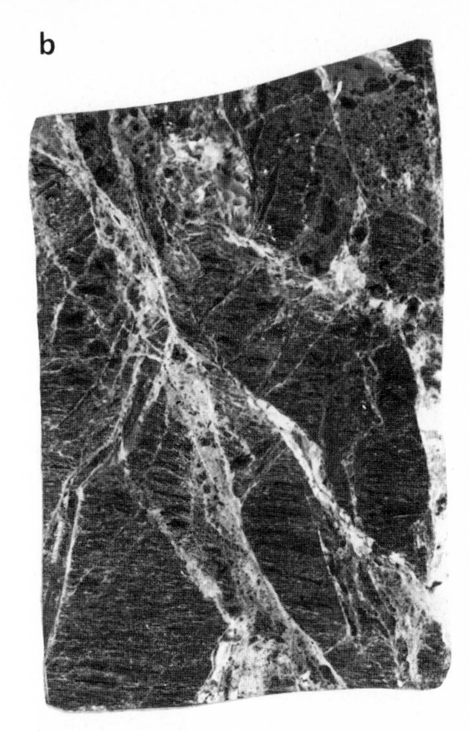

**Fig. 1.65 a, b.** Mineralized shear fissures (**a**) and tear fissures (**b**) in foliated amphibolites at a depth of 6–11 km. Lumps of ore. Natural size

Axinite, actinolite and epidote are present predominantly in igneous rocks (in diabases and gabbro-diabases), and microcline occurs in apodiabase schists. In groups of sedimentary rocks such as phyllites, siltstones, sandstones and tuffs of the Zhdanov suite, veins contain coaly material, pyrrhotite, chalcopyrite and pyrite. Siltstones and sandstones of the Luchlompol suite contain streaks of hematite, magnetite and barite, while sulphides are practically absent. Quartz, calcite, chlorite and sulphides are assigned to the so-called "through" vein minerals which are met in all types of rocks (with rare exceptions). These minerals are met not only in sedimentary rocks of the Luchlompol suite and in volcanites of the Pirttiyarvi suite. However, some features of vertical zonality in the distribution of vein minerals are also observed. For instance, axinite is spread in metabasites of the prehnite-pumpellyite and greenschist facies, but disappears in transition to the epidote-amphibolite facies; biotite appears in the lower parts of the greenschist facies, and is traced down to the bottom of the Pechenga complex, etc.

Passing from the Pechenga complex to the Kola series, the number of fissures decreases. Nevertheless, it is of prime importance that tectonic fissures, associated with zones of retrograde metamorphism, are present in core samples removed at depths between 6 and 11 km (see Fig. 1.65). Even at maximal depths they are represented by shear and tear fissures in amphibolites consisting of epidote, quartz and carbonate. The tectonic nature of these fissures is proved not only on the basis of their known morphological features, but also by taking into account the established rock displacements; these movements, which occurred in the past, are identified in one of the core samples in which a steeply dipping quartz-epidote veinlet obviously displaced the earlier-formed quartz-feldspar streak, as a result of shifting.

On the basis of his own studies carried out in ore fields and deposits Kreiter (1956) has arrived at the conclusion that vertical distribution and internal fabric of ore-bearing fissure structures are considerably affected (other things being equal) by the lithostatic pressure which increases with depth. He estimated that the maximal depth of occurrence of the brecciated and fissured zones, originating as a result of brittle deformations, was 5 km. Sherman's (1978) theoretical calculations serve to confirm this evaluation.

The facts presented above refute these assumptions and demonstrate that during the different stages of geological history of one and the same area, the character of rock deformation changes, depending on the Earth's interior geothermal regime; this, in its turn, results in the occurrence of contrast or more vertically lengthened structural zonality. Correspondingly, in conditions of low-temperature heat flows, fissured structures of the so-called subsurface type can be formed at great depths. It should also be added that a depth interval of 10 – 11 km is the minimal depth at which zones of retrograde fault metamorphism occur. Unfortunately, accurate data on the geological age of the latter and the amount of erosional cutting of overlying rocks are absent. It could be conventionally accepted that erosion has destroyed the upper 3 – 5 km of the sequence. If so, then a vertical range of occurrence of crushed and fissured zones and zones of low-temperature hydrothermal mineralization in the crystalline basement of the Baltic Shield may be lowered to 15 km.

### 1.7.3 Fissuring, Physical Properties and Anisotropy of Rocks

The results of studies of the Pechenga rock fissuring in core samples have been correlated with parameters (measured on a mass scale) such as density ($\sigma$ g cm$^{-3}$), bulk porosity ($P_{bulk}$ %), natural humidity ($H$ %), open porosity ($P_{open}$ %), as well as velocity propagation of longitudinal $Vp$ m s$^{-1}$) and transverse ($Vs$ m s$^{-1}$) oscillations in three mutually perpendicular directions ($Vp_1$, $Vp_2$, $Vp_3$; $Vs_1$, $Vs_2$, $Vs_3$). Correspondingly, two values of $Vp$; $Vp^{x}$ (UKB-1M) and $vp^{xx}$ (UZIS-LOTI) have been obtained.

The whole complex of the analyzing values and their interrelationship is presented in the form of a correlation matrix (Table 1.28). The number of degrees of freedom used in the calculation of this matrix is 53. Correspondingly, the minimal significant correlation coefficient ($r$) is 0.32 at 99% probability, and 0.26 at 95% probability. From this it appears that the number of significant $r$ is 44. It was established that the specific density of fissures ($\Sigma_n$) is the most informative parameter of fissuring, because it is correlated with all other indices of fissuring (total number of significant $r$ is 9). An analysis of the relationship of physical properties between each other and with fissuring parameters shows that the factor of rock anisotropy is characterized by the greatest number of significant $r$ (9); velocity dispersion, $DVp$, is less informative (7 significant values of $r$). The den-

**Table 1.28.** Relationships between fissuring and physical properties of the Pechenga complex rocks

| Correlating value | $\Sigma_n$ 1 | $\Sigma_m$ 2 | $N$ 3 | $\Delta Vp$ 4 | $DVp$ 5 | $AVp$ 6 | $D\sigma$ 7 | $A' Vp$ 8 | $T$ 9 | $AVs$ 10 | $T'$ 11 | $\Delta Vs$ 12 | $P_{bulk}$ 13 | $P_{open}$ 14 | $H$ 15 |
|---|---|---|---|---|---|---|---|---|---|---|---|---|---|---|---|
| 1. $\Sigma_n$ | 1.00 | | | | | | | | | | | | | | |
| 2. $\Sigma_m$ | 0.65 | 1.00 | | | | | | | | | | | | | |
| 3. $N$ | 0.35 | 0.15 | 1.00 | | | | | | | | | | | | |
| 4. $\Delta Vp$ | 0.38 | 0.20 | 0.14 | 1.00 | | | | | | | | | | | |
| 5. $DVp$ | 0.56 | 0.24 | 0.31 | 0.59 | 1.00 | | | | | | | | | | |
| 6. $AVp$ | 0.52 | 0.15 | 0.55 | 0.63 | 0.73 | 1.00 | | | | | | | | | |
| 7. $D\sigma$ | −0.03 | −0.12 | 0.00 | 0.20 | 0.02 | −0.01 | 1.00 | | | | | | | | |
| 8. $A' Vp$ | 0.40 | 0.11 | 0.64 | 0.52 | 0.61 | 0.90 | 0.01 | 1.00 | | | | | | | |
| 9. $T$ | 0.49 | 0.13 | 0.50 | 0.54 | 0.67 | 0.91 | −0.08 | 0.90 | 1.00 | | | | | | |
| 10. $AVs$ | 0.43 | 0.13 | 0.06 | 0.31 | 0.59 | 0.45 | 0.09 | 0.34 | 0.48 | 1.00 | | | | | |
| 11. $T'$ | 0.50 | 0.12 | 0.57 | 0.57 | 0.63 | 0.88 | 0.01 | 0.89 | 0.85 | 0.40 | 1.00 | | | | |
| 12. $\Delta Vs$ | 0.05 | 0.02 | 0.07 | 0.28 | 0.00 | 0.05 | 0.11 | 0.12 | −0.01 | −0.09 | 0.15 | 1.00 | | | |
| 13. $P_{bulk}$ | 0.12 | −0.19 | 0.07 | −0.19 | −0.06 | −0.04 | −0.10 | −0.02 | −0.02 | −0.02 | −0.01 | −0.14 | 1.00 | | |
| 14. $P_{open}$ | 0.08 | −0.01 | 0.35 | −0.06 | 0.11 | 0.17 | −0.15 | 0.14 | 0.06 | −0.05 | 0.07 | 0.09 | −0.15 | 1.00 | |
| 15. $H$ | −0.11 | −0.10 | −0.03 | −0.14 | −0.18 | −0.15 | −0.14 | −0.17 | −0.15 | −0.17 | −0.12 | 0.18 | −0.06 | 0.34 | 1.00 |

sity dispersion ($D\sigma$) is absolutely non-informative. This could have been expected in view of the fact that lithologically uniform intervals were selected for the rock analysis; the result confirms the correctness of the selected subdivisions.

The porosity and natural humidity of the Pechenga complex rocks change within a narrow range: $0 \leqslant n_{open} \leqslant 2\%$; $0 < n_{bulk} \leqslant 6\%$. This means that the Pechenga rocks differ considerably both from typical sedimentary rocks and from igneous rocks characterized by higher porosity, particularly in subsurficial parts of the Earth's crust and in fault zones (Rozanov 1962).

Thus, in the course of the joint study of cores from the SG-3 well, a special method was evolved, allowing investigation of an internal structure of different shear zones and mineral filling of fissures. This method combines studies of the mineral associations, structures and textures of the deformed rocks, a geometrical analysis of the mineralized fissures and ascertaining their genetical nature and the collection of data on rock fissuring and their correlation with physical parameters.

A close relationship was also revealed between prograde metamorphism and the imbrication-thrust displacements of the Pechenga complex along major conformable ruptures. It was established that when the temperature of metamorphism increases down the section, brittle deformations are replaced by ductile ones. At the same time, the recrystallized and newly formed minerals become spatially oriented and anisotropy of the rock elastic properties appears.

The abundant statistical material has for the first time made it possible to study the distribution pattern of the mineralized fissures down to a depth of 7 km. The main factors responsible for density, orientation and mineral filling of fissures have been established. It was proved that the mineralized zones of crushing, cataclastic deformation, fissuring and low-temperature hydrothermal alterations containing sulphide minerals extend to much deeper horizons (by three to four times) than was previously thought on the basis of general assumption.

As distinct from the near-surface faults, the intensity of the mineralized fissures is directly linked with the anisotropy of their elastic properties and the dispersion of the velocity of transverse waves, but not with rock porosity.

# 1.8 Gases and Organic Matter

The results of studies of gas and organic matter are considered together because the task of finding a relationship between hydrocarbon gases (HCG) and other forms of organic matter in metamorphic rocks when drilling through the Middle Proterozoic and Archean metamorphic complexes was among the most important.

In the present work the term "organic matter" implies the reduced carbon compounds present in rocks in a disseminated form, as well as carbonaceous matter and graphite, irrespective of their genesis. Therefore, the symbol $C_{org}$ used here has a genetical meaning, and can be applied equally for metasedimentary and magmatic and highly metamorphosed rocks. Chloroform and alcohol-benzene extracts from rocks are combined in the term "bituminous matter".

In the SG-3 Kola well section, both free gases of the gas-permeable zones and bound gases sorbed at the surface of rocks and fissures and enclosed in minerals have been studied. Free gases were studied with the help of the drilling mud gasometry method both in the course of drilling and between drilling operations during pauses. Since a borehole is left uncased beginning from a depth of 2 km, there is the possibility of studying the process of gas emission from rocks throughout the entire well section for a long time.

The study of gases in closed pores of the extracted core makes it possible to assess their proportion in the total gas factor of the drilling mud.

## 1.8.1 Well Gasometry

Samples of the drilling mud were taken at the well head after very 2–10 m of drilling. Samples were tied to a particular depth with the help of rubber indicators. The extracted samples were degassed by applying the thermovacuum method with the help of a GBE degasser (by heating up to 60° – 70 °C). The emitted gas was analyzed by means of chromatographic method, with the following gas components being determined: He, $H_2$, $O_2$, $N_2$, $CO_2$, $CH_4$, $C_2H_6$, $C_2H_4$, $C_3H_8$, $C_3H_6$, n-$C_4H_{10}$, iso-$C_4H_{10}$, $C_4H_8$, n-$C_5H_{12}$, iso-$C_5H_{12}^+$, n-$C_6H_{14}$. The technogenic contamination and geochemical background were taken into account by using data on the "entry" and "cyclic" samples during the idle stroke of the drilling instruments.

Results of drilling mud gasometry obtained in the process of drilling are presented in $m^3\,m^{-3}$ of rock (Fig. 1.66).

It was established that down the section, the content of each gas component in some intervals is anomalously high and exceeds the content expected in this section. Sometimes these intervals coincide for all or for almost all components

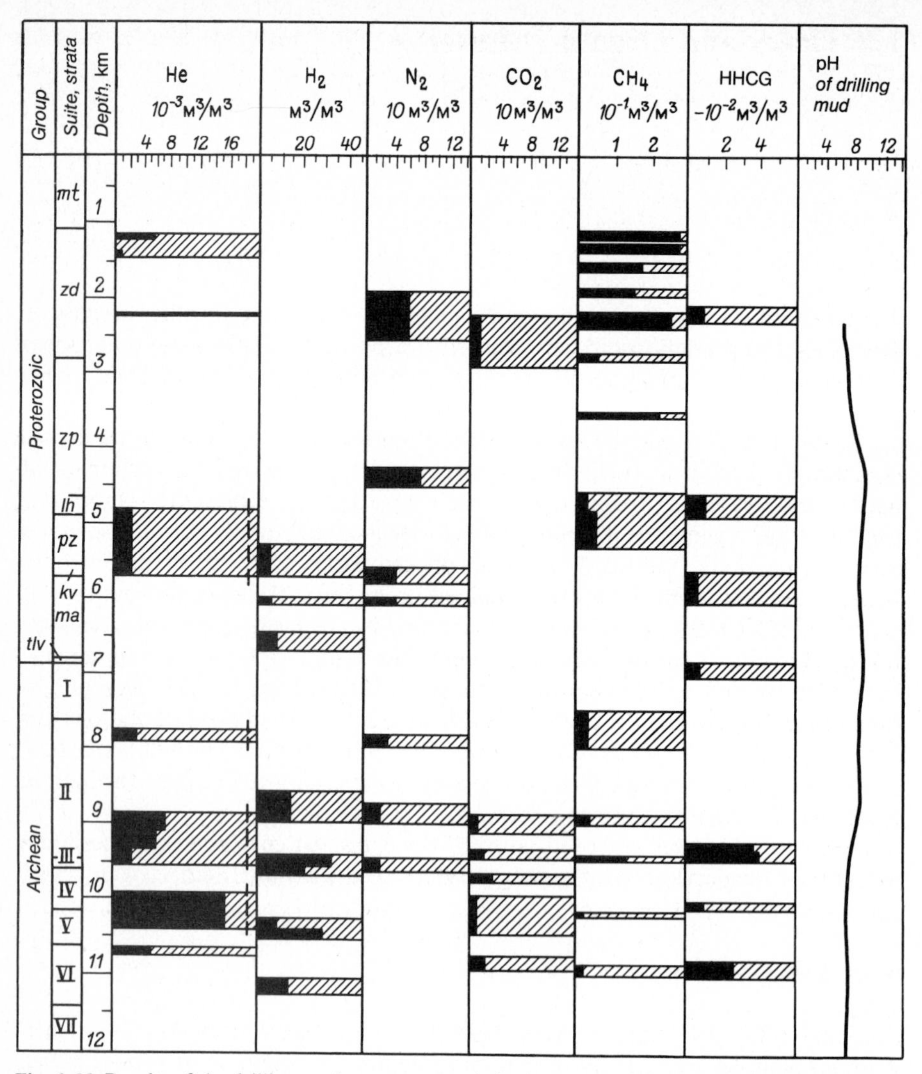


**Fig. 1.66.** Results of the drilling mud gasometry study in the course of the SG-3 well drilling in intervals with anomalously high gas content (*hatched*). *Black solid* mean values for individual intervals; *dotted lines* intervals in which helium was discovered during repeated gasometry studies in the course of drilling operations and inbetween these operations. The following general background content values (in $m^3$ of gas per 1 $m^3$ of the drilled-out rock) have been chosen for individual components: He $0.5 \times 10^{-3}$; $H_2$ 1.0; $N_2$ 10.0; $CO_2$ 0.2; $CH_4$ $1 \times 10^{-2}$; HHCG $0.2 \times 10^{-2}$

(2180 – 2370 m, with the exception of $H_2$: 8700 – 9700 m), but in most cases they coincide only for two to three components or do not coincide at all. The coincidence of intervals for most of the components proves that there was an abrupt change in the fluid system, this change cannot have been due to technogenic causes, particularly if anomalously high contents of helium were identified in the course of drilling operations and in between. The helium content recorded may serve as controlling data for the localization of the fissured gas-permeable zones.

Anomalous values of gas indications are caused by the particular lithological composition and the structural-tectonic features of rocks (fissuring, foliation, sealing factors). It was shown that lithology controls gas indications only in the Zhdanov volcanoclastic-sedimentary suite of the Middle Proterozoic complex. Increased contents of methane are observed here in metasedimentary and volcano-sedimentary rocks (intervals: 1050–1140, 1200–1340, 1440–1570, 1850–2000, 2700–2770 m). In an interval of 2140–2340 m, considered to be a contact zone between coarse-grained gabbro-diabases and an overlying member of silty conglomerates and phyllites, increased contents of heavy hydrocarbon gases (HHCG), nitrogen, carbon dioxide, helium, not only methane, have been established. Rocks bear traces of cataclasis.

Metasedimentary rocks from these intervals are characterized by a maximal content of bound methane (Fig. 1.70); however, this amount can account for no more than 20% of its total quantity in drilling mud. It should moreover be noted that this portion decreases (as facies of metamorphism replaced each other) to 0.05% in the Proterozoic complex (3480–3540 and 4800–5300 m) and to 0.03–0.01% in those intervals of the Archean complex which are characterized by the anomalous content of methane (7500–8000, 8900–9020, 9460–9500, 10,240–10,300, 10,900–11,050 m). In intervals characterized by ordinary values of methane content, they are as follows: 10% in the Matertin and Zhdanov suites and up to 0.5% in rocks below the Zhdanov suite.

The remaining methane appears to come from cataclastic zones and foliation fissures which form isolated or poorly connected systems of limited size, because any other permanent methane inflows were not established by testing carried out in between drilling operations. In drilling through the upper parts of the Proterozoic complex, the drilling mud became oversaturated with gas components; this was expressed in the spontaneous emission of gas from the drilling fluid (intervals: 1100–1140, 1177–1265, 1540–1570, 1760–1807, 2312–2232, 4565–4570 m). In the most representative sample, taken at the 1100–1140 m interval, the proportion of gas (without oxygen) per vol. is as follows (%): He 1.48, $H_2$ 20.6, $N_2$ 37.7, $CH_4$ 40.2, $C_2H_6$ 0.195, $C_3H_8$ 0.006, iso-$C_4H_{10}$ 0.0003, n-$C_4H_{10}$ 0.002, CO and $CO_2$ 0.0. The isotope composition of carbon in methane extracted from this sample ($\delta^{13}C = -4.0\%$, RDB reference, determined by E. M. Galimov) does not differ from the isotope composition of methane in phyllites of this suite ($\delta^{13}C = -4.78$, RDB reference, determined by V. S. Lebedeva).

The anomalous contents of HHCG coincide with those of methane in the following intervals: 2050–2250, 4700–4900, 9250–9500, 10,850–11,060 m, with their absolute and relative values in a sum of hydrocarbon gases (HCG) increasing, particularly beginning from a depth of 8800 m.

Data of repeated gas testing of the drilling mud conducted between drilling operations indicate that some intervals, which were characterized by the increased values of helium content in the course of drilling, may be regarded as intervals composed of fissured, gas-permeable rocks. Some of these intervals are as follows: 4800–5700, 7700–7890, 8800–9250, 9920–10,750 m.

This effect was not established for other gas components, because the minor amount of gas emission did not allow recording the quantity of the emitted gas against fluctuations in its background content and taking analytical errors into account.

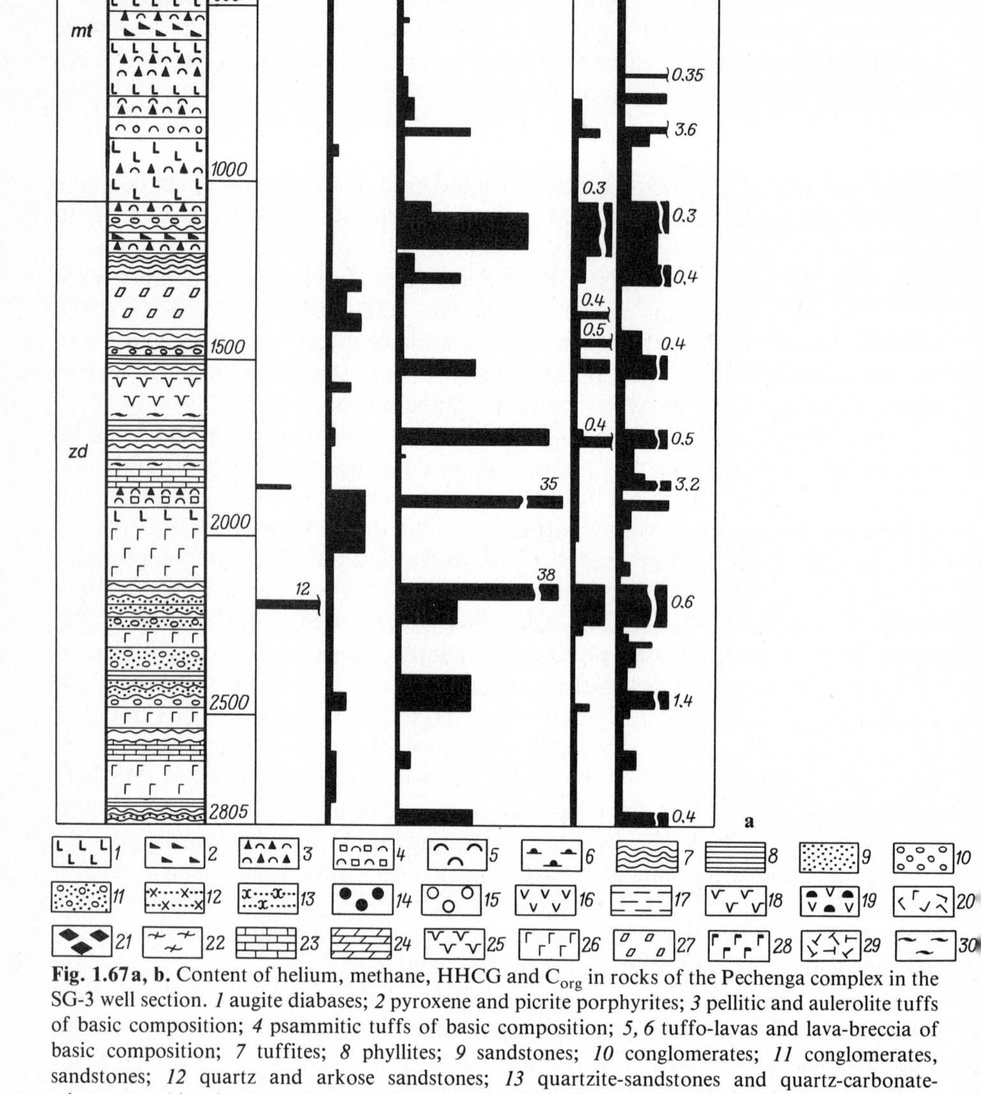


**Fig. 1.67 a, b.** Content of helium, methane, HHCG and $C_{org}$ in rocks of the Pechenga complex in the SG-3 well section. *1* augite diabases; *2* pyroxene and picrite porphyrites; *3* pellitic and aulerolite tuffs of basic composition; *4* psammitic tuffs of basic composition; *5, 6* tuffo-lavas and lava-breccia of basic composition; *7* tuffites; *8* phyllites; *9* sandstones; *10* conglomerates; *11* conglomerates, sandstones; *12* quartz and arkose sandstones; *13* quartzite-sandstones and quartz-carbonate-micaceous schists in them; *14* actinolitic diabases; *15* globular lavas; *16* diabase porphyrites; *17* magnetite-plagioclase-amphibole schists and plagioclase-amphibolite schists in metadiabases; *18* magnetite-plagioclase schists (with amphibole) in andesite porphyrites; *19* altered tuffo-lavas of basic and acid composition; *20* orthophyres; *21* biotite-amphibole-plagioclase schists in andesite-basalts; *22* magnetite-sericite-chloritic schists with carbonate; *23* dolomites; *24* metamorphosed dolomites, marbles, carbonate-tremolite schists; *25* chlorite-serpentine-talc and talc-tremolite rocks; *26* gabbro-diabases; *27* essexitic gabbro-diabase; *28* amphibolites in massive gabbro-diabases; *29* andesite porphyrites; *30* pocket-streaky sulphide ore manifestations

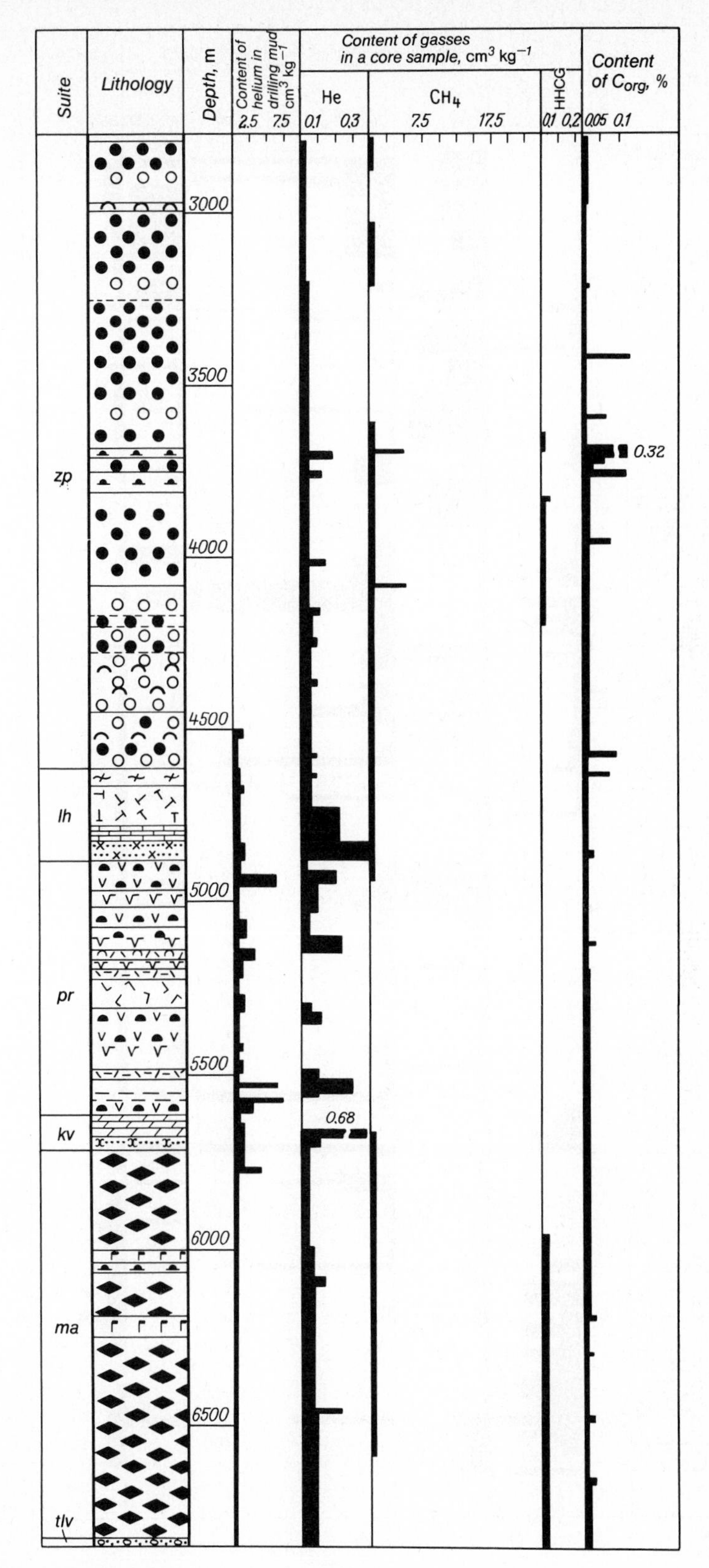
Suite
Lithology
Depth, m
Content of helium in drilling mud cm³ kg⁻¹
Content of gasses in a core sample, cm³ kg⁻¹
He
CH4
HHCG
Content of Corg, %
2.5 7.5 0.1 0.3 7.5 17.5 0.1 0.2 0.05 0.1
3000
3500
4000
4500
5000
5500
6000
6500
0.32
0.68
zp
lh
pr
kv
ma
tlv

Fig. 1.67 b

Stratum
Lithology
Depth, m
Content of helium in drilling mud cm³ kg⁻¹
Content of gases in a core sample, cm³ kg⁻¹
He
CH₄
HHCG
Content of C_org, %
2.5 5.0 7.5
0.1 0.3
0.02 0.06 0.1 0.14
0.002 0.004
0.05 0.1
I
II
III
7000
7500
8000
8500
9000
9500
11.0
15.7
10.2
15.1
0.48
1.12
0.48
0.44
0.57
0.49
2.25
0.25
0.28
0.54
0.35
0.008
0.021
0.019
0.025
0.012
0.027
0.018
0.008

Fig. 1.68 a

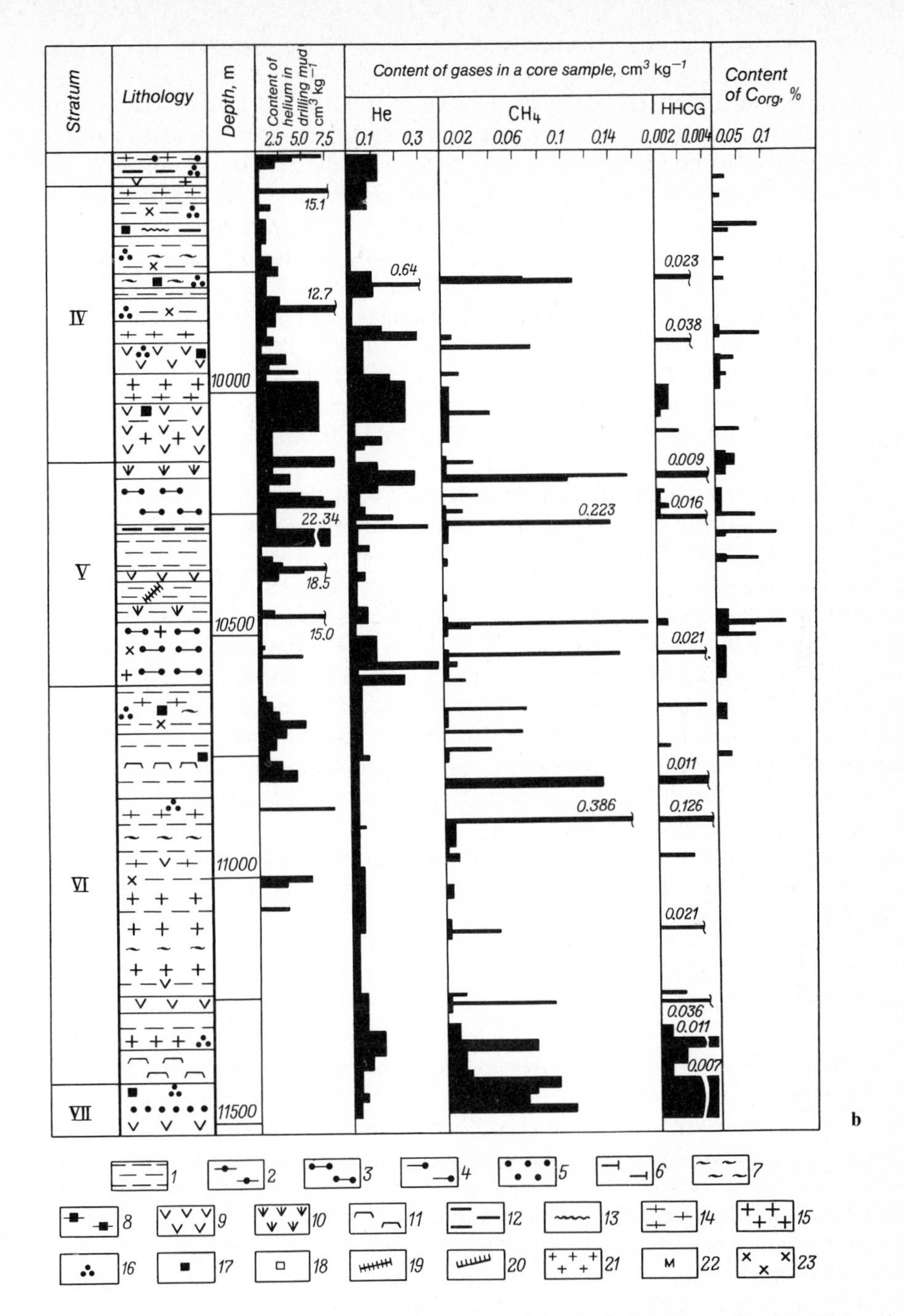


**Fig. 1.68 a, b.** Content of helium, methane, HHCG and $C_{org}$ in the Archean rocks of the Kola series in the SG-3 well section. *1* biotite-plagioclase gneisses; *2* sillimanite-biotite-plagioclase gneisses; *3* bimicaceous gneisses with HAM; *4* bimicaceous gneisses; *5* garnet-biotite-plagioclase gneisses; *6* epidote-biotite-plagioclase gneisses; *7* epidote-biotite-plagioclase schists; *8* magnetite-amphibole schists; *9* amphibolites; *10* amphibolites with cummingtonite; *11* banded amphibolites (leucocratic); *12* biotite-amphibole schists; *13* phlogopite rocks; *14, 15* intrusive rocks (*14* plagiogranites and shadow migmatites; *15* plagiomicroclinic granites, pegmatites); *16–18* ore manifestations (*16* sulphide; *17* magnetite; *18* ilmenite); *19* quartz veins; *20* chloritization; *21* microclinization; *22* muscovitization; *23* zones of migmatization

Figures 1.67 and 1.68 depict the helium content in cores as well as that in rocks calculated on the basis of data obtained in the course of gasometry of the drilling mud; the latter information may be used for determining a portion of the bound helium emitted from crushed rocks with respect to a total quantity of helium fixed in the drilling mud. In passing through fluid-permeable zones of the Proterozoic complex (2180 – 2370, 4800 – 5700 m), this portion of helium does not exceed 10%, and is equal to 30 – 100% in massive beds; in gas-permeable zones of the Archean complex this portion ranges from 0.4 to 4%, and in relatively impermeable blocks it is 10 – 40%. The inflow of helium into the fissured zones is so negligible that it does not affect the values of gas indications obtained in examining the drilling mud during subsequent drilling operations; this is expressed in zero values of helium content recorded in drilling through the underlying rocks.

Considerable amounts of $CO_2$ (see Fig. 1.66) appear in the drilling mud beginning from a depth of 8800 m. Since sodium carbonate is added in the drilling mud beginning from a depth of 8700 m, the question of whether a geochemical component of natural gas mixtures such as $CO_2$ can give any useful information or not, remains open. Nevertheless, it may be assumed that in intervals of 8900 – 9150, 9400 – 9500 and 9900 – 10,500 m, where anomalously high contents of helium, nitrogen, hydrogen, methane and HHCG are fixed, the $CO_2$ is not an artifact caused by technogenic factors, or at least, it is caused not only by these factors. Anomalously high contents of hydrogen and nitrogen, associated with the anomalous content of helium, are also of natural origin. It is characteristic that, beginning from a depth of 8500 m, the content of hydrogen in gas mixtures increases considerably.

It must be emphasized that zones of gas saturation, detected by using the gasometry data, appear to be wider and more "obscure" than zones of true gas saturation.

### 1.8.2 Gases in Rocks

The study of gases in the closed pores (gases of a gas phase or only gases of rocks) has been undertaken to solve the following tasks: finding out regularities in the distribution of gases in the oldest crystalline rocks of different genesis and their composition, depending on their depth of occurrence and geological position; the determination of the role played by gas in processes of endogenic ore formation and metamorphism from the prehnite-pumpellyite to amphibolite facies with intensively manifested granitization; the determination of the proportion of closed pore gases in the total amount of gas contained in the drilling mud in order to mark permeable-to-fluid zones in crystalline massifs and to trace rocks with distinctive gas content.

In the rocks penetrated by the SG-3 well, the following main distinctive features in the distribution of helium and HCG have been established.

1. Decrease in the total content of HCG down to a depth of 4884 m (i.e. down to the bottom of the Luchlompol suite, which approximately coincides with a boundary between the greenschist and epidote-amphibolite facies of metamor-

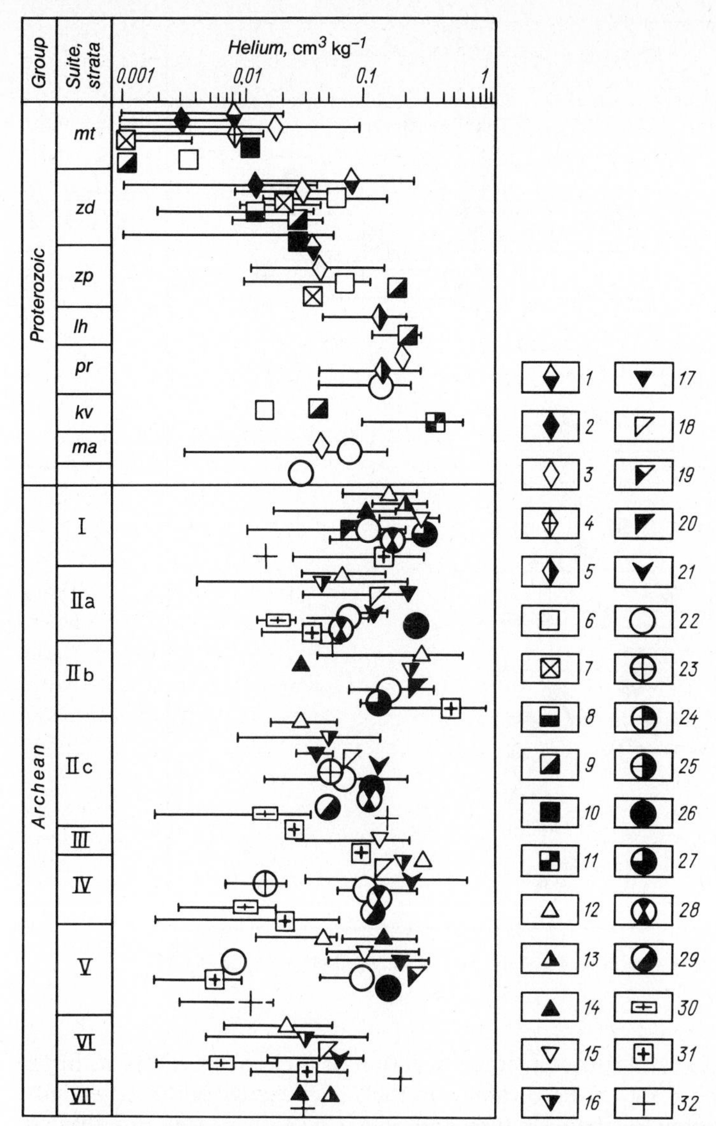


**Fig. 1.69.** Content of helium in rocks of the SG-3 well. *Middle Proterozoic rocks: 1* gabbro-diabases, gabbro-differentiated intrusions; *2* peridotites; *3* diabases; *4* picritic porphyrites; *5* andesite porphyrites, albitophyres, plagioporphyrites; *6* tuffs and tuffobreccia; *7* psammitic metatuffites; *8* pelitic metatuffites; *9* metapsammites; *10* metapellites and auleropellites; *11* carbonate rocks; *12* biotite-plagioclase gneisses, schists; *13* biotite-plagioclase gneisses, schists with HAM; *14* bimicaceous schists, gneisses; *15* bimicaceous gneisses, schists with HAM; *16* epidote-biotite plagioclase gneises; *17* biotite-amphibole-plagioclase gneisses; *18* epidote-biotite-amphibole-plagioclase gneisses; *19* garnet-epidote-biotite-plagioclase gneisses; *20* biotite-amphibole-plagioclase schists; *21* epidote-biotite-amphibole-plagioclase schists; *22* amphibolites; *23* banded coarse- and irregularly-grained amphibolites; *24* leucocratic amphibolites; *25* coarse- and irregularly-grained amphibolites; *26* amphibolites with cummingtonite; *27* amphibole schists with cummingtonite; *28* biotite-amphibole schists; *29* talc-actonilitic schists; *30* plagiogranites and shadow migmatites; *31* microclinic granites; *32* pegmatites. Mean helium contents are confined to central parts of rock symbols: ranges of content are shown by *lines*

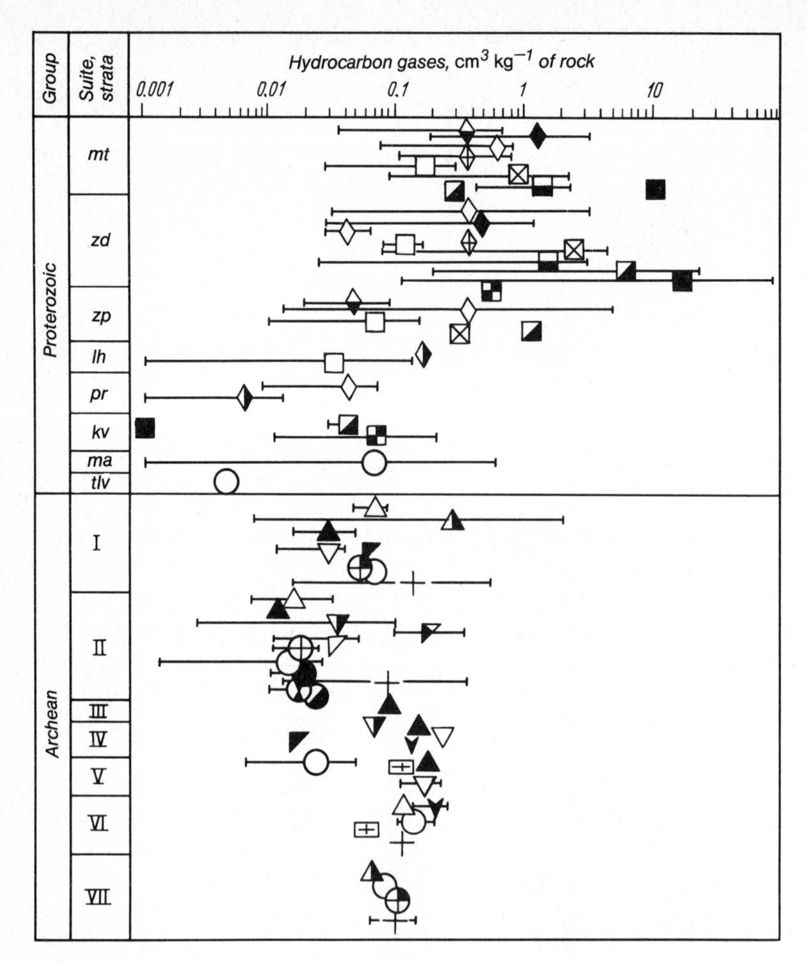


**Fig. 1.70.** Content of HCG sum in rocks of the SG-3 well. Symbols as in Fig. 1.69

phism in metabasalts) and subsequent stabilization of their mean values within a range of $0.01-0.2\ cm^3\ kg^{-1}$; these values, in fact, may change within a second-order range, depending on the rock type and on the particular suite or strata to which this rock belongs. For instance, metasedimentary rocks of the Zhdanov suite differ sharply from the Proterozoic effusive and intrusive rocks as well as from metasedimentary rocks of other suites by their high HCG content ($40-80\ cm^3\ kg^{-1}$); metasedimentary rocks are generally characterized by a considerable decrease of HCG in transition from the muscovite-chlorite to biotite-chlorite zone of greenschist facies metamorphism of meta-aulerolites which occur at a depth of 3180 m; in the Archean sequence, minimal contents of HCG are fixed in strata composed of biotite-plagioclase, biotite-amphibole-plagioclase gneisses and amphibolites (7622–9456 m), they are observed in almost all types of rocks, plagiogranites and migmatites being the only exceptions; the composition of HCG in pegmatites, plagiogranites and migmatites is stable, and is considered to be relatively high for the Archean rocks; maximum contents of HCG in the Archean strata were determined in the biotite and two-mica-gneisses with HAM. A distribution pattern of the HCG in the well section is shown in Figs. 1.67 – 1.70.

2. The helium concentration profile for the SG-3 well illustrates a situation opposite to the distribution pattern of HCG (see Figs. 1.67 – 1.69). The content of helium increases in all types of Proterozoic rocks down to the Luchlompol suite (4884 m). At deeper levels there are considerable variations in content (between 0.01 and 0.50 $cm^3 gm^{-1}$) and beginning from a depth of 10,600 m, the content of helium in rocks tends to decrease. It is characteristic that the maximal helium content occurs in different rocks of those Archean strata and members (see Fig. 1.69) which are particularly enriched in dark-coloured or highly aluminiferous minerals, i.e. amphibole schists with cummingtonite, biotite and two-mica-gneisses with HAM, epidote-biotite-amphibole-plagioclase gneisses, etc. However, in addition to melano- and mesocratic rocks, pegmatites and even granites are also notable for their high helium content in some intervals.

For instance, in an interval of 8320 – 8465 m, in addition to epidote-biotite-amphibole-plagioclase gneisses and magnesium amphibolites, there are potassium-sodium granites and plagiogranites which are also characterized by a high helium content. In Fig. 1.69, a corresponding interval has been recognized by taking this feature into consideration. High helium contents here were apparently, a result of rock recrystallization in conditions of tectonic deformations and inflow of the helium-bearing fluids. Part of the helium in these rocks thus is an impregnation. This caused remarkably high coefficients of helium preservation in rocks calculated according to the law of radioactive decay for the age of $3 \times 10^9$ years (Fig. 1.71). The coefficient of helium preservation is the ratio of the helium found in rocks to the helium which, in the case of its being fully preserved, must have accumulated in rocks as a result of uranium and thorium decay during a certain geological period. Uranium and thorium are considered to be immobile components, and their decrease is caused only by radioactive decay. It is worth noting that in one instance this coefficient is equal to 1, which may be explained by two reasons. First of all, it might be that surplus helium supplied by fluids was captured by rocks. Secondly, the age of rocks might be more than has been accepted ($3 \times 10^9$ years). Judging by geological data, the latter explanation seems to be hardly probable.

Nevertheless, the major part of the granites is in an area characterized by low and medium helium content. In almost all Archean strata, minimal helium contents are inherent to migmatites and plagiogranites.

3. Down to a depth of 5 km, the distribution of helium in rocks largely depends on the depth of their occurrence. At deeper levels, the textural and structural peculiarities of rocks, their mineral composition, their content of $^{238}U$ and $^{232}Th$ (main sources of radioactive helium), the age of rocks and their genesis, and the history of metamorphism became predominating factors.

On diagrams depicting a ratio of helium to uranium and thorium content in rocks of the SG-3 well (see Fig. 1.71), the results of the analysis obtained for the Archean amphibolites are grouped in bands stretching in two directions: the first band, illustrating the direct relationship between the contents of the components

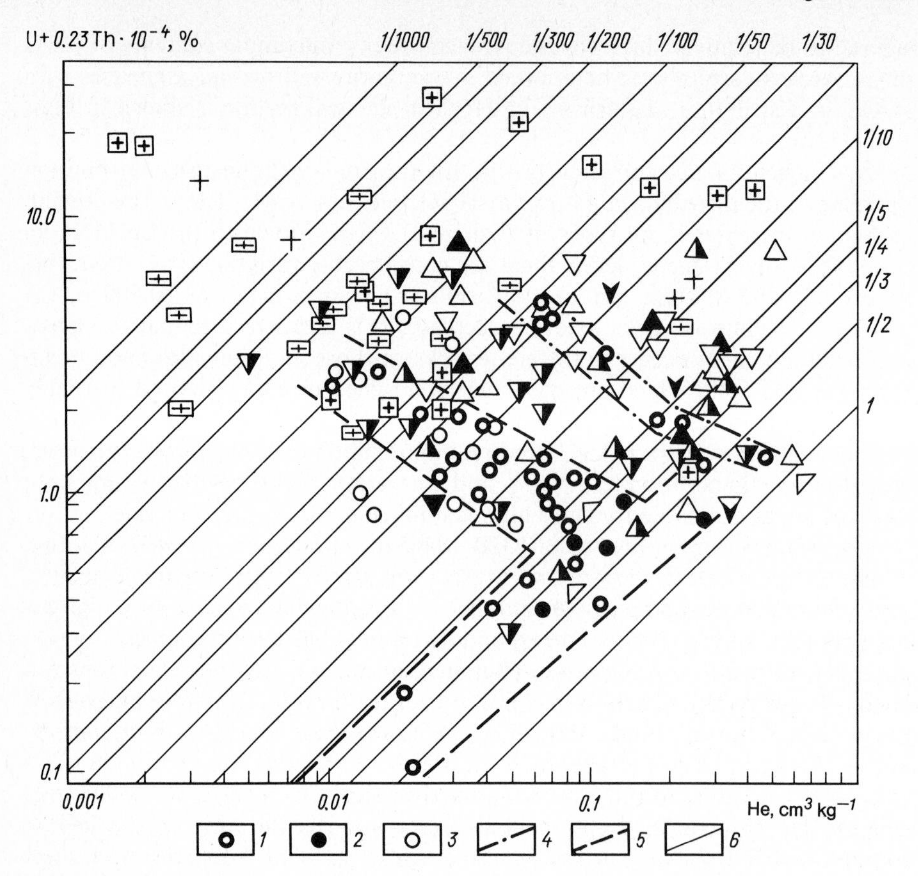


**Fig. 1.71.** Helium content versus uranium and thorium content in amphibolites of the Middle Proterozoic complex and in rocks of the Archean complex of the SG-3 well. *1, 2* amphibolites of the Archean complex, including amphibolites with cummingtonite (*2*); *3* amphibolites of the Proterozoic complex; *4* contours of the amphibolites field of the I stratum (6842 – 7622 m) of the Archean complex; *5* contours of the amphibolites field of the II-VII strata of the Archean complex; *6* isolines of helium preservation coefficients. Other symbols of rocks as in Fig. 1.69

discussed, coincides with isolines of the helium preservation coefficient in an interval of 0.2 – 0.8; two other bands, illustrating the inverse relationship between helium and radioactive elements, extend a 70° to the first band. One of these two latter bands, notable for higher values of component content, mainly comprises analytical data on amphibolites from the upper aluminiferous strata; these data are absent in other bands. Data on the Middle Proterozoic amphibolites (predominantly on the Mayarvi suite), as well as data on amphibolites from other Archean strata, tend to approach the second band, although data on the Middle Proterozoic amphibolites occupy more isometric area.

Values of helium preservation coefficients in amphibolites decrease as the content of uranium and thorium in amphibolites increases. This can be caused by the inflow of these elements into amphibolites by means of granite-forming solutions. It may be supposed that the initial contents of radioactive elements in am-

**Table 1.29.** Mean content of HCG, nitrogen ($cm^3\ kg^{-1}$) and organic carbon (%) in metasedimentary rocks of the Zhdanov suite, depending on conditions of their formation

| Number of samples | Facies | $N_2$ | $CH_4$ | $C_2H_6+C_3H_8$ | $C_{org}$ |
|---|---|---|---|---|---|
| 5 | Lagoonal | 9.11 | 58.45 | 0.67 | 2.14 |
| 3 | Shallow marine, stagnant | 4.11 | 21.76 | 0.35 | 0.30 |
| 8 | Shallow marine | 3.92 | 17.57 | 0.118 | 0.42 |
| 4 | Turbidity flows | 2.50 | 13.77 | 0.028 | 0.69 |
| 3 | Coastal | 5.23 | 11.30 | 0.04 | Not determined |
| 3 | Coastal flows | 1.33 | 4.68 | 0.019 | 0.26 |

phibolites of the first aluminiferous strata were higher. It should be said that highly aluminiferous amphibolites with cummingtonite are the most resistant to the action of granitization processes and, therefore, the degree of helium preservation in them is high. Granitization affects gneisses of all kinds in the same, although less pronounced, manner because during all stages of metamorphism they undergo more intensive internal reconstruction than amphibolites. Plagiogranites and migmatites lose helium in greater amounts, because they are the most vulnerable to the action of granite-forming potassium-bearing fluids, and due to the specificity of their mineral composition. For this reason, results of their analyses are grouped in the left quarter of the diagram.

4. Nitrogen was found in the Middle Proterozoic metasedimentary rocks of the Zhdanov suite. Its content reaches 20 $cm^3\ kg^{-1}$ and it is correlated with the content of HCG and $C_{org}$; the values of the correlation coefficients are rather high: $r = 0.69 + 0.90$ (under $r$ critical$^{0.01} = 0.487$).

5. The presence of a direct correlation between the content of $C_{org}$, HCG and $N_2$, as well as the light carbon isotope composition of HCG ($\delta^{13}C = -4.7$‰, RDB standard), are indicative of the biogenic nature of these components. This conclusion is confirmed by an observed relationship between the content of HCG, $C_{org}$, $N_2$ in metasedimentary rocks and facial conditions of their formation (Table 1.29). Carbon of methane from the Proterozoic gabbro intrusions is characterized by a heavier isotope composition ($\delta^{13}C = -1.0$‰, RDB standard) which is similar to that of the carbonaceous chondrite mantle.

6. Results of gas analysis, represented as average values by rock types, are grouped in a system of coordinates in a manner shown in Fig. 1.72, 1.73, i.e. $C_{org}$ and HCG form bands of varying width and direction or isometric areas; all of them illustrate either the presence or absence of main correlation trends.

A positive relationship between the average contents of HCG and $C_{org}$ and helium in the Matertin and Zhdanov sedimentary rocks, i.e. in a zone of the muscovite-chlorite metamorphism facies, confined to meta-aulerolites, proves that HCG and most of the helium occur together in a single whole pore system. Helium partial pressure in a system of interconnected pores is left unchangeable within individual members during the early stages of sedimentary rock metamorphism, and is preserved unchanged later when pores become sealed. Helium, gen-

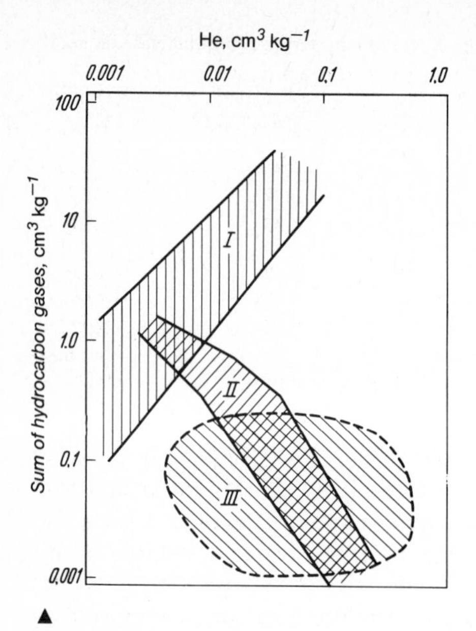


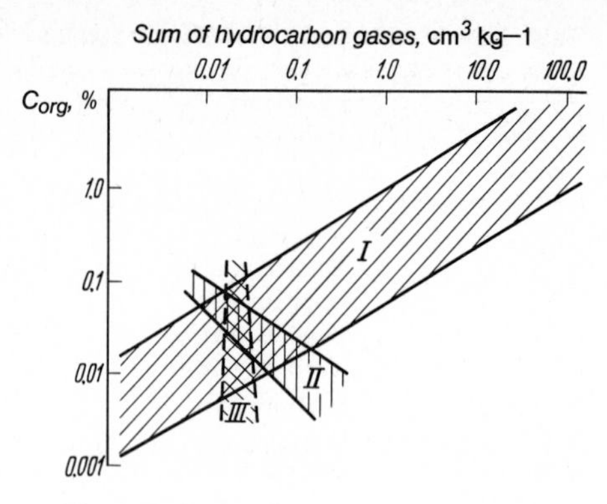


**Fig. 1.73.** Ratio of HCG sum to $C_{org}$ in rocks of the SG-3 well. *I* a field of metasedimentary rocks of the post-Proterozoic complex; *II* a field of rocks of the I, III and V strata of the Archean complex; *III* a field of rocks of the II and IV strata of the Archean complex

▲ **Fig. 1.72.** Ratio of HCG sum to helium in rocks of the SG-3 well. *I* a field of metasedimentary rocks of the Matertin and Zhdanov suites; *II* a field of volcanogenic rocks of the Proterozoic complex; *III* a field of the Archean rocks

erated later by uranium and thorium, whose content in sedimentary rocks is roughly proportional to that of $C_{org}$, is accumulated in pores containing sealed water solutions. There is thus a relationship between the amount of HCG and helium, i.e. the amount of these gases is in interrelation with the general porosity and $C_{org}$ content.

In older sedimentary rocks of the Proterozoic section there is a positive relationship between $C_{org}$ and the total HCG, although any relationship between total HCG and helium is not established. This may be caused by a differential behaviour of gas components during degassing of rocks, when they pass into facies of a higher grade of metamorphism, which leads to a sharp decrease in their porosity. Such factors as helium preservation, which depends on the depth of occurrence, and helium inflow from the Archean basement appear to play a significant role.

7. In the Proterozoic volcanic rocks, any relationship between the average contents of the sum of HCG and $C_{org}$ is absent, while a negative correlation exists between total HCG and helium. The absence of any relationship between $C_{org}$ and the sum of HCG, is apparently indicative of their different genetic nature. With the increase of the grade of metamorphism, the content of HCG decreases (degassing). However, different from sedimentary rocks, the increase of the grade of metamorphism of igneous rocks, characterized by a monotonous composition and texture, leads to a more regular decrease in HCG content. The degree of helium preservation in relatively uniform volcanic rocks systematically increases down to a depth of 4800 m, thus resulting in the appearance of a negative correlation.

8. In the I, II and III Archean strata, characterized by the presence of HAM, a negative relationship exists between the content of the total HCG sum and $C_{org}$. In rocks of the II and IV strata, no correlation between these components was observed. All Archean rocks are characterized by the absence of correlation between the content of helium and total HCG. This is because these gases have originated in different disconnected sources, and because their modes of occurrence in rocks are also different.

9. $CO_2$ and hydrocarbons (90% methane) are main gas components of the ubiquitous quartz (quartz is met throughout the well section) which is the rock-forming mineral of hydrothermal veins and pegmatites. The total content of these components in quartz of the synmetamorphic hydrothermal veinlets, existing in volcanic and metasedimentary rocks of the upper part of the section, changes from 0.14 to 20.5 $cm^3\,kg^{-1}$, with metasedimentary rocks containing higher amounts. HCG predominate in quartz veins of metasedimentary rocks; the HCG content is 5 – 15 $cm^3\,kg^{-1}$. In quartz of magmatic rocks, HCG are met in predominant quantities only if the amount of gases is small; in all other cases $CO_2$ predominates, while the content of HCG is not less than 3 $cm^3\,kg^{-1}$. This indicates that the fluid regime of metamorphism in sedimentary rocks was notable for its more reducing character than that in volcanic rocks.

$CO_2$ is the predominant component of a gas phase in quartz from hydrothermal veins and pegmatites of the Archean strata, its content reaching 78 $cm^3\,kg^{-1}$; HCG are present in negligible amounts ($<0.5$ $cm^3\,kg^{-1}$).

The process of granitization is accompanied by an increase of the $CO_2$ portion in a fluidal phase which participates in the formation of hydrothermal quartz-carbonate and carbonate veins, in both the Archean and Proterozoic complexes. However, if carbonate veins are formed in the Proterozoic complex, part of the carbon is extracted from the organic matter present in considerable amounts in metasedimentary rocks. Because of this, the carbon of carbonates becomes lighter ($\delta^{13}C = -1.2 + 1.7‰$) as compared to the carbon isotope composition of carbonates from the Archean complex ($\delta^{13}C = -0.7‰$; RDB standard; determined by V. S. Lebedeva) which is similar to the carbon composition of the juvenile $CO_2$.

For the relationship between the gaseous phase of rocks and ore genesis, the following facts must be emphasized.

1. A helium inflow zone was established in an interval of 9950 – 10,500 m; it was detected by the gasometry studies carried out after the well dead time. Gasometry studies performed during a process of drilling have revealed that drilling mud contains an anomalously high amount of helium; core samples are characterized by a somewhat increased helium content as compared with that typical of rocks of these types. The same zone is also characterized by the increased content of HCG in rocks. It is apparent that in this interval the well has crossed a permeable-to-fluids and ore-controlling zone of rock slackening.

2. Rocks of the Zhdanov ore-bearing suite differ considerably by the presence of increased amounts of HCG from rocks of the same composition and age developed in other areas of the Kola Peninsula. For instance, in metasedimentary rocks of the Imandra-Varzug complex, analogous to the Pechenga complex, HCG are absent despite high amounts of $C_{org}$. This token may be used in assess-

ing prospects of sulphide mineralization of volcano-sedimentary strata of a similar type.

3. The inverse relationship established between the content of uranium + thorium in monotypic rocks of the Archean complex and helium preservation, calculated according to equations of radioactive decay, indicates that a certain part of the radioactive components flowed into these rocks later (see Fig. 1.71).

### 1.8.3 Geochemistry of Organic Matter

Geochemical studies of organic matter in the SG-3 well section included quantitative determinations of $C_{org}$ and of the chloroform bitumen extract (CHB) of rocks. The extract was analyzed for its elemental, compound class and molecular composition by electronic paramagnetic resonance spectroscopy (EPR), low resolution nuclear magnetic resonance spectroscopy (NMR) and gas chromatography (GC).

More than 500 $C_{org}$ determinations were made for the part of the well section penetrated. On the basis of these data, regularities in $C_{org}$ distribution, depending on the genetic type of the host rocks, were revealed (Fig. 1.74). The content of $C_{org}$ in the Proterozoic volcano-sedimentary complex varies from "traces" to 3%. Maximum $C_{org}$ content is characteristic of metasedimentary rocks (phyllites, siltstones, sandstones) of the Zhdanov suite. For instance, the $C_{org}$ content in phyllites is 0.2 – 0.8% on the average, but in many samples it reaches 1.2 – 2.5%, and in sandstone-siltstone rocks it varies from 0.03 to 3.0%; these changes are controlled by the ratio of pelitic to psammitic material in a studied sample. In tuffs and tuffites the content of $C_{org}$ decreases to 0.1%, and in diabases, gabbro-diabases and porphyrites it falls to 0.01%. In the Archean complex content of $C_{org}$ in most of the samples analyzed changes from 0.01 to 0.03%.

EPR and NMR spectroscopy were used to evaluate the degree of organic matter transformation. The first method provides information about its paramagnetic properties which become more intensive as low grades of catagenesis (protocatagenesis) reach medium stages (mesocatagenesis); paramagnetic properties completely disappear when stages of graphitization are reached. The second method allows the characterization of a degree of the organic matter enrichment by hydrogen and movable water-bearing compounds. The EPR and NMR spectra have not been obtained on analyzing phyllites, siltstones and sandstones, which is considered to be the main indication of the high degree of organic matter transformation corresponding to stages of graphitization. Strong catagenesis has led to the destruction of organic matter; this was accompanied by a loss of paramagmatic properties and the main hydrocarbon mass. The EPR and NMR spectrum of igneous rocks cannot be interpreted because of the presence of terrigenous minerals.

Semi-quantitative evaluation of bitumen content in rock was performed with the help of the bitumen-luminescence analysis; however, it must be noted that at higher stages of catagenesis the strength of physicochemical bonds between bituminous matter and the host minerals increases, and, consequently, the data obtained cannot be reliable. This conclusion was confirmed by comparing data ob-

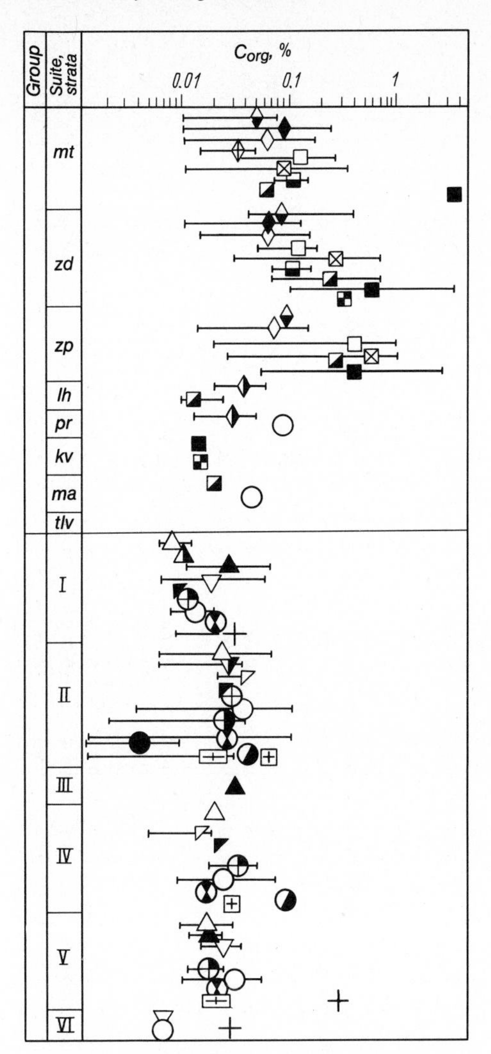


**Fig. 1.74.** Content of $C_{org}$ in rocks of the SG-3 well. Symbols are given in Fig. 1.69

tained by the bitumen-luminescence studies and by rock extraction with the help of the soxhlet apparatus, which differed from each other by one to three orders of magnitude.

CHB, containing hydrocarbon compounds, is the most interesting component of organic matter to be used in geochemical studies. There are two basic characteristics in the quantitative distribution of CHB (Table 1.30) in the well section. The first characteristic is that the Proterozoic rocks possess higher contents of CHB (mainly hundredths of a percent of dry rock) as compared with those established in the Archean rocks (thousandths of a percent). The second characterizes CHB distribution in the complex Proterozoic section in which maximal values (up to 0.2%) were determined in metasedimentary rocks, while in volcano-sedimentary rocks these values decreased to hundredths and thousandths of a percent. In genetically pure rocks, a direct correlation exists between

**Table 1.30.** Results of the chemical analysis of bituminous matter from rocks of the Kola superdeep well

| Depth, m | Rocks | $C_{org}$ | Insoluble residue | CHB, % | Compound class composition of CHB, % | | | | |
|---|---|---|---|---|---|---|---|---|---|
| | | | | | Oils | Petroleum ether | Benzene resins | Alcohol-benzene resins | Asphaltenes |
| 1 | 2 | 3 | 4 | 5 | 6 | 7 | 8 | 9 | 10 |
| Proterozoic complex | | | | | | | | | |
| 765 – 885 | Tuffite | 0.04 | 66.7 | 0.008 | 49.4 | 6.8 | 28.3 | 12.7 | 2.8 |
| 887 – 965 | Diabase | 0.08 | 85.8 | 0.085 | 1.11 | 1.48 | 81.29 | 15.52 | 0.6 |
| 1108.2 – 1228.2 | Phyllites, siltstones, sandstones | 0.26 | 79.5 | 0.028 | 35.2 | 9.4 | 11.4 | 31.7 | 12.3 |
| 1182 – 1496 | Phyllites with siltstones | 0.60 | 82.0 | 0.007 | 42.6 | 10.8 | 14.7 | 22.4 | 9.5 |
| 1289.4 – 1305.0 | Essexitic gabbro-diabase | 0.04 | 71.6 | 0.084 | 8.88 | 8.29 | 58.9 | 23.74 | 0.19 |
| 1284 – 1413 | Gabbro-diabase | 0.02 | 83.4 | 0.004 | 22.7 | 9.2 | 14.3 | 39.3 | 14.4 |
| 1757.8 – 1767.6 | Hyperbasites | 0.13 | 76.3 | 0.161 | 1.74 | 17.6 | 76.2 | 4.46 | Traces |
| 1954.6 – 1958.3 | Diabase | 0.03 | 85.5 | 0.091 | 4.89 | 6.13 | 84.21 | 4.42 | 0.35 |
| 2154 – 2291 | Alternation of phyllites and siltstones | 0.01 | 80.4 | 0.005 | 41.5 | 9.9 | 17.4 | 23.0 | 8.2 |
| 2994 – 3201 | Diabase | 0.03 | 88.3 | 0.068 | 2.49 | 3.38 | 84.94 | 8.8 | 0.39 |
| 3645.8 – 3650.3 | Diabase with carbonaceous material | 0.06 | 89.2 | 0.066 | 3.60 | 3.87 | 79.26 | 12.96 | 0.31 |
| 3752.1 – 3757.4 | Tuffs | 0.27 | 74.0 | 0.095 | 4.65 | 4.62 | 81.54 | 9.02 | 0.17 |
| 3983.8 – 3987.3 | Diabase | 0.06 | 94.4 | 0.065 | 2.25 | 2.02 | 75.00 | 20.19 | 0.54 |
| 4336 – 4389.0 | Diabase | 0.05 | 94.9 | 0.051 | 2.43 | 4.96 | 86.05 | 5.68 | 0.88 |
| 4630.5 – 4784.2 | Schist in tuffaceous-sedimentary rocks | 0.09 | 88.0 | 0.185 | 1.78 | 3.65 | 75.56 | 19.04 | Traces |
| 4749.15 | Andesite porphyrite | 0.04 | 93.8 | 0.099 | 3.28 | 4.25 | 64.91 | 27.56 | Traces |
| 4847 – 4859 | Arkose sandstone | 0.04 | 96.0 | 0.059 | 4.40 | 7.48 | 79.11 | 9.01 | Traces |
| 6569 – 6577.9 | Biotite-amphibole-plagioclase schist | 0.06 | 94.0 | 0.025 | 4.78 | 2.58 | 87.15 | 6.49 | Traces |
| 6757.3 – 6767.3 | Amphibole-plagioclase schist | 0.04 | 96.0 | 0.016 | 2.05 | 2.59 | 82.06 | 10.51 | 2.76 |
| Archean complex | | | | | | | | | |
| 6933.0 – 7048.3 | Two-mica-plagiogneisses | 0.07 | 96.5 | 0.014 | 1.45 | 2.51 | 81.16 | 13.29 | 1.59 |
| 7642 – 7656 | Amphibolite | n.d. | n.d. | 0.010 | 26.1 | 21.6 | 48.3 | 3.4 | Traces |
| 7937 – 7944 | Amphibolite | n.d. | n.d. | 0.011 | 16.3 | 22.0 | 54.1 | 7.6 | Traces |
| 8313.1 – 8324.0 | Biotite gneiss | 0.03 | 95.3 | 0.009 | 59.99 | 16.25 | 12.64 | 10.10 | 0.02 |
| 9600 – 9989 | Epidote-biotite-plagioclase gneisses | 0.01 | 96.4 | 0.009 | 21.92 | 12.95 | | 65.08 | 0.05 |
| 10144 – 10181 | Two-mica-gneisses | 0.03 | 98.0 | 0.011 | 37.75 | 11.85 | | 53.37 | 0.03 |

the content of CHB and $C_{org}$, which indicates the syngenetic nature of the bituminous matter. All types of Middle Proterozoic rocks encountered in the well section down to a depth of 4860 m are characterized by similar carbon isotope composition of CHB ($\delta^{13}C = 2.9\permil$).

The values of CHB elemental composition (Table 1.31) illustrate that hydrogen content in the Proterozoic samples (irrespective of the genetic type of rocks) reaches 8 – 9%, the ratio of C/H does not exceed 7%, and the sum of heteroatoms (O + S + N) is 20 – 35%; the content of hydrogen, the ratio of C/H, and the sum of heteroatoms in the CHB of the Archean gneisses and amphibolites reaches 10 – 12%, not more than 7%, and 20 – 22%, respectively. This means that the CHB composition of the Proterozoic and Archean rock differ from each other in both the distribution of their water-bearing compounds and their structural peculiarities. The difference in distribution patterns of the water-bearing compounds is reflected by the value of the C/H ratio: it seems most probable that at lower values of this ratio compounds with aliphatic structures will predominate, and at higher values cyclic compounds will constitute a considerable part of CHB.

The compound class composition of CHB changes both in the stratigraphic sequence and from one to another genetic type of rocks in the Proterozoic section (see Table 1.30). In the studied portion of the Archean complex, a basic part of CHB is represented by an oil fraction (oils + petroleum ether resins) and benzene resins; asphaltenes are practically absent and alcohol-benzene resins are present in amounts of not less than 13%, i.e. high-molecular weight heterosystems are present only in limited quantities. In the Proterozoic complex, the compound class composition of CHB changes in accordance with genetic types of the host rocks. For instance, metasedimentary rocks are notable for the increased content of asphaltenes (8 – 12%), alcohol-benzene resins (up to 39%) and oil fraction (40 – 50%), whereas in effusive and intrusive rocks CHB is mainly represented by benzene resins; the amount of asphaltenes decreases to "traces", and the amount of an oil fraction falls to 6%. Some exceptions are observed mainly for effusive rocks, which are probably related to the presence of a considerable amount of metasedimentary rocks in the analyzed samples.

A structural analysis of CHB components performed by infrared spectrometry (IRS) has made it possible to establish that, in spite of the fact that the well section contains multigenetic and multi-lithological types of rock, three types of CHB can be distinguished in effusive and intrusive Proterozoic rocks, metasedimentary Proterozoic rocks and metamorphic Archean rocks.

The structure of all components of CHB extracted from the Proterozoic effusive and intrusive rocks is characterized by the presence of predominantly oxygen-bearing functional groups of phthalate type (esters of aromatic dicarboxylic acids) which show intense absorption bands at 1700 – 1750 $cm^{-1}$ and in the range of 900 – 1300 $cm^{-1}$ (C=O, C – OH); aromatic structural units are in close conjunction with the oxygen-bearing groups, and are found by the presence of a "double tooth" in the range of 1600 – 1640 $cm^{-1}$. Paraffinic structures ($CH_2$, $CH_3$ groups) are represented mainly by radicals. Their absorption bands (1460, 1380 $cm^{-1}$) are poorly pronounced. The phthalate composition of CHB is characteristic of both the poorly transformed organic matter present in the oxidizing

**Table 1.31.** Results of the analysis of the elemental composition of CHB from rocks of the Kola superdeep well

| Interval of sampling, m | Rocks | Content of elements, % | | | |
|---|---|---|---|---|---|
| | | C | H | S | N / O, others |
| 1 | 2 | 3 | 4 | 5 | 6 |
| Metatuffites | | | | | |
| 771.2 – 771.9 | Tuffitic aleuropsammite | 65.49 | 6.71 | 1.35 | 0.00 / 26.45 |
| Metapelites and aleuropelites | | | | | |
| 1199.0 – 1199.7 | Layered greywacke silty schists and aleurolite schist | 71.19 | 10.21 | 0.65 | 0.00 / 17.95 |
| 1199.8 – 1200.8 | Layered greywacke silty schist | 60.41 | 8.66 | 0.0 | 1.18 / 29.86 |
| 1710.0 – 1712.2 | Thin-layered greywacke aleuropelite | 64.81 | 9.79 | 1.22 | 0.00 / 24.18 |
| 2163.4 – 2165.9 | Greywacke aleuropelite | 55.85 | 7.03 | 0.002 | 0.00 / 37.12 |
| 2166.9 – 2167.2 | Greywacke aleurolite | 67.68 | 9.40 | 0.95 | 0.00 / 21.67 |
| Metapsammites | | | | | |
| 2459.9 – 2460.5 | Aleuropsammite | 55.82 | 8.65 | 0.00 | 0.00 / 35.53 |
| 2606.3 – 2608.2 | Arkose psammite | 69.90 | 10.53 | 0.00 | 0.00 / 19.57 |
| 2608.8 – 2610.2 | Arkose psammite | 68.87 | 9.86 | 0.61 | 0.00 / 20.60 |
| 4855.6 – 4862.9 | Arkose psammite with hematite-sericite matrix | 55.94 | 8.11 | 0.00 | 0.00 / 35.95 |
| 4863.8 – 4866.1 | Arkose psammite with hematite-sericite matrix | 56.44 | 8.14 | 0.00 | 0.00 / 35.42 |
| Magmatic rocks of basic composition | | | | | |
| 2030.5 – 2031.2 | Gabbrodiabase | 61.00 | 8.23 | 0.68 | 0.00 / 30.07 |
| 2032.2 – 2032.5 | Gabbrodiabase | 70.08 | 7.46 | 0.00 | 0.00 / 22.46 |
| 2657.1 – 2658.7 | Amphibole gabbro | 69.93 | 10.37 | 0.00 | 0.00 / 19.70 |
| Metamagmatic rocks of ultrabasic composition | | | | | |
| 370.3 – 373.8 | Peridotite | 68.82 | 9.73 | 0.86 | 0.00 / 20.59 |
| 902.8 – 903.6 | Serpentinite | 70.29 | 8.13 | 1.28 | 0.34 / 20.19 |
| 1581.3 – 1582.6 | Serpentinite | 66.98 | 8.14 | 0.00 | 0.00 / 24.87 |
| Amphibolites | | | | | |
| 6475.6 – 6485.7 | Schistose amphibolites in diabase | 69.86 | 9.60 | 0.00 | 0.00 / 20.54 |

**Table 1.31** (continued)

| Interval of sampling, m | Rocks | Content of elements, % | | | |
|---|---|---|---|---|---|
| | | C | H | S | N / O, others |
| 1 | 2 | 3 | 4 | 5 | 6 |
| 7647.6 – 7664.1 | Para-amphibolites | 69.90 | 9.60 | 0.00 | 0.00 / 20.40 |
| 7670.0 – 7677.5 | Amphibolites with sphene | 69.41 | 8.22 | 0.00 | 0.00 / 22.37 |
| 8460.4 – 8460.8 | Amphibolites | 63.60 | 11.00 | 0.00 | 0.00 / 25.40 |
| 8941.8 – 8942.5 | Schistose amphibolites | 63.60 | 11.20 | 0.00 | 0.00 / 25.20 |
| Gneisses | | | | | |
| 7120.1 – 7134.0 | Garnet-staurolite two-mica-gneisses | 74.00 | 11.90 | 0.00 | 0.00 / 22.50 |
| 7771.4 – 7778.9 | Muscovitized epidote-biotite gneisses | 68.89 | 8.60 | 0.00 | 0.00 / 29.50 |
| 8993.9 – 8997.9 8998.0 – 8999.7 | Biotite gneisses | 60.70 | 10.20 | 0.00 | 0.00 / 29.50 |
| 8625.6 – 8628.2 8637.3 | Epidote-biotite gneisses | 63.60 | 11.30 | 0.00 | 0.00 / 25.10 |
| 9352.4 – 9355.9 | Biotite gneisses | 70.60 | 11.20 | 0.00 | 0.00 / 18.20 |

environment and the organic matter from zones of high-temperature metamorphism in which organic matter is being destroyed at molecular and atomic levels, and free hydrogen and oxygen appear. This results in the formation of local areas characterized by an oxidizing environment favourable to the preservation of phthalates. It is apparent that similar conditions existed during the formation of effusive and intrusive Proterozoic rocks.

CHB components from metasedimentary rocks possesses other structural properties (Fig. 1.75). Oils are represented by the aliphatic and cyclic hydrocarbon functional groups with the former predominating. Cyclic aromatic structures (750, 820, 1610 $cm^{-1}$) are mainly of a bicyclic type. In the fraction of petroleum-ester resin, which, by its composition, is intermediate between oils and highly molecular resins, a high content of hydrocarbon structures is preserved, while structures represented by polycondensed systems (distinguished by the presence of a "triple-tooth" peak in the range of 750 – 850 $cm^{-1}$) change. In addition to hydrocarbon groups, the fraction contains oxygen groups of the phthalate type. In the phthalate fraction benzene and alcohol-benzene resins exhibit signs of destruction; this is expressed in a change of the "double-tooth" configuration at 1600 – 1640 $cm^{-1}$ (asymmetry of absorption bands appears), as well as in the increase of amounts of paraffinic structures (1380, 1460 $cm^{-1}$).

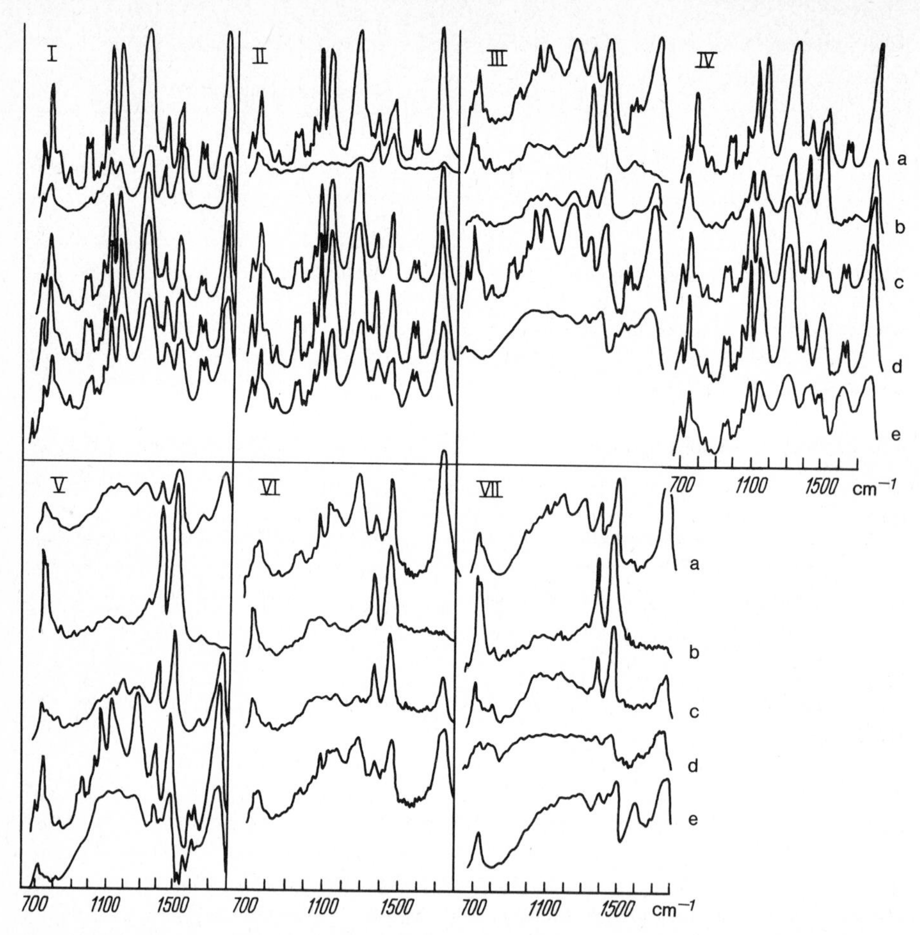


**Fig. 1.75.** IR-spectra of CHB and its components. *Middle Proterozoic rocks: I* diabase (depth 1954.7 – 1958.4 m); *II* tuffs (depth 3752.1 – 3757.4 m); *III* tuffites (depth 765 – 885 m); *IV – V* phyllite with siltstone (depth 1182 – 1496 m). *Archean rocks: VI* epidote-biotite-plagioclase gneiss (depth 9600 – 9980 m); *VII* biotite gneiss (depth 8313.3 – 8324.0 m); *a* CHB, *b* oils; *c* petroleum-ether resins; *d* benzene resins; *e* alcohol benzene resins

An independent structural type of CHB was discovered in the Archean rocks (see Fig. 1.75). The oil fraction mainly consists of high molecular-weight paraffins, and this conclusion is corroborated by the high intensity and deep splitting of the absorption band at 720 $cm^{-1}$, corresponding to a $CH_2$ group in long chains; in petroleum ether resins paraffinic structures also predominate, but oxygen-bearing structures of an acid type also appear (the latter are main components of the benzene and alcohol-benzene resins).

On the basis of the chromatographic analysis of normal alkanes (n-alkanes) contained in CHB, the following geochemical indices can be determined: the extension of the n-alkane series, the position of a concentration maximum in this series and the content of "liquid compounds" (up to $C_{16}$ inclusively), the value of an odd-to-even ratio (the ratio of the sum of n-alkanes with an odd number of carbon atoms to the sum of those with an even number of atoms). In agreement with these indices, the principle differences in the n-alkane distributions have been established for CHB oils of the Archean and Proterozoic rocks.

Oils of CHB from the Archean rocks were studied by gas chromatography, which is notable for high accuracy and sensitivity; this method also reveals the distribution pattern not only of alkanes but also of iso-alkanes, beginning with $C_9$. In the Archean rocks high-molecular-weight compouns ($C_{23-25}$) comprise a series of n-alkanes extending from $C_{14-17}$ to $C_{33-40}$ in maximal quantities (see Table 1.32, Fig. 1.76); the amount of "liquid" n-alkanes does not exceed 5%, the value of the odd even ratio is 1.1. Alkanes of an isoprenoid type are also represented by high-molecular-weight compounds (iso-$C_{18-24}$), with phytane being the predominant component (iso-$C_{20}$; the ratio of total n-alkanes to total isoprenoids is 0.07. It should be noted that in a gneiss sample (depth 8313 – 8324 m) isoprenoids are practically absent in the CHB oil fraction. By correlating results of the chromatographic and structural analyses, it may be concluded that n-alkanes are the predominating components in CHB oils of the Archean rocks.

Oils of CHB, extracted from intrusive and effusive rocks of the Proterozoic complex, were studied by the chromatographic method by applying an attached column. These rocks contain negligibly small amounts of n-alkanes of the series $C_{15,16}-C_{29-33}$ in which high-molecular compounds are characterized by maximal concentrations: "liquid" n-alkanes are either absent, or are present in amounts of not more than 3.5% of the total quantity; the odd-even ratio changes from 0.9 to 1.1.

Oils of CHB from the Proterozoic metasedimentary rocks, analyzed by the same method, are characterized, first of all, by the increased amount of n-alkanes in their total composition, and secondly, by the peculiar distribution pattern of individual compounds in the series. It was established that the series was represented by n-alkanes from $C_{11}$ to $C_{29-30}$, with a concentration maximum at $C_{12}$. The presence of the second maximum at $C_{17}$, $C_{18}$ is caused by the applied method of analysis (another type of chromatographic column is used), because with this method $C_{17}$ and $C_{18}$ peaks are obtained as a sum of coeluting n-alkanes and isoprenoids. The amount of "liquid" n-alkanes reaches 65%. This genetic type of rock was studied in detail in satellite wells in which analogous results on the distribution of n-alkanes and other geochemical indices have been obtained.

While studying rocks of the mixed volcano-sedimentary genesis (tuffites with siltstones, schists in volcano-sedimentary rocks, etc.), the chromatographic indices were characterized by intermediate values (see Table 1.32).

The results of the integrated geochemical investigations allow the conclusion that the quantitative distribution, composition, structural peculiarities and chromatography characteristic of organic matter on the whole, and of the CHB, depend on the conditions in which different genetic types of host rocks were formed.

The geochemical environment of the sedimentation of metasedimentary rocks was conductive to the accumulation and preservation of organic matter which in the process of transformation generated bituminous components containing aliphatic and alicyclic hydrocarbons. It is apparent that the major part of the bituminous matter did not leave the parent rock. It appears most probable that

**Table 1.32.** Distribution of n-alkanes (%) in oils of CHB from the Kola superdeep well

| Depth, m | Lithology | $C_{11}$ | $C_{12}$ | $C_{13}$ | $C_{14}$ | $C_{15}$ | $C_{16}$ | $C_{17}$ | $C_{18}$ | $C_{19}$ | $C_{20}$ | $C_{21}$ | C |
|---|---|---|---|---|---|---|---|---|---|---|---|---|---|
| 1 | 2 | 3 | 4 | 5 | 6 | 7 | 8 | 9 | 10 | 11 | 12 | 13 | 1 |
| 765 – 885 | Tuffite | n.d.[c] | 2.2 | 3.4 | 2.6 | 4.3 | 7.1 | 10.8 | 12.9 | 8.6 | 6.6 | 6.5 | 6 |
| 887 – 965 | Diabase | n.d.[c] | n.d. | n.d. | n.d. | n.d. | n.d. | 2.3 | 4.8 | 3.0 | 1.1 | 6.2 | 7 |
| 1182 – 1496 | Phyllite with aleurolite | 15.1 | 29.1 | 14.6 | 0.1 | 2.1 | 4.2 | 5.3 | 5.5 | 3.1 | 2.1 | 2.1 | 2 |
| 1289 – 1305 | Essexitic gabbro-diabase | n.d. | n.d. | n.d. | 1.4 | 2.0 | 2.6 | 4.5 | 5.8 | 4.3 | 15.4 | 2.2 | 5 |
| 1757 – 1767 | Hyperbasites | n.d. | n.d. | 0.4 | 0.3 | 1.4 | 3.0 | 6.2 | 6.3 | 5.4 | 9.1 | 3.5 | 4 |
| 1954 – 1958 | Diabases | 1.2 | 4.2 | 4.2 | 2.3 | 2.1 | 11.7 | 4.8 | 5.3 | 7.9 | 2.1 | 3.7 | 5 |
| 1284 – 1413 | Gabbro-diabases | n.d. | n.d. | n.d. | n.d. | n.d. | 1.4 | 7.1 | 9.7 | 8.9 | 6.8 | 7.7 | 9 |
| 2154 – 2291 | Alternation of phyllites and aleurolites | 8.6 | 10.3 | 5.5 | 0.4 | 4.1 | 4.7 | 10.6 | 11.1 | 5.9 | 5.7 | 4.9 | 4 |
| 2994 – 3201 | Diabases | n.d. | n.d. | n.d. | n.d. | n.d. | 2.1 | 3.4 | 5.6 | 5.5 | 3.9 | 10.7 | 6 |
| 3645 – 3650 | Diabases with carbonaceous material | n.d. | n.d. | n.d. | n.d. | n.d. | 6.5 | 4.0 | 6.8 | 6.0 | 6.5 | 8.0 | 6. |
| 3752 – 3757 | Tuffs | n.d. | 0.6 | 0.6 | 0.9 | 2.2 | 3.6 | 8.8 | 9.3 | 8.0 | 7.6 | 5.2 | 5. |
| 3983 – 3987 | Diabases | n.d. | n.d. | n.d. | n.d. | n.d. | 3.5 | 3.4 | 6.2 | 4.6 | 6.1 | 7.5 | 6. |
| 4360 – 4389 | Diabases | n.d. | n.d. | n.d. | n.d. | 0.7 | 1.1 | 4.2 | 4.2 | 3.4 | 1.9 | 4.2 | 5. |
| 4630 – 4748 | Schists in volcano-sedimentary rocks | n.d. | 1.4 | 2.3 | 1.7 | 1.2 | 1.6 | 3.7 | 4.2 | 3.7 | 3.2 | 4.6 | 5. |
| 4749 | Andesite-porphyrite | n.d. | n.d. | 0.3 | 0.4 | 0.5 | 0.8 | 1.8 | 1.6 | 1.7 | 3.7 | 9.1 | 9. |
| 4847 – 4859 | Arkose sandstone | n.d. | n.d. | n.d. | n.d. | n.d. | 0.8 | 4.7 | 4.3 | 2.9 | 3.2 | 6.1 | 4 |
| 6757 – 6767 | Schists | n.d. | n.d. | 1.1 | 3.0 | 3.7 | 5.4 | 13.3 | 15.1 | 5.5 | 2.5 | 3.9 | 6 |
| 8313 – 8324 | Biotite gneiss | n.d. | n.d. | n.d. | n.d. | n.d. | n.d. | 0.7 | 1.1 | 1.9 | 2.8 | 6.3 | 8. |
| 9600 – 9989 | Epidote-biotite plagioclase gneisses[a] | n.d. | n.d. | n.d. | 0.5 | 0.8 | 1.2 | 1.6 | 1.9 | 2.9 | 3.8 | 5.2 | 6 |
| 10144 – 10181 | Two-mica-gneisses[b] | n.d. | n.d. | n.d. | 0.8 | 1.4 | 2.3 | 3.6 | 3.8 | 3.6 | 3.8 | 4.9 | 5. |
| Oil-addition of $C_9$, $C_{10}$ in drilling mud – 54, 5.4 | | 6.5 | 5.5 | 4.1 | 4.7 | 3.8 | 4.3 | 3.2 | 3.4 | 2.8 | 2.7 | 2.7 | 2. |

[a] The following alkanes were added for the total sum: $C_{34}$ – 2.4; $C_{35}$ – 2.2; $C_{36}$ – 1.8; $C_{37}$ – 1.7; $C_{38}$ 1.6; $C_{39}$ – 0.4%.
[b] The following alkanes were added for the total sum: $C_{34}$ – 2.1; $C_{35}$ – 2.0; $C_{36}$ – 1.9; $C_{37}$ – 1.6; $C_{38}$ 1.4; $C_{39}$ – 0.4%.
[c] Not detected.

intrusive bodies, which emplaced sedimentary strata, assimilated organic compounds and transformed them into distinctive organic systems under the action of high temperatures. This appears to explain the geochemical regularities revealed for rocks of the Proterozoic complex.

The part of the Archean complex penetrated is mainly composed of gneisses and amphibolites; the major part of these rocks is, in fact, deprived of organic matter. Nevertheless, individual samples contain micro-concentrations of CHB; oils of this CHB are characterized by a hydrocarbon composition similar to that of paraffinic petroleum.

| 3 | $C_{24}$ | $C_{25}$ | $C_{26}$ | $C_{27}$ | $C_{28}$ | $C_{29}$ | $C_{30}$ | $C_{31}$ | $C_{32}$ | $C_{33}$ | $C_{9-40}$ | $C_{9-16}$ | $C_{11-35}$ | $C_{11-6}$ | Odd ratio / Even ratio |
|---|---|---|---|---|---|---|---|---|---|---|---|---|---|---|---|
| | 16 | 17 | 18 | 19 | 20 | 21 | 22 | 23 | 24 | 25 | 26 | 27 | 28 | 29 | 30 |
| 2 | 3.3 | 5.6 | 4.6 | 2.4 | 2.4 | 2.3 | n.d. | n.d. | n.d. | n.d. | – | – | 97.6 | 20.1 | 1.3 |
| 5 | 10.0 | 10.2 | 13.6 | 5.9 | 5.9 | 4.3 | 2.4 | 1.5 | 1.1 | n.d. | – | – | 90.9 | | 1.0 |
| 3 | 2.6 | 2.7 | 1.8 | 1.7 | 1.5 | 1.4 | 0.5 | n.d. | n.d. | n.d. | – | – | 100.0 | 65.2 | 1.3 |
| 6 | 7.9 | 7.3 | 6.2 | 5.0 | 2.9 | 3.3 | 1.1 | n.d. | n.d. | n.d. | – | – | 80.6 | 7.3 | 1.2 |
| 0 | 9.0 | 8.6 | 9.4 | 5.4 | 4.0 | 3.2 | 1.5 | 1.5 | 0.8 | 0.3 | – | – | 91.0 | 5.6 | 1.0 |
| 4 | 6.5 | 9.1 | 8.1 | 4.3 | 2.8 | 1.4 | 0.6 | n.d. | n.d. | n.d. | – | – | 93.1 | 27.6 | 1.1 |
| 4 | 7.9 | 8.9 | 6.0 | 5.3 | 4.1 | 3.8 | 2.0 | 1.7 | 1.0 | n.d. | – | – | 100.0 | 1.4 | 1.3 |
| 0 | 3.7 | 4.5 | 3.9 | 3.6 | 0.9 | 2.3 | n.d. | n.d. | n.d. | n.d. | – | – | 98.8 | 34.0 | 1.4 |
| 0 | 10.4 | 9.3 | 9.1 | 6.0 | 3.5 | 2.6 | 1.9 | n.d. | n.d. | n.d. | – | – | 90.5 | 2.3 | 1.1 |
| 1 | 11.0 | 6.7 | 7.6 | 5.5 | 2.7 | 6.0 | n.d. | n.d. | n.d. | n.d. | – | – | 93.1 | 7.0 | 1.0 |
| 0 | 9.3 | 8.2 | 6.4 | 5.3 | 4.1 | 3.2 | n.d. | n.d. | n.d. | n.d. | – | – | 97.1 | 8.1 | 1.1 |
| 7 | 11.5 | 11.5 | 11.3 | 8.4 | 6.5 | 5.0 | n.d. | n.d. | n.d. | n.d. | – | – | 99.6 | 3.5 | 0.9 |
| 5 | 13.4 | 11.9 | 12.0 | 6.9 | 5.0 | 5.5 | 2.1 | 1.1 | 0.2 | n.d. | – | – | 92.0 | 2.0 | 1.0 |
| 9 | 12.0 | 12.1 | 12.3 | 6.3 | 4.4 | 3.8 | 2.1 | 0.9 | 0.3 | n.d. | – | – | 96.8 | 8.5 | 1.0 |
| 1 | 11.5 | 10.6 | 8.9 | 7.1 | 5.7 | 4.0 | 2.2 | 1.4 | 0.9 | 0.8 | – | – | 94.8 | 2.0 | 1.2 |
| 1 | 11.8 | 12.2 | 8.0 | 6.9 | 3.6 | 3.5 | 1.4 | 5.3 | 3.6 | 4.9 | – | – | 95.8 | – | 1.3 |
| 2 | 6.6 | 7.4 | 5.3 | 3.2 | 2.9 | 4.7 | 1.3 | n.d. | n.d. | n.d. | – | – | 97.2 | 13.5 | 1.3 |
| 6 | 12.7 | 11.5 | 10.9 | 9.6 | 7.3 | 5.9 | 4.7 | 2.2 | 0.9 | 0.4 | – | – | 98.1 | – | 1.1 |
| 2 | 7.4 | 7.0 | 6.5 | 6.4 | 4.6 | 3.0 | 3.4 | 3.1 | 2.8 | 2.5 | 89.2 | 2.6 | – | – | 1.1 |
| 2 | 6.1 | 5.0 | 5.7 | 5.5 | 4.1 | 3.5 | 2.9 | 2.8 | 2.4 | 2.2 | 87.3 | 5.1 | – | – | 1.1 |
| 5 | 1.1 | 0.9 | 0.6 | 0.3 | | | | | | | 61.0 | 65.0 | – | – | 1.3 |

Before proceeding to a discussion of results of the geochemical studies, it must be emphasized that in the course of drilling crude oil was added to the drilling mud, beginning from a depth of 7 km. To determine the effect of technogenic core contamination, special studies were undertaken by gas chromatography. It was established that the distribution of n-alkanes and isoprenoids in crude oils differs basically from that in oils of CHB from the Archean rocks. Crude oil is of low density, the amount of n-alkanes (above $C_{16}$) in it does not exceed 35% of the total quantity, and the series ends with $C_{28}$; the amount of isoprenoid alkanes ($\Sigma iso-C/\Sigma_n-C=0.3$) in crudes is higher by almost one order of magnitude than that in CHB oils. The results allow the supposition that the technogenic contamination of rocks did not significantly affect the CHB composition.

The possibility of forming micro-concentrations of hydrocarbons in intrusive rocks cannot be ruled out; this question is, however, of principle importance,

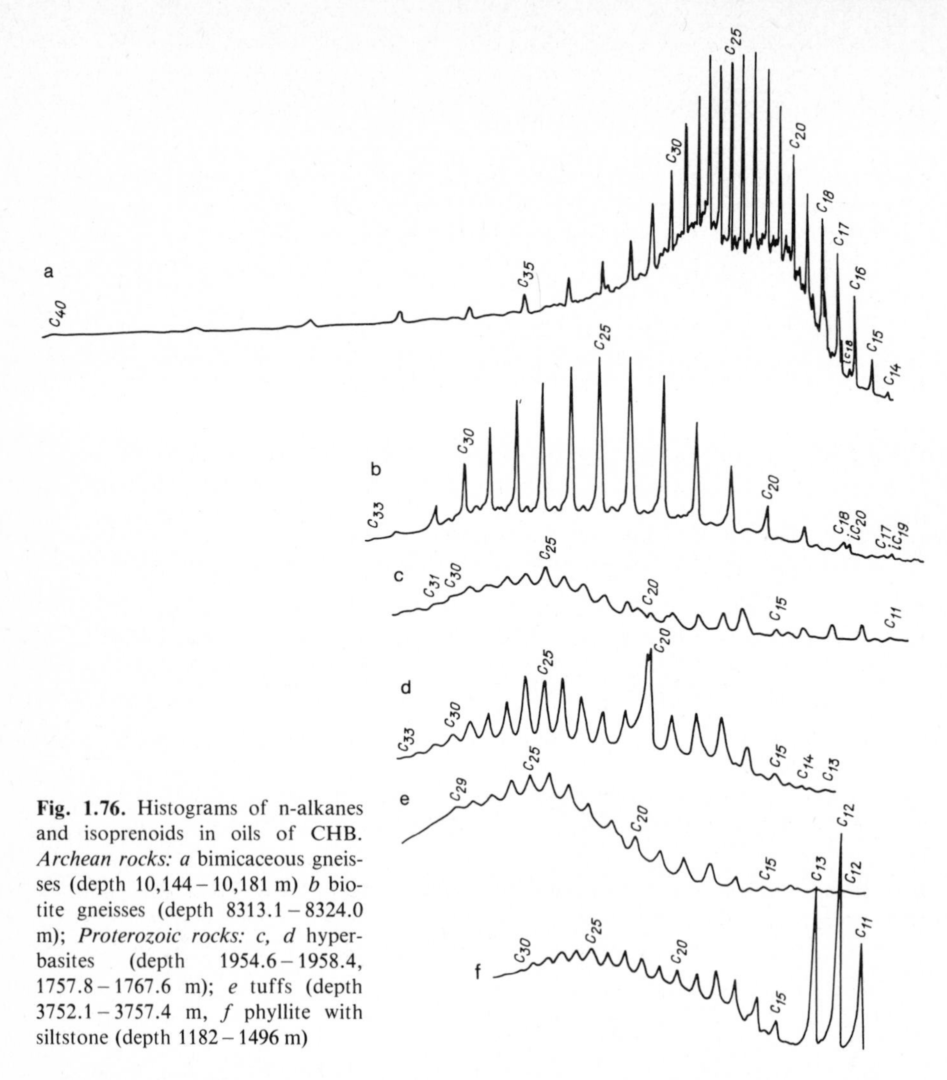


**Fig. 1.76.** Histograms of n-alkanes and isoprenoids in oils of CHB. *Archean rocks: a* bimicaceous gneisses (depth 10,144 – 10,181 m) *b* biotite gneisses (depth 8313.1 – 8324.0 m); *Proterozoic rocks: c, d* hyperbasites (depth 1954.6 – 1958.4, 1757.8 – 1767.6 m); *e* tuffs (depth 3752.1 – 3757.4 m, *f* phyllite with siltstone (depth 1182 – 1496 m)

and it can be resolved to a certain extent by investigating such objects as superdeep wells, providing contamination is avoided during drilling process.

Let us now present the main conclusions which can be arrived at as a result of the studies of gases and organic matter in the well section down to a depth of 11,600 m.

1. Helium gasometry studies (in the upper parts of the section HCG, helium and nitrogen gasometry studies were also carried out) made it possible to recognize the fissured fluid permeable zones in rocks. The maximum gas inflows were identified on drilling through the fissured zones of the Zhdanov suite. It is of interest that the presence of the fluid permeable zones in the Archean rocks at depths of 9400 – 9500 and 9950 – 10,500 m, which control the ore-formation process, is now an established fact.

2. The increased contents of HCG in the closed pores of rocks, reaching some tens of cubic centimeters per kg of rock, are characteristic only of metasedimentary rocks of the Zhdanov suite. The existing direct relationship between HCG content and amount of $C_{org}$, as well as isotope data of HCG carbon and $C_{org}$, are indicative of the biogenic nature of these organic components of metasedimentary rocks.

As a whole, the content of HCG in the closed pores of rocks of the well section studied, particularly of pores in the Archean rocks, is low, and it approaches values inherent to an analytical background.

3. The content of helium in rocks does not exceed 1 $cm^3\ kg^{-1}$; it gradually increases from the surface down to a depth of about 5 km, and then at lower levels it is in fact left unchanged; it exhibits only slight variations, which are caused by the mineral composition of rocks, their textural-structural features, their age, the time of the metamorphic changes and their intensity and content of uranium and thorium. The diffusion of helium into the atmosphere is the main factor which determines the helium distribution pattern in the upper part of the section down to a depth of 5 km.

4. The distribution pattern of organic carbon in metasedimentary rocks of the Middle Proterozoic complex is controlled by the facial conditions of sedimentation. The maximal contents of $C_{org}$ in metasedimentary rocks reaches 3%. The Archean rocks are characterized by a comparatively regular distribution of $C_{org}$ whose content, as a rule, does not exceed some hundredths of a percent. In the Archean gneisses, schists and amphibolite zones of organic carbon accumulation were not met.

5. The Middle Proterozoic rocks in the well section contain CHB in quantity from hundredths to tenths of a percent, with maximal values being identified in metasedimentary rocks in which the content of CHB is correlated with the content of $C_{org}$ and HCG. In the Archean rocks the content of CHB decreases to thousandths of a percent.

The study of CHB has revealed considerable variations in the composition of bitumens in three lithologically and genetically different rock complexes constituting the SG-3 well section, i.e. in the Proterozoic metasedimentary and magmatic rocks and in the Archean metamorphic rocks.

In the Middle Proterozoic metasedimentary rocks, CHB is mainly represented by oils (up to 42%), alcohol-benzene resins (up to 39%) and asphaltenes (8 – 12%). Oils of CHB are characterized by the presence of functional groups of the aliphatic and cyclic (to a lesser degree) structure. A series of alkanes in the oil fraction is represented by alkanes from $C_{11}$ to $C_{29-30}$, with the concentration maximum being confined to $C_{12}$; the quantity of "liquid" alkanes reaches 65%. In the fraction of petroleum-ether resins, in addition to hydrocarbon groups, oxygen groups of the phthalate type are also present. In benzene and alcohol-benzene resins of the phthalate type, indications of the destruction of the phthalene complexes have been discovered.

In the Middle Proterozoic magmatic rocks and benzene resins (up to 85%) constitute the bulk of CHB. The oil fraction is 6 – 18%, asphaltenes are present as traces. All CHB components are characterized by the presence of a phthalate group. An ore fraction of CHB contains a negligible amount of n-alkanes of a series $C_{15,16}-C_{29-33}$, "liquid" alkanes are either present or make up no more than 3.5% of the total content.

In most of the analyses, oils contained in CHB make up 20 to 60%, resin content is 48 – 50%, asphaltenes are practically absent. High-molecular-weight paraffins constitute the bulk of the oils, they also predominate in petroleum-ether resins, while in benzene and alcohol-benzene resins oxygen-bearing structures of the acid type predominate. In the alkane series of the fraction from $C_{14-17}$ to $C_{33-40}$ a maximum is confined to highly molecular compounds ($C_{23-25}$), the amount of "liquid" n-alkanes does not exceed 5%.

6. The negligible contents of bitumen in the Middle Proterozoic metasedimentary rocks identified at the relatively high values of general saturation of rocks with organic matter (HCG, $C_{org}$) may be considered as the surviving traces of a process of biogenic decomposition of matter, which proceeded under the action of metamorphism. For Middle Proterozoic magmatic rocks, the possibility may not be ruled out that hydrocarbons have been synthesized in an abiogenic way. The specific and monotopic composition of bitumens of different Archean rocks may be explained as follows: they were formed from hydrocarbons and brought together with fluids of deep origin during a period following processes of main metamorphic transformation.

## 1.9 Hydrogeological Essay

Hydrogeological investigations in superdeep wells are associated with great technical and methodological difficulties. For instance, high temperatures and pressures encountered in the well 1-Ernest R. Beiden (see Fig. 1.1), drilled to a depth of 9159 m, did not allow the application of a formation tester. Nevertheless, studies performed in superdeep wells provide interesting and unexpected results. For instance, in the Minibayev well No. 20000, an inflow of strongly gas-containing mineralized water (283 $g\,l^{-1}$) of 120 $m^3\,day^{-1}$ yield was produced in testing an interval of 4703 – 5099 m; water was saturated by gas of hydrogen-hydrocarbon composition.

The main aims of hydrogeological investigations carried out in the SG-3 well are as follows; the study of conditions of the subsurface water formation and the character of the water distribution in metamorphic rock of the deep crustal interior and the assessment of the role of water in the course of different geological processes.

In accordance with the intended aims, hydrogeological investigations were carried out in the SG-3 well and in other wells of the district. Additional material on other areas of the Baltic Shield was also used. For reliable interpretation of geological data, the results of geophysical well logging studies, laboratory core examination (petrophysical, geochemical, isotope, etc.), analyses of fissuring, studies of gas-liquid inclusions, thermophysical investigations, etc. were also taken into consideration. Hydrogeological and helium surveys along main major fault zones was also carried out.

A method of operative supervision (OS) for physicochemical parameters of the drilling mud was one of the most frequent modes of investigations conducted in the SG-3 well. This method was applied in the SG-3 well as an independent method for the first time, and was elaborated especially for tasks and conditions of superdeep drilling.

The formation of the drilling mud properties is mainly controlled by the following two factors: (1) the injection of various additions [CMAD, OKSIL, USCHR (trademarks), potassium dichromate, sodium carbonate, etc.]; (2) the inflow of subsurface waters into a borehole, which causes changes in the physical and chemical parameters of the drilling mud.

While conducting the operative supervision for the drilling mud parameters, a series of special laboratory investigations was undertaken to include the effect of technogenic factors. By these studies changes caused by subsurface waters could be distinguished. Quickness, simplicity and the possibility of performing mass measurements are the advantages of the OS method; the impossibility of conducting accurate quantitative determinations of properties of fluids encountered by the well is considered to be a drawback of this method. A special method has been invented to study hydrodynamical parameters of aquiferous zones penetrated by the well (SPR method). This method made it possible to obtain approximate values of formation pressure, the coefficient of water conductivity and some characteristic coefficients of chemical composition without running special equipment. However, this method could not be used on a broad scale, because accurate information about the amount of the "entry" and "outlet" drilling fluids was not known, due to a peculiar construction of the drilling unit. Since the amount of subsurface waters which flow into a borehole is rather small, the water discharge can be measured with high accuracy. The feasibility of applying this method in the ultra-deep well was predetermined by high formation pressure, which at great depth exceeded the pressure of the drilling mud column. Unfortunately, high pressure excluded the possibility of using formation testers attached to a drilling instrument. The application of testers in such conditions may lead to serious failures.

### 1.9.1 Subsurface Waters of the Pechenga Subartesian Basin

The Kola well runs through a section characteristic of the ancient Proterozoic artesian basins of superimposed type. By its structure, this section represents artesian basins of intermontane depressions, and two stages are distinctly recognized. The upper stage (0 – 6835 m) is composed of metamorphosed sedimentary and volcanic rocks of the Pechenga complex (the artesian basin proper), the lower stage is made up of the strongly metamorphosed Archean rocks (basement of the artesian basin). Maximal values of open porosity of rocks do not exceed 2 – 3%, mean values are less than 1%. Hence, the spatial distribution of fluids is controlled by fracturing of the water-bearing rocks. A zone of exogenous fissuring occurs at a depth between 500 and 2000 m, with a level of 800 m being considered average for this zone. Subsurface waters of this zone form a layer of frontal circulation, which controls regional drainage on the Kola Peninsula. These waters of atmospheric origin are usually poorly mineralized.

The intensity of the exogenous fissuring is rather weak, which is accounted for by the character of the rocks developed in the area under study. Coefficients of water conductivity rarely exceed 1 – 2 $m^2 day^{-1}$, and only in zones of tectonic faults are they somewhat higher.

It has already been stated that a lower boundary of the exogenous fissuring zone occurs at a depth of about 800 m. In the SG-3 well section (according to Y. P. Smirnov's data) fissures of different directions are met down to a depth of 2.5 – 3 km, vertical and inclined fissures predominate in an interval of 2.5 – 4.3 km, horizontal fissures are widely developed at lower levels. Beginning from a depth of 8.2 km, steeply inclined fissures again appear in the section. Almost all fissures are filled with vein material.

It should be noted that zones of more intensive fissuring are irregularly distributed in the well vertical section, and that their thickness changes abruptly. Down to a depth of 4.5 – 4.6 km, the thickness of fissured zones changes from 30 to 80 m. These zones are met comparatively rarely, i.e. at a distance of

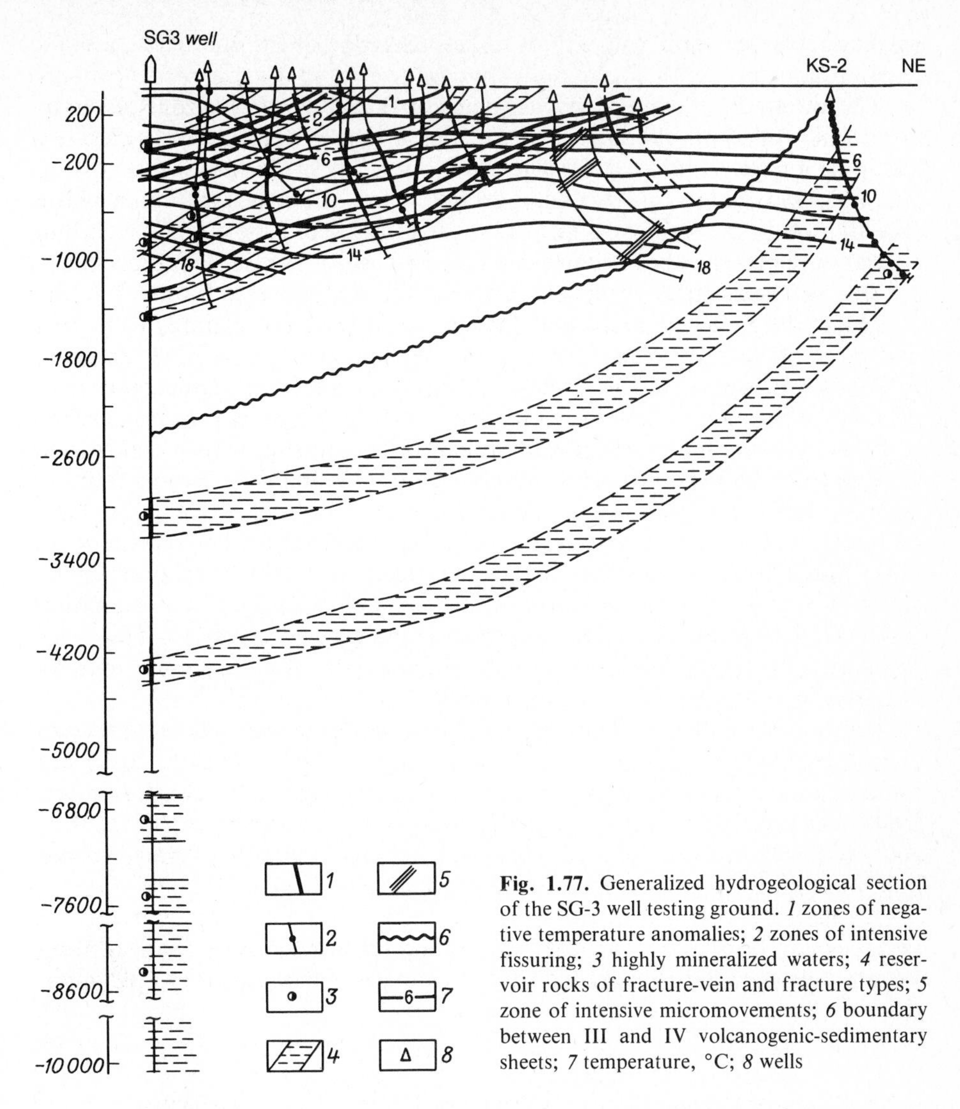


**Fig. 1.77.** Generalized hydrogeological section of the SG-3 well testing ground. *1* zones of negative temperature anomalies; *2* zones of intensive fissuring; *3* highly mineralized waters; *4* reservoir rocks of fracture-vein and fracture types; *5* zone of intensive micromovements; *6* boundary between III and IV volcanogenic-sedimentary sheets; *7* temperature, °C; *8* wells

500 – 1500 m from each other. Beyond these zones rocks are practically impermeable ($10^{-5} - 10^{-6}$ mD). Intervals of increased fissuring are grouped in large zones 500 – 800 m thick, each containing two or three local maxima of fissuring. Waters of a fissured-vein type are confined to these zones (Fig. 1.77). However, judging by the intensity of water manifestations, the permeability and formation pressures of these zones are comparatively low.

Down to a depth of 4.5 km, three such zones have been recognized in intervals of 300 – 620, 1050 – 1840 and 230 – 2870 m. Below a depth of 4.5 km, the thickness of individual fissure zones does not increase, and the distance between them shortens. Foliation becomes the predominate feature of the rocks, and the main direction of the fissures changes. All these facts are evidence of the different nature of the aquiferous fissure zones occurring above and below a depth mark of 4.5 km.

The lower part of the section was named by Bezrodny (1979) a zone of regional tectonic foliation, although it seems that a zone of regional disaggregation would be a more suitable name. The lower boundary of this zone descends to a depth of 9 km, and its total thickness exceeds 4 km. The zone is saturated by free (gravity) water; the permanent inflow of water into a borehole during the drilling process supports the above conclusion (on the basis of the OS data processing).

The hydrogeological phenomenon described was discovered for the first time in drilling the Kola superdeep well. According to results of hydrodynamic testing, a coefficient of rock filtration in the aquiferous zone at a depth of 6170 – 6470 m is equal to $10^{-7}$ m $day^{-1}$; this value exceeds the permeability value determined in a core sample[10] by one or two orders. This fact is of primary importance, because in the past metamorphic rocks occurring at such depths were considered to be impermeable to waters beyond the limits of tectonic faults in conditions of the actual existing gradients of the water head, which were identified in the course of field investigations. In the present case, the gradient of the water head was raised to 100 m $m^{-1}$ in the course of field investigations and to 10 – 40,000 m $m^{-1}$ during laboratory studies. Simultaneously, it was established that the formation pressure here approaches the geostatic pressure. This is the main cause for the inflow of a considerable amount of subsurface waters into the borehole, despite generally low rock permeability.

Below a depth of 8.9 – 9 km, the number and thickness of fissure zones containing free subsurface waters abruptly decrease. These thin (10 – 20 m) and rather intensive zones are most probably of local occurrence; this conclusion proceeds from the observation that water inflow ceases very rapidly as soon as these zones are penetrated by the well. However, the great intensity of water inflows indicates that the permeability of these zones is rather high.

All fissure zones discovered in the well section contain free (gravity) water, and they can be readily recognized on drilling through them by changes in the various indices of the drilling fluid which are analyzed in the course of the operative supervision.

### 1.9.2 Chemical Composition of Subsurface Waters

Chemical and gas composition of subsurface waters from a zone of exogenous fissuring of the Pechenga subartesian basin have been studied in detail. It was established that mainly ultra-fresh and fresh waters of calcium hydrocarbonate composition circulate in fissured rocks of varying genesis. Their mineralization rarely reaches 0.5 g $l^{-1}$. Waters of calcium sulphate composition, with mineralization reaching 2 g $l^{-1}$ and higher, are confined to areas composed of rocks containing sulphides. These waters usually occupy a part of the section, but at places they fill the entire zone of exogenous fissuring. The formation of these waters is

[10] In laboratory conditions permeability was determined in a vertical direction and in field conditions in horizontal direction, i.e. in the direction of the predominant fissure development.

associated with the oxidation of sulphide minerals. The composition of the dissolved gases is indicative of their meteoric origin.

Exceptions to the composition described are noted only in the zone of "opened" major faults. Such conformable major faults as the Poryitash, Lottin and Kolosiyok are morphologically well expressed (drainage hollows, swamps), yet not noticeably deep hydrogeological manifestations within their limits have been discovered. At the same time, water and gas discharge at deep horizons may start if a conformable major fault is crossed by a radial fault of even small extension.

As compared with the zone of exogenous fissuring, the deep parts of the Pechenga subartesian basin were studied in less detail. In the SG-3 well section the fissure-vein subsurface waters in zones of conformable major faults were established in intervals of 463 – 470, 580 – 610, 1135 – 1170, 1760 – 1828 and 3317 – 3448 m. The first two zones were drilled and tested in a satellite well, SW-1. Subsurface waters in these wells are characterized by subreducing properties ($Eh = -20$ mV) and high alkalinity (pH = 9.4), chlorine-hydrocarbonate sodium-calcium composition and mineralization which reaches 3.7 g $l^{-1}$.

Two aquiferous zones have been encountered in one of the exploratory wells drilled 250 m north of the SG-3, at depths of 900 and 1200 – 1350 m, respectively. Testing of these zones showed that subsurface waters are characterized by the chlorine calcium-sodium composition, low alkaline reaction (pH = 8.5) and mineralization of 24 and 51 g $l^{-1}$, respectively.

Results of the OS show that the chemical composition of subsurface waters of the lower zone (3317 – 3448 m) differs only slightly from that of waters in previous intervals, although their mineralization is considerably higher; this is supported by the presence of sulphates and halogen.

A zone of "regional disaggregation", in which water flow was encountered in an interval of 4.5 – 9 km, was not met in exploratory wells; the only exception is the topmost part of this zone (4565 m), which was met at a depth of 1520 m at a distance of 3 km north of the SG-3 well. From this depth, a mixture of technical and formation waters was produced. These mixed waters are characterized by the chlorine-calcium composition of neutral reaction (pH = 6.8). Kurlov's formula is as follows:

$$M_{19.7}\frac{Cl_{99}HCO_1^3}{Ca_{84}Mg_{12}Na_4}.$$

It should be noted that for the purpose of well-flushing, in the course of drilling, ultra-fresh water ($M \leqslant 1$ g $l^{-1}$) was used; this water could only dilute formation water. The described zone is not hydrologically linked with the overlying zones. This conclusion results from the observation that as soon as the zone was drilled, the water level immediately rose to 80 m.

Studies carried out in exploratory wells allowed greater accuracy and confidence in the interpretation of results of OS studies in deeper intervals of the SG-3 well; in fact, this method, used for determining the chemical composition of subsurface waters, proved to be the only one to produce the necessary information.

A water-bearing interval of 5325 – 5845 m subdivides a zone of disaggregation into two parts notable for different hydrochemical characteristics. The upper part is remarkable for the comparatively low salt concentrations of subsurface waters.

If a high degree of the filtrate (as compared with subsurface waters) is taken into account, then one may arrive at the conclusion that the chemical composition of subsurface waters is not subject to considerable changes down to this depth. It is likely that only some variations in chemical composition of waters take place here.

At lower depth levels the geochemical situation changes abruptly. Down to a depth of 7 km, practically pure chlorine-sodium (possibly hydrocarbonate-chlorine) waters entered the borehole. The following formula of a filtrate, produced from a depth of 6720 m, may be cited as an example:

$$M_{10.3}\frac{HCO_{66}^{3}Cl_{33}}{Na_{99}}.$$

In drilling through this zone, the pH decreased to 7.5 with respect to the background values (8.5 – 9).

By taking into account the most probable values of general mineralization of subsurface water, it may be supposed that the degree of dilution in this sample is about 100 – 120.

Below a depth of 7050 m, the hydrochemical situation again changes abruptly. Judging by the intensity of changes of the physicochemical parameters of the flushing fluid, it may be said that highly mineralized chloride calcium and calcium-sodium waters are present here under high pressure.

The absence (due to technical reasons) of full analyses of the filtrate for depths below 8 km makes it difficult to identify the chemical composition of the inflowing brines, although according to data on macrocomponent analysis their changes are insignificant down to a depth of 8.9 km. At lower levels, intensive inflows of subsurface water of different composition were identified for short periods of time.

In spite of the fact that the hydrochemical characteristics of waters from lower intervals of the SG-3 well are rather approximate, some regularities in the change of their chemical compositions with depth may be noted.

1. In an interval of 0.8 – 5 km chlorine sodium-calcium waters are present; they are characterized by pH = 8.5 – 9, and by higher concentrations of J, Br, K, and, possibly, B. In the lower part of this interval, waters are transformed into a chlorine-calcium type. Water mineralization increases with depth, in some intervals it reaches maximal concentrations. In the deepest part of this zone the relative content of Na decreases, and waters acquire a chlorine-calcium type.

2. In an interval of 5.3 – 7 km, neutral or poorly acid ($pH \ll 8$) chlorine-sodium waters are present.

3. Below a depth of 7 km, neutral or poorly acid chlorine-calcium and highly mineralized sodium-calcium waters are present.

As a whole, the upper and lower zones contain waters which by their chemical composition greatly resemble "metamorphosed" sedimentation waters of the Paleozoic rocks of ancient platforms, intermontane and foredeep depressions.

Compositional studies of dissolved gases meet considerable difficulties, because gases were analyzed in the drilling mud and in rocks at the surface. It is natural that, as gases rise to the surface in conditions of decreasing pressure, a considerable part is diffused. In some instances a drilling fluid, flowing out of the well, was "boiling" with hydrogen. It should also be noted that the atmospheric gases and hydrocarbons of technological additives may also contaminate the drilling mud. Other factors could also be mentioned; however, the summary is that all these factors cannot be quantitatively assessed.

In a zone of exogenous fissuring, atmospheric gases predominate among gases dissolved in waters. At lower levels, down to a depth of 2 – 2.5 km, hydrocarbon gases and nitrogen play a significant role. The content of hydrogen increases with depth, and in an interval of 2.8 – 3 km this gas predominates. Down the section, the amount of dissolved helium considerably increases. Beginning from a depth of 6 km, its content in water is commensurable with the content of hydrogen and even exceeds the latter in some places. Since helium is an inert gas, it is the best indicator of water inflow. Carbon dioxide present in the drilling mud may be produced by different sources. Nevertheless, the correlation of data of the drilling mud analysis and data on the composition of gas contained in intercrystal pores and intracrystal fluidal inclusions shows that the role of carbon dioxide in the composition of dissolved gases becomes significant from a depth of 5.5 – 6 km. On the whole, the degree of saturation of subsurface waters with gas is rather low.

### 1.9.3 Subsurface Waters and Hydrophysical Zonality

The results of hydrogeological studies carried out in the Kola deep well have shown that free gravity subsurface waters are spread throughout the entire zone of the strongly metamorphosed older rocks of the Pechenga subartesian basin (more than 10 km thick). It was the first time that manifestations of subsurface waters were observed in such conditions during direct field investigations.

It is important to note that within the limits of the major part of the section, subsurface waters are similar in chemical composition to the metamorphosed marine sedimentary (formation) waters of deep horizons of the ancient platforms (including the Eastern European Platform), foredeep and intermontane depressions. It can thus be assumed that at least some amount of the waters, syngenetic to the initial deposits, were preserved in the metamorphosed rocks, although later these waters passed through very complex evolutionary changes over a long geological period. This assumption is in good agreement with results of investigations conducted in many regions of the world (Taylor 1977; Fyfe et al. 1981; Henley and Sheppar 1975).

In order to elucidate the genesis of water participating in the process of metamorphic rock transformation, six core samples of different rocks (from an interval of 7455 – 10,190 m), containing minerals with hydroxyl groups, have been analyzed to determine the isotope composition of oxygen and hydrogen (Table 1.33). The correlation between deuterium (D) and $^{18}O$ subdivides the samples into two groups: samples containing more than 10‰ of $^{18}O$ are assigned to the first

**Table 1.33.** Results of isotope analysis of OH′-groups in minerals

| Test Nos. | Sample Nos. | Depth of sampling, m | Rock | Rock | | Water | | Genesis |
|---|---|---|---|---|---|---|---|---|
| | | | | D ‰ | $^{18}O$ ‰ | D ‰ | $^{18}D$ ‰ | |
| 1 | 24199 | 7455 | Aluminiferous gneiss | – | +10.6 | – | +8.0 | Sedimentary |
| | 24265 | | | | | | | |
| 2 | 26681 | 7932 | Biotite gneiss (nebulitic migmatites) | −78 | +10.4 | −40 | +8.0 | Sedimentary |
| | 26686 | | | | | | | |
| | 26688 | | | | | | | |
| | 26721 | | | | | | | |
| | 26722 | | | | | | | |
| 3 | 27282 | | | | | | | |
| | 27284b | 8028 | Talc-biotite-tremolite schist | −82 | +8.8 | −50 | +7.0 | Magmatic, worked by seawater |
| 4 | 28748 | 8360 | Biotite-amphibole-plagioclase schists | −80 | +10 | −40 | +8.0 | Sedimentary |
| | 28767a | | | | | | | |
| | 28768a | | | | | | | |
| 5 | 31569 | 8862 | Amphibolites | −92 | +6.9 | −60 | ±6 | Magmatic |
| | 31570 | | | | | | | |
| 6 | 37804 | 10,182−10,190 | Two-mica melanocratic gneisses | −82 | +11.8 | −35 | +9 | Sedimentary |
| 7 | 37804 | 10,182−10,190 | Muscovite from gneiss | −7.8 | +9.4 | −30 | +6.4 | – |

group, and those containing less than 9‰ to the second one. The first group comprises all gneisses studied (samples 1, 2, 6) and biotite-amphibole-plagioclase schist (sample 4), the second group includes amphibolites from a depth of 8862 m (sample 5) and talc-biotite-tremolite schist (sample 3). Numerous investigations show that the mean value of $\delta^{18}O$ for basic and ultrabasic rocks changes from +5 to +7.5‰ and in fact never exceeds the given values. Some increase of $\delta^{18}O$ may be caused by the action of the secondary, mainly low-temperature, processes. However, marine sediments are characterized by a high content of $^{18}O$ (up to 25‰ for clayey and siliceous muds). The value of $D$ relative to content also shows that rocks of the first group are heavier by about 10‰; it should be noted that this value is constant.

Available data on the isotope equilibrium make it possible to reconstruct the isotope composition of the water which participated in the metamorphic transformations of the rocks. The first group is characterized by the presence of $\delta^{18}O$ ranging from +8 to +9‰ and $\delta D$ ranging from −30 to −40‰; the second group is characterized by the presence of $\delta^{18}O$ ranging from +6 to +7‰ and $\delta D$ ranging from −50 to 60‰.

These data allow the following conclusions:

1. By isotope ratio, rocks of the first group are most likely to be primary-marine sedimentary rocks. Hydrated minerals were formed in them by participation of "metamorphic" water which was released during the dehydration of highly hydrous minerals formed at lower grades of metamorphism; it seems most probable that this "metamorphic" water is of marine origin. The increased values of $^{18}O$ in bimicaceous melanocratic gneisses can be explained by the relative predominance of pelitic fractions in the primary deposits.
2. Amphibolite (sample 5) was formed when igneous rocks underwent metamorphism in the presence of magmatic waters.
3. Ultrabasic rocks, which produced the talc-biotite-tremolite schist, were subject to the action of metamorphic, primarily marine, waters; this resulted in some increase in $^{18}O$ and D content. It is likely that the water came from the enclosing sedimentary strata. This explains the intermediate position of this sample between rocks of the first group and the amphibolites mentioned above.

Of particular interest is the isotope composition of water from a depth of 10,190 m. Muscovitization is considered here as a late process, i.e. a process of retrograde metamorphism. In the course of electron-microscopic core studies, it became evident that some muscovite crystals grew in the opened rock hollows, and later, fully or partially, filled these hollows (Fig. 1.78). The result shows that muscovite originated with the assistance of water whose isotope composition is similar to its oceanic standard and, correspondingly, is far from the isotope ratio inherent to "magmatic" waters (see Table 1.33). This water is undoubtedly of metamorphic origin. It contains a lesser amount of oxygen and is enriched in deuterium, because a considerable part of water may have been used earlier, in the process of metamorphism, to form other hydroxyl-bearing minerals. Taking the hydrogeochemical data into account, it seems most probable that these waters are of primary-sedimentary origin; however, they have passed through long and complex evolutionary changes.

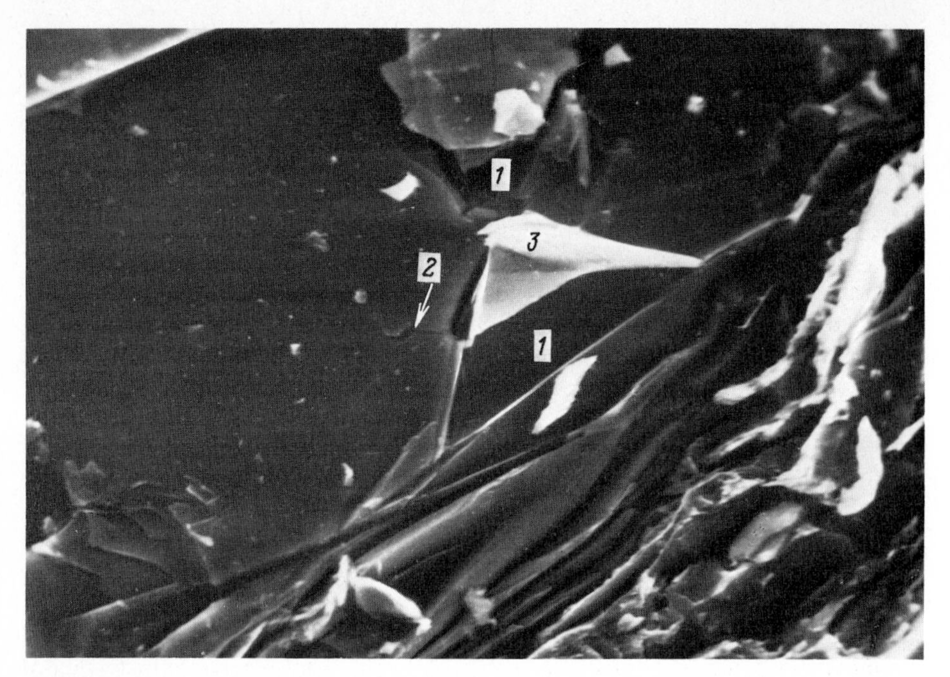


**Fig. 1.78.** Bimicaceous gneiss (depth 10,190 m). Thin section, x 200. *1* intercrystal pores; *2* fissures of hydrofissuring; *3* secondary filling

As was stated earlier, the SG-3 well penetrated through a section typical of the ancient Proterozoic basins. The present data recognize several hydrogeological zones in this section: (1) a zone of regional underground drainage (zone of exogenous disaggregation); (2) an upper zone of the joint waters (zone of metamorphic compaction); (3) a zone of regional tectonic foliation and hydraulic disaggregation of rocks; (4) a lower zone of joint waters (lower zone of compaction).

The upper zone is the result of the formation of exogenous fissuring in the process of the general uplift of the area under study. Its average thickness is 800 m. The zone is underlain by practically impermeable rock in an interval of 0.8 – 4.5 km. The formation of the water-proof zone is explained by the presence of rocks which underwent regional metamorphism of predominantly greenschist facies; as a result of this, a part of the remaining free water was utilized to form highly hydrous minerals which led to strong rock compaction. The established high ratio of the bound to free waters indicates that even slightly mineralized waters (including those notable for the lowest values of mineralization) are characterized by the maximal level of salt concentration; consequently, salt can precipitate in such conditions. This state may be considered as a state of equilibrium for a given stage of metamorphism. The composition of the gas-liquid inclusions and the presence of the sulphate and haloid admixtures in minerals are indicative of a high concentration of metamorphic solutions [11]. Somewhat lower concentra-

[11] Experimental studies on rock leaching in autoclaves have showed that a considerable amount of the readily soluble salts is still preserved in "dry" Proterozoic rocks.

tion of salts in waters of major faults may be associated with two reasons: (1) it seems likely that processes of dehydration and hydration of vein rocks and minerals replace each other in different epochs, thus controlling the amount of solvents present in gravity waters of major faults; (2) a zone of joint waters is a transitional zone with respect to solutions, which are formed in deeper zones.

The lower boundary of the water-proof zone corresponds to a boundary at which the greenschist facies of metamorphism is replaced by the epidote-amphibolite facies.

In the process of prograde metamorphism, a free fluid is released from mineral hydrates (chlorite, epidote, mica, etc.) as a result of their reduction. For dehydration reactions, leading to transformation of rocks of the greenschist facies into rocks of the amphibolite facies, the silicate crystalline lattices release $H_2O$ in an amount of 2 vol.% (Liou 1974) in an interval of 2 – 3 km.

The chemically bound water in this zone is associated with low hydrous minerals (epidote, mica, etc.). At the main stage of metamorphism, transformation of highly hydrous minerals into low hydrous minerals has led to the generation of solvents and the increase of the gravity water volume. This water, together with the fossil sedimentary water, was non-uniformly distributed inside the pore space and in a system of micro-fissures along grain boundaries.

The presence of water, remobilized inside the metamorphosing rock, predetermines the progress of processes which lead to the disaggregation of rocks and to an increase in hollowness. Firstly, high pore pressures result in the formation of microhydrofractures in rocks at those points where the quantity of the regenerated water is small. Secondly, this water prevents rocks from full compaction in the process of rock recrystallization, because in conditions of low permeability water cannot be easily removed from rocks. Thirdly, metamorphic waters, being very aggressive and possessing high leaching ability due to their origin, can dissolve some components of rocks, thus resulting in an increase in rock hollowness.

Hydrodynamical investigations performed at a depth of 6370 m have shown that the coefficient of rock filtration here is $10^{-7}$ m day$^{-1}$, which means that permeability is very low. In spite of this, the results are of prime importance, since previously it was thought that in the past, metamorphic rocks at these depths were impermeable beyond the limits of the tectonically disturbed zones.

The rather low permeability of the metamorphic rock and the presence of a great number of isolated hollows (microfissures) are the factors promoting preservation of water in this rock. Studies carried out in the SG-3 well have showed that water can be preserved over a very long time (more than 1 Ga). A zone of disaggregation is also preserved for this period.

Thus, on the basis of analysis of the distribution pattern of aquiferous zones and the character of fissuring in the well section, it may be concluded that liquid fluids inflow into a zone below 4.5 km as a result of mineral dehydration during their metamorphic transformation.

Since primary rocks are represented by approximately one and the same type throughout the entire Proterozoic sequence of the Pechenga depression, it may be surmised that the initial rocks constituted an interval of 4 – 6.8 km (at a stage of greenschist metamorphism), also contained about 4 vol.% of the bound water. Calculations show that the total loss of $H_2O$ in this interval is

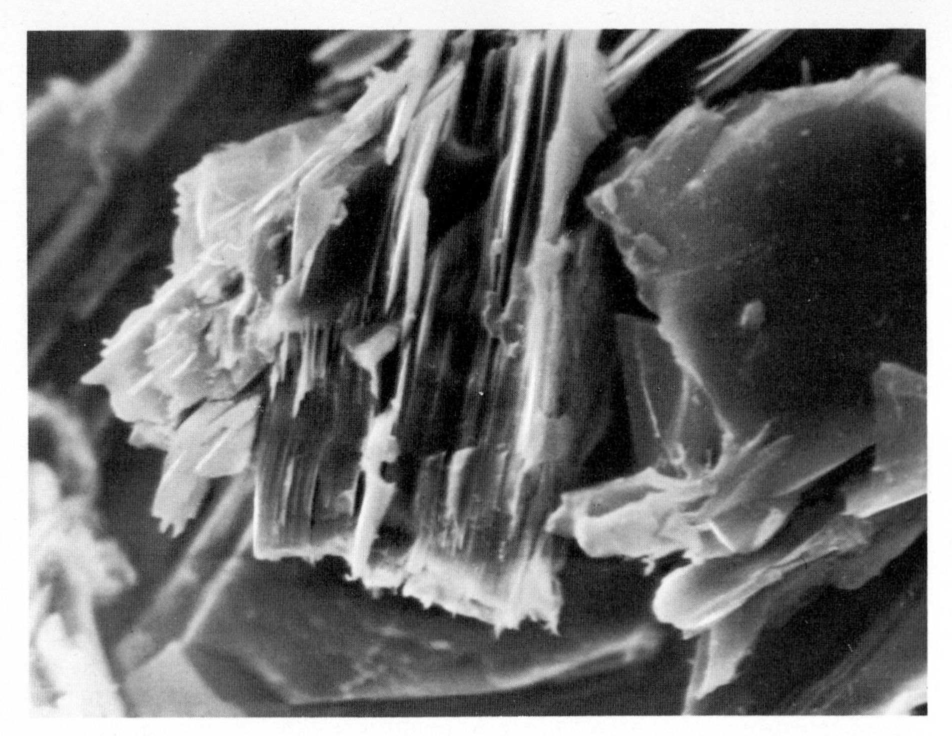

**Fig. 1.79.** Talc-biotite-tremolite schist (depth 8028 m). Thin section, × 600. Intercrystal microfissures are well seen

$5.5 \times 10^7$ gm m$^{-2}$, or 6.7% of the initial rock volume on the average (by taking into account real temperatures and pressures of transformation).

Analysis of the mean mineral composition shows that in the present case the total volume of the transformed minerals passing from the greenschist to epidote-amphibolite facies is 0.96 of its initial volume, and 1.027 of the initial volume if taken together with the released water; this means that it exceeds the initial volume by 2.7%. Water should be compressed by 1.7 times in order to preserve the initial volume of rocks; this requires a pressure of about 30 GPa. Such pressure corresponds to the mantle pressure at a depth of 90 – 100 km (in conditions of "crust" temperature preservation). The volume can be increased only as a result of a rupture in rock continuity, because the rupture strength is less by some orders than the stresses arising during the process of water release. As is shown in Fig. 1.79, mineral crystals in rocks were strongly deformed and hollows were formed (rupture of continuity); this can be explained by the action of high pressures of the releasing fluids. Otherwise, a reaction of dehydration could not result in such volume.

Hollows formed as a result of the hydrofracturing are first developed in the form of intercrystal and intracrystal microfissures filled with water; microfissures may originate only later.

In forming the secondary hydrous minerals (muscovitization), a part of the free water became bounded, and this immediately led to a compaction of the rocks due to loss of porosity. The mean density and velocity of longitudinal waves also increased in spite of the fact that the density of the newly formed minerals was less than that of the primary minerals. It is noteworthy that this process could proceed only providing a certain reserve of free water fluid was available; the same conditions have ensured transformation of the amphibolite facies of metamorphism into the epidote-amphibolite facies. It is of interest that water inflow into the well, fixed by OS, also decreased approximately in the same interval.

It is known that dehydration reactions take place within a rather narrow temperature range. Nevertheless, one facies usually gradually passes into anohter one during the process of regional metamorphism.

In the present case, the decrease in $H_2O$ content in rocks and the corresponding changes of rock physical properties, and parameters of the drilling mud take place very rapidly, i.e. within a time interval corresponding to a rock section length of only some tens of meters. The degree of dehydration in the upper part is higher than that in the lower part of the Proterozoic strata.

An impression can be created that dehydration proceeded very rapidly against the background of an abrupt temperature increase, and that the differential stresses that appeared, caused by high pressure of the releasing fluids, were so great that they led to deformation of the crystalline lattices of some minerals [12].

All these facts can be readily explained if the theory proposed earlier (see Chap. 1.4) is accepted, that the rock was heated by the lava sheets of the III volcano-sedimentary strata in the direction from top to bottom. In this case, the temperature would have risen rapidly and the necessary pressure could have been built up due to the so-called autoclave effect. The subsequent evolution of rocks and their compaction depends on the possibility of water being released from the metamorphosing rock.

A number of recently published papers is devoted to this question (Fyfe et al. 1981; Norris and Henley 1976; Price 1975). In one of these papers, the possibility of water release from rocks during their subsidence due to temperature increase is substantiated. Such a release results in water expansion and hydraulic rock splitting [13]. Other papers discuss the mechanism of rock splitting associated with the decrease of the minimum main stress, which can take place only if some specific tectonic processes are involved in this event.

The first mechanism is considered to be the more real one. This is due, in particular, to the great loss of water in transition from the amphibolite to granulite facies. Decrease in the lithostatic load in a period of uplifting during the process of orogenesis leads to the same consequences and seems to explain the fact of the predominant development of disaggregated zones in synclinal structures (Galdin 1981). The orientation of hydrofissures is controlled by the spatial position of a stress ellipsoid at the moment when stresses arise. Fissures become oriented in a direction perpendicular to the minimum main stress according to the laws of me-

[12] Data of X-ray structural studies of cores in the SG-3 well performed by V. V. Ponomarev.

[13] The effect of rock disaggregation in heating is confirmed by the experimental data (Zaraisky and Balashov 1978).

chanics. In relatively quiet tectonic conditions predominantly subvertical fissures are formed.

If horizontal stresses are considerably greater than the geostatic load (this situation is typical of the Baltic Shield), then subhorizontal fissures arise; these fissures can be left open for a long time. The formation of these fissures is associated with the phenomenon of regional rock foliation in the upper part of the disaggregated zone (Bezrodnov 1979) established by the SG-3 well. Foliation extends in a subhorizontal direction, and does not depend on the position of stratigraphic boundaries. Since great horizontal stresses are common for deep crustal parts, a phenomenon of regional foliation may appear rather frequently.

Despite the generally low permeability of rocks, waters can filtrate throughout a great distance, particularly during the tectonically active geological periods; this is provided by high pressure gradients existing over a long period. When waters are brought into another hydrodynamical environment, for instance, in a zone of the open major fault, soluble components may be precipitated, thus initiating the process of ore mineralization.

Due to the presence of free water in the zone of disaggregation, minerals can leave rocks and enter waters, be transported, redeposited and accumulated. Higher concentrations of the respective components in the parent rocks are not necessary conditions for these processes to take place.

In this connection, one may note that in some intervals of the mentioned zone the content of individual components, for instance rubidium, in rocks decreases as compared with the background content typical of a given rock type (see Chap. 1.4). Simultaneously, an abrupt increase in subsurface water concentration is also registered here.

Judging by the ratio of free to bound water at different stages of metamorphism, one may expect that the phenomenon of hydraulic disaggregation will be most strongly expressed in rocks of the epidote-amphibolite and amphibolite facies.

In conclusion, it should be noted that the seismic boundary, which was previously interpreted as the boundary of a basaltic layer, in fact occurs at some distance below the zone of disaggregation, and it seems that this boundary identifies changes in hydrogeological conditions, rather than changes of rock lithology. The phenomenon of hydraulic disaggregation of metamorphic rocks poses questions on the formation of rock secondary hollows and on the evolution of metamorphic rock from new positions.

As was noted above, a considerable reserve of readily soluble salts, including chlorides, is preserved at low grades of metamorphism. When dehydrated, these salts are again dissolved, thus causing high mineralization of the metamorphic fluids. This facilitates the formation of the readily dissolved chloride complexes of heavy metals and their migration in this form over a great distance; when thermodynamical conditions are changed, they precipitate and form ore deposits. Since waters are widely spread throughout the region, a considerable bulk of the metamorphosing rocks may be involved in the process of mobilization.

This phenomenon may also lead to the decrease in shear strength of rocks and promote the formation of faults of different morphology.

In the first place, this is associated with high pressure of a pore fluid. Experimental data show that if the fluidal pressure is equal to the load pressure, then the rock strength becomes equal to its strength in conditions of uniaxial compression, i.e. blocks can be displaced at small differential stress.

The slow progress of geochemical study provides no possibility of drawing sound conclusions as to conditions of formation and chemical composition of subsurface waters in the zone of disaggregation. However, in view of the almost complete coincidence of boundaries of hydrochemical zones with those of the stratigraphic units, and because these boundaries define certain grades of metamorphism, one may say that such factors as the primary character of rocks, their grade of metamorphic transformation, and the present trend of metamorphism play a leading role in the formation of subsurface waters.

Below a depth of 9 km, subsurface waters are localized in zones of small thickness. However, water manifestations are very intensive, although they are of short duration (depths: 9920 and 10,020 m). It may be assumed that joint waters of unusual composition are developed here. It is not impossible that their formation is related with deeper crustal zones.

Returning to the problem of the hydrodynamic zonality of the Pechenga subartesian basin, it should be noted that each of the defined hydrodynamic zones is distinguished not only by the qualitative and quantitative characteristics of its fissuring, certain ranges of change in filtration properties and values of hydrostatic pressures, but also by ranges of change in geodynamic pressure, temperature, indices of gas and chemical composition of subsurface waters, each zone having its own peculiarities. As a matter of fact, the observed zonality is of a complex hydrophysical character, and every individual zone differs from the underlying and overlying zones by conditions of subsurface water formation; this zonality is very tightly linked with the evolution of rocks of the enclosing strata.

In the zone of regional drainage, the main mass of subsurface water is formed under the action of active water exchange. Contrary to this, in lower zones the formation of subsurface waters displays a definite endogenic character. A zone of regional drainage and a zone of foliation are distinguished by the absolute predominance of free gravity water. On the contrary, a zone of vein subsurface waters is characterized by the predominance of chemically bound water (with the highest absolute content). The chemically bound water in all zones is considered to be a passive reserve which cannot be extracted by natural processes under existing pressures and temperatures. By taking into account the total amount of chemically bound water and the character of this bounding (strongly hydrous minerals in the two upper zones and poorly hydrous in the lower zone), it becomes possible to reconstruct the hydrological evolution of the strata. In an epoch of rock epigenetic transformation two main processes took place: squeezing out of solutions at greath depths and removal of solvents which were used for the formation of the secondary hydrous minerals. Both processes led to the hydration of rocks, shown with greater or less intensity in different geological periods. At the main stage of metamorphism only less than 1 vol.% of water (on the average) was preserved in a volume of hydrous minerals.

In view of the fact that the dehydration of the Earth's interior proceeded by squeezing out towards the Earth's surface, rather than by water bounding,

**Table 1.34.** Hydrophysical zonality in the SG-3 well

| Hydrophysical zone | Depth interval, m | Main type of water in rocks | Type of reservoir rock | Metamorphic facies | Water composition | Gas composition (predominating) |
|---|---|---|---|---|---|---|
| Zone of regional underground drainage | 0–800 | Gravity | Fissured | Greenschist facies | $M<1$ $HCO_2$-Ca $SO_4$-Ca | Atmospheric |
| Upper zone of joint waters | 800–4500 | Chemically bounded | Fissured-veined | | $M_{50-150}\frac{Cl}{Ca\text{-}Na}$ | $(N_2)$, $CH_4$, $H_2$ |
| Zone of regional tectonic foliation and hydraulic disaggregation of rocks | 4500–9200 | Gravity | Fissured | | | |
| Upper subzone | 4500–5850 | | | Epidote-amphibolite facies | $M_{200-300}\frac{Cl}{Ca}$ I, Br, Sr, pH < 8 | $H_2$, He |
| Middle subzone | 5850–6900 | | | Epidote-amphibolite facies | $M>300\frac{Cl}{Na}$ B, F, Rb, pH < 7 | $H_2$, He, $CO_2$ |
| Lower subzone | 6900–9200 | | | Epidote-amphibolite facies | $\frac{Cl}{Na\text{-}Ca}$ Br, I, Rb, B | $CO_2$, $H_2$, He |
| Lower zone of joint waters | 9200 | Chemcially bounded | Vein | Epidote-amphibolite and amphibolite facies | $\frac{Cl}{Ca\text{-}Mg\text{-}Na}$ | (?) |

it becomes evident that the ratio of bound water to gravity water increased in the course of time, solvents were redistributed and metamorphic solutions became highly concentrated.

On the basis of the material presented above, a preliminary model of hydrophysical crustal zonality in the area of the SG-3 well may be proposed (Table 1.34).

This model not only explains the nature of hydrothermal fluids unrelated with intrusions, but also the mechanism of rock foliation in zones of disaggregation and the formation of some types of tectonic faults, etc., but it also radically changes the theory of the process of water circulation in the continental crust.

## 1.10 Evolution of the Continental Crust in the Precambrian

Every year new evidence confirms the theory that the Earth's crust originated at the very beginning of geological history, and that in the Archean-Proterozoic period it passed through complex evolutionary events. The swift progress in the study of the Early Precambrian period was accompanied by the appearance of new hypotheses which caused heated discussions.

The question of the origin of the primitive Earth's crust is the most actively discussed. Some researchers distinguish fragments of "moon" anorthosites and basalts among the oldest crystalline schists. Others believe that the primitive crust is represented by metamorphosed basites-ultrabasites and tonalites which have been formed as a result of the mantle matter melting. Others again are of the opinion that the leading role in the formation of the Earth's crust belongs to processes of sedimentation and weathering which fully destroyed the earlier existing cover of magmatic origin. Actively debated are also such problems as the specificity of the Archean and Lower Proterozoic geotectonic regimes and structures, the impact of tectonic movements, metamorphism and magmatism on the earlier consolidated masses, applicability of the actualistic principles and mobilistic hypotheses of the Early Precambrian time, regularities in distribution of ore deposits remarkable for their unique origin and size (Correlation of the Precambrian Rocks 1977, Early History of the Earth 1980, Archean Geology 1981).

The section of rocks penetrated by the Kola superdeep well does not comprise the full Precambrian sequence. Nevertheless, factual data obtained in the process of the well drilling make it possible to discuss these general questions from new positions by taking a three-dimensional model of the Pechenga ore-bearing region as a basis. The appearance of the third reliable coordinate opens additional possibilities for the reconstruction of the geological processes which have determined the present geological setting of the region and for the assessment of alternative hypotheses on the structure and development of the Earth's crustal zones composed of Precambrian metamorphic rock.

### 1.10.1 Volumetric Geological Model of the Pechenga Region

It should be recalled that prior to the drilling of the Kola superdeep well in the Pechenga structure, the relationships between its southern and northern flanks and the character of the productive strata at depth were interpreted in different ways. For instance, one group of researchers, viz. G. T. Makiyenko, L. Y. Kharitonov, V. G. Zagorodny, S. N. Suslova, and others, were of the opinion that the Pechenga structure was a graben-syncline with a broad trough-like northern flank and steep overturned southern flank. The similarity of the rocks composing the northern and southern flank served as a basis to support the above idea. M. A. Gilyarova and other researchers believed that rocks of the southern flank did not belong to the Pechenga rocks, because they were strongly altered; they considered the Pechenga structure as a cup-like monocline bounded by a major fault in the south.

Recently, in mapping the southern flank, a somewhat unexpected theory as to the structure of the southern junction zone of the Pechenga block has been advanced. According to the opinion of L. I. Uvadyev, P. Z. Skufyin, A. A. Predovsky, E. D. Chalykh, and others, the upper, fifth, sedimentary-volcanogenic strata were formed in the southern part of the Pechenga graben-syncline at the final stages of their development. They differ from the underlying strata mainly by the presence of medium-acid tuffs and layers of dacite-andesite porphyrites. An independent South-Pechenga suite (even a series), assigned to the tuffaceous-picrite-andesite formation, is distinguished by the researchers.

The thickness of the Pechenga complex, assessed conventionally, is from 5 to 18 km.

On the basis of data from the superdeep well, material from deep gravimetric-seismic investigations and results of redocumentation of all outcrops and wells known in the vicinity of the SG-3 well, it became possible to build a reliable three-dimensional model of the whole Pechenga structure and the surrounding area (see Fig. 1.80). An analysis of the integrated data allows the following conclusions about the Pechenga complex:

1. The main seismic marker horizons, corresponding to the lower contact zones of the Nickel and Luostari series (according to drilling data) can be distinctly traced in the northern and central parts of the Pechenga structure, thus indicating the centroclinal dipping of all lineaments of this structure in southern directions. Reverse, northward dips of the sedimentary-volcanic strata are not observed in any of the areas, including the area of the southern flank;
2. The southern flank, a comparatively narrow linear zone, bounded from the south and north by steep major faults, is represented at the surface by volcano-sedimentary rocks (amphibolitized diabases and schists in them, tuffs of different basicity with predominance of medium-acid phyllites, siltstones and sandstones. There is a very close similarity betwen the lithopetrographic composition of rocks of this zone with those of the Nickel series. They are characterized by rather steep dips in predominately southern directions. A dense net of major faults has resulted in intensive rock crushing and the wide occurrence of intrusive formation bodies of dacite-andesite porphyrites, which mark lines of major faults in an area of Proterozoic sedimentary-volcanic rock development. For instance, a formation body of the dacite-andesite porphyrites of 110 m thick was met in the SE-well section. This body is also confined to a large intercalcated tectonic zone (Luchlompol major fault) located between the third sedimentary and volcanogenic strata;
3. Data of seismic and gravity investigations, adjusted with information obtained in the SG-3 well, do not confirm the theory that the Pechenga complex is built up from its top part; moreover, according to these data, any traces of newly discovered strata are absent;

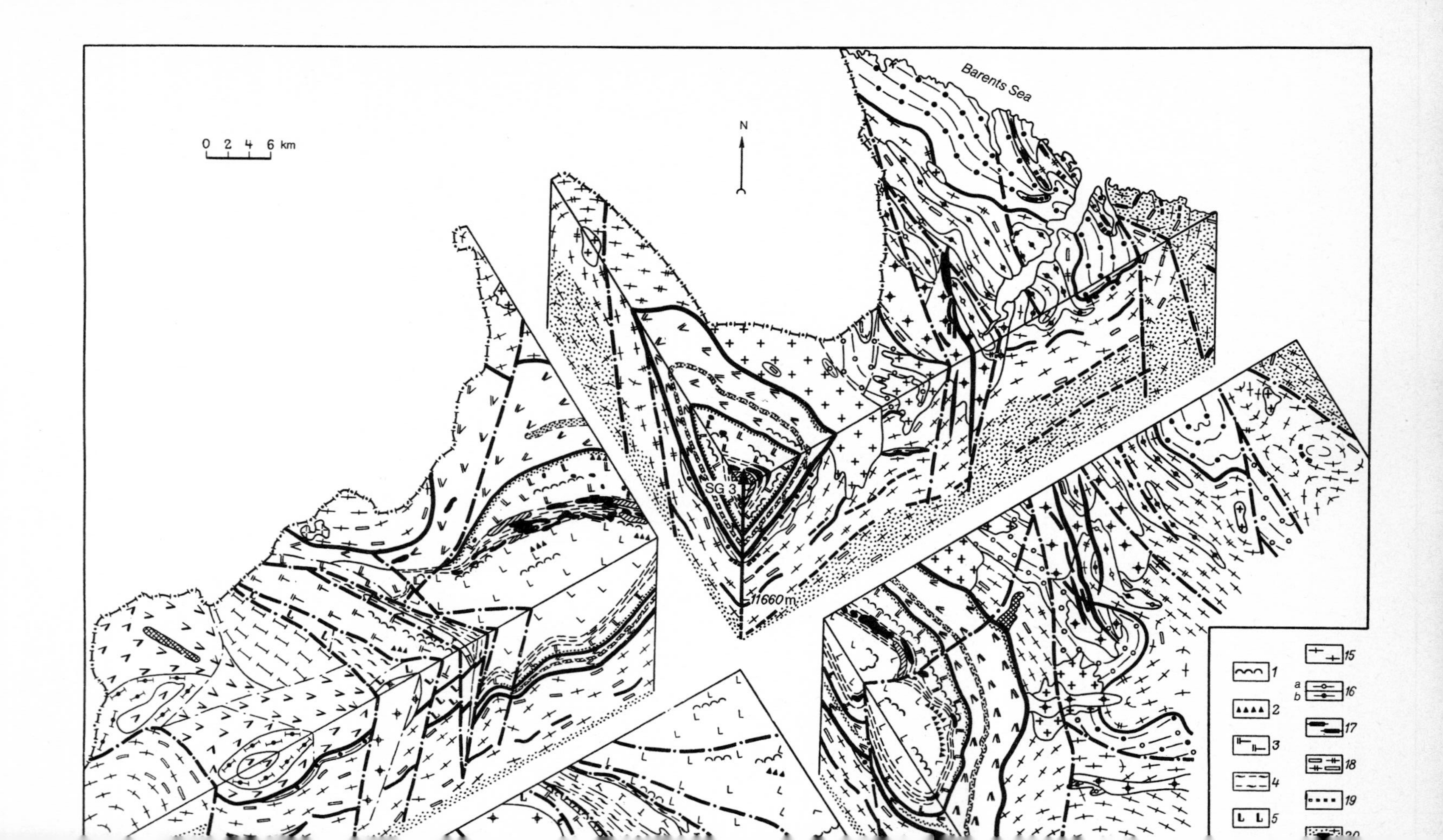
0 2 4 6 km
N
Barents Sea
SG 3
11660 m
1
2
3
4
5
15
a
b
16
17
18
19

**Fig. 1.80.** Geological block-diagram for the area of the SG-well drilling. (After V. S. Lanev and M. S. Rusanov). *Pechenga complex (Lower Proterozoic): 1* tuffs and sedimentary-tuffogenic rocks of different composition and schists in them; *2* volcanites of the basite-hyperbasite and hyberbasite composition and schists in them. *Southern zone, Poryitash stratum* – sutural tectonic block; *3* andesite, andesite-basalt, basalt metaporphyrites – schists and actinolitic amphibolites in them; *4* coaly tuffites, phyllites, siltstones, sandstones and schists in them. *Ruossel stratum* – block of granitoid domes: *5* amphibolites and amphibole-plagioclase schists in tholeiitic basalts; *6* various gneisses and schists with coaly matter in sedimentary and tuffogenic-sedimentary rocks. *Northern zone, Nickel series: 7* metamorphosed tholeiitic basalts (*a* diabases; *b* metadiabases and greenschists; *c* amphibolites and amphibole-plagioclase schists), *8* coaly phyllites, tuffites, siltstones, sandstones. *Luostari series: 9* metamorphosed trachybasalts, trachyandesites, andesite-basalts (*a* metadiabases meta-leucodiabases, metaandesites, green schists; *b* amphibolites, amphibole-biotite-plagioclase and biotite-plagioclase schists); *10* metasandstones, metaconglomerates, quartzite-sandstones, dolomites; *Tundra series (Lower Proterozoic – Upper Archean): 11* Talyin suite – micaceous-chlorite-actinolite-feldspar and micaceous-quartz-feldspar schists and gneisses, often with garnet; *12, 13* Kashkam suite (*12* amphibolites, often with garnet; *13* biotite-garnet-amphibole-feldspar schists and gneisses). *Kola-Belomor complex (Archean), Kola series, upper stratum: 14* biotite-plagioclase gneisses with HAM (*a* without cordierite; *b* with cordierite); *middle stratum: 15* biotite-plagioclase and amphibole-biotite-plagioclase gneisses and migmatites in them; *16* biotite-plagioclase gneisses with aluminiferous minerals (*a* without cordierite; *b* with condierite); *17* amphibole-biotite-plagioclase gneisses with pyroxene and pyroxene-amphibole-plagioclase crystalline schists; *18* amphibolites and amphibole crystalline schists; *19* quartz-magnetite schists; *lower strata (basement complex): 20* amphibolites, pyroxene-amphibole-plagioclase schists, gneisses, granitogneisses, charnockites, enderbites. *Intrusive rocks: 21* andesite-dacites (diorites): *22* diabases and gabbro-diabases, *23* differentiated basite-hyperbasites (*a* gabbro-wehrlites; *b* gabbro-norites); *24* microclinic granites; *25* microcline-plagioclase granites, granodiorites and diorites; *26* plagiogranites and migmatites; *27* geological boundaries (*a* of complexes; *b* of strata and *c* of series); *28* main tectonic dislocations; *29* points and boundaries of passing waves exchange (*PS*)

4. A step-like decrease in gravity gradients in the linear zones of major faults fully refutes the theory of the presence of the overturned southern flank of this structure;
5. It seems necessary to consider that the Pechenga rock complex constitutes a graben-syncline bounded in the south by a zone of regional major faults with their roots, according to seismic data, reaching 20 km. These major faults, judging by the step-like decreasing gravity gradients, must be considered as a system of parallel-contiguous reverse faults. The zone was "healed" by hypabyssal intrusions of andesite composition;
6. With data from gravity and seismic investigations, the maximal thickness of the volcano-sedimentary rocks of the Pechenga complex has been established in the central part of the Pechenga graben-syncline, in immediate proximity to an area due north of the Poryitash major fault, and is estimated to be 8 – 9 km;
7. Since volcano-sedimentary rocks abruptly decrease in thickness south of the Poryitash major fault (following the step-like decrease of the gravity gradients) and even almost disappear beyond the southern boundary of the fault zone, it may be surmised that a total vertical amplitude of major faults and reverse faults here is not less than 8 km.

Such great vertical displacements, which explain the absence of Proterozoic rocks in the southern parts of the Colour Belt of the Kola Peninsula, are indicative of the rather intensive denudation processes which took place over a period of a hundred million years in the north-eastern part of the Baltic Shield. This is considered to be the reason for the great thickness of the Phanerozoic sedimentary rocks in the Barents Sea shelf, which is adjacent to the shield in the north.

The Kola superdeep well has also established some facts of great significance concerning the underlying granite gneisses of the Kola series. They are as follows:

1. The well has confirmed the absence of the so-called Tundra series in the central part of the Pechenga structure, and consequently rejected the theory as to its conformable occurrence with the Pechenga complex in the form of a synclinal fold;
2. As against the previous interpretation of gravity and seismic data, the well has encountered a thick layer of granite gneisses filled with amphibolite bodies in an interval of 6.8 – 11 km; this layer is correlatable with the Kola series;
3. The well crossed strata of the primary sedimentary gneisses with HAM at different levels of the Kola series; as a result of this, the existing theory on the stratigraphy of this series has been changed;
4. An analysis of the orientation of the banded structures in core samples has revealed that granite gneisses and amphibolites occur at steep angles throughout the section drilled; this fact rejects the supposition of the dying out of folding at depth;
5. Data of geological investigations carried out in the SG-3 well, analyzed together with geophysical data, showed that the so-called Conrad discontinuity can be explained neither by transition from more acid to more basic rocks nor by the abrupt appearance of the metamorphism front and rock basification, nor by the presence of the subhorizontal tectonic zones.

## 1.10.2 History of Tectonic Development of the Pechenga Region

The general history of the geological development of the continental crust of the Pechenga region can be considered on the basis of the joint geological petrological-geochemical and radiological study of deep metamorphic complexes of the SG-3 well and by comparing these results with data on analogous formations developed at the surface. This history, embracing a considerable part of the Precambrian, is subdivided here into two major cycles, i.e. Archean and Proterozoic. The Archean cycle includes two stages: (1) sedimentation and volcanism, (2) folding, metamorphism and ultrametamorphism. The Proterozoic cycle includes four stages: (1) the initial formation of the intracontinental mobile belt, (2) andesite-basalt volcanism, (3) picritic-basalt volcanism, (4) folding, faulting and metamorphism. The structural-metamorphic evolution of rocks in this area ended in a stage of platformal activation.

The Archean cycle (>2.6 Ga)[14] is subdivided into two major stages. At the first stage, sedimentation and volcanism took place. At the second stage, folding, metamorphism and ultrametamorphism occurred.

The Archean metamorphic complex penetrated by the SG-3 well at a depth of 6842 – 11,662 m is of varying composition and rhythmically layered structure. It is composed of alternating thick (from 571 to 1834 m) strata of biotite plagiogneisses and strata of two-mica-plagiogneisses with HAM of less thickness (117 – 780 m). As a whole, these rocks occupy 66% of the described part of the section; these rocks are associated everywhere with sporadic bodies of amphibolites (30%) 60 m thick, and granitoids (4%). The question of the primary premetamorphic nature of the rocks is complex. Since unaltered or slightly migmatized and metasomatized metamorphic rocks were preserved in almost every layer of the deep section, it became possible to reconstruct their primary composition with the help of the petrochemical diagram $K_2O - SiO_2 - CaO$ (Fig. 1.81); trends to discriminate of this diagram were established in advance on the basis of studying more than 1000 analyses embracing the full compositional spectrum of magmatic and sedimentary rocks of a normal series. Changes in composition of $K_2O$ and CaO are shown on the diagram, depending on the distribution pattern of $SiO_2$, and they form two diametrically opposite general trends of magmatic and sedimentary activity. Symbols denoting the Archean rocks in the SG-3 well and the surrounding surface are confined to isolated areas of the diagram and are subdivided into three following groups:

1. Metavolcanic rocks which are distributed in accordance with the trend of magmatic differentiation; they include amphibolites, meta-ultramafites and part of the biotite-plagioclase gneisses;
2. Metaterrigenous rocks; the distribution of the corresponding metamorphites is in distinct agreement with the trend of sedimentary differentiation. These rocks are represented in the section by biotite-plagioclase gneisses, including gneisses with HAM; leucocratic rocks correspond in chemical composition to

[14] In accordance with recommendations of the All-Union meeting on the subdivision of the Precambrian rocks (Ufa 1977), the Archean strata comprise all formations whose age exceeds 2.6 Ga.

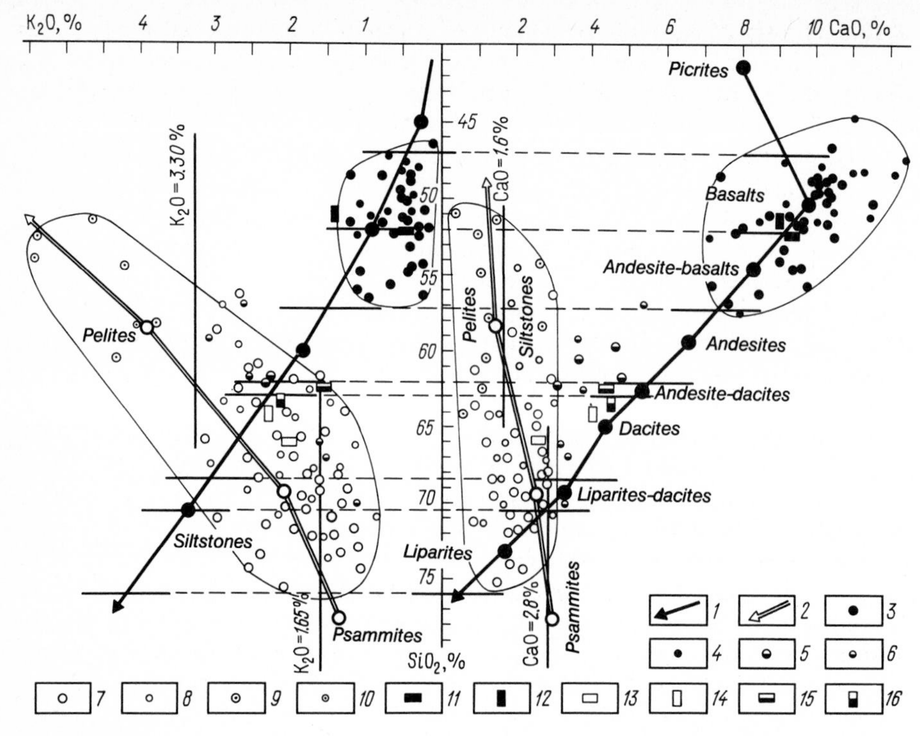


**Fig. 1.81.** Reconstruction of the Archean rock protocomposition in the SG-3 well section (open symbols) and of surrounding surface (*filled symbols*). *1* trend of magmatic differentiation; *2* trend of sedimentary differentiation; *3–4* amphibolites; *5, 6* hornblende-biotite plagiogneisses; *7–8* biotite-plagioclase gneisses; *9–10* biotite-plagioclase gneisses with HAM; *11–16* average compositions; (*11, 12* rocks of the amphibole group; *13, 14* rocks of the plagioclase group; *15, 16* rocks of the Archean complex)

psammites, mesocratic plagiogneisses correspond to siltstones and melanocratic rocks correspond to pelites;

3. Metavolcanic-terrigenous rocks occupy an intermediate position between trends of sedimentary and magmatic differentiation. It is the most scarce (by the extent to which it is spread out) group of the hornblende-biotite plagiogneisses (4% of the complex bulk).

On the strength of geological-petrographical and mineralogical-geochemical indications (see Chap. 1.4), amphibolite rocks are subdivided into two series: Fe-Mg-amphibolites → Fe-amphibolites (including porphyroblastic amphibolites) → Si-amphibolites; Al-Mg-amphibolites → actinolitic schists → talc-actinolitic schists. The first series of amphibolites corresponds to the volcano-plutonic association of theoleiite type; it includes, respectively, olivine basalts, normal basalt (including gabbro) and andesite-basalts; it is not unlikely that part of the rocks of this series (Si- and Fe-amphibolites) can be paralleled with metabasites of the Pechenga complex. The second series of amphibolites-meta-ultramafites is represented by a subvolcanic association (hypersthene-basalts pyroxenites and olivine pyroxenites), and it corresponds to the rocks of the Kola series. By composition and content of rare-earth minerals, amphibolites are identified with the volcano-plutonic associations of basic rocks of the Archean greenstone belts (Fig. 1.82). The primary-mantle genesis of the initial amphibolites of the Kola complex is proved by the low value of the primary ratio $^{87}Sr/^{86}Sr = 0.7040$ (see Table 1.18).

The biotite-plagioclase gneisses, characterized by an alternating mineral composition, belong to rocks of a normal series ($Na_2O + K_2O = 7\%$) and are notable for their high mean content of $SiO_2 = 68\%$; insignificant changes in chemical composition with depth are identified by the increase of $SiO_2$ up to 72% and decrease of $Na_2O$ down to 4%. On the petrochemical diagram they are confined to an area which is overlapped by fields of acid effusive rocks and feldspar sandstones. In a geochemical formula of the biotite-plagioclase gneisses elements such as Sc, Be, Zn, Ba, Mo and V, Co, Ga, Ni, Cu, Cr and Sc are isolated and form independent groups; it is known that such grouping is typical of magmatic rocks (Burkow 1973). The distribution pattern of REE and the low value of the primary ratio $^{87}Sr/^{86}Sr$ (0.70.70) in these rocks draws a parallel between them and the Archean tonalite series.

Metaterrigenous rocks in the SG-3 well section are represented by leucocratic and melanocratic plagiogneisses with HAM. A characteristic feature of this series is the presence of a single whole trend of successively increasing mean concentrations of $\sum$ FeO, MgO, $K_2O$, Rb, Li, Cr, V, Co, Sc and Zn and amounts of the accessory ilmenite, sphene and orthite; on the petrochemal diagram their distribution pattern is in conformity with the trend of sedimentary differentiation. The latter is also proved by the type of REE distribution, the high value of the primary ratio $^{87}Sr/^{86}Sr = 0.7240$ and the presence of the regenerated fragmental zircon grains in these rocks. The geochemical formula of the biotite-plagioclase gneisses with HAM (Sr, Cu, Mo, Sc, Zr, Cr, Be, Ni, Co, V, Ba and Ga) also indicates that they belong to the primary sedimentary rocks similar in composition to the hydromicaceous-montmorillonite clays which are associated in the initial section with pelitic and sandy rocks.

Thus, it appears that the first stage of Archean rock formation was characterized by the accumulation of thick sedimentary strata (clayey siliceous rocks) and active volcanic processes accompanied by the eruption of liparite, dacite and andesite lavas and subordinate basalts and ultramafites (basalt-andesite formation).

During the following stage, these rocks underwent folding, metamorphism and ultrametamorphism under granulite-facies conditions; this is seen by the occurrence of steep contacts and by the presence of gneissose rocks in the SG-3 well, the presence of relicts of mineral associations in amphibolites and biotite-plagioclase gneisses (diopside, hypersthene, labradorite-anorthite and high-titanium biotite with $F = 50-60\%$), as well as associated migmatite-granites (high-temperature plagioclase with non-ordered distribution of Al in a crystalline lattice (see Chap. 1.3). It should also be noted that at the surface the granulite facies is met in the form of relict paragenesis of low and moderate pressures throughout the north-eastern (Belyayev et al. 1977) and south-western (Kremenetsky 1979) fringes of the Pechenga structure. By analogy with the Archean

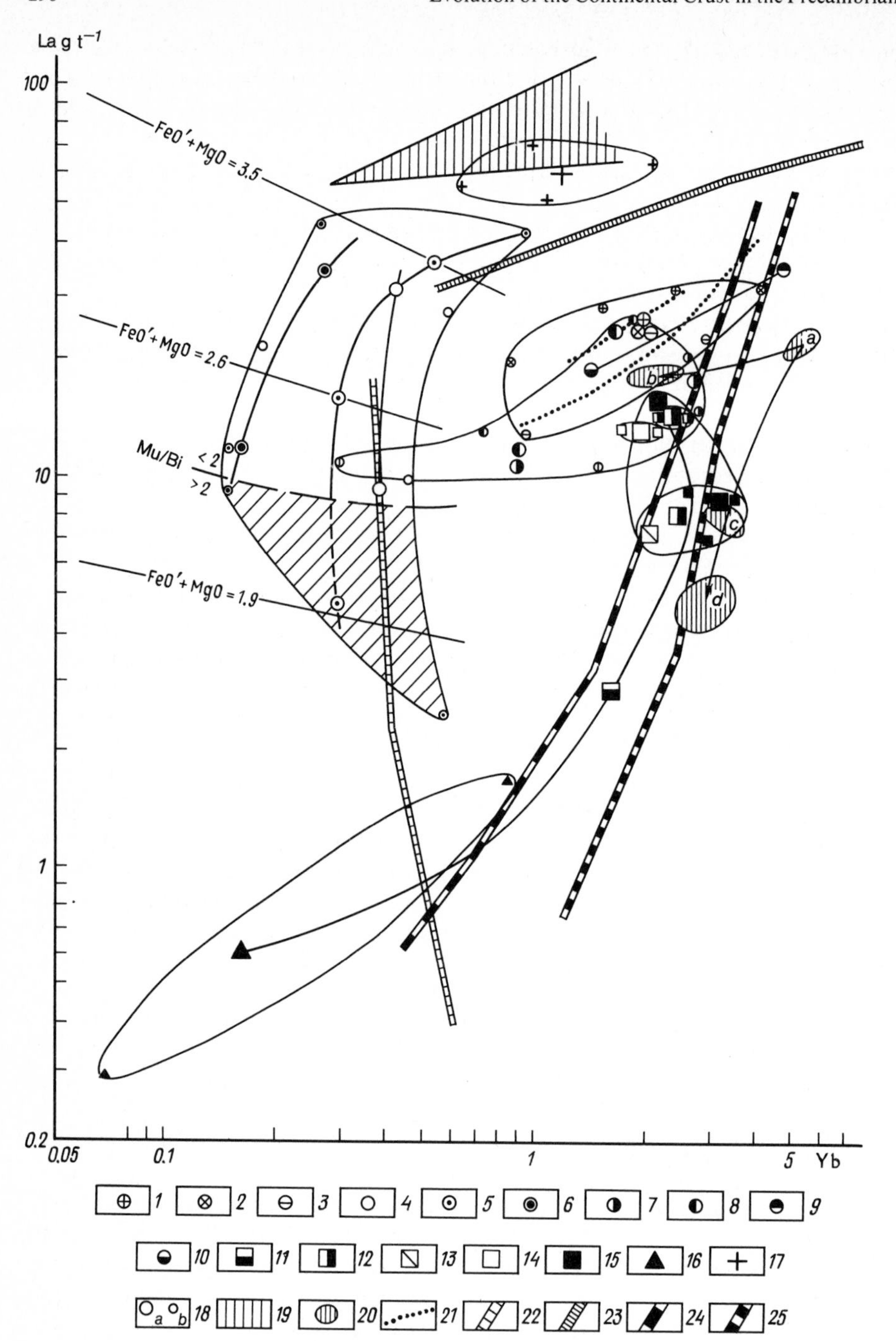
La g t⁻¹
100
10
1
0.2
0.05
0.1
1
5
Yb
FeO'+MgO = 3.5
FeO'+MgO = 2.6
FeO'+MgO = 1.9
Mu/Bi
< 2
> 2
a
b
c
d
1
2
3
4
5
6
7
8
9
10
11
12
13
14
15
16
17
18
19
20
21
22
23
24
25

Fig. 1.82

complexes developed at the surface, one may surmise that tectonic structures of this stage were represented by synclinorial zones or by asymmetric synclines in combination with dome-like structures (Zagorodny 1980). Moreover, surficial observations indicate a multi-recurrent character of the folding and strike-slip faults of the Kola series rocks during processes of metamorphism and ultrametamorphism (Dobrzhinetskaya 1978), although these tectonic elements of different age cannot be distinguished from each other in cores from the SG-3 well.

The radiological age of the culmination of regional metamorphism for this stage is $2754 \pm 4$ Ma. Similar datings of the uranium-lead age, determined by accessory zircon ($2700 \pm 50$ Ma) have been obtained by Tugarinov and Bibikova (1980) for the regional metamorphism stage of the Kola plagioclases which outcrop in the areas of Pulozero, Chudzyarvi, Olenegorsk and Murmasha. It was established that zircons from the amphibolite and granulitic facies are of the same age (2800 – 2700 Ma); products of the earliest granitoid magmatism are also of similar age. This indicates that prograde metamorphism and ultrametamorphism of the Kola complex belong to a single whole zonal series.

The Archean cycle has ended by the consolidation of supracrustal rocks (cratonization) and deep erosion.

The Proterozoic cycle (2.6 – 1.1 Ga) is characterized by the formation of the intracontinental Pechenga-Varzug mobile belt (on the Archean basement) extending throughout the Kola Peninsula in north-western direction from the Varanger-Fiord (Norway) to the White Sea. We distinguish four stages in the development of this belt: (1) the initial formation of the belt, (2) andesite volcanism, (3) picritic-basalt volcanism, (4) strike-slip faults and metamorphism.

The first stage is distinctly identified at the surface in the southern fringe of the Pechenga-Varzug belt by the presence of areas composed of Tundra rocks. The available radiological data do not allow an accurate determination of the time span between the Kola complex consolidation and the beginning of the intracontinental belt development. The presence of conglomerates and metamorphosed crusts of weathering at the base of the Pechenga complex (Televin suite) makes it possible to suppose that the Proterozoic sedimentary-volcanic strata were undoubtedly deposited on the rigid stable base, although they inherited the general trend of the ancient structures. It can also not be ruled out

**Fig. 1.82.** Trends of La and Yb distribution in the Archean rocks of the SG-3 well section and their identification with main structural-genetic Precambrian and Phanerozoic series. *1* melanocratic plagiogneisses with HAM; *2* mesocratic bimicaceous plagiogneisses with HAM; *3* leucocratic bimicaceous plagiogneisses with HAM; *4* biotite plagiogneisses (Qz>10%); *5* biotite plagiogneisses (Qz>10%); *6* biotite gneisses; *7* leucocratic epidote-biotite plagiogneisses; *8* mesocratic epidote-biotite plagiogneisses; *9* hornblende-biotite schists; *10* hornblende-biotite plagiogneisses; *11* Fe-Ti-amphibolites; *12* biotite schists; *13* Al-Mg-amphibolites; *14* amphibolites; *15* gabbro-amphibolites; *16* talc-actinolitic schists; *17* post-folded porphyry granites; *18* medium (*a*) and boundary (*b*) contents; *19, 20* granitoids of the Litsk-Araguba complex (*a* metabasalts; *b* metaandesite-basalts of the continental-rift stage; *c* metagabbrodiabases; *d* metabasalts of the oceanic-rift stage); *21 – 25* trends (*21* of the sedimentary differentiation – Phanerozoic sands-clays and $PR_2$ metapsammites-metapellites, after Balashof (1976); *22* of the Early Archean gabbro-plagiogneiss series, after Simmons et al. (1980); *23* of the Early Archean primary-crust granitoids of the Kola peninsula, after Belkov and Yelina (1979); *24* of intrusive rocks; *25* of the Early Archean effusive rocks of the greenstone belts and Phanerozoic effusive series

that abrupt changes in thickness of the conglomerates (which at places even fully wedge out) may indicate their alluvial origin. Paleotectonic conditions existing in the initial stage of the belt development can be described only in general, because rocks of the Tundra series were strongly foliated and underwent zonal metamorphism in a sequence from the sericite-chlorite zone of the greenschist facies to the sillimanite-muscovite zone of the epidote-amphibolite facies (for instance, in the metaterrigenous rocks of the Talya-Talypvyd suites developed on the south-western and south-eastern fringes of the Pechenga structure).

A transitional period from the first to the second stage was marked by an abrupt increase in volcanic activitiy in subplatformal conditions. This is evidenced not only by the great thickness of the lava flows (Mayarvi and Pirttiyarvi suites) but also by the remarkable uniformity (along the strike and down the dip) of thin horizons of sedimentary rocks (Kuvernerinyok and Luchlompol suites). The composition of these rocks, found in the SG-3 well (Fig. 1.83), and that of rocks developed at the surface, are remarkably similar to each other, thus characterizing a successive replacement of the arkosic terrigenous deposits by the terrigenous-carbonate deposits up the section (content of calcite and dolomite reaches 70%); it is known that such changes are inherent to the continental shallow lagoons. The petrological-geochemical data (see Chap. 1.4) indicate that volcanism commenced with the eruption of uniform non-differentiated magma (andesite-basalt formation); the subsequent evolution of this magma (melts were enriched with alkaline metals and volatile components) and its differentiation have resulted in the eruption of volcanites of successively rhythmic trachybasalt formation (Fig. 1.84). On the whole, the formation mechanism of these rocks can be defined as syntectic (Dobretsov 1981). According to this mechanism, a process of basite magma intrusion into the "cold" acid crust was accompanied by additional melting and assimilation of the crustal material, and consequently, the primary magma composition became "more acid". The subsequent differentiation of such melt led to the segregation of andesites.

The volcanic activity of this stage was accompanied by the formation of a comagmatic complex of the gabbro-diabase intrusions which were injected into rocks of the underlying Mayarvi suite.

As a result of this, the andesite-basalt and trachybasalt formations, genetically associated with each other, form one single successive volcano-sedimentary association together with sedimentary rocks of the Kuvernerinyok and Luchlompol suites. This association may be assigned to a carbonate-terrigenous trachybasalt-andesite-basalt formational series (Luostari series).

The interval between the first and second stages is not well defined, yet its presence is proved by the following facts: the crust of weathering is developed at the top of the Pirttiyarvi suite; fragmented magnetite is present in sandstones of the Luchlompol suite; volcanic formations replace each other abruptly. For instance, the beginning of the third stage is distinctly fixed by the presence of products of the erupted non-differentiated tholeiite magma, which forms the rhythmic tholeiite-basalt formation (Zapolyarny suite) of 1.9 km apparent thickness. The subsequent evolution of mantle hearths and their successive sinking has led to the appearance of the so-called komatiitic magma which can produce varieties from ultrabasic to basic, and even of acid composition (see Fig. 1.84).

As a final result of this process, a successive-rhythmic picrite (komatiite)-tholeiite formation (Matertin suite) and associated intrusions of the gabbro-diabase and nickeliferous gabbro-wehrlite (Synmatertin) complexes have been formed. Volcanites of the picrite-basalt stage of volcanism are generally similar in chemical composition to rocks of the tholeiite-komatiite association, which is widely spread in the Precambrian greenstone belts on the Baltic (Vetryany belt) and other shields (Ukrainian, South-African, Canadian and Australian); this is indicative of the global character of magmatic activity at this stage (and, consequently, of the global character of metallogeny) in the Precambrian time.

The specificity of the third stage is reflected in the character of sedimentation. For instance, the Zhdanov suite is characterized by a heterogenous sequence which comprises two major rhythms: the lower rhythm is represented by gravel-sandstone-siltstone rocks and the upper rhythm is composed of siltstone-pelite rocks with a considerable admixture of tuffaceous material (see Fig. 1.83). These facies and the composition of the salts dissolved in subsurface waters of this suite (see Chap. 1.9) are indicative of the marine conditions of sedimentation. Tuffs and tuffites present in the section are similar in composition to volcanites of the Matertin suite, and their wide development indicates an active role of volcanism in the process of sedimentation; due to volcanic activity considerable masses of hydrogen sulphide entered marine basins. In addition to this, sediments are notable for the presence of organic matter contained in increased amount; organic matter even forms isolated layers in the sulphide-carbonaceous schists.

Thus, the volcano-sedimentary rocks of the third stage possess features of different formations (greywacke, flysch, black-shale), but nevertheless, do not fully correspond to any of them. These formations, together with the underlying tholeiite-basalt and overlying picrite-tholeiite formations, we combine into a single whole tuffaceous-terrigenous picrite-basalt formational series (Nickel series).

There is no doubt that the stage of formation of the folded and ruptured structures and metamorphism which closed the Proterozoic cycle lasted for a long time and was multi-phased. It is impossible to reconstruct in detail the full sequence of tectonic events. Nevertheless, data presented above offer the possibility of a new interpretation of the Pechenga area structure. According to this interpretation, deformations and metamorphism of the Pechenga complex are linked not with the final stage of the Middle Proterozoic geosyncline formation but with the protoactivation of the Early Precambrian structures of the Kola Peninsula (Kazansky 1982). In fact, the Pechenga complex was formed in the subplatformal domain, and Proterozoic volcanism is antidromal in character on the whole. The geological material available at present is most compatible with the theory of the overthrust occurrence of the Pechenga complex, or to be more precise, of the northern flank of the Pechenga graben-syncline. Regular changes of structure and texture of metamorphic rocks with depth, the appearance of a certain orientation of quartz, carbonates, mica and amphiboles in these structures and textures indicate that the underthrust movements took place synchronously with processes of metamorphism of the Pechenga complex whose intensity was increasing from the prehnite-pumpellyite to amphibolite facies. These processes also affected granite gneisses and amphibolites of the Kola series

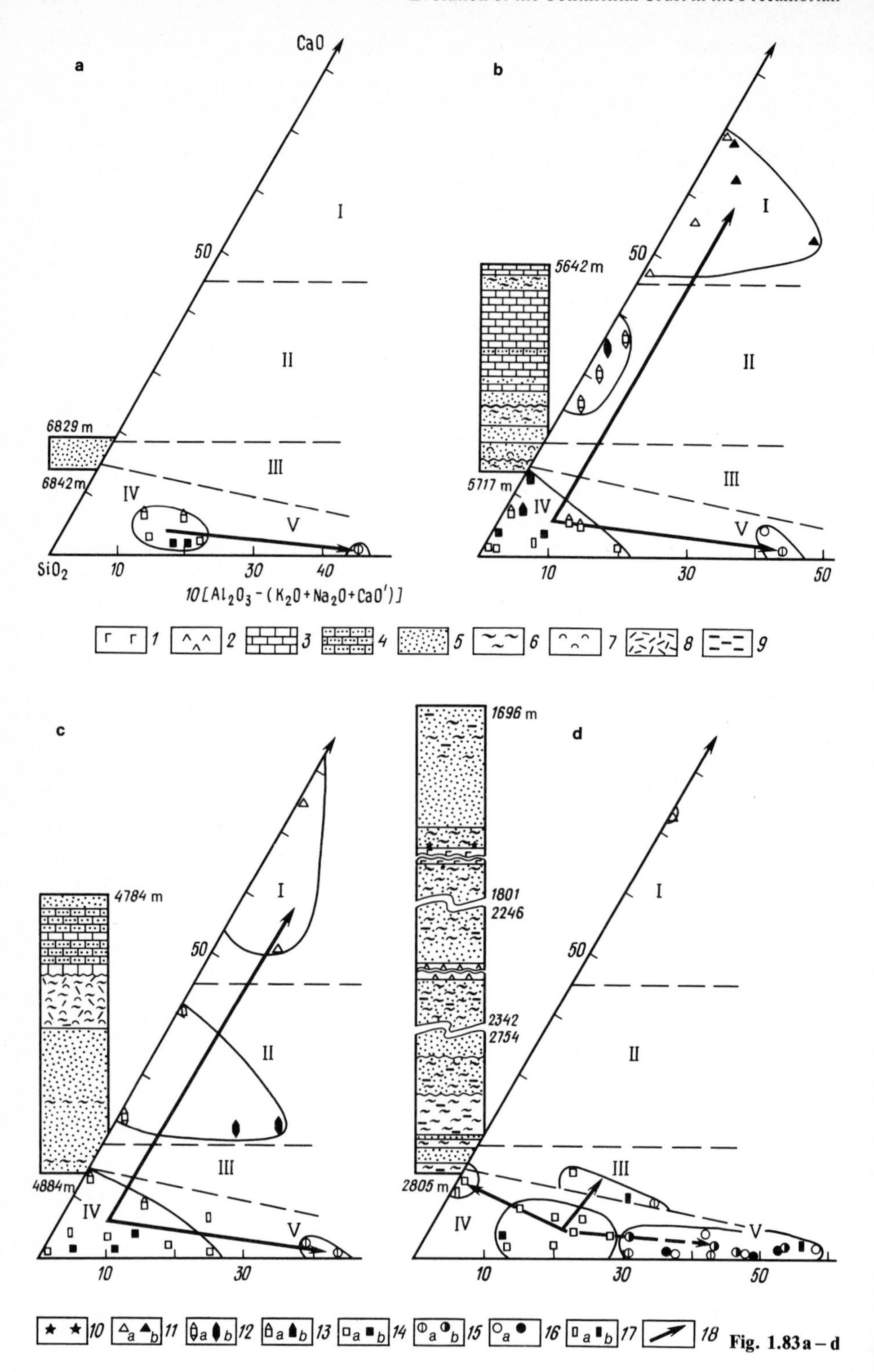
a
b
c
d
CaO
$SiO_2$
10[Al2O3 - (K2O+Na2O+CaO')]
I
II
III
IV
V
6829 m
6842 m
5642 m
5717 m
4784 m
4884 m
1696 m
1801
2246
2342
2754
2805 m
10
30
40
50
1
2
3
4
5
6
7
8
9
10
11
12
13
14
15
16
17
18

Fig. 1.83 a – d

and almost fully destroyed mineral associations of the granulite facies which previously existed in this series.

Thrusting was accompanied by the appearance of the metamorphic rock anisotropy (in internal structure and in elastic properties) in zones of foliation. Such a sharp transition from isotropic to anisotropic rocks in the zone of the Luchlompol major fault coincides with the boundary between the Luostari and Nickel series. It appears most probable that the eruption of effusive rocks of the Mayarvi and Pirttiyarvi suites was accompanied by low-temperature autosomatic and metamorphic changes and the formation of hydroxyl-bearing minerals. During the subsequent prograde metamorphism of the whole Pechenga complex, a temperature increase up to 500° – 600 °C could have caused the pressure increase according to the autoclave effect; as a result of this, synmetamorphic fluids, enriched in chlorine, iodine, barium, strontium and potassium, were released. This effect facilitated the deformation of rocks and caused their hydraulic disaggregation (see Chap. 1.).

The zone of the Luchlompol major fault was undoubtedly active over a long period of time. A body of dacite metamporphyrites is confined to this zone in the SG-3 well section. These porphyrites may be correlated with the volcano-plutonic complex of the southern part of the Pechenga structure if their composition and age are taken into account. Subvolcanic elements of this complex mark the axis of the tectonically active zone between the northern and southern parts of the structure (Poryitash major fault), and effusive elements constitute the basalt-andesite formation of the orogenic type in its southern part. One may suppose that, at depth, the Luchlompol major fault is joined with the steeply dipping Poryitash tectonic zone, and that these fractures build the framework of the reverse fault-thrust dislocations of the Pechenga structure. It should also be recalled that near this fault the foliated and metamorphosed sedimentary and volcanic rocks contain anomalously high concentrations of helium and argon which might have been introduced by fluids.

The intrusion of granites into the Litsk-Araguba complex in the area immediately adjacent to the Pechenga region also occurred at this stage. The age of granites is 1810±50 Ma if dated by the U/Pb technique, and 1720±25 Ma if dated by the Rb/Sr technique (Gorokhov et al. 1982). In the SG-3 well section, vein bodies of porphyritic granites, assigned to this complex, dissect granite gneisses of the Kola series. The ratios of porphyritic granites to muscovite-bearing aplites, pegmatites and quartz veins in cores of the well were not determined, but the radiological data show that their ages are similar to each other.

**Fig. 1.83a – d.** Litho-petrochemical characteristic of metasedimentary rocks of the Pechenga complex. *Suites:* **a** Televin; **b** Kuvernerinyok; **c** Luchlomphol; **d** Zhadonov. *Rocks: 1* metaultrabasites; *2* metagabbro-diabases; *3* dolomites and limestones; *4* dolomitic and calcareous metasandstones; *5* metasandstones and metaaleuro-sandstones; *6* metaaleurosandstones and metapellites; *7* admixture of carbonate material; *8* metatuffs and metatuffites; *9* carbonaceous matter; *10* sulphides. Compositions of rocks from the SG-3 well section (*a*) and surrounding surface (*b*); *11* sandy dolomites and limestones; *12* dolomitic and calcareous metasandstones; *13* calcareous metasandstones; *14* metasandstones; *15* metasiltstones; *16* metapellites; *17* metatuffites; *18* trends of sedimentary differentiation (*I* sandy dolomites and limestones; *II* dolomitic and calcareous sandstones; *III* calcareous sandstones; *IV* sandstones; *V* siltstones and aleuropellites)

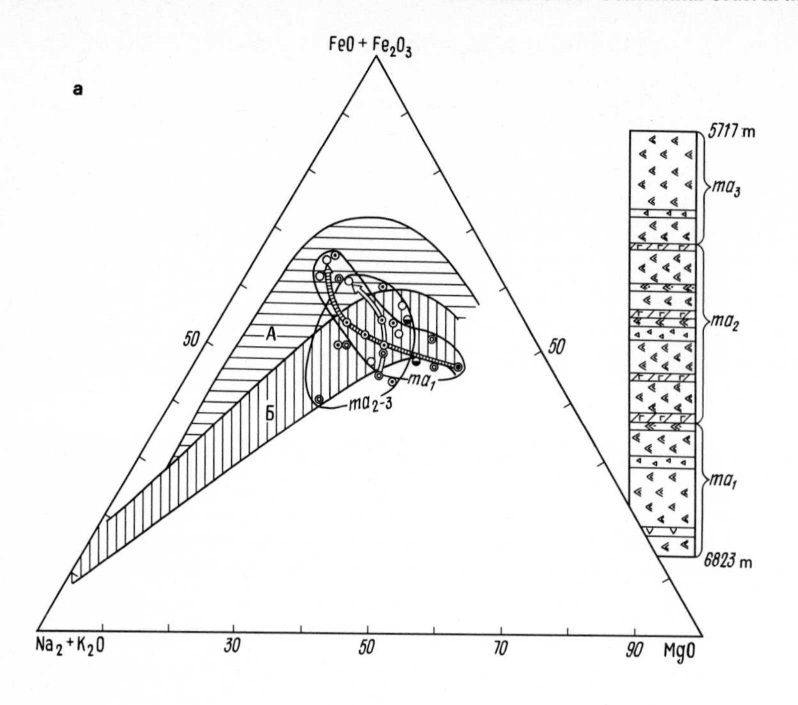
a
FeO + $Fe_2O_3$
50
50
A
Б
$ma_1$
$ma_{2-3}$
$Na_2 + K_2O$
30
50
70
90
MgO
5717 m
$ma_3$
$ma_2$
$ma_1$
6823 m

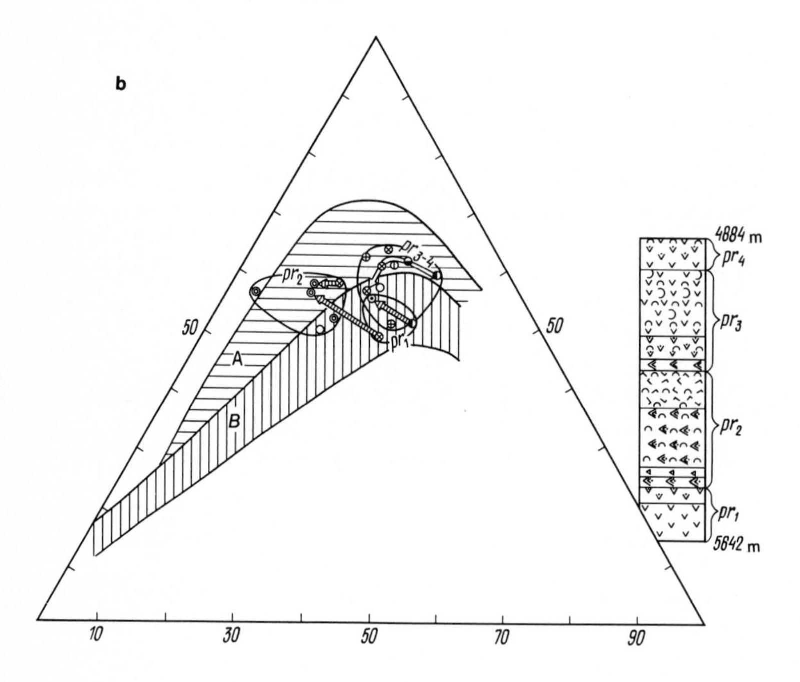
b
50
50
$pr_2$
$pr_{3-4}$
$pr_1$
A
B
10
30
50
70
90
4884 m
$pr_4$
$pr_3$
$pr_2$
$pr_1$
5642 m

**Fig. 1.84 a, b**

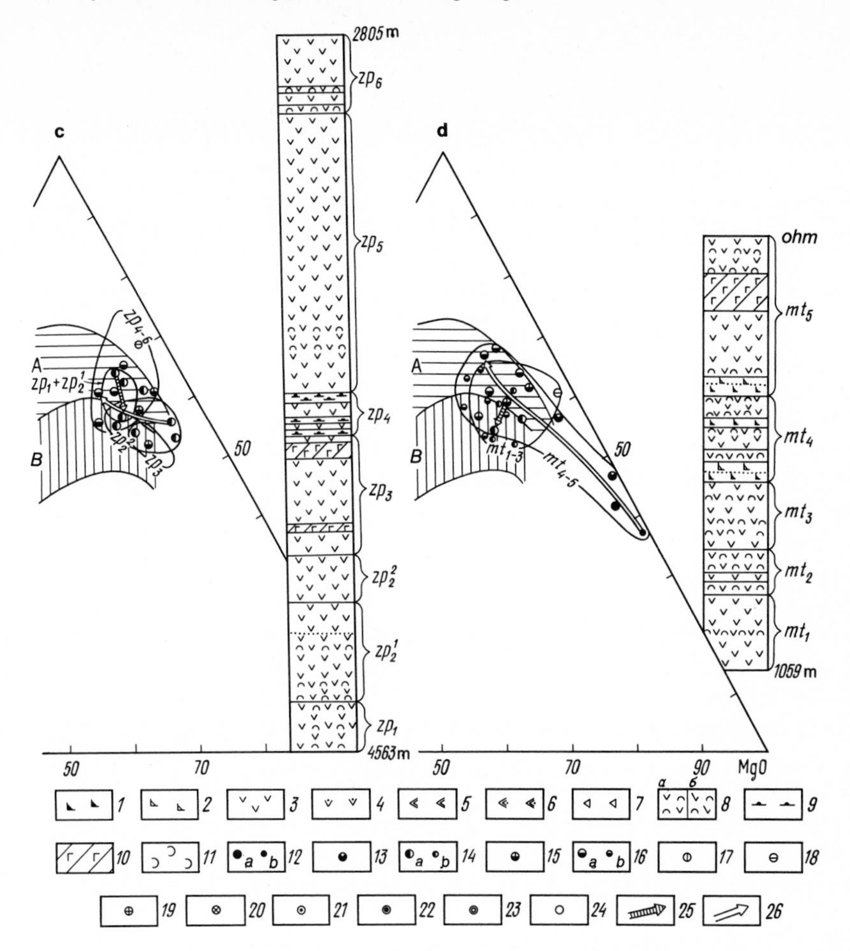


**Fig. 1.84 a – d.** Geological-petrochemical characteristic of metavolcanogenic rocks of the Pechenga complex. *Suites:* **a** Mayarvi; **b** Pirttiyarvi; **c** Zapolyarny; **d** Materin. Main types of rocks on geological sections: *1* metapicrites; *2* metapicrite-basalts; *3* metabasalts; *4* metatrachybasalts; *5* meta-andesite-basalts; *6* metatrachyandesite-basalts; *7* metaandesites; *8* metatuffs of basic (*a*) and medium (*b*) composition; *9* metatuffites; *10* metagabbro-diabases; *11* spilitization; *Compositions of meta-effusive rocks* (*a* mean; *b* individual values); *12* metapicrites; *13* metapicrite-basalts; *14* olivine metabasalts; *15* differentiated metabasalts; *16* low-alkaline metabasalts; *19* spilites; *20* metatrachybasalts; *21* meta-andesite-basalts; *22* pyroxene-olivine meta-andesite-basalts; *24* meta-andesites. Differentiation trends of metaeffusive rocks: *25* in lower and *26* in upper parts of volcanogenic formations. Series: *A* hypersthene; *B* pigeonite

For the zones of retrograde fault metamorphism, superimposed over the thrust structure of the Pechenga region, it may be said that their age status is not clear. Two theories are equally admissible: (1) they were formed in the very end of the Proterozoic cycle; (2) they belong to the next stage of platformal development and subsequent tectonic-magmatic activitation of the Baltic Shield. A description of the final stage is beyond the scope of this work.

### 1.10.3 Evolution and Vertical Zonality of Mineralization

The study of the Kola borehole core has for the first time presented factual material about the vertical zonality of mineralization in deep zones of the Earth's crust and proved the theory of the polychronous origin of this zonality. Manifestations of all ore-forming processes associated with sedimentation, volcanism, intrusive magmatism and hydrothermal activity have been established in this unique section.

Ferruginous quartzites, encountered by the well among metamorphic rocks of the Kola series, were formed at the stage of sedimentation and volcanism of the Archean cycle. Analogous ferruginous quartzites are known in outcrops of the Pechenga region. Beyond the limits of this region, at the Olenegorsk deposit, these quartzites are of commercial importance. The geochronological study of rocks developed at the Olenegorsk deposit, accomplished with the help of the Rb/Sr and K/Ar technique, has established that the minimal age of ferruginous quartzite formation is 2600 Ma (Gorokhov et al. 1981). Results of examination of core samples from the SG-3 well do not contradict this assessment.

Ferruginous titanium and sulphide copper-nickel mineralization, encountered by the SG-3 well in the Archean rocks, is associated with bodies of basic and ultrabasic composition and is probably of magmatic origin. However, this mineralization was affected by processes of metamorphism under conditions of amphibolite facies which were also manifested (as already mentioned above) in the Pechenga complex. When titanium-bearing gabbroids were subjected to the action of metamorphic processes, titanomagnetite was recrystallized; as a result of this, high-grade ilmenite-magnetite ores were formed. Nickeliferous ultrabasites were transformed into biotite-actinolitic schists. During this transformation, copper and nickel sulphides partially migrated into enclosing rocks. Sulphur isotope ratios of these rocks, similar to ratios of meteoric standard, were left unchanged. From the genetic point of view, copper-nickel mineralization encountered by the SG-3 well in the Kola series is similar to copper-nickel ores of the Allarechinsk deposit, which also underwent intensive metamorphism under conditions of amphibolite facies.

The question of the position of the iron-titanium mineralization in the general scheme of the geological evolution of the Pechenga region is still open after studies conducted at the Kola superdeep well.

Formerly, some proofs were adduced in support of a theory about the thrust structure of the Pechenga ore field and to demonstrate that sedimentary and volcanic suites of this region were formed under subplatformal conditions. Data on the copper-nickel intrusion do not contradict this conclusion, but, in fact, confirm it. Additional arguments in favour of this conclusion are as follows: (1) the tectonic nature of the most nickeliferous intrusions both on the lower and upper sides and along the strike; (2) the inclined occurrence of the primary foliation of intrusions and boundaries of the impregnated copper-nickel orebodies; (3) the absence of direct correlation between the thickness of copper-nickel ores and thickness of basic-ultrabasic bodies; (4) the exclusive uniformity of copper-nickel mineralization down the dip. All these arguments, considered together, assume that the formation of sulphide copper-nickel deposits of the Pechenga ore field is associated with a peculiar geotectonic process, i.e. with a process of activation of Early Precambrian structures of the Earth's crust. This process commenced in a time span between the Archean and Proterozoic eras, and reached its maximum in the Middle Proterozoic when continental blocks were dissected from the stabilized Earth's crust by major fault, along which mantle melts and fluids have ascended to the surface (Kazansky 1980).

With this in view, nickeliferous intrusions of the gabbrowehrlite formation, together with the picrite-basalt formation, may be considered to be the latest single whole volcano-plutonic association of the Pechenga complex. The antidromal character of Proterozoic volcanism indicates the successive subsidence of those hearths in which magma melts originated. Nickeliferous ultrabasites were formed out of the mantle matter melts originating at depths of 90 – 150 km (Dobretsov 1981). Before the picrite-basalt formation was formed, thick sediments of the Zhdanov volcano-sedimentary suite had accumulated in conditions of a shallow basin; these sediments contained a great amount of organic matter and massive sulphide mineralization. It is likely that the initial nickeliferous intrusions occurred almost horizontally. With this in view, it becomes understandable why boundaries of magmatic differentiates, poor- and rich-impregnated ores, are parallel to the inclined contacts of intrusions, etc. If the stratigraphic succession and thicknesses of sedimentary and volcanic rocks of the Nickel series are taken into consideration, then it appears that the minimal depth of formation of the basic-ultrabasic intrusions and magmatic impregnated copper-nickel ores is 3 – 5 km.

Present progress in the geological study of the Pechenga region rules out the possibility of accurately assessing a magnitude of changes of the initial size and shape of nickeliferous intrusions, which have been caused by thrust faulting developed in the Pechenga complex and determining the distance from the surface at which these intrusions underwent prograde metamorphism. However, one may assert that this metamorphism took place together with the development of intensive heat flows, with their gradients being equal to 50° – 70° per km. Ultrabasites and copper-nickel ores encountered by the Kola superdeep well were metamorphosed under conditions of greenschist facies, i.e. at medium temperatures and pressures ($T = 350° - 450°C$, $P = 0.2 - 0.3$ GPa). Metamorphism was accompanied by the remobilization of the ore matter. It is known that rich breccia ores, which occupy a certain position in the general vertical metamorphic zonality are associated with zones of fault metamorphism, and that sulphide minerals are tightly intergrown with silicate minerals in foliated meta-ultrabasites and metasedimentary rocks. On the contrary, sulphide minerals of the primary magmatic impregnated ores are intensively heated with hydrosilicates. All these facts are the most important arguments in evidence of the epigenetic origin of rich breccia ores. Nevertheless, the question of the genesis of rich breccia ores is still greatly open to debate.

The pattern of vertical distribution of these types of ore mineralization is fully compatible with the Archean and Early Proterozoic history of the Pechenga region. The Kola series comprises the intensively metamorphosed ferruginous quartzites, iron-titanium ores, copper-nickel mineralization; the Pechenga complex contains the less metamorphosed massive sulphide mineralization and sulphide copper-nickel deposits. However, this ordered scheme is obviously disturbed and complicated by the presence of a low-temperature hydrothermal mineralization which was encountered for the first time in the lower part of the Pechenga complex, and later traced down to a depth of 11 km. Although this mineralization is small in size, it is obvious that the discovery of this mineralization at previously inaccessible depths increases the depth limit down to which ore-bearing fissured structures are formed by three to four times. The great persistence in vertical occurrence of mineral associations in zones of retrograde fault metamorphism, the low temperatures of sulphide formation and the results of sulphur isotope analyses indicate the existence of a juvenile source for hydrothermal solution. This source is likely to be of subcrustal origin.

### 1.10.4 Evolution of the Earth's Crust in Precambrian Time

The elaboration of the theory of the deep structure and chemical composition of the Precambrian continental crust is one of the fundamental problems of modern geology, serving as a starting point for the objective assessment of the global trend in evolution of the matter of the Earth's crust in the process of its development, particularly at early stages. A sound solution to this problem should help in conceiving the evolution of the primary composition of the oldest complexes, beginning from the protocrust and ending with the recent crust, revealing the mechanism of their development, establishing the specific character of the fluidal regime and *P-T* conditions of their metamorphism in Archean and Proterozoic times and, consequently, in determining the sources and evolutionary trends of the ore-forming systems.

It is known that about 60% of the whole Earth's crust is concentrated on the continents. According to calculations of many researchers (The Earth's Tectonosphere 1978), the chemical composition of the Earth's crust (without sedimentary cover) corresponds to the composition of medium rocks[15] and is as follows: $SiO_2$ – 57.4–63.8 ($\bar{x} = 60.6$); $TiO_2$ – 0.6–1.2 ($\bar{x} = 0.9$); $Al_2O_3$ – 15.0–16.1 ($\bar{x} = 15.5$); $Fe_2O_3 + FeO$ – 6.1–9.1 ($\bar{x} = 7.1$); MnO – 0.1–0.2 ($\bar{x} = 0.1$); MgO – 3.1–5.6 ($\bar{x} = 4.0$); CaO – 3.8–7.4 ($\bar{x} = 5.6$); $Na_2O$ – 2.2–3.9 ($\bar{x} = 3.2$); $K_2O$ – 2.0–3.2 ($\bar{x} = 2.8$); $P_2O_5$ – 0.2–0.3 ($\bar{x} = 0.2$). The maximum limits of mean values identified for $SiO_2$ on the one hand, and for $Fe_2O_3 + FeO$, MgO and CaO, on the other, are caused by different ratios of granites to basalts (from 1:2 to 1:1) in the calculated systems; these ratios were calculated on the basis of the existing theory of the primary basalt composition of the protocrust and the degree of subsequent granitization of its upper part.

The strongly metamorphosed Early Archean basite rocks, and the associated gabbro-norite, considered as the oldest ophiolites (Nalivkina 1977), as well as gabbro-anorthosites, are usually taken to be direct evidence of the basic composition of the protocrust. These associations are regarded as the outcropping near-mantle horizons of the primitive crust which were formed during the moon stage

[15] Here and in the forthcoming parts of the text data of analyses have been recalculated for dry matter (in %).

of the Earth's evolution. The ensuing logical contradiction between the basalt composition of the primary crust and the distinct (more than 80%) predominance of the Precambrian acid rocks on the continents might have been caused by a process of superimposed granitization at the expense of intratelluric and transmagmatic flows of the mantle fluids, which saturated the primitve crust and transformed it radically on all continents at one and the same time.

A two-layered model of continental structure, proposed by Harold Jeffreys, served until recently as an indirect basis for the theory of the basalt composition of the protocrust. However, the new interpretation of the geophysical data available, as well as the latest material from deep seismic sounding on the Canadian and Anabar shields, on the Eastern European and other platforms, have served as cogent arguments to show that the theory of the two-layered structure of the continental crust should be rejected because the Conrad discontinuity, subdividing the crust into "granite" and "basaltic" layers, is absent on the continents.

This was confirmed by the factual material obtained in the course of the Kola superdeep well drilling. The well also did not meet the Conrad discontinuity. Moreover, it has showed that the real deep section, to describe it with an image, is "turned inside out" with respect to a supposed seismic section, i.e. the well proved to have encountered metamorphic rocks of predominantly basic composition instead of the anticipated "granitic" layer, and, on the contrary, a migmatite-plagiogneiss complex, extending down to a depth of 11.6 km, was met instead of the "basaltic" layer.

In recenty years, the existing theory of the chemical composition of the oldest formations of the Earth and their role in the structure of the continental crust have been considerably re-evaluated. According to recently published data, there is support for the opinion that more acid magmatic rocks such as tonalite-plagiogranites or andesites (melted out of the mantle), but not basites, are the rocks analogous to the primitive crust. However, although this theory has been actively and fruitfully developed, the question of the chemical composition of the primitive continental crust is still open to discussion. The results of deep crustal drilling obtained in the Kola superdeep well and in a series of deep wells drilled through the crystalline basement in the eastern part of the Russian plate (Minibayev reference well No. 20,000, Ulyanovsk key well No. 663 and Tuimaza key well No. 2000) allowed for the first time on the basis of actual material, a discussion of this problem. Deep sections, constructed with the help of these wells, exhibit many features common to the structure and composition of metamorphic complexes (Fig. 1.85c), and they may be considered as key components of the conventional idea of one single deep Precambrian sequence of the Eastern European platform; within the limits of this platform the total vertical thickness of the sequence is 15 – 25 km (see Fig. 1.85b). In view of the new seismic model of the Earth's crust (1980), elaborated for old platforms (see Fig. 1.85), this sequence was extrapolated down to the M-discontinuity (see Fig. 1.85b) and was subdivided by us into three layers. At present, owing to superdeep and deep drilling, the first two layers are within the reach of direct study and, as will be demonstrated later, they have been identified according to composition as granite gneiss (0 – 15 km) and granulite-gneiss (15 – 30 km) layers. The third (lower) layer of the crust, notable for the sharply increased velocities of the

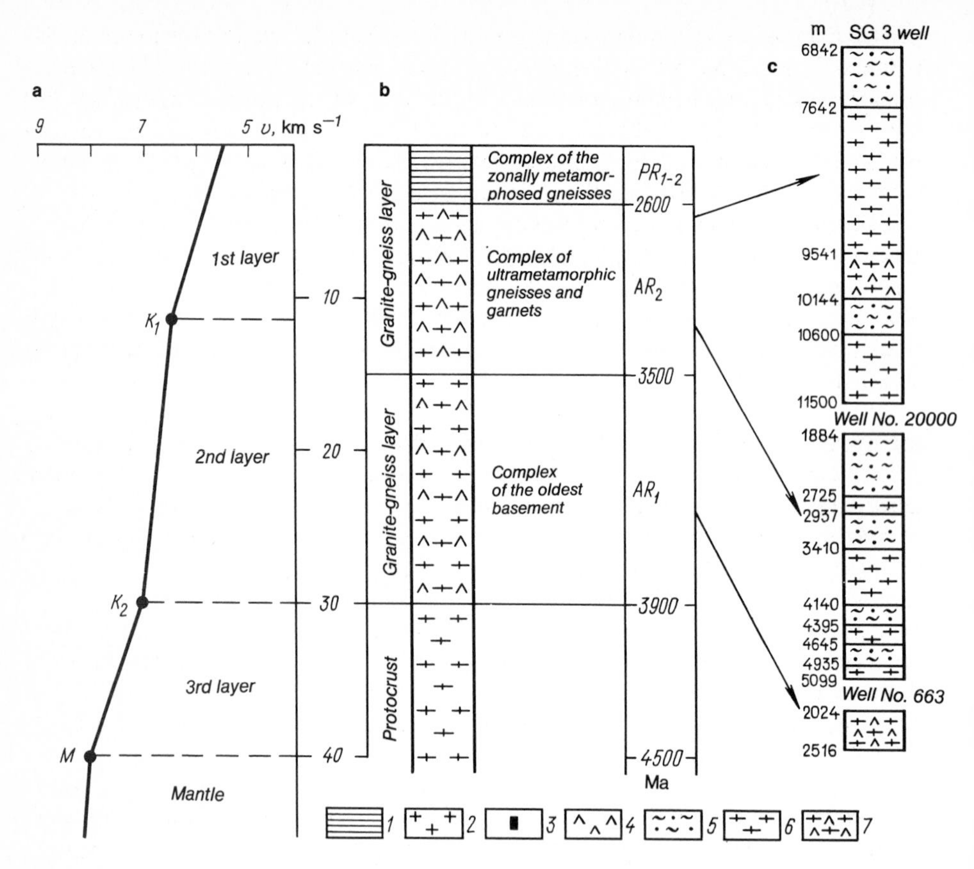


**Fig. 1.85 a – c.** Seismic model of the ancient platform Earth's crust (**a**) after (1980); conjectured (**b**), and studied (**c**) deep sections of the East European platform. *1* volcanogenic and volcanogenic-sedimentary metarocks; *2* granites; *3* metaultrabasites and talc-actionolitic schists; *4* amphibolites and bipyroxene basic crystalloschists; *5* biotite plagioschists; *6* plagiogneisses and enderbites; *7* biotite-amphibole plagiogneisses and hypersthene diorite-gneisses

longitudinal elastic waves (6.8 – 7.6 km $s^{-1}$) and the significant gradient of their increase with depth, represents the protocrust (or its autochthonous relics).

The granite gneiss layer is non-uniform in structure. A seismic boundary between the Proterozoic and Archean rocks is established at depths from 5 to 18 km. In the SG-3 well a contact between them was fixed with a high degree of certainty at a depth of 6842 m; for instance, a sharp transition between main Proterozoic (Pechenga) volcanites and acid Archean (Kola) plagiogneisses was encountered. The Kola complex was penetrated by the well in an interval of 6842 – 11,662 m (see Fig. 1.85c); it is rhythmically layered in structure and mainly represented by plagiogneisses (70% of the apparent thickness of the section), which obviously dominate over amphibolites (30%). By composition and type of the petrogenic oxides (see Fig. 1.81) and REE (see Fig. 1.82), amphibolites correspond to continental tholeiites; plagiogneisses conform to trends of effusive and sedimentary differentiation, and their mean composition is indicative of a primary rock composition corresponding to that of dacites. Typomorphic features of accessory zircon from plagiogneisses with HAM clearly indicate that metaterrigenous strata have been formed as a result of the destruction of the initial dacites (Table 1.35, columns 1 – 3). The primary mantle genesis of the initial tholeiites and dacites is proved by the relatively low values of the primary ratio of $^{87}Sr/^{86}Sr$ (0.7040 in amphibolites and 0.7073 in plagiogneisses).

The polymetallic complex in the section of the Minibayev well No. 20000 is traced at depths from 1.9 to 5.1 km (Fig. 1.86) and characterized by the rhythmically layered structure and even by the more pronounced predominance of aluminiferous gneisses and crystalline schists (96% of the section's apparent thickness) over metabasites (4%). The following two tectonic-metamorphic stages are recognized in the evolutionary history of the complex (Sitdikov et al. 1980): the Archean (granulitic facies, $T = 750° - 800°C$, $P = 0.4 - 0.5$ HPa) and the Early Proterozoic (amphibolite facies, $T = 640° - 660°C$, moderate pressure). Geological-geochemical indications only prove that predominantly aluminiferous (including highly aluminiferous) gneisses and crystallogneisses are of primary-metaterrigenous nature, and the typomorphism of accessory zircon indicates that acid effusive rocks, similar in composition to dacite, served as source rocks for the metaterrigenous deposits (see Table 1.35, columns 4 – 6).

The granulitic-gneiss layer is notable for its uniform structure and rock composition. At the surface of shields it is commonly represented by ancient (more than 3.0 Ga old) monozonal granulitic complexes composed of large rhythmic strata of the hypersthene plagiogneisses and enderbites which alternative with obviously subordinate pyroxene basic crystalline schists.

In the eastern part of the Baltic Shield, the largest relics of the oldest enderbite series are represented by the Kanetiyarvi and Volnavolok complexes (Vinogradov and Vinogradova 1979); Sviridenko 1980). By chemical composition (low potassium content, moderate iron content, relatively high sodium content – see Table 1.35, column 8) enderbites of the first complex are assigned to a particular geochemical type of sodium salts which are named primary crust rocks; analogous rocks are not known in the Phanerozoic sequence. Similar rocks have been encountered in drilling through the crystalline basement of the Russian plate. Well No. 663 penetrated a layer of 1.5 km thick, which is mainly (82%) composed of hypersthene plagiogneisses and enderbites (including biotitized and amphibolitized enderbites); hypersthene and bipyroxene crystalline schists are present in very subordinate (4%) amounts (see Fig. 1.85c). The uniform composition of the hypersthene plagiogneisses present in the section and the very low values of the initial ratio of $^{87}Sr/^{86}Sr$ in these rocks (0.7016) indicate that they are similar to primary-mantle magmatic rocks of dacite composition (see Table 1.35, column 11).

Thus, it appears that metamorphic complexes, representing the granite gneiss and granulite-gneiss layers of the continental crust in the model are characterized by many common features in their structure and composition. The coarse-rhythmical character of strata, the very distinct predominance of plagiogneisses and enderbites of dacite-rhyodacite composition over metabasites, the isochemical regime of early (pregranite) metamorphism and inflow of potassium during subsequent tectonometamorphic (metasomatic) stages are the most characteristic

**Table 1.35.** Mean chemical composition of the Archean plagiogneisses and enderbites which constitute the basement of the eastern part of the Eastern European Platform, and the calculated composition of the primitive continental crust (recalculated for dry matter, %)

| Oxides | Granitic-gneiss layer | | | | | | | Granitic-gneiss layer | | | | | | | | Proto-crust |
|---|---|---|---|---|---|---|---|---|---|---|---|---|---|---|---|---|
| | Baltic Shield | | | Russian Plate | | | Mean value for a layer | Baltic Shield | | | Russian Plate | | | | Mean value for a layer | |
| | 1 | 2 | 3 | 4 | 5 | 6 | 7 | 8 | 9 | 10 | 11 | 12 | 13 | 14 | 15 | 16 |
| $SiO_2$ | 65.05 | 68.18 | 67.27 | 65.27 | 68.34 | 66.68 | 66.94 | 71.29 | 65.49 | 67.23 | 66.16 | 70.74 | 65.18 | 66.23 | 66.45 | 66.10 |
| $TiO_2$ | 0.68 | 0.32 | 0.43 | 0.60 | 0.55 | 0.59 | 0.52 | 0.29 | 0.54 | 0.46 | 0.58 | 0.30 | 0.56 | 0.51 | 0.53 | 0.50 |
| $Al_2O_3$ | 16.55 | 16.86 | 16.77 | 17.19 | 14.37 | 15.91 | 16.28 | 15.55 | 16.47 | 16.19 | 15.30 | 15.78 | 16.94 | 16.71 | 15.98 | 16.20 |
| $Fe_2O_3$ | 1.02 | 0.64 | 0.75 | 1.39 | 1.05 | 1.24 | 1.03 | 0.83 | 1.43 | 1.25 | 1.94 | 0.72 | 1.40 | 1.27 | 1.55 | |
| FeO | 6.17 | 2.17 | 3.33 | 5.63 | 4.87 | 5.28 | 4.44 | 1.59 | 3.90 | 3.21 | 4.60 | 1.60 | 3.33 | 2.90 | 3.73 | 4.50 |
| MnO | 0.07 | 0.03 | 0.04 | 0.08 | 0.07 | 0.07 | 0.06 | 0.03 | 0.10 | 0.08 | 0.06 | 0.03 | 0.08 | 0.07 | 0.07 | 0.08 |
| MgO | 2.33 | 1.41 | 1.68 | 1.99 | 1.57 | 1.79 | 1.74 | 0.83 | 1.87 | 1.56 | 2.10 | 1.76 | 1.91 | 1.88 | 1.89 | 2.20 |
| CaO | 1.88 | 2.54 | 2.35 | 2.75 | 3.59 | 3.12 | 2.79 | 3.39 | 4.59 | 4.23 | 4.91 | 2.98 | 4.65 | 4.32 | 4.55 | 5.10 |
| $Na_2O$ | 3.11 | 5.40 | 4.73 | 2.96 | 3.08 | 3.02 | 3.75 | 5.18 | 3.92 | 4.30 | 3.19 | 4.55 | 4.45 | 4.47 | 3.88 | 4.70 |
| $K_2O$ | 3.05 | 2.38 | 2.58 | 2.07 | 2.33 | 2.18 | 2.35 | 0.92 | 1.69 | 1.46 | 1.05 | 1.74 | 1.50 | 1.50 | 1.30 | 0.62 |
| $P_2O_5$ | 0.08 | 0.07 | 0.07 | 0.07 | 0.18 | 0.12 | 0.10 | 0.11 | n.d. | 0.11 | 0.16 | n.d. | n.d. | n.d. | 0.16 | 0.10 |
| Number of samples | 69 | 92 | 161 | 25 | 25 | 50 | 211 | 6 | 14 | 20 | 13 | 6 | 41 | 47 | 80 | n.d. |
| Portion in the section | 18.7 | 46.2 | 64.9 | 47.4 | 39.1 | 86.5 | 100 | n.d. | n.d. | n.d. | 82.0 | 12.3 | 50.4 | 62.7 | 100 | n.d. |

1 – 3 – biotite plagiogneisses of the Kola series from the SG-3 well (1 – meso and melanocratic, 2 – leucocratic, 3 – weighted average composition); 4 – 6 – rocks of the Cheremshan series from the Minibayev well No. 20000 (4 – granite-sillimanite-biotite crystallogneisses, including highly aluminiferous ones, 5 – hypersthene plagiogneisses, 6 – weighted average composition of plagiogneisses); 7 – mean composition of the plagiogneiss complexes of the granite gneiss layer; 8 – enderbites of the Kannetiyavr massif on the Kola Peninsula (Vinogradov and Vinogradova 1979); 9 – enderbites of the Volnavolok complex, Kareliya (Sviridenko 1980); 10 – mean composition of enderbites of the Baltic Shield; 11 – enderbite-gneisses of the Prikazan series (Nurlat complex) from the Ulyanovsk well No. 663; 12 – 14 – plagiogneisses of the Iksk series from the Tuimaza well No. 2000 (12 – biotite plagiogneisses, 13 – hypersthene plagiogneisses, 14 – weighted average composition); 15 – mean composition of the enderbite-plagiogneiss complexes of the granulitic layer; 16 – calculated composition of the protocrust

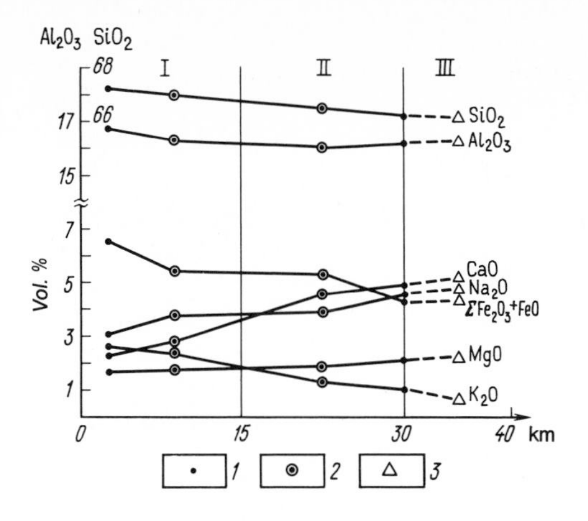


**Fig. 1.86.** Scheme of changes in chemical composition of the enderbite-plagiogneiss complexes along the section of the continental crust. *1* average compositions of metamorphic complexes; *2* average compositions of granite-granulite gneiss layers; *3* extrapolated average composition of the protocrust (*I* granitogneiss layer; *II* granulite-gneiss layer; *III* protocrust)

features of these complexes. Moreover, according to the deep drilling data, a percentage of metabasites tends to decrease in transition from the granite gneiss layer (25 – 40%) to the granulite-gneiss layer (5 – 15%). It is also observed that the composition of the second layer inherits the composition of the plagiogneisses of the first layer when the destroyed products of the second layer are replaced; this is clearly expressed by the decrease in bulk of the metavolcano-plutonic associations of the dacite-tonalite series with the low value of the initial ratio of $^{87}Sr/^{86}Sr = 0.7016$ and metaterrigenous rocks = 0.7240 (5 – 15 km) as depth increases. All these observations, as well as experimental data (Bayuk et al. 1982), indicate the increase of longitudinal elastic wave velocity in acid granulites and enderbites from 5.50 km $s^{-1}$ at normal pressure ($P = 0.1$ GPa) to 6.96 km $s^{-1}$ at higher pressure ($P = 1$ GPa); this allows the conclusion that the primitive crust of continents (third layer in Fig. 1.90a, b) is not basite-hyperbasite but more acid in composition.

For the purpose of determining the chemical composition of the protocrust, the mean contents of the petrogenic elements in the enderbite-plagiogneiss series, predominating in granite gneiss and granulite-gneiss sections, have been calculated (see Table 1.35). In comparing the values obtained, taking into due account the ultimate contents of the elements in these layers of the crust, it becomes evident that there is a tendency to decreasing $SiO_2$, $Fe_2O_3 + FeO$ and $K_2O$ with depth against the background of increasing MgO, CaO and $Na_2O$. The extrapolation of these trends to a depth of 35 km has allowed the calculation of a composition of the primitive continental crust, corresponding to dacite, and the assumption (Kremenetsky and Ovchinnikov 1983) that the primary magmatic melts, which formed the protocrust and then contributed to the formation of volcano-plutonic associations of the granulite-gneiss layer of the crust, were attributed to the sodium series of tholeiite type.

However, the correlation of materials of deep and superdeep wells drilled through ancient metamorphic rocks confirms the theory of the tectonic layering of the lithosphere. In fact, in addition to the SG-3 well, the Minibayev well No. 20,000 has also discovered thick zones of crushed rocks, cataclasis and mylonitization of ultrametamorphic rocks at great depths, which were accompanied by retrograde metamorphism of the greenschist facies. It is interesting that some of these zones are characterized by comparatively gentle angles of dip (60° – 70° with respect to well axis). There is no doubt that these zones are superimposed over the general vertical zonality of the Precambrian metamorphic complexes and formed in conditions of lower temperature gradients than those that existed in the period of prograde metamorphism of the Early Precambrian volcano-sedimentary strata.

The drilling of the Kola superdeep well has revealed extensive new information on the geological structure and mode of occurrence of the Proterozoic and Archean complexes in the north-western part of the Baltic Shield, and made it possible to assess alternative hypotheses on tectonic structures of this territory, elaborate the first trustworthy volumetric (three-dimensional) model of the Precambrian continental crust (with the Pechenga ore-bearing region as a pattern area) and trace its history over a time interval from 3.0 to 1.6 Ga.

According to this model, the formation of the Proterozoic Pechenga structure is linked not with the development of a geosyncline but with the origination and further evolution of the intracontinental mobile belt of riftogenic type. It is characterized by a successive increase of sedimentation depth, an antidromic character of volcanism with the established transition from basalt formations, and by the manifestation (at the late stage) of horizontal displacements of tectonic blocks and progressive zonal metamorphism, superimposed over the Proterozoic and Archean complexes.

Transitions from one stage of the continental Earth's crust development to others were accompanied by a regular change in the ore-forming processes and formation of the polychrome vertical ore zonality. According to the new interpretation, the original nickeliferous intrusions occurred horizontally, but later were dissected into individual fragments as a result of thrust faults. This conclusion extends the prospects of discovering new sulphide copper-nickel deposits at deep horizons of the Pechenga region.

The correlation of the Kola superdeep well section with sections of the Minibayev well No. 20,000, Ulyanovsk well No. 663, Tuimaza well No. 2000 and other wells which entered the basement of the Eastern European platform offered the possibility of making one single geological-geochemical scheme of the whole Archean continental crust structure consisting of three layers, i.e. granite-gneiss (0 – 15 km), granulite-gneiss (15 – 20 km) and the lower (30 – 40 km) layer which represents the protocrust. Extrapolation of the geochemical trends to a depth of 35 km indicates the possibility of the protocrust having been formed at the expense of rocks similar in composition to dacites of the sodium series.

# 2 Geophysics

## 2.1 Geological and Technical Conditions for Logging the Well

The geological and technical conditions for conducting geophysical investigations in the Kola well are essentially different from those encountered in other superdeep wells drilled before in the USSR and abroad. This was the first investigation of a well that penetrated a very thick section of crystalline rocks with extremely high hydrostatic pressure and temperature and a great length of uncased borehole (more than 9000 m). This required the use of a new approach to creating a set of techniques, apparatus, methods and technology of well logging (WL).

The well section is comprised of Lower Proterozoic and Archean rocks of volcano-sedimentary origin–gneisses, aluminium or biotite granites with epidote, amphibolites and schists of various composition, the rocks being subject to microfissuring, lamination and secondary mineral growth. These processes, in the form of metamorphism, microclinization, granitization, muscovitization, chloritization, etc. have led to the alteration of both the initial composition of rocks and the range of their physico-mechanical properties, which affects the geophysical parameters registered. For instance, the rock density varies from 2.56 to 3.15 $g\ cm^{-3}$ over the entire well section, while the velocity of compressional elastic waves are measured on core samples varies within the 2200–6800 $m\ s^{-1}$ range. The electrical resistivity of these rocks varies in a very wide range, from a few tens of ohm-m to $10^9$ ohm-m. However, some of the rocks, although quite different in composition, may have similar physical characteristics owing to the type and degree of secondary alterations.

The rocks penetrated by the Kola well are extremely heterogenous, which considerably complicates the determination of their lithological mineralogical and elementary compositions from well logging data. The problem of lithological determination cannot be solved by the traditional identification of known and studied rocks from a set of log parameters. Great attention should be paid to studying the chemical and mineralogical compositions of the rocks, including recent mineralization, and to analyzing the character and intensity of the destructive processes (microfissuring, lamination, dispersion of secondary minerals) and their effects. The steep dipping of layers crossed by the Kola well further complicates the well-logging investigations.

The differentiation of rocks in the Kola well section with respect to their physico-mechanical properties is governed by differences in their elementary lithological composition and, additionally, to a considerable extent by the stress affecting the rocks. The evaluation of the stress levels and respective zonation of the section, as well as control of the borehole stability with time, are essentially new problems to be solved by the use of well logs for the lengthy open borehole of the Kola well.

For the first time a combination of geophysical and physico-tectonic measurements was available for studying rock stresses. This study involves the dynamic interpretation of faults of various scales, which has helped to provide a three-dimensional picture of tectonic stresses and to determine the alteration of rock stresses caused by drilling in the borehole vicinity. These techniques are described by Nikolaev (1977).

Because of the unusual physico-geological characteristics of the rocks in the Kola well section and the specific technical involvements, the borehole cross-section has a particular shape, which greatly affected both the well casing programme and the WL technology planning. In the depth interval where the Proterozoic rocks of mostly volcano-sedimentary origin occur (2004 – 5000 m), the conditions for WL operations are favourable, while downwards, especially where the Archean granite-gneisses are found, they are much poorer.

Hole enlargements significantly affect the operational possibilities of some well logs and the quality of their information. The complex configuration of the hole cross-section and the presence of vugs from 7000 m down greatly complicates the passage of geophysical tools. The lengthy uncased hole with walls composed of very hard and abrasive rocks causes very intensive wear of the WL cable. To cope with these complications, the geophysical and drilling services had to develop a special technology for the WL operations. The tools were lowered into the relatively short open intervals of investigation via the drill pipe column, which ensured both protection of the cable from wear and a good passage of the WL tools, and helped to avoid accidents.

Some of the WL curves (the nuclear-physical and magnetic logs) were recorded both in the open hole and through drill pipe walls. In the latter case, the influence of the drill pipe string on the measurements had to be taken into account.

## 2.2 Combinations of Well Logs

**Table 2.1.** WL combinations for solving geological problems

| Problem | WL combination |
|---|---|
| 1 | 2 |
| Stratification of the section, determination of lithology and elementary composition | Acoustic (AL), electrical (LL), spectral gamma (SGL), integral gamma (GL), neutron (NGL, NNL-T, NNL-E), pulsed neutron-neutron (PNNL) and magnetic ML) logs |
| Identifying zones of copper-nickel concentration | Electrode potential (EPL), sliding contacts (SCL) and spectral neutron gamma (SNGL) logs |
| Determination of interval and layer velocities of elastic waves and attenuation coefficients | AL with registration of $\Delta t$, amplitudes, phase correlation diagrams and wave trains (PCD and WT) |
| Determining of electrical characteristics of layers and layer sequences | Lateral electrical log (LL) |
| Rock density determination | Gamma-gamma log (GGL) |
| Determining the effective atomic number | Gamma-gamma log (GGL-S) |
| Determining average life-time of thermal neutrons | PNNL |
| Measuring the geomagnetic field components and magnetic susceptibility, determining geometry of magnetic bodies | ML |
| Determining thermal properties of rocks | Temperature log (TL) |
| Evaluation of axial, radial and tangential stresses in rocks and their differentiation with respect to the stress level; identification of fracture zones | AL, GGL, LL |
| Calculation and construction of geophysical, geothermal and geochemical models of the section | AL, NGL, GL, SGL, ML, LL, GGL, TL, vertical seismic profiling (VSP), mud logging (ML) |

*Geological Problems to be Solved* (Table 2.1). The Kola well section has been studied with a variety of WL techniques including seismic, acoustic, electrical, nuclear-physical, magnetic and thermal ones; the drilling mud has been sampled in the deep to study structural-mechanical properties of rocks, their chemical composition and content of gaseous components. Geochemical analysis for de-

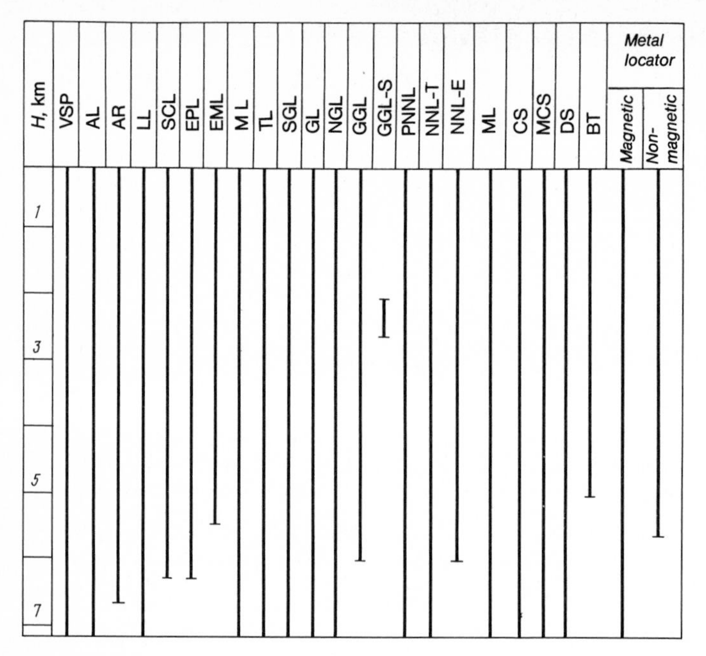


**Fig. 2.1.** WL tools used in the Kola superdeep well at the first stage of drilling. *Seismic-acoustic techniques: VSP* Vertical Seismic Profiling (SSPU-1, SVI-20TS); *AL* acoustic log (SPAK-2, SPAK-D-2, SPAK-4, ZVUK-1,2)[a]. *Electrical surveys: AR* apparent resistivity (7E4, KSP-1, KZ-K); *LL* laterolog (BKR-2, TBK-1, KBK-1,2); *SCL* sliding contact log; *EPL* electrod potential log (sondes): *Electromagnetic logs: EML* electromagnetic log (EML, SM-DT); *ML* microlog (KSM-65SG, MF-3S); *TL* temperature log (TEG-2, T-5, T-7); nuclear-physical methods; *SGL* spectral gamma log (Granat-1,2, SGSL-2); *GL* gamma-log; *NGL* neutron gamma log (DRST-1,2,3, DRST-3D, RKDK-1,2); *GGL* (density) gamma-gamma log (RGP-1); *GGL-S* selective gamma-gamma log; *PNNL* pulsed neutron-neutron log (IGN-4, IGN-D); *NNL-E* neutron-neutron log (thermal); *NNL-E* neutron-neutron log (epithermal) (MNK-KD); *ML* mud log (mud log station). *Technical-technological methods: CS* caliper survey (SKS-4, KSU-1); *MCS* multiarm caliper survey (SKP-1, SKPD-2, SPRUT-2, PTS-2, SPRK-1); *DS* directional survey (ZENIT, KIT-A, IT-200), Metal locator: magnetic (LMM, POT-25, GGOT-50, POLBT-60), non-magnetic (OM-200); *BT* borehole televiewer

[a] Here and elsewhere, the apparatus trade mark is shown in brackets.

termining the percentages of gaseous components in the mud has been carried out to the depth of 11,556 m.

Most attention was paid to studying the elementary lithological composition of rocks and determining the physico-chemical parameters of geological bodies intersected by the well, including the study of stress distribution in the bodies and the calculation and construction of geophysical, geochemical and geothermal models for the Kola well section.

In the first stage of drilling, the well was investigated by relatively more numerous methods, to compare their operational possibilities and select the more successful ones, as well as to solve some specific geological problems, such as the identification of the zones of concentration of copper-nickel sulphide from the EPL and SCL logs (Fig. 2.1).

**Table 2.2.** WL combinations for solving technical-technological problems

| Problem | WL combination |
|---|---|
| 1 | 2 |
| Controlling the three-dimensional configuration of the borehole | Directional survey, magnetic log (LMM – locator of magnetic metal) |
| Controlling the condition of walls of the long open borehole | Multi-arm caliper log, radius measurement, determination of the hole cross-section orientation in space, acoustic log |
| Study of complex zones in detail | Directional survey, multi-arm caliper log with determination of the hole cross-section orientation, magnetic log |
| Identifying and locating metal left in the well | Magnetic log, magnetic metal locator |
| Determining the drill string length | Magnetic log |
| Determination of the drill pipe grade | Acoustic log, spontaneous potential log |
| Determining the ware of removable casing columns | Multi-arm caliper log, induction log, acoustic borehole televiewer (BT) |
| Determining axial deformations of the drill pipe string (while stationary) in the well at any cross-section | Magnetic log |
| Determination of lateral deformations of the drill pipes (while stationary) in the well at any cross-section | Multi-arm caliper log |
| Determination of elevated stress zones in the pipe string in the well (while stationary) | Magnetic and acoustic logs |

An analysis of the results helped to compile an optimum combination of log techniques for the second stage of drilling: the most informative modifications of all the basic geophysical methods were included except for formation density (gamma-gamma) log, because of the too great influence of the rugged hole on measurements made with the side-wall skid tool (Fig. 2.1). The resulting WL combination helped to reduce the logging time and accelerate the drilling.

*Technical and Technological Problems.* The great depths reached by the well have imposed a number of unconventional problems on the geophysicists, such as the evaluation of the long open hole stability, control of the condition of the drill pipe column at high pressure and temperature, etc. Geophysical information was used to optimize drilling practices and find the best ways to avoid or eliminate complications (Table 2.2).

These geophysical surveys supplied the information necessary for drilling operations.

## 2.3 Apparatus, Cable, and Equipment for Logging the Kola Well

At the first stage of drilling, when temperature and hydrostatic pressure (normal drilling muds were used) were relatively low, mainly standard USSR-made logging tools were run in the Kola well. Special instruments, such as *Granat-1,2* were designed for the first time.

**Table 2.3.** Specification of apparatus used to log the Kola well in the second stage of drilling

| Apparatus for | Temperature °C | Pressure MPa | Outer diameter mm |
|---|---|---|---|
| 1 | 2 | 3 | 4 |
| Vertical seismic profiling: | | | |
| ASS-12 | 125 | 100 | 70 |
| Acoustic log | | | |
| SPAK-D-2 | 200 | 150 | 80 |
| SPAK-T-6K | 200 | 170 | 90 |
| Nuclear-physical logs: | | | |
| RKDK-1 | 250 | 150 | 73 |
| RKDK-2 | 250 | 170 | 89 |
| RKDK-3 | 250 | 210 | 89 |
| MNK-KD | 200 | 150 | 90 |
| Granat-2 (SGL) | 200 | 150 | 90 |
| Magnetic log: | | | |
| KSM-65-SG | 200 | 150 | 65 |
| Electric (lateral) focussed log: | | | |
| KBK-2 | 200 | 150 | 73 |
| KBK-3 | 250 | 200 | 90 |
| Electronic thermometer: | | | |
| T-7 | 250 | 150 | 60 |
| Multi-arm caliper: | | | |
| SKPD-2 | 200 | 150 | 73 |
| SKPD-3 | 250 | 200 | 80 |
| Inclinometer: | | | |
| IT-200 | 200 | 150 | 74 |
| KIT-NS (non-standard) | 200 | 170 | 80 |
| Free point indicators and metal locators: | | | |
| POT-25 | 250 | 170 | 25 |
| POT-50 | 250 | 210 | 50 |
| OM-200 | 200 | 170 | 90 |
| Drilling mud samplers: | | | |
| OPN-112 | 200 | 150 | 90 |
| SPS-88 | 200 | 170 | 88 |

The problems in apparatus were not limited to making the available deep well sondes still more temperature- and pressure-proof. It was necessary to construct new instruments that were not available even for medium depths (magnetic logging tool, gamma-spectrometer).

The problem of transmission of reliable information via a very long cable was especially difficult. With increasing cable length and temperature the capacitance and resistivity of cable wires, as well as the leakage of current, also increase, which worsens the power supply to the instruments and weakens the signal transmitted.

Certain difficulties were encountered in adjusting the downhole and wellhead instruments with the long transmitting line, as the cable parameters gradually change on its being lowered into the hole. A number of electronic and engineering improvements had to be introduced to achieve higher accuracy in computing devices, protect the "tracking" systems used, increase the linearity of amplifiers within the dynamic range of the signals, select the sonde length, arrange combination logging tools and stabilize the feeding current, etc.

A general specification of newly constructed instruments and some standard tools (T-7, IT-200) used is shown in Table 2.3.

*Lowering-Hoisting Equipment.* Well-logging and perforating operations were carried out using standard self-propelled hoisters PK-4 and PKS made at the Tuimaza Geophysical Equipment Plant (to a depth of 7943 m) and well-logging winches made at the Gamma Combinate of the Hungarian Peoples' Republic, the winches being of two types: EL-7000 (to 8000 m) and EL-7000 B (to 9800 m).

In the second stage of the Kola well drilling, the conditions for well logging became still more complicated. Abrasion of the cable increased sharply, the friction forces due to the contact of the cable with the hole walls became greater, as well as the danger of the cable or sonde jamming in the open hole key seats (Fig. 2.2).

To protect the cable from wear and wedging in the long open hole with its complex configuration, and to decrease cable tension, technology and technical means were introduced that protected the cable by drill pipes during the hoist-lowering operations (HLO). Prior to running the WL, a drill pipe string is lowered into the wellbore, leaving open only the interval to be logged (700 – 1000 m) above the well bottom. The lower end of the string is equipped with a special funnel.

During HLO the sonde passes through the funnel to the interval under investigation and most of the cable remains inside the drill pipes. Some of the measurements (RL, MK, directional survey) can be made with the sonde remaining within the light metal drill pipes.

All the downhole tools have a diameter smaller than the size of the drill string bottlenecks.

To exclude the possibility of dropping the sonde and cable in the well, a special device called "parachute-stop" is mounted on the cable above its downhole end at a distance greater than the interval to be logged. The diameter of the device is less than that of the pipe bottlenecks, but more than the outlet diameter of the funnel, which stops the cable from passing into the open hole by the exact

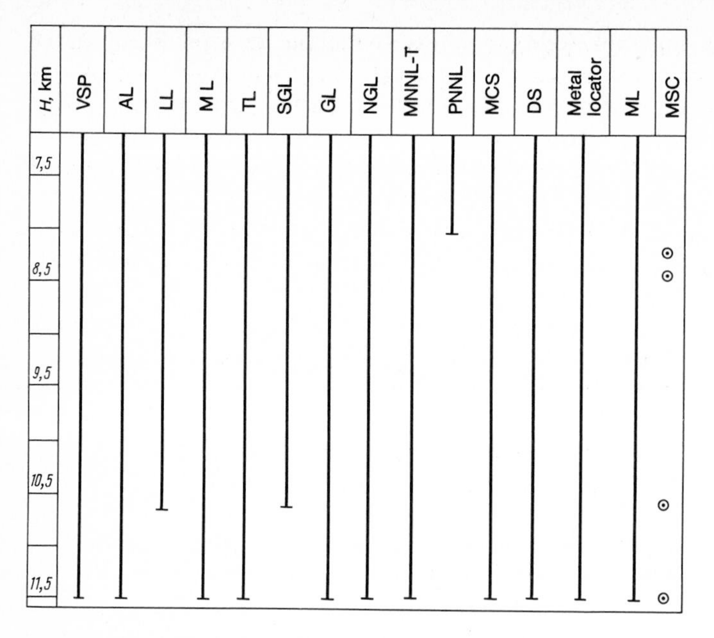


**Fig. 2.2.** WL tools used in the Kola super well at the second stage of drilling. *VSP* (SSPU-1, SVI-20TS); *AL* (SPAKT-6K); *LL* (KBK-2,3); *ML* (KSM-65SG, MF-3S); *TL* (T-5, T-7); *SGL* (Granat 2); *GL* and *NGL* (RKDK-2,3); *MNNL-T* multisonde neutron-neutron log (thermal) (MNK-KD); *PNNL* (IGN-D); *MCS* (SKPD-2,3); *DS* (KIT-A, KIT-NS, IT-200); *metal locator* (LMM, POT-25,50), POLBT-60, OM-200), *ML* mud log station; *MSC* mud sampler lowered with logging scale (OPN-112, SPS-88)

length necessary to do the survey. In the case of cable failure at the most tensed point near the wellhead, all the cable above the parachute-stop will remain inside the pipes, and can be retrieved by pulling out the drill string. A set of parachute-stops and funnels of high-grade steel was used for HLO via drill pipes.

To ease the impact of sondes against pipe joints and smooth their passage so as to avoid the danger of loose cables, the design of the bottom end of the sondes was modified.

The use of several cable sections of varying diameter helped to improve the operation of the cable and the hoisting equipment. The small diameter cable section, together with a sonde attached to it, is lowered to a depth of 6000 – 7000 m and then fixed at the wellhead. The upper end of this section is connected to the next section of the cable, which has a larger diameter. If necessary, some of the cable is wound up to create the desired tension, and then the lowering continues until the sonde reaches the interval to be studied.

HLO were controlled by appropriate sensors and devices to register sudden changes in cable tension (an increase up to 15 – 20 kilonewtons while hoisting, or decrease while lowering), to give optical or acoustical alarm signals to stop the winch smoothly, and to find the cause of any HLO complication and eliminate it.

Magnetic depth marks were put on the cable with the URS-10 device during each hoisting.

To obtain the necessary information on cable behaviour to predict and avoid accidents, a system of control of cable operational characteristics was developed and introduced into WL practice. This includes measuring integral and interval (relative) deformations (both elastic and residual ones) at any point of the cable, its rotation speed, tension and electrical parameters (capacitance, inductivity, resistivity of wires to dc and that of the insulation).

The WL technology described for the Kola well has enable logging operations at greatest depths and has completely excluded complications such as dropping the tools or cable into the well at the second stage of drilling.

## 2.4 WL Operation Procedure

The particular geological and technical conditions in the Kola well necessitated the development of special logging techniques with various WL tools. Such techniques were developed by institutes engaged in devising WL apparatus and methods and by the Kola Experimental-Methodical geophysical crew. The essential elements of these techniques that provide for obtaining informative logs of high quality are described below.

*Determination of Depths for the WL Parameters Registered.* One serious problem encountered in logging the Kola well at depths of 7000 – 8000 m and greater was the determination of accurate depth marks for the logs.

Calculations have shown that the total lengthening of the WL cable $\delta 1$ is caused by at least nine factors, some of which are of physical nature, while others results from errors in measurement.

The calculation of length increments for the standard cable of the KTBF-6 type shows that at the well depth of 11 km its maximum total lengthening can be as great as 19 – 20 m. The probability of length variations of such magnitude necessitates great thoroughness in marking the cable and depth correlation of all the various kinds of logs.

The use of the improved system of cable-length marking and more accurate measurement, as well as the development of some new methods of log-tying, helped considerably to decrease the negative influence of cable length variations.

The well is logged using the same technology of cable marking at each HLO. The technique of successive comparison of log depths is applied, based on mutual correction of the curves registered. First, the data positions are determined on each log, then the depths of all the logs are compared, taking one of the logs as reference for the cycle of investigations. This basic log is usually the one run with a fully stabilized cable. Thus, for example, the results of an electric focussed log compared with the basic log depths can be used as a reference for correcting the depths on the magnetic log and other curves, and consulting the basic log later.

Actual data on cable length increments measured at each HLO were used in depth correction.

*Calibration and Preparing the WL Tools for Operation.* All the WL apparatus used in the well was calibrated according to standard meteorological requirements and checked in the Sputnik SG-3 calibration well and in the test pit.

The procedure used for preparing and tuning the instruments provided quality data. It should be noted that the preparation proved to be time-consuming, which has to be taken into account when planning WL service for superdeep wells.

*Acoustic Log (AL).* The AL apparatus was inspected and calibrated using the field acoustic calibration device (PAUK) and standard meteorological metallic pipes.

The AL was run to a depth of 11,500 m. The information was presented in the form of kinematic and dynamic parameters of compressional and shear waves. The processing and interpretation of the AL data were made taking due account of signal shape and phase correlation results.

Log quality decreases sharply near the well bottom. This is possibly because the high concentration of descending cuttings in the drilling mud absorbs the acoustic energy. After deepening the well and running the AL in the same interval, logs of improved resolution are obtained, provided there are no vugs in the walls. In intervals where vugs filled with thick mud are present, the log quality does not improve; but if such intervals with large cavities are well washed with fresh drilling mud, analogous curves of acceptable quality can be registered.

Generally speaking, the variations of AL quality with depth fully correspond to those of the seismic-geological characteristics of the Kola well section. Taking into account the results of repeated measurements run after prolonged washing to expel the cuttings, the average coverage of the AL for the section investigated is about 70%. For some intervals of the section, the missing kinematic parameters were reconstructed from phase correlation data.

The BT modification of AL was also used to evaluate the technical condition of the open hole and removable casing columns, as well as to study stress levels of the drill pipe string hung free in the borehole.

*Gamma Ray Log (GL).* The GL measurement channel of the apparatus was calibrated with a reference $^{137}Cs$ source. Before and after each run, the channel sensitivity was checked with the special source and the natural background was recorded. To evaluate the operational stability of the apparatus and to control the log quality, repeated measurements were run in the 4673 – 4784-m interval where the reference andesitic porphyrite rocks occurred.

Natural radioactivity was measured in the 1 : 500 and 1 : 200 depth scale both while lowering and hoisting the down-hole tool.

GL was recorded to a depth of 11,440 m, of which the interval from 9300 to 9820 is represented by an open-hole curve. The measurements are noticeably influenced by the drilling mud, the gamma-ray intensity increasing in vuggy intervals. When the GL is run through a drill pipe, this influence decreases by 13 – 16%.

*Spectral Gamma Log (SGL).* The linearity of the spectral gamma-ray registration circuits was inspected; the registration scale was determined by exposing the sonde detectors to radiation from a set of monochromatic gamma-ray sources.

The SGL was run in each 100 – 500-m interval with repeated control measurements while hoisting the sonde. Integral gamma-activity was logged with depth scales 1 : 500 and 1 : 200, and the curves of concentration of $^{40}K$, RaC′ and ThC″ were recorded in the 1 : 200 scale. Energy windows were adjusted so as to exclude the contribution of each isotope to the relatively "hard" domain of the gamma-spectrum.

The SGL readings are affected by the drilling mud in the same way as explained above for GL.

The spectral gamma logs are sufficiently effective enough to evaluate the elementary lithological composition of rocks from the $^{40}$K, RaC′ and ThC″ values.

*Gamma-Gamma Log for Density Evaluation (GGL).* These measurements were performed using two different source-detector spacings (15 and 35 cm) down to the depth of 6000 m. At greater depths where the borehole walls are extremely vugged and the influence of the vugs on the log readings becomes excessive, the GGL was not run at all.

*Selective Gamma-Gamma Log (GGL-S).* This log was run down to 3000 m depth and used in determination of effective atomic numbers for rocks in the intervals containing ore minerals. Later the GGL-S was not included in the log combination.

*Neutron Gamma Log (NGL).* The calibration of the NGL registration channel consisted of determining the relation between the captured gamma-radiation and neutron dose for fixed source-detector spacing, adjusting the channel linearity and defining the recalculation coefficient. The count rate at the highest hydrogen content was taken as a relative unit. The sensitivity of the channel was controlled like that of the GL channel.

The procedure for NGL measurement was the same as for the GL. To correct the NGL readings for pipe joints, the position of the drill pipe string was intentionally changed (by half a pipe length) at each successive run. The NGL was recorded within the drill pipes to the depth of 11,440 m.

NGL readings are considerably influenced by the hole wall cavities, a sharp decrease being noted when the cavity diameter exceeds 40 cm. The drill pipes generally attenuate the influence of the drilling mud by thinning its layer between the sonde and hole walls. However, variations in pipe diameter and wall thickness and the presence of pipe joints and fallen pipes also influence the readings, as was verified by repeated measurements.

*Neutron-Neutron Log (NNL).* Calibration and sensitivity adjustment of the NNL measuring channel were similar to those of NGL.

NNL in the modification of thermal neutrons was run through drill pipes using two source-detector spacings (30 – 50 and 40 – 60 cm) to a depth of 11,510 m. The readings were influenced by the same factors as in the case of NGL.

The pulsed neutron methods, PNNL (double spacing modification, 37 and 87 cm) and PNGL, were run to depths of 8000 and 2000 m, respectively. Errors in parameter values registered at different time gates are also caused by the borehole vugs and drill pipe constructional features.

Information supplied by the neutron logs was used in lithological determination.

*Magnetic Log (ML).* The ML tool was calibrated using the inclinometer table and Helmholtz coil. Prior to magnetic susceptibility measurements in the well, a special calibration device was put against the downhole tool housing to measure the signal level. This operation was repeated after the well investigations were completed.

The ML data were obtained at both lowering and hoisting of the cable in the 1 : 500 and 1 : 200 depth scales. Scales for the parameter registered are as follows: vertical component of the geomagnetic field $Z_a = 100$ nT cm$^{-1}$, horizontal component $H_x - H_y = 1300$ nT cm$^{-1}$, magnetic susceptibility $\varkappa = 300 \times 10^{-5}$ SI units.

The ML was run to the 11,500 m depth. The results were used in the determination of lithological and elementary composition and to study the technical condition of the well. The main purpose was to investigate the variations in magnetite concentration with well depth.

*Electric Log.* The open hole of the Kola well was investigated with various electric logs to a depth of 10,644 m.

The sliding contact (SCL) and electrode potential (EPL) logs run in the interval 0 – 6000 m have proved to be effective for identifying conductive zones including rocks with copper-nickel sulphide minerals. Further on, the well was studied by means of focussed electrical log (Laterolog, LL).

The functioning of the LL apparatus was checked in calibration wells and the recording characteristics were also determined.

The LL curves were registered both on lowering and hoisting the downhole tool in the 1 : 500 and 1 : 200 depth scales.

As the rock electrical resistivities are extremely high, LL is able to characterize only qualitatively of the Kola well geoelectrical section. Among the factors influencing the readings, one should note the large cavities filled with drilling mud.

The ratio of apparent resistivity, $\varrho_a$, to the resistivity of the drilling mud can exceed $10^9$, the average value being $10^3 - 10^4$. The influence of the mud on the LL readings is minimum when the hole diameter is rated or the key seats (if present) are well washed.

*Temperature Measurements.* Inspection and calibration of electric thermometers were carried out with the standard calibration device TS-24 prior to each run.

The temperature was measured in the well both on lowering and hoisting. Registration scales were 0.25° and 0.5 °C cm$^{-1}$ (temperature log) and 0.05 °C cm$^{-1}$ for a differential temperature survey.

Temperature measurements were made to a depth of 11,503 m, under both steady-state and transient conditions. Temperature was also measured with self-reacting instruments during the drill string running in and at mud circulation to the depth of 10,905 m.

Among characteristic features of the geothermal field of the well should be mentioned the thermal anomalies that complicate the static field near the well bottom and in intervals where large cavities are present. The near-bottom anomalies are small when the cavities are minor or completely absent and disappear after thorough washing and elimination of cuttings. If the near-bottom zone includes major cavities, the thermal anomalies are observed for a much longer period of time.

*Vertical Seismic Profiling (VSP).* A VSP survey was carried out when the Kola well had reached a depth of 11,514 m. It supplied information on the velocity distribution along the well section.

*Multi-Arm Caliper Survey.* Before each run the four-arm caliper was calibrated with a cross-shaped device. The stability of the calibration was checked as the tool was lowered.

The four-arm caliper logs were registered using the 1:5 depth scale in the open hole and 1:2 depth scale while inside the drill pipes. The survey was run to the depth of 11,300 m (the well bottom being at 11,514 m). The information on area and shape of the vugs and their orientation was used in evaluating the condition of the open well hole. The results of such evaluations served as a basis for undertaking the technical and technological measures necessary to keep the borehole in a good shape.

The four-arm caliper log registration became complicated due to cable rotation. On some logs the two curves periodically cross each other every 7 – 8 m, the rotation being confirmed by tool orientation sensors. Moreover, this leads to poor repeatability of successive measurements. Use of a freely rotating cable head, VKS-D, helped to avoid the caliper rotation in the open hole.

*Directional Survey.* Inclinometers were tuned and calibrated with the inclinometer table. Instrument operation was checked up at control points during lowering into the hole.

Inclinometer measurements to a depth of 7700 m were carried out in a conventional way in the open (uncased) hole during tool hoisting. Below that depth the measurements were made exclusively during tool lowering within a drill pipe string whose bottom sections are composed of light metal jointless pipes. The bottom pipe of the string was equipped with two constant magnets for determining the depth of the drilling tool with a magnetic metal sensor (coupled with the inclinometer).

A directional survey was run to the 11,500 m depth. The data obtained were used to control the azimuth and inclination of the borehole by applying the appropriate technology of well drift regulation.

*Location of Magnetic Material (LMM).* LMM measurements were conducted both in open hole and drill pipe string at the tool movement speed of 1000 – 2500 m $h^{-1}$. Maximum sensitivity was established when locating the metal in the hole walls, and minimum when taking measurements within the drill pipes.

The magnet locator proved to be a very effective tool for solving many problems. Identification and accurate location of magnetic material in the hole wall were successfully achieved, as well as the prediction of its possible shifts along the well. The length of the drill pipe string was effectively controlled. The non-magnetic metal locator helped to identify and accurately locate metals and differentiate them with respect to their magnetic properties.

LMM data were used to determine the actual length increment of the drill pipe string and evaluate its axial deformations in some intervals of the well. The LMM measurements helped to determine the depths of magnetic data horizons used in comparing the depths on other logs and making perforating and other shooting operations. The elastic deformations of the well logging cable were also studied with LMM.

*Control of the Borehole Position in Three Dimensions.* The WL tools were used successfully in solving many technological problems of drilling.

The history of Kola well drilling and the strict requirements imposed on borehole verticality have created the necessity of systematic routine determination of the 3-D position of the borehole. The data thus obtained included the information necessary for various practical technological situations and even helped to predict borehole deviations more reliably, as well as to judge how trustworthy the initial data were. Isometric images of the borehole were made for its total length (0 – 11.5 km) and for complicated zones these images were shown in more detail, together with the borehole shape.

*The Condition of the Borehole Walls.* Repetition of four-arm caliper logs showed that the borehole in general was stable in time. This is proved by the absence of filtration, well known in the low-porosity crystalline rocks.

*Perforating and Shooting Operations.* These operations were performed in the Kola well mainly to release the jammed bottom sections of the drill column.

Whenever the fact of the drill column jamming became evident, geophysical investigations would be carried out including drill tool gauging with simultaneous LMM measurements. The depth of the jammed pipe is read from the LMM log recorded during rotation of the drill tool under strain. If the jammed point does not appear to be within the interval investigated, the conclusion is reached that the bottom-hole drill assembly is jammed.

Such accidents were eliminated by shaking the drill pipe tool by the explosion of a cordite charge. In some cases, simultaneously with the explosion, the drill pipe string was pulled and rotated both clockwise and counter-clockwise.

In the case of a drill string breaking, the upper end of the part left in the well was located with WL tools. For instance, in the interval of 10,450 and 10,850 m, the combination of logs run with this purpose was as follows: inclinimeter, four-arm caliper, magnetic log and magnetic metal detector.

If necessary, pipe unscrewing was stimulated by exploding the TDSh charge of DShTT-80 cordite. The lowering of the charge into the hole and its ignition by electric current were controlled with the free point indicator POTT-50. The housing of this downhole tool was equipped with a washer 60 mm in diameter to keep it from passing through the fishing apparatus. The length of the bridle with the cordite fastened to it was adjusted so that the charge would be positioned against the joint to be unscrewed when the free point indicator had stopped.

## 2.5 Further Investigations and Improvements in Apparatus and Technology of Superdeep Well Logging

The analysis of geological and technical conditions for WL operations to a depth of 11,514 m helps to predict more reliably the environments to be encountered in the Kola well at greater depths and in other superdeep wells which are to be drilled in massive crystalline rocks to a depth of about 15,000 m. The experience already accumulated helps to estimate the specific features of superdeep drilling technology, propose ways of solving the problems involved and direct scientific investigations so as to ensure normal WL operations to a depth of 15,000 m.

*Prediction of Geological and Technical WL Environments.* Considerable difficulties are expected for WL operation at these depths because of high temperature and pressure. According to a linear extrapolation of the geothermal gradients observed thus far (about 0.019 °C $m^{-1}$), the temperature in the Kola well at 15,000 m depth will be 266 °C. The hydrostatic pressure of drilling mud at this depth can be evaluated from its most probable density of 1.2 – 1.22 g $cm^{-3}$. It is possible, however, that the density of the drilling mud will be different, which should be kept in mind when operating WL equipment.

Considering the further deepening of the well, significant changes in geological and technological conditions of drilling are not to be expected, including the properties of rocks and borehole behaviour known from the previous history of the Kola well. The WL technology is thus not likely to change, and one should be prepared to run the logging via drill pipes or within them.

The most important moment in planning the HLO system is to envisage the possibility of using cable sections of varying diameter and two hoisters, one of them working as an auxiliary winder.

Determination of log depths should be made to include systematically measured cable lengthening.

*Downhole Logging Tools.* To protect the electronics of downhole tools from the expected temperature and pressure, as well as from vibrations and shocks, their housings should be made of superstrong steel. The parts and materials inside the WL sondes must withstand the simultaneous action of high temperature (250° – 270 °C), vibration and shocks. Some of the devices and materials must be highly resistant to drilling mud chemical and must withstand pressure as high as 185 – 190 MPa. All the materials and devices must maintain their characteristics under the repeated action of hostile factors.

To provide for the necessary heat resistance when planning and making the downhole apparatus, it is advisable to proceed as follows: firstly, to use the construction parts and devices available from the local electronic industry with sufficient heat resistance potential; secondly, to employ new materials and technological products. In some cases, if technically possible, passive heat protection systems (downhole thermostats) may be utilized if the limitations imposed on thermostat size and the duration of operation of integral circuits and individual parts can be overcome.

*Transmission of Information.* Analysis of data on the operational history of the cable helps to evalute its possibilities for functioning to 15,000 m in depth. Calculation of electric parameters of the cable of known length and lowering depth, together with experimental data, makes it possible to evaluate the expected inductivity, capacitance, electrical resistivity of wires and isolation of future cable.

*Main Directions for Scientific Research in the Field of Development of WL Equipment for Superdeep Wells.* The superdeep drilling developed in the USSR has noticeably stimulated progress in WL technology. On the basis of the experience gained at the Kola well and the evaluation of geological and technical conditions to be expected in future, the main directions for further scientific research and engineering efforts can be listed as follows:

- improving heat and pressure resistance and operational duration of downhole tools in extreme environments by using the available technological products and new devising items;
- perfection of wireless transmission systems (autonomous instruments installed in the drill pipe string, determination of geophysical and geochemical properties of drilling mud);
- devising multi-wire cable able to withstand high tension forces (the upper section) and elevated temperatures (lower sections);
- construction of powerful hoisting and lowering equipment;
- development of optimum HLO technology to minimize the forces opposing cable movement along the well;
- development of a control system for adjusting WL technological parameters;
- creation of technical means and systems of geophysical control for superdeep drilling.

All these complex problems can be successfully solved by the scientists and technicians within the USSR.

## 2.6 Rock Density, Porosity, and Permeability

A knowledge of the density of rocks and its distribution is indispensable for interpretation of gravity data. Porosity and permeability are closely related to density, therefore all these properties are discussed together.

The density of rock composition was studied on core samples taken from the well (over 46,000 samples) and derived from gamma-gamma and borehole gravimeter logs. For the samples, the density was measured using the method of hydrostatic weighing with a counter-balance, the accuracy being at least 0.01 g $cm^{-3}$.

Total porosity $\phi$, open porosity $\phi_0$ and permeability $K_p$ were studied using standard core analysis techniques: the accuracy of porosity determinations being at least 0.1%. Generalized results are shown in Table 2.4.

Regularities in variations in rock properties in the Kola well section were analyzed using the following criteria: petrographical and mineralogical (including petrochemical), i.e. classifying the rocks according to their elementary composition; structure and texture; the degree of metamorphism and ultrametamorphism; and depth of occurrence.

An analysis of the data obtained shows that rock density is controlled mainly by its mineralogical (elementary) composition and depends considerably less on porosity and fracturing.

**Table 2.4.** Density and permeability of rocks penetrated by the Kola well

| Rocks | Density, g $cm^{-3}$ | Porosity (open), % | Permeability ($mcm^2$) |
|---|---|---|---|
| *Proterozoic, depth interval 0–4500 m* | | | |
| Magmatic (diabase, gabbro-diabase, pyroxene porphyrite, and similar) | 3.0 | 0.4 | 0.03 |
| Volcano-sedimentary and sedimentary (metatuff, metatuffite, volcano-breccia, phyllite, aleurolite, etc.) | 2.9 | 0.45 | 0.044 |
| *Proterozoic, depth interval 4500–6835 m* | | | |
| Magmatic (metadiabase, meta-andesite, etc.) | 2.89 | 0.6 | 0.29 |
| Volcano-sedimentary (various schists) | 2.78 | 0.55 | 1.252 |
| *Archean, depth interval 6835–10,500 m* | | | |
| Gneiss | 2.69 | 1.19 | 16.905 |
| Amphibolite | 2.93 | – | – |
| Ultrametamorphite | 2.98 | – | – |

Mean density weight: thickness calculated for the entire section of the Pechenga (Proterozoic) volcano-sedimentary sequence is 2.95 g cm$^{-3}$, the highest values (2.9 – 3.01 g cm$^{-1}$) being observed in its upper portion above the 4.673 m depth where prehnite-pumpellyite and greenschist facies of metamorphic volcanites and subvolcanic rocks of tholeiite-basaltic and picrite-basaltic composition occur, as well as volvano-sedimentary bodies differently enriched in sulphide minerals. The difference between maximum and minimum average values (for petrophysical intervals) is 0.6 – 0.7 g cm$^{-3}$. The maximum values are characteristic of massive fine-grain pyroxene-porphyrites occurring at the section top (depth intervals 351 – 356.5, 446 – 457, 576 – 584 m). Their average density is 3.11 ± 0.12 g cm$^{-3}$. Minimum densities are observed at the bottom of the section (e.g. $\sigma = 2.61 \pm 0.1$ g cm$^{-3}$ for biotite porphyritic muscovitized granites in depth intervals 9489 – 9499 and 9558 – 9605 m), the minimum values often associating with zones of tectonic faults. Weighted mean densities for mylonitized Proterozoic rocks, sedimentary rocks and tuff-tuffites are 2.88, 2.89 and 2.92 g cm$^{-3}$, respectively. As microscopic analysis of the Pechenga rocks has shown, the higher density values of sedimentary bodies is explained by the presence of ore minerals in significant quantities. These are mainly pyrite and pyrrhotite with relatively high mineral density ($\delta$).

The igneous rocks, as expected, are characterized by the greatest density, 3.01 g cm$^{-3}$.

Density studies have revealed hitherto unknown sharp changes in rock density in the Middle Proterozoic section (the bottom bed of the Zapolyarni series $zp_1$) at about 4500 m in depth and at the border between the Proterozoic and Archean (6842 m). The first interface is associated with manifestation of dislocation metamorphism (V. I. Kazanski et al. 1980), the second one being explained by a change in rock composition. Consequently, the study of laws in density variation helps to identify the boundary between the Proterozoic and Archean, and moreover, reveals an interface within the Lower Proterozoic, thus subdividing the latter (at least in this particular area of the Pechenga synclinorium) into two parts: the upper massive one composed of the Matertin, Zhdanov and Zapolyarnin (excluding the $zp_1$) series, and the lower schistose one (the Luchlompol, Pirttiyarvin, Kuvernerinyok, Mayarvin and Televin series).

The generalized stratification of the Kola well section based on density values consists of three layers (Table 2.4): from 0 to 4500 m of depth there are massive low-porosity diabases, gabbro-diabases, volcano-sedimentary and sedimentary rocks ($\sigma = 2.95$ g cm$^{-3}$) showing the lowest porosity and permeability values of the section; from 4500 to 6842 m are schistose volcano-sedimentary bodies, less dense and more porous and permeable than the rocks of the first interval; and below 6842 m there occur Archean rocks showing (except for the amphibolites) the lowest density and highest porosity and permeability of the whole section.

The weighted averaged density for the part of the Kola amphibolite-migmatite-gneiss sequence of Archean age studied (including ortho-amphibolites of possibly younger age) is 2.73 g cm$^{-3}$. The Archean section is represented by the alternation of contrast (in elementary composition) rocks of epidote-amphibolite and amphibolite metamorphic facies, which is responsible for the greater density variations as compared to the Proterozoic.

The apodiabase amphibolites (with the exception of hornblende schists and iron-silicate rocks with magnetite rarely encountered in the section) have a maximum density of 3.06 g $cm^{-3}$. Higher densities are typical for leucocratic amphibolites (partly with cummingtonite), 2.92 g $cm^{-3}$. The thickness of bodies made of these amphibolites varies as a rule from a few meters to 20 – 30 m, the density variations being negligible (the mean square root deviation not exceeding 0.03 g $cm^{-3}$). The group of apodiabase amphibolites corresponds to the tholeite-basalt petrochemical type, that of leucocratic amphibolites to the andesite-basalt one.

The biotite and two-mica gneisses with a variable percentage of minerals rich in aluminium form a separate group. In the depth interval 6842 – 7622 m, the weighted average density for this group is 2.77 g $cm^{-3}$, the group itself being easily subdividable into three subgroups with mean $\sigma$ values as follows: 2.71, 2.77 and 2.84 g $cm^{-3}$. The two-mica gneiss, together with fibrolites that occur in intervals 9456 – 9541 and 10,144 – 10,600 m have the mean density of 2.72 g $cm^{-3}$ at a very small dispersion.

The density of nebulitic migmatites associated with various gneisses and amphibolites is 2.64 g $cm^{-3}$, or 2.62 g $cm^{-3}$ if the microclinization is intensive. These rocks are the most common of the Archean sequence. Among the Archean rocks some relicts of epidote-biotite and epidote-biotite-amphibolite gneisses ($\sigma = 2.71 - 2.78$ g $cm^{-3}$) and schists (2.87 – 2.97 g $cm^{-3}$) are encountered, as well as those of amphibolites (3.00 – 3.02 g $cm^{-3}$). All these rocks are characterized by high density dispersion and gradual transition from one rock type to another, which is well seen on detail sections (1 : 500).

The transition from prehnite-pumpellyite to greenschist facies (the rock chemical composition being the same) is accompanied by an increase in bulk density (0.05 – 0.06 g $cm^{-3}$) and also in mineral density (0.1 – 0.14 g $cm^{-3}$) (Table 2.5).

From the fact that the bulk density of the crystalline (Precambrian) massive rocks corresponds roughly to their mineral density, it follows that metamorphism has been proved to play an important role in density distribution. Where the well-developed greenschist facies is associated with zones of crushing and mylonitization, the density noticeably decreases, primarily owing to decompaction of the rocks (especially amphibolites) that undergo blastocataclasis and blastomylonitization accompanied by chloritization.

Bulk and open porosity and their interrelation, as well as mineral densities for the core taken from the well, are shown in Tables 2.5 and 2.6. Generally speaking, there is a steplike increase of $\phi$ and $\phi_0$ downwards from depths of 4.5 and 6.8 – 6.9 km (the Proterozoic-Archean interface), respectively.

The rock porosity is mainly controlled by the following factors: depth of the core, rock structure, texture and grade of metamorphism.

Analysis of the $\phi - \phi_0$ interrelation for various rock types shows that the core porosity increase with depth is accompanied by alteration in the pore structure: instead of isolated pores, there appear pore channels and microfractures. In cores taken from depths exceeding 7600 – 7700 m, in addition to the decompressional microfracturing there are evidences of breakage fracturing, which possibly takes place at the well bottom under the action of the drill bit. The samples with

**Table 2.5.** Porosity and mineral density for some rocks from the Kola well section

| Rock | Depth, m | $\phi$ | $\phi_0$ | $\sigma$ | $\delta$ | FG[a] |
|---|---|---|---|---|---|---|
| | | % | | g cm$^{-3}$ | | |
| 1 | 2 | 3 | 4 | 5 | 6 | 7 |
| Diabase | 830.3 | 0.7 | 0.1 | 2.99 | 3.01 | 0 |
| Poikilophytic diabase | 860.8 | 0.5 | 0.2 | 3.09 | 3.10 | 0 |
| Phyllite | 1135.2 | 0.7 | 0.2 | 2.90 | 2.92 | 0 |
| Tuffite | 1142.2 | 1.2 | 0.4 | 2.90 | 2.93 | 0 |
| Blastocataclastic gabbrodiabase | 2318.7 | 0.6 | 0.2 | 2.77 | 2.78 | 0 |
| Metasandstone | 2413.0 | 0.8 | 0.2 | 2.80 | 2.82 | 0 |
| Metadiabase | 3281.8 | 0.6 | 0.3 | 2.91 | 2.93 | 0 |
| Magnetite-sericite schist | 4858.2 | 2.8 | 0.4 | 2.97 | 3.05 | 0 |
| Magnetite-biotite-plagioclase schist | 5111.6 | 2.8 | 0.4 | 2.77 | 2.85 | 0 |
| Amphibolite | 6952.9 | 2.0 | 1.1 | 3.07 | 3.12 | 0 |
| Staurolite-biotite gneiss | 7193.5 | 2.1 | 0.9 | 2.84 | 2.90 | 0 |
| Biotite-epidotic gneiss | 7758.9 | 2.6 | 2.4 | 2.77 | 2.85 | min |
| Nebulitic migmatite | 8118.6 | 3.2 | 1.8 | 2.60 | 2.69 | min |
| Amphibolite | 8446.6 | 2.1 | 0.9 | 3.06 | 3.13 | 0 |
| Nebulitic migmatite | 8625.1 | 2.6 | 1.6 | 2.65 | 2.71 | min |
| Amphibolite | 8733.8 | 2.6 | 1.2 | 3.09 | 3.14 | 0 |
| Nebulitic migmatite | 8811.8 | 2.2 | 0.9 | 2.62 | 2.68 | min |
| Amphibolite | 8939.1 | 2.7 | 0.9 | 3.03 | 3.11 | 0 |
| Nebulitic migmatite | 9050.8 | 2.5 | 1.7 | 2.65 | 2.71 | min |
| Amphibolite | 9091.3 | 2.7 | 2.2 | 3.02 | 3.09 | min |
| Nebulitic migmatite | 9220.5 | 3.5 | 2.4 | 2.62 | 2.72 | max |
| Nebulitic migmatite | 9317.2 | 3.1 | 2.7 | 2.64 | 2.72 | max |
| Nebulitic migmatite | 9418.3 | 2.6 | 1.9 | 2.61 | 2.68 | min |
| Amphibolite | 9921.2 | 3.4 | 0.7 | 2.99 | 3.09 | 0 |
| Amphibolite | 9928.6 | 2.2 | 1.3 | 3.08 | 3.15 | min |
| Blastocataclastic amphibolite | 10035.7 | 1.6 | 0.6 | 3.06 | 3.11 | 0 |
| Blastocataclastic amphibolite | 10043.9 | 1.6 | 0.8 | 2.71 | 2.76 | 0 |
| Amphibolite | 10053.8 | 4.3 | 2.8 | 3.00 | 3.13 | max |
| Biotite gneiss | 10122.5 | 2.5 | 1.8 | 2.70 | 2.77 | 0 |
| Nebulitic migmatite | 10124.0 | 4.8 | 4.6 | 2.59 | 2.72 | max |
| Two-mica gneiss with fibrolite | 10158.8 | 4.5 | 4.4 | 2.63 | 2.76 | max |
| Two-mica gneiss with fibrolite | 10485.3 | 8.2 | 1.9 | 2.69 | 2.37 | 0 |

[a] FG – Fracture Grade for open fractures of a technogenous-decompressional nature according to the three-grade classification: 0, minimum, maximum.

maximum concentration of such fractures have the highest $\phi$ values (3.5 – 4.8%), their density decreasing by about 0.04 g cm$^{-3}$. Porosity of the crystalline rocks as measured at normal $P-T$ conditions is a little higher for rocks of higher grades of metamorphism.

The influence of elevated pressures on rock density is illustrated by Table 2.7, and that of uniform and non-uniform 3-D compression on porosity and permeability by Fig. 2.3.

Analysis of the above data shows that under stresses imitating geostatic pressure, the rock porosity (pore space) and permeability decrease, although the lat-

**Table 2.6.** Permeability ($K_p$) of rocks penetrated by the Kola well

| Sampling depth, m | Rock | Open porosity, % | $K_p \times 10^3 (\mu m^2)$ at pressure (MPa) | | | |
|---|---|---|---|---|---|---|
| | | | 0.01 | 10 | 50 | 100 |
| 3786 | Actinolitic diabase | 1.2 | 0.40 | 0.26 | 0.14 | – |
| 5492 | Albitophyre | – | 0.34 | 0.13 | 0.06 | 0.04 |
| 6140 | Blastoclasite | 2.5 | 0.46 | 0.24 | 0.09 | 0.05 |
| 6300 | Amphibole-plagioclase schist | 2.45 | 0.43 | 0.25 | 0.12 | 0.07 |
| 6325 | Amphibole-plagioclase schist | 1.9 | 0.63 | 0.38 | 0.16 | 0.10 |
| 6621 | Amphibole-plagioclase schist | 1.89 | 0.51 | 0.37 | 0.14 | 0.05 |
| 6371 | Amphibole-plagioclase chlorite schist | 3.17 | 0.76 | 0.46 | 0.12 | 0.05 |
| 6462 | Amphibole-plagioclase chlorite schist | 1.9 | 0.062 | 0.036 | 0.011 | 0.008 |
| 6621 | Amphibole-plagioclase chlorite schist | 4.01 | 0.085 | 0.039 | 0.024 | 0.002 |
| 6739 | Amphibole-plagioclase chlorite schist | 4.19 | 0.047 | 0.026 | 0.019 | 0.008 |
| 6742 | Amphibole-plagioclase chlorite schist | 1.99 | 0.138 | 0.048 | 0.035 | 0.009 |
| 7120 | Staurolite-muscovite-biotite-quartz-plagioclase | 4.9 | 0.022 | 0.010 | 0.004 | – |
| 8710 | Amphibolite | 1.92 | 1.53 | 0.91 | 0.14 | 0.08 |
| 8759 | Biotite-plagioclase gneiss | 4.31 | 6.95 | 3.11 | 1.06 | 0.90 |
| 9003 | Biotite-plagioclase gneiss | 3.18 | 1.51 | 0.88 | 0.11 | 0.06 |
| 9060 | Biotite-plagioclase gneiss | 1.5 | 2.22 | 0.92 | 0.18 | 0.14 |
| 9905 | Amphibolite | 1.94 | 1.59 | 0.86 | 0.25 | 0.20 |

**Table 2.7.** Influence of three-dimensional pressure on mean density ($\sigma_m$) of rocks (g cm$^{-3}$) penetrated by the Kola well

| Depth of sampling, m | Rock | $\sigma_m$ at pressure, MPa | | | |
|---|---|---|---|---|---|
| | | 0.01 | 100 | 200 | 1200 |
| 1 | 2 | 3 | 4 | 5 | 6 |
| 290 – 1000 | Diabase | 3.02 | 3.03 | – | – |
| 2140 | Phyllite | 2.75 | 2.79 | 2.82 | 2.84 |
| 2276 – 4534 | Actinolitic diabase | 2.99 | 3.01 | 3.02 | 3.07 |
| 4640 | Quartz-carbonate-chlorite schist | 2.86 | 2.88 | 2.89 | 2.93 |
| 4742 | Andesite-porphyrite | 2.80 | 2.82 | 2.83 | 2.88 |
| 4844 – 4867 | Arkose sandstone | 2.69 | 2.71 | 2.72 | 2.74 |
| 5540 – 5577 | Amphibole-plagioclase schist | 2.88 | 2.90 | 2.91 | – |
| 5647 – 5688 | Dolomite | 2.77 | 2.80 | 2.81 | 2.84 |
| 5691 | Carbonate-mica-quartz schist | 2.69 | 2.70 | 2.71 | – |
| 5928 – 6337 | Amphibole schist | 2.88 | 2.90 | 2.91 | 2.93 |
| 6329 – 6822 | Amphibole-plagioclase schist with chlorite or biotite | 2.77 | 2.80 | 2.81 | 2.83 |
| 6841 – 9211 | Biotite-plagioclase gneiss | 2.64 | 2.69 | 2.70 | 2.72 |
| 7159 – 9590 | Granite, granitoid rocks | 2.63 | 2.66 | 2.67 | – |
| 7357 – 7760 | Magmatite | 2.77 | 2.79 | 2.80 | – |
| 7469 – 9905 | Amphibolite | 2.95 | 2.99 | 3.00 | 3.02 |

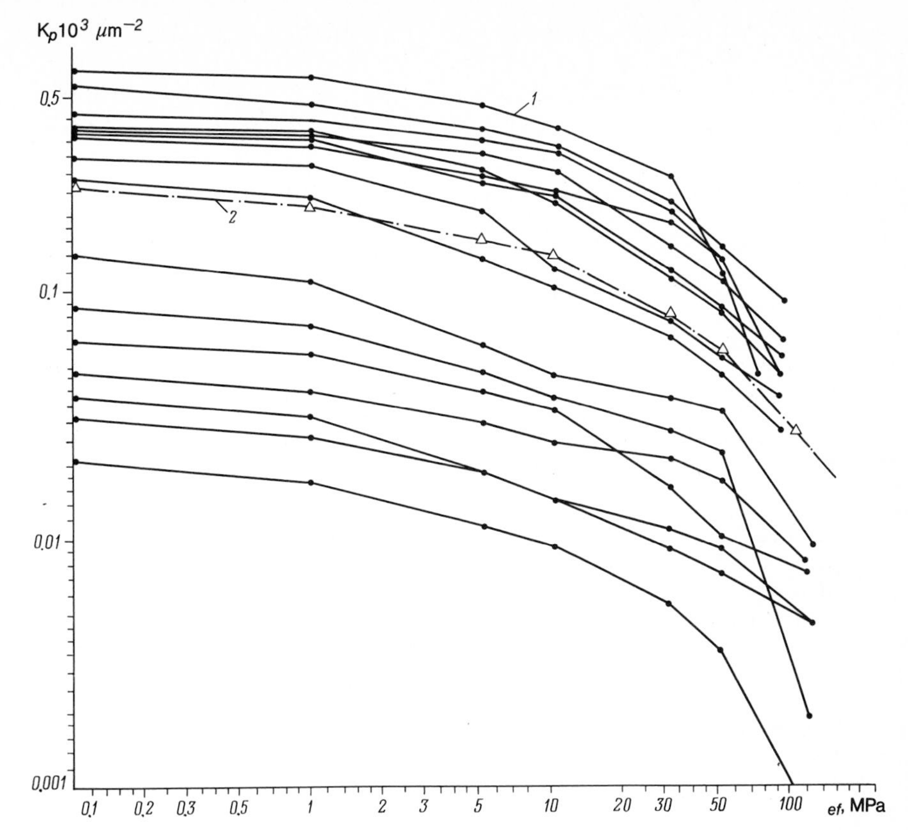


**Fig. 2.3.** Permeability ($K_p$)-effective pressure ($\sigma_{ef}$) relationship for the Kola well rocks. *1* data on individual samples; *2* directive curve

ter remains high enough to pressure the possibility of fluid filtration through the rocks in situ at great depths.

The presence of microfractures and non-uniform deformations in rocks under a uniform 3-D compression proves that they have been affected by tectonic stresses of sufficiently high magnitude.

Experimental reconstruction of stress patterns encountered in tectonically active zones (non-uniform compression) has revealed that the porous space in crystalline rock can noticeably increase under such stresses as a result of microfracturing in some predominant direction. This particular result of rock deformation is, probably, the cause of the somewhat elevated permeability of rocks occurring at greath depths.

Summarizing the results of studying density, porosity and permeability of rocks tested by the Kola well down to the depth of 11,600 m, the following conclusions can be made:

a) the well section is composed of very dense rocks of low porosity. The abnormally high density of the volcano-sedimentary and metasedimentary rocks is explained by their low porosity and the presence of ore minerals;

b) the rock density reflects primarily their elementary (chemical) composition. Regional moderate pressure metamorphism ranging from the prehnite-pumpellyite to greenschist facies leads to an increase in the crystalline rock bulk and mineral density by no more than 2% ($0.05-0.06 \text{ g cm}^{-3}$) and 4% ($0.10-0.14 \text{ g cm}^{-3}$), respectively. The depth of occurrence of a given rock scarcely influences its bulk density, and no density increase with depth is noted;
c) changes in rock composition, together with secondary processes, are responsible for the fact that rock density decreases while porosity and permeability increase with depth;
d) the density values determined for the crystalline rocks can serve as a reliable means of geological and petrographical stratification of the section, provided the accuracy of density measurements is high enough (at least $0.01 \text{ g cm}^{-3}$);
e) from the results of the density, porosity and permeability studies, the section is subdivided into three major layers. The first interface at the depth of 4500 m practically coincides with the border between the greenschist and epidote-amphibolite metamorphic facies, while the second is within the epidote-amphibolite facies and coincides with the Archean-Proterozoic contact. The validity of this stratification is proved by results of studying other physical rock properties.

## 2.7 Acoustic Properties of the Rocks

The ability of rocks to transmit elastic energy is controlled by their acoustic properties, which are classified into two groups, kinetic and dynamic. In the section of the Kola well these properties were studied on core samples taken from the well and by logging. The core was analyzed at atmospheric and higher pressures in order to determine the parameters as follows: compressional wave velocities: $V_p\|$ along the core (well) axis; $V_p\perp$ perpendicular to the core axis; $V_{p1}^d$, $V_{p2}^d$ and $V_{p3}^d$ maximum, medium and minimum velocity respectively in the sample at natural humidity (dry); $V_{p1}^w$, $V_{p2}^w$, $V_{p2}^w$ and $V_{p3}^w$ maximum, medium and minimum velocity in a sample saturated with water; "correlational" velocity (Kuznetsov 1971); $V_p = \sqrt[3]{V_{p1}, V_{p2}, V_{p3}}$; shear wave velocities: $V_s\|$ along the core (well) axis; $V_s\perp$ perpendicular to the axis; coefficient $K_a$ and factor $A$ of rock anisotropy as related to $V_p$:

$$K_a^d = V_{p1}^d / V_{p3}^d, \quad K_a^w = V_{p1}^w / V_{p3}^w, \quad A = V_p\| / V_p\perp ;$$

dynamic Young's modulus = $E_{dyn}$; and dynamic Poisson ratio = $\nu_{dyn}$.

The compressional and shear wave velocities in cores were measured by sonic testing of samples using the standard apparatus UZAS, UZIS, UKB, and DUK.

The pressure measurements (up to 200 MPa = 2 kbar) of $V_p$ and $V_s$ were carried out using a hydrostatic compressional device with a pump station of the NSVD-2500 type. This equipment enables a great number of experiments on samples of relatively large size in a short time (right-angle prisms 20 × 20 mm in cross-section and 35 – 40 mm high were cut from the cores to be tested, their faces thoroughly smoothed). To measure the time of elastic wave propagation through the samples (sonic test), standard ultrasonic registering apparatus was used.

The meteorological service for the investigations was taken over by the Petrophysical Laboratory of the VSEGEI and the acoustic logging of the well using the wide frequency band tool ($AK_{sh}$) was carried out by the Complex Geophysical-Geochemical Expedition of the VNIIYaGG.

### 2.7.1 Acoustic Properties from Core Measurements

*Measurements of Atmospheric Pressure.* Over 30,000 measurements of $V_p$ and over 1000 of $V_s$ were made on the Kola well cores, i.e. the investigation was very comprehensive. Processing of the numerous data was done by employing the mathematical statistics approach. The role of a basic primary unit for data processing was attributed to the "petrographic interval", defined as a depth interval where the dispersion of physical properties did not exceed ±5%.

The analysis of compressional wave velocities shows that the Kola well section can be classified as a high velocity one. For the Proterozoic rocks (to 6842 m in depth) the weighed (by thickness) average velocity $V_p \|$ as measured on samples with natural humidity and saturated with water is equal to 5670 m $s^{-1}$ and 6080 m $s^{-1}$, respectively. The difference between average values of velocities measured along the core vertical axis in different intervals can exceed 3500 m $s^{-1}$. Maximum $V_p$ values are registered in fine-grain massive diabases of the Matertin series occurring in the top part of the section, gabbro-diabases of the Zhdanov series, and actinolitic diabases and globular lavas of the Zapolyarny series. Weighed average velocities in these rocks are from 6600 to 6800 m $s^{-1}$. The lowest values (to 2000 m $s^{-1}$) measured on samples are registered in the Archean section. However, when the samples are saturated with a liquid (kerosene, water) as well as being subjected to external pressure, the velocity in them sharply increases and approaches values determined from well logs. For the up-section massive rocks, the influence of these factors is negligible.

The weighed average $V_p \|$ for the Archean sequence (depth interval from 6842 to 11,000 m) is 3990 m $s^{-1}$. The lowest values (3020 m $s^{-1}$) are found in the two-mica gneisses with fibrolite (depths from 10,114 to 10,600 m), the highest ones (5960 m $s^{-1}$) in the blastocataclastic low-temperature diaphthoritic amphibolites. Unaltered amphibolites, nebulitic migmatites (gneiss-granites) that predominate in the section, and their blastocataclastic diaphthoritic analogues encountered only locally have $V_p = 4750$, $3990-4060$, and 4980 m $s^{-1}$, respectively (Table 2.8).

The $V_p$ distribution over the Kola well section in general is characterized by firstly, a good correlation of velocities with the rock density variations and, secondly, the presence of three abrupt interfaces where $V_p$ decreases are observed. The first interface (depths from 4300 to 4500 m) practically coincides with the boundary between the metamorphic greenschist and epidote-amphibolite facies within the metadiabase sequence. At this interface the velocity-density correlation suddenly becomes poor because of the density increase by 0.03 g $cm^{-3}$, and a sharp increase of velocity anisotropy $K_a^w$ from 1.04 to $1.14-1.17$ caused by rock schistosity takes place. The schistosities being equal, $K_a^w$ depends on the melanocratic index of the rocks. For instance, for the Archean nebulitic migmatites the $K_a^w$ is about $1.05-1.06$, while for amphibolites and two-mica gneisses its value reaches $1.1-1.24$.

A certain linearity resulting from lamination along the hornblende grains or other minerals and their aggregates is typical for the entire downward section, the laminae being superpendicular to the well axis. This is the cause of the $V_p \|$ decrease at a relatively shallow depth of 4300 m, while the "correlational" velocity ($V_p$) begins to decrease sharply at the 4500 m depth only, which naturally cannot be explained exclusively by rock schistosity.

The second abrupt velocity decrease is associated with the Proterozoic-Archean contact where the density also decreases, because the rocks become abruptly more leucocratic, their grade of metamorphism remaining the same. Significant decreases in velocity values below this interface cannot be attributed to petro-

**Table 2.8.** Average values of compressional and shear wave velocities (km $s^{-1}$) measured on Proterozoic rock samples as compared with acoustic log data for homogeneous intervals, at pressures up to 200 MPa)

| Rock | Depth, m | Number of samples | Density g $cm^{-3}$ | Velocity at pressure $P$ (MPa) | | | | | | Velocity from AL |
|---|---|---|---|---|---|---|---|---|---|---|
| | | | | 0.1 | 25 | 50 | 100 | 150 | 200 | |
| 1 | 2 | 3 | 4 | 5 | 6 | 7 | 8 | 9 | 10 | 11 |
| *Matertin series* | | | | | | | | | | |
| Gabbro-diabase | 93 – 186 | 3 | 3.07 | 6.50 | 6.53 | 6.56 | 6.60 | 6.63 | 6.67 | 6.60 |
| | | | | 3.54 | 3.56 | 3.58 | 3.61 | 3.63 | 3.65 | 3.7 – 3.8 |
| Diabase | 186 – 334 | 3 | 3.02 | 6.59 | 6.64 | 6.69 | 6.74 | 6.78 | 6.81 | 6.65 |
| | | | | 3.67 | 3.69 | 3.71 | 3.79 | 3.75 | 3.76 | 3.7 |
| Diabase | 466 – 525 | 1 | 3.00 | 6.52 | 6.56 | 6.59 | 6.65 | 6.67 | 6.70 | 6.7 – 6.8 |
| | | | | 3.60 | 3.62 | 3.64 | 3.68 | 3.70 | 3.72 | 3.8 |
| Metatuff and diabase | 530 – 550 | 1 | 2.83 | 6.00 | 6.06 | 6.11 | 6.15 | 6.18 | 6.20 | 6.3 – 6.5 |
| | | | | 3.36 | 3.38 | 3.41 | 3.45 | 3.47 | 3.49 | 3.7 – 3.8 |
| Gabbro-diabase | 599 – 664 | 2 | 3.04 | 6.81 | 6.88 | 6.93 | 7.00 | 7.04 | 7.07 | 7.0 |
| | | | | 3.64 | 3.68 | 3.71 | 3.75 | 3.79 | 3.82 | 3.95 |
| Metatuff | 762 – 817 | 1 | 2.89 | 6.24 | 6.27 | 6.30 | 6.33 | 6.35 | 6.37 | 6.3 – 6.4 |
| | | | | 3.63 | 3.64 | 3.65 | 3.67 | 3.69 | 3.70 | 3.7 |
| Diabase | 832 – 840 | 1 | 3.03 | 6.79 | 6.86 | 6.95 | 7.04 | 7.08 | 7.12 | 6.6 |
| Diabase | 888 – 1059 | 2 | 3.05 | 6.73 | 6.81 | 6.88 | 6.96 | 7.00 | 7.03 | 6.8 – 6.9 |
| | | | | 3.73 | 3.77 | 3.81 | 3.86 | 3.89 | 3.92 | 3.8 – 3.9 |
| Metatuff | 965 – 971 | 1 | 2.81 | 6.35 | 6.41 | 6.45 | 6.50 | 6.53 | 6.56 | 6.3 |
| | | | | 3.63 | 3.65 | 3.66 | 3.68 | 3.70 | 3.72 | – |
| *Zhdanov series* | | | | | | | | | | |
| Tuff, tuffite | 1059 – 1119 | 1 | 2.81 | 6.10 | 6.14 | 6.17 | 6.23 | 6.26 | 6.30 | 6.3 |
| | | | | 3.49 | 3.51 | 3.53 | 5.57 | 3.59 | 3.62 | 3.6 |
| Essexitic gabbro | 1282 – 1417 | 3 | 2.84 | 5.64 | 5.89 | 6.03 | 6.26 | 6.32 | 6.39 | 6.2 – 6.3 |
| | | | | 3.21 | 3.31 | 3.38 | 3.49 | 3.52 | 3.56 | 3.7 |
| Serpentinite peridotite | 1570 – 1635 | 1 | 3.20 | 6.67 | 6.71 | 6.74 | 6.82 | 6.86 | 6.90 | – |
| | | | | 3.66 | 3.71 | 3.75 | 3.80 | 3.83 | 3.86 | – |
| Serpentinite | 1570 – 1635 | 4 | 2.84 | 6.18 | 6.23 | 6.27 | 6.34 | 6.38 | 6.41 | 6.3 – 6.4 |
| | | | | 3.41 | 3.43 | 6.45 | 3.47 | 3.49 | 3.51 | 3.6 |

**Table 2.8** (continued)

| Rock | Depth, m | Number of samples | Density g cm$^{-3}$ | Velocity at pressure $P$ (MPa) | | | | | | Velocity from AL |
|---|---|---|---|---|---|---|---|---|---|---|
| | | | | 0.1 | 25 | 50 | 100 | 150 | 200 | |
| 1 | 2 | 3 | 4 | 5 | 6 | 7 | 8 | 9 | 10 | 11 |
| Aleurophyllite | 1678 – 1756 | 1 | 2.98 | 5.63 | 5.66 | 5.69 | 5.75 | 5.78 | 5.82 | 5.9 – 6.1 |
| | | | | 3.29 | 3.31 | 3.33 | 3.37 | 3.41 | 3.43 | 3.6 – 3.7 |
| Aleurophyllites | 1941 – 1952 | 1 | 2.80 | 6.12 | 6.16 | 6.19 | 6.25 | 6.28 | 6.32 | 6.1 |
| | | | | 3.53 | 3.55 | 3.57 | 3.61 | 3.63 | 3.65 | 3.7 |
| Gabbro-diabase | 1988 – 2125 | 6 | 3.05 | 6.35 | 6.51 | 6.61 | 6.78 | 6.84 | 6.89 | 6.9 – 7.0 |
| | | | | 3.57 | 3.61 | 3.65 | 3.72 | 3.75 | 3.78 | 4.0 |
| Aleurophyllite and sandstone | 2135 – 2264 | 4 | 2.82 | 5.84 | 5.92 | 5.99 | 6.07 | 6.10 | 6.13 | 5.8 – 6.1 |
| | | | | 3.51 | 3.54 | 3.61 | 3.64 | 3.67 | 3.70 | 3.5 – 3.7 |
| Gabbro-diabase | 2264 – 2325 | 1 | 3.05 | 6.22 | 6.43 | 6.60 | 6.68 | 6.74 | 6.78 | 6.9 |
| | | | | 3.49 | 3.56 | 3.62 | 3.66 | 3.68 | 3.70 | 3.9 |
| Aleurophyllite and sandstone | 2325 – 2487 | 6 | 2.80 | 5.94 | 5.99 | 6.03 | 6.09 | 6.13 | 6.17 | 6.0 – 6.2 |
| | | | | 3.52 | 3.54 | 3.56 | 3.60 | 3.63 | 3.66 | 3.6 – 3.7 |
| Gabbro-diabase | 2637 – 2747 | 4 | 2.98 | 6.16 | 6.42 | 6.56 | 6.65 | 6.68 | 6.71 | 6.8 – 6.9 |
| | | | | 3.44 | 3.51 | 3.57 | 3.61 | 3.63 | 3.65 | 3.9 |
| Aleurophyllite | 2764 – 2805 | 3 | 2.85 | 6.06 | 6.12 | 6.17 | 6.22 | 6.24 | 6.26 | 5.9 – 6.0 |
| *Zapolyarny series* | | | | | | | | | | |
| Actinolitic diabase | 2820 – 2845 | 1 | 3.01 | 6.26 | 6.52 | 6.70 | 6.75 | 6.78 | 6.81 | 6.7 |
| Actinolitic diabase | 2845 – 2975 | 4 | 3.00 | 6.56 | 6.64 | 6.68 | 6.71 | 6.74 | 6.76 | 3.9 – 6.9 |
| Actinolitic diabase | 2975 – 3065 | 5 | 2.97 | 6.29 | 6.45 | 6.53 | 6.64 | 6.67 | 6.70 | 6.5 – 6.6 |
| Actinolitic diabase | 3175 – 3242 | 1 | 3.06 | 6.52 | 6.63 | 6.70 | 6.75 | 6.78 | 6.81 | 6.8 – 6.9 |
| Actinolitic diabase | 3257 – 3460 | 3 | 3.03 | 6.49 | 6.54 | 6.56 | 6.58 | 6.60 | 6.62 | 6.7 – 6.8 |
| Actinolitic diabase and diabase-porphyrite | 3460 – 3589 | 5 | 3.04 | 6.68 | 6.77 | 6.83 | 6.88 | 6.91 | 6.93 | 6.8 – 6.9 |
| Actinolitic diabase and diabase-porphyrite | 3589 – 3602 | 2 | 3.04 | 6.45 | 6.51 | 6.55 | 6.59 | 6.62 | 6.64 | 6.4 |
| Actinolitic diabase | 3625 – 3645 | 1 | 3.04 | 6.40 | 6.48 | 6.52 | 6.58 | 6.61 | 6.64 | 6.6 |
| Diabase-porphyrite | 3645 – 3682 | 1 | 3.01 | 6.15 | 6.48 | 6.60 | 6.64 | 6.67 | 6.70 | 6.8 |
| Schistose tuff-breccia | 3691 – 3707 | 1 | 3.01 | 5.63 | 5.68 | 5.73 | 5.78 | 5.81 | 5.83 | 5.8 – 6.0 |
| Actinolitic diabase | 3790 – 3886 | 1 | 3.00 | 6.23 | 6.48 | 6.58 | 6.77 | 6.80 | 6.83 | 6.7 – 6.8 |

**Table 2.8** (continued)

| Rock | Depth, m | Number of samples | Density g cm$^{-3}$ | Velocity at pressure $P$ (MPa) | | | | | | Velocity from AL |
|---|---|---|---|---|---|---|---|---|---|---|
| | | | | 0.1 | 25 | 50 | 100 | 150 | 200 | |
| 1 | 2 | 3 | 4 | 5 | 6 | 7 | 8 | 9 | 10 | 11 |
| Actinolitic diabase-porphyrite | 3886 – 4082 | 11 | 3.04 | 6.39 | 6.68 | 6.81 | 6.92 | 6.96 | 7.00 | 6.9 – 7.0 |
| Actinolitic diabase-porphyrite | 4082 – 4169 | 2 | 3.05 | 6.25 | 6.47 | 6.60 | 6.67 | 6.70 | 6.73 | 6.9 |
| Actinolitic diabase-porphyrite | 4220 – 4244 | 4 | 3.00 | 6.52 | 6.78 | 6.87 | 6.95 | 6.99 | 7.02 | 6.9 |
| Actinolitic diabase-porphyrite | 4244 – 4268 | 1 | 3.00 | 6.18 | 6.57 | 6.70 | 6.80 | 6.85 | 6.89 | 6.9 |
| Actinolitic diabase-porphyrite | 4268 – 4286 | 3 | 3.02 | 6.05 | 6.83 | 6.48 | 6.62 | 6.66 | 6.70 | 6.9 |
| Actinolitic diabase-porphyrite | 4299 – 4325 | 1 | 3.09 | 6.25 | 6.60 | 6.74 | 6.97 | 7.03 | 7.08 | 6.9 |
| Actinolitic diabase-porphyrite | 4325 – 4435 | 6 | 3.05 | 6.42 | 6.66 | 6.80 | 6.89 | 6.94 | 6.98 | 6.8 |
| *Luchlompol series* | | | | | | | | | | |
| Andesite-metaporphyrite | 4673 – 4784 | 3 | 2.74 | 5.38 | 5.82 | 5.99 | 6.09 | 6.12 | 6.16 | 5.9 – 6.0 |
| Sandstones, schists | 4831 – 4884 | 3 | 2.76 | 4.03 | 5.16 | 5.54 | 5.83 | 5.94 | 6.02 | 5.8 – 6.0 |

*Note*. For each depth interval the $V_p$ and $V_s$ values are shown in the first and second lines, respectively.

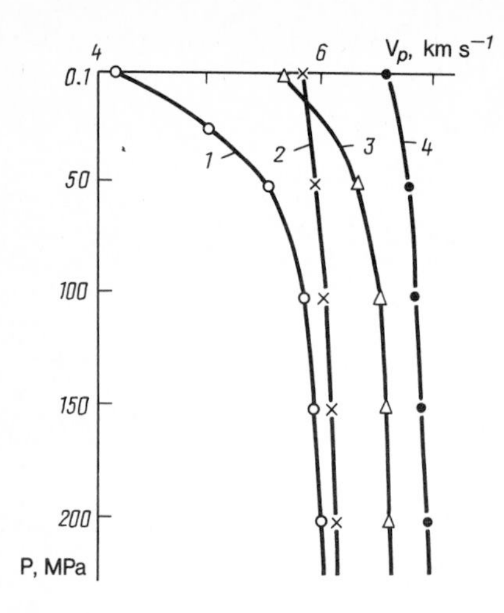


**Fig. 2.4.** Compressional wave velocity ($V_p$)-pressure ($P$) curves obtained from core measurements for different rocks from the Kola well. *1* granite-gneiss; *2* aleurolite; *3* blastoamphibole schist; *4* diabase

graphical and mineralogical factors only, for there is a velocity decrease at about 10,100 m depth without any contrast in rock composition.

Simultaneously with the determination of compressional wave velocity, the shear wave velocity was measured, which enabled the calculation of the dynamic elastic parameters – Young's modulus $E_{\mathrm{dyn}}$ and Poisson ratio $\nu_{\mathrm{dyn}}$. Variations of these parameters along the Kola well section closely follow the $V_p$ variations; the sharp interfaces in the middle of the Proterozoic and at the Proterozoic-Archean boundary are distinctly identified. For instance, Young's modulus drops from 100 – 104 GPa in the top part of the Pechenga sequence to 63 – 71 GPa in its bottom part, while in the Archean sequence this parameter average weighted value equals 27 GPa only. The Poisson ratio respectively drops from 0.23 – 0.27 to 0.14 – 0.21 within the Pechenga sequence and down to 0.11 in the Archean one. The main cause of such a great downward change in the $E_{\mathrm{dyn}}$ and $\nu_{\mathrm{dyn}}$ values is undoubtedly the partial disintegration of rock samples owing to decompressional fracturing of the cores, changes in rock composition playing a subordinate role.

*Measurements at Pressures of up to 200 MPa.* Over 200 measurements of $V_s$ and 302 measurements of $V_p$ were made on the Kola well cores at hydrostatic pressures of up to 200 MPa (see Table 2.8 and Fig. 2.4).

As follows from Table 2.8, $V_p$ and $V_s$ in different rocks are essentially different even within the limits of the relatively homogenous Matertin series. The dense fine-grained and massive rocks of this series show small changes in velocity with pressure; however, the difference between the $V_p$ values at the initial and final pressures is at least 2 – 3%, more often 3 – 4%.

In the Zhdanov series there is considerable differentiation in rock mean densities and velocities. For instance, in the essexitic gabbro, these parameter variations reach 10 – 12% and considerably depend on pressure (see Fig. 2.4). Velocity variations in the gabbro-diabase rocks with pressure are 8% on the average.

The rocks of deeper horizons are highly schistose and altered by secondary processes, therefore they have not been studied well yet (although their properties under elevated pressures are known). It follows from Table 2.8 that the common rocks of the Luchlompol series have relatively low velocities, $V_p$. These rocks comprise the top part of the zone of lower velocities. They consist of grains of a larger size and are characterized by higher porosity (up to 1.5 – 2%), variability and heterogeneity of composition and structure, more or less prominent anisotropy of elastic properties, lower velocities and their more obvious dependence on pressure. Moreover, intensive secondary changes and both inter- and inside-grain cataclasis are common in these rocks, which complicates velocity measurement under pressure.

Saturation of the rock samples with water increases the velocity considerably; however, when pressure is applied to the samples, the velocity in them becomes lower than in dry ones because the liquid "trapped" in their pores creates a back-pressure inside the pores. The data shown in Table 2.8 are from air-dry samples during the first cycle of pressure application. At successive loadings, in a number of cases some deviations of the $V_p = f(P)$ curves from the first one were observed, however at $p \approx 100$ MPa the curves, as a rule, closely approach each other.

### 2.7.2 Results of Acoustic Properties Determination from the Wide Frequency Band Acoustic Log ($AL_w$). Comparison with Data Obtained from Core Samples

To obtain more accurate $V_p$ values for the Kola well rocks in situ, an attempt was made to compare the velocities measured on samples with those derived for the same rocks from the $AL_w$ curve in intervals where the curve quality was good enough.

Using the method described in Leshchuk (1977), the zone of the borehole affecting the rock elastic properties was determined from AL data. For most of the section this zone size does not exceed 0.15 m and consequently the $V_p$ values as derived from AL characterize the untouched rocks. It should be noted that the rocks in question vary relatively little lithologically.

Unlike the Proterozoic, the Archean section is much less suitable for such a comparison, and here it is feasible to derive the $V_p$ average values from log readings taken in different intervals composed of similar rocks. The whole Archean section of the Kola well from 6842 to 11,636 m depths has been stratified into separate beds according to the geological column (see Table 2.9), their total number being 408. To study the general relation between acoustic properties and depth of occurrence of rocks and particular variations of properties of some rocks occurring at different depths, velocity determinations were carried out for three units of the Archean Kola sequence.

It was impossible to derive accurate $V_p$ values for some beds, as the quality of AL was poor in intervals complicated with borehole vugs. The smallest number of determinations is typical of ultramafite beds where plenty of large vugs are found in plenty. Few determinations were made also for the epidote-biotite-amphibole and epidote-biotite schists for the same reason. A great number of determinations is characteristic for rocks such as granite, biotite-plagioclase gneiss and amphibolites, i.e. the most common rock types of the section. Rocks like

**Table 2.9.** Compressional wave velocities (km $s^{-1}$) derived from AL

| Depth interval, m | Rock | Thickness, m | Number of beds | Number of determinations | $V_p$ |
|---|---|---|---|---|---|
| 6842 – 7627 | Biotite-plagioclase gneiss with high-aluminium minerals | 507 | 38 | 31 | $\frac{5.5-6.1}{5.7}$ |
| | Granite, granite gneiss, pegmatite, nebulitic migmatite | 131 | 17 | 12 | $\frac{5.5-6.0}{5.75}$ |
| | Amphibolite and schistose amphibolite | 99 | 13 | 10 | $\frac{6.0-6.7}{6.3}$ |
| | Meta-ultramafite | 55 | 8 | – | – |
| 7627 – 9455 | Biotite-plagioclase gneiss and schist | 399 | 37 | 42 | $\frac{5.4-6.2}{5.91}$ |
| | Pegmatite | 8 | 3 | 2 | $\frac{5.9-6.0}{5.9}$ |
| 9455 – 10,636 | Amphibolite and schistose amphibolite | 219 | 12 | 9 | $\frac{6.0-6.7}{6.4}$ |
| | Meta-ultramafite | 25 | 4 | – | – |
| | Muscovite-biotite-plagioclase gneiss and schist | 75 | 8 | 5 | $\frac{5.3-5.8}{5.5}$ |
| | Two-mica gneiss with cummingtonite (high-aluminiferous) | 132 | 9 | 13 | $\frac{5.0-6.0}{5.38}$ |
| | Biotite-amphibole gneiss | 7 | 1 | 1 | 6.45 |
| | Biotite-plagioclase gneiss and schist | 370 | 30 | 33 | $\frac{5.4-6.1}{5.73}$ |
| | Biotite-amphibole-plagioclase schist | 187 | 26 | 11 | $\frac{5.5-6.3}{5.95}$ |
| | Epidote-biotite gneiss | 82 | 3 | 6 | $\frac{5.6-6.2}{5.97}$ |
| | Epidote-biotite-plagioclase gneiss and schist | 452 | 45 | 53 | $\frac{5.5-6.3}{5.97}$ |

**Table 2.9** (continued)

| Depth interval, m | Rock | Thickness, m | Number of beds | Number of determinations | $V_p$ |
|---|---|---|---|---|---|
| | Amphibolite and schistose amphibolite | 461 | 43 | 29 | $\frac{6.2-6.7}{6.45}$ |
| | Meta-ultramafite | 43 | 7 | 1 | 4.5 |
| | Epidote-biotite-amphibole schist | 105 | 20 | 14 | $\frac{5.7-6.2}{6.04}$ |
| | Epidote-biotite schist | 15 | 4 | 1 | 5.8 |
| | Biotite-amphibole-plagioclase schist | 32 | 5 | 2 | 5.9 |
| | Biotite-amphibole-plagioclase gneiss | 66 | 4 | 3 | $\frac{5.9-6.0}{6.0}$ |
| | Epidote-biotite-amphibole-plagioclase gneiss | 75 | 10 | 16 | $\frac{5.5-6.1}{5.9}$ |
| | Nebulitic migmatite | 134 | 10 | 7 | $\frac{5.8-6.1}{5.92}$ |
| | Biotite granite | 38 | 12 | 5 | $\frac{5.5-6.1}{5.9}$ |
| | Granite and pigmatite | 84 | 13 | 13 | $\frac{4.9-6.1}{5.76}$ |
| 6835 – 10,636 | Kola sequence | 3801 | 408 | 309 | $\frac{4.5-6.7}{5.9}$ |

*Note.* Numerator – the maximum and minimum values, denominator – the average value.

granite, granite-gneiss, pegmatite and nebulitic migmatite are common in all three units and do not differ much from each other with respect to their average velocity, which in all cases varies from 5.0 to 6.1 km $s^{-1}$ (mean value 5.7 – 5.9 km $s^{-1}$). According to AL data there is no velocity increase with depth for these rocks. Comparison of the $AL_w$ readings with velocity measurements made on core samples for the granite type rocks shows that they fit closely enough. However, the granites from the Kola well certainly have lower velocities as compared to the $V_p$ values (both derived from AL and measured on cores) known for granites from other regions, which is explained by the particularities of composition and structure of the rocks penetrated by the Kola well. In the upper unit the rocks of the biotite-plagioclase gneiss with a high aluminium mineral group have relatively low AL velocities, which can be attributed to a high percentage of mica (both muscovite and biotite) and a low content of high-aluminium minerals, relatively high porosity, and strong cataclastic manifestations. Amphibolites and schistose amphibolites are characterized by elevated average velocity (6.3 km $s^{-1}$) with the considerable dispersion of individual values partly explained by the pronounced rock anisotropy. The amphibolites from the upper unit have a somewhat lower velocity than those from the middle unit.

The middle unit of the Kola sequence of the Archean section of the Kola well is confined to the depth range from 7627 to 9455 m and is composed of mainly biotite-plagioclase, epidote-biotite-plagioclase or biotite-amphibole-plagioclase gneisses and schistose amphibolites. Gneiss represents 992 of the 1828 m, i.e. a little more than half of the unit's thickness, its leucocratic and mesocratic species being distinguished in the geological column. At 200 MPa pressure, the leucocratic rocks have average $V_p$ value of 6 km $s^{-1}$ (variations from 5.7 to 6.3 km $s^{-1}$), the mesocratic ones 6.08 km $s^{-1}$ (variations from 5.8 to 6.4 km $s^{-1}$). The considerable variations in velocity observed even in strongly compressed samples cannot be explained by the rock elastic anisotropy only, all the more so that in these rocks the anisotropy is relatively weak. The main reason is the variation in rock composition, the presence of secondary minerals like muscovite, chlorite and saussurite and also a wide development of cataclastic structures. AL shows even greater velocity variations of these rocks which, in addition to the rock influence, is caused by the instrument's high sensitivity with regard to the borehole wall geometry. However, the mean weighted velocities derived from the AL for gneisses of the middle unit (5.9 – 6 km $s^{-1}$) fit well those measured on cores under pressure of 150 – 200 MPa, which proves the reliability of velocity values derived for these rocks in situ. It is worth noting that gneisses of acid and medium composition from the Archean section of the Kola well, as compared to their analogues from other regions, have lower velocities and higher values of $V_p/V_s$ and Poisson ratios.

Schistose amphibolites of the middle unit occupy about 1/4 of its thickness. In Table 2.8 they are given without classification into various species. Comparison of data derived for these rocks from AL and core analysis is of a certain interest. Velocity was measured in various directions with respect to the rock structure, particularly parallel to the laminae and linear features and perpendicular to them, i.e. along the minimum and maximum velocity vectors, which resulted in considerable variations (5.7 – 7.4 km $s^{-1}$ at $P = 200$ MPa).

The variations in velocity derived from AL for these rocks are essentially smaller (6.2 – 6.7 km $s^{-1}$), the highest values in excess of 7 km $s^{-1}$ completely lacking. This is explained by the fact that the acoustic log measures the velocity along the walls (axis) of the well, while the rock linearity is oriented, as a rule, subhorizontally. These rocks are commonly characterized by lamination and banding, the schistosity planes being parallel to each other and crossing the well axis at various angles, more often from 40° to 60°.

Thus the velocity derived from AL has values that are always between the maximum and minimum ones, so that it is less variable than the velocity measured on samples. However, average weighted values of both velocities are sufficiently close, especially if the measurements are made with samples under pressure of 150 – 200 MPa.

The middle unit includes rocks such as nebulitic migmatite, biotite granite and pegmatite, forming 25 beds with a total thickness of 180 m (about 10% of the entire unit thickness). The average weighted values and range of variation of velocity for these rocks correspond well to those measured on their samples at pressures from 150 to 200 MPa (see Table 2.8).

The lower unit (downwards from the 9455 m depth) comprises a great amount of biotite-plagioclase, muscovite-biotite-plagioclase and epidote-biotite-plagioclase gneisses (527 m, or 45%) analogous to such gneisses known in the middle unit regarding their mineral composition and structure. However, the AL velocities appear to be lower in the lower unit, although one could expect them to become higher at greater depths. This reversal can be attributed to more intensive deformations of the rocks that manifest themselves in pronounced schistosity and fracturing.

Granites and pegmatites in the lower unit have also somewhat lower velocities. Of this unit 132 m are composed of two-mica gneiss with sillimanite characterized by low velocities (5.38 km $s^{-1}$), which is evidently caused by intensive rock crushing in the hole walls independently of the rock mineral composition.

Amphibolites from the lower unit have the same properties as their analogues from other units. Meta-ultramafites from four beds, their velocities not having been studied with acoustic logging.

The results of studying the elastic properties of rocks in the Kola well can be summarized as follows:

1. the Earth's crustal section tested by the well is differentially layered. The common belief that velocity in rocks of the same composition increases with depth (the gradient model) has not been proved by the investigations. The changes in rock structure caused by secondary processes are evident from changes in elastic properties. The highest velocities of elastic waves are registered in the top part of the section, i.e. a reverse seismogeological section is revealed with a zone of lower velocities (the depth interval from 4500 to 6835 m);
2. the three-layer structure of the section established from density observations is even more evident from the elastic properties behaviour (see Table 2.9). The first interface (the 4500 m depth) almost coincides with the boundary between greenschist and epidote-amphibolite metamorphism facies. The principal significance of this interface lies in the fact that the primary composition of

rocks is preserved below it in general, though schistosity increases sharply. The anisotropy of rock elastic properties also increases.

At the second interface (the 6842 m depth) the elastic properties do not change as sharply as the density and nuclear-physical parameters, probably because the processes of deformation and metamorphism are essentially common to both the Proterozoic and Archean rocks;

3. the range of variations in elastic properties over the section is rather wide. From this follows the important role of these properties in the section stratification and high efficiency of seismic-acoustic investigations;
4. for the main rock types the correlation between velocities derived from AL and core analysis at pressures from 150 to 200 MPa is good enough, which proves the reliability of velocity determinations for the rocks in situ and the efficiency of the combined velocity measurements on cores and in the well.

## 2.8 Seismic Investigations in the Borehole

### 2.8.1 Velocity Characteristics of the Section

To determine the velocities of seismic wave propagation, time-depth curves were plotted for the first arrivals and other phases of the passing compressional wave ($P$) as well as time curves for the phases of intensive passing shear waves ($S$) registered at greater times. Reduction of the observed time curves to zero-offset ones and plotting of the velocity section were carried out using the *Time* programme of processing time curves for layered media models with flat dipping interfaces, taking into account the interface inclinations, refractions and formation of head-waves. Other methods of processing were also used.

From the VSP data (Fig. 2.5), layers of variable lithology were identified. The section being weakly differentiated, the velocities in these layers for the purpose of kinematic studies can be assumed to be equal to mean velocities of $P$- and $S$-waves, which vary insignificantly with depth. Mean velocities of compressional and shear waves are 6500 m s$^{-1}$ and 3600 m s$^{-1}$, respectively. The $V_p/V_s$ ratio varies from 1.65 to 1.84. These values are characteristic of the crystalline rocks which composed the well section.

To obtain interval velocities for the section in more detail, the time curves were reduced and numerically differentiated. Reduction of the VSP time curves helped to identify and make more definite many velocity contrasts that had been too weakly expressed on the primary time curves.

Numerical differentiation was carried out using the least-squares method with the shifting window size equal to the geophone spacing used in the field observations. From the apparent velocity curves plotted for two unequal bases (160 and 400 m) of differentiation the depth intervals of stable velocity were more accurately determined. In most cases the time curve bending points identified or defined more accurately by the reduction, as well as the bed contacts determined by differentiating the apparent velocity curves, corresponded to geological boundaries such as contacts between geological bodies of different composition or stratigraphic interfaces.

The section was stratified into intervals of different velocities with a resolution from 100 to 200 m, the standard error in velocity determination not exceeding 100 – 200 m s$^{-1}$. In thin beds the velocity was determined with a relatively larger error.

Some intervals of the velocity section could be interpreted not only as layered media, but as gradient ones as well, i.e. characterized by a smooth velocity change with depth. Such a possibility was provided, first of all, by the apparent velocity plots.

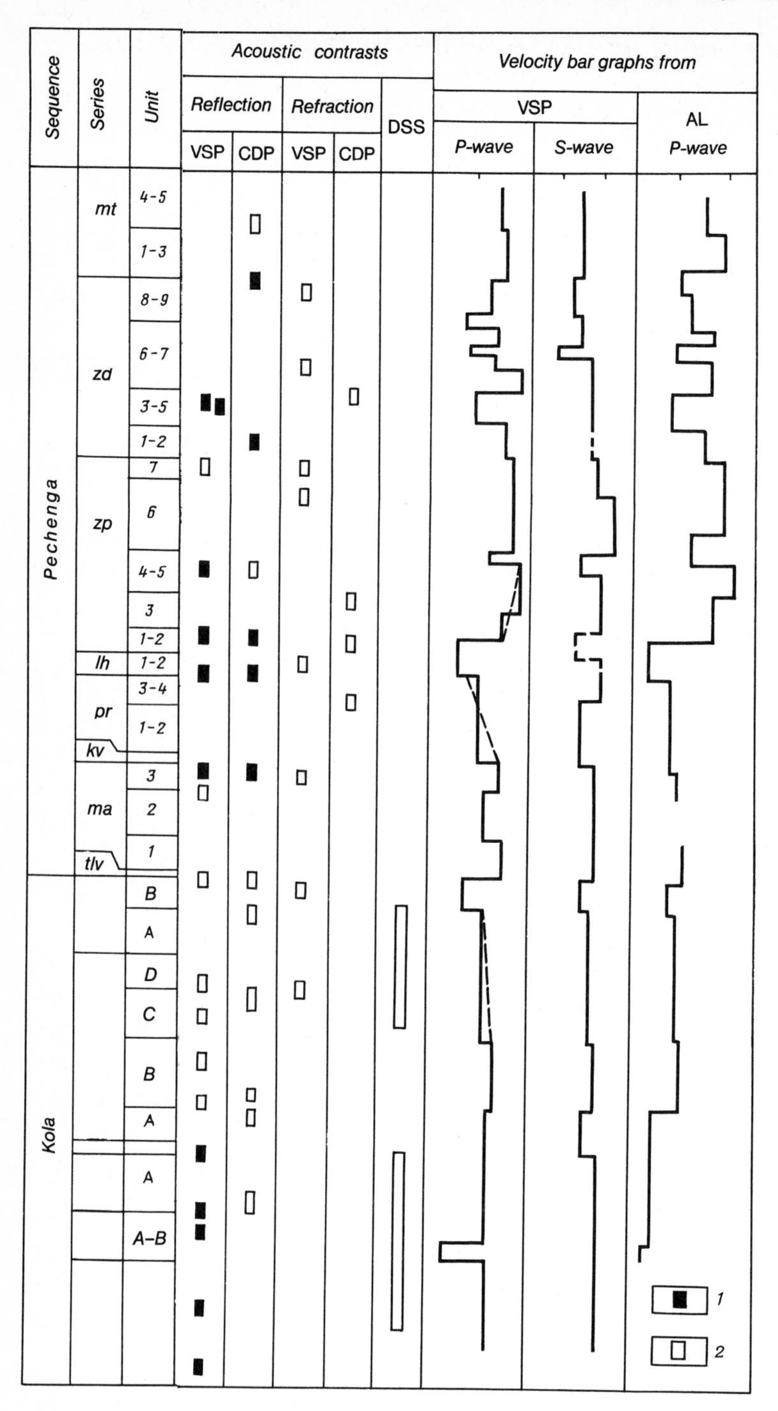


**Fig. 2.5.** Seismic characteristics of the Kola well section from VSP data. Reflection ties up to: *1* strong acoustic contrast; *2* weak one

The velocity sections were verified by combined interpretation of linear geoacoustic models derived from the AL and VSP data. A good correlation between the smoothed acoustic model (at the 0.965 level of smoothing) and the VSP data was obtained (±2%). This has considerably helped to improve the previously available velocity-depth curves and concentrated on studying the nature of reflecting interfaces.

The interval velocities of elastic waves for the well section are as follows:

From the results of studying the Kola borehole and adjacent shallow boreholes, as well as surface seismic observations, the top part of the velocity section (about 150 m thick) is shown as a gradient interval with gradual velocity increase with depth from 5200 to 6400 m $s^{-1}$. Still deeper, the Matertin series is characterized by practically stable interval velocities: 6400 – 6500 m $s^{-1}$ for compressional and 3500 m $s^{-1}$ for shear waves (see Fig. 2.5).

The Zhdanov series is one of the most differentiated *P*-wave velocity intervals of the well section. The velocity varies by 1000 m $s^{-1}$ and more, thus reflecting the heterogeneity of the rock composition in this series. The lower velocities characterize sedimentary rocks, while the maximum ones are typical for intrusive gabbro-diabases. With respect to the shear waves, this series is less differentiated as to both the number of layers (5 instead of 9) and the variations in velocity values (800 m $s^{-1}$). The most prominent $V_s$ anomaly here (3000 m $s^{-1}$ at 1680 – 1810 m) coincides with the anomaly of compressional wave velocity. In this part of the section the wellbore is very vuggy, from which tectonic faults can be assumed.

The Zapolyarny series has maximum interval velocities. In practically the entire section of this series, with the exception of the thin first and fifth units, the $V_p$ values exceed 6500 m $s^{-1}$, and for more than 70% of the section they are higher than 6700 m $s^{-1}$. Such velocities are characteristic of actinolitic diabases. Velocity decrease in the fifth unit (down to 6200 m $s^{-1}$) is caused by the predominance of volcano-sedimentary rocks. Among the Pechenga sequence rocks, the lowest velocities (5500 m $s^{-1}$) are observed in the first unit of the Zapolyarny series. Here the dislocational metamorphism associated with the Luchlompol fault is most pronounced.

In the Zapolyarny series, the $V_s$ values (like the $V_p$ ones) are generally high. Some discrepancy in these two velocities possibly results from the fact that they were derived from VSP measurements with a relatively remote shot point (3000 m), therefore the heterogeneities in rocks far from the well could have influenced the VSP data.

The $V_p$ values derived for the Luchlompol series are also among the lowest within the Pechenga sequence. This series is united with the first unit of the Zapolyarny series into a single interval with $V_p = 5600$ m $s^{-1}$. Such a low velocity is explained by the predominance of sedimentary rocks in the Luchlompol series and, moreover, by strong tectonic faulting. The $V_s$ values vary from 3400 to 3900 m $s^{-1}$.

In the Pirttiyarvy series $V_p = 6000$ m $s^{-1}$, which is some 500 – 700 m $s^{-1}$ lower than the velocity in effusive rocks of the Nickel series. Most of this series also has lower shear wave velocities, down to 3500 m $s^{-1}$.

The interval velocity of the *P*-wave in the Mayarvi series reflects its three-unit stratigraphy. The lower velocity values here as compared to the Nickel series effusions are explained by the intensive schistosity of the rocks.

The thin Kuvernerinyok and Televin series cannot be individually characterized by interval velocities, so they are included in the adjacent intervals of the Pirttiyarvi and Mayarvi series.

The gneiss sequence shows considerably smaller velocity variations than the Pechenga; here several velocity units are often united into a single interval of constant velocity.

The upper series of two-mica plagioclase gneisses includes two intervals: the top one (to the depth of 7160 m) has a lower (5700 m $s^{-1}$) *P*-wave velocity, while the bottom interval has a higher velocity (up to 6100 m $s^{-1}$). The *S*-wave velocity also increases with depth from 3500 to 3700 m $s^{-1}$, respectively.

The underlying series of biotite-plagioclase gneisses and amphibolites includes three intervals of stable $V_p$ and $V_s$. The upper one has $V_p$ and $V_s$ equal to 6100 and 3700 m $s^{-1}$, respectively; it comprises units C and D made of two-mica plagioclase gneiss. The middle interval corresponds to a part of the B unit (see Fig. 2.5) and has the highest $V_p$ value (6400 m $s^{-1}$) of all the gneiss sequence rocks. In the lower interval $V_p$ and $V_s$ decrease to 6200 and 3600 m $s^{-1}$, respectively.

$V_p$ behaviour in the first of the above intervals can be interpreted as a gradient one, like those in the bottom part of the Zapolyarny series and Pirttiyarvi section (see the broken line, Fig. 2.5). For the gneiss interval discussed and the Pirttiyarvi series, the velocity gradient is positive, while for the Zapolyarny rocks it is negative.

The relatively thin layer of two-mica plagioclase gneiss (9460 – 9570 m) and underlying gneiss series as well as the lower part of the series of biotite-plagioclase gneiss with the lower part of the series of biotite-plagioclase gneiss with amphibolite are characterized by a stable interval *P*-wave velocity equal to 6200 m $s^{-1}$. Shear wave velocity for the two-mica plagioclase gneisses in the upper part is 3600 m $s^{-1}$ and downsection becomes equal to 3900 m $s^{-1}$.

The compressional wave velocity-depth plot has a specific interval corresponding to a two-mica plagioclase gneiss unit occurring within the depth range from 10,420 to 10,600 m. Here the velocity has an abnormally low value of 5300 m $s^{-1}$, the lowest of the entire well section.

In the deepest unit of the gneiss series investigated in the interval from 10,610 to 11,500 m the velocities observed are as follows: $V_p = 6200$ m $s^{-1}$ and $V_s = 3900$ m $s^{-1}$.

In general, the velocity section obtained for the gneiss sequence has rather stable $V_p$ values: for 75% of its thickness studied thus far they are within 6100 – 6200 m $s^{-1}$. At this background there are two intervals with abnormally low values – 5700 and 5300 m $s^{-1}$, and one interval with velocity elevated to 6400 m $s^{-1}$.

According to the VSP data there is no $V_p$ and $V_s$ increase with depth for both the Pechenga and gneiss sequences. Consequently, the overburden pressure increase does not seem to play a role.

A clear correlation of velocity with rock structure and texture is established for the Pechenga sequence. High velocities are characteristic of weakly altered effusive rocks in the Matertin and Zapolyarny series as well as of gabbro-diabase sills. Low velocities are registered in sedimentary rocks and intervals affected by schistosity and mylonitization processes.

The gneiss sequence velocities are less differentiated as compared to the Pechenga, which probably reflects a more uniform distribution of elastic properties over the section owing to a wide development of polymetamorphism. The presence of basic rocks within the gneiss section on one hand, and migmatization processes on the other has an opposite effect on the elastic wave velocity. Thus, the same velocity 6100 m $s^{-1}$ is observed in both the 7160 – 7620 m interval where the basic rocks (amphibolites) account for only 18 – 20% of its thickness and migmatization is weak (about 20%) and the 7620 – 8460-m interval where the amphibolite percentage is almost two times greater (34%) but the migmatization manifestation is much stronger. The depth interval 9600 – 10,400 m is also characterized by a single $V_p$ value – 6200 m $s^{-1}$, although it includes units with various basic rock content, in places increasing up to 50% but with a simultaneous increase in migmatization.

## 2.8.2 Wave Patterns

The VSP wave patterns observed in the Kola well differ qualitatively and quantitatively from those obtained when investigating the thick sedimentary cover in platform areas. The complex seismogeological environments (steeply dipping discordant interfaces, numerous tectonic faults, etc.) are responsible for the specific wave fields like the one presented in Fig. 2.6.

The first arrivals on all the VSP traces correspond to the incident *P*-wave of complicated shape. Its dominant frequency depends on seismic energy generation conditions and varies from 50 to 100 cps for different shot points (*SP*). The successive arrivals mark numerous incident waves of differing nature. A wave with a low apparent velocity is seen on the $VSP_1$ over the entire depth range. From its kinematic and dynamic parameters it can undoubtedly be interpreted as the incident shear wave. The amplitude ratio of the shear and compressional waves as a function of registration depth [$As/Ap = f(H)$] from 300 to 1740 m changes from 0.28 to 1.36. The great amplitude of the *Z*-component of the shear wave propagating along an almost vertical ray path proves its high intensity. As follows from data processing, the value of ratio $As/Ap$ is equal to $1.73 \pm 0.42$ (average for the well borehole). Velocity contrasts here being relatively weak, this wave is not likely to be an incident *PS* converted wave. It is rather a pure shear wave coming directly from the source.

Within the time interval between the arrivals of the passing *P*- and *S*-waves there are relatively weak passing waves whose phase correlation (coherent line-up) axes are parallel to the first arrivals. These axes are not traceable up to the surface in the over-critical time domain. From their kinematic parameters they can possibly be classified as incident converted waves of the *SP* type or partly multiple passing *P*-waves. The partly multiple incident waves must have shorter phase correlation axes because of the poor continuity of reflectors and their frequent unconformity.

In addition to the incident *P*-, *S*- and *SP*-waves on the VSP sections, there are incident converted waves of the *PS* type, especially well visible after digital processing of three-component records ($VSP_3$, Fig. 2.7).

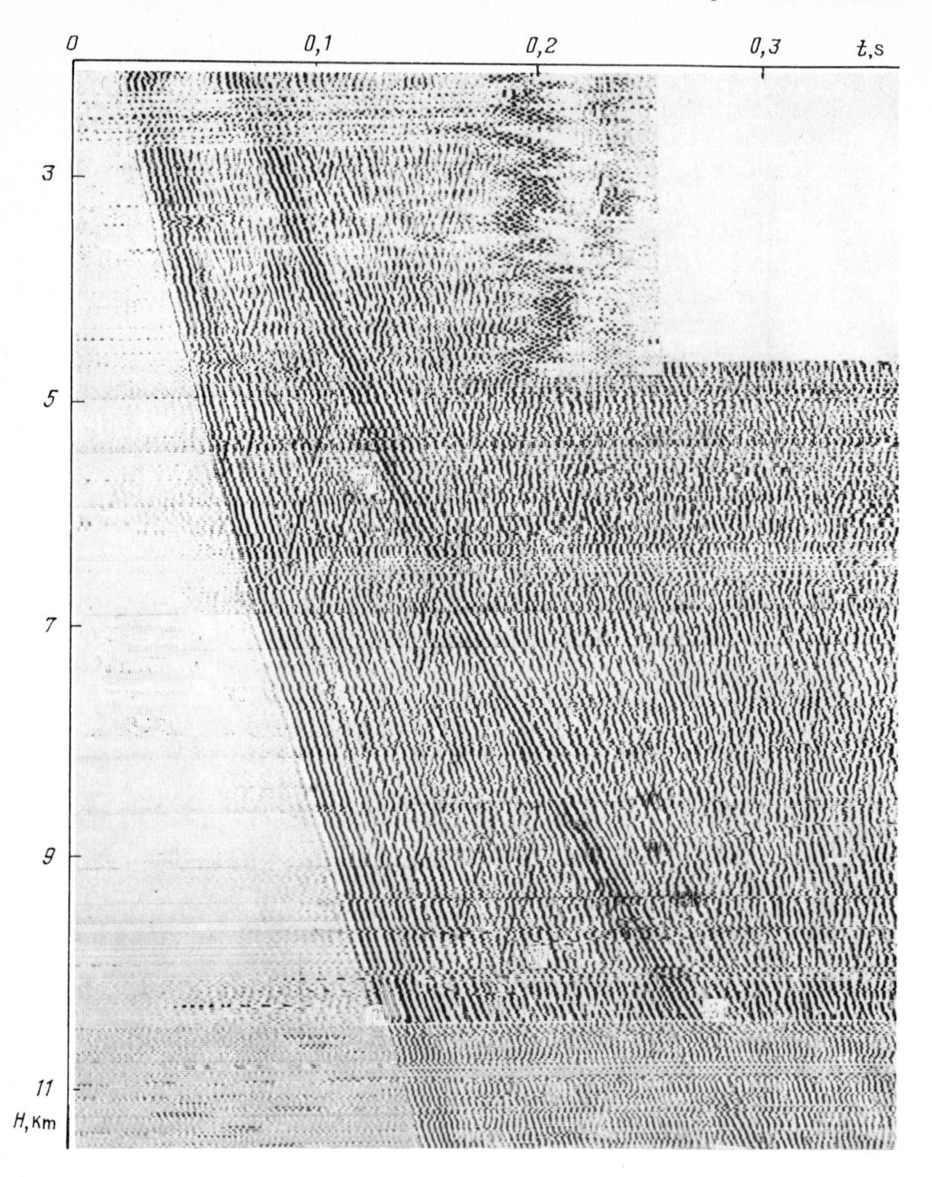


**Fig. 2.6.** Vertical seismic profile $VSP_1$ (*SP* offset is 680 m)

On the *Z*-component profile (Fig. 2.7a) the direct incident *P*-wave of a stable form is clearly seen; incident shear wave *S* is identified with certainty, as well as converted $PS\downarrow$ and $SP\downarrow$ waves. Reflected waves of the $PP\uparrow$ and $PS\uparrow$ types as seen on the *Z*-profile have shorter coherent line-up axes. The $SS\uparrow$ and $SP\uparrow$ reflected waves could not be identified, possibly because of a low intensity of the direct *S*-wave.

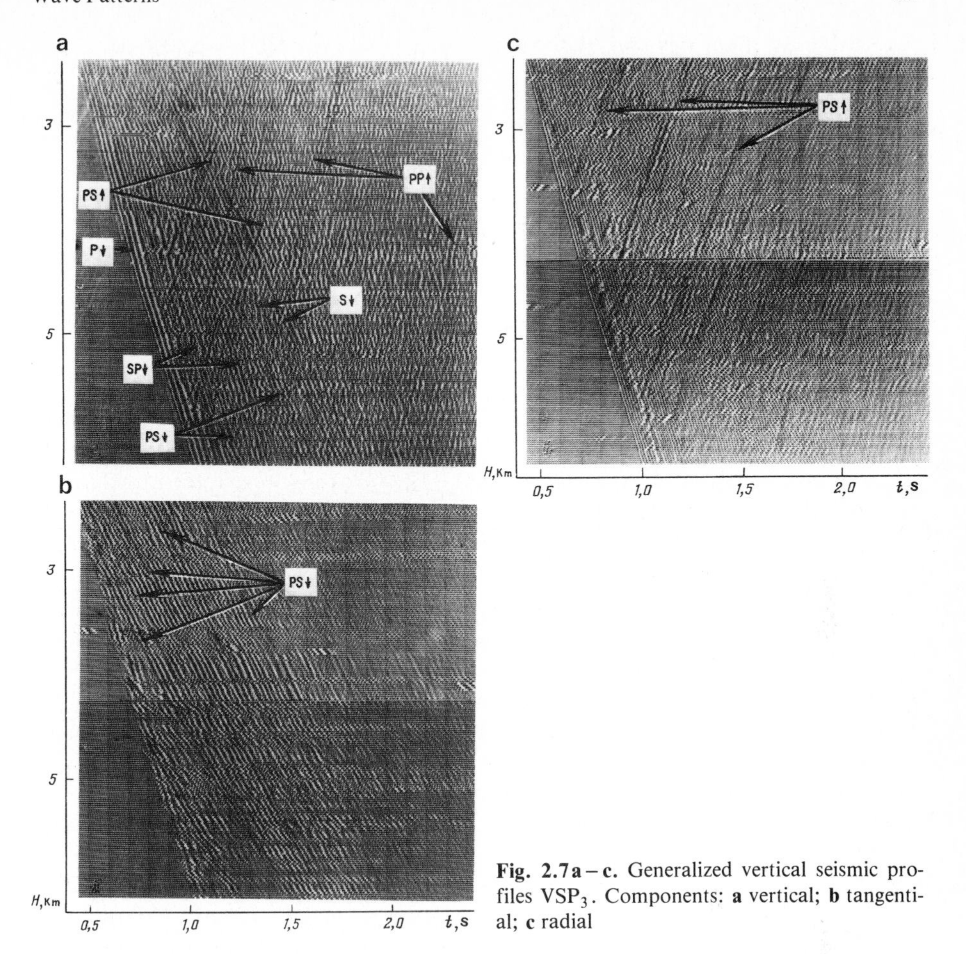


**Fig. 2.7a – c.** Generalized vertical seismic profiles $VSP_3$. Components: **a** vertical; **b** tangential; **c** radial

Polarization transformations gave the possibility of reliably determining the nature of some waves, tracing them for a longer distance and tie up stratigraphically. Improvement of the signal/noise ratio by respective component orientation is best when the directional characteristic minimum is in parallel to the main noise vector. By suppressing the incident waves one can more reliably trace the reflected and converted waves.

Registration of the component oriented tangentially to the direct waves displacement vector (Fig. 2.7) effectively suppresses the incident waves and emphasizes numerous converted waves of the $PS\downarrow$ type.

Reflected waves are very useful for studying the medium structure at a considerable distance from the well. On the images obtained the reflections are traceable for only short distances.

The wide range of apparent velocities (from 4000 to 12,000 m $s^{-1}$) proves the presence of both *P*- and *S*-waves. Because of steeply dipping interfaces, the reflected wave type cannot be defined unambiguously, for at great dips the apparent velocity of shear waves can exceed the interval velocity of compressional waves for the entire length of the reflection or in some places.

Geological identification of seismic reflections in the Kola well environment is complicated by the fact that the reflections remain continuous for relatively short depth intervals. From kinematic and dynamic parameters such short coherent line-ups are interpreted as reflections from small-size reflectors. Short coherent line-ups of the up-coming waves and the complex structure of the geological medium do not permit unambiguous determination of the reflection wave type. Therefore, the interpretation of reflections is to a considerable extent based on qualitative data available on the geological structure, and is made more precise when plotting the reflectors.

A considerable number of up-coming waves shown in the $VSP_1$ are classified as shear waves. There is no doubt about it, for their apparent velocities range from 4900 to 6000 m $s^{-1}$, i.e. they are smaller than the interval velocity of *P*-waves. Because the intensity of the incident *S*-wave is higher than that of the *P*-wave, these reflections can be supposed to be predominantly of the $SS\uparrow$ type. Reflections with an apparent velocity of 6900 – 12,000 m $s^{-1}$ were interpreted as the $PP\uparrow$ waves, which is in good agreement with our knowledge of dips of reflectors.

The $VSP_2$ section differs from the $VSP_1$ and $VSP_3$ ones in that mostly the $PS\uparrow$ and $PP\uparrow$ reflections are visible on it. On the *Z*-component section (see Fig. 2.7a) the *PP* waves are predominant, while on the radial component image (see Fig. 2.7c) the exchange $PS\uparrow$ reflections prevail. Apparent velocities for the $PS\uparrow$ waves are from 6000 to 6500 m $s^{-1}$, while for the $PP\uparrow$ waves they exceed 7500 m $s^{-1}$, which is in agreement with the geological section of the well.

Reflected *SS*-waves are not visible on the seismic sections because of the low intensity of the direct *S*-wave.

The identification of reflections from various images of the complex interferential wave field is possibly feasible only at the beginning of the traces where they can be tied up to definite depths. On the $VSP_2$ and $VSP_3$ profiles some reflections have been correlated with appropriate geological boundaries, for instance with those occurring at $H = 3800$ and 4900 m. Other reflections identified from successive arrivals recorded at greater times can be tied up with depth only conditionally, for they correspond to various reflecting features occurring sometimes very far from the borehole.

Weak velocity contrasts in the subsurface and discordance of interfaces do not stimulate the generation of strong multiples, therefore the waves identified at later arrivals were interpreted as deep reflections. On the $VSP_1$ and $VSP_3$ records they can be traced to a depth of about 10,000 m.

Experimental reflected waves are considerably more intensive than reflections computed for a thick-layer model. From this one can suppose that in the real subsurface there are units made up of thin layers well differentiated with respect to velocity and lense-shaped inclusions. Such a structure of the subsurface provides for a great variety of frequency spectra of the reflected waves.

### 2.8.3 Elastic Wave Absorption

The determination of absorption coefficients for compressional and shear waves in rocks of the Pechenga depression is complicated by a number of factors: low values of the coefficients conditioned by the rock petrography; difficulties in accounting for the wave geometrical spreading because of the complex geological structure; dispersion of seismic energy on the transmitting medium heterogeneities and its influence on effective attenuation of individual waves; instability of the explosive energy source and complicated characteristics of outgoing signals; and numerous wave-conversion events and variations in wave polarization in the process of its propagation which complicate the control of the source instability.

An attempt was made to calculate the effective absorption coefficients for compressional waves by investigating the *P*-wave amplitude behaviour with depth in the Kola and some exploration wells. The total displacement vector amplitude was reconstructed from records made with downhole vertical geophones accounting for the angle that raypaths made with the well axis. The angle was evaluated assuming the raypath to be linear. The plot of approximate amplitudes of the *P*-wave first phases against depth thus obtained for the superdeep well shows a complex behaviour of amplitude with depth. Starting from the Earth's surface and down to about 4500 m there is a general decrease in amplitude. In the depth interval from approximately 4500 to 7400 m an abnormal amplitude increase is noted. The reason for this should possibly be sought in the significant velocity changes known to appear here, like the presence of low velocity layers which distort the wave front and probably have particular properties.

The data obtained do not contradict the general knowledge of the absorption coefficient.

Reflecting features were plotted using all the time-depth curves available for *SS*, *PS*, and *PP* reflections seen on the VSP sections, assuming the medium to be homogeneous and all the elastic wave velocities constant. The substitution of the real subsurface with a single medium is justified as follows: the wave incidence being close to normal in a medium with low velocity contrasts, the velocity values as measured along the raypaths do not differ much from interval ones.

Modelling of thin-layer units with curved interfaces was not used, for it would have complicated the inverse problem solution and could lead to significant errors. The mean velocities derived from vertical profiles were used.

Reflecting features were plotted from the time-depth curve data using the intersection method. In some cases the inverse problem was solved using special techniques of converted wave interpretation (Recommendations on Procedure 1983).

Results of determination of reflector dips from the VSP data are shown in Fig. 2.8. Although the inverse problem solution obtained for such a complex region is two-dimensional and somewhat rough, it does characterize the structure of the near-borehole subsurface. Because of the poor continuity of reflections continuous reflectors could not be plotted, however, considering the complexity of the tectonic environment, here they are likely not to exist at all. The most extensive interfaces may be traced from one reflecting feature to another using respective stacking procedures.

The dips obtained are from 30° to 40°, which agrees with results by other methods.

The results of seismic observations in the Kola well are summarized as follows:

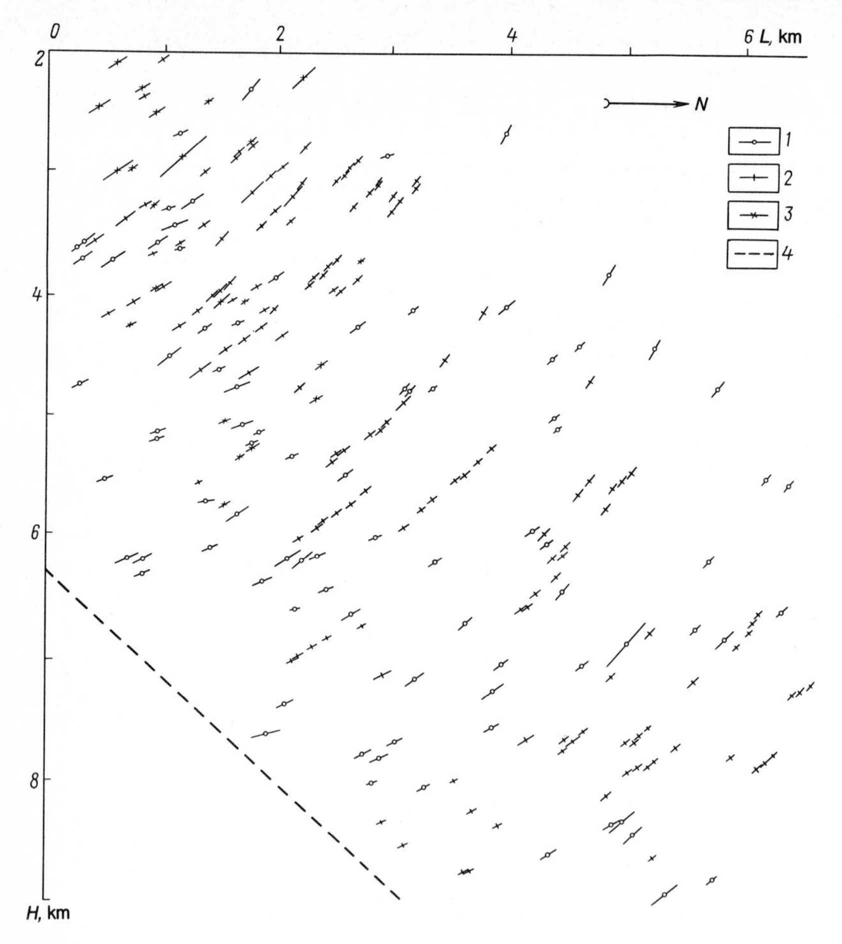


**Fig. 2.8.** Reflecting features. According to the: *1* $VSP_3$; *2* $VSP_2$; *3* $VSP_1$; *4* outlines of the domain studied; *L* distance from the well head along a horizontal profile

- high intensity of shear waves produced by the source provides for studying the well section using both *P*- and *S*-wave velocities, as well as more effective application of shear wave analysis to exploration;
- dominant frequencies of *P*- and *S*-waves (both incident and reflected) do not change much with depth, which demonstrates the weak attenuation of the waves in the Pechenga sequence rocks, and is favourable for investigating great depths and applying high frequency modifications of seismic methods with higher resolution;
- the VSP observations help to trace deep reflections characterized by short coherent line-ups, variable frequency spectrum and relatively high intensity, which indicates their interferential nature. The reflection wave field is formed of both monotype and converted waves;
- the subsurface medium is not favourable for the generation of multiples, consequently the upcoming waves can be interpreted as simple reflections;

- on all the VSP sections the three reflection wave types are predominant: $SS\uparrow$, $PS\uparrow$, and $PP\uparrow$. Predominance of the $SS\uparrow$ or $PS\uparrow$ waves on the records depends mostly upon the conditions of energy generation;
- analysis of the wave field together with a geoacoustic model of the subsurface in detail has resulted in the identification of several types of seismic reflector in the Kola well vicinity: (a) lithology contrasts at certain depths, like those at 2000, 2800, 5750 m; (b) lens-shaped bodies or discontinuous areals of thin layering (these features generate strong reflections of the *SS* and *PP* types with short coherent line-ups); (c) zones of abnormal stresses, like the one at depth of 3800 m; (d) tectonic boundaries associated with rock cataclasis and schistosity, for instance the one at the 4500 m depth.

In further more detailed studies of the wave field, dynamic and kinematic wave parameters as well as deep structure, it is necessary to use orientable downhole three-component geophone assemblages and stable seismic energy sources, particularly non-explosive ones.

## 2.9 Rock Electrical Properties

Electrical resistivity is one of the most important parameters measured in wells under study ($R_L$) or on core samples ($R$). The $R_L$ was determined for the rocks in situ from the LDS and LL readings and the $R$ was measured on samples with single- and two-electrode devices. Variations of $R$ and $R_L$ with depth are similar. Maximum values are noted in diabases, globular lavas and sandstones, while minimum ones are characteristic of aleurolites and phyllites of the Pechenga series. In the high-resistivity rocks the $R$ and $R_L$ magnitudes may differ by a factor of $10-10^2$, e.g. in metadiabase and globular lava at depths from 3200 to 3600 m the $R_L$ values reach $6\times10^4$ ohm-m, while $R=4.5\times10^5-5.6\times10^6$ ohm-m. For low resistivity rocks the values of $R_L$ and $R$ sometimes almost coincide, as in sedimentary rocks of the Zhdanov series, where $R$ and $R_L$ are of the same order of magnitude. The Archean rock specific resistivities measured on cores have appeared to be about ten times lower than the $R_L$ values, which may be explained by a high porosity (up to 7%) and permeability observed in the samples and possibly non-existent in situ at the 7000 m depth. An inverse picture is observed more often, when $R$ of the rocks in situ saturated with formational water is 10–50 times as low as measured on completely saturated samples (Physical Properties of Rocks and Minerals 1976).

The results of $R$ measurements within a polished section cut from a phyllite core taken from the depth of 1600 m are presented in Fig. 2.9. Measurements were taken each 1 mm (sampling span) and the readings were processed using the

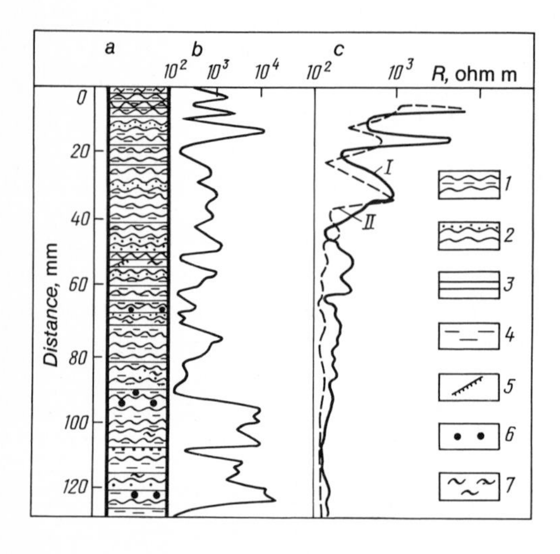


**Fig. 2.9.** Results of studying the polished section of a phyllite sample. *a* petrographic column of the core sample; *b* curve *R*; *c* spectral processing results for: instrumental (*I*) and visual (*II*) observations. *1* aleurite quartzous phyllite; *2* phyllite with sandstone; *3* aleurolite; *4* enrichment in carbon matter; *5* quartz streaks; *6* sulphide impregnation; *7* sulphide ore streaks

frequency-spacing analysis (Bat 1980). The given sample underwent petrographic analysis also with a 1-mm span (Fig. 2.9a).

For further data processing, each petrographic characteristic thus determined (see Fig. 2.9 *1* – *7*) was assigned its own code proportional to a mean value of $R$. Mathematical processing of the cyclogram obtained (see Fig. 2.9b, *11*) has shown good agreement between the results of visual and instrumental observations (as proved by a distribution bar graph reflecting the cyclicity of alternation of various layers in the polished section under study). The $R$ curve spectrum includes two major maxima with amplitdues of $5 \times 10^3$ and $10^3$ ohm-m and spacing periods of 18 and 10.5 mm, respectively. As follows from microscope examination of the cores, low-conductivity streaks alternate with syngenetic sulphide impregnations in the rock mass with these same periods. Consequently, the amplitude spectrum contains information on the periodicity of sulphide accumulation during sedimentation. Moreover, the spectrum of the $R$-curve observed is qualitatively similar to the results of visual petrographic analysis of the polished section. Thus, in order to study sulphide syngenetic impregnation it is feasible to measure the distribution of $R$ in core samples using a single-electrode circuit and to derive the predominant cycles of sedimentation from the amplitude spectra obtained. This technique enables more complete information on particular rock composition and sedimentation periodicity, needing less time and with lower costs than visual analysis.

Temperature influence on specific resistivity (conductivity, $C$) of some rocks from the Kola well is shown in Figs. 2.10 – 2.15. Experimentally obtained rock $C$ values (in ohm-cm) are shown in these figures as conductivity versus logarithmic-absolute temperature reciprocal plots, each curve (isobar) being plotted for a certain pressure. Such a presentation of the results allows for determination of the activation energy of electrical conductivity $E_0$ and exponential coefficient for the equation $C_t = C_0 \cdot e^{-E_0/kT}$, where $C_t$ and $C_0$ are electrical conductivities at temperatures $t$ and 0, respectively, $k$ is a constant. Moreover, the curve character (rate of sloping, bending points associated with certain temperature and the curve anomalous behaviour) may be used as an indicator of the mechanism of rock conductivity, physico-chemical transformations, dehydration processes, mobility of charge carriers, their diffusion and other phenomena. The activation energy and exponential coefficient values obtained experimentally are used in theoretical calculations of the conductivity of the deep layers of the Earth (Magnitski 1978). The "mantle" (ultrabasic) rocks, represented by peridotite in the Kola well section, are of the utmost interest in the light of this.

Electrical conductivity of serpentinite ultrabasic rocks from the well varies with temperature over a wide range (see Fig. 2.10). For comparison, a relatively narrow dispersion of $\lg C = f(1/T)$ points is shown in this figure for unaltered olivine, olivinite, pyrope-peridotite and dunite. These data demonstrate the effect of rock serpentinization and, consequently, the alteration of its mineral composition on electric conductivity. There is a relation between serpentinite conductivity and the amount of ore minerals (fine impregnation with magnetite) dispersed in it, as well as the type of minerals of which it is composed. The lowest conductivity is characteristic of serpentinites containing up to 10% or more of magne-

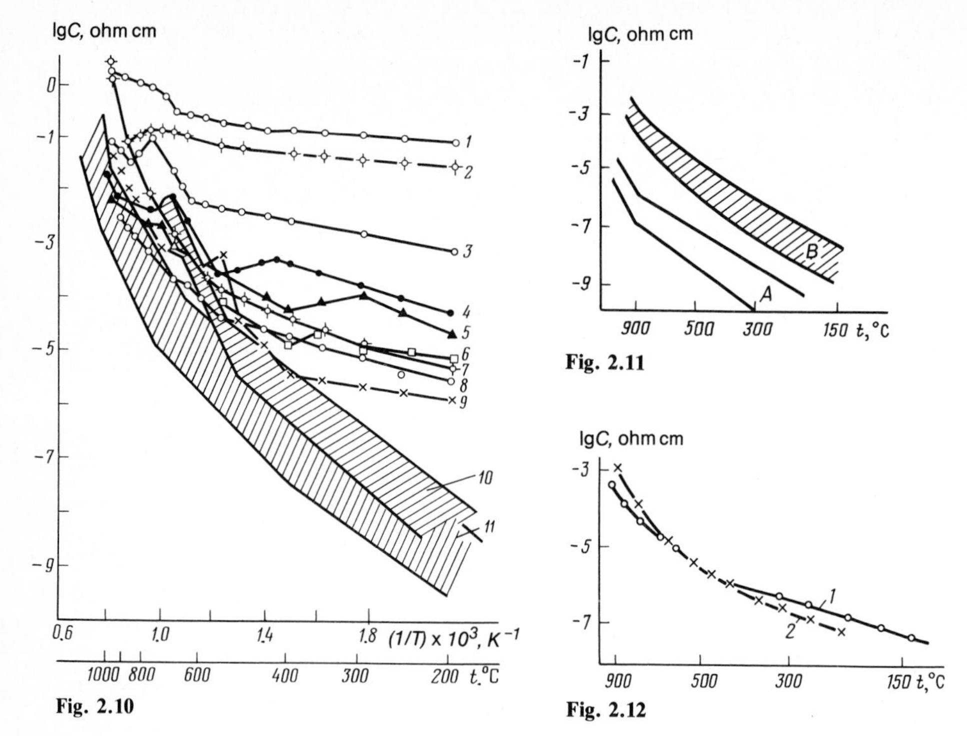


**Fig. 2.10.** Ultrabasic rock conductivity-temperature relationship. *1–9* curves for serpentinite samples from the Kola well; *10* dispersion of conductivity values for the ultrabasic rocks at 1500–2000 MPa pressure; *11* the same at atmospheric pressure

**Fig. 2.11.** Actinolitic diabase conductivity-temperature relationship. Conductivity dispersion for: *A* unaltered gabbro; *B* conductive basalt

**Fig. 2.12.** Conductivity-temperature relation for two samples (*1* and *2*) of globular lava

tite. The variations in ultrabasic rock conductivity at pressures from 1500 to 2000 MPa are illustrated in Fig. 2.10 by the hatched area.

The $\lg C = f(1/T)$ curves shown in Fig. 2.11 for actinolitic diabases are close enough at high temperatures, which indicates the similarity of their mineral compositions. In Fig. 2.11 for comparison purposes, the limits of the most characteristic variations in conductivity of unaltered gabbro and highly conductive basalt are also shown. The diabase conductivities come within these limits. High conductivity of the diabases studied as compared to other diabases is explained by their particular mineral composition (plagioclase, augite, actinolite) and texture. The fine-grained structure of these rocks and ore mineral inclusions are also likely to augment their electrical conductivity (as compared to other common basic rocks).

The presence of actinolite is responsible for certain specific features of the $\lg C = f(1/T)$ curves, such as gently sloping segments corresponding to low activation energy. This is explained by the effect of dehydration, which causes the

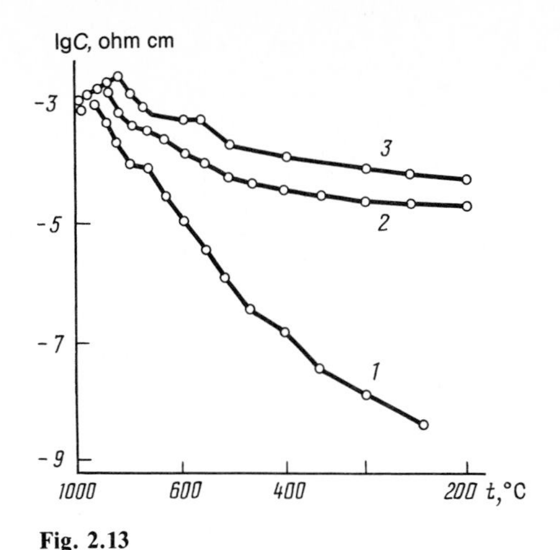


Fig. 2.13

Fig. 2.14

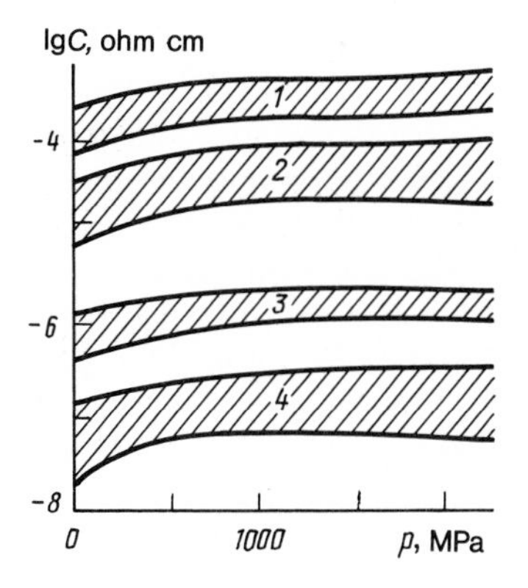


**Fig. 2.13.** Conductivity-temperature relation for phyllite-carbonate aleurolite alternation. *1* normal to laminae; *2* along the laminae; *3* along the core axis

**Fig. 2.14.** Conductivity-temperature curves for some samples (*1*–*4*) of Kola well actinolitic diabase obtained at various pressures

◄ **Fig. 2.15.** Conductivity-pressure relation for actinolitic and hornblende diabase at different temperatures, °C: *1* 600; *2* 400; *3* 300; *4* 200

complex anions (OH) with low activation energy to play a leading role in rock conductivity at such temperatures. The activation energy of the actinolitic diabase within the self-conductivity domain is relatively low: for the majority of samples investigated its value is 1.35 eV, sometimes lower, and does not exceed 2.2 eV.

A comparison of conductivities measured on samples of slightly actinolitic globular lavas of the Kola well taken from different depths shows good agreement for all the temperature intervals (Fig. 2.12), which shows the similarity of their chemical and mineralogical compositions and structures as well.

Among the rocks investigated the magnetitic andesite porphyrite and cyclic aleurite-phyllite alternation have the lowest and highest electrical conductivities, respectively. The latter is characterized by a strong conductivity anisotropy that, however, decreases with temperature (see Fig. 2.13). The maximum conductivity is measured parallel to the alternating layers, and the minimum one perpendicular to them. These two conductivities differ from each other by five orders of magnitude, from which it follows that the anisotropy is not of crystallographic nature but simply results from the alternation of highly conductive (due to the ore mineral presence) layers with poorly conductive ones.

The effect of high (up to 2000 MPa) quasi-uniform compression on electrical conductivity at temperatures from 150° to 600 °C was studied for actinolitic and hornblende diabases, picritic porphyrite and phyllite schists from the 2276 – 4534 m depth interval, where they are represented most fully.

It follows from the experimental data that at high pressure and temperature the conductivities of actinolite and hornblende diabases are close or equal to each other and vary within a rather narrow range (see Fig. 2.14). These diabase conductivities increase by 40 – 70% with pressure increase from 50 to 2000 MPa, and with temperature rising from 200° to 600°C they increase about 10 times each 100 °C. For some diabases, the increase in $C$ as a function of $P$ is accompanied by a drop in activation energy $E_0$ of charge carriers controlling the mechanism of electrical conductivity and thus determining its absolute value (see Fig. 2.14). The activation energy has been calculated from the isobars. At pressures from 50 to 2000 MPa and temperatures from 150° to 650 °C it is equal to 0.5 – 0.7 eV.

An analysis of generalized results (see Fig. 2.15) shows that at pressures of 50 – 2000 MPa and temperatures of 150° – 600 °C the actinolitic and hornblende diabases have maximum conductivity values, i.e. lowest resistivities. Lower conductivities are found for diabases with globular structural features (a sample from the 2937.2 m depth). At pressures from 1000 to 2000 MPa and within the above temperature range the andesite porphyrite (more acid rocks) conductivity is three orders of magnitude lower than that of the actinolitic diabases.

A change in the character and value of conductivity anisotropy of schistose tuffogenic aleurites and phyllites with temperature is evident from Fig. 2.13. The highest conductivity (at temperatures from 200° to 600 °C) is measured along the lamination, the lowest normally to it, and the intermediate one along the core vertical axis.

The anisotropy factor of the specific resistivity as calculated (in %) using formula $[(R\| - R\perp)/R\perp]100$, where $R\|$ and $R\perp$ are resistivities along and normal to the schistosity direction, increases considerably at temperatures exceeding 400 °C (Table 2.10), probably because of a sharp rise in conductivity accompanying the dehydration processes known to progress in a laminated rock with different rates in different directions.

The study of electrical properties of rocks from the Kola well by well logging and core analysis can thus be concluded as follows:

- the specific resistivity of rocks varies from $1 \times 10^{-6}$ to $1 \times 10^{9}$ ohm-m, the rocks with high $R = 10^5 - 10^6$ ohm-m predominating in the well section;
- on the background of high-resistivity rocks (diabase, gabbro-diabase, globular lava, sandstone, etc.) there are sharply anomalous intervals of elevated conductivity resulting from graphitization of aleurolites and phyllites and the presence of sulphide streaks. This provides for a high efficiency of electrical logs in stratification and lithological studies of the Proterozoic volcano-sedimentary sequence of the Pechenga structure;
- the Archean rocks have the highest $R$ values and their electrical properties practically do not vary.

**Table 2.10.** Pressure and temperature effect on anisotropy factor for phyllites

| T, °C | Anisotropy factor (%) at pressures (MPa) | |
|---|---|---|
| | 50 | 1000 |
| 200 | 810 | 857 |
| 300 | 434 | 404 |
| 400 | 540 | 380 |
| 500 | 605 | 589 |
| 600 | 608 | 1360 |

The method of frequency-spacing analysis of $R$ values obtained from core samples with a one-pole measurement circuit can be used in determining the dominant frequencies of these parameter variations and studying the cyclicity of sedimentation of metamorphosed volcano-sedimentary rocks on the basis of their electrical properties. Increase in pressure has a lower effect on rock electrical conductivity than increase in temperature. Besides the change in electrical conductivity, the rocks studied demonstrate also the change in activation energy of ions controlling rock conductivity.

These conclusions are very important for studying the deep structure of the Earth (Magnitski 1978), as they give grounds for assuming that a local increase in the electrical conductivity of crystalline rocks in the Earth's crust and subcrustal layers can be caused by processes of rock dehydration.

## 2.10 Magnetic Properties of Rocks

Magnetic parameters reflect the concentration and composition of ferromagnetic minerals, their structural features and interrelation with paramagnetic iron minerals.

The magnetic properties of rocks from the Kola well were studied both in situ and on core samples. For rocks in situ, the components of the magnetic field vector $\boldsymbol{H}$ were measured, as well as the magnetic susceptibility ($\varkappa$). Magnetic parameters derived from the Kola well cores are as follows: magnetic susceptibility $\varkappa$ and anisotropy $A_x$; value and direction of residual magnetization $\boldsymbol{J}_n$; Koenigs-

**Table 2.11.** Quantitative magnetic characteristics for the Kola well section

| Rock type | Rock group | $\varkappa \times 10^5$, SI units | $Q$ | $J_n \times 10^5$, $A/m$ |
|---|---|---|---|---|
| *Proterozoic, depth interval 0–4586 m, zone of sulphide mineralization* | | | | |
| Igneous, tuff | Diabase | 88 | 3.05 | 231 |
| | Gabbro-diabase | 105 | 4.1 | 153 |
| | Ultrabasite | 6160 | 2.23 | 10612 |
| | Basic tuff | 151 | 1.79 | 278 |
| Weighted average | | 561/94[a] | 3.13/3.2[a] | 1017/218[a] |
| Tuffitic | Tuffitic | 108 | 1.94 | 50 |
| sedimentary | Clastic, chemogenic | 129 | 7.21 | 825 |
| Weighted average | | 125 | 6.05 | 654 |
| *Proterozoic, depth interval 4586–5642 m, zone of oxide mineralization* | | | | |
| Igneous, tuff | Apodiabase schist | 4760 | 1.2 | 1440 |
| | Andesite-porphyrite | 66 | 4.1 | 243 |
| | Ultrabasic, tuff | 7067 | 1.5 | 7639 |
| Weighted average | | 4542 | 1.4 | 1507 |
| Sedimentary | Clastic, chemogenic | 86 | 1.02 | 88 |
| *Proterozoic, depth interval 5642–6842 m, zone of mixed mineralization* | | | | |
| Igneous | Apodiabase schist | 43 | 0.4 | 23 |
| | Andesite-porphyrite | | | |
| | Ultrabasic | 56 | 0.2 | 7 |
| Weighted average | | 45 | 0.4 | 22 |
| *Archean, depth interval 6842–10,500 m, zone of mixed mineralization* | | | | |
| Gneiss | | 20 | 2.0 | 22 |
| Amphibolite | | 70 | 9.1 | 680 |
| Ultrametamorphite | | 40 | 1.4 | 39 |
| Weighted average | | 46 | 2.5 | 70 |

[a] Numerator shows the values calculated for all the rocks, the ultrabasic included; denominator without the ultrabasic ones.

berger ratio $Q = J_n/0.5\,\varkappa$; saturation magnetization $J_s$, residual saturation magnetization $J_{rs}$; demagnetizing field $H'_{cs}$ for the residual saturation magnetization, and parameter $N_t$ (see Chap. 2.10).

These parameters were measured using an astatic magnetometer and ROK-generator ION-1, thermomagnetometer and vibration magnetometer. Respective techniques of core parameter measurements and data processing are described in the literature (Gontcharov et al. 1982).

The results of the magnetic properties study for cores from the Kola well are listed in Table 2.11.

From the well cores the most representative data (over 35,000 readings) were obtained for the magnetic susceptibility, natural residual magnetization and Koenigsberger ratio.

### 2.10.1 Magnetic Susceptibility

This is a very important magnetic characteristic of rocks, showing their ability to be magnetized by a constant magnetic field $\boldsymbol{H}$. In the Kola well section the $\varkappa$ values vary considerably.

Computer processing of data from one petrophysical interval to another showed that for 80% of the intervals the $\varkappa$ values differ significantly from each other with probabilities from 99 to 99.9%, the latter limit prevailing.

In the Proterozoic Pechenga sequence the greatest average $\varkappa$ values are observed for apotrachybasalt schists of the Pirttiyarvin series (4884–5619 m) as $65 \times 10^{-3}$ SI units and serpentinitic peridotites of the Zhdanov series (1541–1678 m) as $124 \times 10^{-3}$ SI units. Relatively high $\varkappa$ is characteristic of talcose and serpentinitic picrite-porphyrites of the Matertin series, $(19-38) \times 10^{-3}$ SI units, and andesite-dacite-metaporphyrites of the Luchlompol series, $5 \times 10^{-3}$ SI units. All these rocks have a magnetic susceptibility of ferromagnetic nature. In metavolcanites of the tholeiite-basalt and andesite-basalt petrochemical types, the $\varkappa$ varies within a narrow range (from $0.5 \times 10^{-3}$ to $1.0 \times 10^{-3}$ SI units) independently of their grade of metamorphism, and is of ferroparamagnetic nature. The lowest $\varkappa$ values are observed in dolomites and metapsammites of the Luchlompol and Kuvernerinyok series. Consequently, magnetic anomalies in the Pechenga rock sequence are related to either ultramafic bodies represented by the greenschist and prehnite-pumpellyite metamorphic facies or metavolcanites of the alkaline petrochemical type independently of their grade of metamorphism.

The magnetic susceptibility of Archean rocks in the Kola well is generally lower than in the Proterozoic sequence or the Archean rocks outcropping at some distance from the well (Fig. 2.16). Ferruginous silicates have the greatest values of $\varkappa = (0.6-1.0) \times 10^{-3}$ SI units. The ferromagnetic nature of $\varkappa$ is also characteristic of epidote-biotite-amphibole gneisses, schists and amphibolites migmatized with plagiomicrocline granites, whose $\varkappa = (4.6-9.6) \times 10^{-3}$ SI units. These rocks are generally the subordinate ones of the section and concentrate in its lower part (depth interval 9541–11,000 m). The background value of the magnetic susceptibility is given by nebulitic migmatites and amphibolites. The nebulitic migmatites show the increase in $\varkappa$ downsection from $0.09 \times 10^{-3}$ to

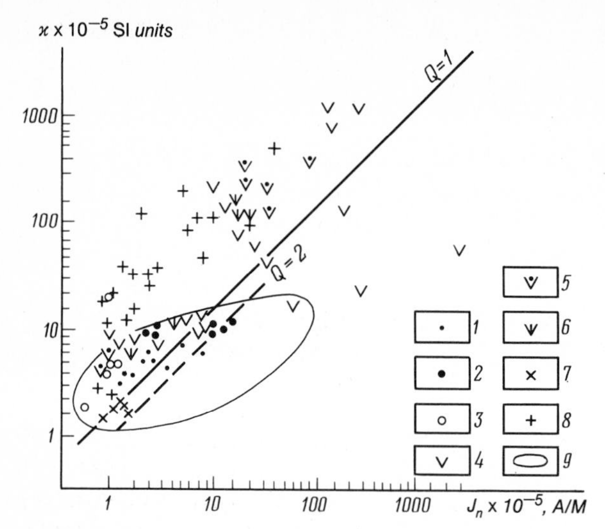


**Fig. 2.16.** Magnetic susceptibility-residual magnetization relation for types of Archean rocks from the Kola well and distant outcrops. From outcrops: *1* biotite gneiss with garnet; *2* high-aluminium gneiss; *3* biotite gneiss with epidote; *4* crystalline schist and amphibolite; *5–7* rocks of the I granitization series, diorite, tonalite and plagiogranite variously diaphthoritic; *8* rocks of the II granitization series of plagiomicrocline granite-granodiorites. From the well: *9* domain for the rocks from depth interval 7 to 10 km (for 80% measurements $Q<2$)

$0.54\times10^{-3}$ SI units and in amphibolites the $\varkappa$ varies from $0.53\times10^{-3}$ to $1.57\times10^{-3}$ units, the highest $\varkappa$ values and their greatest dispersion being observed in ancient pre-migmatic amphibolites. These are characterized by a log normal distribution of $\varkappa$, which is not the case with their apodiabase dyke analogues. Considerable dispersion of $\varkappa$ values for these rocks proves the superimposed nature of the magnetite minerogenesis probably associated with granitization.

Results of petromagnetic investigations of Archean rocks outcropping to the North-east from the Pechenga structure have shown that in this region there is a regular relation between the $\varkappa$ values and the degree of preservation of granulitic paragenic mineral associations. High $\varkappa$ is characteristic of enderbites and charnockites, i.e. nebulitic migmatites of the granulite facies. Diaphthoresis of the epidote-amphibolite and amphibolite facies leads to the complete disappearence of magnetite in these rocks or a sharp decrease in its concentration to the level observed in similar rocks from the Kola well, where the Archean rocks are known to have been subjected to the Lower Proterozoic metamorphism-diaphthoresis of peak intensity.

Comparison of $\varkappa$ measured on core samples with that derived from downhole magnetic survey data has shown their perfect agreement, i.e. the Kola well investigations do not prove the supposedly considerable influence of pressure on rock $\varkappa$, as was believed by some investigators (Lebedev and Poznanskaya 1970). Consequently, the in situ conditions (at least in the upper part of the Earth's crust) do not affect $\varkappa$, therefore it is right to measure this quantity at atmospheric environments on cores taken from wells and use the values obtained as true ones.

Statistical processing of numerous measurements of magnetic susceptibility helped to establish the regularities in its distribution over the well section (see Table 2.11) as follows:

– downsection from Proterozoic rocks to Archean gneisses the $\varkappa_{av}$ regularly decreases from $10^{-3}$ to $10^{-4}$ SI units with two exceptions: peridotites of the

Zhdanov series (sedimentary unit IV, $\varkappa > 10^{-2}$) and zone of magnetite enrichment in the Pirttiyarvi series (volcanic unit II, $\varkappa > 10^{-3}$);
- difference in mean magnetic susceptibilities of igneous and sedimentary rocks diminishes with depth. In unit IV (Zhdanov and Matertin series) it persists still, while starting with unit III (Zapolyarny and Luchlompol series) and deeper on the mean $\varkappa$ values for these rocks are equal to each other ($\varkappa_{av} = 8 \times 10^{-5}$ SI units) with one exception (magnetite zone in volcanic unit II).

These regularities are in agreement with the prograde metamorphism down from the prehnite-pumpellite and greenschist facies to the epidote-amphibolite one. Deep metamorphism is usually accompanied by the destruction of magnetic minerals on one hand, and equalization of rock chemical composition on the other.

A comparison of average $\varkappa$ values calculated for the petrophysically defined intervals proved the high efficiency of this parameter in the identification of rock lithology. The Student's *t*-test was used as a statistical measure of distinction.

Anisotropy of magnetic susceptibility and residual magnetization varies over a wide range ($\varkappa_{max}/\varkappa_{min}$ and $J_{rs_{max}}/J_{rs_{min}}$ are as high as 2.9 and 2.4, respectively) and reaches its peak in schistose intervals of the magnetite zone of the Pechenga sequence and also in Archean gneisses. Considerable dispersion of the anisotropy ratio values is partly explained by a somewhat casual choice of core sample orientation with respect to the direction of stresses in situ. There are some laws in anisotropy behaviour: (1) general growth of anisotropy with depth; (2) within a single unit the sedimentary rocks are more anisotropic than the igneous ones (the ratios are 1.21 and 1.05 respectively). This is evidently caused by more intensive schistosity of sedimentary rocks as compared to igneous ones. This difference is smoothed with depth, which agrees with the increase in intensity in metamorphism.

It follows that magnetic anisotropy is paragenetically associated with elastic anisotropy of the rocks, i.e. increases with depth and grade of regional metamorphism.

### 2.10.2 Natural Residual Magnetization

Because azimuths were not determined for the Kola well cores and, consequently, the behaviour of the $\boldsymbol{J}_n$ vector could not be fully analyzed, only its absolute value $J_n$ and the angle that the vector makes with the horizontal ($I$, i.e. the inclination of vector $\boldsymbol{J}_n$) were studied.

Quantity $J_n$ has a log normal distribution even in the most homogeneous rocks and depends first of all on magnetite concentration and superimposed pyrrhotite mineralization of metamorphic nature.

The maximum value of $J_n$ equal to $21{,}610 \times 10^{-3}$ A m$^{-1}$ was measured for serpentinite peridotites of the Pechenga sequence enriched in magnetite and pyrrhotite. Rather high values are characteristic of phyllites and aleurolites of the productive unit ($J_n = 2101 \times 10^{-3}$ A m$^{-1}$) rich in pyrrhotite as well as alkaline

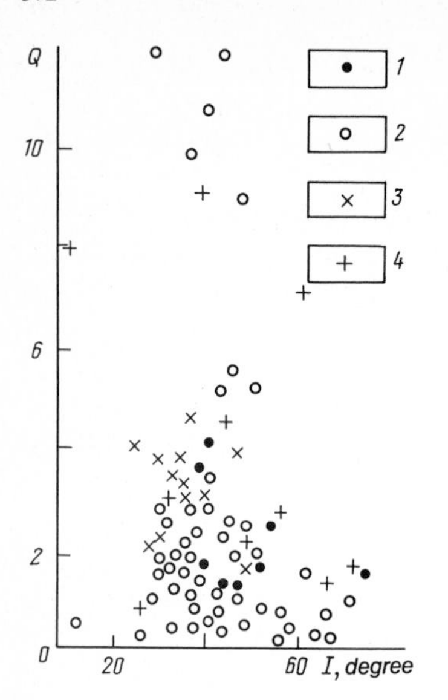


**Fig. 2.17.** Relation between the average values of inclination of the natural residual magnetization vector and the $Q$ factor for the series as follows: *1* Matertin, Zapolyarny; *2* Zhdanov, Luchlompol and Pirttiyarvi; *3* Kuvernerinyok, Mayarvi; *4* Archean

metavolcanites of the Pirttiyarvin series enriched in magnetite ($J_n = 1650 \times 10^{-3}\ \mathrm{A\,m^{-1}}$). In metavolcanites of the tholeiite-basalt type, the transfer from prehnite-pumpellyite to greenschist facies is accompanied by a certain increase in $J_n$ with successive decrease in this quantity still deeper where the epidote-amphibolite facies is encountered. This behaviour is in agreement with the distribution of pyrrhotite mineralization in the facies of regional metamorphism.

In the Kola well the Archean rocks have generally lower $J_n$ values than Proterozoic ones; however, the contribution of the residual magnetization to the total magnetization for these (Archean) rocks is often considerably more than for their analogues outcropping to the north-east from the Pechenga structure, which again is explained by the pyrrhotite mineralization characteristic of the most of the Kola well section.

Normal polarity of the $\boldsymbol{J}_n$ vector prevails in the Kola well section. As determined from results of the investigation, mean values of inclinations $I$ for rocks of relatively high $Q$ (enriched in pyrrhotite) are confined to the range of 30° – 50° (Fig. 2.17). For lower $Q$ values the inclination shifts to greater angles, i.e. approaches the character of the recent geomagnetic field (about 80°). Consequently, less stable rocks become partly or completely remagnetized, being affected by the present geomagnetic field. Such magnetically unstable rocks are most common in the first volcanic unit (the Matertin series). The stability of average inclinations of the $\boldsymbol{J}_n$ in rocks of a high $Q$ all over the section, their distinction from the inclination of the present geomagnetic field and similarity with inclinations measured for reversely magnetized rocks prove that the high $Q$ value rocks preserve their ancient magnetization.

From the insignificant scattering of the inclination values, the age of the residual magnetization $\boldsymbol{J}_n$ can be assumed to be approximately the same for all the well section.

### 2.10.3 Koenigsberger Ratio ($Q$-Factor)

The $Q$-factor is equal to the ratio of the natural residual magnetization to the induced one. It is very important when interpreting anomalous magnetic fields. Along the Kola well section the $Q$ varies from 0.1 to 29, and correlates with the magnetic mineral composition: in the zone of pyrrhotite concentration high $Q$ values are observed, while in the magnetite zone $Q \leqslant 1.5$. It should be noted that in sedimentary rocks of the pyrrhotite zone the $Q_{av}$ are, as a rule, greater than in igneous ones, because in the former a more "strong" magnetic mineral (antiferromagnetic pyrrhotite) is relatively more common. Relatively high $Q$ values measured for such ancient rocks prove either the thermoresidual or chemical nature of $J_n$.

Normal polarity of the $J_n$ vector prevails in the section. The net thickness of reversely magnetized rocks comprises about 10% of the section. Reverse magnetization is encountered more often in the lower part of the section studied. It is characteristic of a variety of rocks (independent of their petrography, genesis and composition of magnetic minerals); consequently, the reverse polarity of $J_n$ can be supposed to result from reversals of the geomagnetic field.

Zones of pyrrhotite and magnetite mineralization manifest themselves differently on downhole magnetic logs. The sulphide (pyrrhotite) concentrations give sawtooth-shaped sharply differentiated anomalies of the magnetic field (up to $5 \times 10^3$ nT), the anomaly amplitude of the horizontal component considerably exceeding that of the vertical one, which proves the predominance of the horizontal magnetization over the vertical one (about 1.5 times). No prominent anomalies of magnetic susceptibility were noted in these zones ($\varkappa$ does not exceed $2.5 \times 10^{-3}$ SI units).

According to the downhole magnetic survey, the $Q$ factor measures up to 40 units.

The oxide (magnetite) concentrations are marked by less fluctuating peaks of high amplitude (up to $10-20 \times 10^3$ nT) on the magnetic field logs which very closely correlate (Fig. 2.18) with intensive anomalies of the magnetic susceptibility (up to $500 \times 10^{-3}$ SI units). The relative contribution of the residual magnetization to the total one is smaller for this zone as compared with the zone of sulphide mineralization. The $Q$ values here vary from 0.1 to 1.

Dips and azimuths for magnetic rock layers as derived from magnetic log readings are shown in Table 2.12.

The listed data demonstrate angular and azimuthal discordances in the occurrence of magnetically active bodies. Sharp interfaces identified in the well section earlier from density, velocity and other rock properties are also evident from the behaviour of magnetic parameters.

#### Results of Rock Magnetization-Mineralogy Analyses

Saturation magnetization controlled by the composition and percentage of ferromagnetic minerals in rocks varies over a wide range (from 0.05 to 15.9 A $m^{-2}$ kg $-1$) even for similar rocks represented by small-size samples. This points

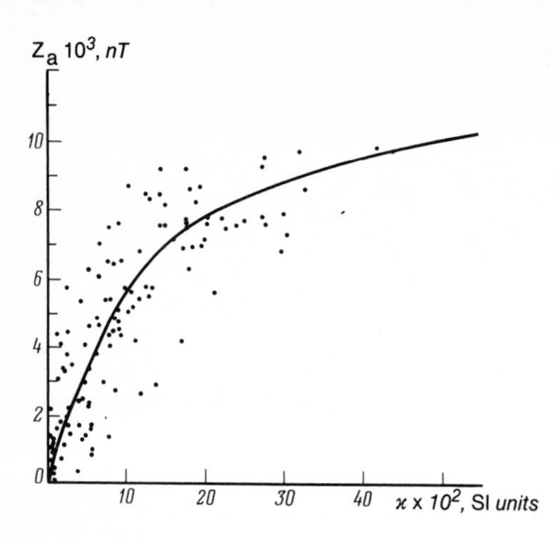


**Fig. 2.18.** Correlation of magnetic field vertical component ($Z_a$) with the magnetic susceptibility ($\varkappa$) for zones rich in magnetite from magnetic log data

out a very irregular distribution of the ferromagnetic material that could result only from a superposed mineralization. In general, low values of $J_s$ are predominant ($<0.2\ \mathrm{A\,m^{-2}\,kg^{-1}}$), particularly in the lower part of the section. The latter observation, together with thermomagnetic analysis results [a hyperbolic shape of the $J_s = f(T)$ curve], indicate a low percentage of ferromagnetic minerals and a significant contribution of paramagnetic ferruginous compounds to the rock magnetic behaviour. Elevated values of the $J_s$ are measured for strongly sulphidic rocks ($0.7-1.0\ \mathrm{A\,m^{-2}\,kg^{-1}}$) and rocks enriched in magnetite (up to $15.9\ \mathrm{A\,m^{-2}\,kg^{-1}}$). The pyrrhotite concentration is below 0.2% except for the zones of a high sulphide content.

It follows from analysis of thermomagnetic data and coercivity spectra that the rocks studied contain both ferromagnetic (monoclinic) and antiferromagnetic (hexagonal, or $\lambda$-type) pyrrhotite types. Curie points for the pyrrhotites are distinguished statistically: for the monoclinic pyrrhotite average $T_c = 350\,°\mathrm{C}$. A rather distinct tendency is observed as follows: the monoclinic pyrrhotite with relatively high $T_c$ is much more common for igneous rocks, while the hexagonal one prevails in sedimentary ones. Magnetic "strengths" of these pyrrhotites are also different. For instance, the $H'_{cs}$ as measured on samples enriched in the $\lambda$-type pyrrhotite varies from 13 to 250 mT or more, coercivity spectra obtained for these samples are widespread and their peak values ($H_{max}$) exceed 15 mT at depths shallower than 7.5 km. At greater depths the material becomes "softer" and its $H_{max}$ varies from 9 to 13 mT to the depth of 10.5 km. For the samples rich mainly in the monoclinal ferromagnetic pyrrhotite these characteristics are as follows: $H'_{cs} = 8-20$ mT, coercivity spectrum is less spread and its peak is within the range of $10-14$ mT. The samples containing the different pyrrhotite types are more distinctly distinguished by the ratio of demagnetizing field $H'_{cs}$ to peak value $H_{max}$ of the coercivity spectrum. This ratio approximately reflects relative percentage of single- and multi-domain grains. For samples with the ferromagnetic pyrrhotite it is equal to $0.9-1.2$, while for those with the $\lambda$-type pyrrhotite $H'_{cs}/H_{max} = 1.0-1.1$ to the depth of 4.5 km, then increases to $1.3-1.4$, and below 9 km it becomes as high as 1.5, i.e. the distinct general decrease in pyrrhotite concentrations with depth is accompanied with a decrease in the grain size.

**Table 2.12.** Results of magnetic log interpretation

| Depth interval | Dip of layers Degree | Azimuth of the dip Degree |
|---|---|---|
| 4600 – 4616 | 35 | 200 |
| 4880 – 5640 | 65 | 150 |
| 5717 – 5720 | 35 | 150 |
| 7628 – 7633 | 70 | 130 |
| 9640 – 9655 | 50 | 170 |

In the thick magnetite zone of unit II where pyrrhotite is practically absent the magnetite concentration as derived from the $J_s$ value varies from 0.5 to 17%. The Curie point varies from 580° to 600 °C and higher. This is connected with a one-phase oxidation of magnetite and development of its cation deficiency which is proved by the X-ray phase analysis. In the rock unit studied, the magnetite grains are relatively large, which is supported by respective parameters: $H'_{cs} =$ 7 – 15 mT, $J_{rs}/J_s = 0.04 - 0.08$. In some samples, besides magnetite hematite is also present ($H'_{cs} = 60$ mT, $T_c \geqslant 650$ °C). Below the magnetite zone in the Archean rocks at depths of 7.2 – 10.5 km there is (besides pyrrhotite) some magnetite spread all over the section, its concentration not exceeding 0.01% as evaluated from the $J_s$ value and $T_c = 580° - 620$ °C according to thermomagnetic measurements. Coercivity spectra obtained for the samples with magnetite have shown that the $H'_{cs}$ and $H_{max}$ values are close. Their ratio does not vary to the depth of 7 km and at greater depths becomes equal to 1.3 – 1.4. The temperature of magnetization (or formation) of ferromagnetic mineral can be estimated from parameter $N_t = H_x/H_0$ (where $H_0$ is the maximum constant field of the Rayleigh domain of the sample, and $H_x$ is the horizontal distance between the straight-line segments of coercivity spectra derived for the natural state and zero state of the sample). When $N_t \geqslant 0.25$, the temperature of sample magnetization is equal to the mineral Curie point or exceeds it. On average, the rocks from the top part of the Pechenga sequence have $N_t = 0.22$, i.e. the temperature of pyrrhotite magnetization (and possibly crystallization) was close to its Curie point (300° – 350 °C). Downsection in unit III the $N_t$ decreases and then drops down to 0.07 at the top of the magnetite zone. High values of $Q$ in combination with very low $N_t$ are proof of a low-temperature chemical origin of rock magnetization in the transition zone. Rocks from the bottom part of unit I have a similar low-temperature magnetization; in the top part of the Archean sequence (depths from 7.3 to 8.2 km) the $N_t$ changes from 0.04 to 0.3 and still deeper it again drops down to $N_t \geqslant 0.16$. In the magnetite zone of unit II the $N_t$ changes from 0.16 to 0.3. Consequently, the magnetization temperature was generally close to its Curie point, i.e. over 500 °C.

Side by side with the general increase in the temperature downwards from unit IV to unit II, or from the greenstone metamorphism to the amphibolite one, there are zones of low-temperature magnetization present at the bottoms of unit III and unit I as well as in the Archean. This irregularity is most probably indicative of a repeated pyrrhotite minerogenesis (or recrystallization), particularly a low-temperature one.

The $\lambda$-type pyrrhotite is known to undergo a transformation at about 250 °C when being heated. Analyzing the results of a step-up heating of samples [the $J_s(T)$ curve features like the peak intensity, width and temperature, the changes in $J$ and $H'_{cs}$, etc.], one can determine whether the pyrrhotite has already been transformed or not, i.e. has been heated in situ and if so, to what temperatures. The core samples of rocks containing pyrrhotite were studied in this manner in detail. The results have shown that for most of the section the pyrrhotites either were not heated at all or the repeated heating did not exceed 200 °C. Pyrrhotites heated to higher temperatures are encountered only in exocontact zones of some intrusive bodies. From the presence or absence of the pyrrhotite heating indications in near-contact zones of the intrusive bodies, the age of various intrusions can be evaluated and the pyrrhotite metallogenesis of different periods distinguished.

Using the above indications, it is easy to identify the undoubtedly superposed processes that have affected the bottom of gabbro bodies (depth from 1282 to 1421 m) and distinguish the peridotite bodies (550 – 570 m and 1540 – 1670 m depths) with adjacent surrounding rocks. The most ancient pyrrhotite generation I thus identified is found in the units occurring at depths of 500 m and shallower (gabbro to 1282 m). The younger pyrrhotite generation II is superposed on the gabbro bottom zone (1421 m and deeper) involving the underlying rocks. Changes in pyrrhotite magnetic properties towards the peridotite body indicate a younger age of this body (1540 – 1670 m) as compared to the age of generation II and the gabbro, respectively. The pyrrhotite metallogenesis that has affected the bottom zone of this peridotite body together with underlying rocks is identified as the next generation, III. It includes also the pyrrhotite found in the peridotite body occurring at the 550 m depth and its surrounding rocks. The IV pyrrhotite generation is recognized within the rocks overlying the gabbro body occurring at the 1987 m depth. Towards the contact of these rocks with the body top the pyrrhotite magnetic properties do not change. Extremely narrow and intensive $\lambda$-peaks characteristic of these pyrrhotites give ground to suppose that it is the best preserved and possibly youngest mineralization.

From the study of the magnetic characteristics of the Kola well rocks by core analysis and well logging, the following conclusions can be drawn:

1. a great dispersion of the $\varkappa$ and $J_n$ values measured on small rock samples is observed, which indicates the superimposed nature of the mineralization;
2. the difference in average $\varkappa$ and $J_n$ values for sedimentary and igneous rocks diminishes with depth; the mean $\varkappa$ values for these rocks become equal in Proterozoic unit III and remain the same still deeper;
3. Proterozoic peridotites of unit IV and volcanites of unit II where high magnetite concentrations are registered have anomalous magnetic properties;
4. analysis of the magnetic properties helped to stratify the section into three zones as follows: depth interval 0 – 4586 m corresponds to the zone of sulphide mineralization. The rocks here generally behave as weak magnetics, their mean magnetic susceptibility (except for the ultrabasic inclusions) not exceeding $4 \times 10^{-3}$ SI units. The presence of a strong magnetic field in this zone is related to the residual magnetization of pyrrhotite and magnetite. The principal mag-

netic mineral for this zone is pyrrhotite. The ultrabasites with considerable amounts of magnetite are an exception;

– depth interval 4586 – 5642 m comprises the zone of oxide mineralization and is characterized by a higher magnetic susceptibility reaching 0.2 – 0.3 SI units; magnetite, magnetite deficient in cations and hematite are present here;
– depth interval 5642 – 10,500 m is the zone of magnetically weak rocks. Magnetic susceptibility is as high as $(2-4) \times 10^{-3}$ SI units at some places only; low pyrrhotite and magnetite concentrations are common;

5. behaviour of the inclination of the $J_n$ vector (with relatively high $Q$ values) is considerably different from that of the recent geomagnetic field, and magnetic reversals in rocks are independent of their composition and magnetic mineralization. All this proves an ancient origin for the residual magnetization. There are about 20 zones of different geomagnetic polarity within the section. Average inclinations of the $J_n$ vector measured for various rocks are close to each other and do not correlate with the magnetic anisotropy values. Consequently, the age of magnetization of rocks containing pyrrhotite is close to that of the rocks rich in magnetite;
6. analysis of some magnetic properties helps to distinguish a few types of magnetic mineralization differing from each other with respect to the time and conditions of generation as follows: (a) dispersed pyrrhotite mineralization in sedimentary rocks of units III and IV; (b) dispersed pyrrhotite mineralization in basic igneous rocks of the same units; (c) a younger (as compared to the first two) pyrrhotite "ore" mineralization associated mainly with the peridotite body endomorphic zone and fractures within sedimentary rocks of units IV and III; (d) dispersed pyrrhotite mineralization in units I and II as well as in Archean gneisses. This is probably the youngest (low-temperature); (e) enrichment of the unit IV peridotites with magnetite that has accompanied their serpentinization. With respect to time (age) it matches the pyrrhotite "ore" mineralization; (f) hematite-magnetite mineralization of unit II, whose timing is not clear, but which probably took place prior to the pyrrhotite mineralization of units I, II and Archean rocks. From the magnetization temperature of the magnetite (over 500 °C) it follows that it was formed earlier than or contemporaneously with the amphibolite metamorphic facies;
7. crystallization, transformation and magnetization of pyrrhotite and magnetite are closely related to rock metamorphism.

## 2.11 Rock Radioactivity

### 2.11.1 Natural Radioactivity

Natural radioactivity of rocks penetrated by the well was studied both on cores and in situ. Core measurements were aimed at determining concentrations of naturally radioactive elements (NRE) such as uranium, thorium and potassium. The methods used are as follows: gamma-spectrometry; scintillation gamma-spectrometry; instrumental neutron-activation analysis for uranium and thorium, X-ray spectral and X-ray fluorescence analyses, delayed neutron technique. Mean NRE concentrations were determined for the rock types identified by petrochemical methods.

Rock radioactivity in situ was derived from gamma-log readings. The log was run in both the integral (GL) and the spectral (SGL) modifications.

Proterozoic

From natural radioactivity data these rocks are subdivided into four groups.

The first group includes the rocks characterized by the lowest natural radioactivity level (NRE concentrations less than their clarkes). This group is comprised by rocks whose natural radioactivity as derived from the integral GR readings (intensity of gamma-radiation $1_{gL}$) is $(7.9-15.8)\times10^{-14}$ A kg$^{-1}$ on average. The second and third group rocks have average total radioactivities (from the GL data) of $(14.3-32.3)\times10^{-14}$ A kg$^{-1}$ and $(40.8-79.5)\times10^{-14}$ A kg$^{-1}$, respectively. The fourth group is characterized by the highest GL readings for all the Middle Proterozoic, $(79.6-133.4)\times10^{-14}$ A kg$^{-1}$.

The rocks of groups I and IV can be unambiguously distinguished on gamma logs. Their range of radioactivity variations of the II and III groups $(14.3-79.6)\times10^{-14}$ A kg$^{-1}$ is characteristic of a number of rocks of varying composition and origin. Therefore the parameter in question is not suitable for differentiating the rocks of these two groups.

The efficiency of natural radioactivity as a criterion for differentiating the Middle Proterozoic rocks greatly improves if the gamma logs are used in combination with data on individual concentrations of the NRE in rocks when analyzing certain intervals of the section. For this purpose it is advisable to determine the natural radioactivity for each series.

The Matertin series is composed of mainly magmatic rocks of basic composition. Dispersion of the GL (and also SGL) readings is not great here. However, the character of the curve varies from rock to rock. For instance, all the gabbro-diabases have about same GL reading. In basalts and diabases, whose thicknesses are also considerable, there are sharp peaks (up to $14.3-21.5 \times 10^{-14}$ A kg$^{-1}$) against certain intervals, caused probably by the presence of thin beds of tuffitic sediments within the diabases.

The GL readings for basic composition of tuffs, picrite and pyroxene porphyrites do not differ greatly from those for diabase and gabbro-diabase, although contacts between these rocks are quite clearly seen on the logs.

The efficiency of the natural radioactivity studies in stratifying the Zhdanov series is sufficiently high.

A group of basic and ultrabasic rocks (group a) is distinguished with certainty in this series. Their radioactivity level is essentially different from that of the rest of the series.

To study ultrabasites, other techniques must be employed including core and cutting measurements. The gamma-spectral (SGL) data in combination with other logs provide additional information for identifying the zone of significant ultrabasite alteration (the serpentine-talc-chlorite-tremolite rocks) where an increase in the uranium and thorium content is noted.

Differentiation of sedimentary rocks (group d) is hard to establish, as they are composed of alternating macro- and microlayers of sediments with various grain sizes. However, the predominantly phyllitic rocks are characterized by higher thorium and potassium content than the essentially aleurolitic ones.

In the Zapolyarny series, GL helps to identify the volcano-sedimentary rocks (group b) occurring among diabases and basic composition tuffs (group a). The latter can be identified from SGL readings; the tuffs are relatively enriched in thorium and uranium.

The rocks of the Luchlompol series all contain almost the same amounts of radioactive elements and are the most radioactive in the Proterozoic section. Integral values of gamma activity for these rocks are also very close to each other; however, SGL helps here to distinguish the andesite-dacite porphyrites as having a higher content of uranium and thorium and a relatively lower percentage of potassium than the arkosic sandstones.

Differentiation of the Pirttiyarvi series rocks can be effected using the total radioactivity values derived from gamma logs.

In the sedimentary metamorphic section of the Kuvernerinyok series, the gamma log helps to distinguish the chemogenic sedimentary rocks from the primarily chemogenic ones with admixtures of sedimentary clastic material.

The rocks of the Mayarvi series, known as the most highly metamorphosed of the whole Proterozoic, have other radioactivity values than unaltered rocks of the same composition.

### 2.11.2 Nuclear-Physical Parameters

Nuclear-physical characteristics of rocks in the composition of the Kola well section were determined in situ using GGL-S and PNNL techniques (the latter including thermal and epithermal neutron modifications).

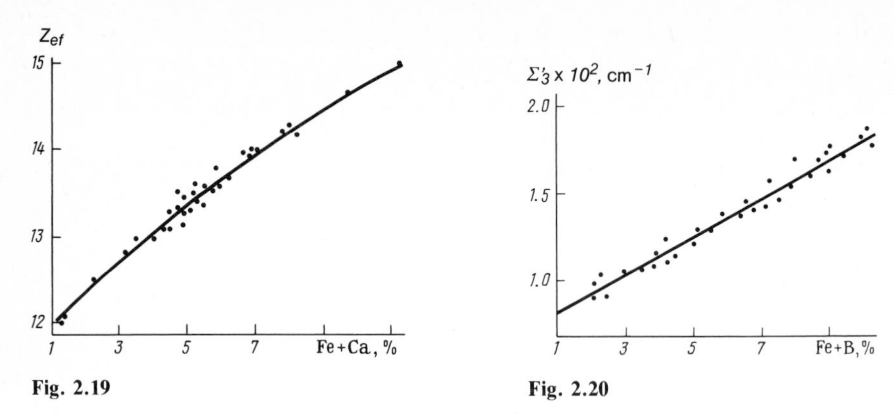


**Fig. 2.19.** Effective atomic number – total amount of iron and calcium (in % Fe) cross-plot for crystalline rocks

**Fig. 2.20.** Thermal neutron capture macro-cross-section as function of total amount of iron and boron (in % Fe) in the rocks

To run investigations of crystalline rocks at great depths, new PNNL and GGL-S apparatus was inverted, made and adjusted for the Kola well on the basis of standard instruments IGN-4 and DRST-2.

For the first time in world practice, measurements with the double-spacing (40 and 90 cm) PNNL tool were effected to a depth of 8000 m at high temperature (to 140 °C) and pressure ($\approx$100 MPa), recording radiation intensity simultaneously in three preselected time intervals (800, 1000 and 1300 $M_s$) and in the interval-to-interval mode for each spacing. The mean lifetimes ($L$) of thermal neutrons have been thus obtained.

Nuclear-physical parameters of crystalline rocks were derived from core chemical analysis results.

The combined investigations helped to establish relationships between $L$, effective atomic number $Z_{ef}$ and elementary composition of rocks, and to develop operational procedures for PNNL and GGL-S interpretation.

The relation between $Z_{ef}$ and the total amount of iron and calcium (Fe + Ca) in crystalline Archean rocks of the Kola well (at depths greater than 6842 m) is shown in Fig. 2.19. Thermal neutron capture macrocross-section ($\Sigma'$) as a function of total amount of iron and boron (Fe + B) expressed in %-equivalent of Fe content is shown in Fig. 2.20.

On the basis of the $Z_{ef}$ and $L$ calculations, the Proterozoic and Archean rocks of the Kola well are classified into three groups (Table 2.13) taking into account their acidity and the relation between the silicium and iron percentages (Fig. 2.21).

The first group includes diabase, gabbro-diabase, essexitic gabbro, diabase and actinolitic gabbro-diabase, amphibolized rocks; porphyroblastic, plagioclase-amphibole, magnetite-biotite-plagioclase-amphibole schists and other rocks; pyroxene porphyrite, pyroxene metaporphyrite, serpentinitic peridotite, serpentine-chlorite-talcous rocks; amphibolite, plagioclase-amphibole, magnetite-biotite-amphibole-plagioclase rocks and schists.

**Table 2.13.** Nuclear-neutron parameters for general rock types

| Rock group | Rock | $Z_{ef}$ | $\tau$, μs |
|---|---|---|---|
| 1 | Basic, ultrabasic | 15 – 17 | 110 – 260 |
| 2 | Medium | 13.5 – 15 | 260 – 380 |
| 3 | Acid | 12 – 14 | 380 – 500 |

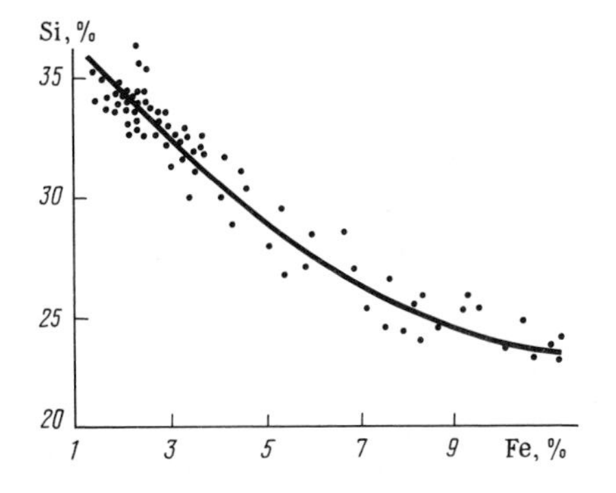


**Fig. 2.21.** Relation between silicium and iron mass percentages for the crystalline rocks

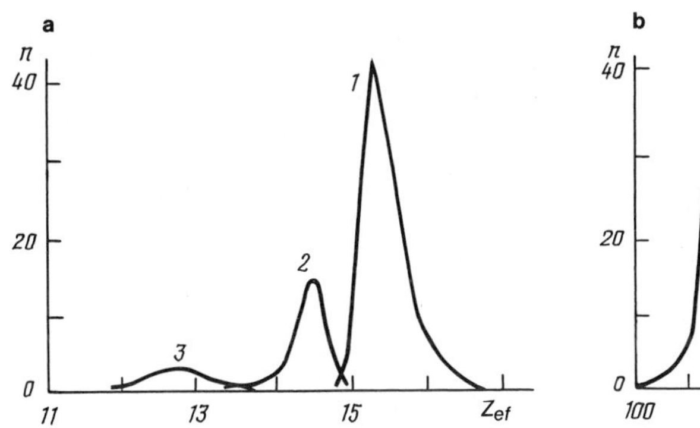


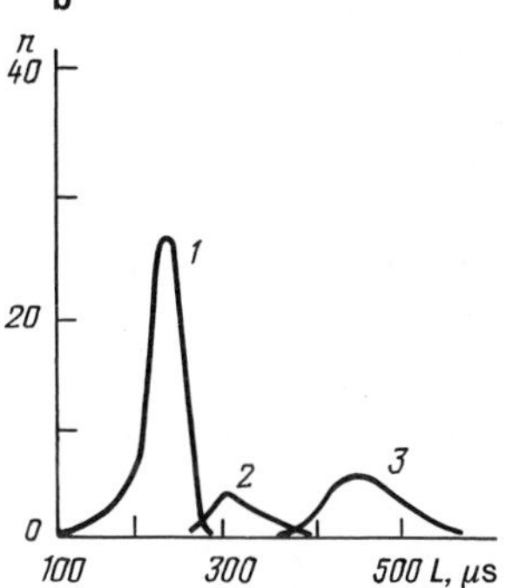


**Fig. 2.22 a, b.** Statistical distribution of the $Z_{ef}$ (**a**) and $L$ (**b**) for rock groups 1 – 3 (according to Table 2.24)

The second group comprises moderately acid rocks, quartz-plagioclase schist with amphibole, biotite-amphibole-plagioclase schist and leucocratic and melanocratic gneiss with high-aluminium material, actinolite-talcose rocks and oligoclasite.

Andesite-dacite porphyrite, potassium and sodium granite and granite-gneiss are included in the third group.

The statistical distribution of the $Z_{ef}$ and $L$ parameters is shown for the rock groups in Fig. 2.22. Experimental values of the $Z_{ef}$ and $L$ determined from the GGL-S and PNNL readings were also presented in the form of continuous bar graphs for the Kola well section.

The study of natural radioactivity and nuclear-physical properties of the Kola well rocks on cores and by well logging has thus resulted in: determining the content of natural radioactive elements (NRE) in certain rock types and achieving better stratification of the well section; identifying zones of a higher NRE concentration associated with secondary changes in rocks, particularly with granitization; discovering the regular increase in natural radioactivity with depth; employing nuclear-physical measurements in determining rock elementary composition.

## 2.12 Rock Thermal Properties

A study of thermal properties on rock samples from the superdeep well is necessary to obtain reliable information on the thermal field in the region and predict temperatures in the deep using the heat conductivity equation, as the thermal parameters appear in the equation as coefficients determining the rate of heat transfer to the surface. The methods used in measurements of rock thermal properties, such as thermal conductivity $\lambda$ ($W\,m^{-1}\,K$), thermal diffusivity $K$ ($m^2\,s^{-1}$) and heat capacity $C$ ($J\,kg^{-1}\,K$), are as follows:

1. flat instantaneous source for measuring $K$, $C$ and $\lambda$;
2. comparison technique (Smirnova 1979) employing plates 40 mm in diameter and 6 – 10 mm thick; this method was used in measuring $\lambda$ on small samples or when the $\lambda$ values were extremely high;
3. split rod technique used for small samples (plates 21.5 mm in diameter and 3 – 6 mm thick);
4. contactless determination of heat conductivity with a mobile point-source;
5. thermal comparator with a two-pointer probe for the sample surface.

The last four techniques are used only in thermal conductivity measurements. An analysis of control measurements (over 40) on reference core samples has shown that comparability of the results obtained by methods 1 – 4 is within the 3 – 15% range for $\lambda$ values from 1.5 to 4.5 $W\,m^{-1}\,K$.

Thermal properties of the Pechenga sequence rocks were measured by methods 1 – 4 on core samples representing all the main rock types to the depth of 6842 m. Air-dry samples were investigated. At shallow depths (to 1700 m) the samples were also taken from other wells, besides the superdeep one. The thermal conductivities obtained are shown in Fig. 2.23.

As follows from Table 2.14, among igneous rocks the andesite porphyrites (group IV) and various schists (group VII) have the greatest thermal conductivities ($\lambda = 4.27 - 5.23$ $W\,m^{-1}\,K$), which can be explained by the high percentages of highly conductive minerals in the samples, such as: up to 50% for chlorite ($\lambda = 5.2$ $W\,m^{-1}\,K$) or carbonate ($\lambda = 3.6 - 5.8$ $W\,m^{-1}\,K$) in the schists and up to 60% for quartz ($\lambda = 7.7$ $W\,m^{-1}\,K$) in andesite porphyrite, the albite ($\lambda = 1.9 - 2.3$ $W\,m^{-1}\,K$) content being low (to 10 – 20%). The lowest conductivities are characteristic of amphibolite sampled at depths in excess of 5 km (group III, $\lambda_{av} = 1.38$ $W\,m^{-1}\,K$[16]) and serpentine ultrabasic rocks from the Zhdanov series (group IV, $\lambda_{av} = 1.84$ $W\,m^{-1}\,K$).

---

[16] $\lambda_{av}$ – average value.

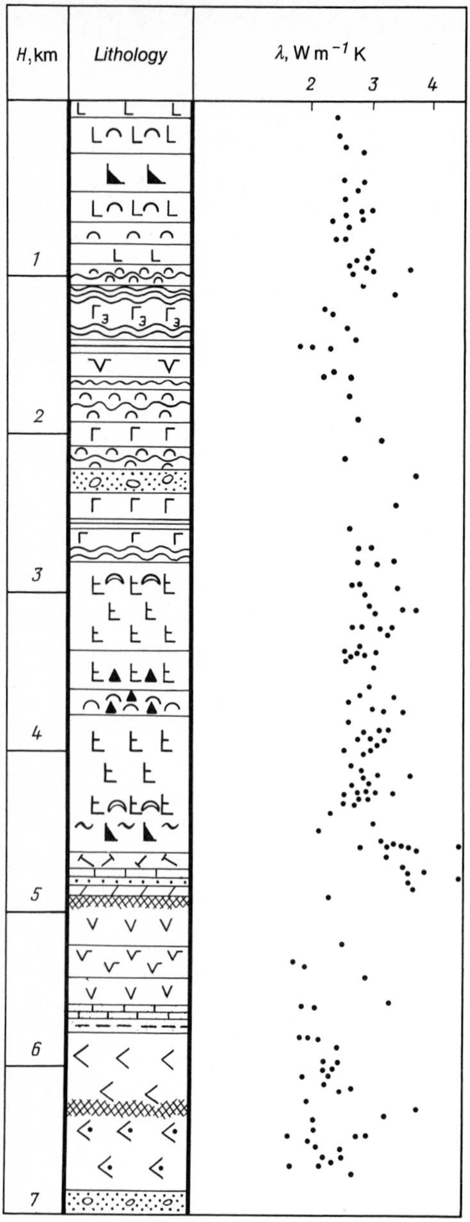

**Fig. 2.23.** Thermal conductivity in the Proterozoic section

For group II rock samples ($H = 2805 - 4500$ m), the range of variation and average values of thermal conductivities ($\lambda_{av} = 3.8$ W m$^{-1}$ K) are a little greater than for the weakly altered rocks sampled from shallower depths($\lambda = 2.7$ W m$^{-1}$ K). The difference in values measured along the sample axis and perpendicular to it does not exceed 4%. Below the 4500 m depth is the border between the greenschist and amphibolite metamorphism facies, the top section massive rocks changing for schistose ones.

**Table 2.14.** Variation ranges and mean values of thermal properties and density for main rock groups of the Pechenga sequence

| Rock | $\delta$ g cm$^{-3}$ | | K $\times 10^6$, m$^2$ s$^{-1}$ | | C, J kg$^{-1}$ K | | Number of samples | $\lambda$ W m$^{-1}$ K | | | Number of samples |
|---|---|---|---|---|---|---|---|---|---|---|---|
| | Variation range | Average | Variation range | Average | Variation range | Average | | Min. | Max. | Medium | |
| Gabbroids | 2.86 – 3.12 | 3.00 | 0.88 – 1.43 | 1.09 | 630 – 880 | 800 | 10 | 2.22 | 3.01 | 2.70 | 14 |
| | 2.80 – 3.27 | 3.04 | | | | | | | | | |
| Metagabbro-diabase | 2.92 – 3.13 | 3.03 | 0.83 – 1.43 | 1.07 | 630 – 1700 | 840 | 41 | 2.22 | 3.18 | 2.70 | 45 |
| Amphibolite | 2.85 – 3.13 | 2.92 | 0.98 – 1.24 | 1.09 | 670 – 1090 | 840 | 24 | 2.42 | 3.68 | 3.00 | 58 |
| | | | | | | | | 1.38 | 3.14 | 2.29 | 24 |
| | | | | | | | | 2.05 | 3.89 | 2.97 | 15 |
| Serpentinite ultra-basic rocks | 2.84 – 3.10 | 2.93 | 0.84 – 1.18 | 1.03 | 670 – 920 | 800 | 5 | 1.84 | 3.60 | 2.55 | 7 |
| Tuff | 2.80 – 3.07 | 2.92 | 1.01 – 1.41 | 1.15 | 670 – 920 | 840 | 5 | 2.55 | 2.93 | 2.74 | 6 |
| Andesite porphyrite | 2.69 – 3.05 | 2.78 | 1.52 | | 840 | | 5 | 3.26 | 4.27 | 3.50 | 9 |
| Schist | 2.96 – 3.01 | 2.99 | 1.62 | | 920 | | 4 | 3.0 | 5.23 | 3.88 | 6 |
| Sedimentary and | 2.72 – 2.91 | 2.84 | 1.03 – 1.77 | 1.32 | 750 – 1000 | 920 | 10 | 2.66 | 3.91 | 3.22 | 14 |
| tuffitic-sedimentary rocks | 2.72 – 3.17 | 2.83 | – | | – | | 44 | 2.13 | 3.93 | 3.00 | 48 |

The thermal conductivity of metabasites occurring below this border is sharply different from that of the basic rocks of the top section. For instance, the rocks of group III (below the 5000 m depth) show a sharp decrease in $\lambda_{av}$ (from 3.0 to 2.3 $W m^{-1} K$), its values as measured along the axis differing from those measured perpendicular to it by 12% on average. This decrease in thermal conductivity is possibly caused not only by transformation of actinolite ($\lambda = 3.8$ $W m^{-1} K$) into hornblende ($\lambda = 2.5$ $W m^{-1} K$), but also by textural and structural changes associated with metamorphism and accompanied with elastic property changes.

Among sedimentary rocks the greatest $\lambda$ values are measured in mica-bearing arkosic sandstones (3.8 $W m^{-1} K$) and quartz-mica aleurolites (3.2 – 3.5 $W m^{-1} K$).

The mean thermal conductivities in rocks of the Proterozoic sequence thus vary from 2.3 to 3.9 $W m^{-1} K$ (see Table 2.14). The sample wetness correction for volcanic rocks of this sequence is negligible.

Considerable anisotropy of thermal conductivity has been established (both for dry and water-wet samples). Maximum $\lambda$ values were observed when measuring when the conductivity perpendicular to the schistosity or linearity, and the minimum ones when parallel to them. The thermal conductivity anisotropy of gneiss and schist generally agrees well with their elastic anisotropy.

Generally speaking, the results of the investigation of thermal properties for all the rock types of the Kola well that has penetrated the major structural and formational zones of the Baltic shield (the Pechenga paleocaldera of Proterozoic age and the underlying Kola sequence of the Archean) have shown the zonal character of the $\lambda$ distribution.

# 2.13 Geothermic Investigations

## 2.13.1 Thermal Field and Its Characteristics

The most important characteristics of the thermal field of the Earth's crust are as follows: geothermal gradient (GTG), thermal conductivity ($\lambda$), heat flux (HF) density in the near-surface zone ($q$), and distribution of radiogenic heat sources ($A$). The geothermal gradient is derived from temperature measurements in a well after having kept it under static conditions until the thermal field becomes stationary. The heat flow (for the case of horizontal strata and stationary heat transfer) is determined as a product of GTG and thermal conductivity, $q = \gamma \times \lambda$. Consequently, the reliability and accuracy of heat flux calculations depend upon the accuracy of determination of these two quantities. The accuracy of the GTG and $\lambda$ measurements for the real well is now about 5% and 7–10%, respectively. The relatively low accuracy of the $\lambda$ evaluations is explained by the fact that this parameter is measured on cores, not in situ. When evaluating $\lambda$ for subsurface environments (especially at great depths), the effect of water saturation has to be accounted for. At 5–10 km depths, a considerable portion of pores and fractures becomes closed. Real values of thermal conductivity for low-porosity crystalline rocks in situ usually exceeding those measured in laboratories.

Prior to studding the superdeep Kola well, 10 wells from 470 to 1675 m deep located on the Pechenga structure were investigated. The GTG values were determined for these wells after they had been kept under static conditions for 1 year and longer. Thermal conductivities from core sample measurements and heat flux densities were also determined (Earth heat flows 1972). All these wells have penetrated Proterozoic rocks, represented mainly by fine- and very fine-grained diabase, tuff, phyllite, aleurolite, sandstone, serpentinite and partly by more coarsely grained rocks like gabbro-diabase. Water saturation does not greatly affect the thermal conductivity of these rocks, therefore it is permissible to use the results of air-dry sample measurements in the $\lambda$ calculations. Average thermal conductivities for the ten wells vary from 2.6 to 3.6 W m$^{-1}$ K, most of the values being within the range of 2.8–2.9 W m$^{-1}$ K. The average GTG of these wells ranges from 1° to 1.4 °C/100 m, the 1.2°–1.3 °C/100 m gradients being predominant. Consequently, average $q = (3.5-3.6)$ M W m$^{-1}$. For the Kola well the most reliable GTG values were derived from a temperature log recorded in 1976 to the depth of 7 km, the well having remained under static conditions for about 1.5 years prior to measurement. Down to 9 km and deeper the temperature was measured under transient conditions, therefore the gradient values obtained are not very accurate. In the Proterozoic section of the Kola well (to the depth of 6842 m), averaged GTG values rise (from up to down) from 1° to 1.8 °C/100 m, the step-like changes in the GTG roughly correlating with interfaces between the series.

At the 2.8 km depth, a sharp local decrease in the GTG is noted within a 50-m-thick interval, below which a stable GTG (about 1.6 °C/100 m) is observed to the depth of 4.3 km. Still deeper there is a zone of abnormal behaviour of the GTG: it rises from 1.65 at the 4.3 km depth to 1.9° – 2.0 °C/100 m at 5 km. From seismoacoustic data (Kazanski et al. 1980), this interval is known as a zone of tectonic faulting. The gradient behaviour is probably affected by the increase in fracturing and permeability, as evident from the seismic-acoustic data beginning with the 4.5 km depth.

In the 5 – 6.6 km depth interval, the GTG is stable and remains as high as 2.2 °C/100 m. Further on at the depth of 6.8 km, there is a local increase in the GTG up to 2.5 °C/100 m, then it decreases to 1.5 and still deeper again rises to 2.3 °C/100 m at depths of 7 – 7.2 km. The GTG values obtained for the great depths should be regarded as approximate, for the static conditions of the well were maintained for a long enough time interval.

Thermal conductivities of the Pechenga sequence rocks were measured on cores using the comparison method, the data obtained for 34 samples having been compared with the results provided by the contactless measurements.

In the top section of the Matertin series to the depth of 1 km, the average $\lambda$ is equal to 2.9 W m$^{-1}$K, then in the Zhdanov series the $\lambda_{av}$ decreases a little to 2.7 W m$^{-1}$K, but starting with the depth of 1.8 km, it rises to 3.4 W m$^{-1}$K. At the 2.4 – 2.6 km depth the values are from 2.9 to 3.1 W m$^{-1}$K.

The transition from the Zhdanov to the Zapolyarny series is not marked by the sharp change in thermal conductivity values measured on samples in the laboratory, although GTG does change sharply enough. The Luchlompol series is known to correspond to a faulted zone. For this series the $\lambda$ values (corrected for water saturation) are as high as 3.7 – 4.1 W m$^{-1}$K, but according to discontinuous measurements become lower (down to 2 W m$^{-1}$K) starting at the depth of 5 km, which corresponds to a general increase in GTG to 2.2 °C/100 m.

At the bottom of the Pirttiyarvi series (depths from 5600 to 5750 m), $\lambda$ increases to 3.75 W m$^{-1}$K, which corresponds well to the highly conductive Kuvernerinyok series. In the Upper Mayarvi series, where amphibole-plagioclase schists are common, the $\lambda$ decreases to 2.2 W m$^{-1}$K. Downwards from the 6 km depth, the $\lambda$ values are from 2.2 to 2.7 W m$^{-1}$K. At the Proterozoic-Archean interface at the bottom of the Televin series there is a sharp drop in $\lambda$ to 2.0 W m$^{-1}$K, which is in accordance with GTG behaviour at this depth.

The $q$ values were derived by interval-to-interval averaging of the mean values of GTG and $\lambda$. Down to the depth of 1000 m, the $q_{av}$ is equal to (26 ± 2) W m$^{-2}$, while for the 1000 – 2800 m depth interval it is (36 ± 2) W m$^{-2}$. In the zone of transition from the Zhdanov to the Zapolyarny series, there is a step-like change in the heat flux, corresponding to a similar change in the gradient (to the depth of 4300 m). Since the rocks occurring in the 2800 – 4300 m depth interval are massive enough and homogeneous with respect to their composition, their GTG and $\lambda$ values can be regarded as very reliable and $q = (49 \pm 1)$ W m$^{-2}$ be used as a reference.

In the 4300 – 4900 m depth interval there is a zone of abnormally high heat flow, its maximum reaching (67 ± 7) W m$^{-2}$. At the depth of 5000 m the average value of $q$ drops sharply and becomes equal to 48 – 56 W m$^{-2}$. It should be noted that $q$ only begins to decrease of the depth of 6800 m. Very low values of flux in the upper part of the section (to the depth of about 1000 m) can be explained by the effect of such factors as filtration, recent glaciation and denudation.

### 2.13.2 Radioactive Heat in the Thermal Balance of the Earth's Crust and Upper Mantle

The heat flux constituent generated by radioactive elements (RE) decaying in the Earth's crust and its constituent that flows up from the upper mantle are determined on the basis of the commonly known model of the Earth's structure and composition, as well as the distribution of uranium, thorium and potassium in it. The distribution of average RE concentrations for rock types and section strata is of specific interest, as well as the radioactive heat distribution in the Proterozoic and Archean rocks.

*Pechenga Sequence.* The total number of determinations of uranium, thorium, and potassium content in the Pechenga sequence rocks is 532. The lower series were tested in less detail than the upper ones because of relatively poor core recovery at depths exceeding 4500 m. However, all the main rock species have been tested, the determinations of mean RE concentrations for each series having taken the GL and GL-S readings into account.

The lowest U, Th and K content is revealed in the Matertin and Zapolyarnin series comprised of mainly diabase and metabasite, while the maximum one is characteristic of the Zhdanov and Luchlompol series. The high RE percentages for the latter are explained by the presence of an andesite porphyrite body known to contain elevated concentrations of uranium, thorium and potassium. The low RE content in the Kuvernerinyok series results from the great amount of low-radioactive carbonate-tremolite schists present in it.

A comparison of mean RE concentrations measured in volcanic sheets shows their gradual decrease with depth from the Matertin to Zapolyarny series down to the depth of 4673 m. Below this depth an increase in RE content is noted. The relatively high RE content in the Pirttiyarvi series is explained by its complex composition: here in addition to the metabasites, there are many effusive bodies made of acid and median rocks (meta-andesite and meta-albitophyre) characterized by elevated RE concentrations. The Mayarvi series, mainly composed of basic effusions, contains relatively high concentrations of RE, which is explained by superposition of such processes as lamination, metamorphism and granitization.

*Kola Sequence.* Stratification of the Kola Archean sequence was carried out by the KGRE and VSEGEI specialists using mainly lithological criteria and accounting for the superposed polymetamorphism processes. The criteria used in stratification are as follows: the presence of rocks with high-aluminium minerals; relative amount of biotite-plagioclase gneisses and amphibolites; and extent of granitization.

It was assumed that in spite of the great influence of superimposed metamorphism, the primary character of the rocks had been at least partly preserved, particularly with respect to their elementary composition. For example, the presence of high-aluminium minerals (sillimanite, staurolite, andalusite) in the rocks clearly indicates the primarily pelitic nature of the metasediments. Six units were identified using these criteria.

The mean RE concentrations obtained for these units depend upon the extent of rock granitization and the relative amounts of granites and pegmatites or acid and basic rocks.

### 2.13.3 Heat Generation

As follows from the vertical distribution of heat generation values down to the 11,400 m depth (Fig. 2.24), its character is mainly controlled by the lithology of the Pechenga and Kola sequence rocks. The maximum heat generation is characteristic of the tuffitic-sedimentary rocks ($1.0-2.5\ \mathrm{M\,W\,m^{-3}}$) and essexite gabbro-diabase ($0.7-1.7\ \mathrm{M\,W\,m^{-3}}$) occurring mostly in the top portion of the series to a depth of about 2.5 km. Downsection in the zone of predominant development of highly metamorphosed basic volcanites (represented by the greenschist and epidote-amphibolite facies), the heat generation level drops to $0.04-0.08\ \mathrm{M\,W\,m^{-3}}$ on average. A sharp increase in the heat generation at some narrow depth intervals from 3700 to 4700 m is related with thin layers of tuffites and andesite-porphyrites. The heat generation peak at the 4700 m depth can be caused (besides respective lithological changes) by the thick highly permeable zone of the Luchlompol fault, as also the sharp increase in heat generation at about 9500 m depth, where a high permeability decompressed zone is found. Inflow of some elements (including the radioactive ones) into this latter zone is assumed on the basis of primary helium concentration analysis.

It is worth noting that the basic volcanic rocks of the Pechenga sequence known to be close to each other in chemical composition but differently metamorphosed show great differentiation (ten times as great and more) in the RE concentrations, which is possibly caused by the particular behaviour of uranium in the process of prograde metamorphism. The heat generation levels as derived for the Pechenga sequence volcanic and sedimentary units are about 0.4 and $1.3\ \mathrm{M\,W\,m^{-3}}$, respectively.

In the Kola Archean sequence the heat generation level derived for the two-mica gneisses with high-aluminium minerals is close to that registered for sedimentary units ($1.2-1.7\ \mathrm{M\,W\,m^{-3}}$); in units composed of more basic gneisses (amphibole-plagioclase ones) with amphibolite interlayers, this parameter becomes essentially lower and is equal to about $0.8\ \mathrm{M\,W\,m^{-3}}$. Consequently, the investigation has not shown any regular decrease in heat generation with depth. The vertical distribution of heat generation has a step-like character as controlled mostly by the rock elementary composition.

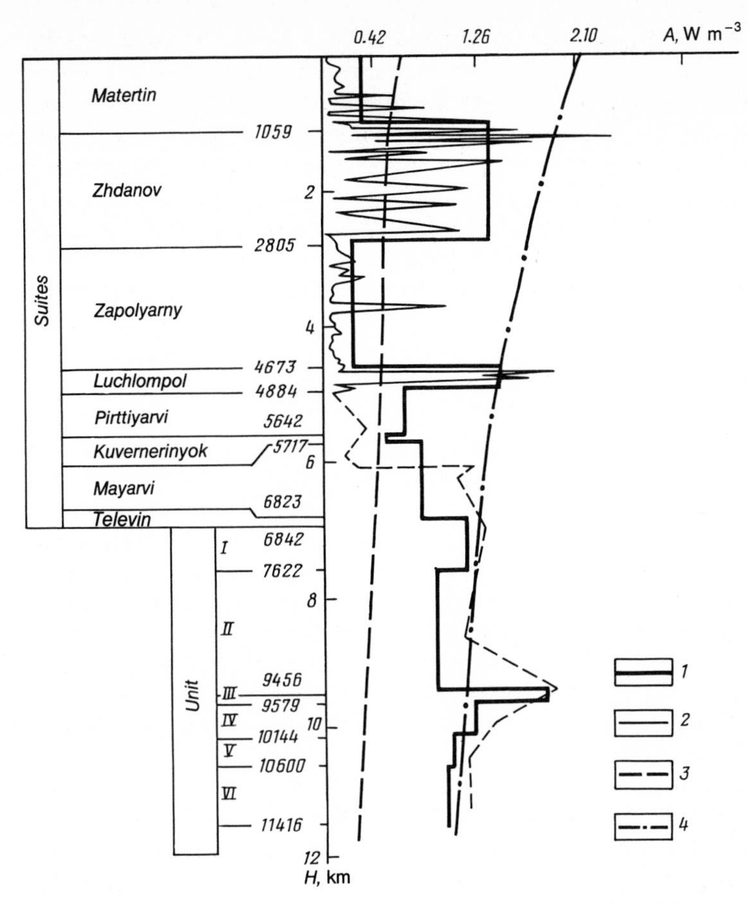


**Fig. 2.24.** Heat generation variations in the Kola well section. *1* Heat generation distribution for series and units; *2* heat generation variations within the series; distribution of heat generation with depth for: *3* Ukrainian shield; *4* Sierra Nevada batholiths

For the sake of comparison, the averaged curves of heat generation evolution with depth are shown for the Ukrainian Precambrian shield and the Sierra-Nevada Mesozoic batholiths. Absolute values of heat generation for these regions are within about the same range ($0.2 - 2.0$ M W m$^{-3}$) as those calculated for the Kola well. It should be also noted that the vertical distribution of heat generation in lithologically identical rocks (separately for the metabasites and tuffitic-sedimentary rocks of the Pechenga sequence) is close to the exponential distribution.

## 2.13.4 Heat Generation Models

According to modern knowledge on the structure of the Earth's crust in the Pechenga region, two models of heat generation must be considered as follows:

*Model I.* This model assumes the bottom of the Earth's crust to be a "granite-gneiss" layer occurring at the depth of 12.5 km, where a seismic interface is identified by VSP data. The interpretation of this interface is still ambiguous. On one hand, it may be the transition from the amphibolite-metamorphic facies to the granulite one, and on the other hand it may correspond to the lower boundary of the granitization zone of the Earth's crust.

According to this model, the "granite-gneiss" layer conditionally comprises both the Pechenga volcano-sedimentary sequence of the Proterozoic and the Kola sequence. The latter (as the Kola well drilled to the depth of 11,612 m has shown) includes rocks of varying composition (both acid and basic). The section penetrated by the well is stratified into six units of different composition. The lithology, similar to that of the section already studied, is supposed to persist to the interface mentioned above (12.5 km deep). Below this interface down to the depth of 40 km, a "granulite-basite" layer is likely to occur, as to the composition of which there are different opinions.

This model is based on the assumption that the layers in question is composed of metabasites represented in various metamorphic facies (amphibolites, pyroxene granulites, eclogites). Depending on relative amounts of "dry" granulites and water-rich amphibolites, the heat generation in this layer can vary from 0.2 to 0.6 M W $m^{-3}$. The average heat generation value used in the model is 0.4 M W $m^{-3}$. Calculations made for this model show that radioactive heat sources dispersed in the Earth's crust generate a heat flux equal to 23 M W $m^{-2}$ (the rest is contributed by the mantle heat flow).

*Model II.* This model assumes that the rocks similar to those of the Kola gneiss sequence tested by the superdeep well persist to the depth of 15 km, below which (down to 20 km in depth) there are rocks whose composition corresponds to that of rock species encountered in the Murmansk block (the so-called basement of the Kola sequence). This uppermost part of the Earth's crust (to the depth of 20 km), conditionally including also the Pechenga sequence, is believed to be its "granite-gneiss" layer. Below it, down to 40 km in depth, the "granulite-basite" layer occurs. Unlike Model I, this model assumes the "granulite-basite" layer to be composed of two sublayers as identified from DSS data, their respective depth intervals being 20–30 and 30–40 km. Concentrations of radioactive elements (and heat generation capacities) in these sublayers are evaluated from their abundance ratios (clarkes) known for respective rock types (mainly gabbro-amphibolites for the upper sublayer and granulite-eclogite for the lower one). The radioactive constituent of the heat flow of the Earth's crust calculated from this model does not differ greatly from the result obtained with model I and is equal to 26 M W $m^{-2}$, that is 45–55% of the total near-surface heat flux as measured in the Kola well (49 M W $m^{-2}$).

The results of geothermic investigations can be summarized as follows:

1. Geothermal gradient increases with depth from 1°–1.1 °C/100 m in the top part of the Pechenga sequence to 1.7°–2.0 °C/100 m in its lower part (at depths of 6000–7000 m). The GTG variation with depth seems to be caused by changes in rock thermal conductivity, heat generation and water saturation.

2. In the Proterozoic section the heat flux intensity increases with depth owing to the increase in concentration of radioactive elements.
3. Vertical distribution of heat generation (according to the model assumed) varies with depth in a step-like manner, being controlled by the rock lithology.
4. The contribution of radioactive heat to the total heat flow of the Earth's crust is about 45 – 55% (as measured for the near-surface zone in the Kola well).

## 2.14 Physico-Mechanical Properties

These properties of rocks from the Kola well were measured on core samples only and then used mainly in solving various drilling technology problems. The parameters that had to be determined are as follows: static Young's modulus $E_{st}$; static Poisson ratio $\nu_{st}$; compressive strength $G_c$; tensile strength $G_t$; indentation hardness $P_i$; plasticity $K_{pl}$; and abrasivity $z$.

$P_i$, $K_{pl}$ and $E_{st}$ (die indentation) were determined using instruments UMGP-3 and UMGP-4 according to the GOST-12288-66 (meteorological standards). $E_{st}$ and $\nu_{st}$ were also measured with these instruments operating in a quasi-static single-axis loading mode. The deformations of cubic samples were measured along their entire length. Abrasivity $Z$ was determined by grinding a steel rod against a flat surface of the rock studied.

The most common parameters characterizing rock strength are its strength limits as measured using the simplest modes of loading, i.e. single-axis compression and tension. These parameters were determined by repeatedly splitting the rock plates $15 \times 300 \times 100$ mm in size and subsequently crushing the rock (cubic samples about 15 mm in the side size). For each rock at least 15 splitting and 10 crushing tests were done. The results of the tests were then used to derive the rock tensile and compression strength, respectively. The numerous values obtained for each rock were then processed statistically to study their distribution (Turtchaninov and Medvedev 1973).

It should be noted that the conventional methods of rock strength evaluation require the use of rock plates (to be tested) at least 30 mm thick and rock cubes with a 40 mm side (Turtchaninov and Medvedev 1973). Because the amount and size of cores from the Kola well do not provide for the preparation of such large samples, the plate-splitting and cube-crushing methods was tried on smaller sized samples.

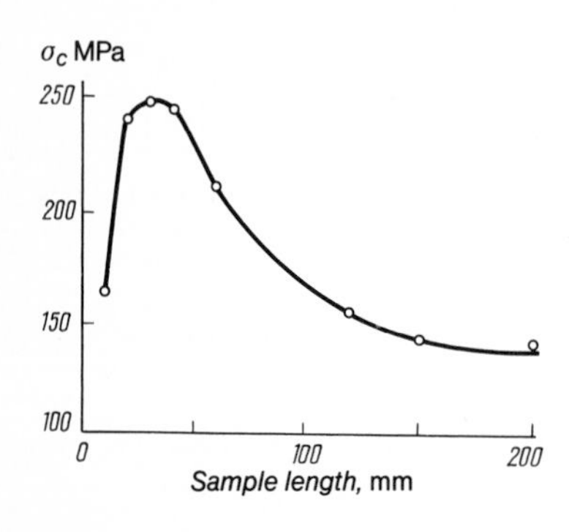


**Fig. 2.25.** Rock compressive strength as a function of the sample length

Experiments with optically active materials have proved the major role of tensile strains in breaking plates 10 – 15 mm thick by driving a sharp wedge into them. A comparison of $\sigma_c$ values obtained for homogeneous pyroxenite samples of very small and very large sizes shows their close agreement (Fig. 2.25).

The rock strength measured for the Kola well cores varies over a wide range, depending upon a number of factors such as mineral composition, structure and texture of rocks, as well as the presence of fractures (including healed) and microfractures.

In the top portion of the Pechenga sequence to the depth of 4500 m, the rock strength variations are controlled mainly by petrographical features and the changes in mineralogical composition. The $\sigma_c$ in this part of the section varies from 114 MPa (talcose picrite and peridotite) to 219 MPa (diabase), the latter being the highest value for the entire section of the Kola well (Table 2.15). The tensile strength varies from 22 MPa (finely laminated metasedimentary rocks) to 45 MPa in diabase. The layered metasedimentary rocks manifest a distinct anisotropy of strength properties. For instance, the $\sigma_t$ values measured in phyllites along their lamination and perpendicular to it are 40 and 32 MPa, respectively.

In the bottom part of the Pechenga sequence (to the depth of 6800 m) the average $\sigma_c$ values (160 MPa) do not differ greatly from those in the top part. The main difference consists in relatively lower $\sigma_t$ and much more pronounced anisotropy of rock strength owing to more prominent rock schistosity. Thus, the leucodiabase amphibole-plagioclase schists have $\sigma_t$ equal to 15 and 24 MPa when the tensile stress is directed across and along laminae, respectively. The strength of these rocks (the most common ones in the 4900 – 6800 m depth interval) appears to be two to three times as low as that of the diabase and metadiabase known to compose most of the Upper Pechenga subsequence. Abnormally low strength is characteristic of chloritic and talc-chloritic apodiabase and apopicrite schists occurring at the 4500 – 4673 m depth. Their $\sigma_c = 29$ MPa, while the tensile strength across and along the lamination is 10 and 64 MPa, respectively.

In the Archean sequence, the rock compressive strength varies from 65 MPa for phlogopite-talc-tremolite schists found in subordinate quantities to 182 – 194 MPa for shadow migmatites (granite-gneiss) and amphibolites known as the most abundant rocks in the section. In the Archean rocks the $\sigma_t$ values vary from 5 MPa for phlogopite-talc-tremolite schists (as measured across the lamination) to 23 MPa for amphibolites (along the laminae), i.e. they are essentially lower than $\sigma_t$ obtained for the Lower Proterozoic rocks.

A good correlation between elastic properties and tensile strength measured on the Kola well cores is evident from Fig. 2.26. There is also a correlation between the elastic and strength anisotropies of these rock samples. These correlations are believed to result from a partial disintegration of cores caused by decompression. Therefore, the strength parameters measured on cores cannot be regarded as characterizing the true strength of rocks in situ, at least of those occurring at great depths.

When the pilot well was drilled, the information on rocks being penetrated had to be obtained continuously, particularly the information on rock strength. Therefore, besides the time-consuming measurements of $\sigma_c$ and $\sigma_t$, a proximate method of evaluating the rock strength had to be used. The determination of

**Table 2.15.** Mean values of physico-mechanical properties for the principal rock types from the Kola well

| Rock | $\sigma$, g cm$^{-3}$ | $V_p$ km s$^{-1}$ | $V_s$ km s$^{-1}$ | $K_a^w$ | $E_{dyn} \times 10^{-4}$ MPa | $\nu$ | $\sigma_t$, MPa | | $P_i$ | | $K_{pl}$ |
|---|---|---|---|---|---|---|---|---|---|---|---|
| | | | | | | | $\sigma_t \perp$ | $\sigma_t \parallel$ | MPa | MPa | |
| 1 | 2 | 3 | 4 | 5 | 6 | 7 | 8 | 9 | 10 | 11 | 12 |
| *Pechenga sequence* | | | | | | | | | | | |
| *Matertin series (0 – 1059 m)* | | | | | | | | | | | |
| Diabase | 3.02 | 6.80 | 3.66 | 1.02 | 10.4 | 0.27 | 45 | 45 | 219 | 27.8 | 2.3 |
| | (1987) | (63) | (60) | (63) | (61) | (61) | (7) | (7) | (7) | (24) | (24) |
| Tuff | 2.92 | 6.48 | 3.50 | 1.05 | 9.1 | 0.28 | 41 | 41 | 153 | 29.6 | 2.7 |
| | (582) | (34) | (31) | (34) | (31) | (31) | (5) | (5) | (5) | (6) | (6) |
| Komatiite rocks | 3.03 | 6.28 | 3.23 | 1.04 | 8.8 | 0.29 | 26 | 26 | 114 | 17.3 | 2.0 |
| | (369) | (23) | (23) | (23) | (23) | (23) | (4) | (4) | (4) | (6) | (6) |
| *Zhdanov series (1059 – 2805 m)* | | | | | | | | | | | |
| Phyllite, aleurolite | 2.90 | 6.03 | 3.41 | 1.06 | 8.0 | 0.20 | 32 | 40 | 183 | 15.7 | 2.3 |
| | (1457) | (54) | (50) | (54) | (49) | (49) | (7) | (7) | (7) | (16) | (16) |
| Alternation of sandstone, | 2.84 | 5.91 | 3.59 | 1.10 | 8.8 | 0.20 | 22 | 26 | 122 | 14.9 | 2.0 |
| aleurolite, phyllite | (1297) | (48) | (42) | (48) | (44) | (44) | (10) | (10) | (10) | (31) | (31) |
| Tuff, tuffite | 2.88 | 6.20 | 3.48 | 1.08 | 8.7 | 0.27 | 32 | 36 | 116 | 11.9 | 2.4 |
| | (325) | (18) | (17) | (18) | (17) | (17) | (4) | (4) | (4) | (4) | (4) |
| Amphibolized and chlori- | 2.98 | 6.24 | 3.56 | 1.03 | 9.3 | 0.24 | 29 | 30 | 179 | 27.7 | 2.0 |
| tized gabbro-diabase | (1948) | (52) | (50) | (52) | (50) | (50) | (8) | (8) | (8) | (20) | (20) |
| Essexite gabbro-diabase | 2.84 | 5.92 | 3.32 | 1.02 | 8.1 | 0.27 | 28 | 28 | 175 | 35 | 2.0 |
| | (552) | (15) | (13) | (15) | (13) | (13) | (4) | (4) | (4) | (3) | (3) |
| Serpentinite peridotite | 2.88 | 6.29 | 3.49 | 1.02 | 9.0 | 0.27 | 30 | 36 | 190 | 17.7 | 2.3 |
| | (627) | (17) | (15) | (17) | (15) | (15) | (5) | (5) | (5) | (8) | (8) |
| *Zapolyarny series (2805 – 4673 m)* | | | | | | | | | | | |
| Metadiabase, greenschist | 3.02 | 6.58 | 3.76 | 1.04 | 10.6 | 0.23 | 30 | 31 | 175 | 35.3 | 1.6 |
| | (5630) | (214) | (216) | (214) | (216) | (216) | (35) | (35) | (35) | (112) | (112) |
| Talc-chlorite apopicrite | 2.93 | 5.46 | 3.35 | 1.35 | 8.2 | 0.22 | 6 | 10 | 29 | 3.8 | 2.1 |
| schists with hematite | (192) | (8) | (6) | (8) | (6) | (6) | (1) | (1) | (1) | (11) | (11) |
| Tuff, tuffite | 2.88 | 6.18 | 3.66 | 1.14 | 9.6 | 0.24 | 10 | 46 | 126 | 24.4 | 1.7 |
| | (130) | (4) | (3) | (4) | (4) | (3) | (3) | (1) | (1) | (4) | (1) |

**Table 2.15** (continued)

| Rock | $\sigma$, g cm$^{-3}$ | $V_p$ km s$^{-1}$ | $V_s$ km s$^{-1}$ | $K_a^w$ | $E_{dyn} \times 10^{-4}$ MPa | $\nu$ | $\sigma_t$, MPa | | $P_i$ | | $K_{pl}$ |
|---|---|---|---|---|---|---|---|---|---|---|---|
| | | | | | | | $\sigma_t \perp$ | $\sigma_t \parallel$ | MPa | MPa | |
| 1 | 2 | 3 | 4 | 5 | 6 | 7 | 8 | 9 | 10 | 11 | 12 |
| *Luchlompol series (4673 – 4884 m)* | | | | | | | | | | | |
| Dolomite and sandy dolomite | 2.77 | 5.25 | 3.41 | 1.08 | 6.6 | 0.05 | – | – | – | 13.4 | 1.4 |
| | (55) | (2) | (1) | (2) | (1) | (1) | – | – | – | (1) | (1) |
| Arkose, sandstone | 2.78 | 5.22 | 3.20 | 1.12 | 6.6 | 0.15 | 22 | 22 | 161 | 21.8 | 1.8 |
| | (214) | (11) | (10) | (11) | (9) | (9) | (1) | (1) | (1) | (10) | (10) |
| Andesite-dacite metaporphyrite | 2.75 | 5.47 | 3.39 | 1.11 | 7.5 | 0.16 | – | – | – | 21 | 1.9 |
| | (259) | (9) | (9) | (9) | (8) | (8) | – | – | – | (7) | (7) |
| Serpentinite peridotite | 2.88 | 6.29 | 3.49 | 1.02 | 9.0 | 0.27 | 30 | 36 | 190 | 17.7 | 2.3 |
| | (627) | (17) | (15) | (17) | (15) | (15) | (5) | (5) | (5) | (8) | (8) |
| *Pirttiyarvi series (4884 – 5619 m)* | | | | | | | | | | | |
| Magnetite-biotite-amphibole-plagio-clase schists in trachybasalt | 2.97 | 5.80 | 3.50 | 1.13 | 7.8 | 0.18 | – | – | – | 28 | 2.3 |
| | (315) | (18) | (13) | (18) | (11) | (11) | – | – | – | (21) | (21) |
| Magnetite-biotite-plagioclase schists in trachyandesite-basalt | 2.86 | 5.04 | 2.96 | 1.14 | 6.1 | 0.20 | – | – | – | 20.5 | 2.3 |
| | (244) | (20) | (5) | (20) | (4) | (4) | – | – | – | (2) | (2) |
| *Kuvernerinyok series (5619 – 5717 m)* | | | | | | | | | | | |
| Tremolitic dolomite | 2.85 | 6.10 | 3.12 | 1.11 | 7.1 | – | – | – | – | 12.2 | 3.1 |
| | (55) | (3) | (2) | (3) | (6) | – | – | – | – | (3) | (3) |
| Carbonate and micabearing meta-psammite | 2.74 | 5.32 | 3.22 | 1.10 | 6.5 | 0.22 | – | – | – | 21.6 | 2.0 |
| | (87) | (6) | (5) | (6) | (6) | (6) | – | – | – | (5) | (5) |
| *Mayarvi series (5717 – 6835 m)* | | | | | | | | | | | |
| Amphibole-plagioclase schists in leucodiabase | 2.90 | 5.28 | 2.91 | 1.14 | 6.1 | 0.21 | 15 | 34 | 160 | 20.4 | 2.2 |
| | (1895) | (79) | (86) | (79) | (44) | (44) | (5) | (5) | (5) | (78) | (78) |
| Quartz-biotite-plagioclase schists in andesite | 2.78 | 5.28 | 2.88 | 1.12 | 5.7 | 0.23 | – | – | – | 29.4 | 1.6 |
| | (15) | (1) | (1) | (1) | (1) | (1) | – | – | – | (1) | (1) |
| Apodiabase amphibolite | 3.03 | 5.76 | 3.26 | 1.08 | 7.6 | 0.23 | – | – | – | 16.4 | 1.8 |
| | (122) | (4) | (6) | (4) | (4) | (4) | – | – | – | (7) | (7) |

**Table 2.15** (continued)

| Rock | $\sigma$, g cm$^{-3}$ | $V_p$ km s$^{-1}$ | $V_s$ km s$^{-1}$ | $K_a^w$ | $E_{dyn} \times 10^{-4}$ MPa | $\nu$ | $\sigma_t$, MPa | | $P_i$ | | $K_{pl}$ |
|---|---|---|---|---|---|---|---|---|---|---|---|
| | | | | | | | $\sigma_t \perp$ | $\sigma_t \parallel$ | MPa | MPa | |
| 1 | 2 | 3 | 4 | 5 | 6 | 7 | 8 | 9 | 10 | 11 | 12 |
| *Archean group* | | | | | | | | | | | |
| *Unit I. Gneiss with high-aluminium minerals (6842 – 7622 m)* | | | | | | | | | | | |
| Biotite gneiss with high-aluminium | 2.77 | 4.29 | 2.54 | 1.18 | 3.4 | 0.01 | – | – | – | 12 | 3.3 |
| minerals | (241) | (12) | (1) | (12) | (12) | (1) | – | – | – | (10) | (10) |
| Migmatite-granite, pegmatite | 2.64 | 4.34 | – | 1.97 | 3.5 | – | – | 11.5 | 117 | 22.3 | 2.7 |
| | (223) | (5) | – | (5) | (3) | – | – | (1) | (1) | (7) | (7) |
| *Unit II. Shadow migmatite, gneiss and amphibolite (7622 – 9456 m)* | | | | | | | | | | | |
| Nebulitic migmatite, gneiss-plagio- | 2.65 | 3.99 | 1.84 | 1.05 | 2.02 | 0.17 | 7 | 11 | 194 | 27.8 | 2.2 |
| granite | (5245) | (40) | (15) | (40) | (15) | (15) | (8) | (8) | (9) | (42) | (42) |
| Plagiomicrocline gneiss-granite | 2.62 | 3.50 | 1.83 | 1.05 | 1.6 | 0.20 | – | – | – | 375 | 2.0 |
| | (241) | (4) | (4) | (4) | (4) | (4) | – | – | – | (4) | (4) |
| Epidote-biotite, epidote-biotite-am- | 2.78 | 4.34 | – | 1.08 | – | – | 9.5 | 21 | 151 | 23.1 | 2.8 |
| phibole gneiss and schist | (328) | (4) | – | (4) | – | – | (1) | (1) | (1) | (4) | (4) |
| Amphibolite | 3.02 | 4.62 | 1.89 | 1.13 | 2.3 | 0.13 | 5 | 20 | – | 25.8 | 2.0 |
| | (417) | (10) | (10) | (10) | (10) | (10) | (1) | (1) | – | (7) | (7) |
| *Unit III. Two-mica gneiss with fibrolite (9456 – 9541 m)* | | | | | | | | | | | |
| Two-mica gneiss with fibrolite | 2.72 | | | | | | | | | | |
| | (98) | | | | | | | | | | |
| *Unit IV. Shadow migmatite, gneiss and amphibolite (9541 – 10144 m)* | | | | | | | | | | | |
| Nebulitic migmatite, gneiss-plagio- | 2.68 | 4.08 | 2.55 | 1.06 | 1.3 | 0.05 | – | – | – | 22.8 | 2.2 |
| granite | (1118) | (3) | (1) | (3) | (3) | (3) | – | – | – | (5) | (5) |
| Epidote-biotite, epidote-biotite-am- | 2.67 | | | | | | | | | | |
| phibole gneiss and schist migma- | (318) | | | | | | | | | | |
| tized with plagiomicrocline | | | | | | | | | | | |
| granites | | | | | | | | | | | |
| Amphibolite | 3.00 | 4.24 | – | 1.24 | – | – | – | – | – | 17.6 | 2.4 |
| | (411) | (3) | – | (3) | – | – | – | – | – | (6) | (6) |

**Table 2.15** (continued)

| Rock | $\sigma$, g cm$^{-3}$ | $V_p$ km s$^{-1}$ | $V_s$ km s$^{-1}$ | $K_a^w$ | $E_{dyn} \times 10^{-4}$ MPa | $\nu$ | $\sigma_t$, MPa | | $P_i$ | | $K_{pl}$ |
|---|---|---|---|---|---|---|---|---|---|---|---|
| | | | | | | | $\sigma_t \perp$ | $\sigma_t \parallel$ | MPa | MPa | |
| 1 | 2 | 3 | 4 | 5 | 6 | 7 | 8 | 9 | 10 | 11 | 12 |
| Blastocataclastic and chloritized | 2.65 | 4.98 | – | 1.05 | – | – | – | – | – | 28.2 | 1.8 |
| shadow migmatite | (219) | (2) | – | (2) | – | – | – | – | – | (5) | (5) |
| Blastocataclastic and chloritized | 2.96 | 5.96 | – | 1.12 | – | – | – | – | – | 25 | 2.0 |
| amphibolite | (183) | (3) | – | (3) | – | – | – | – | – | (6) | (6) |
| *Unit V. Two-mica gneiss with fibrolite (10,144 m – 10,600 m)* | | | | | | | | | | | |
| Two-mica gneiss with fibrolite | 2.70 | 3.02 | – | 1.22 | – | – | – | – | – | 13.4 | 2.1 |
| | (1683) | (2) | – | (2) | – | – | – | – | – | (13) | (13) |
| Amphibole-biotite gneiss, grano- | 2.66 | 3.30 | – | 1.06 | – | – | – | – | – | 14.3 | 2.0 |
| diorite | (427) | | – | (1) | – | – | – | – | – | (2) | (2) |
| Gneiss-granite and plagiomicrocline | 2.62 | – | – | – | – | – | – | – | – | 20.9 | 1.9 |
| pegmatite | (93) | – | – | – | – | – | – | – | – | (3) | (3) |
| *Unit VI. Migmatite, gneiss and amphibolite (10,600 – 11,000 m)* | | | | | | | | | | | |
| Nebulitic migmatite | 2.64 | – | – | – | – | – | – | – | – | 19.5 | 1.8 |
| | (594) | – | – | – | – | – | – | – | – | (2) | (2) |
| Epidote-biotite-amphibole gneiss, | 2.97 | 3.63 | – | 1.11 | – | – | – | – | – | 18.9 | 1.9 |
| schists | (43) | (3) | – | (3) | – | – | – | – | – | (1) | (1) |
| *Lower Proterozoic-Archean (?)* | | | | | | | | | | | |
| *(Dike?) sequence of orthoamphibolites and ultramafic orthoschists (6842 – 11,000 m)* | | | | | | | | | | | |
| Apodiabase-amphibolite | 3.06 | 4.75 | 2.25 | 1.18 | 3.48 | 0.20 | 13 | 18 | 182 | 26.5 | 2.2 |
| | (1209) | (13) | (4) | (13) | (6) | (4) | (2) | (2) | (2) | (22) | (22) |
| Leucodiabase amphibolites including | 2.92 | 4.42 | 2.65 | 1.12 | 3.40 | 0.04 | 11 | 23 | 190 | 22.8 | 2.2 |
| those with cummingtonite | (1120) | (15) | (3) | (15) | (3) | (3) | (3) | (3) | (3) | (12) | (12) |
| Phlogopite-talc-tremolite schists in | 2.94 | 4.27 | 1.91 | 1.21 | 2.6 | 0.19 | 5 | 11 | 64 | 7.5 | 2.1 |
| ultrabasic rocks | (77) | (4) | (1) | (4) | (1) | (1) | (2) | (2) | (2) | (5) | (5) |

*Note.* Number of measurements is shown in brackets.

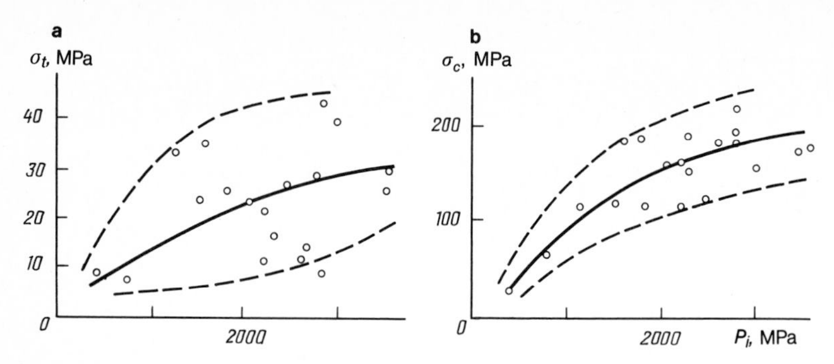


**Fig. 2.26 a, b.** Correlation between strength and its indentation hardness of rocks. **a** tension; **b** compression. Confidence intervals are shown by *broken lines*

rock mechanical properties by forcing a flat-based die into a stressed sample is the most successful method of the kind.

The weighted average value of indentation hardness for the Pechenga sequence rocks is 2520 MPa (Table 2.16), the variations being from 380 MPa for apopicritic talc-chloritic schists (4578 – 4673 m) to 3530 MPa for metadiabases. The $P_i$ values for metamorphic volcano-sedimentary rocks (1190 – 2180 MPa) are considerably lower than those for metamorphosed magmatic bodies (1640 – 3530 MPa), with the exception of blastomylonites of the greenschist facies enriched in chlorite, mica and talc and developed in basic and ultrabasic rocks. These blastomylonites have abnormally low $P_i$ values.

The weighted average $P_i$ for the Archean sequence is 2230 MPa, which is just a little below that shown above for the Proterozoic. Low $P_i$ is characteristic of phlogopite-talc-tremolite schists (750 MPa) and biotite-two-mica gneiss with high-aluminium minerals (1200 MPa) commonly developed at depths from 6842 to 7622 m. Nebulitic migmatite and amphibolite known as the most common rocks in the Archean section have $P_i$ equal to 2780 and 2570 MPa, respectively.

As the correlation analysis has shown (see Fig. 2.26), the indentation hardness is closely related to quantity $\sigma_c$ and not so closely to quantity $\sigma_t$, correlation coefficients ($r$) being equal to 0.89 and 0.35, respectively (confidence bound at a 1% level of significance, $r = 0.56$). Moreover, the analysis has proved that core decompression does not greatly influence its $P_i$ characteristic, while rock mineral composition plays a major role. However, the role of the decompression factor cannot be excluded, for the indentation hardness of some rocks sampled from outcrops appears to exceed that of their analogues sampled in the well from 7 – 11 km depths by 20 – 30%.

The ratio of rock compressive and tensile strengths is known to characterize its brittleness (plasticity) as follows: the higher ratio $\sigma_c / \sigma_t$ is, the higher brittleness and, respectively lower plasticity the rock has. The most brittle rocks are characterized by $\sigma_c / \sigma_t = 15 - 20$.

The brittleness of rocks penetrated by the well to the depth of 6842 m, i.e. to the Archean basement, is generally low. Thus, the $\sigma_c / \sigma_t$ ratio is 5 – 6 for diabase, gabbro-diabase and metadiabase, while it is 3 – 4 for tuffs, serpentinite and talc-bearing ultramafites and talc-chloritic schists. Some increase in rock brittleness is

**Table 2.16.** Weighted average (by thickness) values of rocks physico-mechanical properties for series, units and sequences of the Kola well section

| Series, unit, sequence | $H$, $m$ | $\sigma$, g cm$^{-3}$ | $V_p$ km s$^{-1}$ | $V_s$ km s$^{-1}$ | $K_a^w$ | $E_{dyn} \times 10^{-3}$ | | $\sigma_t$, MPa | | $\sigma_c$ MPa | $P_i$ MPa | $K_{pl}$ |
|---|---|---|---|---|---|---|---|---|---|---|---|---|
| | | | | | | MPa | MPa | $\sigma_t \perp$[b] | $\sigma_t \parallel$[b] | | | |
| 1 | 2 | 3 | 4 | 5 | 6 | 7 | 8 | 9 | 10 | 11 | 12 | 13 |
| *Series:* | | | | | | | | | | | | |
| Matertin | 0–1059 | 3.00 | 6.68 | 3.56 | 1.03 | 10.0 | 0.27 | 42 | 42 | 195 | 26.9 | 2.3 |
| Zhdanov | 1059–2805 | 2.90 | 6.09 | 3.50 | 1.05 | 8.7 | 0.23 | 28 | 32 | 162 | 20.4 | 2.1 |
| Zapolyarny | 2805–4673 | 3.01 | 6.50 | 3.73 | 1.05 | 10.4 | 0.23 | 28 | 31 | 164 | 33 | 1.7 |
| Luchlompol | 4673–4884 | 2.76 | 5.36 | 3.35 | 1.11 | 7.1 | 0.14 | 22 | 22 | 161 | 20.1 | 1.8 |
| Pirttiyarvi | 4884–5619 | 2.92 | 5.47 | 3.26 | 1.13 | 7.0 | 0.19 | – | – | – | 24.8 | 2.3 |
| Kuvernerinyok | 5619–5717 | 2.80 | 5.79 | 3.16 | 1.11 | 6.9 | 0.22 | – | – | – | 16.5 | 2.6 |
| Mayarvi | 5717–6835 | 2.91 | 5.33 | 2.94 | 1.14 | 6.3 | 0.21 | – | – | – | 20.4 | 2.2 |
| Pechenga sedimentary-volcanic sequence as a whole | 0–6842 | 2.94 | 6.08 | 3.45 | 1.07 | 8.7 | 0.23 | – | – | – | 25.2 | 2.0 |
| *Unit:* | | | | | | | | | | | | |
| I – gneiss with high aluminium minerals[a] | 6842–7622 | 2.75 | 4.29 | 2.54 | 1.16 | 3.4 | 0.01 | – | – | – | 140 | 2.9 |
| II – nebulitic migmatite, gneiss and amphibolite | 7622–9456 | 2.73 | 4.12 | 1.85 | 1.07 | 2.4 | 0.16 | 77 | 150 | 1920 | 283 | 2.2 |
| III – two-mica gneiss with fibrolite | 9456–9541 | 2.72 | – | – | – | – | – | – | – | – | – | – |
| IV – nebulitic migmatite, gneiss and amphibolite | 9541–10144 | 2.77 | 4.26 | 2.65 | 1.09 | 2.8 | 0.08 | – | – | – | 226 | 2.1 |

**Table 2.16** (continued)

| Series, unit, sequence | $H$, $m$ | $\sigma$, g cm$^{-3}$ | $V_p$ km s$^{-1}$ | $V_s$ km s$^{-1}$ | $K_a^w$ | $E_{dyn} \times 10^{-3}$ | | $\sigma_t$, MPa | | $\sigma_c$ MPa | $P_i$ MPa | $K_{pl}$ |
|---|---|---|---|---|---|---|---|---|---|---|---|---|
| | | | | | | MPa | MPa | $\sigma_t \perp$[b] | $\sigma_t \|$[b] | | | |
| 1 | 2 | 3 | 4 | 5 | 6 | 7 | 8 | 9 | 10 | 11 | 12 | 13 |
| V – Two-mica gneiss with fibrolite | 10144 – 10600 | 2.72 | 3.20 | – | 1.20 | – | – | – | – | – | 153 | 2.1 |
| VI – nebulitic migmatite, gneiss and amphibolite | 10600 – 11000 | 2.65 | 3.32 | – | 1.06 | – | – | – | – | – | 195 | 1.8 |
| Archean amphibolite migmatite gneiss sequence (including orthoamphibolites) as a whole | 6842 – 11000 | 2.73 | 3.99 | 2.17 | 1.10 | 2.7 | 0.11 | – | – | – | 223 | 2.2 |

[a] Calculated accounting for orthoamphibolites.

[b] Tensile strength was measured in two directions as follows: parallel ($\sigma_t \|$) and perpendicular ($\sigma_t \perp$) to lamination of schistosity.

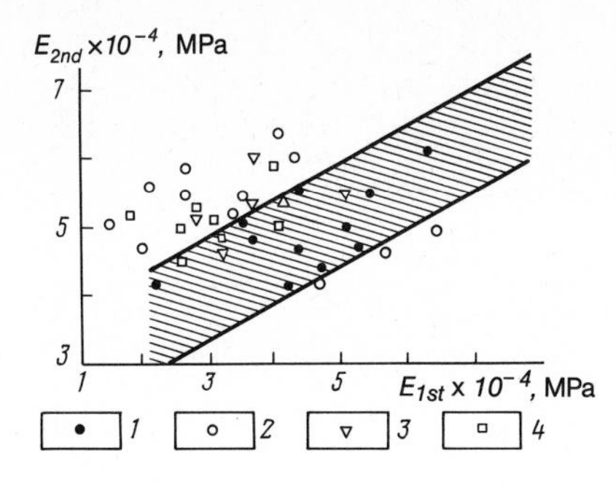


**Fig. 2.27.** Correlation between modulus of elasticity $E_{1st}$ determined by die forcing and modulus of deformation $E_{2nd}$ derived from single-axis loading. *1* diabase and gabbro-diabase; *2* peridotite; *3* tuff and tuffite; *4* sandstone and phyllite

noted in the bottom portion of the Pechenga sequence, e.g. this ratio is equal to 7 for amphibole-plagioclase schists. Brittleness of rocks taken from depths in excess of 6842 m is still higher; the $\sigma_c/\sigma_t$ values for talc-tremolite schists, gneiss with amphibolite and nebulitic migmatite are 8, 10–12 and 11–21, respectively. In this part of the section, the rock physical properties are greatly affected by decompression, their mineral composition playing a subordinate role.

The plasticity coefficient derived from the indentation analysis results (see Table 2.16) is theoretically reciprocal to rock brittleness. However, there is no strict correlation between rock plasticity and brittleness as measured by the indentation method, because this method yields parameter values that depend more on the core mineral composition than the extent of its disintegration.

To widen the knowledge on the physico-mechanical properties of rocks from the Kola well, the results of $E_{dyn}$ evaluation were compared with the $E_{st}$ values derived from measurements of deformations along the entire sample length, the samples being under a quasistatic single-axis load. The results of the comparison (Fig. 2.27) showed that the $E_{st}$ modulus is smaller than the $E_{dyn}$ quantity for all the rock types, but the character of variation with depth is practically similar for both these parameters, their highest values being observed in the upper part of the section (diabase and gabbro-diabase IV and III of the volcanogenic sheets) where $E_{st} = (5.0-7.5) \times 10$ MPa.

Ratio $R_{dyn}/E_{st}$ appeared to be different for different rock types: for diabase and gabbro-diabase it is equal to 2, while for sedimentary rocks it is about 1.7, which is in agreement with changes in the rock plasticity.

For solving technological problems it is important to know rock abrasivity $Z$.

Mean abrasivity values obtained for the Kola well rocks vary from 3 to 48 mg. The abrasivity of crystalline rocks (with respect to the steel) is known to depend on the hardness of the minerals of which the rocks are composed. However, during the friction of steel against polymict rocks an additional roughness is acquired owing to different wear of different minerals, which enhances the abrasivity of rocks, especially clastic ones. The lowest abrasivity (as was expected) is characteristic of phyllite and chlorite-sericitic or highly talcous rocks ($Z = 3-16$ mg). Because of the additional roughness effect, an elevated abrasivity is noted for tuff, tuffite and sandstone ($Z = 24-44$ mg).

According to the conventional classification of rocks with respect to their abrasivity, the Kola well rocks can be grouped as shown in Table 2.17.

**Table 2.17.** Abrasiveness of rocks from the Kola well

| Rock | Class | Grade of abrasiveness |
|---|---|---|
| Phyllite | I | Lower than medium |
| Chlorite-sericitic and talcose rocks | II | – |
| Porphyrite, peridotite, diabase, gabbro-diabase, sandstone, gneiss | IV | Medium |
| Tuff, granite, crystalline schist | V | Higher than medium |

The results of study of physico-mechanical properties of the Kola well rocks can be summarized as follows:

- the well section comprises mainly elastic-plastic rocks of high strength and medium abrasivity;
- the rock strength parameters vary with depth very irregularly, which causes serious drilling complications and necessitates the evaluation of these parameters by proximate methods for better adjustment of the drilling operations to the changing conditions;
- determination of rock hardness with the flat-die indentation method is the most successful technique; the results of hardness evaluation correlate well with compressive and tensile strength values;
- rock physico-mechanical properties (like the acoustic properties) are controlled by a number of factors as follows: petrographical and mineralogical, metamorphic and decompressional factor (the latter prevailing in the lower part of the section).

The rock tensile strength is most informative for evaluating the borehole wall stability. This parameter decreases step-wise with depth and the anisotropy of rocks with respect to this parameter were established.

# 2.15 Vertical Zonation in Rock Physical Properties and of the Crustal Structure from the Results of the Present Study

The results of the superdeep drilling, well logging, core analysis and geological and geophysical investigations of the nearby territories provide the basis for a new approach to the problem of the general structure of the Earth's crust.

Among the most interesting but least studied, and therefore still questionable, problems facing the Earth Sciences is the problem of the behaviour of rock physical properties at great depths. An unambiguous solution of this problem is only possible based on deep drilling information and petrophysical modelling[17] of the subsurface. The modelling is indispensable for improving the accuracy and reliability of surfacial geophysical surveys.

## 2.15.1 Elastic-Density Model

The principal rule derived from the study of the elastic characteristics and density of the Kola well rocks is as follows: the greatest densities ($\sigma$), velocities of compressional ($V_p$) and shear ($V_s$) waves and lowest porosities are observed in the upper portion of the section. A sharp decrease in values of $V_p$, $V_s$ and $\sigma$ and increase in rock anisotropy is noted at 4500 m depth within a homogeneous (in composition) unit of rocks of the Zapolyarnin series (Fig. 2.28). The first interface (at 4500 m) almost coincides with the boundary between the greenschist and epidote-amphibolite metamorphic facies, the second with the boundary between the Archean and Proterozoic. Below the first interface rock schistosity is common.

It should be noted that at the contact between the upper and lower groups of the Lower Proterozoic there is a sharp increase in the anisotropy factor of elastic wave velocities: within a single unit it increases from 1.06 to 1.37. Downwards from 4500 m, the anisotropy factor varies over a wide range, reaching 1.6, but nowhere being as low as in the upper group of the Middle Proterozoic, i.e. not dropping below the 1.05 – 1.09 limit.

The Kola well did not prove the commonly believed increase in elastic wave velocity with depth for rocks of the same composition.

Such factors as prograde and dynamic metamorphism leading to changes in rock structure appeared to influence the elastic properties greatly; the maximum elastic wave velocity is observed in the upper part of the section, i.e. we have here an inverse seismogeological section (velocity inversion).

---

[17] According to Karus and Shkerina (1974) under "petrophysical model" one understands the distribution of rock physical characteristics (as determined in a single well) in the subsurface.

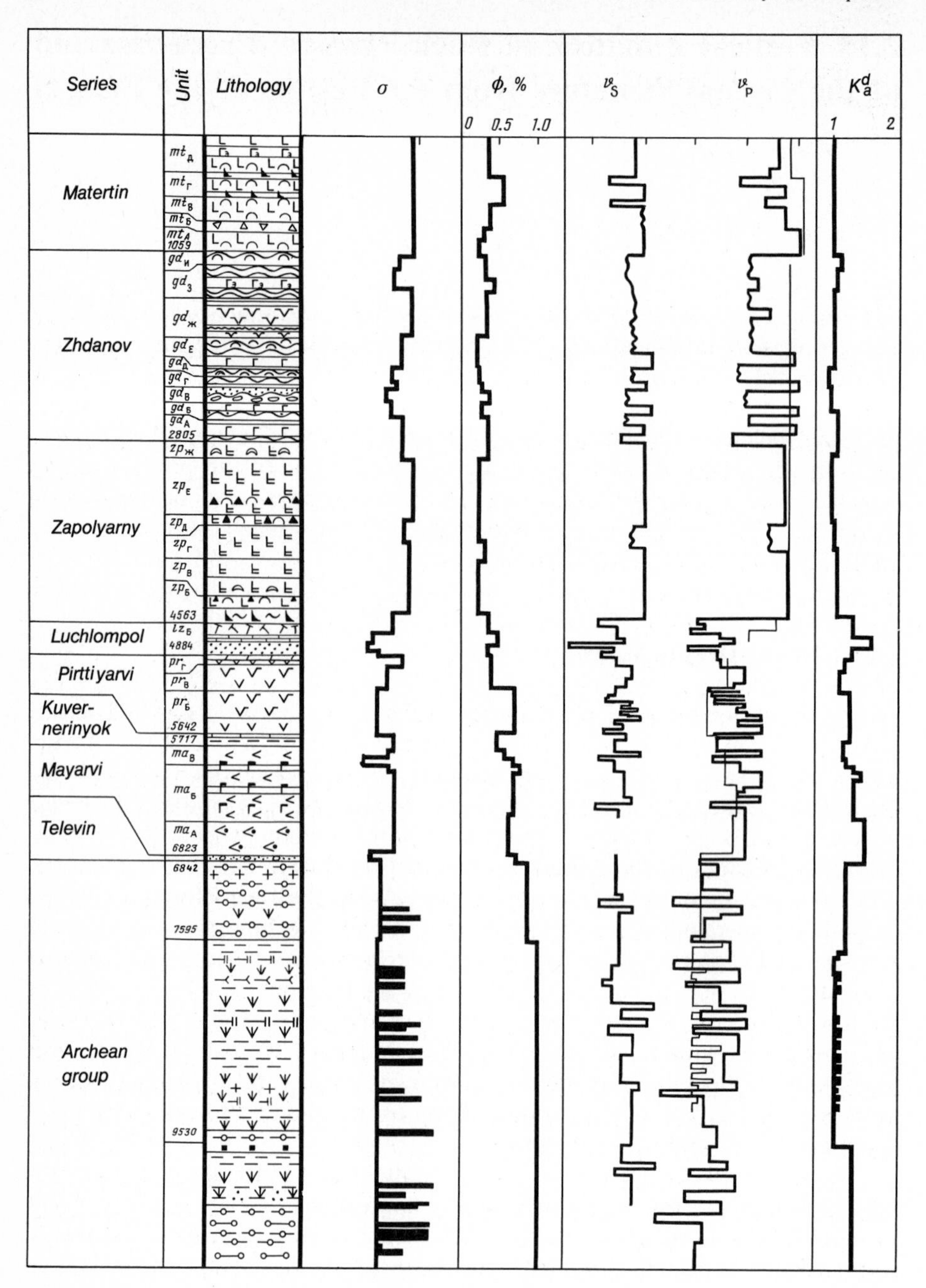


**Fig. 2.28.** Elastic properties and density of rocks in the Kola well section. Legend to the lithology column see Appendix

In the Archean section there is good density-velocity contrast between amphibolites and gneisses, the upper unit of high-aluminium gneisses being more dense than the lower one. The anisotropy of rock elastic properties does not increase with depth in this part of the section.

### 2.15.2 Seismic-Acoustic Model

Use of modern AL technology in the Kola well has helped to compile detailed velocity sections for both compressional and shear waves, plot vertical time (time-depth) curves, to determine effective attenuation and derive elastic modulus values (taking the rock density into account) for various layers. These data, together with the results of vertical seismic profiling, were used in constructing the seismic-acoustic model of the section (Fig. 2.29). The main characteristic feature of this model is as follows: there is no velocity gradient for both $V_p$ and $V_s$ down to 4500 m in depth; below this a local negative gradient is observed, thus showing a certain decrease in elastic wave velocity (according to AL and VSP data). Comparison of time-depth curves independently obtained from the results of these two different techniques has shown their complete agreement for practically the entire section with the exception of its deepest part, where the AL quality is much poorer than in the upper part of the section.

It should be noted that the distinct zone of inversion (decrease) of the $V_p$ and $V_s$ velocities identified from geophysical data is just as clearly evident from core measurements. However, because of the influence of in-situ conditions on the $V_p$, its values derived from AL readings exceed those measured on cores even when they are saturated with water (see Fig. 2.29). For dry samples this difference is still greater. According to the AL interpretation, the velocity decreases sharply at a depth of 4500 m from 6.8 – 7 to 5.5 – 5.8 km s$^{-1}$.

The investigations have thus shown that there is a zone of inversion of rock elastic properties (and also of density).

The seismic-acoustic model of the subsurface was used in the experimental determination and calculations of stress distribution within the upper part of the Earth's crust penetrated by the Kola well. The main objective of these investigations was to evaluate the conditions of rock occurrence important for solving geological and technological problems of drilling. The irregular distribution of rock strain with depth thus revealed, and the presence of a thick zone (about 3 km) of considerable decrease (to 40 MPa) in the stress level (Fig. 2.30) are important from the geological and geophysical point of view, particularly for evaluating the possible nature of deep seismic horizons. The information on temperature gradient, heat flow, stress levels and secondary changes in rocks penetrated by the Kola well is generalized in Fig. 2.30. An analysis of these data has shown that abrupt changes in the rock stress level (curves II and IV) correlate with step-like variations in the temperature gradient (curve I) and heat flow (curve III) and correspond to depth intervals where the borehole crosses metamorphic and seismic interfaces, the contrasts becoming less prominent with depth for all the curves.

The majority of investigators believe that a great number of factors is responsible for redistribution of stresses in the Earth's crust. Results of tectonophysical

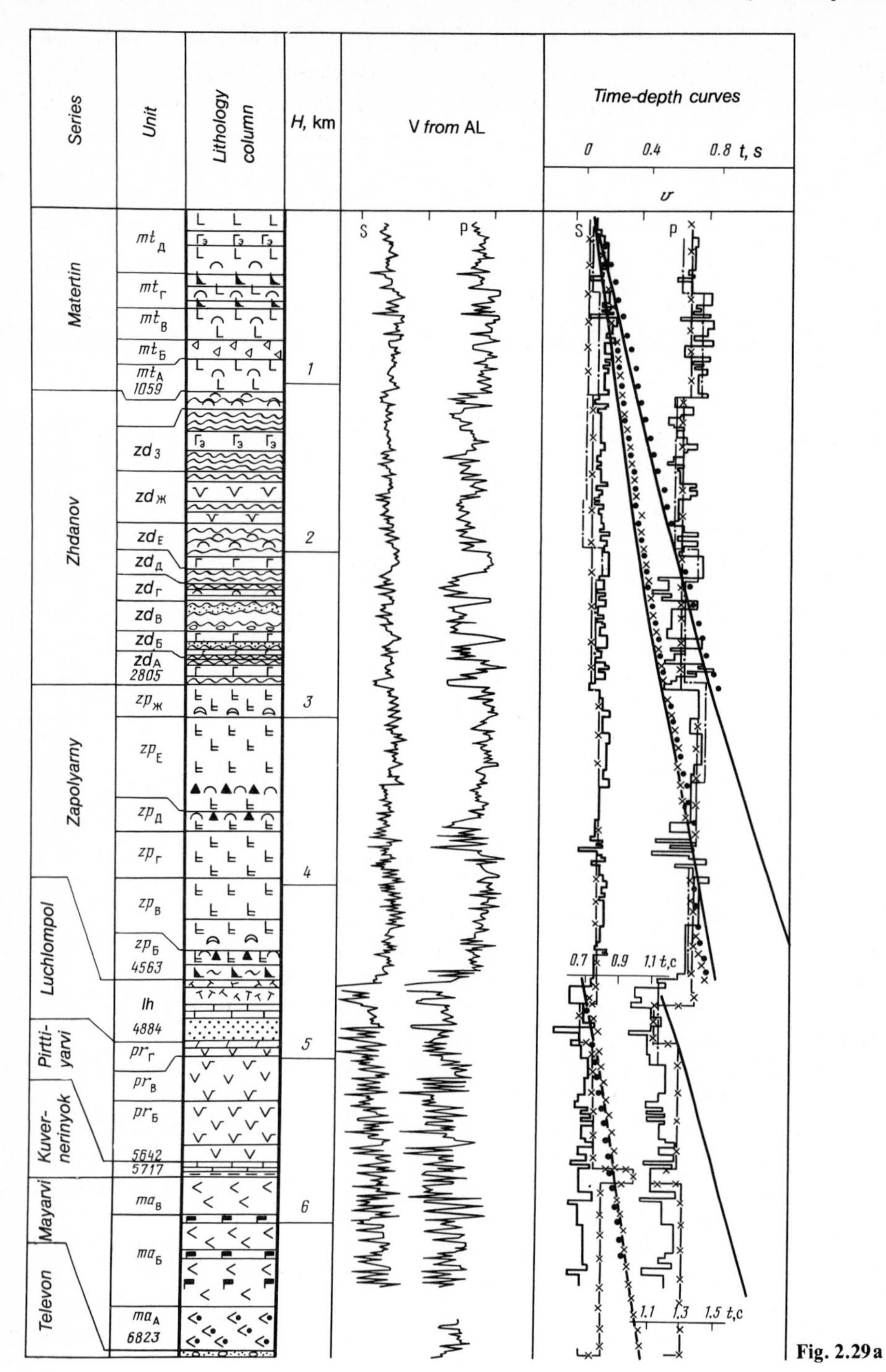


**Fig. 2.29 a – d.** Seismic-acoustic model of the Kola well section. *AL data: 1* reliable; *2* not reliable; *VSP data: 3* by VNIIYaGG; *4* by LGI; *5* Legend to the lithological column see Appendix; *6* Continuation of Fig. 2.29

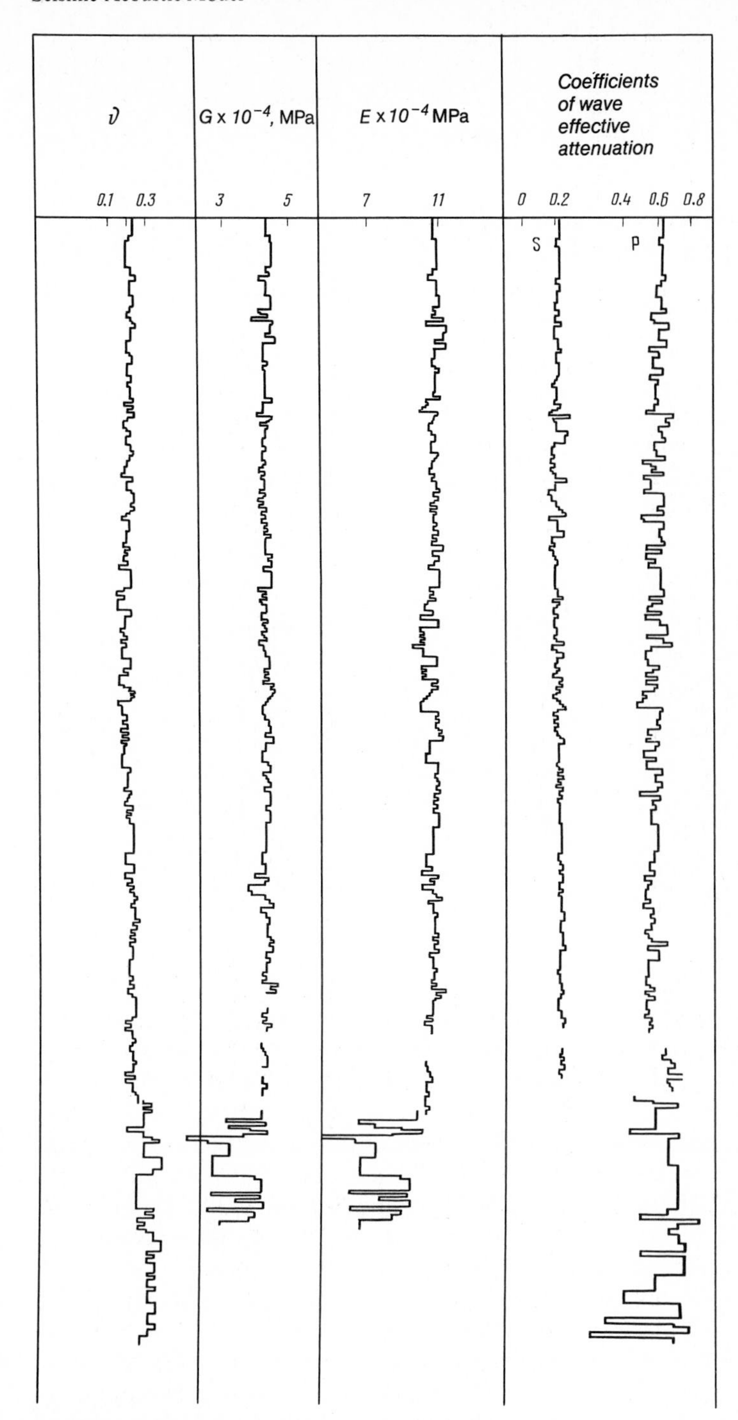


**Fig. 2.29 b** (Figs. 2.29 c and d see page 410 and 411)

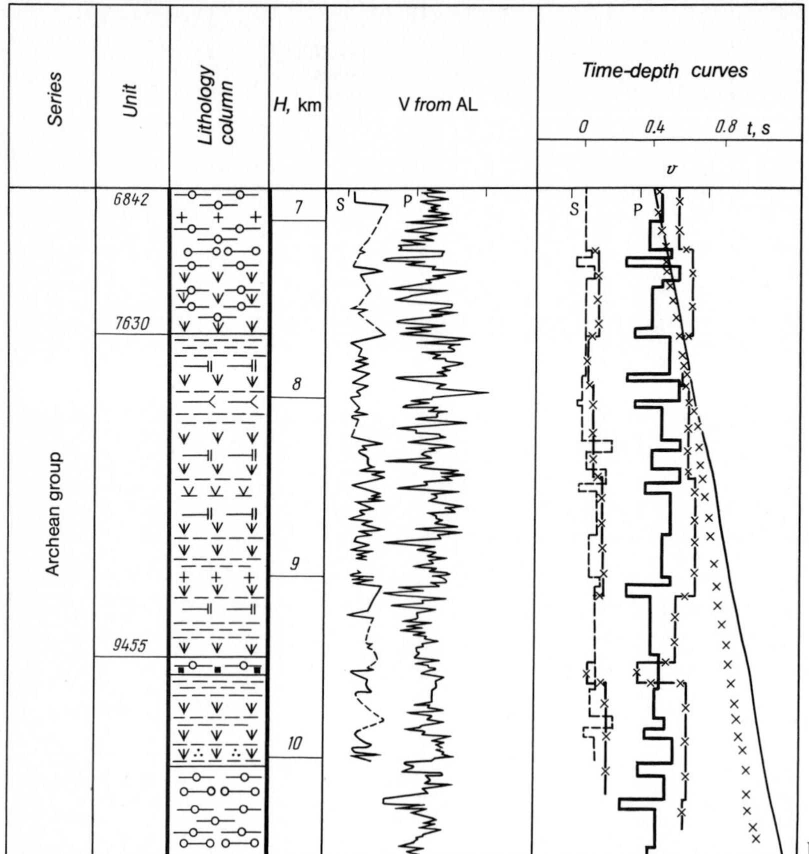


**Fig. 2.29 c**

modelling seem to prove this opinion (see Fig. 2.30). However, as shown in Markov (1977) and Panasyan (1981), the horizontal tectonic forces are active in the upper part of the Earth's crust only and do not contribute to the vertical component of stress ($\sigma_z$). It is the variations of this component with depth that are shown in Fig. 2.30 in two versions: calculated theoretically with due account taken for weakened zones (curve IV) and measured experimentally (curve II). It follows from analysis of these curves that $\sigma_z$ does not change linearly with depth: there are zones of both unloading and overloading (see Fig. 2.30). This redistribution of stresses is likely to result from processes that take (or have taken) place within the Earth's crust, such as rock metamorphism primarily.

The changes in petrophysical properties with progress in metamorphism lead to an alteration of the rock reaction to the stresses, the alterations being most evident in the zone of transition from greenschist to epidote-amphibolite facies, where the massive metabasites change for schistose rocks. An intensive granitization of rocks of the biotite-muscovite schist facies is noted in the bottom part of the section. It is with these features that the geophysical anomalies (especially seismic reflections) are associated (see Fig. 2.29).

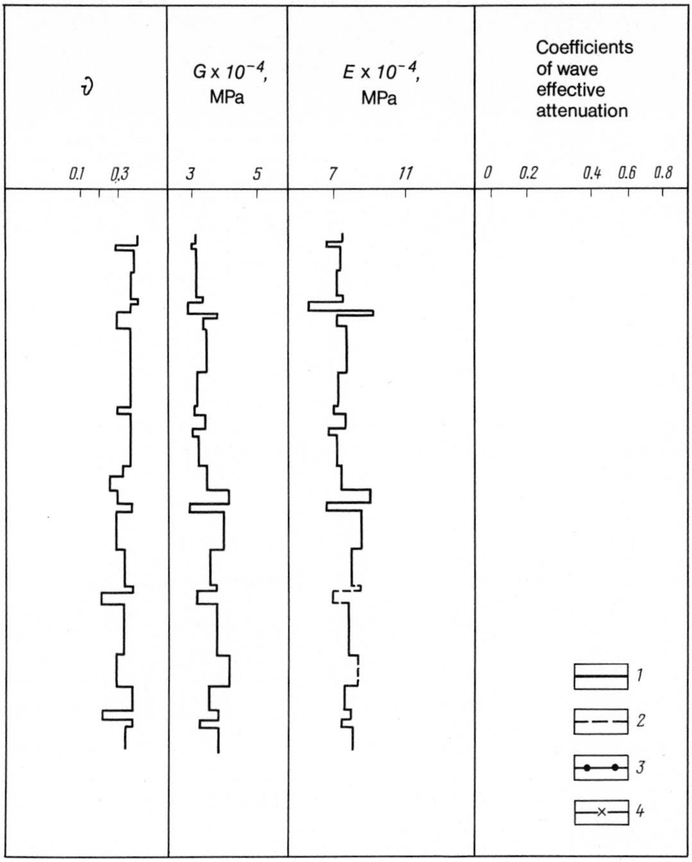

Fig. 2.29d

Since the metamorphic alterations in rocks include their recrystallization and changes in volume and heat flow, these alternations should be accompanied by a redistribution of stresses in the subsurface. For a better explanation of the observed change in stresses, it is necessary to employ the laws of physical properties (mainly elastic properties and density) as revealed by rock studies, particularly the three-layer structure of the well section.

Calculations made with the real elastic modulus and density values known for the rocks composing the well section show that the presence of the 40 MPa unloading zone can be explained by a relative change in the volume of the middle (unloaded) layer, the change being as small as 0.08% only. Such a minute volume change can persist for a long geological period.

Consequently, the irregular distribution of stresses in the subsurface will also be maintained. The relative increase in the middle layer volume (dilation) can also be associated with rocks fracturing and their irregular thermal expansion (caused by heat flow variations). The observed phenomenon is probably caused by factors such as secondary changes in rocks accompanied by changes in physical (especially elastic) properties of the rocks. The zones of sharp decrease (or in-

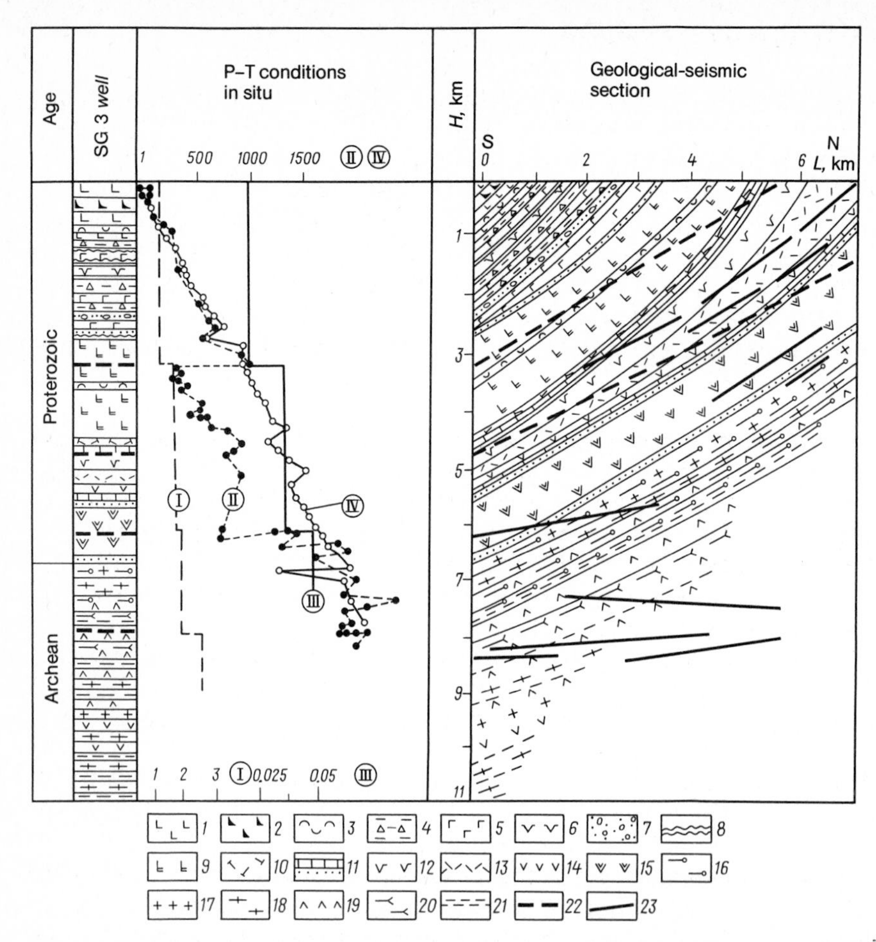


**Fig. 2.30.** Correlation of P-T conditions in situ with outlines of metamorphic facies (subfacies) and seismic horizons for Precambrian rocks of the Baltic shield penetrated by the Kola well. Number of curve and respective scale: *I* temperature gradient, °C/100 m; *II* vertical component of the stress tensor (physically measured), MPa; *III* heat flow, W $m^{-2}$; *IV* vertical component of the stress (analytically calculated), MPa; *1* diabase; *2* picrite porphyrite; *3* tuff; *4* tuffite; *5* gabbro-diabase; *6* peridotite; *7* breccia; *8* aleurolite, phyllite, sandstone; *9* diabase, globular lava, actinolite; *10* plagioporphyrite; *11* dolomite, sandstone; *12* metaandesite; *13* orthophyre, albitophyre; *14* metadiabase; *15* apobasalt-amphibole schists and amphibolite; *16* high-alumina gneiss; *17* plagiogranite; *18* biotite-plagioclase gneiss; *19* amphibolite; *20* talc-tremolite schist; *21* biotite-amphibole gneiss; *22* boundaries between metamorphic facies and subfacies; *23* seismic reflection horizons

crease) in the stress level must inevitably affect the geophysical parameters being measured, including the seismic ones. This is what happens in reality (see Fig. 2.30).

Thus, in this particular case the deep seismic reflections are generated not by lithological and stratigraphic interfaces, but rather by zones of changes in rock physical properties or stress level caused by secondary processes.

### 2.15.3 Tectonic-Physical Model

As shown above, the rock stresses and their distribution in the subsurface play a great role in the behaviour of physical properties and the formation of geophysical interfaces within the Earth's crust. In order to obtain more information on rock stresses, tectonic-physical investigations were carried out in the Kola well region. The goal of these investigations was to reconstruct the recent and present fields of tectonic stresses for the superdeep drilling region and extend these fields to greater depths. To achieve this goal the decomposition method[18] was used, which enables the determination of the stress field constituents by a systematic approach. They are related to tectono-dynamic systems of different ranks. Under a tectono-dynamic system, one understands a systematic genetic model of interrelated tectonic movements, deformations and strains generated within the lithosphere whose properties can be assumed to be uniform at a given scale of investigation. The tectono-dynamic systems from a hierarchy closely related to the hierarchy of tectonic deformations (structures). A relationship between the area of averaging (size of the measurement basis) and the depth of roots (extention) of the field of stresses is established for a given tectono-dynamic system. This approach, commonly used in geophysics (in sounding), is now being applied to studying rock stresses in mines (Markov 1977).

The relationship between the orientation of principal normal stress axes and the geometry of faults and tectonic fractures as established theoretically and proved by field observations, seismological data and modelling (Nikolaev 1977) was used to reconstruct the stress fields.

When reconstructing recent and present stress fields for the region studied, the main difficulties were encountered in dating the fractures. An analysis of air- and space photographs, as well as the topography of the studied region (including the entire north-west part of the Kola peninsula), has helped to recognize some fractures distinctly expressed in the topographic relief. These are thought to be those renewed for recently formed. Statistical analysis of variations in the length of these fractures with respect to their average spacing (performed using the method proposed by S. I. Scherman) has revealed a discrete distribution of this parameter, i.e. the possibility of establishing a hierarchy of faults for the region.

Various faults were thus found to belong to different ranks of tectono-dynamic systems. Azimuth frequency diagrams plotted for the strikes of these faults (mostly with pronounced dips) have shown that the faults of different ranks differ from each other both in general direction of the strike and predominant direction of dispersion. This helped to assume a displacement character of the stress field in the Kola well region. The compression axis of this field is directed to north-north-east, while the relative extension axis strikes from NW to SE (Fig. 2.31).

To date this field of stresses more precisely the mechanism of earthquake foci was analyzed over the entire Arctic area (see Figs. 2.32 and 2.33a, b). The analy-

[18] This method was elaborated in the Neotectonic Laboratory of Geological Faculty of the Moscow State University named after M. V. Lomonosov.

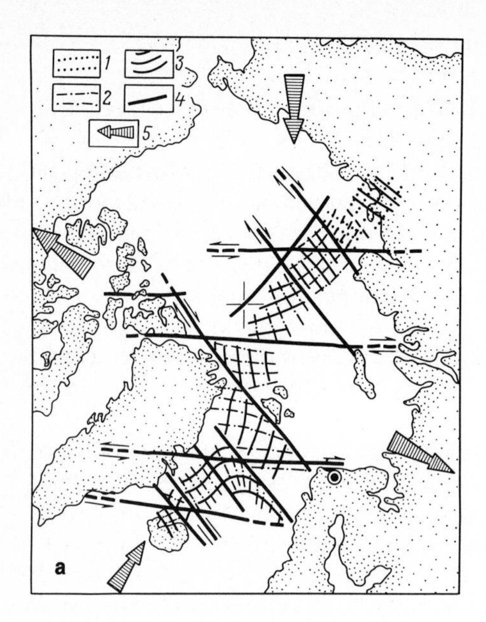


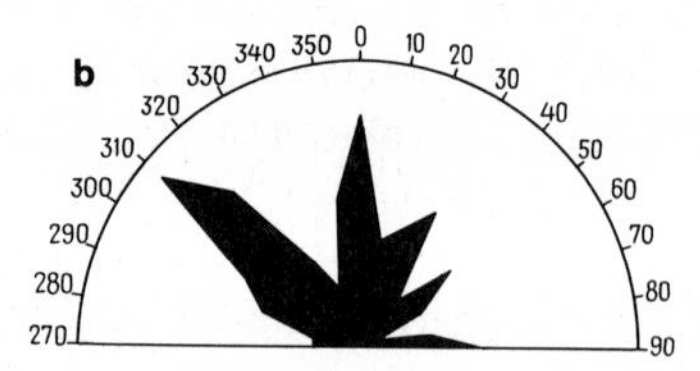


**Fig. 2.31 a, b.** Diagram of tectonic stresses for the Arctic rift area **(a)** and star-diagram for major faults of the Kola peninsula from geological survey, air- and space-photo data **(b)**. Axes of: *1* extension; *2* intermediate; *3* compression; *4* assumed fractures; *5* regional extension and compression directions

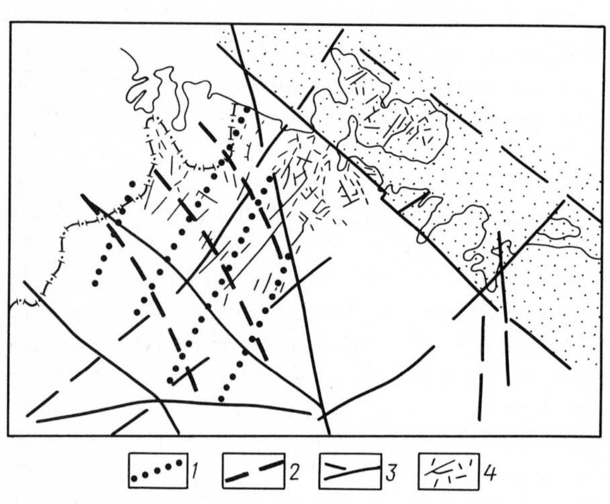


**Fig. 2.32.** Field of tectonic stresses in the north-west of the Kola peninsula. Axes of: *1* extension; *2* compression; *3* major faults; *4* minor faults

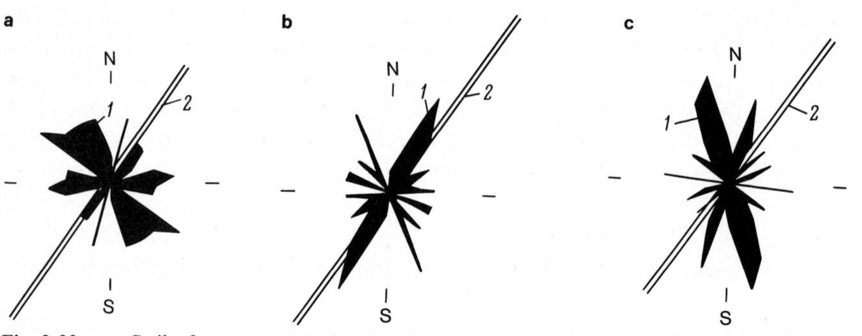


**Fig. 2.33 a – c.** Strike frequency plots for the recent tectonic fractures (*1*) in the zones of ancient faults (*2*). Fractures: **a** and **c** superimposed neotectonic ones; **b** "inherited" present-day ones

sis has helped to predict major faults in this region that coincide with fractures independently identified by geological, geophysical and remote surveys and additionally to confirm the earlier assumed orientation of axes of the main normal stresses. Consequently, the present-day orientation of the stress field (seismological data) essentially inherits the orientation of the recent axes of the main normal stresses (at least the Holocene ones) in the Kola well region.

These preliminary results served as a basis for detailed studies of the stress field by numerous (over 5000) measurements of tectonic fractures within the limits of the Pechenga structure and adjacent areas. Several systems of fracturing filled with various minerals have been revealed in the Pechenga sequence rocks (Kazanski et al. 1980). These fractures correspond to ancient fields of tectonic stresses. In addition to these, there are thin hair-shaped or open cracks without mineral filling widely developed in the region. They form fracturing systems clearly expressed in the present topography in some places. They are often characterized by friction planes and displacements crossing more ancient fractures filled with minerals. These fracture systems appear to be parallel to the lineaments easily seen on large-scale space imageries. These features of fracturing favour its being of young (or even very young) in age.

Fracture measurements were carried out using a more or less regular network of profiles study all the rocks of different lithology and age of the Pechenga sequence uniformly. For each point of observation (where about 100 measurements were made on average), the local stress field was reconstructed. The points were grouped for outcrops of individual series, associations of series and, finally, for the entire area. Characteristic features of the three-dimensional position of the fractures were derived using various averaging charts. Analysis of the data set obtained helped to identify several different stress fields corresponding to different tectonodynamic events of various ranks, i.e. having their roots at various depth levels of the crust.

It was clear that the lithology of individual units, minor faults, dikes, boundaries between differently metamorphosed rocks, etc. essentially affected the local character of tectonic stress fields. The relative roles of the factors vary greatly depending upon their structural position in and orientation to the external stress field of a lower rank. Thus, the recent fractures either follow the ancient structural pattern perfectly, or cross it as if "ignoring" the structural features formed earlier (see Fig. 2.33). A more detailed analysis shows that the ancient fractures "revive" when parallel to the main normal stresses, and are not renewed if their direction differs considerably from the usual direction of the most recent stresses.

It should be emphasized that the recent fractures appear to "neglect" even major ancient faults (the orientation of the earlier structural features being unfavourable), particularly the layer-to-layer overthrusts (see Fig. 2.33).

A separate study of recent and more ancient fractures was carried out at some observational points. It was noted that the ancient stress fields differed sharply from the recent ones (even when a renewal of displacements along ancient trends took place), which indicates a decisive rebuilding of structures and a basic change in the mechanism of the deformation of the Earth's crust during the neotectonic stage of its history.

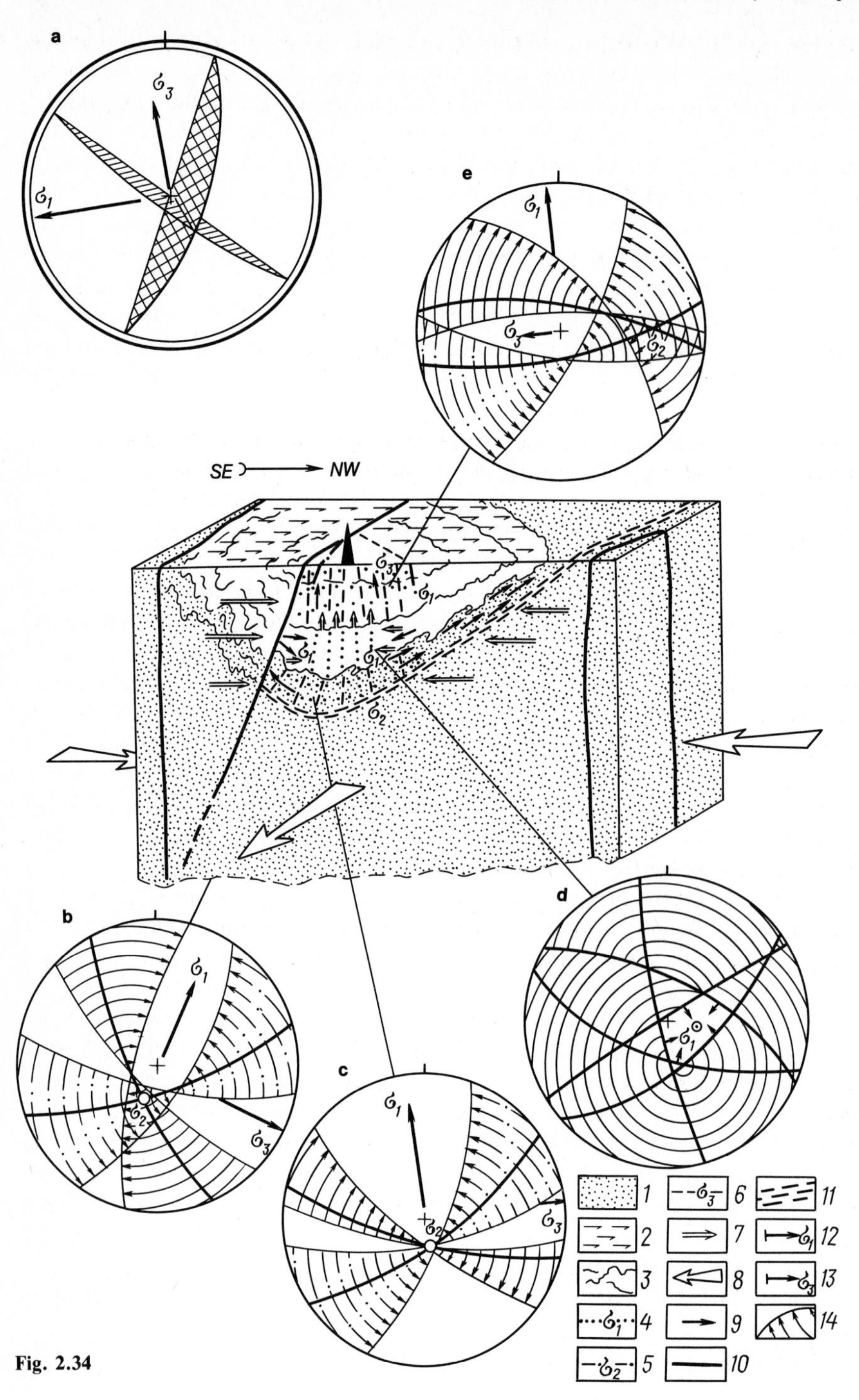


Fig. 2.34

Besides the reconstruction of stress fields from the data on neotectonic fracturing, a dynamic microstructural analysis of oriented samples was made for some points, as well as a study of microsections prepared from the well cores. In most of the cases, the analysis resulted in the reconstruction of ancient stress fields. The data on twinned carbonates (including those from individual streaks) and (in some cases) plagioclase helped to reconstruct the most recent field of stresses.

These investigations led to reconstructing this regional stress field for the entire Pechenga structure and also the more local fields corresponding to tecto-no-dynamic systems of the three ranks. The stress field of the third rank known to have the shallowest roots and form the uppermost tectonophysical layer is characterized by a vertical axis of compression and subhorizontal orientation of the other two axes. The underlying second tectonophysical layer has a subvertically oriented expansion strain pattern. The deepest layer is characterized by a shearing tectonic stress field and corresponds to the tectono-dynamic system of the first rank.

The results have been used to compile a tectono-physical model of the Kola well region (Fig. 2.34).

In the Lower Proterozoic rocks there are stress fields of the three different ranks, each having its own size of averaging area and characterizing a particular depth level in the crust. Analysis of these fields has shown that the "lens" of the more rigid rocks of the Pechenga sequence, whose petrophysical properties differ from those of the surrounding Archean rocks, undergoes a flattening deformation that is accompanied by a quasi-plastic upward squeeze of the material and characterized by a particular local stress field. This field is characterized by a subvertical relative expansion axis and more or less uniform horizontal compression (see Fig. 2.34d), which should be accompanied by a certain increase in volume of the rocks being deformed. At the top and bottom of the Lower Proterozoic "lens" the boundary conditions change abruptly, which leads to a change in the mechanism of rock deformation. This is proved by the change in rank of the tectonic stress field, as shown by results of the stress reconstruction using both shorter and longer bases.

The squeezed-out rocks of the middle and lower parts of the subsurface act as a "soft die" with respect to more brittle deposits in the upper part of the Lower Proterozoic section, making them laterally bend upwards, the respective additional local field of stresses having a vertical compression axis and subhorizontal tension axes (see Fig. 2.34e). The rank analysis of the stress field shows that the interface between these tectono-physical layers must be at depths from 3500 to 3800 m.

**Fig. 2.34 a – e.** Tectonophysical model of the Kola well region. Block-diagram. *Outcrops: 1* Archean; *2* Lower Proterozoic; *3* reference horizons; *strain orientation: 4* relative expansion; *5* intermediate; *6* shrinkage; *orientation of external forces: 7* local scale; *8* regional scale; *9* direction of the deformed material shift; *10* suture faults; *11* schistosity zones. Stereograms (*dash points* to the North). *Direction of stress axes: 12* tension; *13* compression; *14* fracture planes (arcs), dispersion in their orientation (*arrows*) and model orientation of fractures (*heavy arcs*). Stress field stars for: **a** Arctic basin; **b** Pechenga structure as a whole; **c** Archean part of the subsurface; **d** zone of unloading; **e** upper part of the section

This mechanism of rock deformation can be valid only if there is some slippage of the bottom of the Pechenga "lens" being deformed with respect to the more ancient rocks caused by horizontal compression. This is in fact the case, as the investigations have shown that there is another change in the tectonic stress field at the lower Proterozoic-Archean border (depth ca. 6800 m). The new field is essentially a shearing one and is characterized by compression and relative expansion axes, subhorizontally oriented (Fig. 2.34c), which indicates a layer-to-layer shift of the material along the contact of the rock sequences of different ages. A zone of slippage and rock schistosity shown in this figure (a series of upthrusts from the kinematic point of view) is expected to occur at depths from 13 to 15 km. Structural petrology studies of the Archean rocks from the lower part of the section have shown the possibility of identifying two fields of stresses, one replacing the other with time. One of the fields corresponds to the period of formation of schists, and its compression axes are normal to schist laminae. The second (younger) field is directed so that the schistosity planes appear to be parallel to the greatest tangential stresses, i.e. the supposed layer-to-layer displacement of the material is confirmed by factual data.

Analysis of stress fields of various ranks shows that there is well-developed horizontal layering in the Earth's crust, not generated by wide-scale displacement of rock masses, but caused by changes in local conditions and mechanisms of deformation from one deep horizon to another.

Thus, from the tectono-physical point of view, the Pechenga structure is a specific element of the Earth's crust characterized by an elevated compression in the deep caused by some increase in volume of the middle layer rocks and their squeeze-out. At the top and bottom of this layer there exist particular boundary conditions which are responsible for the reorientation of the tectonic stress field and subhorizontal lamination of the Earth's crust which is evident from seismogeological observations.

The geophysical investigations of the Kola well section and numerous systematic measurements of rock physical properties on core samples have thus resulted in compiling geophysical and petrophysical sections (models) of the upper part of the Earth's crust penetrated by the well, including seismic-acoustic, magnetic-electric (magnetic and electrical properties measured both in the borehole and on samples), natural radioactivity (based on well logging and core measurements) and elastic-density models. Analysis of the vertical zonation in rock physical properties and the structure of the Earth's crust in the Kola well region led to the following observations:

The Kola well seismic-acoustic model reflects in detail the vertical distribution of elastic wave velocity, elastic modulus and effective attenuation. The quality of acoustic logs registered below 4500 m in depth becomes much poorer because of a more vugged borehole and a larger average borehole diameter. Below this depth there is a low velocity zone comprising the rocks of the Pechenga (to depth 6842 m) and Kola (to the well bottom) sequences. This zone is characterized by lower velocities as compared to the uppermost high-velocity part of the Pechenga sequence (Proterozoic) and clearly demonstrates a sharp decrease in elastic wave velocity with depth.

## Appendix

Legend to lithology columns

| Symbol | Rock |
|---|---|
| | **Proterozoic group** |
| 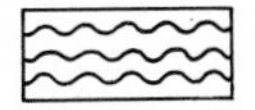 | Phyllite |
| 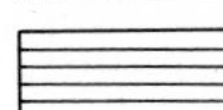 | Aleurolite |
| 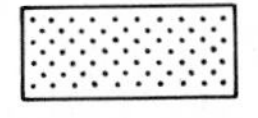 | Sandstone, metamorphic sandstone |
| 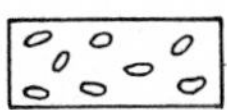 | Sandstone, gritstone, conglomerate |
| 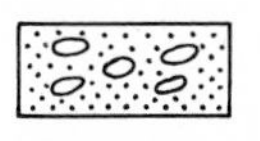 | Metamorphosed sandstone, gritstone-sandstone, fine-clastic conglomerate |
| 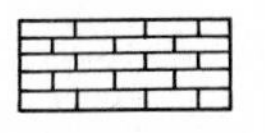 | Sandy limestone, dolomite, including metamorphosed ones |
| 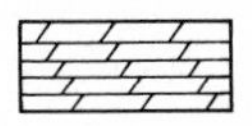 | Calcareous, calcite-dolomite and talc-tremolite marble |
| 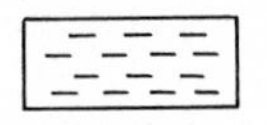 | Quartz-mica-carbonate and quartz-mica schist, metamorphic quartzite and sandstone |
| 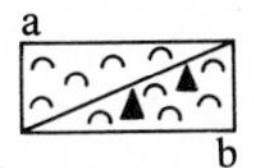  | Tuffs of basic composition: (a) pelitic, (b) psammitic, including metamorphosed ones |
| 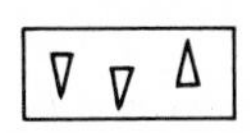 | Lava-tuff breccia and fine-clastic tuff of basic composition |
| 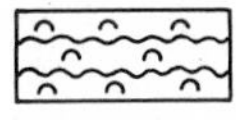 | Tuffitic phyllite, aleurolite, sandstone |
|  | Actinolitic lava breccia, tuff-lava and tuff breccia of basic composition |
|  | Talc-chlorite and carbonate-chlorite schists in ultrabasic and basic tuff-sand effusions |
| 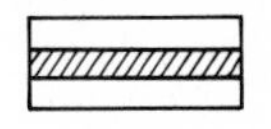 | Sericitic schist, metamorphosed crust of weathering |
| 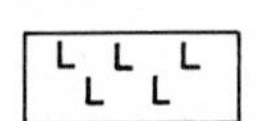 | Diabase, actinolitic diabase |
| 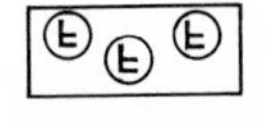 | Actinolite diabase of globular lava |
| 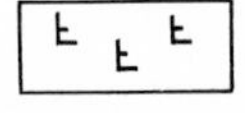 | Porphyroblastic actinolite diabase |
| 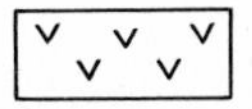 | Apodiabase magnetite-amphibole-plagioclase schist |
| 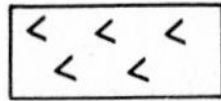 | Apodiabase amphibole-plagioclase schist |
| 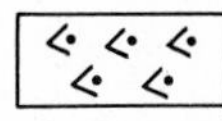 | Apodiabase biotite-amphibole-plagioclase schists |
| 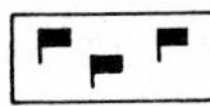 | Melanocratic plagioclase-amphibole rocks |
| 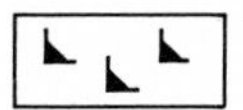 | Picrite porphyrite |
| 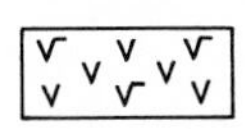 | Apoandesite magnetite-biotite-plagioclase schists with interlayers of apodiabase magnetite-amphibole-plagioclase schists |
| 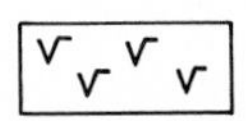 | Apoandesite magnetic-biotite-plagioclase schist and metamorphosed andesite and trachy-andesite |
| 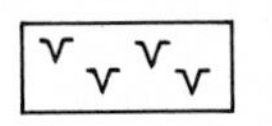 | Serpentinite peridotite, serpentinite, talc-serpentinite rocks |
| 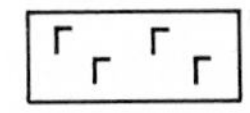 | Gabbro |
| 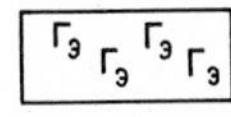 | Essexite gabbro |
| 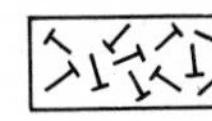 | Dacite-andesite porphyrite |
| | **Archean group** |
| 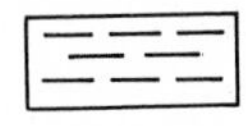 | Biotite-plagioclase gneiss |
| 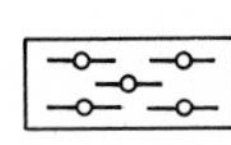 | Biotite-plagioclase gneiss with high-alumina minerals (garnet, staurolite, andalusite, sillimanite) |
| 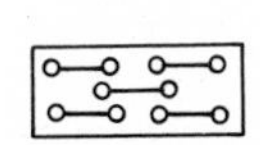 | Two-mica gneiss, schist with high aluminium minerals |
| 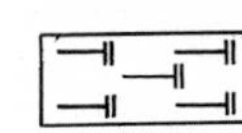 | Biotite-amphibole-plagioclase gneiss |
| 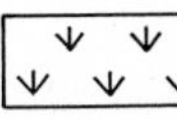 | Amphibolite |
| 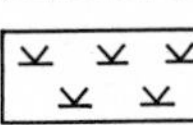 | Amphibolite with cummingtonite |
| 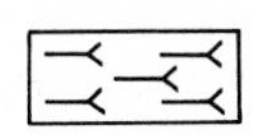 | Meta-ultramafites and biotite-amphibole schist |
| 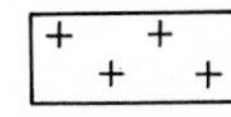 | Biotite bearing porphyric granite |
| 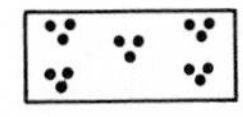 | Sulphide impregnation |
|  | Magnetite impregnation |
| 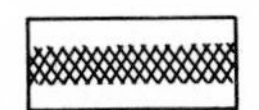 | Cataclastic zone |

# 3 Drilling

## 3.1 Technical and Technological Problems of Drilling. The Scientific Basis for Choice of Drilling Equipment and Tools

World drilling practice is an indication of the considerable difficulties in drilling super-deep wells. Present-day capabilities permit the construction of several unique research holes (under standard geological environments) to a depth of 15 km. This depth is believed to be the maximum possible for science and technology for the next 30 – 35 years.

This section deals with the brief results of the research work aimed at solving basic technical problems of drilling a super-deep well in crystalline rocks with continual core recovery. These data have been used to devise equipment for SG-3 well and to prepare project documentation for its construction and drilling.

### 3.1.1 Drilling Parameters

The technical and economic characteristics of the drilling process and its duration are determined from the basic parameters that include: footage per run $h_{ave}$, overall round trip speed $\vartheta_{ort}$ and overall drilling rate $\vartheta_{odr}$. Theoretical determination of these parameters consists in deriving the analytical relationship $T_{pr} = f(h_{ave} \vartheta_{ort} \vartheta_{odr})$ and in searching for an optimum combination of these parameters to provide the minimum time expenditure for drilling a superdeep hole.

Productive drilling time $T_{pr}$ is the sum of the overall drilling time $T_{dr}$, round trip time $T_{rt}$, hole casing time $T_{cas}$, and the time for auxiliary $T_{aux}$ and repair $T_{rep}$ operations.

$$T_{pr} = T_{dr} + T_{rt} + T_{cas} + T_{aux} + T_{rep} . \tag{3.1}$$

The overall drilling time

$$T_{dr} = L_{max} / \vartheta_{odr} \tag{3.2}$$

($L_{max}$ is the maximum project depth of the hole in metres).

The drilling results indicate that the dependence of the well-deepening on the number of runs varies. As is shown in Vorozhbitov et al. (1975), this dependence is usually expressed by parabolic curves

$$L = A \beta^m ,$$

where $A$ is the proportionality factor, $\beta$ is the number of runs; $m$ is the parabola index characterizing a change of the footage per run in the course of drilling; in crystalline rocks $m > 0.7$, and in sedimentary rocks $m < 0.3 \div 0.6$.

The total length of the drill string being lowered and raised during the whole period of drilling is

$$\sum_{i=1}^{i=\beta} L_i = \frac{1}{m+1} \beta L_{max} \quad \text{or} \quad \sum_{i=1}^{i=\beta} L_i = \frac{L_{max}^2}{(1+m)\,h_{ave}}, \tag{3.3}$$

where $L_i$ is the depth of the hole at a given run, in metres.

The total time consumption for lowering and raising a drill pipe string for the whole period of drilling

$$T_{rt} = 2 \sum_{i=1}^{I=\beta} L_i / \vartheta_{ort} = \frac{2\,L_{max}^2}{3{,}600\,(1+m)\,h_{ave}\,\vartheta_{ort}}, \tag{3.4}$$

where $\vartheta_{ort}$ is the average overall round trip speed in m s$^{-1}$ [a ratio between the total length of the lowered and hoisted pipes and the total time of the round trip operations (RTO)].

Based on deep drilling experience, the time required for the hole casing may be taken as being equal to 0.4 $T_{dr}$.

$$T_{cas} = 0.4\,L_{max}/\vartheta_{odr} \tag{3.5}$$

The total time for auxiliary operations

$$T_{aux} = L_{max}\,t_{aux}/h_{ave}, \tag{3.6}$$

where $t_{aux} = \sum_{i=1}^{i=\beta} t_i = 0.52\,h_{ave} + 1.7$ is an empirical formula obtained at the Heavy Machine-Building Research Institute.

Substituting Eqs. (3.2), (3.4) – (3.6) into expression (3.1) and knowing that $T_{rep}$ does not exceed 10% of $T_{pr}$, we obtain

$$T_{pr} = 1.1\,L_{max}\left[\frac{1.4}{\vartheta_{odr}} + \frac{2\,L_{max}}{(1+m)\,h\,\vartheta_{ort}\,3{,}600} + \left(\frac{1.7}{h_{ave}} + 0.52\right)\right]. \tag{3.7}$$

Below are the data characterizing the effect of the basic parameters on the productive time of drilling a 15-km-deep well. The nature of the effect of $\vartheta_{odr}$, $h_{ave}$ and $\vartheta_{ort}$ on $T_{pr}$ is shown in Fig. 3.1. The curves have been constructed for the case where one of the above parameters varies, while the others remain constant.

As follows from Fig. 3.1, an increase of the overall drilling rate over 2.5 – 3 m s$^{-1}$ does not cause any essential change of $T_{pr}$.

Figure 3.2 demonstrates the effect of $h_{ave}$ on the productive time $T_{pr}$ $h_{ave}$ as changes from 0.1 to 1 m s$^{-1}$ and $\vartheta_{odr} = 2.5$ m h$^{-1}$. Analysis of the curves indicates that in order to reduce the productive time to be within the limits of the real rig depreciation period (4 – 5 years), it is necessary to estimate the level of the $h_{ave}$ and $\vartheta_{ort}$ values which would allow us to drill a 15-km-deep hole for this period. Had a standard drill rig that is capable of providing $\vartheta_{ort} = 0.1 + 0.15$ m s$^{-1}$ been used for drilling to such a depth, it would have been possible to drill a hole for the given period, provided that the average footage per run is not less than 33 m, which is hardly possible in the case of continuous coring. With an increase of $\vartheta_{ort}$ to 0.35 m s$^{-1}$, the average footage per run must be not less than 12 – 15 m. A further increase of $\vartheta_{ort}$ to 0.6 – 0.7 m s$^{-1}$ would reduce $h_{ave}$ to 7 – 8 m, which is quite achievable.

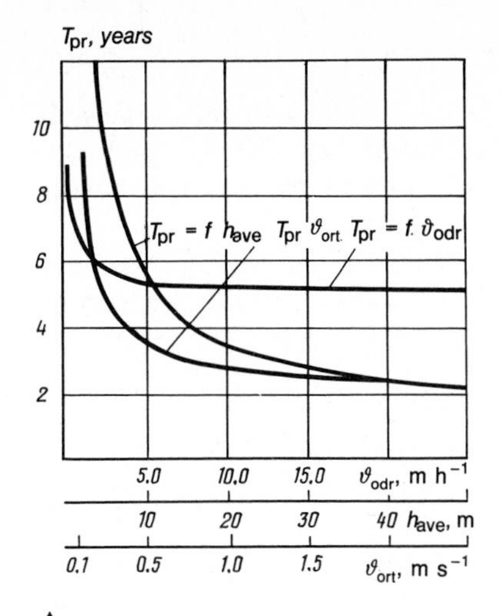


▲ **Fig. 3.1.** Dependence of $T_{pr}$ on the basic parameters

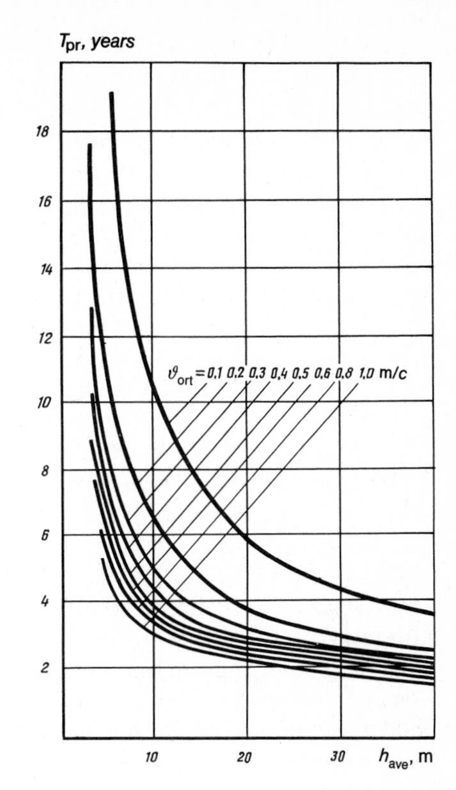


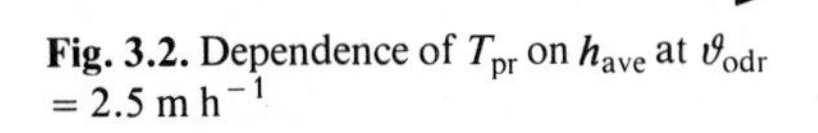

**Fig. 3.2.** Dependence of $T_{pr}$ on $h_{ave}$ at $\vartheta_{odr} = 2.5$ m h$^{-1}$ ►

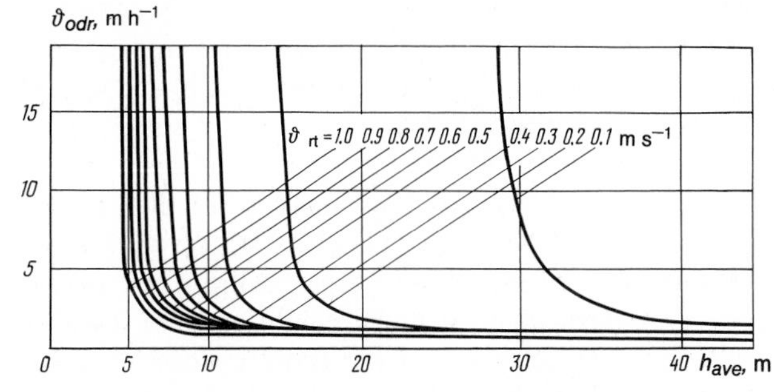


**Fig. 3.3.** Graph showing choice of the basic parameters on condition that $T_{pr} = 5$ years

A graph-nomogram in Fig. 3.3 is constructed for a 15-km-deep well and permits determination of two basic parameters, the third parameter being specified in advance.

Analysis of the graph gives the following minimum values of the basic parameters of the drilling process: $\vartheta_{odr} \geqslant 2.5$ m h$^{-1}$; $\vartheta_{ort} \geqslant 0.6$ m h$^{-1}$; $h_{ave} \geqslant 7$ m. An estimate of the parameters obtained allows us to choose a drilling method, and to define the main trends in improvement drilling equipment and superdeep drilling technology.

### 3.1.2 Choice of Well Design

One of the major problems of deep and super-deep drilling in the areas with poorly known rock succession is the correct choice of well design, which will meet the geological and technical requirements of its drilling and its aim. Inadequately planned well design may lead to serious complications.

The information about geology and potential fields in the drilling site of superdeep well obtained by indirect methods of investigation often appears to be rather unreliable. This is confirmed by the comparison between the inferred and actual data on all the known super-deep holes. For example, the Miocene formations in the Dzharly Well (Azerbaijan SSR) of 7000 m depth were expected at a depth of 3500 m. In fact, however, these rocks were exposed at a depth of 2875 m. The top of the Mesozoic succession appeared to be at a depth 3111 m and not at 4300 m, as was expected.

More serious errors were revealed when we compared the actual stratigraphic levels of the succession with the project ones in the well SG-1 (Aral-Sor), where the top of the Jurassic rocks was intersected at a depth of 2150 m instead of the project depth of 1450 m, and the Permo-Triassic strata were encountered at a depth of about 2800 m instead of 2000 m, while the Carboniferous strata, expected at a depth of 4700 m, was not even met at 6000 m. Drilling of a number of other deep wells showed that the project data on them were also of approximate character, which led to errors increasing with depth. Moreover, it becomes clear that the depth of stratigraphic units exceeds even the length of the casing below the shoe of the previous casing.

The information about anticipated complications is even more vague, since no direct methods exist to define these complications without exposing the section, while the use of data from other wells, even from those sited in the neighbourhood, leads to serious errors. Thus, only a number of assumptions as to rock succession and its variation with depth provide rough initial data for substantiating and projecting the designs of deep exploratory wells. In this, a well design implies a ratio between lengths and diameters of casing strings and other parameters derived from the results of many years' practice of drilling and casing of wells under the most typical geological and technical conditions. Because of the inadequate information about rock succession at the drilling site, difficulty arises in choosing an optimum well design. The problem is that no initial data are available to substantiate the number of casing strings and the depth to which they must be sunk. In this case, the overestimating or underestimating of the project rock succession may lead to lower efficiency in the process of well construction.

The most reliable solution in this respect seems to be through the development of a drilling technology that could make it possible to introduce corrections to the well design directly in the course of drilling by incorporating the actual geological and physical characteristics of the rocks as they are exposed. Such an approach has been developed at the "problem laboratory" for drilling to the mantle (the All-Union Research Institute of Drilling Technique) to solve the problem of optimum well design with only inadequate data on rock succession. The principle of the method lies in the fact that for planning we include not the entire well design, but only its upper part, for which the initial data are most reliable. These usually give the depth of the first (occasionally the second) casing. The diameter of the casing is chosen so as to allow for lowering sufficient casing strings in the case of a complicated formation being exposed in the well.

After sinking and cementing of the casing, the hole is not deepened. A removable casing is lowered into the stationary cemented casing to prepare for drilling a pilot hole. It is secured at the surface in such a manner that it can be subsequently turned and recovered. Then a pilot hole is drilled, with a diameter chosen to ensure the best technical and economic factors.

The pilot hole is drilled after thorough study of the actual state of the section already drilled and the complications encountered. If troublesome zones have been exposed, the pilot casing is raised, the hole is widened and cased in a conventional way. The diameters of the widened hole and casing strings are chosen appropriately to the depth achieved, the interval to be drilled and its presumed degree of complication.

This approach has the following advantages:

- it greatly simplifies a well design;
- it helps to standardize rock-crushing tools, bottom-hole assembly and drill pipes;
- it provides protection of the stationary casings against wear, by the application of a removable casing;
- it leads to a reduction of material consumption and productive time;
- it provides better drill mud washing and pressure control in a well by means of a hydraulic channel between the removable and stationary casings, etc.

This drilling technology is most effective in areas with poorly studied geological and mining conditions, great depths of drilling, and zones of drilling not compatible with formation pressures. The crystalline rock section corresponds most fully to the above factors, which predetermined the selection of this technology for drilling the Kola well.

### 3.1.3 Stability of the Borehole Wall Rock Mass

As has been already mentioned, drilling of a deep well is undertaken with a minimum of information about the rock mass to be exposed. Successful drilling to the projected depth is determined by optimun well design, the major parameter of which is the uncased portion of the advance hole. It is very important in this case to retain the strength (stability) of the sidewall, to predict zones of possible complications associated with a decrease in density of the rocks, as well as the possibility of controlling these over the processes within the drilling site.

There are many different factors which affect the stability of the well walls. They may be tentatively subdivided into two groups, geological and technological. The former include the strength characteristics of rocks, their structure, mode of occurrence, gravitational and tectonic stresses; the latter involves geometric parameters of a well, type and density of the washing agent, manner of drilling, depth of sinking of intermediate casings, and the peculiarities of round trip operations (RTO).

At a first approximation, an increase in complications within a continuous drill site boundary in one and the same rock types, also caused by caving in of rocks in the interval drilled, may be taken as a criterion of the strength of the hole walls. In the process of drilling the equilibrium of the gravitational forces acting in the rock mass is disturbed, resulting in a concentration of stresses near the surface of the walls and bottom of the hole. A certain ratio between these stresses and rock strength may lead to destruction of the hole outline. In these conditions, the problem of controlling the process of preservation of the hole walls in crystalline rocks is reduced to controlling the stresses arising in the hole, since there are as yet no direct methods to control the strength characteristics of the rocks being exposed.

A gravitational field is associated with the overlying rock mass and with the specifics of its structural pattern, whereas a stress vertical component may be written as

$$\sigma_z = \Sigma h_i \gamma_i , \tag{3.8}$$

where $h_i$ is the thickness of the $i$-th interval, $\gamma_i$ is the specific weight of the rock of the interval.

Horizontal components $\sigma_x$ and $\sigma_y$ in an arbitrary point of the elastic isotropic medium are found from the expression

$$\sigma_x = \sigma_y = \xi \Sigma h_i \gamma_i , \tag{3.9}$$

where $\xi = \nu/(1-\nu)$ is a coefficient of the radial stress showing which part of the active load acting on the rock section is the reactive stress, given that the possibility of rock deformation in a plane perpendicular to the action of the active load is absent; $\nu$ is the Poisson ratio.

According to many investigators (Baklashov and Kartoziga 1976; Krupennikov et al. 1972), even the crystalline rocks of very high strength properties change with depth (tens of kilometres) and when exposed to by high pressure and temperature to a plastic state ($\nu$ approaches 0.5) when the nature of the stress state becomes close to hydrostatic. In the boundary case, where $\nu = 0.5$ and $\xi = 1$, a hydrostatic distribution of stresses takes place, i.e. $\sigma_x = \sigma_y = \sigma_z$.

The tectonic component of the stress field differs from the gravitational component in its appreciably more complex nature. The acting forces are assumed, as the first approximation, to be horizontal; the direction of their action theoretically depends on the orientation of tectonic structures, particularly of deep faults. In a general case, the major stresses caused by the action of tectonic force $T_{st}$ may be written in the following form:

$$\sigma_x = T_{st}; \quad \sigma_y = \Psi T_{st}, \quad \sigma_z = \chi T_{st}, \tag{3.10}$$

where $\Psi$ and $\chi$ are the coefficients of the horizontal and vertical repulsions in a field of tectonic forces.

In this, $\xi > \chi \geqslant 0$ and $\xi > \Psi \geqslant \nu$.

The appearance of the tectonic stresses is established experimentally. These stresses are often found at shallow depths to considerably exceed the gravitational ones, in other words, $\Psi T_{st} > \xi \gamma H$ ($H$ is the depth of the well). The nature of the tectonic stresses greater than gravitational ones, the mechanism of their generation and their retention in the course of geologically long periods have not been studied adequately.

Assumptions made in Turchaninov et al. (1978) indicate that the excess horizontal stresses in the upper crustal layers occur as a secondary effect of vertical uplifting of tectonic blocks. It has been established that the Baltic Shield rises roughly by 11 mm annually, which means that an interrelation probably existed between the crustal uplifts and manifestations of excess stresses over the entire territory of Finnoscandinavia.

Ascending movements on the Kola Peninsula involve the Khibiny and Lovozero massifs, where excess horizontal stresses, four times as great as the vertical component at a depth of 500 – 600 m, are observed. The Pechenga massif, however, where the Kola well is sited, lies in a zone of present-day subsidence of the crust or in a zone intermediate between uplifts and subsidences. The investigations carried out in this area at the Pechenga and Nickel mines at shallow depths have failed to detect any tectonic stresses. The horizontal stress components amount to 0.7 of the vertical ones (Turchaninov et al. 1978). It was also found that folding of the massif resulted in a field of stresses represented by successive alternation of zones, showing the values of vertical components to be increased and decreased respectively in comparison with the values of $\gamma H$.

To check and specify more exactly the values of vertical components of the stress field in the vicinity of the Kola drill hole, the Laboratory for Rock Mechanics at the Moscow State University has attempted an analytical calculation of the static field of stresses using a variation-difference method for an elastic medium under plane deformation environments. Account was also taken of the thickness and dip of beds composed of rocks with different properties. The plane with the well in its centre measures $6000 \times 8000$ m in size. The minimum thickness of the identified rock layer was taken to be 200 m and was equal to a subunit, being of similar length both vertically and horizontally. The average density of rocks for all identified layers was taken to be constant and equal to 2.9 g cm$^{-3}$, the Poisson ratio 0.25, and Young's modulus corresponded to the average values obtained by the analysis of the rock specimens collected from the exploratory wells in the area of the SG-3 hole.

The following regularities in a distribution of stresses may be noted:

- values increase in low-elasticity and decrease in high-elasticity rocks;
- in front of the weakened layer or near the fault there is a local rise of the stress vertical component;
- in an interlayer between the low-elasticity rocks or in blocks between tectonic deformations, a general level of the stresses decreases;
- the presence of weakened interlayers in the lower part of the succession where the rock beds display a monoclinal dip leads to a deviation of contour lines of the stress vertical component from the horizontal; a general trend of the contour lines is across the dip of beds.

As is shown by the analysis, the value in an interval of 0 – 9000 m corresponds to a hydrostatic distribution of stresses (Fig. 3.4).

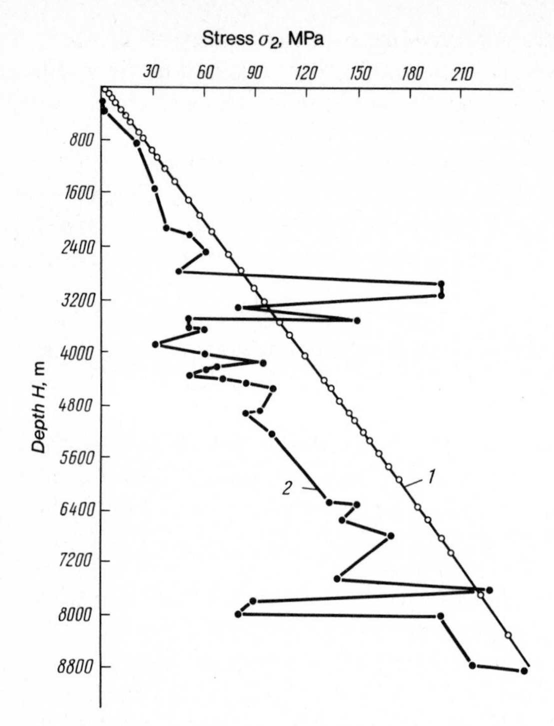


**Fig. 3.4.** Dependence of stresses $\sigma_z$ on the hole depth $H$: *1* calculation based on the geostatic law of stress distribution; *2* calculation based on elastic properties of rocks

We considered a model for the stress distribution in the rock mass of a vertical mine. The formation of any cavities in the rock mass causes the initial natural field of stresses to change. Near the outline of a cylindrical vertical cavity filled with fluid, the stress field is described by the following system of equations (Turchaninov et al. 1971).

$$\sigma_z = \gamma_r H;$$

$$\sigma_r = \xi \gamma_r H \left(1 - \frac{r^2}{a^2}\right) - \gamma_l H \frac{r^2}{a^2}; \tag{3.11}$$

$$\sigma_\theta = \xi \gamma_r H \left(1 + \frac{r^2}{a^2}\right) + \gamma_l H \frac{r^2}{a^2},$$

where $\sigma_z$, $\sigma_r$, $\sigma_\theta$ are the axial, radial and tangential components of the stresses; $\gamma_r$ and $\gamma_l$ a specific weight of the rock and fluid respectively; $H$ is the depth, $a$ is the distance from the axis of the cavity to a measurement point; $r$ is a radius of the cavity.

This system of equations characterizes a field of stresses in the isotropic mass near the walls of the hole due to the action of the rock stratum and taking the backpressure of the drill mud into account. Since the transverse section of the hole is taken to be circular in shape, the components $\sigma_z$, $\sigma_r$ and $\sigma_\theta$ are major stresses, in other words, are equal to $\sigma_1$, $\sigma_2$, and $\sigma_3$ respectively; in the annulus part of the rock mass only compressing forces are active. However, the rocks composing the SG-3 hole section are characterized by a varying degree of anisotropy, and therefore expressions (3.11) cannot reflect a true distribution of the stresses. Anisotropy may be accounted for by introducing the coefficient that characterizes the change of the stress state.

From Hooke's law, we obtain

$$\sigma_z = \frac{E}{E_1} \frac{\nu}{1-\nu_1} \sigma_r + \frac{E}{1-\nu} \varepsilon_0 . \tag{3.12}$$

$E$ and $E_1$ are the elasticity moduli respectively in the longitudinal and transverse directions; $\varepsilon_0$ is the relative deformation of rock.

Many investigators (Baklashov and Kartoziya 1976; Katsaurov 1972) believe that a hypothesis of the absence of linear horizontal deformations is most probable. In this case

$$\sigma_z = \frac{E}{E_1} \frac{\nu}{1-\nu_1} \sigma_r .$$

From here

$$\lambda = \frac{E}{E_1} \frac{\nu}{1-\nu_1} \leqslant 1 .$$

Comparing the obtained values of the components of stress and displacement in the homogeneous anisotropic mass ($E/E_1 = 1.5$; $\nu = 0.3$; $\nu_1 = 0.2$) with respective components in the isotropic mass ($E = E_1$; $\nu = \nu_1 = 0.3$), we see that although there are no changes in the mechanical processes, taking the anisotropy of rocks into account, even with their horizontal bedding, introduces appreciable quantitative corrections. For example, in the above-mentioned case, the values of the stress components were reduced roughly by 20%.

According to Markov (1977), the rocks of the Kola Peninsula (Khibiny massif) will not show failure in drilling provided that between the acting stresses $\sigma_{act}(\sigma_\theta, \sigma_r)$ and strength characteristic of the rock under uniaxial compression in the "piece" $\sigma_{cmp}$ there exists a relationship

$$\sigma_{act} < 0.3 [\sigma_{cmp}] . \tag{3.13}$$

Types of deformation such as exfoliation, sheeting and cracking of rocks take place when stresses reach

$$\sigma_{act} = (0.5 \div 0.8)[\sigma_{cmp}] . \tag{3.14}$$

Intensive "cracking" and shocks in rocks are observed on the condition that

$$\sigma_{act} > 0.8[\sigma_{cmp}] . \tag{3.15}$$

Hence, the stresses corresponding to a range of $(0.5 \div 0.8)$ $[\sigma_{cmp}]$ may be assumed to be sufficient for the loss of stability of the well outline. An increase in the intensity of these stresses must be accompanied by an instantaneous breaking of crystalline rocks. At $\sigma_{act} \leqslant 0.3$ $[\sigma_{cmp}]$ the hole will be stable, while at

**Table 3.1.** Tangential stresses versus depth

| Tangential stresses | Well depth, (m) | | | | | |
|---|---|---|---|---|---|---|
| | 4000 | 6000 | 8000 | 10,000 | 12,000 | 15,000 |
| For isotropic mass $\sigma_\theta$ in MPa | 39.0 | 59.0 | 78.5 | 98.0 | 117.5 | 147.0 |
| For anisotropic mass $\sigma_\theta$ (at $E/E_1 = 1.5$, $\nu/\nu_1 = 1.5$), MPa | 31.2 | 47.2 | 60.8 | 78.4 | 94.0 | 117.6 |

$0.3\,[\sigma_{cmp}] \leqslant \sigma_{act} \leqslant 0.5\,[\sigma_{cmp}]$ transitional processes will occur resulting in exfoliation and sheeting in rocks.

These factors may be applied with a sufficient degree of certainty to the behaviour of the annulus rock mass in the conditions of the complex-stress state. In this, the $[\sigma_{cmp}]$ may be substituted for the ultimate strength of the rocks under the action of rock pressure and hydrostatic pressure, and may be estimated as a difference between them, while the acting stress $\sigma_{act}$ is taken to be maximum of those acting on the well outline, i.e., $\sigma_\theta$.

Experimentally derived values of $\sigma_{ult}$ for the crystalline rocks studied are shown below (in MPa):

| | |
|---|---|
| Schists | 103 – 231 |
| Amphibolites | 140 – 287 |
| Biotite gneisses | 150 – 269 |
| Epidote-biotite gneisses | 194 – 268 |
| Granites | 215 – 298 |
| Porphyrites | 160 – 244 |
| Basalts | 217 – 309 |

Calculations made for the case of drilling when water is applied as drill mud in the crystalline rocks with a density of 3 g $cm^{-3}$ show that in the well outline tangential stresses $\sigma_\theta$ will arise with depth ($H$) (Table 3.1).

Apart from the action of the stresses caused by static loads, in the drill hole outline there arise also dynamic loads due to round-trip operations and washing of the hole. The strength characteristic of the rocks is strongly affected by the temperature of the surrounding mass and its variations caused by the drill mud. In this case, a cyclic repetition of temperature contrasts, like that of hydrostatic stresses, is capable of causing fatigue destruction of wall rocks.

Experiments (Timofeyev et al. 1971) have helped to clarify the nature of the fatigue failure, to determine the number of cycles capable of causing a loss in strength of the hole walls on a certain combination of rock and hydrostatic pressures and at a current value of the hydraulic pulse. Timofeyev et al. (1971) have shown that the least amplitude value of the hydrostatic pressure capable of causing an irreversible process is $\sigma_{cmp} = 5.0$ MPa.

The formula derived for probable intervals of fatigue destruction as a function of the technological process of drilling and physicomechanical properties of rocks is:

$$L_{\text{fat}} = \frac{\sigma_{\text{ult}}(\sigma_{-1} - \sigma_{\text{amp}})}{\sigma_{-1}(\gamma_r - \gamma_{\text{rep}})}, \tag{3.16}$$

where $\sigma_{\text{ult}}$ is the ultimate strength of rocks under static load and omnidirectional compression $\sigma_{\text{ult}} = (1 \div 3)\,\sigma_{\text{cmp}}$; $\sigma_{-1}$ is the endurance limit at a symmetrical cycle of loading; $\gamma_r$ and $\gamma_{\text{rep}}$ the specific weight of the rock and drill mud, respectively.

The ratio between the static $\sigma_{\text{ult}}$ and hydrodynamic strength $\sigma_{\text{hyd}}$ depends on rock type, and characterizes a change in strength of the sidewall under the action of temperature, cyclic temperature or cyclic hydrodynamic stresses.

$$\sigma_{\text{ult}}/\sigma_{\text{hyd}} = 1.20 \div 1.35\,. \tag{3.17}$$

We may treat the hydrothermodynamic strength of the wall of the well as another parameter of the mechanical properties of rocks characterizing a decrease in rock resistance to ultimate stresses. According to experiments, the ultimate strength of the majority of rocks lies within a range of 100 – 300 MPa. This means that the boundary values of the above rock strength at different depths are characterized by ratios of the acting maximum stresses $\sigma_\theta$ and strength indices for isotropic ($K = 1$) and anisotropic ($K_1 = 0.8$ K) rocks when a well is filled with water ($\gamma_{\text{rep}} = 1$ g cm$^{-3}$). The expected behaviour of rocks with different values of $\sigma_{\text{hyd}}$ (100, 200, and 300 MPa) in the function of well depth is shown in Table 3.2.

Figure 3.5 demonstrates a plot constructed from the data given in Table 3.2. This plot characterizes the expected behaviour of rocks of the sidewall at the moment of exposure depending on their depth of occurrence and strength characteristics.

The described model for stressed rocks is not unique. Once the actual drilling conditions appear to differ from the assumed ones, it is desirable to correct the controlled parameter – the density of the drill mud, which can be changed depending on the indications of stability (or instability) of the open hole evident in drilling. In any case, it is wise from the technical standpoint to use drill mud with a minimum density, that provides sufficient stability of the sidewalls to carry on drilling with continuous core recovery without any essential complications.

Based on general concepts about the behaviour of brittle crystalline rocks in complex-stress environments, it is possible to predict with some certainty evidence of the stability loss of the sidewall. With depth, this becomes clear in the progressive destruction of the continuity of the drill hole outline, in disking the core and in decreasing the thickness of disks to their full disintegration (slime) and in improved drillability of rocks, all other conditions being equal.

A general estimate of the sidewall stability shows the need of a universal technology of superdeep drilling which would permit a more exact specification of well design and changing drill mud density, etc. during drilling, when actual data on drilling conditions are required.

### 3.1.4 Admissible Parameters of a Spatial Trajectory of the Borehole

Implementation of drilling technology in the deep well and the intensive wear of the hole walls in a drill casing depend largely on the parameters of the spatial trajectory of the drill hole. The total sweep angle (inclination) $\varphi$ is the criterion for

**Table 3.2.** Probable behaviour of rocks with different $\sigma_{hyd}$ versus depth

| Stress and stability parameters of rocks | Depth of well, (m) | | | | | |
|---|---|---|---|---|---|---|
| | 4000 | 6000 | 8000 | 10,000 | 12,000 | 15,000 |
| $\sigma_{hyd} = 100$ MPa | | | | | | |
| $K$ | 0.39 | 0.59 | 0.78 | 0.98 | 1.17 | 1.47 |
| Nature of stability loss | Exfoliation, sheet jointing | Sheet jointing, cracking | "Cracking", shocks in rocks | | | |
| $K_1$ | 0.312 | 0.472 | 0.608 | 0.784 | 0.940 | 1.176 |
| Nature of stability loss | Exfoliation | Exfoliation, sheet jointing | "Cracking", shocks in rocks | | | |
| $\sigma_{hyd} = 200$ MPa | | | | | | |
| $K$ | 0.2 | 0.3 | 0.4 | 0.5 | 0.6 | 0.7 |
| Nature of stability loss | No destruction is observed in rocks | | Exfoliation, sheet jointing | | Sheet jointing, "cracking" | |
| $K_1$ | 0.15 | 0.23 | 0.3 | 0.35 | 0.47 | 0.6 |
| Nature of stability loss | No destruction is observed in rocks | | | Exfoliation, sheet jointing, "cracking" | | |
| $\sigma_{hyd} = 300$ MPa | | | | | | |
| $K$ | 0.13 | 0.2 | 0.26 | 0.32 | 0.4 | 0.5 |
| Nature of stability loss | No destruction is observed in rocks | | | Exfoliation, sheet jointing | | |
| $K_1$ | | | | | | |
| Nature of stability loss | No destruction is observed in rocks | | | | Exfoliation, sheet jointing | |

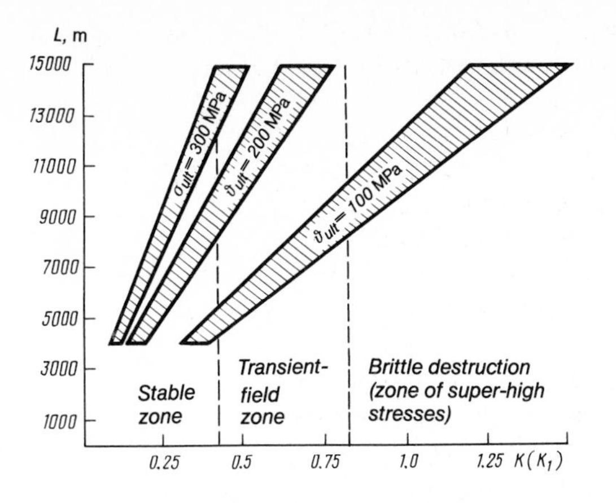


**Fig. 3.5.** Dependence of the stability of annulus rock mass on $H$ and $\sigma_{ult}$

the degree of hole inclination. An increase of this angle leads to intensifying the forces of resistance to movement of a drill pipe string, to wearing the drill tools, to key-seating, and to increasing the tensile loads while drill pipes are being pulled up; it also becomes difficult to transfer the axial load to the bottom hole in the process of drilling. At a certain value of the angle $\varphi$, the tensile forces arising from additional axial loads may exceed the maximum admissible loads on the pipes, and resistance forces plus the weight of the drill string may exceed the load capacity of the drill rig.

The sweep angle (inclination) $\Delta\varphi$ for each interval including the hole drift angle and azimuth is determined by the formula

$$\Delta\varphi = 2 \arcsin \sqrt{\sin^2\frac{\Delta\alpha}{2} + \sin^2\frac{\Delta\psi}{2} + \sin^2\alpha_{ave}}\,, \tag{3.18}$$

or by the formula

$$\cos\Delta\varphi = \cos\alpha_1\cos\alpha_3 + \sin\alpha_1\sin\alpha_3\cos\Delta\psi\,, \tag{3.19}$$

where $\alpha_1$, $\alpha_3$ are inclination angles at the upper and lower limits of the interval drilled; $\Delta\psi$ is the variation in azimuthal angle within a section; $\Delta\alpha$ is the variation of inclination angle within a section; $\alpha_{ave}$ is the average inclination angle of the section.

The angle $\varphi$ is the sum of all the angles of inclination $\Delta\varphi$ admissible in the course of drilling. It is known that, given equal values of the angle $\varphi$, the closer the inclination to the well mouth, the greater the resistance forces, and as a consequence, the tensile load on the hook. Therefore, when designing and drilling deep wells it is necessary to calculate and provide an optimum distribution of the total angle of sweep $\varphi$ over separate hole intervals $\Delta\varphi$.

This problem was solved methodologically by Vorozhbitov and Golubev (1976). The basic condition for the solution is the limitation of the friction forces between the drill string and hole walls by the present value $T$, and the necessity of adhering to the principle of uniformity in the distribution of friction forces over the entire depth of the hole, i.e. $\Delta T = T/n$, where $\Delta T$ is the friction force at a hole interval; $n$ is the number of intervals of the equal length $l$.

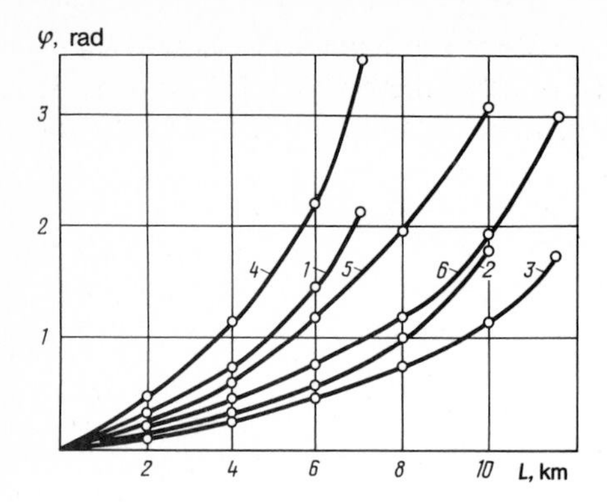


**Fig. 3.6.** Dependence of allowable variation of the total sweep angle $\varphi$ on the hole depth: *1*, *2* and *3* wells with depth being respectively 7, 10 and 11.5 km on condition that friction forces between the drill string and hole wall are limited to $T = 300$ kN; *4*, *5* and *6* the same, with $T = 600$ kN

Hence, the sweep angle $\Delta\varphi$ (in rad) for each of the intervals may be found by the formula (Vorozhibitov and Golubev 1976):

$$\Delta\varphi_n = \frac{\Delta T}{f(Q + ngl + n\Delta T}, \qquad (3.20)$$

where $f$ is the friction coefficient of the drill string against the hole walls; $Q$ is the weight of the load concentrated at the lower end of the drill column (drill-collar string, turbodrill) including the lightened drilling mud; $g$ is the weight of a length unit of the drill string including the lightened drilling mud.

The total angle of sweep for the entire well, $\varphi$, is determined as a sum of angles $\Delta\varphi$ of sweep of all drilled intervals $n$ of the equal length

$$\varphi = \sum_1^n \Delta\varphi .$$

Examples showing the rational distribution of parameters of the spatial trajectory of the holes at 7, 10 and 11.5 km depth are illustrated in Fig. 3.6.

Analysis of the calculated data (fig. 3.6) indicates the very strict requirements for inclination parameters of deep wells, especially in their upper intervals. This leads to the conclusion that particular emphasis must be placed in the drilling programme on the observance of admissible (calculated) parameters of the spatial trajectory of deep wells.

### 3.1.5 Drilling Method and Practices. Prediction of Drilling Indices

The choice of the scientifically based drilling method is important in determining the efficiency and reliability of drilling a superdeep well.

Drilling practices must provide:

- average footage per run not less than 7 m at a penetration rate of 2.5 m $h^{-1}$;
- maximum core recovery;
- retaining of stability of the sidewalls;
- transfer of energy to the bit at great depths without any significant loss;
- possibility of maintaining the hole inclination and correcting it.

These requirements are met by the two most conventional drilling methods – turbine and rotary drilling; in other words, rotation methods of drilling causing destruction of rocks by bits of the crushing, grinding and cutting types. The so-called novel drilling methods, including pulse action on the bottom hole, melting of rocks etc., have not yet found extensive use in drilling operations.

Despite the fact that the deepest wells in the world have been drilled by the rotary methods (USA), calculations indicate (for depths of over 10,000 m) that a turbine drilling method (more exactly, a drilling method based on the application of bottom motors) has indisputable advantages over rotary drilling. Turbine drilling involves a lower intensity increase of the stressed state in the drill pipe string with depth. In rotary drilling, the intensity of the stressed state of the drill pipe metal in the upper sections of the string exceed the strength capacities of current pipe steel at a depth over 10,000 m. In comparison with rotary methods, the turbine drilling method permits the use of light-alloy drill pipes, with greater possibilities of reaching great depths. The turbine drilling method in combination with a light-alloy drill string appreciably decreases the wear of the casing strings and walls of the open hole due to decrease in the normal forces between the rock-metal interfaces.

Turbine drilling also incorporates a wide range of drilling bit rotations per minute, including the range of the rotary drilling method. This is realized on the basis of the known technical solutions for designing thermoresistant gear mechanisms and downhole volumetric motors. In this respect and especially at great depths, the rotary drilling method cannot compete with the turbodrill, because it is unable to provide the higher rotations per minute such as were required, for example, when using a diamond bit.

Drilling based on the application of the downhole drill motor and light-alloy drill pipes (LBT type) faces a number of new vital problems concerned with securing the reliable operation of the turbodrill and LBT pipes in conditions of heavy drilling muds, increased temperature of the surrounding medium and the presence of mineral corrosion.

The efficiency of drill bits in hard abrasive rocks may be determined by the durability of a bit bearing and cutting structure, depending on their working conditions.

To estimate in advance the wear factor of the rock-crushing tool, as applied to the conditions of the Kola well, a special technique was developed (Bergshteyn and Vugin 1974), enabling us to determine duration, footage per run and penetration rate of drilling. The technique is based on the following formulas:

$$H_{\mathrm{eff}} = KB/[\Delta\vartheta_{\mathrm{smp}}\, p_{\mathrm{st}}\, n(\alpha+\theta)]\,; \tag{3.21}$$

$$\vartheta_{\mathrm{ord}} = (c\, p_{\mathrm{act}}\, n)/p_{\mathrm{st}}\,; \tag{3.22}$$

$$T_{\mathrm{str}} = A/[\Delta\vartheta_0\, n^2 p_{\mathrm{act}}(\alpha+\theta)]\,; \tag{3.23}$$

$$T_{\mathrm{smp}} = T - a\, p_{\mathrm{act}} - b\, n\,, \tag{3.24}$$

where

$$B = \Delta\, v_{smp} p_{st.\,ref} H_{ref}(\alpha_{ref} + \theta_{ref}) = \text{const}\ ;$$

$$C = (p_{st.\,ref}\, \vartheta_{ref})/(p\, n_{ref}) = \text{const}\ ;$$

$$A = \Delta\, \vartheta_{0\,ref}\, n^2_{ref}\, p\, T_{rev}(\alpha_{ref} + \theta_{ref}) = \text{const}\ ;$$

$H_{eff}$ is the effective footage per drill bit; $K$ is the empirical coefficient accounting for drillability variation with depth (predicted as 1 for SG-3); $\Delta\,\vartheta_{smp}$ is the wear of the specimen (abrasion coefficient); $p_{st}$ is the standard hardness of the rock; $n$ is the rotations per minute; $\alpha$, $\theta$ are respectively an angle between the hole axis and axis of the bottom-hole assembles (BHA) lower semi-wave and a zenith angle; $\vartheta_{ord}$ is the penetration rate; $p_{act}$ is the load on a drill bit; $T_{str}$ is duration of drilling (resistance of drilling bit cutting structure); $T$ is duration of drilling (resistance of drilling bit bearing); $T_{smp}$ is some constant value for a given type of drilling bit (for 2V-K214/60TKZ) the $T = 22$ h); $a$, $b$ are the constants for all types of drilling bit bearings, $a = 0.45$, $b = 0.08$ (Bergshteyn and Vugin 1974); index "ref" corresponds to the reference run.

Calculations of wear factors for the drilling bit 2V-K214/60TKZ applied to rocks expected in the SG-3 section indicate that with rotations of the drilling bit decreased to 50 – 60 rpm, the rate of wear of the bit cutting structure becomes lower than that of the bit bearing. Hence, in order to provide the maximum footage per drill bit, the drilling practice at which the resistance of the structure and resistance of the bearing will be equated must be chosen. These calculations are valid for the following boundary conditions: rotations in the range of 60 – 1000 rpm; load on the drilling bit from 0 to 270 kN.

In every actual case, to predict in advance the efficiency of drilling bits by drilling practice, the maximum penetration rate per run must be ensured. The study of the function $\vartheta_{ref}$ for the maximum at $p_{act} = \text{const}$ (for 2V-K214/60TKZ) has shown that drilling by means of bits with a hard-alloy cutting structure in high-abrasive rocks, where $d\vartheta_{ref}/dn = 0$, must comply with the conditions of $T_{dr} = T_{ort}$, i.e. time at the bottom must be equal to round trip time.

The tentative nature of these recommendations and calculations should be stressed, since it is hardly possible to take into account all the factors affecting the wear of drilling bits. These include: changes in rock abrasiveness and hardness during one bit run the possible trapping of the core in the downhole device during a run, and the ensuing change in the work of drilling bits; a change of the angle of hole inclination and skewness of the drill bit caused by an irregular supply of drill pipes. The results are also affected by the characteristics of the drilling mud, the dynamics of the drilling process leading to splitting-off of the hard-alloy cutting structure, and jamming of the bearings of drilling bits and downhole motors. Additionally, the calculations have been performed without information on the variation of rock drillability with depth.

However, a number of these factors become less important when we drill the basement crystalline rocks, and they can therefore may be neglected in pre-planning. In this case, the results of the calculation may be treated as average technical-economic indices to an accuracy of 20%.

The following conclusions may be drawn from analysis of the calculation data:

1. In abrasive rocks (where failure of drilling bits is due to wear of the core-forming structure) drilling accompanied by core recovery using a hard-alloy insert rolling-cutter bit does not depend on the load. For this reason, the load on the drilling bit must correspond to the load determined by the procedure introduced by L. A. Shreiner

$$p_{\text{act.eff}} \leqslant a p_{\text{st}} \sum_{i=1}^{n} S_c , \tag{3.25}$$

where $a$ is the proportionality factor; $S_c$ is an area of the contact of the drilling bit cutting structure with the bottom.
2. For drilling bits 2V-K214/60TKZ in which the wear intensity is inversely proportional to rotations per minute; depending on the physico-mechanical properties of rocks, the maximum penetration rate is within a range of 100 – 200 rpm; the higher the abrasiveness of the rocks and the greater the load, the narrower the range of rotations at which the maximum penetration is achieved by a drilling bit.
3. The greater the hole inclination angle and skewness of the drilling bit, the more intense is the wear of its cutting structure at deeper penetration.

Based on the analysis of the calculated drilling practices in hard abrasive rocks, we may estimate the existing downhole motors in view of the possibility of using them for drilling wells in the crystalline basement. Rocks composing the crystalline basement are mostly elastic-brittle varieties with a low plasticity factor ($K_{\text{pl}} < 3$). When drilling these rocks with a 215 mm-diam. bit, the specific torque varies in a range from 3 to 9 (N · m)/kN. Table 3.3 demonstrates the torques on a shaft of the downhole motor determined from the effective load values $p_{\text{act.eff}}$ of these rocks.

Loads corresponding to the volumetric area of destruction are optimum, i.e. $p_{\text{act.eff}}$. For drilling heads with a 60 mm diam. core, the effective load is 20 – 30% less than for a continuous drilling bit (Table 3.3). However, the actual torque requirements of the drilling process will be much higher than is necessary for the destruction of the rocks. This can be explained thus: the application of core-barrel downhole devices suspended from the shaft of the bottom motor with full-size reamer stabilizers ROP-9V, TRS-9, KLS-9 increases the torque requirements of the drilling process by 1500 – 3000 N · m. Anisotropy of the rocks in the section, their probably increased fracturing and the stressed state of the walls may, during the drilling process, cause an inclination of the well hole of the ellipsoid section. Drilling in such a hole is accompanied by an increase in torque requirements due to blocking of the drill string assembly during rotation. In this, the absolute magnitude of this torque depends on local bends of the well hole, on the weight of 1 m of the assembly and on the ratio between radial dimensions and diameter of the bottom motor.

According to the specified drilling practices in basement crystalline rocks, the optimum rotation per minute of drilling bits, depending on the physico-mechanical properties of the rocks being drilled, is 100 – 200 rpm. Thus, the bottom-hole motor must remain stable with a working torque on the shaft of 2000 – 4000 N · m at a rotation speed of 100 – 200 rpm and a temperature of the surround-

**Table 3.3.** Energy parameters of failure for rocks of different hardness and plasticity

| Hardness, MPa | Plasticity coefficient, $K_{pl}$ | Energy parameters of failure | | |
|---|---|---|---|---|
| | | Effective load, kN | Specific torque, (N · m)/kN | Breaking moment, N · m |
| 2000 | 2.5 – 3.1 | 90 | 6 – 9 | 540 – 810 |
| 3500 | 1.1 – 1.5 | 170 | 5 – 7 | 850 – 1190 |
| 4500 | 1.1 – 1.3 | 230 | 4 – 6 | 920 – 1380 |
| 6000 | 1.0 – 1.1 | 300 | 3 – 4 | 900 – 1200 |

**Table 3.4.** Technical data on A7Sh and A7GTSh turbodrills

| Turbo-drill | Number of sections of turbine | Drill mud flow rate, $l\,s^{-1}$ | Speed of rotation at $N_{max}$, rpm | Torque at $N_{max}$, N · m | Pressure drop at $N_{max}$, MPa |
|---|---|---|---|---|---|
| 7Sh | 236 | 30 | 520 | 2300 | 10 |
| 7GTSh | 279/39 | 30 | 300 | 2300 | 7.5 |

*Note:* Drilling mud density is 1.2 g $cm^{-3}$.

ing medium of up to 200° – 250 °C. These requirements are most fully met by a thermo-resistant reduction gear turbodrill. A screw volumetric downhole motor and its modifications capable of providing the required torque and speed of rotation may also be recommended for the hole intervals showing temperatures reaching 150 °C. Serial turbodrills of the A7Sh and A7GTSh-TL type whose technical data are shown in Table 3.4 are not suitable for the recommended drilling practices.

However, the core-barrel device installed on the shaft of the downhole motor decreases the rotation speed of the shaft of the turbodrill. By changing the length of the bottomhole assembly and using the A7Sh and A7GTSh type turbodrills, we may obtain acceptable rotation speeds by decreasing to some extent the level of the axial load.

Table 3.5 gives data of the direct experiment involving the change of the speed of rotation of the shaft of the A7Sh/A7N4Sh type turbodrill depending on the load on the drilling bit and length of the core-barrel device. These data were obtained in the course of drilling a well in the crystalline rocks of the Ukrainian Shield at a depth of 800 – 1000 m. As follows from Table 3.5, the rotation speed at idle decreases by two to four times; with an increase of the length of the suspended assembly and with a transfer of a given load on the bit, the rotation speed was reduced to 120 – 300 rpm. At the same time, while reaming the hole and washing it, as well as while working with decreased loads to combat the hole inclination, we have to keep the rotation speed at a high level, which may certainly affect drilling, particularly during core recovery operations. Therefore, this method of decreasing the rotation speed of the bit should be considered as being constrained and must be applied on a limited scale only in cases where reduction-gear turbodrills and volumetric downhole motors are not available.

**Table 3.5.** A7Sh turbodrill shaft rpm for core-barrel devices of different length at different weight-on-bit values

| Drill mud flow rate, $l\,s^{-1}$ | Bit load, kN | Rotation speed of the shaft in rpm, with length of core-barrel device, in m | | | |
|---|---|---|---|---|---|
| | | 0 | 10 | 20 | 30 |
| 26 | 0, at the well mouth | 1050 | 850 | 650 | 300 |
| | 0, over bottom face | – | 800 | 500–600 | – |
| | 20 | – | – | 490–570 | – |
| | 40 | – | 650–690 | 480–550 | – |
| | 60 | – | 350–450 | 460–490 | – |
| | 80 | – | 260–300 | 350–470 | – |
| | 100 | – | 200–290 | 325–340 | – |
| | 120 | – | 350–400 | 180–220 | – |
| | 140 | – | – | 180–220 | – |
| | 160 | – | – | 180–220 | – |
| 32 | 0, at the well mouth | 1300 | 1000 | 800 | 500 |
| | 0, above bottom face | – | 850–900 | – | 325–350 |
| | 20 | – | 760–800 | – | 320–330 |
| | 40 | – | 800 | – | – |
| | 60 | – | 775 | – | 225–250 |
| | 80 | – | 700–750 | – | 120–180 |
| | 100 | – | 300–320 | – | 120 |
| | 120 | – | 200–430 | – | 150–160 |
| | 140 | – | 330–340 | – | 160 |
| | 160 | – | – | – | 80–140 |

### 3.1.6 Drilling Tools

For successful drilling of a superdeep well by the turbodrilling method in crystalline rocks, use of aluminium alloy as material for drilling pipes appears to be very effective (Timofeyev et al. 1971b). The use of a reliable light-alloy drill string greatly simplifies the problem of decreasing the load capacity of the drill rig, and markedly reduces the drilling time for a well. Sufficient experience has been accumulated in the USSR for successful application of drill pipes made of aluminium alloy D16T (yield strength 330 MPa, density 2.8 $g\,cm^{-3}$) with steel tool joints. These pipes are, however, applicable only for a depth of 7.5 km (Well 1 Shevchenko).

Two more factors, namely, increased temperature of the surrounding medium and additional forces that must be applied to overcome the resistance forces arising in the course of pulling out drill pipes from a hole must be taken into account when manufacturing lightened drill pipes for deep drilling operations. The physico-mechanical properties of the aluminium alloys depend significantly on the temperature-time conditions under which they are used (Aluminium 1972, Aluminium alloys 1973), while the acting loads, particularly the axial ones, depend on resistance forces.

Calculations show that serial pipes LBT-147 of D16T light alloy with joints ZLK-172 are able, as far as temperature and strength characteristics are concerned, to drill wells in crystalline rocks down to a depth of 7 – 7.5 km. The factors that limit their application in greater depths are: temperature-caused decrease in density of the pipe metal (temperatures exceeding 150° – 160 °C), in other words, a loss of pipe body strength in the bottom-hole section of the well, very high state of stress in the upper part of the drill string, and low strength of the joint-drill pipe connecting thread.

The requirements imposed on pipes for deep drilling were determined by analytical studies. The technique of these investigations was constantly being perfected, using the experimental data. The procedures for calculating drill strings for superdeep wells have been most fully described in Neymark et al. (1979). This work shows that in order to manufacture a reliable drill pipe string, one must account for the intensity of the stressed state typical of three main technological processes which differ from each other under differing circumstances: drilling, round trip operations and failure situations.

Based on the condition of the simultaneous action of the axial load, rotation torque of the rotor (a slow turn of the string) and the reactive torque on the turbodrill, we may determine the static strength for any section of the drill string using the following relationship

$$[\sigma_i] = \sqrt{[\sigma]^2 + A[\tau]^2} \leqslant \sigma_0, \tag{3.26}$$

where $[\sigma_i]$ is the admissible intensity of the stressed state calculated using the norm safety factor; $\sigma_0$ is the strength parameter characterizing serviceability of the material and being generally a function of the force and temperature-time factors (determined experimentally); $[\sigma]$ are the admissible normal stresses ($[\sigma] = \sigma n_\sigma$); $A$ is the anisotropy factor of the LBT pipe material determined experimentally (Neymark 1975); $[\tau]$ are admissible twist stresses ($[\tau] = \tau n_\tau$); $\sigma$ are the total normal stresses ($\sigma = \sigma_1 + \sigma_2$); $\sigma_1$ normal tensile stresses ($\sigma_1 = p/F$); $\sigma_2$ are normal stresses of bending ($\sigma_2 = M_{\mathrm{bend}}/W$); $\tau$ are tangential stresses of twisting ($\tau = M_{\mathrm{rev}}/W_{\mathrm{ref}}$); $n_\sigma n_\tau$ are safety factors with respect to normal and tangential stresses; $F$ is an area of the calculated transverse section of a pipe; $W$ is the calculated section modulus of a pipe when bent; $W_{\mathrm{ref}}$ is the polar moment of resistance of the pipe.

The yield strength $\sigma_y$ was used as a strength parameter characterizing the serviceability of the drill pipe material (Neymark et al. 1979). The specific moment of resistance per length unit of the drill string was changed within a range of 2 – 3 $(\mathrm{N} \cdot \mathrm{m})\,\mathrm{m}^{-1}$, the factor of resistance to drill pipe string movement is 1.2 ÷ 1.4, the safety factor $n_\sigma = 1.3 \div 1.4$; $n_\tau = 1.5 \div 2.0$.

Figure 3.7 shows the results (Tumanov et al. 1972; Neymark 1975) of determining the static strength of the pipe material when subjected to high temperature.

The analytical studies, together with the analysis of a different combination of variable factors, have shown that in order to undertake drilling of a well to a depth of 11 – 13 km in crystalline rocks by the turbine drilling method with slow rotation of the drill string (2 – 6 rpm), we must make use of drill pipes made of different aluminium alloys: in the lower part of the drill column of heat-resistant alloy AK4-1 of 4 – 5 km long, in the middle part of alloy D16T of 1.5 – 2 km in length, and in the upper part of an alloy of relatively high strength 01953,

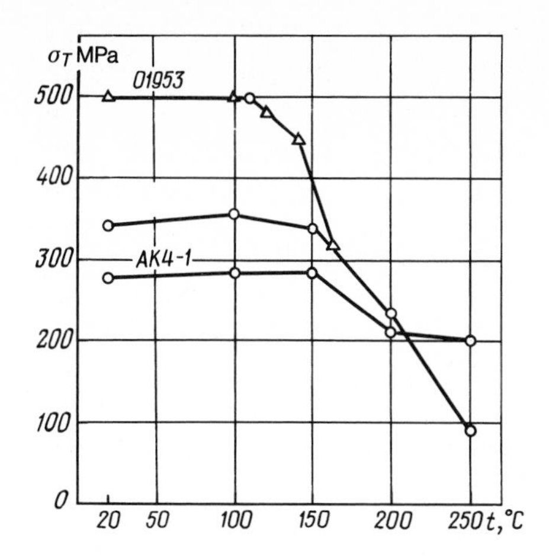


**Fig. 3.7.** Dependence of strength of aluminium alloys on temperature (high temperature curing period is 100 h)

2 – 2.5 km long. It is expedient to use a section of steel drill pipes of 1.5 – 2 km in length for the upper part of the drill column to combat failure situations.

While manufacturing the lightened drill pipes, great emphasis was placed on their protection against wear when the pipes rub against the hole walls composed of crystalline rocks. The use of specially made protectors made of wear-resistant materials was found to be the best protective means for the pipes.

Analytical investigations have revealed a number of improvements in the general technology of drilling a superdeep well. These include: the observance of the vertical alignment of the hole; the prevention of caving and loss of stability; the increase in drill mud density with depth; the improvement of lubricative and anticorrosive properties of the drill mud.

An essential link in the drill column is the joint connection. Table 3.6 (Shcherbyuk and Yakubovsky 1974) demonstrates formulas used to calculate the joint connections in drill strings.

The strength required for tool joints for different depths of the hole was obtained taking into account the stresses generated in these joints when joint connections are screwed together. The calculations revealed basic laws for increasing the strength requirements in the material for joint connections at depth, and optimum conditions for their exploitation. The conclusions made refer to the fact that the lower segment of a drill string made of lightened pipes of 8 – 9 km in length may be furnished with tool joints ZLK-178 of steel 40XH and yield strength not less than 750 MPa. The upper section of the drill string will require joints of steel of higher strength with a yield limit of 1000 – 1000 MPa. An increase at the moment of screwing to 50 kN · m and the necessary technical facilities are needed to put these joints into operation.

When drilling a well to a depth of 13 km and lower with steel drill pipes in its upper section, the material for these pipes must have a yield strength of 1100 – 1200 MPa. These calculated data served as a basis for the programme of designing and furnishing the Kola well with reliable light-alloy drill pipes LBTBK-147 with steel joints ZLK-178 and steel pipes TBVK2-140 for drilling the well by an open pilot hole of 214 mm in diameter using the turbodrill method.

**Table 3.6.** Loading conditions and formulae

| Conditions of loading | Calculation formula |
|---|---|
| Optimum tightening force | $Q_{30} = (0.45 - 0.5)\,\sigma_y F_n$ |
| Tensile stress in a joint nipple | $\sigma_n = (Q_{30} + K_n p_{max})/F_n$ |
| Moment of screwing | $M_{sc} = Q_a;\ M_{max} = F_n \dfrac{\sigma_y}{n} a$ |
| Condition of leak-proof connection | $M_{min} = K_{cl}(p_{max} + p_{in.\,p} F_c)\,a$ |

*Note:* $Q_{30}$ = the optimum tightening force for screwing of tool-joint thread, $H$; $F_n$ = area of dangerous section of the nipple, $cm^{-2}$; $\sigma_n$ = stress in the joint nipple, MPa; $K_n$ and $K_{cl}$ = coefficients of the external loading of nipple and clutch respectively; $p_{max}$ = maximum external loading on the nipple, $H$; $Q$ = weight of the drill pipe string, $H$; $a$ = constant depending on the size of the joint; $n$ = safety margin coefficient; $p_{in.\,p}$ = inner pressure, MPa; $F_{an}$ = area of the joint clutch anvil, $cm^2$.

### 3.1.7 Drilling Fluid

Many results, such as drilling process indices, the condition of the wall rock, the service life of the surface equipment and drill tools, as well as the possibility of performing various operations within a well, particularly geophysical measurements, depend largely on the correct choice of the drilling fluid. The experience of deep drilling in the USSR and abroad and the prediction of the drilling conditions expected in crystalline rocks made it possible to formulate the actual requirements which served as the basis for preparation of the drilling fluid for the Kola well. These requirements were dictated not only by the geological environment in the area of the well, but also by specific details of the drilling technology and equipment.

On the one hand, the load capacity of the *Uralmash-15000* Drill Rig for the Kola well is 4000 kN. This is one of the factors that made it necessary to apply a great amount of light-alloy drill pipes for a drill string, which in its turn affected the parameters of the drilling fluid. On the other hand, the use of turbodrills makes the regulation of the bottom-hole drilling parameters difficult and does not allow the pump pressure to be changed over a wide range.

In drilling the upper interval of the well (0 – 7000 m), standard requirements are imposed on the washing fluids, that is, to provide the maximum rate of drilling and footage per bit, good hole and bottom cleaning, etc. Further deepening of the hole requires stability of the sidewalls, provision for maximum core recovery, counteracting the resistance to movement in the drill string, reducing the aggressive action on the drill string, and preventing the flocculation of clay particles.

Resistance to drill string movement consists of friction and adhesion forces produced by the interaction of pipe surface and rock or filtration cake. These forces depend on physico-chemical properties and the composition of the washing fluid, and also on the trajectory, the shape of the section and length of the open hole, the weight of the drill string and the type of bottom-hole assembly. These factors affect the torque magnitude necessary for rotating the string, and the hook load when the drill string is being pulled out. It is difficult to estimate the degree of the effect of washing fluid on the resistance forces, but it is certain that one of the main requirements for the washing fluid will be an increase in lubricating properties to reduce the friction forces in the course of the movement of the drill string along the hole.

The aggressive impact of the washing fluid on the drill pipe gives rise to surface microfractures, activation of corrosion processes, and to hydroabrasive and cavitation wear. The problem of prevention of corrosion of the drill pipe is considerably simplified because the crystalline rocks being drilled are chemically neutral and the supply of acid gases to the drill mud is hardly possible.

Flocculation of the solid phase is a widespread phenomenon under high-temperature drilling conditions (Anderson 1980; Kister 1972; Paus 1973). Its intensity depends on the composition and content of the solid phase, the temperature and the mode of chemical treatment of the washing fluid. Unless the flocculation of fine particles of the drilled rock is desired [for cleaning purposes, where it is often artificially intensified by the introduction of flocculating agents (Anderson 1980)], the flocculation of clay particles is highly undesirable and should be combatted.

A prediction of the geological and technical conditions of drilling in crystalline rocks permits us to formulate the qualitative parameters of the washing fluid that will lead to successful drilling of an ultra-deep well: thermostability, high lubricating ability, high resistance to flocculation of clay particles, anti-corrosive properties, ability to provide stability of the hole walls and high lifting capacity.

An increase in thermostability of the water-based mud to 220° – 230 °C is achieved by application of new reagents manufactured in the USSR – acrylic polymers (metas, metasol, larkis, etc.). The thermostability of the oil-based mud can now be increased to 260° – 270 °C. It is more difficult to maintain the lubricating properties of the emulsion mud at higher temperatures of the medium. More detailed studies are necessary in this direction to produce special additives, possessing relatively high temperature stability of the lubricating layer and antifriction properties, suitable for diverse drilling conditions (the composition of the dispersion medium, specific friction loads, pH value, etc.).

It is also very important to produce an effective emulsifier for the dispersion and stabilization of the lubricating agent, since it is the emulsifier properties that determine the ability of the emulsion to resist coalescence under the action of high temperature and pressure. The appearance of lubricating and corrosive properties of the mud depends largely on the state of the metal surface and adsorption phenomena on it.

A neutral medium is optimum for aluminium and its alloys which possess amphoteric properties (i.e., their oxides are dissolved both in acidic and alkaline media). For example, at pH $\approx$ 7 (a passive region) a monohydrate oxide-belite of aluminium evolves on the surface (Sinyavsky 1977) and has a protective function. The process is accelerated at temperatures of 60° – 80 °C and higher. Thus, maintaining the value of pH $\approx$ 7 makes it possible to sustain the necessary stability of the surface of light-alloy pipes and to control the corrosion and lubrication effects. To reduce the aggressive action of the mud on the drill string, the latter should be thoroughly cleaned of drilling chips and degassed.

The choice of the dispersion phase is another vital procedure for preparing the washing fluid with desirable properties. When drilling the most important intervals of a deep well, preference should be given to coarse-dispersed clay (kaolinite), despite the obvious advantages of the bentonite clay powder used for conventional shallower drilling. A suspension with the fine-dispersed clay phase exhibits greater sensitivity to concentrations of chemical agents. It is very difficult to maintain this sensitivity beyond a certain level, due to thermo- and mechanical failure, adsorption phenomena on the drilling chips, and large amount of mud in the hole where the working portion of this mud is variable and depends on rheological parameters and pump capacity. Apart from this, the bentonite suspension is very sensitive to contamination of drilling chips which may enter the suspension from the wider dead zones of the open hole, or may be concentrated due to the complicated evacuation of its fine (less than 40 μm) fractions. It should also be remembered that bentonite is more sensitive to high temperatures, which will undoubtedly hamper and complicate the chemical treatment of the mud. A pronounced dependence of the rheological parameters of the bentonite suspension on a displacement rate on the one hand, and the expected high speeds of the ascending flow in the annular space caused by the application of bottom-hole motors on the other, lead to a decrease in the capacity to lift the mud with ensuing negative results in hole cleaning.

Thus, preference should be given to coarse-dispersed clay (kaolinite) as the main dispersion phase. This is also supported by the experience of drilling geothermal wells (Paus 1973). Bentonite fluids may be utilized to decrease a solid phase content and to regulate filtration and rheological parameters.

When using weighted drilling mud, we should try to prevent the high-speed flow of the mud in drill pipes. This is due to the fact that a high content of solid phase enhances the hydroabrasive effect of the mud on the inner surface of the drill pipes, especially in places where a drill string narrows. These phenomena may be dealt with by reducing drill mud flow rate by a simultaneous increase of the energy parameter of the bottom-hole motor and improvement of the lifting capacity of the mud, thorough cleaning of the weighted drill mud from the solid phase, application of light-alloy drill pipes with a varying port diameter and special anti-wear coating. The important condition for using the weighted drilling mud is to prevent high-temperature and electrolytic flocculation of clay particles without a decrease in anti-friction properties of lubricating fluids.

### 3.1.8 Coring

One of the major objectives of drilling the Kola well is maximum core recovery. Coring crystalline rocks at great depths is characterized by: wear of the rolling-cutter drilling head is accompanied by vibrations and intensive wear of the core-forming structure; natural fracturing of rocks encourages core blocking in the core pipe; the rocks are inert with respect to the washing fluid and are not liable to washing-out; the high density of rocks ($2.7-3$ g $cm^{-3}$) causes the pressure to grow rapidly with depth. The probable relief of the stressed rocks when exposed by a hole may lead to self-destruction of the core into disks (the so-called disking of the core).

In the process of coring, the central part of the hole bottom is not subjected to the action of cutting elements of the tool and the rock column thus formed moves in the course of deepening along a core barrel upwards. In the ideal case a length of the recovered core must correspond to footage per run. Under real conditions, however, this does not happen, i.e. linear dimensions of the core do not correspond to the drilled intervals.

A decrease in the linear size of the core may be due to: the dynamic nature of the drilling process resulting in impacts on the contacting surfaces between core columns and the column of the core and hole bottom caught (core-core and core-rock mass contacts); the dynamic effect of the cutting elements of the drilling head on the formed column of the core, causing splitting-off and retrieval of the core from the bottom hole (core-drilling head contact); rotation of the core pipe, and as a consequence, slippage and grinding of core columns and the hole bottom (core-core pipe contact).

However, in all the cited cases we observe only a decrease of the linear size of the core columns in comparison with the length of the drilled interval. The core continues to enter a core pipe and a ring-shaped hole bottom is drilled. If, however, the core material is blocked inside the core receiver device, the supply of the newly cut core columns to the core pipe ceases, and the core is crushed and ground at the bottom hole and in the core receiver pipe. This process may occasionally be accompanied by reduction of penetration rate.

This phenomenon may be brought about by: (1) wear of the cutting elements of the drilling head exceeding the admissible level and resulting in an increase in core diameter, and, as a consequence, in its locking inside a core receiver compartment of the downhole device, (2) a high degree of fracturing of the wall rocks, (3) large angles of dip of beds also causing the blocking of the core inside the core receiver compartment and self-destruction of the core material at a great depth under the complex stressed bottomhole conditions. Although the above factors act permanently in the course of the drilling process, their effect depends on the geological, technological, technical and organizational aspects of drilling.

Based on the analysis of the cause of core losses, a drilling process accompanied by coring the hard abrasive rocks may be subdivided into two stages termed tentatively zones of (a) effective and (b) ineffective drilling (Fig. 3.8). A zone of effective drilling is characterized by the loss of core length in it being due to the

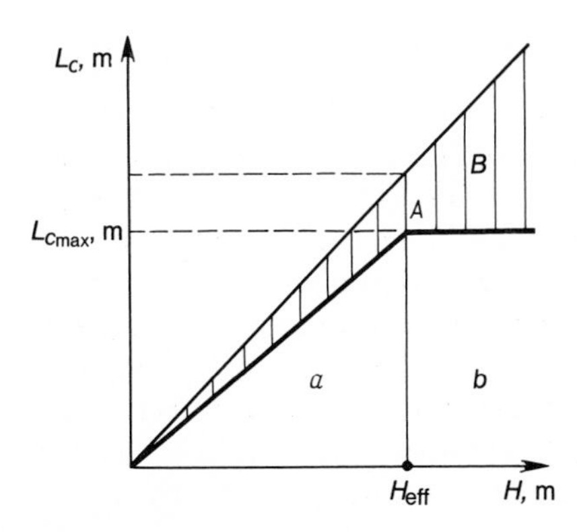


**Fig. 3.8.** Graph showing the core recovery with account of core locked in a core barrel: *a* zone of effective drilling; *b* zone of ineffective drilling; *A* the moment of core locking; *B* core loss

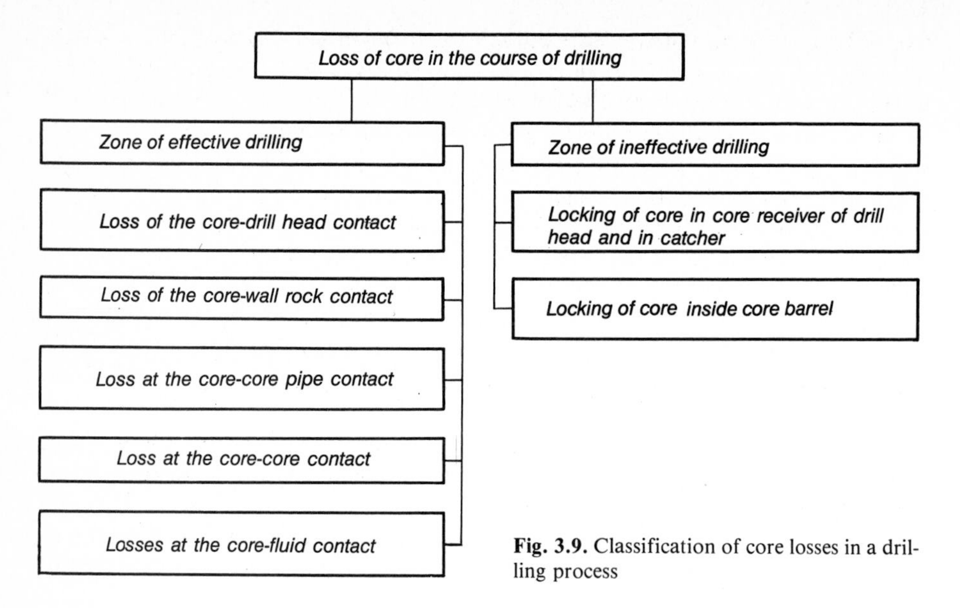


**Fig. 3.9.** Classification of core losses in a drilling process

dynamic effect of the cutting structures of the head and elements of the core catcher upon the core column, to active contacts between the cores along butt surfaces, to contacts between the core and wall rock, between the core and a receiver compartment, and washing fluid. Typical of the zone of ineffective drilling is a further penetration of a borehole after the core has been locked in the core receiver of the drill head, or in a receiver compartment of the core barrel.

Figure 3.9 gives a classification of core material losses in the course of drilling. Below are analytical expressions for determining the length of the core per run and percentage of its recovery

$$L_c = \frac{H_{\text{eff}}}{1 + \frac{\sum_{i=1}^{n} \Delta \bar{l}_c}{\sum_{i=1}^{n} \bar{l}_c}}, \tag{3.27}$$

$$K = \frac{H_{\text{eff}} \bar{l}_c}{H(\bar{l}_c + \Delta l_c)} \cdot 100\%, \tag{3.28}$$

*where* $L_c$ is the core length; $K$ is the core recovery percentage, $H_{\text{eff}}$ is the effective penetration equalling $[V_{\text{cr}}]/\Delta V_{\text{cr}}$; $H$ is the total footage; $\Delta \bar{l}_c$ is the average losses of the length of a core column; $\bar{l}_c$ is the average length of a core column; $[V_{\text{cr}}]$ is the allowable wear of the core-forming structure of the drilling bit; $\Delta V_{\text{cr}}$ is the wear of the core-forming structure of the drilling head per 1 m of drilling.

Thus, the problem of optimization of core recovery is reduced to the identification and realization of the factors leading to an increase of the effective drilling zone $H_{\text{eff}}$ and to a decrease of the linear size of the core in this zone $\Delta \bar{l}_c$. It was found that the core losses during drilling in hard abrasive rock in a zone of effective drilling are negligible (<5%) and may be disregarded. To improve the effectiveness of the coring, it is necessary to exclude jamming of the core in a downhole device (Bergshteyn et al. 1977).

It is known that drilling in the abrasive rocks causes maximum wear of the structure-cutting parts of the rock-crushing tool in places where a high degree of slippage is observed. Hence, maximum wear of the core-forming rows may be anticipated in any drill heads of the rolling cutter type. The wear of these core-forming rows per unit of the friction path (one revolution) may be expressed as follows

$$\Delta V_{cr} = a_0 p_c \vartheta \pi d_c , \tag{3.29}$$

where $a_0 = \Delta V_0 / p_0 \vartheta_0 l_0$ is the proportionality factor; $\Delta V_0$ is the wear of the core sample (abrasivity index); $p_0$ is the contact load on the core; $\vartheta_0$ is the circumferential velocity of the sample; $l_0$ is the friction path of the sample; $p_c$ is the pressure of contact on the core; $\upsilon = \pi d_c n$ is the circumferential velocity; $n$ is the rpm of the drill head; $d_c$ is the core diameter.

The wear of the core-forming rows for the entire penetration path is

$$V_{cr} = \Delta V_{cr} N ,$$

where $N = H/\delta$ is the total number of revolutions per run $H$; $\delta = \Psi p_{act}/p_{st}$ is the penetration of the drill head per revolution, $\Psi$ is the coefficient accounting for the contact surface, the number of inserts and their wear with time; $p_{act}$ is the load on the drill head; $p_{st}$ is the standard rock hardness.

Thus, the wear of the core-forming rows of the drill head of rolling cutters may be written as

$$V_{cr} = \frac{\Delta V_0 (\alpha + \theta) \pi^2 d_c^2 n p_{st}}{p_0 \vartheta_0 l_0 S_c \Psi} H \leqslant [V_{cr}] , \tag{3.30}$$

where $\alpha$, $\theta$ are the angle between the hole axis and axis of the lower half-wave and zenith angle respectively; $S_c$ is the contact area of the core-forming rows.

The wear of the core-forming rows of the truncated-cone rolling cutters may be represented by the following relationship

$$V_{cr} = \frac{\Delta V_0 (\alpha + \theta) \pi d_c p_{st}}{p_0 l_0 S_c \Psi} H \leqslant [V_{cr}] . \tag{3.31}$$

Analysis of the relationships (3.27) – (3.31) shows that in order to exclude locking of the core in the drill head caused by abrasive wear of its cutting structure it is necessary: to maintain the minimum allowable circumferential velocity of the core-forming structure; to use wear-resistant material in places where maximum slippage is observed; to make sure of accurate alignment of the drill string to prevent the string from bending relative to the hole axis. As experience has shown, the locking of the core in a core receiver pipe occurs primarily in the course of drilling fractured rocks. Under these conditions, core recovery drops sharply, and does not exceed 1 m per run, the average footage per run for a drill head attaining 4 m and more. These are parameters three to four times lower than those required for the successful penetration of an super-deep well.

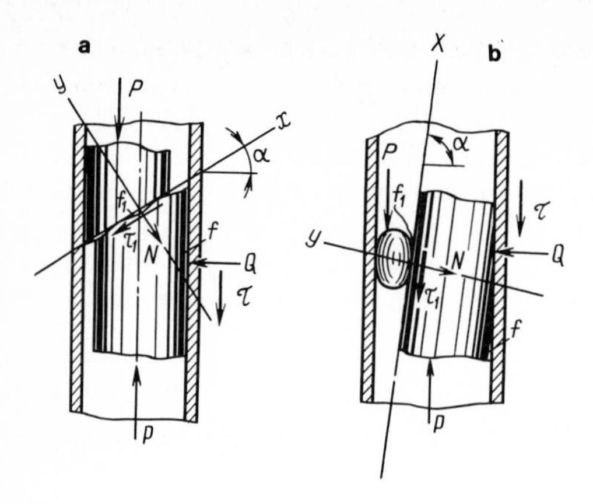


**Fig. 3.10 a, b.** Scheme showing the locking of core in a core barrel: **a** in fractured rocks; **b** in abrasive rocks

Occasionally, the core in a receiver pipe can lock, due to the many changing factors in the course of a run. The probability of the core's locking grows with an increase in core length in the core receiver compartment. The friction forces arising at the contact of the core with the inner wall of the core barrel, self-destruction of the core being formed, radial vibrations of the drill tools, wobbling, the absence of alignment and rotation of the core receiver pipe are factors which aggravate the effect of self-locking. As a result of their joint action, the core entering the core barrel is split along the discontinuities and is blocked inside the core barrel, thus excluding the inner part of the downhole device from access to the newly formed core.

The self-locking of the core in a barrel is represented in Fig. 3.10. Let a core sample be broken at the angle $\alpha$. During downward movement of the drill string, the action of the compression force directed from the upper section to the lower one gives rise to the reaction $N$, and to the normal pressure $Q$ acting on the pipe wall. Apart from this, the friction forces $\tau$ and $\tau_1$ arise at the contact surfaces and prevent the movement of core parts relative to each other and of the core relative to the pipe.

The friction forces arising at the contact surfaces will be equal to:

$$\tau = Qf; \quad \tau_1 = Nf_1, \tag{3.32}$$

where $f$, $f_1$ are the friction coefficients respectively at the core-core receiver and core-core contacts.

While determining the forces operating in the section under discussion, we may conclude that the friction forces provide a bond at the contact core-core receiver pipe. In this, a sliding of the core relative to the pipe is excluded. The force $p$ applied to the butt of the lower fragment of the core changes from the minimum value determined by the weight of the core positioned above to the maximum value determined by the drill head load. Since the bond between the core and the inner surface of the core barrel is provided by the friction forces, the following condition must be met to ensure the movement of the core

$$p > \tau; \quad p > Qf. \tag{3.33}$$

Considering a system of forces determining the conditions of self-locking, we find from the equilibrium ($\Sigma F_x = 0$ and $\Sigma F_y = 0$), that the expansion force $Q$ and the maximum allowable friction coefficient at the core-core receiver pipe contact which prevents the locking of the core in a core barrel may be found from the expressions

$$Q = P\left(\frac{\operatorname{tg}\alpha - f_1}{1 + f_1 \operatorname{tg}\alpha}\right), \tag{3.34}$$

$$f \leqslant (1 + f_1 \operatorname{tg}\alpha)/(\operatorname{tg}\alpha - f_1). \tag{3.35}$$

Equation (3.35) shows that the larger the dip angle of the rocks or angle of chipping, the lower is the friction coefficient $f$ and the less must be the difference between $f$ and $f_1$.

Thus, to improve the effectiveness of the core recovery in the case of self-locking it is necessary to:

1. Decrease the friction coefficient at the contact between the core and core receiver pipe and decrease the value of the expansion force $Q$. This can be achieved by a better preparation of the inner surface of the core receiver pipe; application of the antifriction materials with a low friction coefficient (glass enamel, plastics, metallographitic films, etc.), as well as the use of lubricants in the washing fluid; the change of sliding friction to rolling friction and the application of conical contacting surfaces.
2. Provide a forced transportation of every individual core column from the bottom hole up the core barrel by hydraulic or mechanical means using: the downhole tools with partial inverse flushing of a core receiver compartment; core receiver devices with complete inverse flushing and core storing facility outside the core receiver channel, as well as tools that facilitate mechanical upward transportation of the core (core receiver devices furnished with elastic compartment, auger tools, tools with a mobile "wall", etc.).

Table 3.7 incorporates the basic technological requirements for successful coring and possible ways of fulfilling them.

### 3.1.9 Control of Drilling Parameters

The system of measuring information for drilling the superdeep hole is determined by technological and technical problems, each of which corresponds to a certain set of information. The desired composition of the information system and regulated parameters for solving the technological and technical drilling problems are demonstrated in Table 3.8.

At present the design of a means of control over the surface parameters of drilling encounters no fundamental difficulties. The problem consists only in providing interaction between the information system and the computer facilities that control technological parameters following an optimum model of the drilling process.

The most complex problem is to devise a measurement system on the basis of bottomhole information on drilling parameters. The necessity of devising such a

**Table 3.7.** Basic technological requirements for coring and possible solutions

| Technological requirements | Possible solutions |
|---|---|
| Provision of the maximum footage per run based on the endurance of drill head bearing and cutting structure | Design of insert hard-alloy drill heads with minimum sliding, and hermetically sealed drill head cutter bearing |
| Coincidence of the footage per run with the length of the core being recovered | Application of core receiver devices with sectional modifications |
| Preventing spontaneous hole inclination | Use of coring assemblies based on a double-centralizer scheme |
| Elimination of radial vibrations of the drill string bottom-hole assembly, stabilization of the core recovery assembly | Installation of full-size reamer stabilizers |
| Increase in wear resistance of reamer stabilizers, preventing decrease in hole diameter, reduction of their torque capacity | Design of reamer stabilizers with spherical bearing |
| Elimination of the effect of vertical vibrations on core recovery and possibility of starting bottom hole motors without cutting off the core | Design of floating core catchers with a vertical shaft, with inverse flushing facilities, and installation of intermediate core catchers |
| Provision of the optimum rpm of the drill head in a range of 100 – 200 rpm | Design of turbodrill with reduction gear, volumetric motor |
| Preventing sludging-up of core barrel during drilling, self-adjustment of core barrels with respect to length | Forced partial circulation through a core assembly, telescopic connection of core receiver pipes |
| Preventing self-blocking of cores inside the core receiver downhole device | Hydraulic conveyer for core transportation, installation of centralizing devices, reverse flushing of a core receiver compartment, inner coating and treatment of core receiver pipe, core catcher on a spherical bearing |
| Preventing self-destruction (disking) of core | Increase of mud density and axial load on drill head |

system is motivated by a significant decrease in accuracy of the surface control systems over the working parameters of the drill bit and bottomhole motor at a depth exceeding 5 – 7 km. In this respect a hydraulic communication system appears to be most promising, as being the most advanced of those known.

The hydraulic power consumed by a downhole motor determines the noise-resistant reception of the incoming information. At the same time, the fact that the telesystem consumes considerable power limits the scope of its application with increasing depth. To estimate the required energy conditions of exploitation of telesystems in drilling we must introduce a concept of the coefficient of power consumed $\kappa = (N_r/N) \cdot 100\%$, where $N_r$ is the hydraulic power consumed by a downhole motor and $N$ the hydraulic power of mud pumps.

Depending on the capacity of the drilling equipment, the maximum value of the given coefficient must not exceed 5%, i.e. not more than 5% of the hydraulic power of drill pumps may be utilized for telemetry of bottomhole parameters.

**Table 3.8.** Solutions for technological drilling problems

| Local control over drilling with amendment | Recognition and prevention of failures | Diagnostics of equipment and tools | Specified parameters of drilling practices for optimization of well drilling | Express search for best method |
|---|---|---|---|---|
| Bit load | Hook load | Hook load | Hook load | Hook load |
| rpm of bit | rpm of bit | Shift of pulley block | rpm of bit | rpm of bit |
| Drilling fluid flow rate at inlet | rpm of bit | Drilling fluid pressure | Bottom-hole motor torque | Bottom-hole motor torque |
| Drilling fluid flow rate at the bit outlet | Drill head torque | Number of double strokes of the pump | Drilling fluid flow rate | Drilling fluid flow rate |
| Drilling fluid density | Rotary table torque | rpm of rotary table | Drilling fluid density | Penetration rate |
| Speed of travelling movement | Tongs torque | Rotory torque | Depth of hole | Time of hole drilling |
| Footage per bit | Drilling fluid pressure | Bottom-hole motor torque | Penetration rate | |
| rpm of rotor | Drilling fluid flow rate at inlet | Load on bottom face | Bit run rate | |
| Zenith angle | Drilling fluid flow rate at outlet | rpm of bit | Bit position in hole | |
| Horizontal angle | Level of driling fluid in tanks | Length of unreeled line when restrung | Footage per run | |
| | Drilling fluid level in hole | Drawworks work-measuring device | Bit wear parameters | |
| | Round-trip rate | Mud pumps work measuring device | Drilling fluid parameters | |
| | Number of drill stands in a hole | Rotary work measuring device | Description of rocks in Kola section | |
| | Movement of travelling block | Swivel work measuring device | | |
| | Gas content in drilling fluid | Downhole motor work measuring device | | |
| | Drilling fluid density | Drill bit work measuring device | | |
| | Level of drilling fluid and chemicals in storage tanks | | | |

With the given value of the power coefficient and assumed consumption of hydraulic energy at a depth of 15,000 m, the average power value of the telesystem generator must not exceed $N_r = Qp \approx 60$ kW, where $Q$ is the delivery of slush pumps; $p$ is the pressure in their discharge line.

In view of the fact that the most optimum shape of hydraulic pulses (telesystem type) is sinusoidal, the average value of the generator power may be determined from the expression

$$N_r = \frac{Q}{T} \int_0^T p_{max} \sin \omega t \, dt ,$$

where $T$ is the period of pulse repetition; $p_{max}$ is the maximum pressure increment; $\omega$ is the angular frequency; $t$ is the time.

It follows that the maximum value of the desired signal in the pulse regime, i.e. the maximum pressure increase, is 2.2 – 2.5 MPa. At the same time, the surface apparatus must receive the information at a sensitivity threshold of 0.01 – 0.02 MPa in a discharge line, and 0.001 – 0.002 MPa in the annulus at a stable operation in a frequency range of 0.03 – 0.15 Hz.

All the parts of the bottomhole devices and tools must be of the same material as other components of the bottomhole drill string assembly, so that the reliability of all the elements is commensurable.

### 3.1.10 Drill Rig

Analysis of the scientific ways of solving the technological and technical problems of drilling a super-deep well in crystalline rocks enables us to formulate the basic requirements for surface drilling equipment and tools.

Turbodrilling combined with the use of light-alloy drill pipes, and a strength of the crystalline rocks high enough to permit simplifying the well design have made it possible to reduce the hoisting capacity of the drill rig. Calculations indicated that the weight of the drill string made of light-alloy drill pipes (LBT-147) 15,000 m long is 1500 – 1700 kN, a decrease in weight of the pipes in the mud being also taken into account. It would be erroneous, however, in determining the load capacity of the drill rig, to orient oneself only to the specified weight of a drill string of light-alloy pipes. Under the unique conditions of superdeep drilling, the upper portion of the string may be composed of steel pipes, the lower part of the string (bottomhole section), however, of drill pipes made of titanium alloy, or thin-walled steel pipes whose length and weight will depend on the temperature. Thus, the weight of the drill string may be increased to 2400 – 2600 kN; but even this figure must not be taken as final. The drag forces acting during pulling out the drill string are capable of increasing the hook load by 300 – 600 kN, under unfavourable conditions even by 800 – 1200 kN, a total of some 3000 – 3800 kN.

Results show that a drill rig with a load capacity of not less than 4000 kN should be used to drill a well down to 15,000 m deep in crystalline rocks by the turbodrilling method and using light-alloy drill pipes. Using this rig, we may run in one operation casing strings of 426 mm in diameter to a depth of 3300 M, the

324-mm diameter casing to 4500 m, and the 245-mm diameter to 6000 m. These depths of running the casing strings (in one operation) exhaust practically all the strength potential of casing pipes manufactured at present.

If necessary, casing the well with casing strings of these types at great depths in crystalline rocks may be accomplished by a method of section-by-section running-in or by the specially developed built-in mechanism of large load capacity.

The specific conditions of drilling in crystalline rocks do not allow us to mount more than two preventers at the well mouth. Moreover, an insignificant length of the ditch system, which is constructed and operates in an enclosed space, does not require a very high substructure. Nevertheless, for the sake of convenience in maintaining the wellhead equipment and in performing a number of technological operations, the height of the substructure should not be less than 5 – 6 m.

The requirements of slush pumps were determined for drilling conditions down to 15,000 m by an advance hole of 215 mm in diameter, by a turbodrill and light-alloy drill pipes of the type LBT-147.

On the basis of the calculations for the final period of drilling with an optimum flow rate of drilling fluid of 28 – 30 l $s^{-1}$, the conclusion is drawn that slush pumps providing 30 MPa pressure at a hydraulic power of 736 – 900 kW will be needed.

The practice of long-term exploitation of slush pumps at their allowable maximum horsepower shows that the hydraulic part of the pumps tends to wear out rapidly. Using them, however, at only 2/3 – 3/4 of the hydraulic power provides for stable and continual operation. In this case, the power reserve and respective decrease of load on all the hydraulic parts of the pump (as compared to the allowable loads) are compensated for by more rigid requirements as to reliability under conditions of failure-free exploitation. Thus, when drilling a well with one single pump, its hydraulic power should be 930 – 1100 kW at the maximum pressure of 35 – 40 MPa with a compulsory variation of drilling fluid flow rate by regulating the number of double strokes.

The most important characteristics of the rotary table is a smooth change in rpm, especially in the low frequency range (up to 10 rpm). This is dictated by the necessity for slow rotation of a drill string in turbodrilling where the admissible intensity of the stressed state in drill pipes at a great depth is achieved by a maximum decrease of their rpm down to 2 – 3. This characteristic of the rotary table drive makes it possible to pick up a string of drill pipes smoothly, preventing its destruction, and fulfills the requirements for failure combatting and other operations. The diameter of the rotary table opening must be as large as possible (not less than 760 mm), load capacity not less than 4000 kN, torque up to 50 – 60 kN · m. A built-in power slip must execute the work with drill pipes of 114 – 190 mm in diameter.

The swivel load capacity must provide round trips of drill string with the load on the hook up to 4000 kN.

The rpm range of the kelly is from 2 to 100 (in some cases up to 200 rpm).

Support and sealing of the swivel must work reliably at the different loads listed here:

| | | | |
|---|---|---|---|
| Hook load, kN | 4200 | 2600 | 1500 |
| Rotation frequency, rpm | – | 2–6 | 100 |
| Pressure, MPa | – | 3.0 | 2.0 |

Analysis of the expected time distribution in drilling of the 15,000-m-deep well indicates that the overall drilling time of such a hole depends largely on the round trip velocity (RTV). To reduce the time for round trip operations, the following is required of methods and mechanisms of the round trips:

- drilling practices involving an inserted tool without pulling out the drill string;
- round trip systems of continuous action without stoppage of the drill string for screwing and unscrewing of pipe stands, maximum integration of various operations;
- an increase in power capacity of the hoisting mechanisms and a decrease in the weight of the drill string;
- maximum optimization of the round trip operations and mechanization of work with drill stands.

In designing the drill rig *Uralmash-15,000*, all the above-mentioned considerations were taken into account, except for the system of continuous round trip operations. The time loss for the round trip operations due to a decrease in their speed caused by the application of the high-load capacity mechanisms with a pulley-block system may be compensated for by increase of footage per bit.

Pulling out and running in of a drill string using optimum speed diagrams depending on the weight of the string, automated and mechanized round trip operations and recuperation of electric energy to the mains when braking the drill string in motion are provided by application of dc motors for the drawworks drive. The time loss for the round trip operations depends essentially on the length of a drill pipe stand. For drilling a deep well involving a large number of round trip operations, it is advisable to utilize pipe stands of not less than 37 m in length.

The volume of the hole of 15,000 m deep is 500 – 700 $m^3$. Cleaning such a volume by the circulating drilling fluid from the crushed rocks will require continual pumping at high pressure. To prevent intensive wear on the hydraulic part of the pumps and on the whole circulation system, including the light-alloy drill pipes, at considerable velocities and pressures of the drilling fluid, the latter must be cleaned at the day surface. This task is complicated by the presence of highly abrasive particles in the slime. The cleaning system must provide a three-step cycle of slime removal: a vibrating screen, hydraulic cyclon and a centrifuge. Removal of hard particles of 30 – 40 μm in size is also desirable.

The severe climatic conditions of the area where the Kola well was situated require that the derrick be fully lagged and partially heated. Among other relevant measures are: the heating of the lower ends of drill pipe stands on a rack to prevent their icing, application of a system of purging a discharge line to free it from the fluid in case of a long stoppage at low surrounding temperature, location of all main and auxiliary drilling equipment, including a circulation system, in the heated premises with a temperature level maintained at not lower than +5 °C.

The fact that the Kola well is located in an area far away from industrial centres and localities of active drilling operations and the long period of drilling, necessitate constructing near the drill rig site a heated building of auxiliary services to provide current maintenance, partly also for overhaul and major repairs of equipment and tools, assembling and dismantling of turbodrills and BHA, storing of permanent stock of tools and materials, as well as suitable working conditions for the drilling crew and engineering and administrative personnel.

# 3.2 Surface Drilling Equipment and Facilities

### 3.2.1 Drilling Equipment and Electric Drives

The well construction programme involves two-stage drilling using an improved standard drilling rig *Uralmash-4E* to the depth of 7000 m in the first stage and a recently designed drilling rig *Uralmash 15,000* in the second. The surface drilling facilities were therefore constructed with a view to their future modernization and change of drilling rig within the shortest possible time.

The technical data on the drilling rigs used are given below.

The drilling unit rigged up to drill the well for the first stage differed from a standard *Uralmash-4E* by its improved derrick and higher total hydraulic power of the mud pumps.

| Drilling rig | Improved *Uralmash-4E* | *Uralmash 15 000* |
|---|---|---|
| Total input drawworks power, kW | 640 | 2300 |
| Total input slush pump power, kW | 2550 | 5650 |
| Maximum load capacity, kN | 2000 | 4000 |
| Maximum drilling depth (when using aluminium drill pipes), m | 7000 | 15000 |
| Maximum number of lines strung up | 12 | 14 |
| Maximum running line end pull, kN | 273 | 420 |
| Drilling line OD, mm | 32 | 38 |
| Drive | Electric, ac | Electric, ac |
| Drive type | Independent | |
| Number of motors driving: | | |
| drawworks | 2 | 2 |
| slush pumps | 3 | $3+2=5$ |
| Drawworks model | U2-5-5 | LBU-3000 |
| Drawworks main drum power, kW | 809 | 2646 |
| Slush pump model | U8-7M | UNB-1250, U8-7M |
| Slush pump power, kW | | |
| U8-7M | 825 | 825 |
| UNB-1250 | – | 1250 |
| Total power of rig slush pumps, kW | $3 \times 825 = 2475$ | $3 \times 825 + 2 \times 1250 = 4975$ |
| Total slush pump hydraulic power, kW | $3 \times 700 = 2100$ | $3 \times 825 + 2 \times 1060 = 4220$ |

| Drilling rig | Improved *Uralmash-4E* | *Uralmash 15 000* |
|---|---|---|
| Maximum pump discharge pressure, MPa: | | |
| U8-7M | 32 | – |
| UNB-1250 | – | 40 |
| Maximum pump flow rate, $s^{-1}$: | | |
| U8-7M | 50.2 | – |
| UNB-1250 | – | 51.4 |
| Rotary model | R-560 | R-760 |
| Rotary table opening, mm | 560 | 760 |
| Static rotary table load capacity, kN | 3200 | 4000 |
| Rotary input power, kW | 368 | 368 |
| Swivel model | U6-ShV-14-16OM | UV-450 |
| Static swivel load capacity, kN | 1600 | 4500 |
| Drilling derrick model | VB-53 × 300 (improved) | VBA-58 × 400 |
| Useful derrick height, m | 53.3 | 58 |
| Safe load derrick capacity, kN | 2000 | 4000 |
| Crownblock model | U3-200-2 | UKBA-7-500 |
| Crownblock load capacity (maximum number of lines strung), kN | 2000 | 5000 |
| Hook-block model | U4-200-3 | UTBA-6-400 |
| Hook-block load capacity, kN | 2000 | 4000 |
| Diesel-generator power unit: | | |
| Model | ASDA-200 | – |
| Number | 1 | – |
| Power, kW | 200 | – |
| Compressors: | | |
| Model | KSE-5M | KSE-5M and VSh6/10 |
| Number | 2 | 2 and 2 |
| Maximum air working pressure, MPa | 0.8 | 1.0 |
| Pipe handling devices: | | |
| Stand racking device | – | ASP-6 |
| Power slips | PKR-U7A | PKR-300M |
| Pipe tongs | AKB-3M | AKB-3M-300 |
| Bit feeding device | – | RCDE-3-300 |
| Tool joint breakout device | Tool joint pneumatic breaker | |
| Cranes: | | |
| – for servicing slush pumps | Electrically driven overhead travelling crane | |
| – for handling tools on pipe slide | Travelling gantry crane | Electrically driven overhead travelling crane |
| Method of moving-in and rig-up | In separate assemblies | |

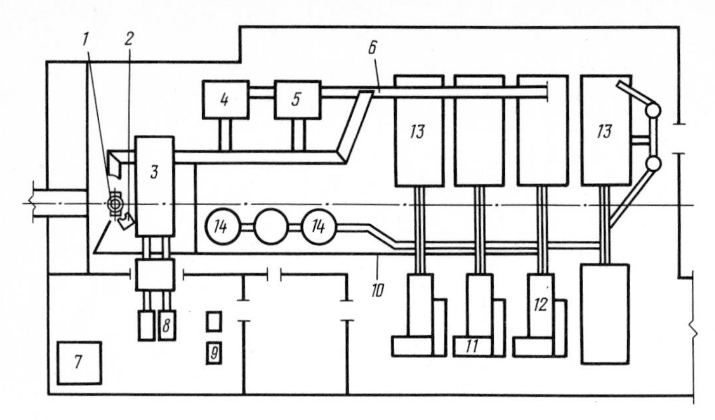


**Fig. 3.11.** Rig layout during the first stage of the SG-3 hole drilling. *1* rotary table; *2* power tong; *3* drawworks; *4* desander; *5* shale shakers; *6* surface circulation system; *7* diesel electric power unit; *8* AC electric power units; *9* compressors; *10* unified manifold; *11* AC electric motors; *12* slush pumps; *13* suction tanks; *14* mud chemical tanks

The drilling rig included a derrick which was an improved variety of the standard VB-53 × 300 drilling derrick. The legs and girths were made of pipes; its racking capacity was increased up to 7000 m of 147 mm OD drill pipes. The derrick was fully covered with wooden panels. The hydraulic power capacity of the rig was enhanced by using three U8-7M slush pumps driven by ac electric motors through one-sided V-belt drive.

The rig layout is shown in Fig. 3.11.

All the equipment was stored in the derrick-drawworks and pump units as well as in the drilling site facilities. A 50-kN travelling gantry crane was used to handle equipment and tools on the pipe ramp. The derrick-drawworks unit substructure accommodated the drilling derrick with a block and tackle system (crown-block, travelling block and hook and 12-line string-up), drawworks with a hydromatic brake, rotary table with pneumatically driven built-in slips, AKB-3 semiautomatic power tongs, driller's control panel etc.

The drilling site facilities accommodated ac electric power units to drive the drawworks and rotary, compressors, dc diesel-electric power units, mud-handling equipment including a desander, shale shaker and surface circulation systems.

U8-7M slush pumps were completely piped with an UMB-320 unified manifold comprising a full set of shut-off, distributing and starting gate valves, pressure relief valves, gauges, remote control devices and discharge line pipe sections.

A mud cleaning system consisted of a SV-2 dual shale shaker, two GC-350 hydroclone desanders, GC-150 desander and GC-75 ten hydrocyclone desilter. This cleaning system provided successive removal of drilling cuttings of different sizes. The total volume of suction tanks was 120 $m^3$. Three 25 $m^3$ tanks were used to store mud chemicals.

The drilling rig power units are able to drive simultaneously three U8-7M slush pumps, drawworks during round-trips and rotary table. The drawworks

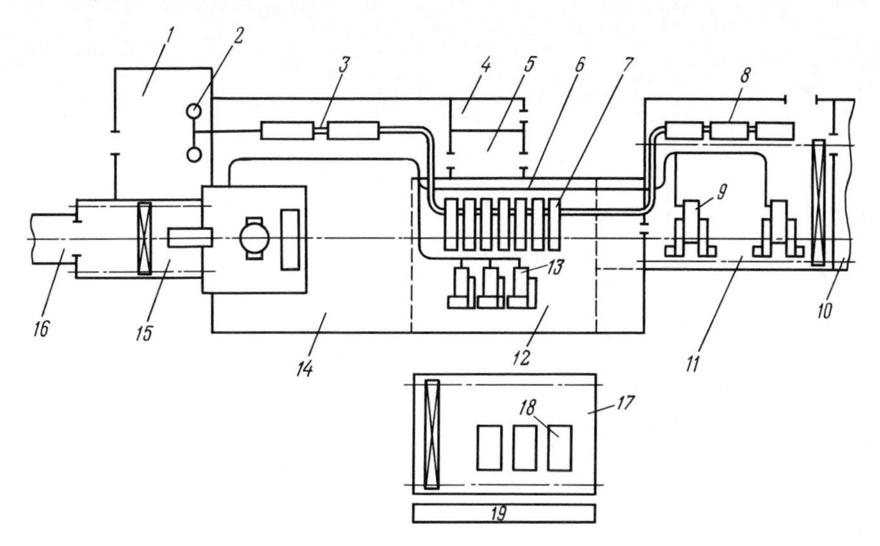


**Fig. 3.12.** *Uralmash-15000* Drig Rig layout. *1* mud and cement dry material storage No. 1; *2* mud mixing unit; *3* cement slurry mixing unit; *4* transformer substation No. 2 (TS-2); *5* hot water heater compartment; *6* discharge line; *7* 24 $m^3$ mud suction tank; *8* 40 $m^3$ suction tank; *9* UNB-1250 slush pump; *10* mud and cement dry material storage No. 2; *11* slush pump unit No. 2; *12* slush pump unit No. 1; *13* U8-7M slush pump; *14* derrick-drawworks unit; *15* drilling compartments; *16* maintenance and repair compartments; *17* power unit compartment; *18* AC/DC converter; *19* 36/6 kV transformer substation

and rotary are powered by two AKB-114-6 320 kW, 980 rpm electric motors. The rotary input power is 368 kW. The slush pumps are powered by three SDB-14-46-8 AC 850 kW electric motors.

The drilling rig *Uralmash-15,000* was used after the hole reached a depth of 7263 m. Among the design features of the rig are the following:

- use of independent adjustable dc drive (generator-motor) of the rotary, slush pumps and drawworks;
- possibility of making round trips using optimized tachograms which are chosen automatically by a computer depending on the drill string weight;
- use of two UNB-1250 slush pumps developing 40 MPa pressure and three U8-7M pumps;
- high input drawworks power (equal to 2646 kW);
- use of drawworks motors as braking machines working in a generator mode, the regenerated power being returned into the supply system during round trips;
- highly mechanized drill pipe stand handling during round trips;
- use of a vertical column drilling derrick of a pylon design;
- simple and easy drilling equipment control, maintenance and repairs;
- use of a heavily instrumented system to control drilling process and performance of all the drilling rig components.

The equipment of the *Uralmash 15000* drilling rig is arranged in four units (Fig. 3.12): a derrick-drawworks unit, power unit, slush pump unit (No. 1) and

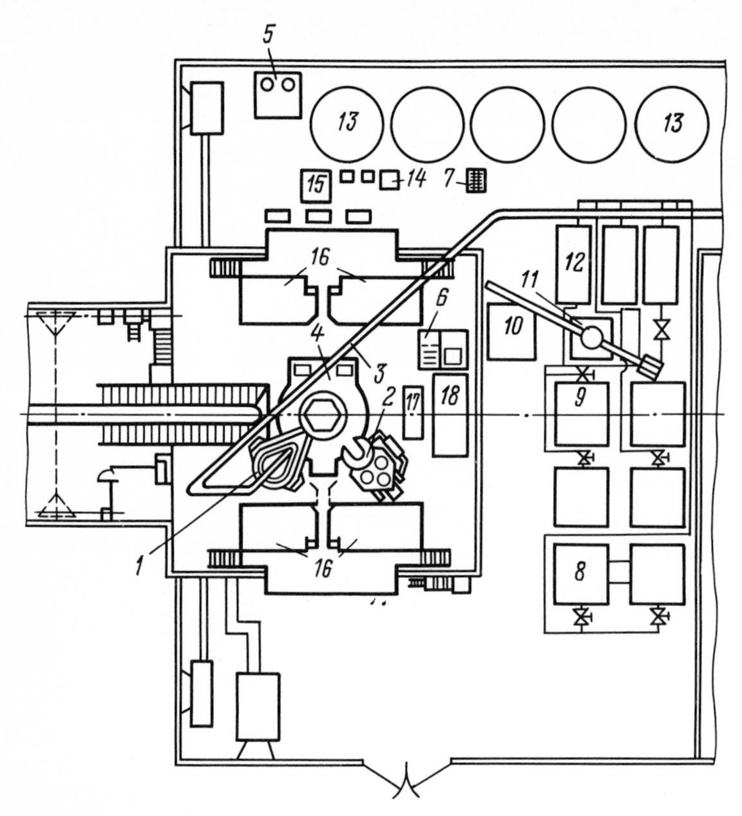


**Fig. 3.13.** Derrick drawworks unit layout. *1* UV-450 swivel; *2* AKB power tongs; *3* manifold; *4* UR-760 rotary with built-in pneumatic slips; *5* VShN-I50 pump; *6* driller's control panel; *7* IG-45 desilter; *8* air dryers; *9* compressor units; *10* drilling tool feed device; *11* pillar rib crane; *12* air receiver; *13* reserve drilling mud tanks; *14* desander; *15* ale shaker; *16* racking boards; *17* auxiliary hoist; *18* LBU-3000 drawworks

auxiliary slush pump unit (No. 2). The derrick-drawworks and slush pump units are placed in a common elongated compartment and directly connected to each other. The power unit is placed in a separate compartment at a certain distance from the derrick-drawworks and slush pump units.

The derrick design is best suited for racking drill stands used in drilling to a depth of 15,000 m. Derrick floor elevation is 6 m. The derrick was erected by the "from-top-to-bottom" method using a special PV-750 frame hoist of 7500 kN capacity. The derrick is fully encased in fluted metal sheets.

The derrick-drawworks unit layout is shown in Fig. 3.13. The hoisting equipment consists of LBU-3000 drawworks and drilling tool feed device and is mounted on a separate foundation. On the left and right sides of the unit substructure are mounted racking boards which provide place for setting back up to 30,000 m of 37-m drill stands using ASP-6 pipe-handling system. Four KSE-5M and VSh-6/10 compressor units connected through air dryers to air receivers are placed at the ground level.

**Table 3.9.** Technical data on mud-cleaning equipment

| Mud cleaning device | Model | Pressure, MPa | Throughput capacity, $l\ s^{-1}$ | Size of solid particles, removed, μm |
|---|---|---|---|---|
| Shale shaker | VS-1 | – | 38 | 160–200 |
| Desander | GC-350 | 0.25 | 44 | 200–250 |
| Desander | GC-250 | 0.34 | 40 | 100–150 |
| Desilter | IG-45 | 0.25 | 45 | 40–50 |

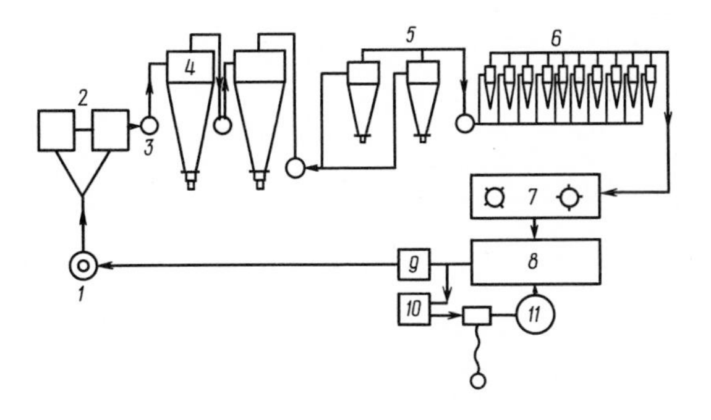


**Fig. 3.14.** Mud cleaning system layout. *1* hole; *2* VS shale shaker; *3* centrifugal pump; *4* GC-350 desander; *5* GC-250 desander; *6* IG-45 desilter; *7* tank furnished with stirring devices; *8* suction tank; *9* slush pump; *10* auxiliary mud pump; *11* bentonite feeding unit

The air supply system operates AKB power tongs, pneumatic drill stand breaker and drawworks aircluthes. The drill pipe stands are assembled, dissassembled and transported to the well axis from the racks and in the opposite direction with the help of the ASP-6 pipe handling system and AKB-3M-300 power tongs. The tong carriage needs much room when being transported. The height of its jaws is rigidly fixed. The independent drive of the rotary is mounted on special support beams under the derrick floor.

Each of two UNB-1250 slush pumps is powered by two dc 800 kW electric motors through two-sided V-belt drives. Theoretical pump capacity and pressure at 60 double strokes per minute are given below.

| | | | | | | |
|---|---|---|---|---|---|---|
| Liner ID, mm | 150 | 160 | 170 | 180 | 190 | 200 |
| Pump capacity, $l\ s^{-1}$ | 26.7 | 31.7 | 38.7 | 40.7 | 45.9 | 51.4 |
| Maximum pressure, MPa | 40.0 | 35.0 | 30.5 | 26.5 | 23.5 | 21.0 |

Two UNB-1250 and three U8-7M slush pumps can work simultaneously. Two AK2-150M high-pressure compressors are used to charge pump pulsation dampeners. The inside diameter of the common discharge line of the slush pump manifold is 125 mm, this line being designed for a pressure of 40 MPa. The inside diameter of lines connecting hydraulic ends of the pumps with the common discharge line is 100 mm.

The main equipment is mounted in an enclosed space. The loads on the receiving platform of the derrick are handled with a travelling crane, its rail track being laid in the enclosed gallery connecting the drilling compartment with the other drilling site facilities.

Technical data on the mud-cleaning equipment are given in Table 3.9. The mud-cleaning system layout is shown in Fig. 3.14.

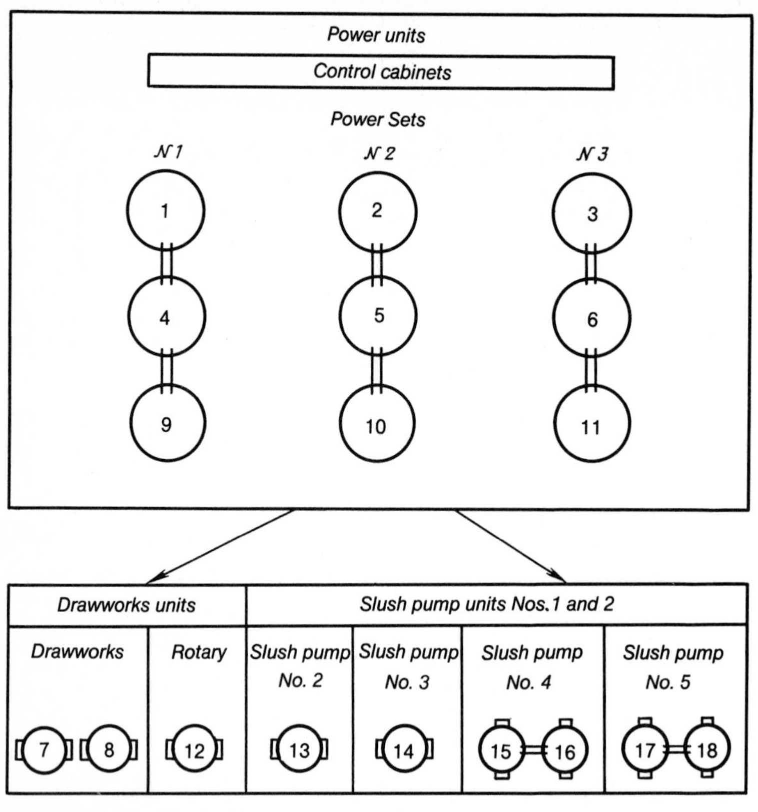


**Fig. 3.15.** Block diagram of the main electric drives. *4, 5, 6* synchronous motors of SDEZ-15-64-6 type (1900 kW, 6000 V, 1000 rpm); *9, 10, 11* main DC generators of GPE-1700-1000 type (1700 kW, 900 V, 1000 rpm); *7, 8* DC electric motors of PS-179-9K type (1150 KW, 660 V, 220/440 rpm); *1, 2, 3* auxiliary DC generators of MPE-800-900 type (400 kW, 230 V, 1000 rpm); *12, 13, 14, 15, 16, 17, 18* DC electric motors of MPE-700-800 type (800 kW, 450 V, 800/1000 rpm)

**Table 3.10.** Main supply circuits

| Operation | Main generators | Auxiliary generators |
|---|---|---|
| I. Drilling | 1. Three main generators power motors of any three slush pumps<br>2. Any of two main generators power motors of any two slush pumps | One or two auxiliary generators power a motor of the draw-works or rotary |
| II. Running and pulling of drill string | Any of two main generators power two drawworks motors | One or two auxiliary generators power rotary motor |
| III. Reaming of borehole | Slush pumps, drawworks and rotary operate simultaneously | |

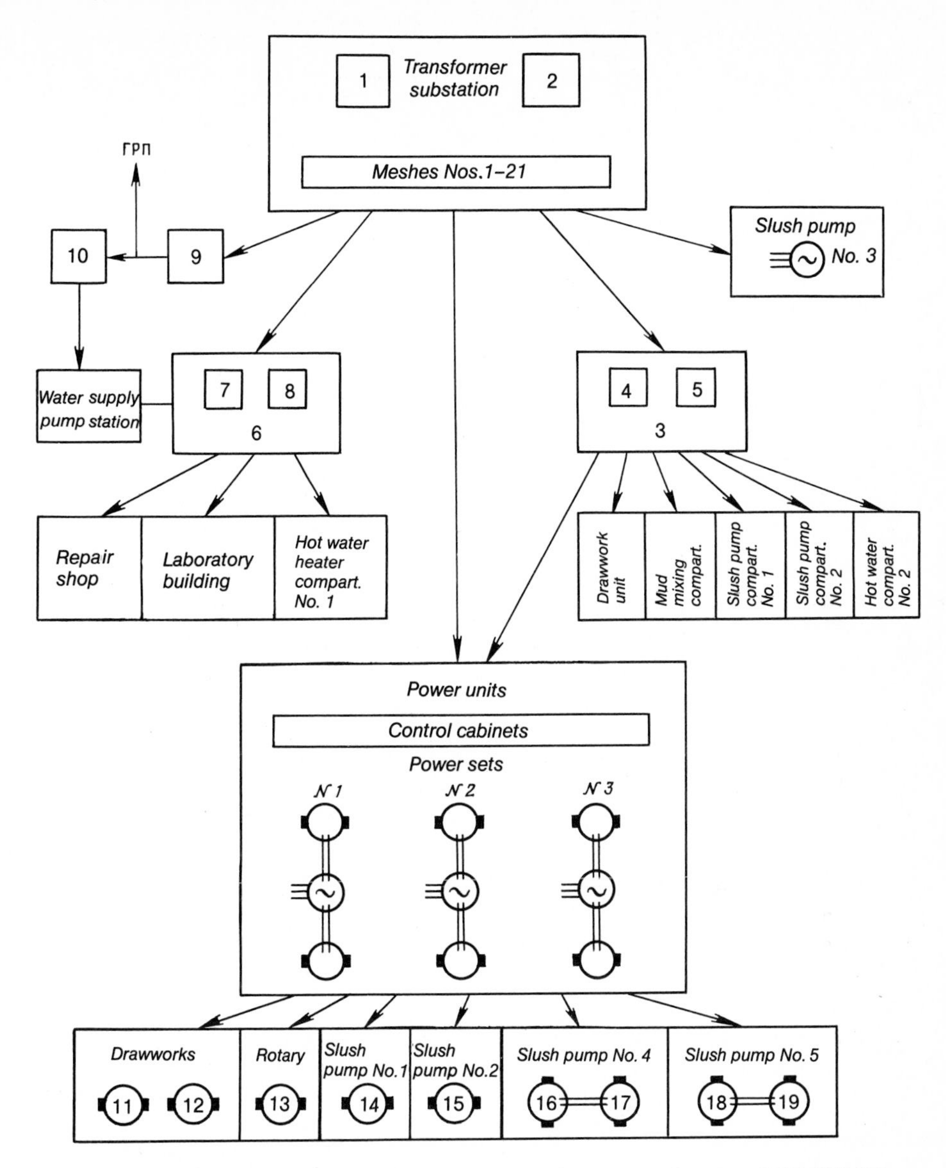


**Fig. 3.16.** Block diagram of the SG-3 well site power supply. *1, 2* transformers of TM-6300/35 type (6300 kVA, 35/6 kV); *3* transformer substation; *4, 5* transformers of TM-1000/6 type (1000 kVA, 6/0.4 kV); *6* transformer substation; *7, 8* transformers of TM-1000/6 type (1000 kVA, 6/0.4 kV); *9* transformer substation with TM-180/6 transformer (180 kVA, 6/0.4 kV); *10* transformer substation with TM-63/6 transformer (63 kVA, 6/0.4 kV); *11 – 19* electric motors

The total volume of the mud tanks installed is 350 $m^3$. The following drilling sludge disposal system is used. The liquid phase is directed by centrifugal pumps into the pits. The solid phase is gathered and used for grading the drill site territory. The volume of the settling pit is 15 $m^3$, it is equipped with two VShN-150 sludge centrifugal pumps. A set of five reserve tanks used for storing special chemicals and filling the well is located adjacent to the mud cleaning compartment. Each reserve tank is connected to the distributing manifold system.

There is heated storage for dry mud and cement materials and a 2BPR mechanized unit for mud and cement slurry mixing on the drilling site.

The main electric drives shown in Fig. 3.15 provide a supply of adequate power during all drilling rig operations including drilling, reaming and round trips. Switching of the main supply circuits is illustrated in Table 3.10.

The main electric drive circuits include three bus systems for the main generators and motors of the slush pumps and drawworks. The circuit is set up in such a way that there is one switch in the line of every main generator-auxiliary generator-slush pump block, which not only provides a connection of the block to the bus system, but also provides disconnection of the block by an apparent break in the main circuit when generators or motors are faulty or being serviced.

The electric power supply for the drilling site is shown in Fig. 3.16. Emergency power supply is automatically switched on if a transformer of a 35 kW line is dead. The electric block diagram chosen provides an uninterrupted power supply for all the services on the drilling site.

### 3.2.2 Drilling Process Instrumentation

A special information system has been used to control the main process parameters in Kola well drilling from a depth of 8000 m on. The process parameters controlled, instrument range, gauges, recording instruments and control action of the system are given in Table 3.11.

The information system is divided into separate subsystems according to the part of the process monitored: hole making, round trip, mud circulation and a subsystem of main drilling equipment operating parameters. The system is arranged on the unit assembly principle, i.e. independent information channels are made of unified elements and blocks, the information being recorded in analogue form. The standard output of each channel makes it possible to connect it to a computer to acquire and process technological information. The system includes a set of monitoring gauges on the driller's control console (Fig. 3.17), remote control gauges in the slush pump compartment, an alarm subsystem which provides control action when a deviation of the monitored parameter from the pre-set value, and a technological control room (Fig. 3.18), where all the gauges and recording instruments necessary to monitor and control hole-making are concentrated.

### 3.2.3 Auxiliary Facilities at the Drilling Site

According to their purpose these facilities may be subdivided into two complexes: one for inspection and servicing of the main drilling tools and processing and storing of all drilling data acquired, the other for handling of all materials and tools delivered to the drilling site.

**Table 3.11.** Controlled process parameters, indicating and recording instruments. Limits of their range and output action of information system

| Process parameter controlled | Instrument range | Type of indicating and recording instrument | | | | Control action |
|---|---|---|---|---|---|---|
| | | Indicator on driller's control console | Indicator in technological control room | Remote indicator | Analogue data recorder | |
| 1 | 2 | 3 | 4 | 5 | 6 | 7 |
| | | | *Hole-making subsystem* | | | |
| Hook load, kN | 0 – 4000 | + | + | + | + | Disconnection of drawworks motors and application of brake when allowable load is exceeded |
| Weight on bit (hook load vernier), kN | 0 – 200 | + | + | – | + | Warning of deviation of weight on bit from set value |
| Rotary torque, kN · m | 0 – 60<br>0 – 6 | + | + | – | + | Disconnection of rotary motor when allowable torque is exceeded |
| Downhole motor rpm | 0 – 1200 | + | + | – | + | – |
| Rotary rpm | 0 – 300<br>0 – 10 | + | + | – | + | Automatic changeover of instrument range |
| Penetration rate, $m\ h^{-1}$ | 0 – 50<br>0 – 5 | + | + | – | – | Automatic changeover of instrument range |
| Number of metres drilled | No limits | – | – | – | + | – |
| Length of kelly under rotary table upper surface, m | 0 – 20 | – | + | – | – | – |
| | | | *Round-trip subsystem* | | | |
| Round-trip rate, $m\ s^{-1}$ | 1 – 0 – 1<br>3 – 0 – 3 | + | + | – | + | Automatic change-over of instrument range |
| Number of stands in the hole | 0 – 400 | + | + | – | – | Warning of dangerous borehole intervals and zones |

**Table 3.11** (continued)

| Process parameter controlled | Instrument range | Type of indicating and recording instrument | | | | Control action |
|---|---|---|---|---|---|---|
| | | Indicator on driller's control console | Indicator in technological control room | Remote indicator | Analogue data recorder | |
| 1 | 2 | 3 | 4 | 5 | 6 | 7 |
| Travelling block height above rig floor, m | 0 – 50 | + | + | – | – | – |
| Length of drilling line discarded during restringing, m | 0 – 2000 | – | – | + | – | – |
| Power tongs torque, kN · m | 0 – 60 | + | + | – | – | – |
| | | | *Mud circulation subsystem* | | | |
| Mud pressure, MPa | 0 – 40 | + | + | + | + | Disconnection of slush pump motors when allowable pressure is exceeded |
| Mud flow rate (riser), $l\ s^{-1}$ | 0 – 100 | + | + | + | + | Warning of deviation of flow rate from steady state value |
| Mud flow rate (mud flow line), $l\ s^{-1}$ | 0 – 100 | – | + | – | – | Warning of deviation of flow rate from steady state value |
| Mud temperature (mud flow line), °C | 0 – 100 | – | + | – | – | Warning of deviation of flow rate from steady state value |
| Mud density (mud flow line), $g\ cm^{-3}$ | 0.8 – 2.6 | + | + | – | – | Warning of deviation of flow rate from steady state value |
| Mud pressure (annulus), MPa | 0 – 6 | + | + | – | – | Warning of deviation of flow rate from steady state value |
| Mud level (suction tanks), m | 0 – 2.5 | – | + | + | – | Warning of deviation of flow rate from steady state value |
| Mud level (storage tanks), m | 0 – 3 | – | + | – | – | Warning of deviation of flow rate from steady state value |

**Table 3.11** (continued)

| Process parameter controlled | Instrument range | Type of indicating and recording instrument | | | | Control action |
|---|---|---|---|---|---|---|
| | | Indicator on driller's control console | Indicator in technological control room | Remote indicator | Analogue data recorder | |
| 1 | 2 | 3 | 4 | 5 | 6 | 7 |
| | | *Subsystem of main drilling equipment operating variables* | | | | |
| Hoisting system and drilling line work metre, kN · km | 0 – 99999 | – | + | – | – | – |
| Swivel work metre, kN · rpm | 0 – 99999 | – | + | – | – | – |
| Rotary work metre, (kN · m) × rpm | 0 – 99999 | – | + | – | – | – |
| Mud pump work metre, MPa × number of double strokes $min^{-1}$ | 0 – 99999 | – | + | – | – | – |
| Mud flow rate during downhole motor testing on wellhead, $l\ s^{-1}$ | 0 – 50 | – | + | – | + | Warning of deviation of flow rate from steady state value |
| Mud pressure during downhole motor testing on wellhead, MPa | 0 – 20 | – | + | – | + | Warning of deviation of flow rate from steady state value |
| Reactive torque at downhole motor housing during its testing on wellhead, kN · m | 0 – 10 | – | + | – | + | Warning of deviation of flow rate from steady state value |
| Downhole motor rpm during its testing on wellhead | 0 – 1200 | – | + | – | + | – |

Note: + = information is available; – no information.

**Fig. 3.17.** Indicating instruments in driller's panel

**Fig. 3.18.** Technological control room

The complex intended to inspect and service the main drilling tools and to process and store information on the drilling process is placed on the drilling site so that it is a natural extension of the derrick-drawworks unit and connected to the latter by a special enclosed gallery. This enables unhampered conveyance of drill pipes, bits, bottom-hole assemblies etc. to the derrick-drawworks unit independently of meteorological conditions.

The layout of the facilities intended to inspect and service the main drilling tools and to process and store data on the drilling process is shown in Fig. 3.19. A two-storeyed building accommodates areas to prepare sets of drill pipes, turbine downhole motors and bottom-hole assemblies, for forging, welding and heat

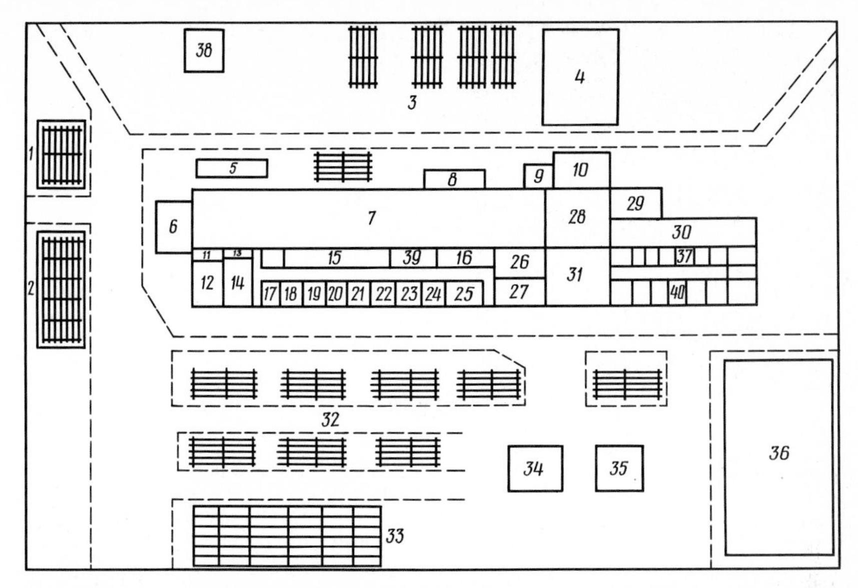


**Fig. 3.19.** Layout of facilities intended to inspect and service the main drilling aids and to process and store data on the drilling process. *1–3, 32, 33* drill pipe racks; *4–6, 9* enclosed storage compartments for everyday reserves; *7* pipe-downhole motor area; *8, 13, 16–19* fault detection, instrumentation and automatic equipment laboratories; *14* electric equipment repair area; *10* forging and welding area and heat-treatment bay; *11, 12* gauging area; *15* electric hot-water heater compartment; *20–22* primary core analysis laboratory; *23–24* mechanical engineer office; *25* site control department (control board); *26* metal strength testing laboratory; *27* tool preparation area; *28* fitting area; *29* transformer substation of the complex; *30* ready-for-service tool gallery; *31* metal working area; *34* fuel and lubricant storage; *35* mud chemical storage; *36* power unit; *37, 38* research and methodological geophysical party; *39* core storage; *40* research and methodological party for processing geological data obtained by drilling; drilling mud laboratory

treatment for metal working and metal strength testing, for electric equipment repairs, adjustment and alignment, for tool preparation and fitting. The building also has a drilling site control compartment (control board) equipped with an industrial television screen.

In addition to preparation for drilling, this complex supplies all the power for the drilling site, supports all logging in the borehole, provides primary core analysis and comprehensive processing of all geological and geophysical data obtained in the well. All research in the sphere of mud composition and drilling technique is also carried out in this building.

Sanitary facilities and shower-rooms, a first-aid station, a canteen to cater for staff day and night, a meeting hall and rooms for preventive medical aid provide normal living conditions for the operating personnel of the rig. The building is furnished with ventilation, sewerage, heating, and hot and cold water supply.

The complex is a compact drilling and research unit. Efficiency in the drilling process is directly dependent on the punctual and efficient operation of this complex due to the importance and diversity of the problems solved with its help.

The facilities shown in Fig. 3.20 are intended for loading and unloading expendables, preparing sets of tools, moving in and storing expendables, main-

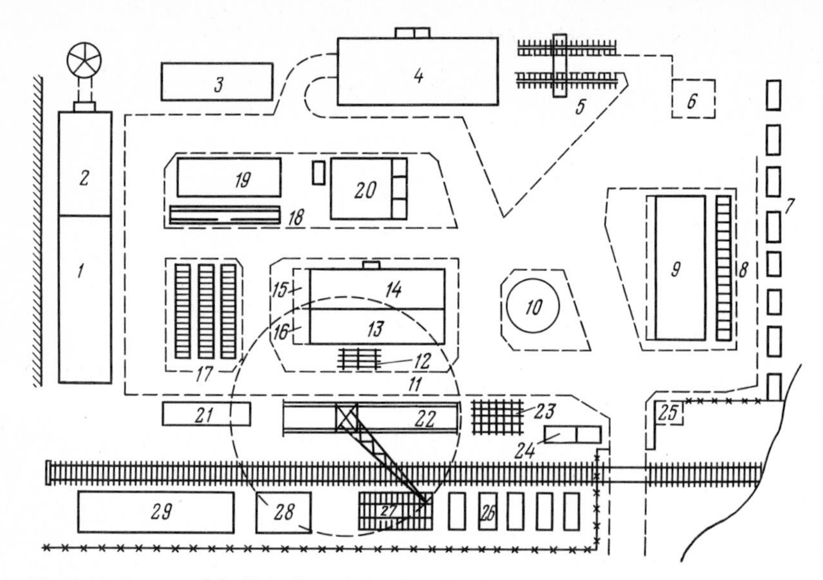


**Fig. 3.20.** Layout of facilities intended to handle all materials and tools delivered to the drilling site *1* repair boxes; *2* parking garage; *3, 15, 16* open parking lot; *4, 5* drill tool selection area; *6, 12, 17, 23* open store areas; *7* storage wagons for building materials; *8, 26, 27* rack areas; *9* main core storage; *10* fire tank; *11, 22* radius of tower crane section, crane rails; *13, 14* central storage; *18* washing rack; *19* area for reserves; *20* main fuel and lubricant storage; *21* loading and unloading rack; *24* loading and unloading area; *25* check; *28, 29* roofed racks

taining and repairing approach roads to the drilling site, drilling and living facilities, vehicles and cranes. The main core storage building is located in the territory of this complex. The area of shed and open storages is quite sufficient to preserve all the expendables and equipment in a good condition. Loading and unloading equipment and areas make it possible to prevent delays in the wagons and hence drilling rig shut-downs.

The living conditions of the operating personnel of this complex are equipped in the same way as those of the first complex with sanitary facilities and shower-rooms, recreation hall etc. The building is furnished with ventilation, sewerage, heating and hot and cold water.

## 3.3 Drilling Tools and Aids

Since the beginning of the preparatory stage of the Kola well drilling project and during its realization, the All-Union Research Institute for Drilling Technology and All-Union Research Institute for Oil-Field Tubulars, together with expedition personnel, have been continuously developing and improving different drilling tools. Prototype tools were first tested in a full-scale test well in the Ukrainian State Project Planning and Research Institute of Oil Industry and then in a satellite SG-3 well and in the Kola well.

A short description and the characteristic features of the main tools developed and used in drilling of the SG-2 well are given here.

### 3.3.1 Drill Bits

As the main purpose of the Kola well drilling was core recovery and analysis, core drill heads, the main type of drill bit, have received primary attention. Some types of reamers and above-bit calibrators are also described.

*A KC-212.7/60 TKZ (2B-K212.7/60 TKZ) core head* is designed for high rpm coring of 212.7-mm borehole in hard rocks interbedded with extremely hard rocks, the diameter of core being 60 mm (Fig. 3.21). The body of this four-cutter core head with crushing action is formed of four arms welded together. Each roller cutter is mounted on the axle of its arm using two ball and two journal bearings. One of the ball bearings serves as a retainer and prevents the cutter from axial displacement. All four cutters take part in forming both a borehole and core. The core head cutting structure consists of hard-alloy inserts with a wedge-shaped cutting surface. The inserts are fitted by the pressing-in method instead of the brass-brazing method used earlier.

A special bushing is located in the inner space of the core head. It is intended to lower the level of core entry and to protect it from erosion by drilling mud.

The core head is developed to substitute for a serially produced 2B-K 214/60 TKZ core head. The new model features stricter tolerances for its outside and inside diameters, as well as for cone height difference and play.

*A 21N-K212.7/80 TKZ core head* is designed for coring hard abrasive rocks interbedded with extremely hard rocks using the rotary method or low-rpm downhole motors (maximum 350 rpm) (Fig. 3.22).

The core head has six roller cutters. Three outer cutters cut the peripheral part of the bottom face and form the borehole, three inner cutters form the core and cut the bottom face part adjacent to the core. The welded core head consists of

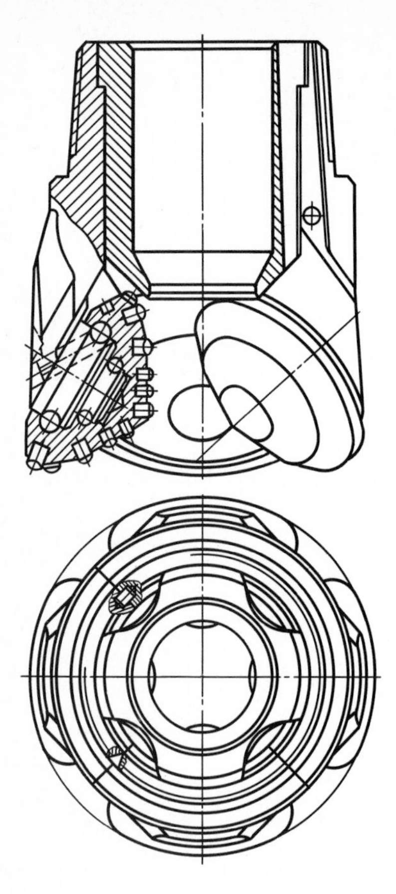

**Fig. 3.21.** KS-212.7/60TKZ (2V-K212.7/60TKZ) core head

an adapter, an inner section with three axles for the inner cutters, three plates and three outer sections with axles for the outer cutters.

The core head roller cutters are mounted on cylindrical hard-alloy faced journal bearings. The core head cutting structure consists of hard-alloy inserts with a wedge-shaped cutting surface.

*A KC-212.7/60 TKZ (25N-K212.7/60 TKZ) core head* differs from existent core heads for core barrels with retrievable core receivers. It has six cutters and axles of restrained design. Its roller cutters are mounted on journal bearings (Fig. 3.23). The core head is designed to core hard abrasive rocks interbedded with extremely hard rocks using the rotary method or downhole motors operating at a speed of not more than 350 rpm.

Core head roller cutters are mounted on hard-faced journal bearings. The outer roller cutters have a tapered cylindrical journal bearing, the journal bearings of the inner ones are cylindrical in form, their hard-faced axles being of restrained design. The core head cutting device consists of two types of hard-alloy insert. The inserts of the main rows have a wedge-shaped cutting surface, those of the rows which form a borehole and core a hemispherical cutting surface. This new design made it possible to use the annulus to best advantage and to place within clearance limits bigger roller cutters with more efficient

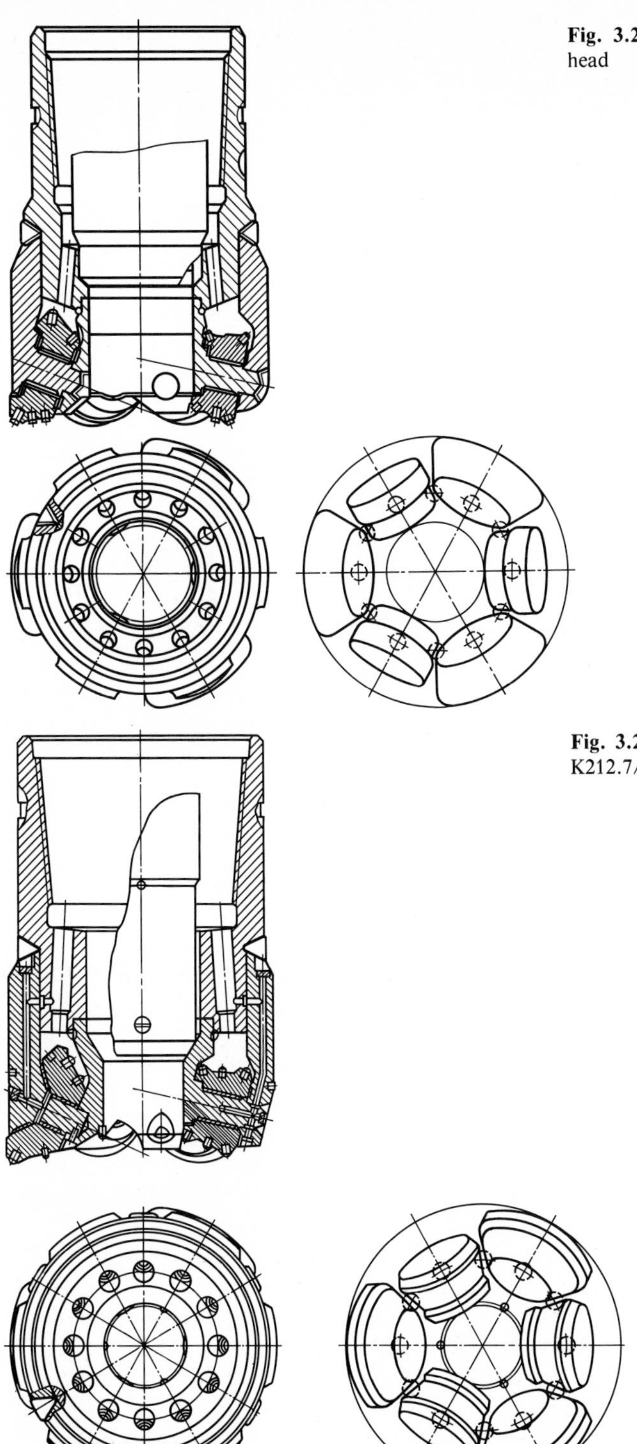

**Fig. 3.22.** 21N-D212.7/80TKZ core head

**Fig. 3.23.** KS-212.7/60 KZ-N (25N-K212.7/60TKZ) core head

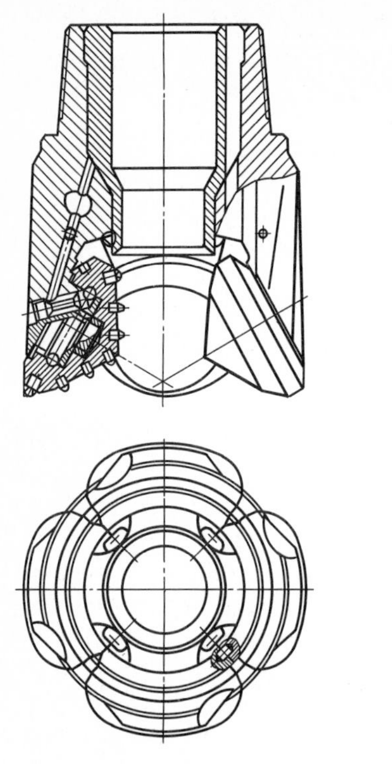

**Fig. 3.24.** KS-212.7/60 TKZ-NU core head

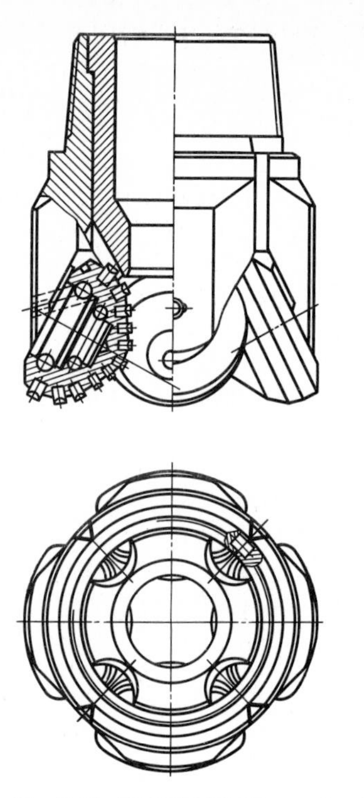

**Fig. 3.26.** 1N-K214/60 TZ core head

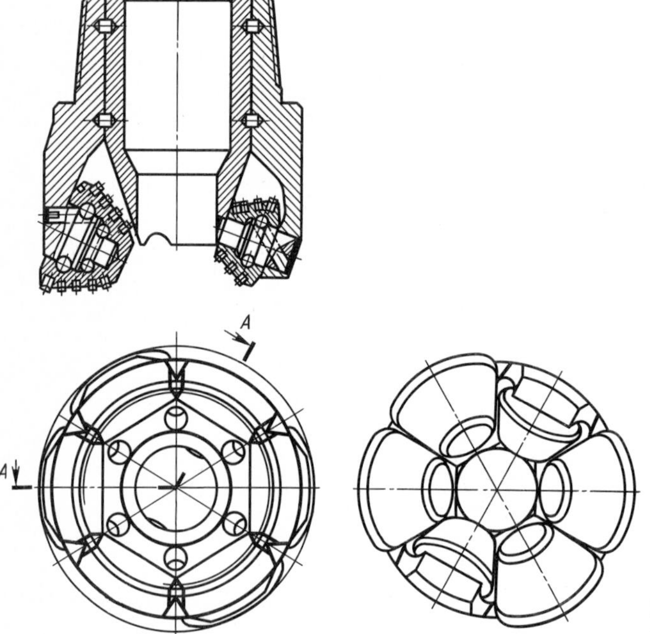


**Fig. 3.25.** 15N-K214/60 KZ core head

cutting structure and bearings of higher load-carrying capacity. In the core head, provision is made for cooling the bearings with the aid of a system of channels through which drilling mud is fed to the sliding surfaces.

*A KC-212.7/60 TKZ-NU core head* is designed for low rpm drilling in hard rocks interbedded with extremely hard rocks. It consists of four arms with cantilever axles. Roller cutters are mounted on these axles, using one locking ball bearing and two journal bearings. All four cutters simultaneously form a borehole and core (Fig. 3.24).

The inner space of the cutters is sealed with a cup. The arms are welded to each other to form the core head body. A core-receiving bushing is welded into the inner space of the body. The bushing has longitudinal channels on its outer surface to feed a drilling fluid to the bottom face. The cutting structure of the cutters consists of T-shaped hard-alloy inserts. The sealing and pressure-compensating system in the core head bearings is similar to those used in serially produced GAU-type roller-cutter drilling bits.

*A 15N-K214/60 KZ core head* is designed to drill extremely hard rocks. This is a core head with six truncated-cone roller cutters of true roll (Fig. 3.25). Four outer cutters cut the peripheral part of the bottom face, two inner opposite cutters cut the bottom face part adjacent to the core. The outer cutters are mounted on the arms, whose axles are directed from the periphery to the center, the inner ones mounted on inserted axles. The outer cutters are fitted on double-row ball bearings, the inner ones are fitted on single-row ball bearings and two radial end journal bearings. The cutting device consists of hard-alloy inserts. A central bushing mounted into the core head inner space provides a low level of core entry. The bottom hole is flushed through the channels arranged in the core head body. A drilling fluid is fed to the bottom face between the roller cutters.

*A 1N-K214/60 TZ core head* is designed to drill in abrasive rocks interbedded with extremely hard rocks. It was developed using the design of the 1V-K core heads for core barrels with retrievable core receivers as the base. The core head (Fig. 3.26) consists of four cone roller cutters mounted on the axles of four arm sections using antifriction bearings. The sections welded together form the core head body, the upper end of which represents a pin-type shank with a Z-161 tool joint thread. The bearing assembly of the cutters consists of a double-row ball bearing and two radial end journal bearings. A special bushing is fitted into the inner space of the core head to provide low core entry. Axles of the arms are inclined to the axis of the core head at an angle of 63°. All four cutters simultaneously take part in forming a borehole and core. The cutting structure of the cutters consists of wedge-shaped tungsten carbide inserts.

*A 20N-K214/60 K core head* is designed to drill in extremely hard abrasive rocks using core barrels with retrievable core receivers (Fig. 3.27). It differs from existing core heads by the axles of restrained design. The core head has six roller cutters; three of them form a borehole and three others form a core.

The core head body consists of an inner section with axles for core-forming cutters and three outer sections with axles for hole-forming cutters and plates. A bearing assembly of each hole-forming cutter consists of radial thrust and radial antifriction bearings and two end radial journal bearings. The bearing assembly

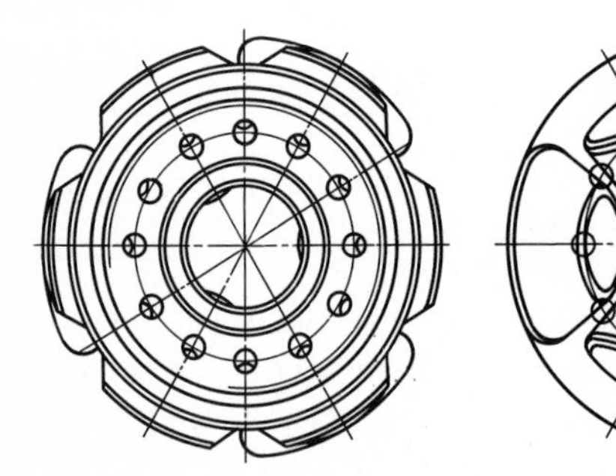

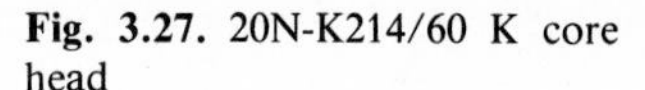

**Fig. 3.27.** 20N-K214/60 K core head

of each core-forming cutter consists of a double-row radial antifriction bearing and end journal bearings.

Drilling fluid is directed to the bottom face through twelve channels of round cross-section. The core head cutting device consists of hemispherical hard-alloy inserts.

*An ISM-214.3/60T core head* is designed to drill in hard and extremely hard highly abrasive rocks using the turbine drilling method (Fig. 3.28). It consists of a body and core-forming central part fitted into the body. The core head cutting structure consists of six segments and is furnished with cutting elements secured in segments by the brazing method. The cutting elements are mounted in the segments with the bottom face coverage factor of 0.35. The core head is 300 mm high, 214.3 mm in diameter. The core-receiving part diameter is 60 mm.

The end face of the core head is fitted out with hemispherical rock-cutting elements projecting 3 mm out of the body. The gauge surface is fitted out with flat-top wear-protecting elements flush-mounted into the body. The rock cutting and wear protecting elements are made of the ultrahard composite diamond-containing material *Slavutich*.

The weight on the head ranges from 80 to 100 kN, drilling fluid flow rate being equal to $25-30\,\mathrm{l\,s^{-1}}$. The well bottom is flushed with the fluid coming out of three mud-discharge ports 14 mm in diameter.

*A III-215.9TKZ-GNU three cone drill bit* is designed to drill hard and extremely hard abrasive rocks. It consists of three arm sections welded together.

Each arm is fitted out with a roller cutter mounted on the axle, bearing seal, pressure compensator and jet assembly with replaceable nozzles (Fig. 3.29). The

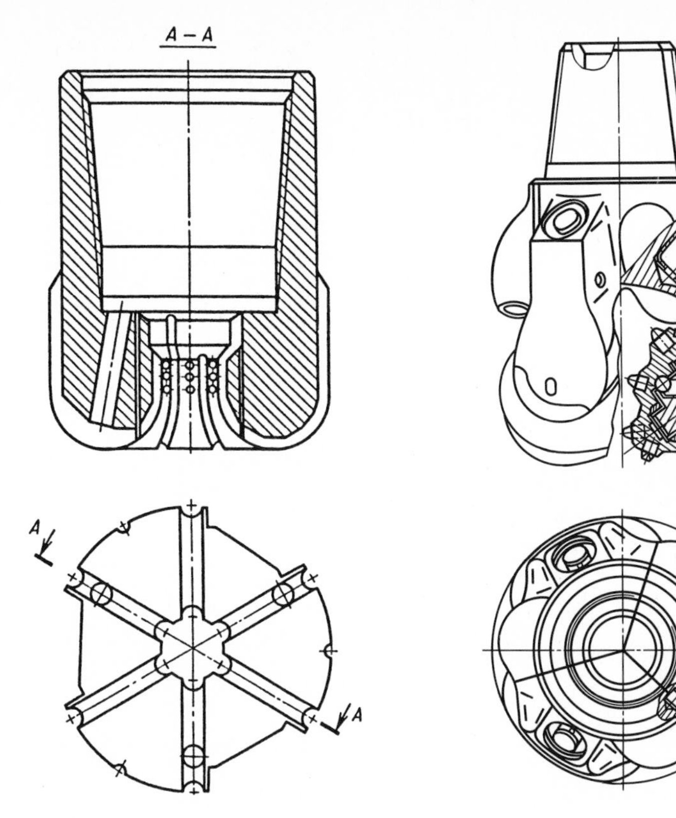


**Fig. 3.28.** ISM-214.3/60 T core head

**Fig. 3.29.** III-215.9 TKZ-GNU three cone drill bit

bit is supplied complete with three replaceable sealing and locking elements, the standard outlet diameter of the nozzles being 8.1, 9.5 and 11.1 or 12.7 mm.

Three arm sections with the cones mounted on their axles using bearings are welded together to make up the bit, the body of which is threaded to form a shank with a tool joint thread. The bit is an integral non-repairable product. Its rock-cutting structure consists of hard-alloy inserts pressed into the outer surfaces of the cones. Its cone-bearing assembly enables a free rotation of the cone mounted on it and consists of two antifriction bearings (radial and radial thrust ones) located at the root and in the middle of the axle, one radial journal bearing and one journal thrust bearing.

The bearing assembly is filled with a special grease and sealed with a rubber and metal cup. An elastic diaphragm, a working element of the compensator, enables equalization of pressure in the inner space of the bearing assembly and well-bottom pressure.

When using the bit, drilling variables should be as follows: the weight on the bit should not exceed 310 kN at 60 rpm and 220 kN at rpm higher than 60. The pressure differential should not exceed 15 MPa.

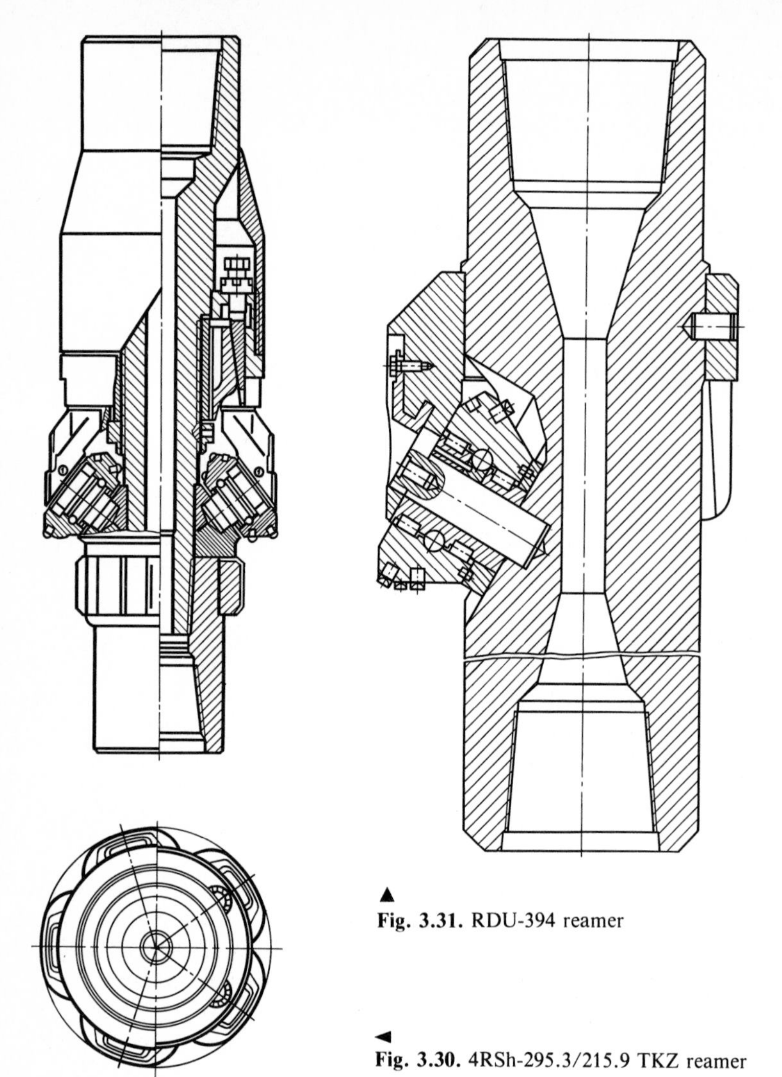

▲
**Fig. 3.31.** RDU-394 reamer

◄
**Fig. 3.30.** 4RSh-295.3/215.9 TKZ reamer

*A 4RS-295.3/215.9 TKZ reamer* consists of a body, split sleeve and thrust carrier welded to the body. Five rock cutting assemblies are mounted into the thrust carrier (Fig. 3.30). The rock cutting assembly consists of an arm and cone roller cutter. Shank parts of the arms are inserted into the recesses of the thrust sleeve and pressed to the body from the top through the outer tapered surfaces by a socket. Axle ends are inserted into the recesses in the thrust carrier. The weight on the reamer is transmitted from the body to the arms through intermediate half-rings located between the body and arm ends and through the socket welded to the body and arms. The torque is transmitted to the arms by star-like projections of the thrust sleeve.

The cone roller cutter rotates on one locking ball bearing and one roller bearing. The rows of cone cutting structure consists of alternately pressed-in

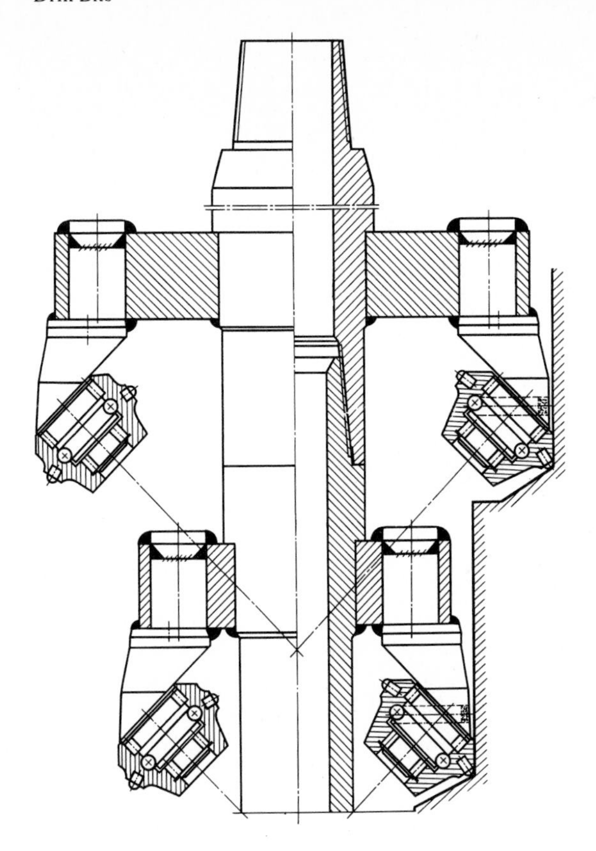

**Fig. 3.32.** RD-445/640 reamer

wedge-shaped and hemispherical hard-alloy inserts. To prevent the body from abrasive wear in the lower part of the thrust carrier, wings are welded with their thrust ends to the carrier. The wings are protected with hard-alloy inserts. To change worn-out rock-cutting assemblies, the welds on the socket and intermediate half-rings are removed and they are dismantled in succession. Then new rock-cutting assemblies fastened with a socket are welded to the body and together.

*Reamer technical data*

| | |
|---|---|
| Reamer size, mm | 295.3 |
| Minimum allowable hole size to be reamed, mm | 215.9 |
| Maximum reamer height, mm | 740.0 |
| Maximum cone height difference, mm | 1.0 |
| Maximum cone play in relation to thread axis | 1.5 |
| Connection thread (State Standard of the USSR 5286-75): | |
| Upper | Z-171 |
| Lower | Z-147 |
| Central bore diameter, mm | 50 |
| Allowable weight-on-reamer, kN | 300 |
| Maximum mass, kg | 170 |

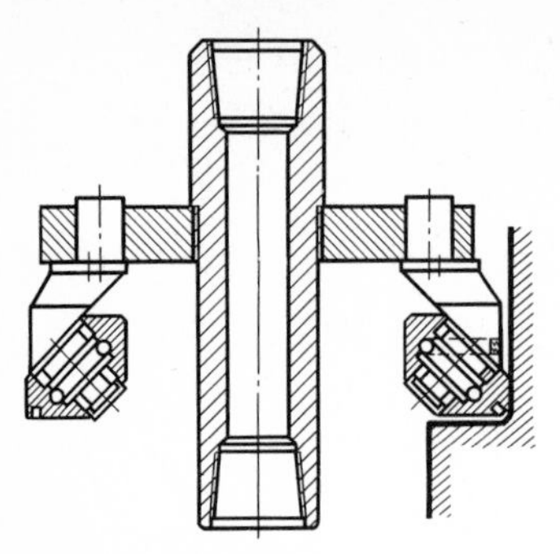

**Fig. 3.33.** RD-920 reamer

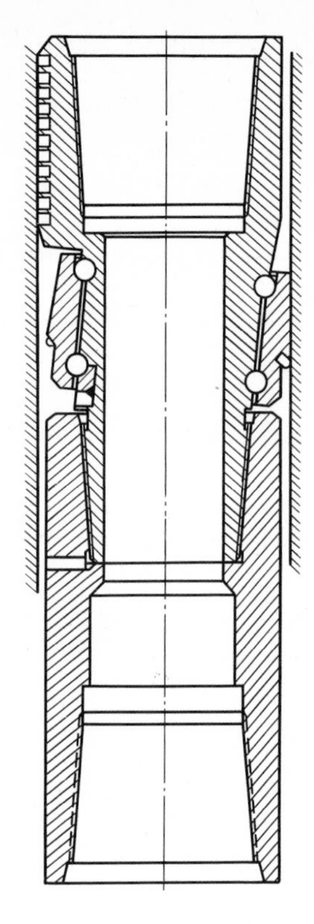

**Fig. 3.34.** ROP-9V reamer ►

*An RD(RDU) roller cutter reamer* is designed to ream a borehole in extremely hard abrasive rocks by the rotary method. It provides successive reaming of the borehole from 295 to 394 mm size (Fig. 3.31), from 394 to 445 mm size, from 445 to 640 mm size (Fig. 3.32) and from 640 to 920 mm size (Fig. 3.33). Simultaneous reaming of the borehole is possible with a set of several RD reamers of different sizes.

Among design features of the reamer are: a thread to connect a pilot bit or another reamer at the lower end; true roll (crushing) action of the rock cutting elements; replaceability of RDU-394 reamer roller cutters; unification of roller cutter assemblies in sections of the RD-445 and RD-920 reamers. Good results are obtained when using the reamers in extremely hard abrasive rocks, i.e. in cases where the reactive-turbine method of drilling large-sized holes is found to be ineffective due to the high rate of cutting structure wear caused by its sliding on the bottom face.

The roller cutter reamer is developed to be used in the upper conductor section of the SG-3 well and in the intermediate string section of 325 mm diameter (RDU-394).

*A ROP-9V one-cutter pilot reamer* is especially designed to be used in turbine downhole motor drilling (Fig. 3.34). The high rpm of roller cutters in turbine downhole motor drilling usually shortens conventional roller-cutter reamer life due to quick failure of bearing assemblies, which may result in roller cutter loss and a fishing job.

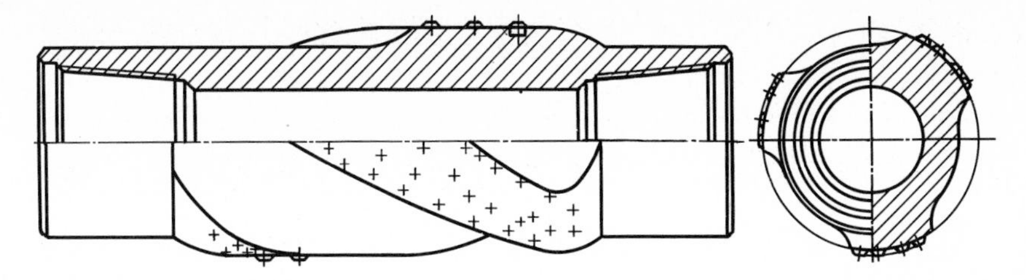

**Fig. 3.35.** CS-212.7 ST stabilizer

These disadvantages are avoided in the ROP-9V design, which has one roller cutter of big diameter, its axis being inclined at an angle of 5° to the reamer body axis. The angle and eccentricity of the roller cutter relative to the borehole axis are selected so that the cutter crushes a rock by true roll action while reaming.

The holes drilled with a 190 – 215-mm bit can be reamed with the ROP-9V reamer, but the latter is mainly used as a calibrator. While drilling, the ROP-9V reamer is aligned with the hole axis by means of hard-alloy inserts located in the reamer body on the side opposite to the maximum roller cutter offset. The speed of the roller cutter rotation about its axis is relatively low due to its large diameter. It provides better operating conditions for the roller cutter bearing assembly and long reamer mean time failure.

*A CS-212.7 ST stabilizer* is used as a part of a bottomhole assembly when coring in rocks of medium hardness interbedded with hard ones in ultradeep wells (Fig. 3.35). It consists of a box with three spiral blades protected with hard-alloy inserts. The ends of the blades are ground to a radius of 106 mm, which makes the form of blade ends very close to their natural wear form and extends their life.

*Technical data*

| | |
|---|---|
| Connection box thread | Z-147 |
| Stabilizer length, mm | 690 |
| Minimum ID, mm | 110 |

### 3.3.2 Coring Tools

Three coring tools and their modifications have been developed and used to core in the Kola well: a KTD4S-195-214/60 – 80 sectional core turbine downhole motor, KDM-195-214/60 core barrel and MAG-195-214/60 core barrel with core recovery by reverse circulation.

*A KTD4S-195-214/60 – 80 sectional core turbine downhole motor* is a downhole turbine motor with a hollow shaft along its full length. In addition to its main purpose (to transmit torque from the turbine and weight-on-core-bit from the drill collars) the shaft serves to accommodate a special core receiver. The latter consists of a core catcher, core receiving tube, vent valve, extension

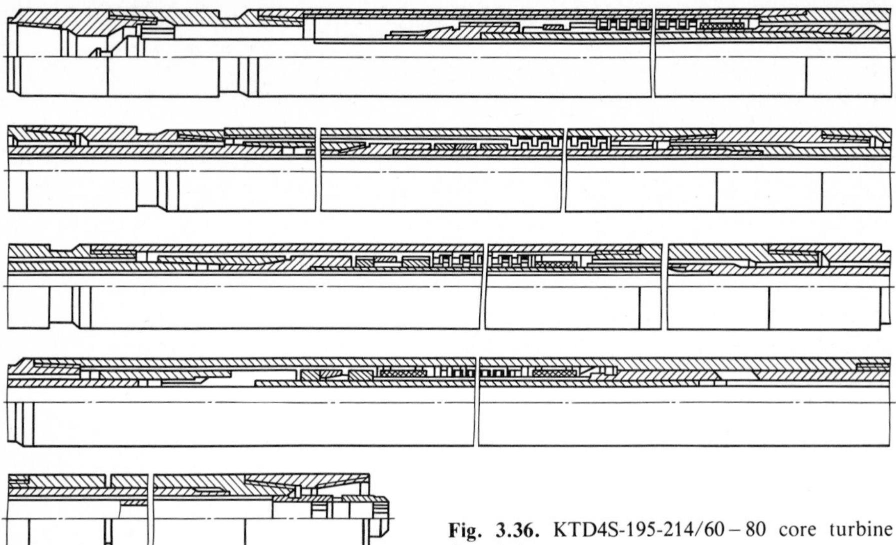

**Fig. 3.36.** KTD4S-195-214/60 – 80 core turbine downhole motor

tube and adjusting head. To fix the core receiver in the inner hollow of the rotating turbine shaft, a special bearing assembly is used upon which the core receiver is suspended by means of the adjusting head. The core receiver may be retrievable or non-retrievable.

*A KTD4S-195-214/60 – 80 three-section spindle core turbine downhole motor* (Fig. 3.36) is similar in design to a 3TSSh-195 three-section spindle turbine downhole motor. Provision is made in the design of the core turbine downhole motor to use a non-retrievable core receiver for 80 mm core or a retrievable core receiver for 60 mm core. All three sections of the motor are alike and have 105 axial turbine stages with blades cast to shape. An open rubber and metal or radial axial ball bearing is placed into a separate spindle connected to the bottom turbine section with cone splined clutches and a sub. The sections are connected to each other in the same way. A core receiver bearing is mounted in the upper sub which serves to connect the motor to the drill string. The retrievable core receiver is made according to the drawing given in Fig. 3.36.

The non-retrievable core receiver consists of hollow shafts of the turbine sections, spindle and a special tube screwed into the spindle extension. A spring-loaded adjusting assembly is connected to the tube. An assembly consisting of a spring finger and toggle-type core catchers is freely inserted into the adjusting assembly. In the hollow shaft of the upper section an ejector valve is mounted, which serves to avoid direct drilling fluid flow onto the core and to clean core-receiving space of drill cuttings.

Serially produced KS 212.7/60 ST and KS 212.7/60 TKZ core heads are used with the core turbine downhole motor fitted out with the retrievable core receiver. 6V-K-214/80 SZ, 21V-K 214/80 TKZ core heads and their modifications are used when the core turbine downhole motor is fitted out with the non-retrievable core receiver.

*Technical data on the KTD4S-195-214/60–80 core turbine downhole motor*

| | |
|---|---|
| Length, m | 25.92 |
| Number of turbine sections | 3 |
| Section length, m | 7.955 |
| Spindle length, m | 4.415 (4.922) |
| Number of turbine stages | 315 |
| Number of turbine stages in one section | 105 |
| Number of thrust bearing stages | 30 |
| Number of intermediate bearings | 11 |
| Number of intermediate bearings in one section | 3 |
| Number of intermediate bearings in spindle | 2 |
| Housing diameter, mm: | |
| Outside | 195 |
| Inside | 165 |
| Core receiver length, m | 25.835 |
| Outside diameter of core tube, mm: | |
| Retrievable | 83 |
| Non-retrievable | 105 |
| Inside diameter of core tube, mm: | |
| Retrievable | 70 |
| Non-retrievable | 86 |
| Connection threads: | |
| to drill string | Z-147 |
| to 60 mm ID core head | Z-161 |
| to 80 mm ID core head | MK150 × 6 × 1 : 8 |
| Core turbine downhole motor mass, kg | 4450 |

*Operating characteristics of the core turbine downhole motor*

| | |
|---|---|
| Drilling fluid flow rate, $l\,s^{-1}$ | 28–36 |
| Power, kW | 84–178 |
| rpm | 480–600 |
| Shaft torque, N · m | 1450–2380 |
| Pressure drop, MPa | 6.6–10.8 |

*A KDM-195-214/60 core barrel* is a set of tools connected to each other to core a vertical borehole in hard abrasive rocks. It is a universal assembly designed to be used both in rotary and downhole motor drilling methods (Fig. 3.37). In downhole motor drilling it is used as a suspension connected to the downhole motor shaft.

The assembly may include one, two or more sections. A core-receiving part of one section is 5.8 m long. The sections are connected to each other with cone clutches and Z-171 threads. Core receivers have telescoping connections to eliminate the necessity for their adjustment during the assembly of the sections. A set of the ROP-9V and TPS-9 calibrating reamers is installed between the upper sub and section. Each section consists of a housing 195 or 203 mm in diameter, an 83 mm OD core receiver is placed into the housing and suspended

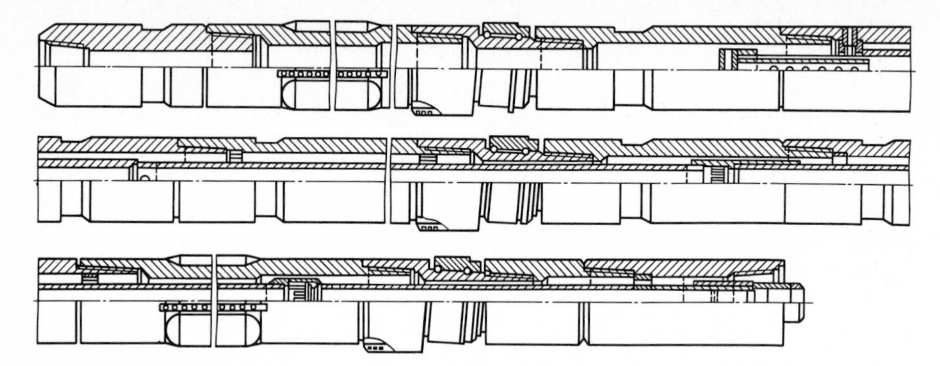

**Fig. 3.37.** KDM-195-214/60 core barrel

on the upper end face of it. The ROP-9V or TPS-9 calibrating reamers are installed at the lower end of the housing.

The core receiver is centered in the housing by means of a ring and furnished with a toggle-type core catcher. In the core barrel there is a special section of reduced length consisting of a PM-3-171/Z-161 sub and ROP-9V calibrating reamer. This section accommodates a short core receiver with a core catcher assembly mounted on a ball bearing. A vent valve or subhydraulically connecting core barrel inner space with the annulus may be mounted in the upper part of the core-receiving hollow of the core barrel.

Thus, the KDM-195-214/60 core barrel assembly differs from conventional coring tools used in deep drilling by having full-size centering devices (ROP-9V and TPS-9 reamers), floating core catchers mounted on the ball bearing and separated from the rotating core receiver, a forced core-receiving hollow drainage device by means of reverse or direct circulation. Additionally, optional core tubes with low friction core – tube contact may be supplied to complete the core barrel assembly.

*Technical data on the KDM-195-214/60 core barrel*

| | |
|---|---|
| Drilling fluid flow rate, l $s^{-1}$ | 20 – 40 |
| Allowable weight-on-head, kN | 160 – 200 |
| rpm | 60 – 600 |
| Connection threads: | |
| Upper | Z-117 |
| Lower | Z-161 |
| Diameter, mm, of: | |
| Core head | 212 – 214 |
| Calibrating reamer | 212 – 214 |
| Section housing | 195 |
| Core | 60 |
| Two-section core barrel length, m | 18.58 |
| Maximum core length (two-section barrel), m | 15.5 |

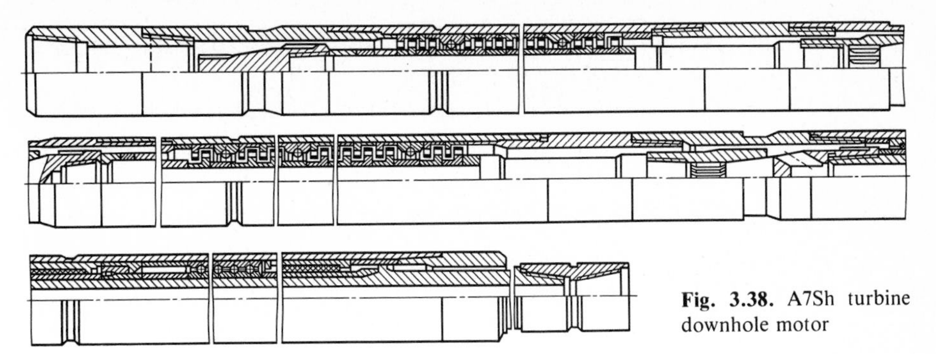

**Fig. 3.38.** A7Sh turbine downhole motor

### 3.3.3 Downhole Motors

Great efforts in research and development to create the appropriate downhole motors were necessary to put into practice the turbine downhole motor method of coring chosen for the superdeep Kola well drilling at the preliminary research stage of the project. The main efforts to refine conventional technique were aimed at reduction of downhole motor shaft rpm while preserving high torque motor characteristic, improvement of thermal stability of bearing and sealing downhole motor elements, achievement of failure-free performance, longer service life and reliability of the motor.

Considerable advances have been made in solving the problems posed. This is evidenced by drilling the well to a record depth at temperatures up to 200 °C with the downhole motors made in the USSR.

The super deep Kola well was drilled to a depth of 8000 m with high torque spindleless A7N4S and spindle A7S turbine downhole motors with inclined pressure curve. Low-rpm high-torque A7GTSh and TRM-195 turbine downhole motors furnished with turbine and spindle sections of the A7Sh turbine motor were tested in the 8028 – 8914 m depth interval.

*The A7Sh (Fig. 3.38) turbine downhole motor* has a multistage sectionalized turbine and multiple row thrust-radial bearing mounted in a spindle section. One hundred and eighteen turbine stages and five radial ball bearings are mounted in each turbine section of the A7Sh downhole motor. Standard radial rubber and metal plain bearings may be installed, if necessary, instead of the ball ones. The multiple row thrust-radial ball bearing takes up the axial load. The radial rubber and metal bearings and sealing device take up the pressure drop across the core barrel connected to the motor shaft.

The A7N4S and A7Sh turbine downhole motors can be sectionalized to reduce their rpm and increase their torque. Three-section modification of the A7Sh downhole motor (3A7Sh type) has lower working rpm (500 – 600), but drill bits and core heads with sealed bearings cannot be used with it. Core recovery is also reduced when using the 3A7Sh turbine motor. To eliminate this disadvantage a system of hydrodynamic braking may be used, i.e. A7GTSh turbine downhole motors.

*Three-section A7GTSh turbine downhole motors* do not differ in their design from the A7Sh downhole motors. A certain number of turbine stages are substi-

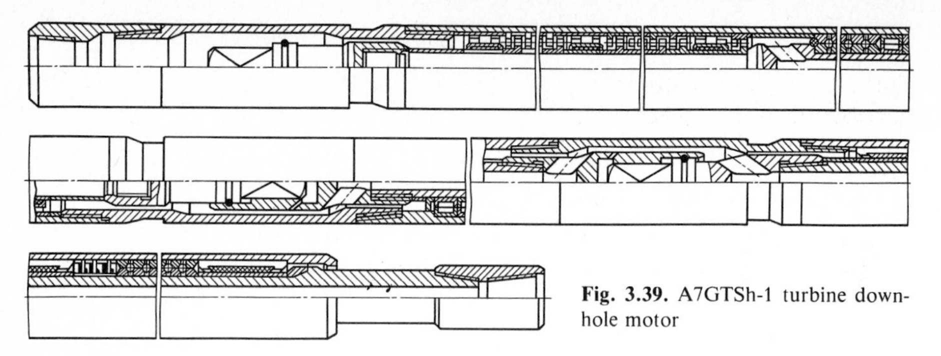

**Fig. 3.39.** A7GTSh-1 turbine downhole motor

tuted for the hydrodynamic braking stages. The flow of the drilling fluid through the rotating stages of hydrodynamic braking sets up a counteracting torque at the turbine shaft. The more the shaft's rpm, the higher is the counteracting torque. This results in starting and operating turbine motor rpm reduction, the torque curve of the motor falling steeply. The braking and driving torques are dependent on the number of the motor turbine stages. The load carrying capacity of the downhole motor is increased due to the steeply falling torque curve.

To ascertain the performance characteristics of the turbine downhole motor with hydrodynamic braking necessary for drilling under specific geological conditions, an appropriate ratio between the number of turbine stages and hydrodynamic brake stages should be ensured. Operating rpm of the downhole motor is defined from the following formula:

$$n_2 = 0.9\, n_1 \frac{z_t}{z_t + z_{hb}} ,$$

where:

$n_1$ = operating rpm of A-type turbine downhole motor;
$z_t$ = number of turbine stages;
$z_{hb}$ = number of hydrodynamic brake stages.

In addition to rpm reduction, the design of an A7GTS-1 downhole motor enables the axial vibration generated by the rock cutting tool to be reduced, which results in an increase of the tool life and footage per run.

*The A7GTSh-1 turbine downhole motor* (Fig. 3.39) is designed to drill at 250 – 300 rpm.

This motor embodies the following design features: high-torque low-rpm turbine sections of low volume capacity with by-pass ports, two systems of hydrodynamic braking full-flow system mounted in succession with the turbine stages and restricted-flow system mounted under the section thrust bearing; an axial vibration suppression system which consists of hydraulic dampers made integral with clutches of square cross-section connecting turbine and spindle sections; a multiple-stage system of labyrinth seals which ensures high pressure drops across the core barrel and core head; an independent mounting of turbine section shafts on ball bearings which substantially increases overhaul period of the downhole motor.

*Three-section downhole motor technical data*

| Turbine downhole motor type | 3A7Sh | 3A7GTSh |
|---|---|---|
| Number of turbine stages | 354 | 135 |
| Number of hydrobrake stages | – | 270 |
| Flow rate, l $s^{-1}$ | 30 | 30 |
| rpm: | | |
| Operating | 520 | 200 |
| Starting | 1200 | 400 |
| Torque, N · m | | |
| Operating | 2850 | 1700 |
| Braking | 5700 | 3400 |
| Pressure drop, MPa: | | |
| Braking | 7.9 | 6.5 |
| Operating | 12.2 | 7.0 |
| Starting | 18.8 | 7.5 |

*Notes.*

1. Performance curves were recorded using water as a working medium.
2. 3A7GTSh performance curves were recorded at the Kola expedition test stand. The motor had a A7PZ turbine.

The performance curves and operating characteristics of these turbine downhole motors correspond to the super deep drilling requirements. Antifriction bearings ensure an easy start and trouble-free operation of the motor during flushing, reaming and drilling, which is impossible when using turbine downhole motors with rubber-metal plain bearings at high bottomhole temperatures in highly abrasive mud environment. The A and AGT turbine downhole motors can be successfully used at mud temperatures up to 200 °C. They are also suitable for drilling with muds of high oil or other hydrocarbon content.

The A turbine downhole motor develops an increased torque and, which is of particular importance for superdeep drilling, has an inclined pressure curve, i.e. the pressure drop across its turbine is maximum at no-load operation and gradually decreases until it stalls. The changes of pressure depending on the motor shaft rpm make it possible to control the motor operation which is very important from the technological point of view. These downhole motors are equipped with the clutches of square cross-section between the motor sections instead of cone splined clutches, which allows for relative axial displacement of the shafts of the turbine and spindle sections (Fig. 3.40).

Such a substitution has some important advantages. The turbine and spindle sections can be replaced directly on the drilling site. The section which has to be repaired or replaced due to necessary change in downhole motor rpm can be removed above the rotary table and replaced without complete disassembly of the downhole motor. The turbine downhole motor overhaul period is increased. When the upper limit of the axial spindle bearing play is reached, an adjustment ring is installed between the lower turbine section and spindle section shafts, its height being equal to the value of the axial bearing wear. After this, the rotors return to their original position relative to the stators and the spindle bearing may be used again.

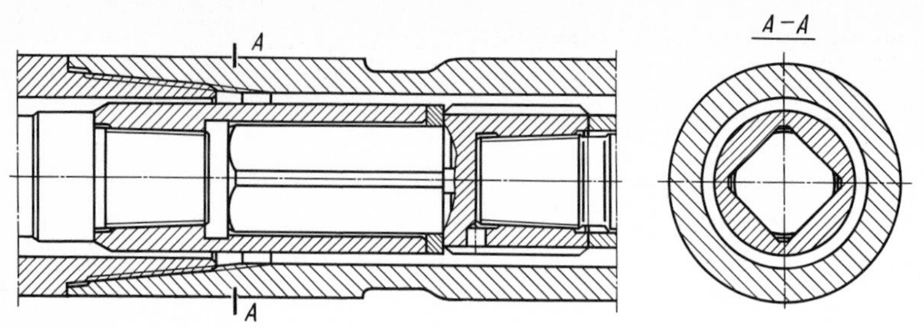


**Fig. 3.40.** Connection of shafts of turbine and spindle sections

Working hours of adjustment of the downhole motor sections of the turbine are reduced. The adjustment is attained by inserting adjusting rings into the half-couplings, the size of the rings depending on the protrusion or depth of the half-couplings relative to the end face of the housing parts. The serviceability of the turbine downhole motor housings and subs is increased by cutting the number of the screwed threads by half. The latter becomes possible due to elimination of the turbine and spindle section adjustment.

The strict demands imposed upon the turbine downhole motor used in combination with long coring tools causing higher loads necessitated the substitution of the Z-117 spindle shaft connecting the thread of the A7Sh motor to the Z-147 thread. Some complications in the assembly of the spindle are more than offset by the higher reliability of this troublesome connection.

*A D2-172M screw downhole motor* falls into the category of positive displacement (hydrostatic) machines. The screw downhole motors have some advantages over the other hydraulic downhole motors used to drive the drill bit. Low rpm and high torque developed by its shaft enable higher footage per run to be attained. Low pressure drop across the motor makes it possible to use it in combination with jet drill bits. The operation of the motor and weight-on-bit can be observed by the slush pump discharge pressure. Repairs and operation of the motor are much easier due to the small number of parts of the motor. All these advantages permit the screw motors to be successfully used in deep well drilling, especially in combination with low rpm sealed-bearing and jet drill bits.

The screw downhole motor is designed as a planetary and rotary hydraulic machine of volumetric type with internal helical gearing of working parts. The main parts of the motor are a rotor and a stator (Fig. 3.41). The stator 1 is a steel housing with threaded ends. The bore of the housing is rubber plated, the vulcanized plating having helical left-hand blades on its inner surface.

The steel rotor has outer helical left-hand blades, the number of which is that of the stator less one. The rotor axis is offset relative the stator axis, the eccentricity being equal to a half of the blade height. The pitches of the helical surfaces of the rotor and stator are proportional to the number of blades of these parts. Such a design of rotor and stator blades provides a constant contact of the parts and formation of separated working chambers closed within the length of the stator pitch.

The drilling fluid supplied into the motor by the drilling rig slush pumps can pass to the bit only if the motor rotor rotates within the stator plating revolving

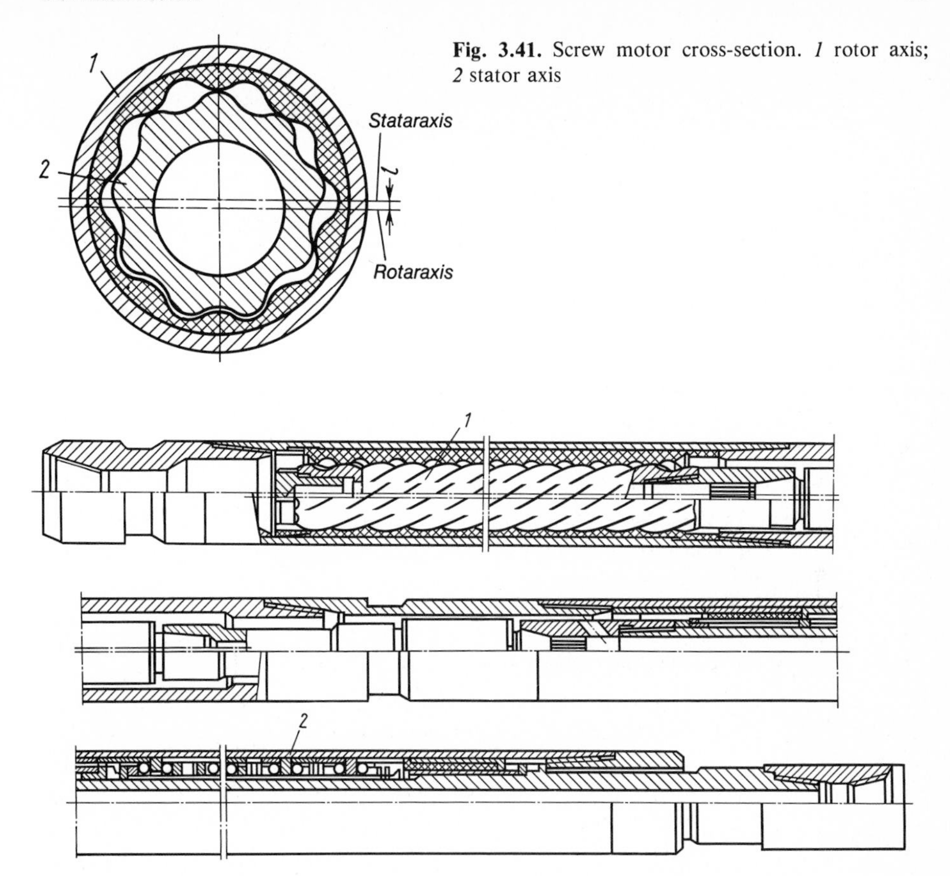


**Fig. 3.41.** Screw motor cross-section. *1* rotor axis; *2* stator axis

**Fig. 3.42.** D2-172M screw downhole motor. *1* screw (motor); *2* spindle section

around its blades under the action of unbalanced hydraulic forces. The rotor thereby executes a planetary motion, its axis rotating counter-clockwise around the stator axis (a transportation motion) and the rotor itself rotating clockwise (an absolute motion). Because of the difference in the number of blades between the rotor and stator, the transportation motion is reduced into the absolute one with the gear ratio equal to the number of rotor blades, which results in low rpm and high torque of the downhole motor.

The rotor planetary motion is transformed into a coaxial rotation of a spindle shaft driven by a cardan shaft transmitting the torque and axial load from the rotor. The cardan shaft consists of two double-toothed joints filled with grease and an intermediate tube. The Cardan joints are connected with the tube by means of mated cones in the motor, and to the rotor and spindle sleeves with cone splined couplings. The motor spindle includes an axial multistage anti-friction bearing and radial rubber and metal bearings.

Figure 3.42 shows the general arrangement of the D2-172 motor. The motor has the usual design of the downhole motors and includes: a screw section and a spindle section. The D2-172M is equipped with a ShShOI-172 spindle with ball

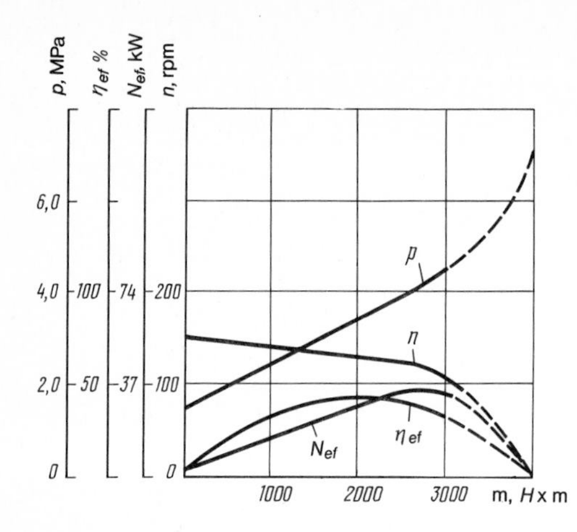


**Fig. 3.43.** D2-172M screw downhole motor characteristics. $p$ pressure; $n$ rotation speed; $N_{ef}$ power; $\eta_{ef}$ efficiency (washing fluid-water, flow rate Q = 23 l s$^{-1}$)

multistage damped bearing, a variety of a double-row thrust-radial ball bearing. The motor is equipped with a fill-and-dump valve to fill and dump the drill string during round trips.

*Technical data on the screw downhole motor*

| | |
|---|---|
| Motor displacement volume, l | 9 |
| Mud flow rate, l s$^{-1}$ | 23 – 36 |
| Shaft rpm, at maximum power | 115 – 220 |
| Torque, at maximum power, N · m | 2900 – 4150 |
| Maximum power, kW | 33 – 92 |
| Pressure drop across motor, at maximum power, MPa | 4.5 – 6.0 |
| Outside diameter, mm | 172 |
| Length, mm | 6875 |
| Mass, kg | 878 |

*Note.* The performance data were recorded during tests at maximum power using water as a drilling fluid.

Figure 3.43 shows the main operation variables as functions of the torque developed by the motor. It should be noted that as wear of the working parts proceeds and their play increases, the performance curves become "quiet", i.e. the $n$-$M$ relationship approaches a linear one, $n$ and $M$ decrease. The motor preserves its serviceability until the torque developed by it exceeds that necessary to rotate the bit at the given axial load. To determine performance characteristics at another drilling fluid flow rate it is possible to use the approximate relationships: at the maximum motor power its rpm, torque and pressure drop vary directly with the flow rate, and the motor power varies in proportion to the square of the flow rate.

The D2-172M screw downhole motor is designed to drill vertical and directional wells with 190 – 215.9 mm drill bits (preferably with low rpm sealed-bearing and jet bits), to core in combination with a core barrel using water, mud

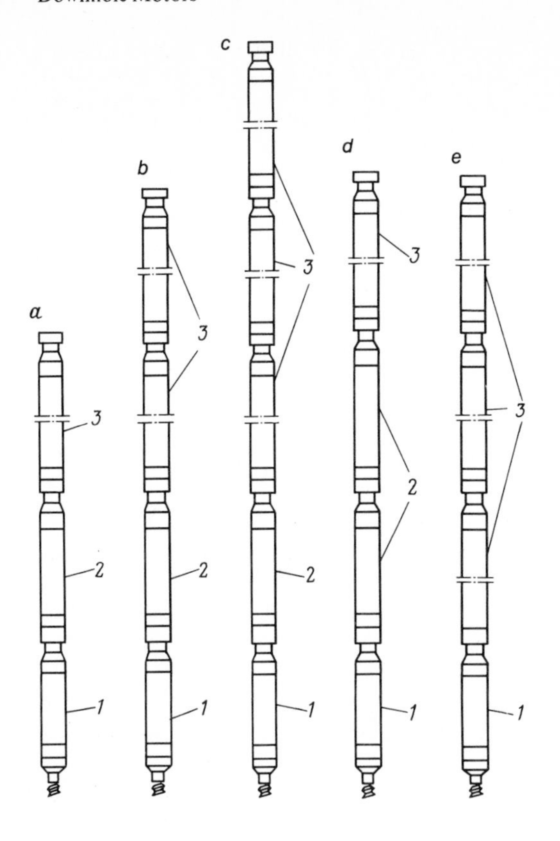


**Fig. 3.44.** Different downhole assemblies consisting of motor and RM-195 reduction gear section. *1* spindle section; *2* reduction gear section; *3* turbine section

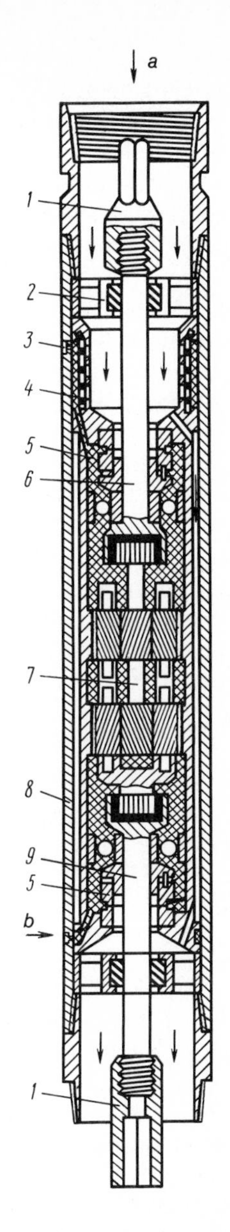


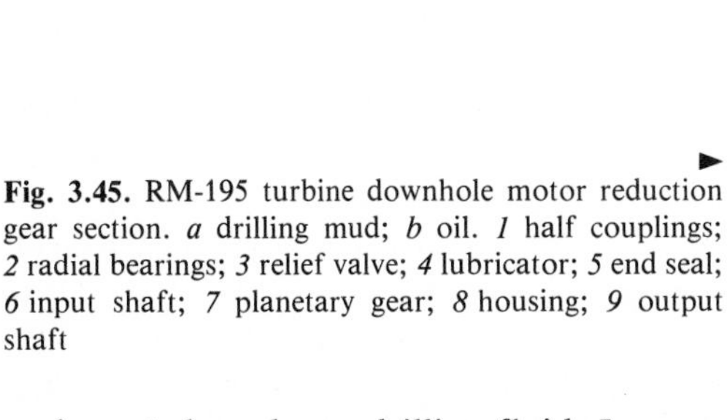

**Fig. 3.45.** RM-195 turbine downhole motor reduction gear section. *a* drilling mud; *b* oil. *1* half couplings; *2* radial bearings; *3* relief valve; *4* lubricator; *5* end seal; *6* input shaft; *7* planetary gear; *8* housing; *9* output shaft

and aerated mud as a drilling fluid. In exceptional cases it is allowed to use the motor in combination with bits up to 269 mm in diameter.

*A TRM-195 turbine downhole motor* with a reduction gear section ensures rock cutting tool rpm in the range of 80 – 250. It becomes possible due to the selective method of motor assembly using reduction gear sections. The required performance characteristics are obtained by means of combining only three main assemblies: turbine section, reduction gear section and spindle section (Fig. 3.44). The turbine downhole motor is assembled directly on the drilling site using Z-171 tool joint threads. Depending on drilling conditions, the motor is assembled of one or several turbine sections, one or two reduction gear sections. If the reduction gear section is not necessary, for example when using diamond drill bits, the motor is assembled only of standard turbine sections and the spindle section.

The turbine motor reduction gear (Fig. 3.45) is a changeable assembly placed in a housing of 195 mm outside diameter (OD). It consists of a planetary gear, output and input shafts, radial bearings, oil protection system and relief valve. Double-row planetary Novikov helical gear is capable of transmitting torque in excess of 10000 N · m and distinguished for its wear resistance and strength in spite of the fact that it has limited radial dimensions. Of great importance is reduction gearing protection from overloading and load variation incorporated in its design.

The reduction gearing is connected to the turbine and spindle section shafts with the input and output shafts mounted on bearings through half-couplings. The oil protection system consists of an end seal and lubricator. The lubricator is designed to compensate for oil losses during operation and to stabilize pressure in the oil-filled reduction gear chamber. The design and production technology of the seals ensure their reliable operation in abrasive and corrosive medium, under intensive axial and radial vibrations and drilling fluid pressure surges. Maximum allowable operating temperature of the reduction gear section is brought up to a level of 250 °C.

The main characteristics of turbine and reduction gear combinations are given in the Table 3.12. Drill bits of 212 – 270 mm in size are recommended for use with the TRM-195 motor.

### 3.3.4 High-Strength Drill Pipes

*LBTVK-147 internal upset light-alloy drill pipes* with tapered stabilizing surfaces and ZLK-178 steel tool joints for them are designed to drill deep and superdeep wells using the turbine downhole motor and rotary method. A pipe tool joint connection consisting of a TT-138 taper buttress thread in combination with smooth taper stabilizing surfaces eliminates thread shear and fatigue failure, which considerably increases light-alloy drill pipe reliability and useful life. Interference fit of taper stabilizing surfaces and metal-to-metal contact of inner shoulders make pipe tool joint connection very tight.

A buttress thread with an 1 : 32 taper, 5.08 pitch and 30° profile angle developed by the All-Union Research Institute for Drilling Technique (VNIIBT) is used in this connection. The use of the thread mating on its inside diameter and one profile side enables the connection to be assembled precisely with a given fit in the thread. The stabilizing surface also has a 1 : 32 taper. The fatigue limit of the LBTVK pipes is substantially higher than that of standard LBT pipes.

The ZLK-178 tool joints are screwed on LBTVK-147 pipes with shrink fit or in a "cold" condition. When shrink fit is used it is necessary to take measures not to overheat the pipe in the zone of contact with the tool joint (it is provided by cooling the pipe on the inner side) to maintain the initial mechanical properties of

**Table 3.12.** Main characteristics of turbine and reduction gear combinations

| Downhole motor assembly (see Fig. 3.44) | Number of | | | Assembly length, m | Assembly mass, kg | Gear ratio | Drilling fluid flow rate, $l\ s^{-1}$ | rpm | Torque, $N \cdot m$ | Maximum power, kW | Pressure drop, MPa | Efficiency |
|---|---|---|---|---|---|---|---|---|---|---|---|---|
| | Turbine sections | Turbine stages | Reduction gear sections | | | | | | | | | |
| a | 1 | 109 | 1 | 13.7 | 2600 | 3.69 | 24 – 28 | 150 – 175 | 2200 – 3020 | 35 – 55 | 2.2 – 2.9 | 64 |
| b | 2 | 218 | 1 | 21.2 | 4000 | 3.69 | 20 – 28 | 125 – 175 | 3080 – 6040 | 41 – 110 | 3.1 – 6.1 | 64 |
| c | 3 | 327 | 1 | 28.6 | 5400 | 3.69 | 18 – 20 | 115 – 125 | 3740 – 4620 | 45 – 61 | 3.7 – 4.6 | 64 |
| d | 1 | 109 | 2 | 17.2 | 3260 | 11.76 | 20 – 28 | 35 – 48 | 5230 – 10250 | 20 – 52 | 1.6 – 3.2 | 59 |
| e | 3 | 327 | – | 25.9 | 4740 | – | 24 – 30 | 560 – 700 | 1960 – 3060 | 115 – 225 | 6.5 – 10.0 | 70 |

*Note.* Power characteristics of the turbine downhole motor are given for a drilling fluid of 1 g $cm^{-3}$ density.

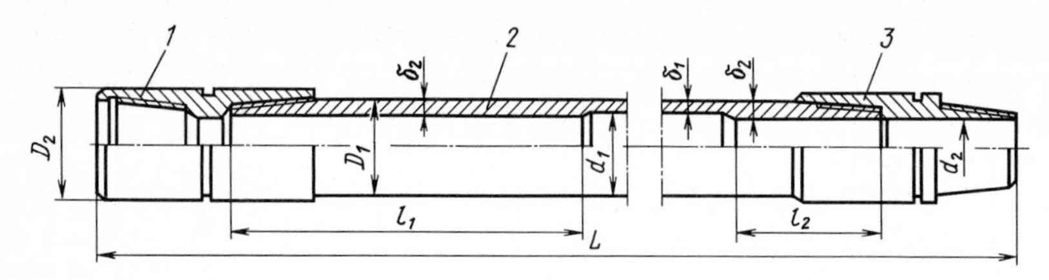


**Fig. 3.46.** LBTVK-147 pipe with ZLK-178 tool joint. *1* box; *2* pipe; *3* pin

the aluminium alloys. The design of the LBTVK-147 pipe with the ZLK-178 tool joint is given in the Fig. 3.46.

*LBTVK-147 pipe dimensions and characteristics*

| | | | | |
|---|---|---|---|---|
| Wall thickness $\delta_1$ | 11 | 13 | 15 | 17 |
| Pipe diameter, mm | | | | |
| Outside, $D_1$ | 147 | 147 | 147 | 147 |
| Inside, $d_1$ | 125 | 121 | 117 | 113 |
| Upset portion wall thickness, $\delta_2$, mm | 17 | 20 | 22 | 24 |
| Nominal pipe length, *m*: | | | | |
| without tool joint | 12 | 12 | 12 | 12 |
| with tool joint, *L* | 12.4 | 12.4 | 12.4 | 12.4 |
| Upset portion length, mm: | | | | |
| Box side, $l_1$ | 1750 | 1750 | 1750 | 1750 |
| Pin side, $l_2$ | 250 | 250 | 250 | 250 |
| Cross-section area, cm: | | | | |
| Pipe body | 47.0 | 54.7 | 62.2 | 69.4 |
| Pipe bore | 122.7 | 114.9 | 107.4 | 100.2 |
| Upset portion bore | 100.2 | 89.8 | 83.3 | 76.9 |
| Mass of 1 m of pipe, kg: | | | | |
| – including upset portion mass | 13.9 | 16.1 | 18.2 | 20.1 |
| – including upset portion and tool joint mass | 18.8 | 21.0 | 23.1 | 25.0 |

*ZLK-178 tool joint dimensions*

| | |
|---|---|
| Diameter, mm: | |
| Outside, $D_2$ | 178 |
| Inside, $d_2$ | 101 |
| Assembled tool joint length, mm | 573 |
| Thread: | |
| Tool joint | Z-147 |
| Pipe | TT-138 × 5.08 × 1 : 32 |
| Tool joint mass, kg | 61 |

Standard light-weight drill pipes with ZL-172 light-weight tool joints were used to drill SG-3 well to depth of 8000 m.

LBTVK-147 light-weight drill pipes made of different aluminium alloys (D-16T, 01953 and AK4-1) with steel ZLK-178 tool joints (which were earlier called ZLK-172 tool joints) made of 40XN and 40XM1FA steels are used at greater depths in SG-3 well. The use of different aluminium alloys to make drill pipes is dictated by different operating conditions in the well intervals, i.e. by acting stresses and downhole temperatures.

On the basis of the studies which had been carried out earlier, the following alloys were selected and developed: (1) DT16T is medium-strength alloy of Al-Cu-Mg group produced in the USSR; it is thermally hardened and retains corrosion and heat resistance up to a temperature of 150 °C; the pipes made of this alloy can be used to a depth of 9500 m; (2) 01953 is a high-strength alloy of Al-Zn-Mg-Cu group specially developed to make drill pipes and mastered by the industry; it is thermally hardened, has a medium corrosion resistance and retains heat resistance up to a temperature of 100 °C; the pipes made of this alloy can be used to a depth of 6000 m; (3) AK4-1 is a medium-strength alloy of Al-Cu-Mg-Fe-Ni group widely used to make pressure forged products; it is thermally hardened, has medium corrosion and heat resistance up to a temperature of 200 °C; pipes made of this alloy can be used to a depth of 13,000 m.

*Main physical and mechanical properties of drill pipe materials*

| Alloy | D16T | 01953 [a] | AK4-1 [b] |
|---|---|---|---|
| Yield strength, $\sigma_T$, MPa, no less than | 330 | 490 | 280 |
| Tensile strength, $\sigma_B$, MPa, no less than | 450 | 540 | 400 |
| Elongation, $\delta_0$, % | 11 | 7 | 12 |
| Reduction of area, $\psi$, % | 20 | 15 | 26 |
| Brinell hardness number | 120 | 150 | 130 |
| Density, $\varrho$, g cm$^{-3}$ | 2.8 | 2.8 | 2.8 |
| Modulus of elasticity, MPa: | | | |
| modulus of elongation, $E$ | $0.72 \times 10^5$ | $0.70 \times 10^5$ | $0.73 \times 10^5$ |
| shear modulus, $G$ | $0.26 \times 10^5$ | $0.275 \times 10^5$ | $0.275 \times 10^5$ |
| Poisson ratio | 0.33 | 0.31 | 0.31 |
| Coefficient of linear thermal expansion, $\alpha$, 1/°C | $22.5 \times 10^{-6}$ | $23.8 \times 10^{-6}$ | $23.8 \times 10^{-6}$ |

[a] Physical and mechanical properties of the 01953 alloy are given according to STU-94-4-68.
[b] Physical and mechanical properties of the AK4-1 alloy are given according to the data from the handbook *Aluminium alloys*, Metallurgia, 1972.

*Mechanical properties of ZLK-178 tool joint materials*

| Steel grade | 40XN | 40XM1FA |
|---|---|---|
| Yield strength, $\sigma_T$, MPa, no less than | 750 | 800 |
| Tensile strength, $\sigma_B$, MPa, no less than | 900 | 900 |
| Elongation, $\delta_0$, %, no less than | 10 | 14 |
| Reduction of area, $\psi$, %, no less than | 45 | 50 |
| Impact strength, $10^5$ J m$^{-2}$ | 7 | 8 |
| Brinell hardness number | 285 – 341 | 285 – 363 |

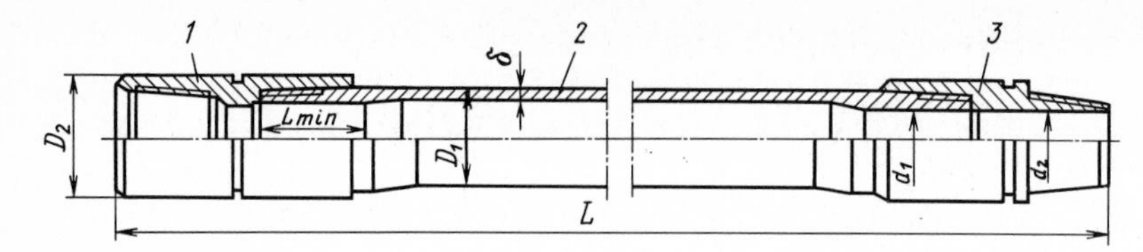


**Fig. 3.47.** TBVK-140 pipe with ZShK-178 tool joint. *1* box; *2* pipe; *3* pin

TBVK-140 internal upset steel drill pipes with tapered stabilizing surfaces and ZShK-178 tool joints are designed to drill deep and ultra-deep wells under severe and troublesome conditions using rotary and turbine downhole motor methods. The pipe-tool joint connection consisting of a taper buttress thread in combination with smooth taper surfaces eliminates its fatigue failures. Interference fit of tapered stabilizing surfaces and metal-to-metal contact of inner shoulders make the pipe-tool joint connection very tight.

The tool joints are screwed onto the drill pipes with shrink fit at a temperature up to 400° – 450 °C. A buttress thread with an 1 : 32 taper, 5.08 pitch and 30° profile angle developed by the VNIIBT is used. The use of the thread mating on its inside diameter and one profile side enables the connection to be assembled precisely with a given fit in the thread. The stabilizing surface also has an 1 : 32 taper. The fatigue limit of the TBVK pipes is much higher than that of conventional pipes.

The design of the TBVK-140 with the ZShK-178 tool joint is given in the Fig. 3.47.

*TBVK-140*[a] *drill pipe dimensions*

| | | | | |
|---|---|---|---|---|
| Wall thickness, mm | 10 | 11 | 12 | 13.5 |
| Upset end inside diameter, $d_1$, mm | 100 | 100 | 100 | 98 |
| Upset end length, $L_{min}$, mm | 155 | 155 | 160 | 160 |
| Standard | All-Union Standard 631-75 | | TU 14-3-1002-81 | |

[a] *T* pipe; *B* drill; *V* with internally upset ends; *K* with tapered stabilizing surfaces; 140 size in mm.

*ZShK-178 tool joint dimensions*

| | |
|---|---|
| Diameter, mm | |
| Outside, $D_2$ | 178 |
| Inside, $d_2$ | 101 |
| Assembled joint length, mm | 573 |
| Thread: | |
| Tool joint | Z-147 |
| Pipe | TT-132 × 5.08 × 1 : 32 |
| Mass, kg | 61 |

The high-strength TBVK-140 drill pipes made of 30XGCNM steel (TU 14-3-1002-81) and the high-strength ZSK-178 tool joints made of 40XN steel (All-Union Standard 4543-71) and 40XM1FA steel (TU 14-1-2634-78) were used in drilling of the SG-3 well.

*Mechanical properties of 30XGCNM steel*

| | |
|---|---|
| Yield strength, $\sigma_T$, MPa, no less than | 900 |
| Tensile strength, $\sigma_B$, MPa, no less than | 1000 |
| Elongation, $\delta_0$, %, no less than | 12 |
| Reduction of area, $\psi$, %, no less than | 40 |
| Impact strength, $10^5$ J m$^{-2}$, no less than | 3 |
| Brinell hardness number | – |

At present TBVK-140 drill pipes of P grade with 13.5, 12 and 11 mm wall thickness in combination with the ZShK-178 tool joints made of 40XM1FA steel are used to drill the SG-3 well.

### 3.3.5 Drilling Fluids

The 0 – 81 m interval of the UD-3 well was drilled using clay drilling mud treated with CMC-350 and soda ash. The mud had a bentonite content of 10 – 15% and the following properties:

| | |
|---|---|
| Density, g cm$^{-3}$ | 1.14 – 1.20 |
| SPV-5 apparent viscosity, s | 30 – 60 |
| 30 min filtration loss, cm$^3$ | 15 – 40 |
| Filter cake thickness, mm | 2 – 3 |

The 81 – 1875 m interval was drilled according to the drilling programme using untreated water. But unsatisfactory cuttings lifting and frequent preventive flushing out of the borehole necessitated substitution of water for mud. Lignin-alkaline agent was added to the mud which had been used in drilling 0 – 81 m interval in amounts of 0.2 – 1.0 mass percent. The mud acquired the following properties:

| | |
|---|---|
| Density, g cm$^{-3}$ | 1.08 – 1.14 |
| SPV-5 apparent viscosity, s | 20 – 30 |
| 30 min filtration loss, cm$^3$ | 10 – 15 |
| Filter cake thickness, mm | 1.5 – 3.0 |
| Sand content, % | 0.5 – 1.5 |
| Gel strength after 1/10 min, Pa | 2/5 – 10/14 |
| pH | 7 – 9 |

This mud was used in the 1875 – 5295 m interval, and then it was changed for a low clay content emulsion mud. Owing to use of a high-quality modified bentonite clay content was reduced to 5%. A lubricating agent *Smad-1* in amounts of 1 – 3 vol% was introduced into the mud system. The emulsion obtained effectively protected the drill string from corrosion. It had the following properties:

| | |
|---|---|
| Density, g cm$^{-3}$ | 1.04 – 1.06 |
| SPV-5 apparent viscosity, s | 25 – 50 |
| 30 min filtration loss, cm$^3$ | 20 – 40 |
| Filter cake thickness, mm | 1 – 2 |
| Gel strength after 1/10 min, Pa | 2/3 – 12/14 |
| pH | 8.5 – 9.0 |

Further improvement of the composition led to a change of CMC and lignin-alkaline agent for *Metas* (methacrylamid-methacrylic acid copolymer). Thus, *Metas* (in amounts of 0.25 – 0.3 mass %) and *Smad-1* (in amounts of 1 – 3 vol%) were used to treat the drilling fluid in the 7065 – 7952 m interval, and *Metas* not only effectively stabilized clay solids, but also acted as a selective flocculating agent for fine drilling cuttings which led to reduction of hydraulic losses.

The drilling fluid had the following properties:

| | |
|---|---|
| Density, g cm$^{-3}$ | 1.03 – 1.04 |
| SPV-5 apparent viscosity, s | 50 – 100 |
| 30 min filtration loss, cm$^3$ | 8 – 12 |
| Filter cake thickness, mm | 1.0 – 3 |
| Gel strength after 1/10 min | 0.5/3 – 14/16 |
| pH | 8.5 – 9.8 |
| Solids content, % | 7 – 10 |

However, low efficiency of mud cleaning equipment, high sensitivity of the bentonite mud to drilling and cutting contamination and difficulties in maintaining optimum *Metas* concentration in 400 – 500 m$^3$ of the drilling mud resulted in considerable fluctuations of practically all properties of the mud which presented difficulties in the drilling of the well. In addition, the well-bottom zone was sludged up and plugs of drill solids were formed due to the insufficient viscosity of the drilling fluid in this zone caused by the nature of the polymer-bentonite mud. Periodic flushing-outs of the borehole helped greatly in eliminating these difficulties. As the well became deeper, the frequency of flushing-outs had to be drastically increased. To avoid these negative phenomena and to increase mud density, a drilling mud on the basis of Druzkovsky clay was used from a depth of 7959 m. The composition of the drilling mud (on a percentage basis) is as follows:

| | |
|---|---|
| Kaolinite clay | 25 – 30 |
| CMC-400 (500) | 0.3 – 0.5 |
| Nitrolignin | 0.2 – 0.3 |
| Potassium bichromate | 0.05 – 0.1 |
| Smad-1 | 2 – 4 |
| Graphite | 1.5 – 2 |

The mud had the following properties:

| | |
|---|---|
| Density, g cm$^{-3}$ | 1.15 – 1.20 |
| SPV-5 apparent viscosity, s | 40 – 60 |
| 30 min filtration loss, cm$^3$ | 3 – 7 |
| Filter cake thickness, mm | 0.5 – 1.5 |
| pH | 8.5 – 9.2 |

Borehole cleaning was improved and the range of hydrodynamic pressure surges was decreased by the use of this mud, but the drill string drag continued to grow to a depth of 8800 m. The rather higher consumption of CMC-500 (600) which was presumedly ascribed to thermal degradation of the agent caused *Metas* to be used as the main stabilizer. Optimization of the content-to-size distribution ratio made it possible to abandon nitrolignin as a thinning agent.

As the light-alloy pipes constituted 80% of the drill stem mud, alkalinity (pH) reduction to 7.5 – 7.9 improved operating conditions for pipes in the borehole. This was accomplished mostly by treating the mud with powdered *Metas*. The torque required was reduced by 15 – 20%. To improve *Smad* emulsification under extreme conditions anionic surfactants such as alkyl benzene sulphonate and NP-3 were used. This resulted in lower torque and drag and better drilling efficiency in 8900 – 9600 m interval.

The 9350 – 9750 m interval was drilled using a weighted emulsion mud of the following composition (on the percentage basis):

| | |
|---|---|
| Kaolinite clay | 15 – 18 |
| Barite | 15 – 30 |
| Metas | 0.2 – 0.3 |
| Nitrolignine | 0.2 – 0.3 |
| Potassium bichromate | 0.02 – 0.05 |
| Smad-1 | 3 – 4 |
| Graphite | 2 – 2.5 |

The use of barite as a weighting material gave rise to hydrophobic flocculation of the solids which resulted in intensive hydroabrasive wear on the inner surface of the light-alloy drill pipe contracted sections due to high working pressure and turbulent mud flow. High sedimentation stability of the mud did not prevent pressure surges ranging from 2 to 4 MPa. The mud had the following properties:

| | |
|---|---|
| Density, g cm$^{-3}$ | 1.30 – 1.43 |
| SPV-5 apparent viscosity, s | 40 – 120 |
| 30 min filtration loss, cm$^{-3}$ | 4 – 8 |
| Filter cake thickness, mm | 1 – 2 |
| Gel strength after 1/10 min, Pa | 6/8 – 16/20 |
| pH | 8.2 – 8.8 |
| Solids content, % | 32 – 45 |

Drilling with unweighted emulsion mud was resumed from a depth of 9910 m. Sodium tripolyphosphate (0.01 – 0.05%) was added to the mud to prevent and eliminate high temperature clay flocculation and to improve the chemical composition of the mud water phase. The kaolinite clay content was reduced to 15 – 22%, and that of *Metas* was reduced to 0.15 – 0.2%. The mud with the following properties was made:

| | |
|---|---|
| Density, g $cm^{-3}$ | 1.12 – 1.15 |
| SPV-5 apparent viscosity, s | 30 – 100 |
| 30 min filtration loss, $cm^{-3}$ | 5 – 8 |
| Filter cake thickness, mm | 2 – 3 |
| Gel strength after 1/10 min, Pa | 2/4 – 14/16 |
| pH | 7.2 – 7.6 |
| Solids content, % | 20 – 25 |

In spite of high bottomhole temperature the consumption of the materials per 1 m of hole was sharply decreased (particularly *Smad, Metas,* potassium bichromate and graphite) owing to the optimization of the materials-to-solids ratio. The thermal stability of this mud was in excess of 200 °C, although its lubricity was considerably impaired at temperatures above 160 °C.

### 3.3.6 Telemetering System of Downhole Motor rpm Indication

Specialists of the Kola geological expedition developed and introduced in SG-3 well drilling from a depth of 9745 m a telemetering system of turbine downhole motor rpm indication based on the experience gained in the development and operation of GTN-type downhole telemetering systems which had been designed for use only to a depth of 5000 m. The system consists of a GIZ.03 downhole pulse generator and UNP-1 surface receiving equipment. The most important assembly of the telemetering system which must satisfy stringent reliability requirements is a transmitting device (a downhole generator) working in the well-bottom zone under conditions of high temperature, high hydrostatic pressures, corrosive medium and substantial vibration.

The GIZ.03 downhole pulse generator (Fig. 3.48) includes an oil-filled planetary reduction gear, an input shaft of which is connected to the downhole motor shaft through a drive end, and an output one is connected with a cam mechanism transforming downhole motor rotation into reciprocation movement of the stem of the valve. The valve seat is mounted in a "floating" piston movable relative to the housing. During drill string running-in the piston is lifted by upward flow of the drilling fluid passing into the drill string. During generator operation the piston comes down on its seat, providing a specified clearance between the valve and its seat. When coarse drilling cuttings are entrapped between the valve and the seat surfaces the piston is lifted, enlarging the clearance between them, thus affording protection of the reduction gear and cam mechanism from overloading.

The inner space of the reduction gear is filled with high-density fluid (1.9 g $cm^{-3}$) which is heavier than the drilling mud used. The space is sealed from the

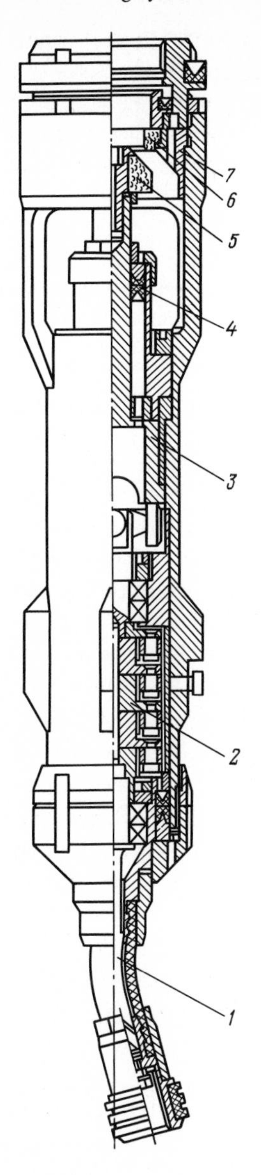


**Fig. 3.48.** Hydraulic transducer of downhole motor shaft rpm. *1* drive end; *2* planetary reduction gear; *3* cam mechanism; *4* rubber element; *5* valve; *6* valve seat; *7* floating piston

mud with a movable rubber element. Such a design affords protection for the reduction gear from the drilling mud entry and equalization of pressure of the two media.

Pulse generator kinematics provides reduction of the clearance between the valve and its seat to the specified value after every 100 revolutions of the downhole motor shaft, and a pressure pulse is fed into the hydraulic telemetry link. This signal is received and processed by a UNP-1 low-frequency receiver, block diagram of which is given in the Fig. 3.49. Blocks I and II are located respectively at the drilling site and in the technological control room.

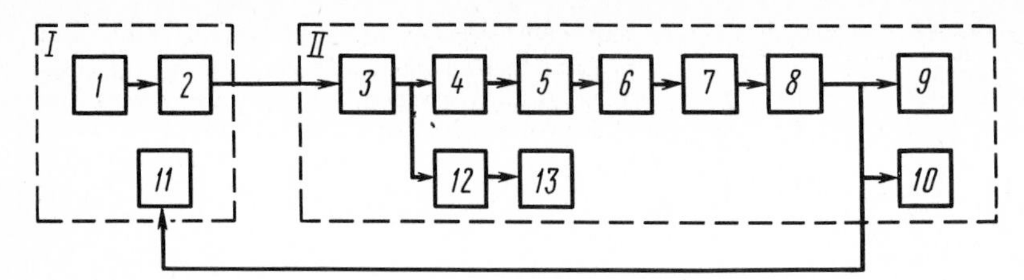


**Fig. 3.49.** UNP-1 block diagram. *1* presure transducer; *2* pulse-width converter; *3* demodulator; *4* frequency filter; *5* comparator; *6* pulse converter; *7* linearizer; *8* scaler unit; *9* analogous read-out; *10* control board indicator; *11* driller's console indicator; *12* filter; *13* recorder

A MED-600-type pressure transducer of the receiver is furnished with a pulse-width converter. A modulated signal is transmitted to a demodulator in the control room through a cable. An infrasonic frequency filter enables a legitimate signal to be developed. A comparator forms a sequence of pulses, the duration of their period being inversely proportional to the turbine downhole motor rpm. A pulse converter transforms the time of pulse duration into a proportional voltage, a linearizer performing piecewise linear inverse transformation of the $1/x$ function into voltage directly proportional to the turbine downhole motor rpm. A scaler unit enables an analogous read-out, control board indicator and driller's console indicator to be connected. The signal from the demodulator is applied to a recorder through a filter to check the telemetry link and downhole generator. The recorder makes it possible to determine the amplitude of the legitimate signal and noise level in the telemetry link.

The telemetering system provides reliable indication of the downhole motor rpm using the hydraulic telemetry link.

### 3.3.7 Wellhead Equipment

The Kola well is drilled in crystalline rocks where flows of formation fluids are not expected, formation pressure does not exceed the hydrostatic one, and therefore wellhead equipment is of simple design (Fig. 3.50), taking into account specific features of pilot drilling and reverse circulation, turning and change of temporary casing string, as well as carrying out of auxiliary operations associated with experimental drilling.

A special pedestal (see Fig. 3.50) is screwed onto the top pipe of the 325 mm OD casing string (landed and cemented at a depth of 2000 m). A 245 mm OD temporary casing string (landed at a depth of 1995 m to protect the 325 mm OD casing string from wear and to provide a uniform borehole cross-section) is freely suspended upon the pedestal. Two – three pup joints are screwed into the upper box of the temporary casing string. These joints are used to reciprocate and turn the drill string and to prevent the thread of the box from possible wear during these operations.

An experimental universal rotating blowout preventer of UVP-230 × 210 type made by the Schmidt oil-field equipment works (the city of Baku) is mounted on the upper pup joint.

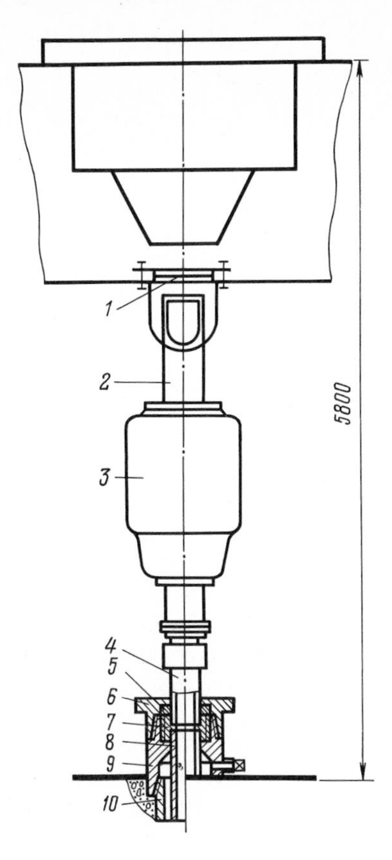


**Fig. 3.50.** Wellhead equipment layout. *1* UPP-324 device; *2* drain pipe; *3* universal BOP; *4* pup joint; *5* sealing set; *6* special nut; *7* upper box; *8* temporary casing string; *9* pedestal; *10* top pipe

*BOP technical data*

| | |
|---|---|
| Minimum inside diameter, mm | 230 |
| Working pressure, MPa: | |
| without rotation | 21 |
| allowable at 10 rpm | 10 |
| of hydraulic control system, maximum | 6.5 |
| Maximum rpm | 100 |
| Maximum diameter of drill string elements packed, mm | 225 |
| BOP maximum outside diameter | 990 |
| BOP height, mm | 1465 |
| BOP mass, kg | 492 |

The blowout preventer is designed to pack the wellhead and to turn and reciprocate the drill string with the packed wellhead. A BOP packing element seals against any member of the drill string and plugs the well when the drill string is pulled out of it. The force with which the packing element is pressed against the drill string is controlled. The preventer is remotely operated with the help of a hydraulic system.

An UPP-324 device developed by the Poltava division of the Ukrainian Geological Research Institute is mounted in a special tank on the drain pipe to

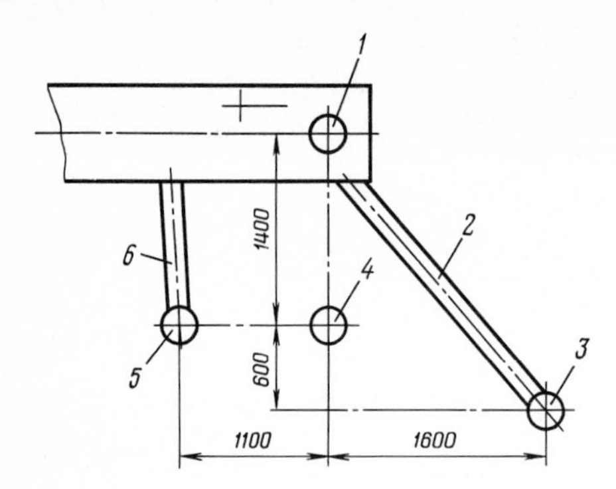


**Fig. 3.51.** Layout of technological holes on drilling site. *1* well axis; *2, 6* lines; *3* kelly rathole; *4, 5* technological holes

prevent foreign objects from falling into the well. The annulus between the strings is sealed by tightening a sealing set with a special nut. Landing and sealing arrangements allow for reverse mud circulation at a pressure of 15 MPa, which is specified according to the strength of the 325 mm OD casing string. A pressure gauge and discharge line of a centrifugal pump are connected to the annular space. The pump serves to flush this space after every run and to fill up the borehole during drill string pulling-up.

The wellhead equipment layout enables an unhampered change of the temporary casing string and periodical turning of it which materially increases its service life.

Figure 3.51 shows arrangement of auxiliary technological holes and a kelly rathole and their reference to the well axis. The holes are connected with the mud ditch system by lines. The depth of the technological holes corresponds to the length of the drill collars and bottomhole assemblies used. The holes may be useful both for carrying out different experiments and testing geophysical devices.

### 3.3.8 Bottomhole Assemblies

Different bottomhole assemblies (BHA) were used in the Kola well drilling depending on specific aims, geological and technical conditions. The assemblies were improved as the borehole was deepened and experience was gained. Maximum core recovery and keeping the borehole vertical were the most important requirements.

#### Bottomhole Assemblies for Coring

Three types of core barrel were predominantly used for coring. They are: (a) the KDM-195-214/60 core barrel, the design of which is shown in Fig. 3.37; (b) the MAG-195-214/60 core barrel; (c) a KN-2 core-receiving extension for a turbine downhole motor. The KN-2 core-receiving extension was developed by specialists of the expedition to core in troublesome intervals. Its housing is connected to the spindle housing, its hollow shaft being connected to the spindle shaft.

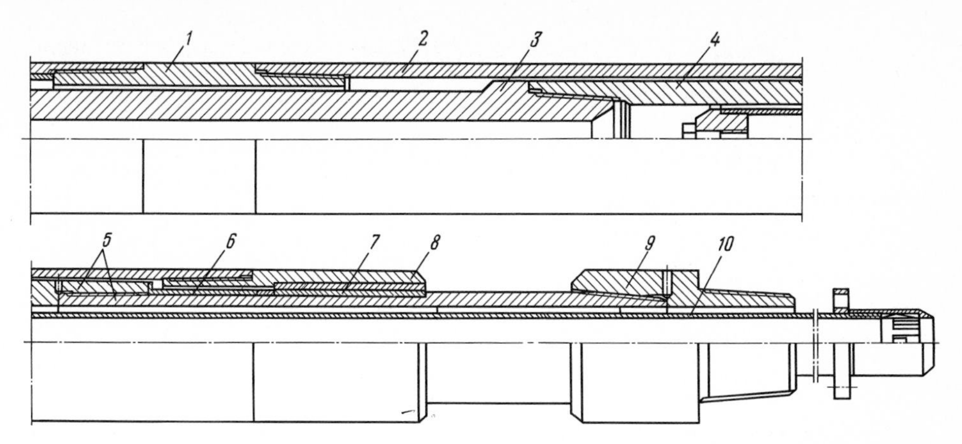


**Fig. 3.52.** KN-2 core-receiver extension for a turbine downhole motor. *1* housing sub; *2* housing; *3* upper shaft section; *4* intermediate shaft section; *5* lower shaft section; *6* thrust ring; *7* bearing sleeve; *8* pin; *9* shaft adapter; *10* core tube

The core-receiving extension (Fig. 3.52) consists of the following assemblies: a housing sub, housing, upper, intermediate and lower shaft sections, thrust ring, bearing sleeve, pin, shaft adapter and core tube.

The lower shaft section is connected to a 2V-K214/60 TKZ core head with a sub. The length of the rotating part is less than 1 m.

## Well-Bottom Assemblies for Hole Deviation Control

Well-bottom assemblies which included a bent sub, or turbine downhole motor with a ribbed spindle or OT turbine deflecting tool were used to control hole deviation in drilling to a depth of 5340 m. A schematic of an active-type assembly for hole deviation control is shown in Fig. 3.53.

The deflecting tool working plane was oriented by means of a STT bottomhole telemetering system which consists of a bottomhole transmitter, cable telemetry link, cable box, cable winch and surface equipment. Altogether more than 100 runs of the STT system transmitter in the SG-3 well were made, their aim being the straightening-out of its deviating borehole.

The most effective deviation correction was carried out in the 3125 – 3150 m interval (1° per 10 m) using the OT-195 turbine deflecting tool. At depths exceeding 5340 m, the use of this active-type bottomhole assembly proved to be ineffective because orientation work became much more complicated and time-consuming. A pendulum well-bottom assembly was used as a means of correcting borehole deviation on further deepening of the borehole. The design of the assembly is shown in Fig. 3.54. The main trends in increasing efficiency of the pendulum assembly are maximum increase of “pendulum” mass, improvement of drilling bit gaging action, reduction of penetration rates in crooked hole sections.

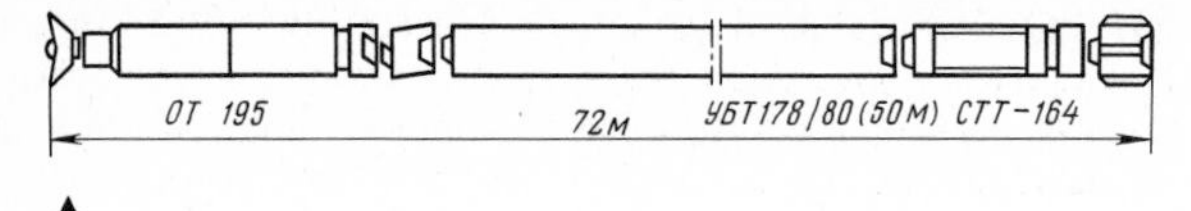


▲ **Fig. 3.53.** Active-type bottomhole assembly for hole deviation control

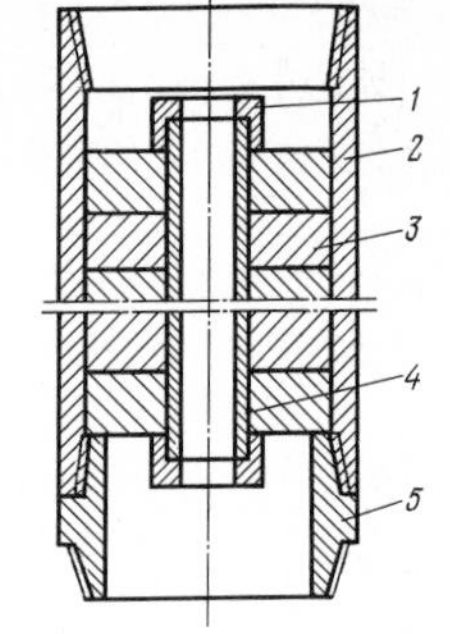


**Fig. 3.54.** Pendulum-type bottomhole assembly. *1* hold-down nut; *2* housing; *3* lead disk; *4* circulation tube; *5* lower sub ►

## Bottomhole Assemblies for Drilling in Intervals of Sticking Hazard

To decrease the probability of BHA sticking in troublesome intervals, well-bottom assemblies with a minimum number of tools suspended from the turbine downhole motor shaft were used in the following combinations: three-cone bit-ROP-9V reamer-TPS reamer; three-cone bit-TPS reamer; three-cone bit-ROP-9V reamer; three-cone bit.

Drilling in such intervals is performed at a limited rate of penetration, turbine downhole motor rpm being constantly monitored and the bit being periodically picked up above the bottom.

## Well-Bottom Assemblies for Sidetracking

The use of conventional well-bottom assemblies for sidetracking with an oriented deflecting tool proved to be impossible at great depths. Sidetracking with the help of bridging cement plugs was very difficult because the cement particles contaminated the drilling mud deteriorating its properties (drag was increased by 200 – 300 kN, torque was built up by 13 – 15 kN · m and up).

A well-bottom assembly employing the principle of plumb bob (Fig. 3.55) was developed to sidetrack without orienting a deflecting tool and placing a cement plug. Sidetracking begins with forming a shoulder in the borehole in four – six runs depending on the specified vertical deviation angle value in the sidetracking interval. The drill pipes are selected so that the bit is at the depth of the shoulder formed when it finishes to drill ahead the length of the kelly. The shoulder is formed from its bottom upwards. The overall length of the key seat formed above the shoulder comes to 30 – 35 m. The rate of forming the shoulder is reduced from its top to bottom. The presence of the shoulder is checked up by unloading the drill string with the standing drill bit.

When the shoulder is formed, the bent sub is removed out of the assembly and the first 80 – 100 m below the shoulder are drilled at a limited penetration rate and then drilling is performed using normal drilling practices. Sidetracking at deviation angles from 2° to 13° was performed with such bottomhole assemblies according to the technology described.

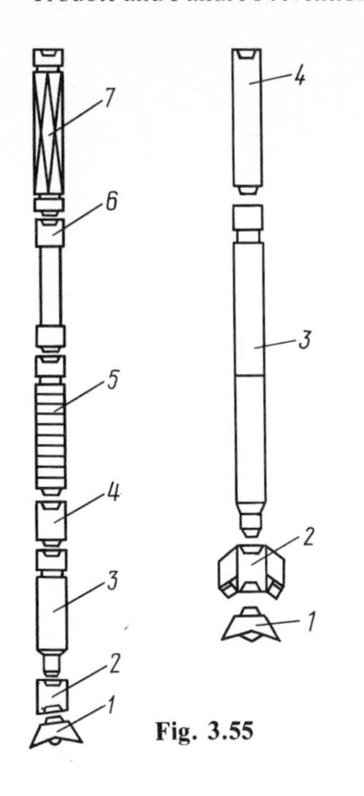


Fig. 3.55

Fig. 3.56

**Fig. 3.55.** Bottomhole assembly for sidetracking. *1* 215.9 mm three-cone drill bit; *2* 5° 30′ bent sub; *3* A7Sh turbine downhole motor (one section); *4* turbine shaft rpm transducer; *5* lead pendulum 14 m long; *6* steel drill pipe 114×10 (flexible link); *7* square cross-section drill collar

**Fig. 3.56.** Bottomhole assembly for borehole reaming. *1* drill bit; *2* reamer; *3* turbine downhole motor; *4* drill collar

Well-Bottom Assemblies for Reaming a Hole to a Diameter of 394 mm

The bottomhole was reamed from 214 to 394 mm diameter to a depth of 2000 m in two stages. At the first stage the hole was reamed to a diameter of 295 mm with K-type hard-alloy drill bits using the A7N4S turbine downhole motor. At the second stage it was reamed from 295 to 394 mm diameter with the bits tailor-made for the SG-3 well and RShP-394 and RDU-394 reamers (Fig. 3.56).

### 3.3.9 Trouble and Failure Prevention and Control Aids

One of the characteristic features of superdeep drilling grows with depth: difficulty in timely prevention and control of dangerous situations and trouble. Conventional aids prove ineffective or fully inadequate. The great volume of the well makes specific borehole flushing-out after every run hardly feasible.

Considerable difficulties are encountered in running fishing tools to the bottom when a danger arises of borehole plugging and blockage of circulation; much remains to be done in the sphere of junk removal from the well bottom.

A most frequently encountered and very dangerous situation at great depth arises when it becomes necessary to disconnect the drill string from the stuck bottomhole assembly. Aids especially designed and introduced into superdeep drilling proved to be very effective.

*A fill-up valve* has been designed for permanent filling of the drill string during running-in to avoid downhole motor or fishing tool plugging with drilling cuttings. This valve also enables the borehole to be flushed with the shut down downhole motor after a run is finished. It is installed into the drill string above the downhole motor or fishing tool with a small flow section (magnetic junk basket, fishing tap etc.).

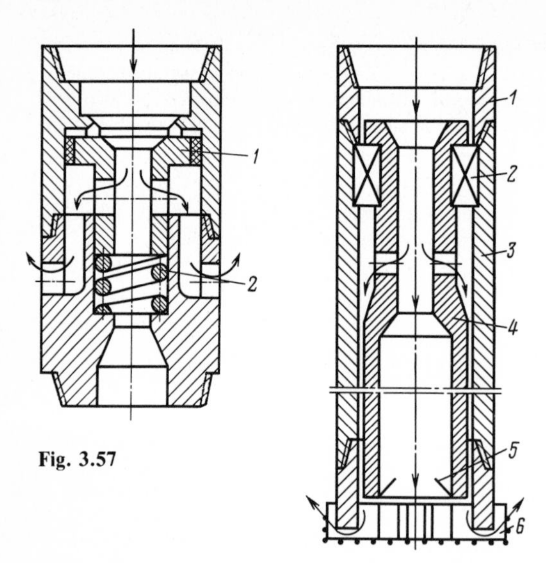


Fig. 3.58. Junk basket. *1* sub; *2* seal; *3* housing; *4* shell; *5* spring finger; *6* washover shoe

Fig. 3.57. Drill string fill-up valve. *1* piston; *2* calibrated spring

The following principle of operation is applied in the valve (Fig. 3.57). The drill string is filled with a drilling mud from the drill string-borehole annulus through the connected hollows of the valve. For a short time the mud flow rate is increased by 4–6 l $s^{-1}$ as compared to the usual one. A piston moves into its lowermost position overcoming elastic force of a calibrated spring thereby shutting off the hydraulic channel communicating the valve hollows with the annulus. This moment is evidenced by pressure build-up. Subsequently, mud flow may be decreased to the specified level. After the run, mud circulation is stopped for a short time. The piston moves into its uppermost position under the action of the spring and the borehole may be flushed at the specified mud flow rate until the mud is sufficiently clean. The fill-up valve described is especially effective when used in combination with the D2-172 hydraulic screw downhole motors.

A great length of highly cavernous open borehole is responsible for the accumulation of large pieces of clastic rocks on the bottom. This impedes lowering conventional magnetic junk baskets to the metal fragments left on the bottom (roller cutters etc.). These baskets do not ensure mud circulation at increased flow rates. Their efficiency is also limited by temperature conditions in the borehole. The problem of cleaning the bottom from metal fragments was solved by the introduction of special junk baskets into ultra-deep drilling.

*A junk basket* (Fig. 3.58) consists of a housing connected to the downhole motor shaft with a sub. The lower end of the junk basket housing is furnished with a replaceable washover shoe faced with hard-alloy inserts. Its shell may be cylindrical or conical in form, the former being furnished with spring fingers. The inside diameters of the washover shoe and shell should correspond to the size of a metal fragment fished. The junk basket works on the bottom like a drill bit.

At the moment when a core-like rock is pressed into the shell pressure build-up is observed in the pump discharge line, which indicates that further penetration should be stopped.

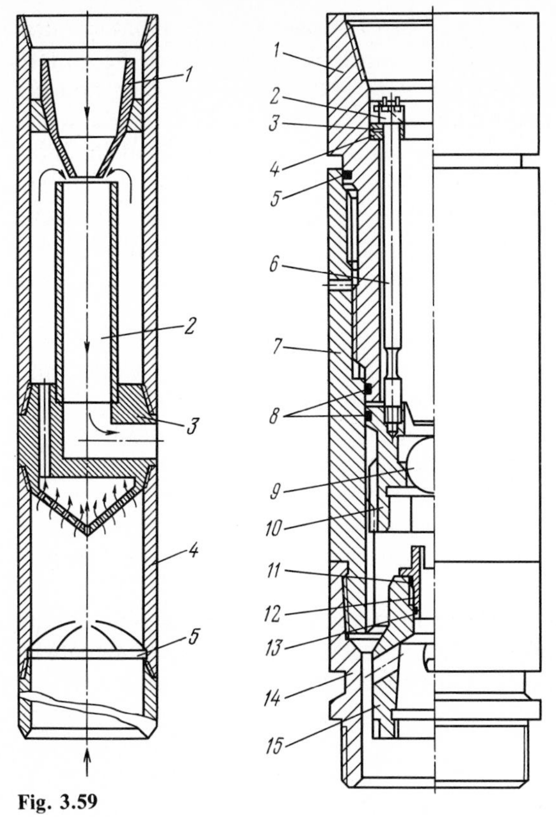


Fig. 3.59

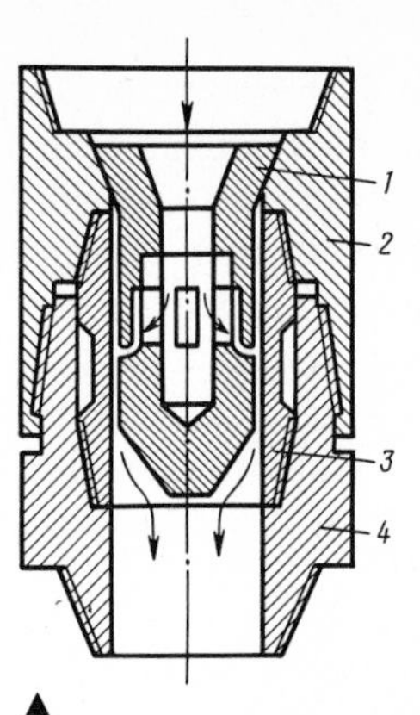


▲ **Fig. 3.61.** Hydraulic disconnector of drill string GRBK-195. *1* cutter; *2* sub; *3* short pipe; *4* sub

◄ **Fig. 3.60.** RBK-195 drill string disconnector. *1* pin; *2* nut; *3* angle piece; *4* bearing ring; *5* ring; *6* stud; *7* housing; *8* rubber ring; *9* ball; *10* driven element; *11* O-ring; *12* thrust block; *13* O-ring; *14* sub; *15* clutch

**Fig. 3.59.** Jet junk basket. *1* jet; *2* diffusor; *3* sub; *4* basket; *5* wire catcher

*A jet junk basket* is shown in the Fig. 3.59. The principle of its operation is based on bottomhole reverse circulation. Metal and rock fragments are thrown into a basket furnished with a wire catcher. Appreciable lifting force is provided by high speed flow in a jet-diffusor system placed in a special sub. The jet junk basket efficiency depends to a large measure on the quality of the product, its assembly and adjustment.

Built-in and go-devil jets may be used in the junk basket. Continuous rotation of the drill string by the rotary is necessary when operating the jet junk basket on the bottomhole.

*A RBK-195 drill string disconnector* is installed directly above the 3TSSh-9 or A7Sh turbine downhole motor. It is designed to disconnect the drill string when the rock cutting tool or downhole motor housing is stuck. The disconnector is fully controlled from the surface and operated by clockwise rotary rotation. The RBK-195 is used in deep, superdeep and deviated wells where there are many borehole deflections resulting in excessive drag and torque forces. Due to these forces after a backoff shot explosion or in the presence of right-hand releasing subs in the string, disconnection of the latter may take place in any threaded connection when the string is rotated counter-clockwise, which complicates further releasing of stuck pipes.

The RBK-195 consists of a pin (Fig. 3.60) connected to a housing using an MK $150 \times 6 \times 1:32$ left-hand thread, a sub connected to the turbine downhole motor shaft with a clutch screwed on the shaft. The upper clutch end has an outer surface of hexagon cross-section. A thrust block is screwed into the clutch. A driven element is connected to the sub with end cams and to the housing with splines which prevents the left-hand threaded connection from unscrewing. The driven element is held in its initial position with two tension-broken studs, nuts and bearing ring. The nuts are locked with angle pieces. A pressure differential between the RBK-195 disconnector inner space and drill pipe-borehole annulus is maintained and the driven element is sealed thanks to rubber rings. Rings prevent the thrust block from unscrewing. A ring holds grease in the MK $170 \times 6 \times 1:32$ thread.

When the drill bit or turbine downhole motor are stuck, a ball is dropped into the drill pipes to disconnect the string. The ball pushed down by drilling mud flow plugs the channel in the driven element. In this way mud high pressure acting on the driven element sealed by the ring is developed. The resulting downward force breaks the studs and pushes the driven element down to the stop thereby releasing left-hand thread and throwing hexagonal surfaces of the element and clutch into engagement. As a result, the downhole motor shaft is prevented against rotation, the ball seats on the thrust block and leaves a passageway for the drilling mud.

After the drill string is disconnected and free pipes are pulled out, fishing tools with a jar are run and connected to the stuck tools by means of the MK $170 \times 6 \times 1:32$ thread and recovering of the stuck tools begins.

*Technical data on RBK-195*

Outside diameter, mm . . . . . . . . . . 195
Minimum (driven element) inside diameter, mm . . . . . . . . . . 58
Go-devil ball diameter, mm . . . . . . . . . . 70
Length (transport position), m . . . . . . . . . . 1.44

Connection threads:

Upper . . . . . . . . . . Box, Z-171
Lower . . . . . . . . . . Pin, RKT $177 \times 5.08 \times 1:16$
Mass, kg . . . . . . . . . . 300
Maximum allowable axial force, kN . . . . . . . . . . 400

Torque applied, kN · m:

– when screwing on, not in excess of . . . . . . . . . . 40
– when disconnecting, no less than . . . . . . . . . . 4
Pressure differential necessary for breaking studs, MPa, not in excess of . . 7.5
Downhole temperature, °C, not in excess of . . . . . . . . . . 150

*A GRBK-195 jet pipe disconnecting cutter* is designed for operation at great depth (Fig. 3.61). It is used for disconnecting the drill string from the bottomhole assembly. It is used when the RBK-195 is ineffective due to the excessive torque necessary to rotate the string. A special sub with an MK 150 taper fishing thread is screwed with a shrink fit on a lower end of an aluminium short pipe with a TT $138 \times 5.08 \times 1:32$ pipe thread on both ends. A special sub with a cone seat of a hydraulic go-devil cutter is screwed on the short pipe upper end. The latter sub protects the fishing thread.

When it is necessary to release the drill string of the stuck bottomhole assembly above the downhole motor the go-devil pipe cutter is dropped into the drill string and forced down to its seat in the sub of the disconnecting cutter by the slush pump pressure to increase its travelling speed. The seating of the cutter is accompanied by mud pressure build-up. The slush pump provides a specified mud pressure for effective action of the hydraulic jet ring through a cutter slot 0.7 – 1 mm wide. During jet cutting, a surplus tension load of 100 – 150 kN is applied to the string. When the operation is finished, the drill pipes are pulled out of the well with the upper sub and hydraulic jet cutter. The bottomhole assembly left in the well is ready for further fishing with the help of a jar and by connecting it to the GRBK-195 thread.

## 3.4 An Integrated Interpretation of Technical and Economic Drilling Results

This section deals with the main laws of drilling depending on well depth, in other words, an integral index that determines the natural change in the geological and technical conditions of penetration. The section also contains data describing improvements introduced by advanced technology. An analysis of the drilling results reveals the quantitative interrelation of the basic drilling indices and geological and technical conditions at depth, making models of drilling at great depths more accurate, and suggesting methods of technological and technical progress in drilling the SG-3 well and other wells in similar environments.

### 3.4.1 Technical and Economic Evaluation of the Drilling Process

Some difficulty in carrying out a technical and economic analysis of the SG-3 drilling process cames from the new approach to designing the well, and the use of two types of drill rig with the technological limitations caused by the great depths.

The first drilling stage in an interval down to 7263 m was performed by the drill rig *Uralmash-4E*. The total time necessary for these operations was 51.5 months. During this period the well design was corrected by enlarging the open pilot hole to a depth of 2000 m; then it was cased by a casing string of 325 mm in diameter with the subsequent setting of withdrawable string of 245 mm in diameter to the same depth. The general time expenditure for these operations made up 8.1 rig month. The next stage started from a depth of 7263 m with the use of the *Uralmash-15000* drill rig. The total time consumption for these operations, including the time for geophysical survey of the borehole at this depth was 68.3 rig per month.

Prior to the beginning of the second stage of drilling, the *Uralmash-4E* drill rig was dismantled and the drill rig *Uralmash-15000* was installed accompanied by construction of production and research facilities for drilling to the project depth.

The subdivision of the hole section into separate depth intervals was dictated by the specific technological problems to be solved, differences in rock lithology of the section, the type of the drill rig and other aspects. Such an approach permitted, a somewhat tentative differentiation of the section into the following intervals:

0 – 2000 m – preventive measures against hole inclination, hole enlargement and casing;

| | |
|---|---|
| 2000 – 4000 m | – productive zones and the assumed end of the Pechenga Series (has not been confirmed); |
| 4000 – 6000 m | – abrupt alternation of rock strength properties; the interval is represented by four formations of the eight which characterize the entire section and by 12 geological and geophysical units of the 24 identified in the pre-Archean rocks; |
| 6000 – 7263 m | – transition to the Archean rocks and the end of the *Uralmash-4E* drill rig operation; |
| 7263 – 9000 m | – zones of intensive formation of the ellipse-shaped section of the pilot hole, adjustment and installation of the *Uralmash-15000* drill rig; |
| 9000 – 10000 m | – zones of large faults, introduction of low-revolution turbodrills and bit rotation frequency meter; |
| 10000 – 11500 m | – zones of large faults causing complications in drilling and requiring special technology to prevent trouble. |

The basic technical and economic indices for productive time in drilling are the drilling rate, footage per run, overall round trip velocity and the time of the auxiliary operations.

The on-bottom time (Table 3.14) comprises from 21.7% of the total time in an interval of 0 – 2000 m and up to 3.1% in an interval of 10,000 – 11,500 m. The penetration rate changes with depth from 1.3 to 2.5 m $h^{-1}$ (Table 3.13). Such a low drilling rate observed in the major portion of the hole section (down to 7263 m) is due to the technological conditions of drilling that make it necessary for the well to be held vertical in zones of frequent change in the rock strength. Moreover, below a depth of 7263 m, the necessity arose to reduce the speed of drilling tool feed so as to facilitate evacuation of the crushed rocks. This is one of the measures of preventive maintenance in the case of jamming.

As is known, the on-bottom time does not markedly affect the effectiveness of the drilling process at depths of 1500 – 2000 m, and the penetration rate of 2.5 m $h^{-1}$ has been found to be minimum for superdeep wells. Therefore, the rate for the Kola well may be taken as satisfactory.

Round-trip operations constitute up to 27.4% drilling time by the *Uralmash-4E* (Table 3.14), the average overall round trip velocity being 0.24 – 0.29 m $s^{-1}$ (Table 3.13). Utilization of the *Uralmash-15000* drill rig reduced the round trip time to 13.7 – 17.3% of the total while the average round trip velocity is up to 0.34 m $s^{-1}$. It is of interest that at a depth of 11,500 m the overall round trip velocity remained practically unchanged (0.33 – 0.34 m $s^{-1}$).

A limitation of the round trip velocity at these depths was introduced purposely to lower the hydrodynamic surge in the open well and to decrease the tensile forces acting on the drill tool and arising as a result of the increase of friction forces in the course of round trip operations. Because of this, the round trip velocity given does not represent the capacity of the *Uralmash-15000* drill rig, which according to specifications could provide an overall round trip velocity amounting to 0.55 m $s^{-1}$. Obviously such a velocity can be achieved only in the cased hole. It is believed that the round trip velocity in SG-3 hole is rather high, though it has not achieved the possible maximum.

**Table 3.13.** Drilling performance of *Uralmash-4E*

| Indices | Total in interval | *Uralmash-4E* drill rig | | | | *Uralmash-15 000* drill rig | | |
|---|---|---|---|---|---|---|---|---|
| | 0 – 11 500 m | 0 – 2000 m | 2000 – 4000 m | 4000 – 6000 m | 6000 – 7263 m | 7263 – 9000 m | 9000 – 10 000 m | 10 000 – 11 500 m |
| Penetration in m | 11 500 | 2000 | 2000 | 2000 | 1263 | 1737 | 1000 | 1500 |
| Overall drilling rate, m/rig month | 96 | 313 | 175 | 103 | 99 | 62 | 69 | 55 |
| Penetration rate, m $h^{-1}$ | 1.8 | 2.0 | 1.4 | 1.3 | 2.0 | 2.0 | 2.5 | 2.5 |
| Penetration per run, m | 7.2 | 7.9 | 5.8 | 6.8 | 7.2 | 6.9 | 8.6 | 9.7 |
| Single round trip, h | 10 | 2 | 6 | 9 | 14 | 14 | 15 | 17 |
| Overall round trip velocity, m $s^{-1}$ | 0.32 | 0.24 | 0.29 | 0.29 | 0.28 | 0.33 | 0.34 | 0.34 |

**Table 3.14.** Time distribution in Kola well drilling

| Indices (per 1 m of drilling) | Total in the interval of 0–11 500 m | *Uralmash-4E* drill rig | | | | *Uralmash-15 000* drill rig | | |
|---|---|---|---|---|---|---|---|---|
| | | 0–2000 m | 2000–4000 m | 4000–6000 m | 6000–7263 m | 7263–9000 m | 9000–10000 m | 10000–11 500 m |
| Total calendar time | 7.5 | 2.3 | 4.1 | 7.0 | 7.3 | 11.7 | 10.4 | 13.1 |
| of drilling, h % | 100.0 | 100.0 | 100.0 | 100.0 | 100.0 | 100.0 | 100.0 | 100.0 |
| Including net drilling, | 2.0 | 0.8 | 1.7 | 2.2 | 2.5 | 2.5 | 2.2 | 2.2 |
| h % | 26.7 | 34.8 | 41.5 | 31.4 | 34.3 | 21.4 | 21.2 | 16.8 |
| Of which: | | | | | | | | |
| On-bottom, h % | 0.6 | 0.5 | 0.7 | 0.8 | 0.5 | 0.5 | 0.4 | 0.4 |
| | 8.0 | 21.7 | 17.1 | 11.4 | 6.9 | 4.3 | 3.9 | 3.1 |
| Round trip, h % | 1.4 | 0.3 | 1.0 | 1.4 | 2.0 | 2.0 | 1.8 | 1.8 |
| | 18.7 | 13.1 | 24.4 | 20.0 | 27.4 | 17.1 | 17.1 | 13.7 |
| Auxiliary operations, | 3.1 | 1.0 | 2.0 | 3.1 | 2.5 | 5.4 | 3.5 | 5.1 |
| h % | 41.3 | 43.5 | 48.8 | 44.3 | 34.3 | 46.1 | 33.6 | 38.9 |
| Repairs, h % | 0.6 | 0.1 | 0.2 | 0.6 | 0.5 | 0.5 | 1.0 | 1.5 |
| | 8.0 | 4.3 | 4.9 | 8.6 | 6.8 | 4.3 | 9.6 | 11.5 |
| Problem control, h % | 1.0 | – | 0.1 | 0.3 | 1.3 | 2.2 | 1.3 | 2.9 |
| | 13.3 | – | 2.4 | 4.3 | 17.8 | 18.8 | 12.5 | 22.1 |
| Overall drilling time, | 6.7 | 1.9 | 4.0 | 6.2 | 6.8 | 10.6 | 8.0 | 11.7 |
| h % | 89.3 | 82.6 | 97.6 | 88.6 | 93.2 | 90.6 | 76.9 | 89.3 |
| Unproductive time, | 0.8 | 0.4 | 0.1 | 0.8 | 0.5 | 1.1 | 2.4 | 1.4 |
| h % | 10.7 | 17.4 | 2.4 | 11.4 | 6.8 | 9.4 | 23.1 | 10.7 |
| Including: | | | | | | | | |
| Accident elimination, | 0.5 | – | 0.1 | 0.5 | 0.2 | 0.9 | 2.3 | 0.8 |
| h % | 6.7 | – | 2.4 | 7.2 | 2.6 | 7.7 | 22.1 | 6.1 |
| Idle time, h % | 0.3 | 0.4 | – | 0.3 | 0.3 | 0.2 | 0.1 | 0.6 |
| | 4.0 | 17.4 | – | 4.2 | 4.2 | 1.7 | 1.0 | 4.6 |
| Of which: | | | | | | | | |
| Delay in supply of new | 0.1 | 0.4 | – | 0.1 | 0.1 | 0.1 | – | 0.6 |
| equipment and tools, | 1.3 | 17.4 | – | 1.4 | 1.4 | 0.8 | – | 4.6 |
| h % | | | | | | | | |
| Delay in supply of | 0.2 | – | – | 0.2 | 0.2 | 0.1 | 0.1 | – |
| geophysical instrumen- | 2.7 | – | – | 2.8 | 2.8 | 0.9 | 1.0 | – |
| tation, h % | | | | | | | | |

The reduction in round trip time in superdeep drilling by the use of an pilot hole is apparently possible only by of an increase in penetration per run. Based on available mining and geological data on the hole section and technical capacities, an average penetration per run of 6 m was programmed for the Kola well. In fact, however, in the drilled interval of pre-Archean rocks (down to 6840 m) the penetration per run amounted to 5.8 – 7.2 m. This is indicative of the fact that the forecast of the physico-mechanical properties of rocks in the section was basically confirmed. When the well struck the Archean formations the penetration per run increased to 9.7 m.

Without detailed discussion of the mining and geological evaluation of the state of annulus rock mass with depth, we note only that rock drillability, as is confirmed by drilling results, tended to improve with depth. Therefore, in view of the fact that the drilling bit run was carried out throughout the entire drilling interval under unfavourable conditions due to technological limitations, possible well inclination from the vertical and other reasons, the figures of penetration per run obtained cannot be considered as being the maximum possible. This is confirmed by the drilling results in an interval of 9500 – 11,500 m, where low-revolution bottomhole motors were used and information on drill bit rpm was available. Here the penetration per run attained 9.7 m. The assumption can be made that a further decrease of the bit rpm to 80 – 90 rpm at limited axial loads will lead to increased penetration per run up to 12 – 20 m, provided drilling bits equipped with sealed bearings are used.

Auxiliary operations constitute more than 40% of the whole time, and largely determine the profitability of the drilling process. Table 3.15 specifies all expenditures for each type of work. These operations may be tentatively subdivided into those characteristic of superdeep drilling and those for conventional drilling in general.

In the case of conventional exploratory drilling, the auxiliary work does not usually exceed 16 – 18% of the total time. This work does not include experiments to reduce the drilling rate; an insignificant period of time is allocated to preventive maintenance of hole condition, which is commonly done in the complicated zones; geophysical measurements as a rule constitute not more than 5 – 7% of the total time; very little time is spent for maintenance of drill pipes and flaw detection.

Considerable time is allocated for auxiliary operations typical of a superdeep well. For example, in an interval 0 – 2000 m, assembly and dismantling of the BHA took more than 8.6% of the total time (Table 3.15). It was the first time that core downhole devices KDM-195 were used in the Kola well as suspension beneath a shaft of the turbodrill. Preparation and assemblage of this tool were carried out on the drilling site. The time expenditure for this type of job decreased gradually, and, in an interval of 4000 – 6000 m, reached 2.9% as a result of standardization of the bottom hole assemblies.

In a depth range of 6800 to 7263 m, the hole reached the zone of Archean rocks and the KDM assemblies no longer proved effective in core sampling. There was a 4.1% increase in duration stemming from a new testing cycle and selection of appropriate BHA components. In a depth range of 7263 to 9000 m, use was made of new core barrels of simpler design than those of KDM-195 type,

**Table 3.15.** Time distribution in auxiliary Kola well drilling operations

| Indices (per 1 m of drilling) | Total in the interval of 0–11 500 m | *Uralmash-4E* drill rig | | | | *Uralmash-15 000* drill rig | | |
|---|---|---|---|---|---|---|---|---|
| | | 0–2000 m | 2000–4000 m | 4000–6000 m | 6000–7263 m | 7263–9000 m | 9000–10000 m | 10000–11 500 m |
| Auxiliary operations, h % | 3.1<br>41.3 | 1.0<br>43.5 | 2.0<br>48.8 | 3.1<br>44.3 | 2.5<br>34.3 | 5.4<br>46.1 | 3.5<br>33.6 | 5.1<br>38.9 |
| Including | | | | | | | | |
| Assembling and dismantle of BHA, h % | 0.3<br>4.0 | 0.2<br>8.6 | 0.2<br>4.8 | 0.2<br>2.9 | 0.3<br>4.1 | 0.4<br>3.4 | 0.2<br>1.0 | 0.3<br>2.3 |
| Experimental work, h % | 0.3<br>4.0 | –<br>– | 0.3<br>7.2 | 0.1<br>1.4 | –<br>– | 0.8<br>6.8 | 0.1<br>1.0 | 0.6<br>4.6 |
| Preventive conditioning of the hole, flushing the hole, conditioning drilling fluid, regaining circulation, round trip operations, h % | 0.7<br>9.3 | 0.2<br>8.8 | 0.2<br>5.0 | 0.6<br>8.6 | 0.3<br>4.1 | 1.6<br>13.6 | 1.0<br>9.6 | 0.2<br>9.1 |
| Borehole reaming, h % | 0.2<br>2.7 | –<br>– | 0.1<br>2.5 | 0.2<br>2.9 | 0.2<br>2.7 | 0.3<br>2.6 | 0.3<br>2.9 | 0.4<br>3.1 |
| Geophysical measurements, h % | 0.9<br>12.0 | 0.4<br>17.3 | 0.6<br>14.6 | 1.2<br>17.1 | 1.2<br>16.5 | 0.9<br>7.7 | 1.1<br>10.6 | 1.6<br>12.2 |
| Replacement of separate components of the equipment, restringing of drilling line, maintenance of equipment, etc., h % | 0.1<br>1.3 | –<br>– | 0.1<br>2.5 | 0.1<br>1.4 | 0.1<br>1.4 | 0.5<br>4.3 | 0.2<br>1.9 | 0.2<br>1.5 |
| Replacement and pressure testing of turbodrills, h % | 0.1<br>1.3 | –<br>– | 0.1<br>2.5 | 0.1<br>1.4 | 0.1<br>1.4 | 0.1<br>0.9 | 0.1<br>1.0 | 0.1<br>0.8 |
| Assembling, dismantling and flaw detection of drill pipes, h % | 0.2<br>2.7 | 0.1<br>4.4 | 0.1<br>2.5 | 0.2<br>2.9 | 0.1<br>1.4 | 0.3<br>2.6 | 0.3<br>2.8 | 0.5<br>3.8 |
| Other operations, h % | 0.3<br>4.0 | 0.1<br>4.4 | 0.3<br>7.2 | 0.4<br>5.7 | 0.2<br>2.7 | 0.5<br>4.2 | 0.2<br>1.9 | 0.2<br>1.5 |

and the time required for their assembly in a depth range of 9000 to 11,500 m amounted to 2.3% of the total time.

Experimental operations consumed as much as between 6.8 and 7.2% of total time, averaging 4% throughout the entire section.

The time spent on well conditioning (flushing, mud conditioning, regaining circulation), including round trips, was fairly long, reaching 13.6% at a depth of 7263 to 9000 m, as compared with an average of 9.3% throughout the entire section. These operations are most typical of superdeep drilling with an open pilot hole of maximum length.

Reaming of intervals drilled during preceding runs did not take too much time, being more or less stable throughout the entire section.

The geophysical measurements in conventional holes generally account for 5 to 6% of total rig time. However, in the case of SG-3 drilling, these went up to 12%. Well logging data are of great importance, and the overall duration of geophysical studies is likely to remain at a high level.

Assembly, disassembly and flow detection of drill pipes as auxiliary operations consumed on the average between 2.5 to 2.8% of time in all the depth ranges, which is 1.5 to 2% higher than in drilling conventional holes. This stems from complicated drill string assemblies and stringent requirements imposed on drilling tools to be used at large depths.

Largely determined by the use of non-standard drilling tools, the period of time involved in connection with other operations greatly varied (1.5 to 7.2%), exceeding that in conventional drilling by at least 1.5 to 2%.

The overall estimation of time spent on auxiliary operations, excluding BHA assembly, experimental operations, hole conditioning, logging, assembly and flaw detection of drill pipes, shows that these account for not more than 20% of total rig time, which indicates a high level of SG-3 drilling operation management.

It seems possible to further reduce auxiliary operation time by optimizing BHA assembly and disassembly processes using standardized elements, shortening hole conditioning duration, using up-to-date hoists, cables and thermobaro-resistant geophysical equipment, detecting defects of drill pipes and tools in round trips, etc. All this is likely to bring down auxiliary operation time from 40–45% to 20–25% in terms of 1 m penetration.

Repairs consumed 4.3 to 11.5% of total rig time, averaging 8% throughout the entire section, not only for the highly reliable *Uralmash-15 000* rig, but also for routine maintenance.

Special mention must be made of the elimination of geological hazards in drilling below 6000 m. This work consumed 18 to 22% of the time, mostly due to sticking-hazard situations with drill pipes or tools when drilling in the upper Archean (6500 to 9000 m) and in an interval of large faults (10,000 to 11,500 m). Sidetracking was resorted to, whenever recovery of the stuck pipes from troublesome sections was unsuccessful.

Overall time spent on the elimination of such troubles can be reduced by further updating methods making up for the loss of stability of the surrounding rock mass, and by developing special impact tools included into drill string assembly to eliminate its sticking. Already well under way, these improvements are likely to yield positive results.

Productive drilling time reached 97.6% in a number of depth ranges averaging 89.3% throughout the entire section. Unproductive time for the elimination of accidents was fairly low (6.7%), but there was one in an interval of 9000 to 10,000 m that consumed 22.1% of drilling time. Downtime, mostly due to the absence of new drilling and geophysical equipment, as well as cables, reached on the average 4%.

Analysis has shown that the overall drilling rate in the well as an integral value of penetration effectiveness is determined not only by actual drilling time, which tends to decrease with depth, but also by overall time spent on auxiliary operations and elimination of drilling troubles, which represent a total of 54.6%. To further increase overall drilling rate we must, therefore, not only optimize drill bit performance but also reduce the duration of auxiliary operations and the period of time required for the elimination of drilling troubles.

The technical-economic assessment of the drilling process in the Kola well shows that a number of complicated technical and technological problems have been solved, thus ensuring a high level of productive drilling time. All this offers ample opportunities for further increase in effectiveness of geological prospecting in drilling superdeep wells.

### 3.4.2 Major Trends in Variation of Drilling Indices with Depth

Rock Drillability

While drilling through complicated zones with reduced penetration rate and penetration per run, conventional methods as a rule prove inadequate in appraising drillability variation as well as the influence of physical and mechanical properties on drilling parameters. Drillability variation with depth is determined by penetration per 1 revolution of a drilling bit with an almost constant bit load for one and the same type of tool and rock of almost the same (± 300 MPa) standard hardness ($P_s$).

Average penetration per 1 revolution of a drilling bit is $\delta_i = h/Tn60$, where $h$ is penetration per run, mm; $T$ is drilling time; $n$ is drilling bit rpm.

Penetration per 1 revolution of a drilling bit must not change for one and the same type of tool at a constant load and rock hardness. However, this has not been confirmed in working the Kola well.

Variation in drillability with depth is characterized by the factor $K_i = \delta_i/\delta_0$, where $\delta_i$ is penetration per 1 revolution of a drilling bit at a particular hole depth and $\delta_0$ is penetration per 1 revolution of a drilling bit in the upper portion of the hole. The whole section (0 to 11,500 m) is divided into depth ranges, 1000 m each, to facilitate calculations. Penetration per 1 revolution of a drilling bit in the first depth range – 0 to 1000 m ($\delta_0$) is a datum mark for determining $K_i$.

Depth-dependent variation in drillability has been studied only for a 2V-K2I4/60TKZ core bit used most commonly and for A7Sh and A7GTSh downhole motors for rocks with $P_s$ between 2000 and 2500 MPa and two ranges of bit load. The results of these calculations are summarized in Table 3.16 and shown in two diagrams (Fig. 3.62).

**Table 3.16.** Drillability versus depth (for 2B-K214/60TKZ core bit)

| Depth range, m | Average standard hardness, MPa | Number of runs | Average penetration per run, m | Average drilling time, h | Rotation speed, rpm | Penetration per 1 mm | | Factor $K_i$ |
|---|---|---|---|---|---|---|---|---|
| | | | | | | Estimated | Actual | |
| | | | | $P_b = 80 \pm 20$ kN | | | | |
| 0 – 1000 | – | – | – | – | – | – | – | – |
| 1000 – 2000 | 1800 | 7 | 7.6 | 2.44 | 350 | 0.31 | 0.149 | 1 |
| 2000 – 3000 | 1870 | 11 | 6.3 | 3.37 | 300 – 500 | 0.3 | 0.082 | 0.55 |
| 3000 – 4000 | 3260 | 17 | 4.0 | 2.62 | 300 – 500 | 0.179 | 0.083 | 0.55 |
| 4000 – 5000 | 2860 | 11 | 4.2 | 3.01 | 300 – 500 | 0.179 | 0.077 | 0.51 |
| 5000 – 6000 | – | – | – | – | – | – | – | – |
| 6000 – 7000 | 2000 | 24 | 8.0 | 2.34 | 350 – 450 | 0.283 | 0.163 | 1.09 |
| 7000 – 8000 | 2320 | 6 | 8.1 | 1.91 | 350 – 450 | 0.227 | 0.183 | 1.23 |
| 8000 – 9000 | 2370 | 11 | 6.0 | 1.71 | 300 – 400 | 0.227 | 0.203 | 1.36 |
| 9000 – 10000 | 2300 | 3 | 15.0 | 3.2 | 300 | 0.227 | 0.262 | 1.76 |
| | | | | $P_b = 20 \pm 20$ kN | | | | |
| 0 – 1000 | 2390 | 12 | 7.2 | 3.06 | 500 – 700 | 0.057 | 0.068 | 1 |
| 1000 – 2000 | 1710 | 19 | 7.5 | 3.75 | 600 | 0.07 | 0.056 | 0.82 |
| 2000 – 3000 | 2110 | 17 | 6 | 5.38 | 500 – 800 | 0.065 | 0.031 | 0.46 |
| 3000 – 4000 | – | – | – | – | – | – | – | – |
| 4000 – 5000 | 2250 | 10 | 3.6 | 2.92 | 650 | 0.060 | 0.032 | 0.47 |
| 5000 – 6000 | 2210 | 10 | 8.8 | 6.94 | 500 | 0.050 | 0.044 | 0.65 |
| 6000 – 7000 | 2000 | 13 | 8.4 | 4.01 | 500 | 0.065 | 0.078 | 1.15 |
| 7000 – 8000 | 1700 | 10 | 9.6 | 5.57 | 500 | 0.07 | 0.099 | 1.46 |
| 8000 – 9000 | 2510 | 8 | 6.8 | 2.99 | 500 | 0.057 | 0.08 | 1.18 |
| 9000 – 10000 | 2410 | 6 | 8.1 | 2.38 | 600 | 0.057 | 0.094 | 1.4 |

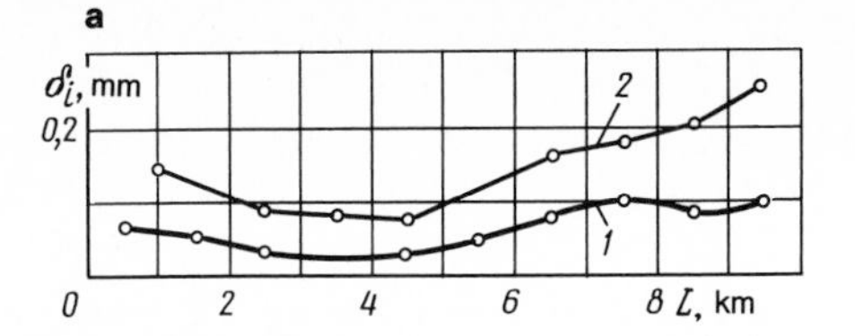


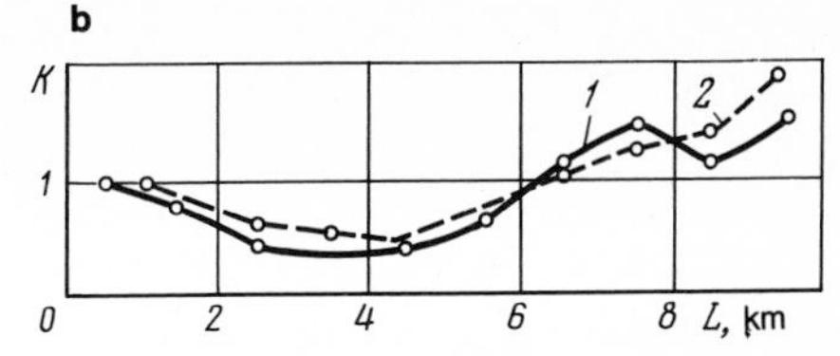


**Fig. 3.62 a, b.** Penetration per 1 revolution of drill bit (drillability) **a** and drillability factor **b** as a function of hole depth: *1* load 20 ± 20 kN at n = 500 ÷ 700 rpm; *2* load 80 ± 20 kN at n = 300 ÷ 500 rpm

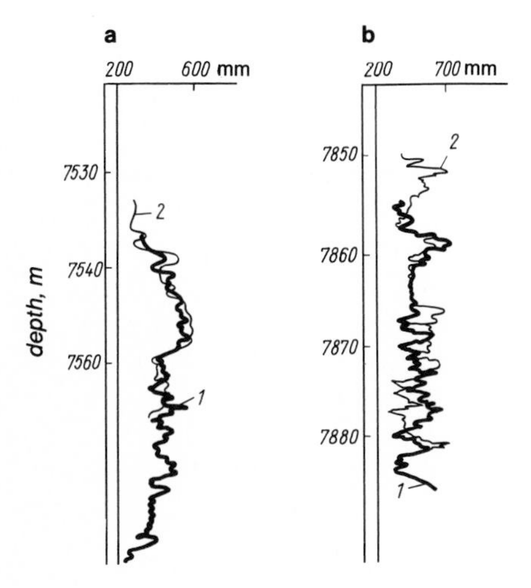


**Fig. 3.63 a, b.** Rock cavity development with time: **a** for interval 7530 – 7560 m; *1* May 8; *2* August 14; **b** for interval 7850 – 7880 m, *1* March 1; *2* July 30

The variation in shaft speed is specified on the basis of both rated performance of a motor and run-in standing measurements, as well as using GIZ-03 turbotachometer directly in the process of drilling.

As is seen from Table 3.16, drillability deteriorated with depth between 0 and 5000 m (drillability factor $K_i$ decreased from 1 to 0.46). In the Archean zone (6000 to 11,000 m), the drillability factor increased between 1.5 and 2 times, as compared with the 0 to 6000 m depth range. Better drillability may be due to the different rock texture with strength characteristics unaffected, lower rock density on account of metamorphism, tectonic effects, stress relaxation during geological periods or higher stresses at the well bottom.

## Borehole Enlargement

Drilling practice has shown that well stability is affected mostly by local sloughing-in of side walls, which gives rise to rock cavities while drilling through unstable formations. The data obtained from section gauge logging and caliper survey indicate that rock cavities do not actually increase in size (Fig. 3.63). This indicates the brittle nature of caving with the stresses in the surrounding rock mass, which is confirmed by results of experiments carried out in the Groznyi Oil

Institute. Holes 5 to 10 mm in diameter have been drilled in cylindrical core samples recovered from the Kola superdeep hole to serve as a well model. A state of composite stress was subsequently imparted to the sample corresponding to actual pressure drop $(\gamma_n - \gamma_x) \times H$ at a given depth.

Another important factor determining sidewall stability is a joint block structure of the crystalline rock mass which is dissected into individual blocks varying in size by different mineral grain boundaries, macro- and microfissuring and boundaries between rock beds. The influence of the interfaces on hole wall stability is determined by the size of these blocks limited by surfaces of weakness. For instance, surfaces of weakness dissecting the rock mass into blocks 10 to 100 mm in size commensurable with a hole diameter are prevalent for holes 250 to 300 mm in diameter. Such surfaces include contacts of alternating beds of different rocks, systems of micro- and macrofissures which are open or filled in with low-strength secondary minerals.

The SG-3 cross-section is characterized, as is noted above, by numerous fissures and vein formations. Mineralized fissures occur throughout the whole section. The infilling minerals include calcite, quartz, chlorite, serpentine, talc, epidote, axinite, feldspar, etc. Strength of specimens in fissures filled with secondary minerals frequently proves to be lower, as is shown by tests, than the strength of the same rocks in a solid block. The structural blocks are represented mostly by solid bodies composed of strong rock-breaking minerals.

Rock strength in a solid block of the SG-3 cross-section is as follows:

| Rock | Gneiss | Gabbro | Sandstone | Amphibolite |
|---|---|---|---|---|
| Strength $\sigma$, MPa: | | | | |
| Tensile strength | 5 – 23 | 10 – 30 | 17 – 30 | 7 – 43 |
| Shear strength | 7 – 32 | 12 – 40 | 20 – 40 | 12 – 50 |

The strength of the same rocks with mineralized fissures is determined by the infilling mineral.

| Mineral | Hydromuscovite, chlorite, mylonite, talc | Carbonate, ore minerals | Quartz, microcline, epidote |
|---|---|---|---|
| Strength $\sigma$, MPa: | | | |
| Tensile strength | 1 – 4 | 4 – 12 | >12 |
| Shear strength | 2 – 6 | 6 – 15 | >15 |

Hence, it follows that tensile strength and shear strength are lower in fissures filled in with hydromuscovite, talc and chlorite than in a solid block. In the crystalline rock mass, tectonic fractures with slickensided surfaces featuring extremely low adhesion may be encountered.

The caliper survey and section gauge logging have shown that the hole diameter is fairly uniform in the upper portion down to 1750 m. There were only some minor local enlargements associated, as a rule, with contacts of different rock units (582 to 600 m, 1120 to 1130 m) and zones of strongly metamorphosed rocks (1379 to 1416 m). The first large cavities were observed in fracture zones at the contact between ultrabasic intrusives and basic rocks (1756 to 1790 m and

1798 to 1812 m). Subsequently, the cavities rapidly increased in size, reaching more than 70 cm in diameter in a couple of years. As a result of caving-in of the well, the drill string was repeatedly stuck. Further penetration proved increasingly difficult, and an intermediate casing 325 mm in diameter was, therefore, landed down to 2000 m. In a depth range of 2000 to 4500 m, the well diameter is also fairly uniform with some minor enlargements at 2392, 2568, 2572 and 2806 m.

Intensity of jointing sharply decreased after the boundary of the 3rd volcanic sequence (2805 m) had been intersected, which agrees well with the fact that there were actually no cavities. It was only at the contact with the 3rd sedimentary sequence (4500 to 4670 m) that a large cavernous zone was observed. The repeated section gauge logging and detailed survey using a *Sprut* downhole device showed a marked increase in the number of cavities.

From 4800 m downwards, the well exhibits numerous cavities associated with large faults, contacts between sedimentary and volcanic rocks and intrusion zones. There are cavities of an irregular shape in the intrusion zone at a depth ranging from 6340 to 6470 m, as found by the *Sprut* device. The cavity narrows at a depth of 6424.5 to 6426 m to form a narrow cross-section of the hole. Depth ranges of 5150 to 5440, 5790 to 6000, 6150 to 6260, 6300 to 6320 m also exhibit cavities of irregular shape in strongly deformed rocks.

From 7200 m downwards, there is an oval cross-section of the well with different maximum distances from the imaginary axis. The large oval axis is arranged in the direction of rock bedding. The average diameter of the well ranges from 25 to 48 cm. The well enlargement is negligible, ranging from 0 to 2.97 $m^3$ per 100 m. The total increase in well volume at a depth of 7200 to 8700 m is 10 $m^3$, i.e. less than 10% of the total section volume, which may be accounted for by measuring errors (Table 3.17). The well is, therefore, mostly enlarged in depth ranges composed of brittle altered rocks, or in areas with intense fissuring, which is supported by theoretical and experimental results.

Analysis of the state of stress and properties of the rock mass, as well as dynamics of the SG-3 borehole, reveals that local caving-in of the hole is likely to become more intense with the growing number of cavities, stemming from higher static stresses. This necessitates further studies of the mechanism of rock deformation and disintegration at high pressures and temperatures.

## Borehole Temperature

Penetration is a complicated process which includes round trips, flushing, drilling and downtime. All these operations affect borehole temperature. The distribution of heat flow in the hole is largely determined by the depth-dependent increase of rock temperature, as well as by variable heat transfer from the rocks to drilling mud and then to the drill string. The drill string in the Kola superdeep hole consists mostly of light-alloy pipes with strength properties as a direct function of temperature. Dynamic temperature variation necessitates a design for the drill string with due regard for limitations of application of a particular alloy.

**Table 3.17.** Well volume versus depth

| Depth range, m | Well volume, $m^3$ | | | |
|---|---|---|---|---|
| | March 23, 1976 | May 8, 1977 | February 4, 1979 | February 12, 1981 |
| 0 – 2000 | 131-drill string volume | | | |
| 2000 – 2100 | 3.59 | 4.26 | 4.31 | 4.32 |
| 2100 – 2200 | 4.89 | 5.31 | 4.87 | 5.55 |
| 2200 – 2300 | 4.78 | 5.1 | 4.83 | 5.04 |
| 2300 – 2400 | 5.14 | 5.47 | 5.19 | 5.56 |
| 2400 – 2500 | 5.26 | 5.6 | 5.24 | 5.56 |
| 2500 – 2600 | 4.82 | 5.18 | 4.87 | 5.3 |
| 2600 – 2700 | 4.56 | 4.6 | 4.37 | 4.78 |
| 2700 – 2800 | 5.32 | 5.76 | 5.39 | 6.04 |
| 2800 – 2900 | 4.52 | 4.56 | 4.81 | 4.99 |
| 2900 – 3000 | 4.15 | 4.26 | 4.25 | 4.27 |
| 3000 – 3100 | 4.15 | 4.3 | 4.25 | 4.18 |
| 3100 – 3200 | 4.18 | 4.33 | 4.35 | 4.43 |
| 3200 – 3300 | 4.23 | 4.48 | 4.37 | 4.14 |
| 3300 – 3400 | 4.12 | 4.41 | 4.29 | 4.09 |
| 3400 – 3500 | 4.09 | 4.34 | 4.26 | 4.17 |
| 3500 – 3600 | 4.01 | 4.67 | 4.27 | 4.18 |
| 3600 – 3700 | 4.30 | 4.83 | 4.57 | 4.66 |
| 3700 – 3800 | 4.37 | 5.1 | 4.49 | 4.57 |
| 3800 – 3900 | 4.25 | 4.79 | 4.56 | 4.5 |
| 3900 – 4000 | 4.1 | 4.48 | 4.25 | 4.33 |
| 4000 – 4100 | 4.1 | 4.45 | 4.23 | 4.14 |
| 4100 – 4200 | 3.85 | 4.45 | 4.14 | 3.89 |
| 4200 – 4300 | 4.13 | 4.71 | 4.32 | 4.05 |
| 4300 – 4400 | 4.03 | 4.87 | 4.64 | 4.38 |
| 4400 – 4500 | 4.32 | 4.95 | 4.67 | 4.51 |
| 4500 – 4600 | 6.56 | 8.29 | 7.08 | 8.37 |
| 4600 – 4700 | 7.62 | 10.81 | 8.38 | 8.3 |
| 4700 – 4800 | 4.14 | 4.87 | 4.47 | 4.23 |
| 4800 – 4900 | 5.66 | 7.26 | 5.49 | 5.87 |
| 4900 – 5000 | 6.56 | 8.19 | 6.82 | 7.18 |
| 5000 – 5100 | 5.11 | 6.24 | 5.19 | 5.46 |
| 5100 – 5200 | 5.39 | 6.24 | 5.57 | 5.62 |
| 5200 – 5300 | 6.14 | 8.76 | 6.63 | 6.29 |
| 5300 – 5400 | 6.78 | 8.39 | 7.35 | 7.58 |
| 5400 – 5500 | 5.02 | 5.85 | 6.01 | 5.69 |
| 5500 – 5600 | 4.39 | 5.10 | 4.75 | 4.52 |
| 5600 – 5700 | 4.48 | 5.35 | 4.74 | 4.6 |
| 5700 – 5800 | 5.09 | 6.78 | 5.43 | 5.31 |
| 5800 – 5900 | 6.72 | 10.40 | 8.74 | 6.14 |
| 5900 – 6000 | 7.37 | 9.78 | 8.3 | |
| 6000 – 6100 | 4.98 | 6.15 | 5.47 | |
| 6100 – 6200 | 5.37 | 6.97 | 6.02 | |
| 6200 – 6300 | 6.39 | 9.56 | 7.59 | |
| 6300 – 6400 | 5.19 | 7.74 | 6.19 | |
| 6400 – 6500 | 6.48 | 8.91 | 8.03 | |
| 6500 – 6600 | 4.39 | 7.21 | 5.10 | |
| 6600 – 6700 | 5.18 | 7.54 | 5.99 | |
| 6700 – 6800 | 5.39 | 8.81 | 6.67 | |
| 6800 – 6900 | | 7.74 | 11.16 | 9.61 |
| 6900 – 7000 | | 6.44 | 10.46 | 7.64 |
| 7000 – 7100 | | 7.54 | 10.57 | 8.63 |
| 7100 – 7200 | | 11.91 | 16.18 | 12.45 |
| 7200 – 7300 | | 9.59 | 10.72 | 9.28 |

| Depth range, m | Well volume, m | |
|---|---|---|
| | July 30, 1978 | Oct 2, 1 |
| 7300 – 7400 | | |
| 7400 – 7500 | | |
| 7500 – 7600 | | 11.7 |
| 7600 – 7700 | | 11.7 |
| 7700 – 7800 | | 9.2 |
| 7800 – 7900 | | 10.2 |
| 7900 – 8000 | 8.99 | 7.1 |
| 8000 – 8100 | 10.78 | 6.4 |
| 8100 – 8200 | 10.29 | 10.2 |
| 8200 – 8300 | 10.63 | 9.5 |
| 8300 – 8400 | 8.86 | 7.7 |
| 8400 – 8500 | | 6.6 |
| 8500 – 8600 | | 10.6 |
| 8600 – 8700 | | |
| 8700 – 8800 | | |
| 8800 – 8900 | | |
| 8900 – 9000 | | |
| 9000 – 9100 | | |
| 9100 – 9200 | | |
| 9200 – 9300 | | |
| 9300 – 9400 | | |
| 9400 – 9500 | | |
| 9500 – 9600 | | |
| 9600 – 9700 | | |
| 9700 – 9800 | | |
| 9800 – 9900 | | |
| 9900 – 10000 | | |
| 10000 – 10100 | | |
| 10100 – 10200 | | |
| 10200 – 10300 | | |
| 10300 – 10400 | | |
| 10400 – 10500 | | |
| 10500 – 10600 | | |
| 10600 – 10700 | | |
| 10700 – 10800 | | |
| 10800 – 10900 | | |
| 10900 – 11000 | | |
| 11000 – 11100 | | |
| 11100 – 11200 | | |
| 11200 – 11300 | | |
| 11300 – 11400 | | |
| 11400 – 11500 | | |

| October 30, 1978 | February 27, 1979 | July 15, 1979 | October 6, 1979 | March 5, 1980 | June 17, 1980 | December 25, 1983 | April 14, 1981 | October 8, 1982 |
|---|---|---|---|---|---|---|---|---|
| | 7.88 | | | | | | | |
| | 8.31 | | | | | | | |
| 13.04 | 12.29 | | | | | | | |
| 13.44 | 11.85 | | | | | | | |
| 11.22 | 10.48 | | | | | | | |
| 12.45 | 10.68 | | | | | | | |
| 8.47 | 8.08 | | | | | | | |
| 10.9 | 8.23 | | | | | | | |
| 10.3 | 10.67 | | | | | | | |
| 9.35 | 9.2 | | | | | | | |
| 8.49 | 7.58 | | | | | | | |
| 6.19 | 6.32 | | | | | | | |
| 10.69 | 10.15 | | | | | | | |
| 5.01 | 5.35 | | | | | | | |
| | 5.76 | | | | | | | |
| | 8.37 | | | | | 7.92 | | |
| | 7.52 | 7.96 | | 9.2 | | 8.08 | | |
| | | 7.72 | 7.11 | 8.21 | | 7.18 | | |
| | | 9.47 | 9.19 | 9.21 | | 8.51 | | |
| | | 7.72 | 7.46 | 8.98 | | 7.31 | | |
| | | 7.73 | 7.54 | 9.2 | | 7.51 | | |
| | | 8.00 | 8.55 | 9.73 | 7.69 | 7.89 | | |
| | | 6.85 | 10.84 | 7.04 | 5.71 | 6.05 | | |
| | | | | 7.83 | 6.4 | 6.61 | | |
| | | | | | 7.9 | 6.73 | | |
| | | | | | 9.7 | 9.24 | | |
| | | | | | 5.84 | 5.89 | | |
| | | | | | 6.82 | 6.49 | | |
| | | | | | 6.63 | 6.55 | | |
| | | | | | 8.28 | 7.96 | | |
| | | | | | | 7.99 | | |
| | | | | | | 7.19 | | |
| | | | | | | 8.07 | | |
| | | | | | | | 9.16 | 9.81 |
| | | | | | | | 8.72 | 9.10 |
| | | | | | | | 5.78 | 8.50 |
| | | | | | | | 7.68 | 10.43 |
| | | | | | | | | 8.28 |
| | | | | | | | | 7.40 |
| | | | | | | | | 6.84 |
| | | | | | | | | 7.24 |
| | | | | | | | | 7.25 |

**Table 3.18.** Shut-in and circulated downhole temperatures versus depth (measurements with GSRT-4)

| Temperature measurement depth, m | Temperature, °C | | Drill string length, m |
|---|---|---|---|
| | Shut-in | Circulated | |
| 890 | 34.6 | – | – |
| 2990 | 45.1 | – | – |
| 3470 | 51.5 | – | – |
| 4000 | 58.6 | – | – |
| 4540 | 67.4 | – | – |
| 4850 | 75.0 | – | – |
| 6015 | 93.2 | 82.5[a] | – |
| 6275 | 97.5 | 90.6[a] | 10600 |
| 6510 | 110.5 | 103.6[a] | 10160 |
| 6950 | 116.6 | 109.5[a] | 10160 |

[a] Drilling mud circulation rate is 32 to 34 $l\,s^{-1}$.

**Table 3.19.** Shut-in and circulated well temperatures versus depth (measurements with GN4)

| Temperature measurement depth, m | Temperature, °C | |
|---|---|---|
| | Shut-in | Circulated |
| 3960 | 56.8 | 43.0[a] |
| 6350 | 98.2 | 89.0[a] |
| 7800 | 123.0 | – |
| 8230 | 131.0 | 123.0[a] |
| 10425 | 167.8 | 161.0[a] |
| 10909 | 181.0 | 147.6 |
| 10909 | 181.0 | 146.6 |
| 10909 | 185.4 | – |
| After 42-h shutdown | | |
| 10909 | 185.4 | 150.6 |
| 10100 | 132.0 | – |

[a] Temperature was measured after intermediate flushings (15 to 20 min); drilling mud circulation rate is 34 $l\,s^{-1}$.

To obtain actual temperature, experimental operations have been carried out directly while circulating, using special self-contained thermometers. Measurements were taken in the drill string at depths of 6015, 6275, 6510, 6950 and 10,909 m.

Results of temperature measurements with GSRT-4 and GN-4 instruments are shown in Tables 3.18 and 3.19 respectively. Figs. 3.64 – 3.66 illustrate graphs constructed from these data.

After analyzing temperature data, one may draw the following conclusions:

– the difference in temperature between the ascending and descending drilling mud flows both during drilling and washing does not exceed 40 °C;
– the distribution of the heat flow with depth is governed by the linear law;

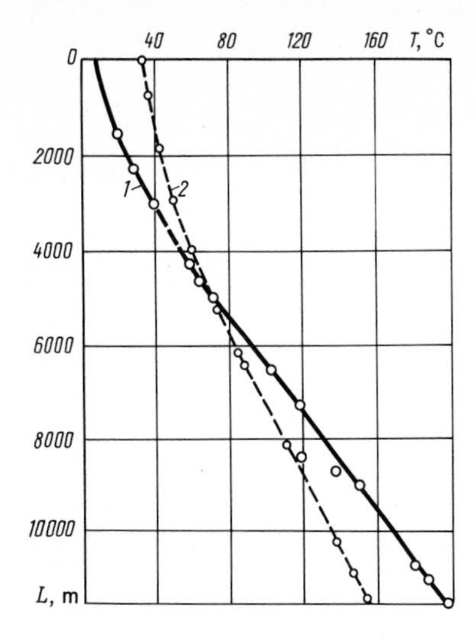


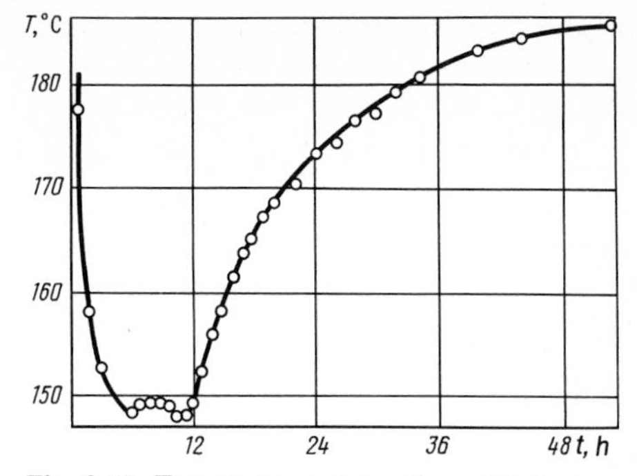


**Fig. 3.65.** Temperature restoration with time

**Fig. 3.64.** Graph showing distribution of temperature in hole: *1* shut-in temperature; *2* flowing temperature at $Q = 32\,l\,s^{-1}$

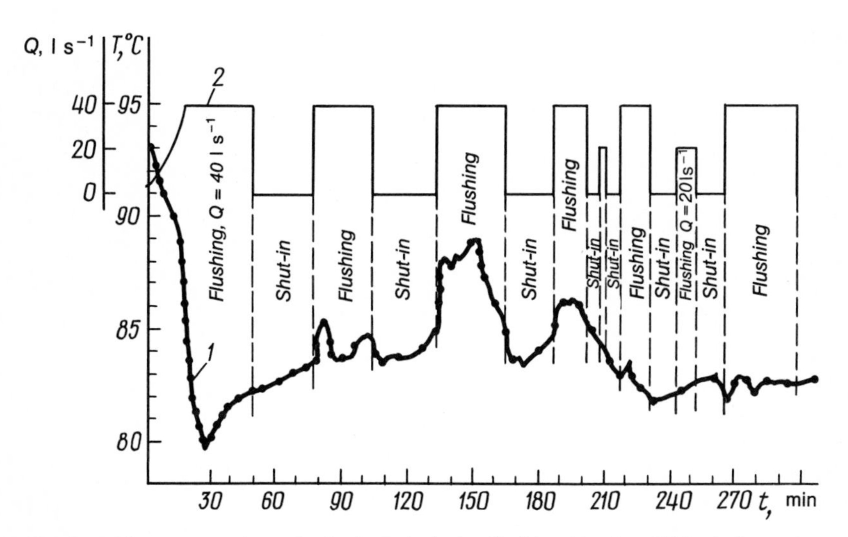


**Fig. 3.66.** Temperature dynamics in the hole during flushing (depth = 6015 m): *1* temperature; *2* circulation rate

– the zone of equilibrium of a temperature gradient in the drilling mud column ($Q = 30-40\,l\,s^{-1}$) is at a depth of 5000 m;
– the period of time required to restore the temperature conditions in the well does not exceed 50 h;
– the rate of temperature field restoration points to minor thermal effects in the lower portion of the well as compared with higher (down to 5000 m) zones, and the temperature measured in the bottomhole region is close to that calculated by geothermal gradient.

## Depth-Dependent Variation of Service Life of Downhole Motors

In the first stage of SG-3 drilling, use was made mostly of high-torque (500 to 700 rpm) spindleless (A7N4S) turbodrills with inclined pressure curve and spindle (A7Sh) turbodrills with bearing assemblies running under severe service conditions. High rotation speed (500 – 700 rpm) of the turbodrills should be noted. Positioned below the shaft were assemblies consisting of three sections of a KDM core sampler with a total length of over 20 m, pendulum drill collar assemblies, lead pendulums, etc.

Radial thrust bearings made of 55SM5FA steel and heat treated to ensure maximum impact strength (HRC 44 – 47) often failed under these conditions. Bearing races and especially balls used to break down. The balls, made of 55SM5FA steel with a hardness of 50 to 62 HRC, had, therefore, to be tempered. The bearing races and balls were soon severely abraded and loosened, if not broken down, under the action of high loads while running in the highly abrasive drilling mud, i.e. a turbodrill had to be repaired due to the high axial play of its bearing assembly. That is why the overhaul period of a turbodrill was from 7 – 10 to 25 – 30 h and the failure interval from 16 to 18 h in the first stage of drilling.

The life of downhole motors largely increased in the second stage, as motors and bearing assemblies of an entirely new design were put to use (see Table 3.20).

The use of 3A7 and 3A7GT turbodrills with ShIP-2 torsional spindles to dampen axial vibration caused by rock cutting tools has made it possible to increase drilling time and penetration per run, as well as to extend the turbodrill overhaul period. Two pre-unloaded axial bearings mounted on the spindle allow their life to be increased from 2.5 to 3 times without turbine malfunction, i.e. without its rotors landing on stators.

The overhaul period of 3A7GTSh and 4A7GTSh low-rpm turbodrills largely increased primarily due to lower (250 to 350 rpm) rotation speed. The use of square couplings and a composite rolling-and-journal bearing has improved bearing performance in terms of both load and protection from abrasive medium. The independently suspended turbine sections with liquid dampers arranged between these sections and spindles have extended the turbodrill service life up to more than 50 h.

As drilling progressed, the temperature conditions of the hole placed increasingly stringent requirements on downhole motors. This is most true of D2-172M positive displacement motors, which ensured far greater penetration per run due to the lower rotation speed with a satisfactory durability of the working pair (the short overhaul period specified in Table 3.20 is a disadvantage of the articulated joint). The rubber lining of the stator used to break and peel off at a shut-in bottom-hole temperature between 140° and 150 °C. Heat-resistant rubber and adhesives used in stator curing did not work at a depth of 9000 m at about 160 °C.

Tests of reduction gear turbodrills (RM) have shown that the design of the reduction gear is inadequate for drilling at depths exceeding 8100 m. The first reduction gear with a failure interval of 290 h in the Tatar region was under operation for 15.4 h only. After examining broken components, we came to the conclusion that these were subjected to intensive transverse, torsional and longitudinal vibrations. The failure interval of the modified version (RM-195) increased up to 84 h, which far exceeded that of downhole motors used in the same depth range.

**Table 3.20.** Downhole motor performance in Kola well

| Motor | Depth range 7263 – 7460 m | | | | |
|---|---|---|---|---|---|
| | Penetration, m | Number of runs | Per run | | Overhaul period, h |
| | | | Penetration, m | Drilling time, h | |
| 3A7NAS | | | | | |
| 3A7Sh | 168.0 | 16 | 10.5 | 4.9 | 27.7 |
| 3A7 + ShIP-2 | | | | | |
| 3A7GT + ShIP-2 | | | | | |
| 3A7GTSh | | | | | |
| 3A7GTSh-I | | | | | |
| 4A7GTSh | | | | | |
| D2-172M | | | | | |
| A7Sh-RM | | | | | |
| A7GTSh-RM | | | | | |

| Motor | Depth range 7460 – 8028 m | | | | |
|---|---|---|---|---|---|
| | Penetration, m | Number of runs | Per run | | Overhaul period, h |
| | | | Penetration, m | Drilling time, h | |
| 3A7NAS | 113.7 | 15 | 7.6 | 2.1 | 20.0 |
| 3A7Sh | 308.3 | 50 | 6.2 | 2.8 | 21.9 |
| 3A7 + ShIP-2 | 80.8 | 11 | 7.4 | 4.7 | 35.1 |
| 3A7GT + ShIP-2 | | | | | |
| 3A7GTSh | | | | | |
| 3A7GTSh-I | | | | | |
| 4A7GTSh | | | | | |
| D2-172M | | | | | |
| A7Sh-RM | | | | | |
| A7GTSh-RM | | | | | |

| Motor | Depth range 8028 – 8914 m | | | | |
|---|---|---|---|---|---|
| | Penetration, m | Number of runs | Per run | | Overhaul period, h |
| | | | Penetration, m | Drilling time, h | |
| 3A7NAS | | | | | |
| 3A7Sh | 296.3 | 54 | 5.5 | 2.7 | 23.4 |
| 3A7 + ShIP-2 | | | | | |
| 3A7GT + ShIP-2 | 48.3 | 7 | 6.9 | 2.8 | 33.6 |
| 3A7GTSh | 189.7 | 31 | 6.1 | 2.9 | 25.2 |
| 3A7GTSh-I | | | | | |
| 4A7GTSh | | | | | |
| D2-172M | 101.1 | 14 | 7.2 | 6.7 | 15.0 |
| A7Sh-RM | | | | | |
| A7GTSh-RM | 174.7 | 17 | 10.3 | 6.3 | 38.3 |

**Table 3.20** (continued)

| Motor | Depth range 8914 – 9711 m | | | | |
|---|---|---|---|---|---|
| | Penetration, m | Number of runs | Per run | | Overhaul period, h |
| | | | Penetration, m | Drilling time, h | |
| 3A7NAS | | | | | |
| 3A7Sh | 41.1 | 7 | 5.9 | 3.5 | 23.7 |
| 3A7 + ShIP-2 | | | | | |
| 3A7GT + ShIP-2 | 178.0 | 22 | 8.1 | 2.9 | 33.1 |
| 3A7GTSh | | | | | |
| 3A7GTSh-I | 200.6 | 21 | 9.6 | 2.7 | 50.8 |
| 4A7GTSh | 220.6 | 22 | 10.0 | 3.5 | 30.9 |
| D2-172M | 29.8 | 6 | 5.0 | 2.4 | 13.4 |
| A7Sh-RM | | | | | |
| A7GTSh-RM | 114.8 | 14 | 8.2 | 4.2 | 25.7 |

| Motor | Depth range 9711 – 10636 m | | | | |
|---|---|---|---|---|---|
| | Penetration, m | Number of runs | Per run | | Overhaul period, h |
| | | | Penetration, m | Drilling time, h | |
| 3A7NAS | | | | | |
| 3A7Sh | 28.7 | 5 | 5.7 | 2.6 | 10.9 |
| 3A7 + ShIP-2 | | | | | |
| 3A7GT + ShIP-2 | 66.1 | 9 | 7.3 | 3.4 | 35.3 |
| 3A7GTSh | | | | | |
| 3A7GTSh-I | | | | | |
| 4A7GTSh | | | | | |
| D2-172M | | | | | |
| A7Sh-RM | 770.9 | 88 | 8.8 | 3.3 | 22.4 |
| A7GTSh-RM | | | | | |

However, as drilling progressed and bottom-hole temperature rose higher than 150 °C, which exceeds the limiting values for materials used in sealing rings and a lubricator diaphragm, the average failure interval of the RM-195 unit shortened noticeably to 42 h. Its design had to be updated again; the diaphragm lubricator was replaced with the one of a piston type, heat-resistant rubber was used for sealing rings and a C-52 cylinder oil substituted the MK-22 or MS-20 aviation oil. As a result, the average failure interval of the RM-195 unit increased up to 107.9 h.

As the bottom-hole temperature increased to 180 °C at a depth of about 10,000 m, almost all the rubberized components of the turbine sections and shafts, radial bearings of turbines (OSI-195/95), spindles (ON-195/120) and axial bearings (3TSSh1-195) used as spindle stuffing boxes proved to be unserviceable. For lack of heat-resistant adhesive it was not possible to use heat-resistant rubber and other mixes specially developed for the Kola well, since the rubber used to peel off the metal parts.

Journal bearings were, therefore, put to use in all the downhole motors with pressed-in bronze bushings instead of rubber linings. With the bronze-steel sliding pair employed, the radial bearings of the turbine sections were running for 20 to 25 h and those of the spindles, for 12 to 15 h before a limiting radial play developed. The low performance of motors and, consequently, continuous repairs and downtime have necessitated application of a number of experimental journal bearings.

Bearings with heat-resistant fluorine rubber and adhesive have shown high performance (up to 60 h running time). Positive results have also been obtained with bearings in which taper slots milled in the metal parts were filled in with heat-resistant rubber that was thus attached to the parts. The life of such bearings developed in the Kola Geological Superdeep Drilling Expedition ranged from 65 to 70 h. Work still remains to be done to ensure higher performance of downhole motors at a temperature of 200 °C and above. For this, oil-filled bearing assemblies, bearings employing hard-alloy and other composite materials, as well as heat-resistant polymers are being developed at present.

## Drill Pipe Wear and Consumption

The wear of a set of drill pipes in drilling is a total of its wear in different depth ranges

$$S = \sum_{1}^{n} LKCa \ ,$$

where $S$ is the wear of a set of drill pipes in a particular depth range, kg; $L$ is the distance covered by a set of drill pipes in this depth range, m; $K$ is the depth factor; $C$ is the coefficient determining penetration per run in the given region; $a$ is the average consumption rate of drill pipes per 1 m of penetration.

The above formula fails, however, to take proper account of actual work performed by a set of drill pipes since wear, as defined above while being dependent on the meters drilled, does not allow for round trip operations which are responsible for major wear of drill pipes in turbodrilling in an deep open well.

It seems to be appropriate to calculate wear with respect to the outer diameter of a set of drill pipes as a function of its location in a drill string, as well as of respective frictional forces. This integral value is determined by the distance covered by a set of drill pipes in round trip operations.

This approach involves drill pipe loads which determine pressing force and the distance covered by a set of drill pipes in the hole, i.e., work performed by resistance forces,

$$A_i = \sum_{1}^{n} L_i(F_{ip} + F_{ir}) \ ,$$

where $A_i$ is the work performed by resistance forces per $n$ runs; $L_i$ is the location of a set of drill pipes in a string (distance from the well head), km; $F_{ip}$, $F_{ir}$ are resistance forces in pulling out and running in of a tool respectively, $N$.

The experimental studies undertaken in serviceability checks of drill pipe sets (Table 3.21) have confirmed a correlation interdependence between the wear of the outer surface of drill pipes and work performed. This interdependence has a parabolic configuration

$$\delta_i = \mathrm{a}A_i^2 + \mathrm{b}A_i + \mathrm{c}$$

(a, b, c are constant coefficients for the hole).

Thus, after determining the work performed by a set of drill pipes to overcome resistance forces, we can estimate its wear.

This approach was used for determining drill pipe wear, as well as frequency and priority of non-destructive tests, relocating drill pipe sets within the string sections to ensure their uniform utilization, etc. The operation of aluminium and steel drill pipes in the first drilling stage has resulted in their commensurable wear, thus making aluminium pipes fairly effective in drilling at great depths. The main reason for which drill pipes have been discarded in the first and second stages is a worn out Z-147 tool-joint thread responsible for removing drill pipes (class 1) out of operation and using them in wells of secondary importance.

Data on drill pipe consumption in the first and second stages of SG-3 drilling, including rejection, are specified in Tables 3.22 and 3.23 respectively.

## Dynamics of Drag Force Variation

The operating conditions of a drill string in the well in impervious crystalline rocks have the following distinguishing features: large hole depth; considerable amount of round trip operations; fairly extensive length of the open pilot hole; intricate hole path; high abrasiveness and vugular porosity of a hole of elliptical cross-section; high sludging-up of a bottom hole region; absence of wall mud cake; absence of pressure differential (rock imperviousness); absence of hydrophilic rocks and adhesive forces between drill pipes and sidewalls.

These features of interaction between a drill string, sidewalls and drilling mud account for greater tool drag forces, i.e. a total of forces responsible for higher load on a hook and in different sections of a drill string in pulling out tools, as well as for its higher torque in rotation. The drag forces are, therefore, dependent mostly on physical and chemical rock properties and mud composition, path, section and length of open well, as well as on drill string cross-radial dimensions and material.

The depth-dependent variation of drag forces in pulling out tools is shown in Fig. 3.67. The depth-dependent variation of drill string torque is similar to that of drag forces.

Both drag forces and torque have been determined stage-wise in pulling a drill string, its design weight, readings of ground monitoring devices and the effects of parameters of the block and tackle system upon hook load measuring accuracy being allowed for.

**Table 3.21.** Drill string wear in Kola well

| Set No. (if counted from hole bottom) | Depth range 7200 – 7700 m (100 runs) | | | Depth range 8500 – 9200 m (120 runs) | | | Depth range 9600 – 10700 m (95 runs) | | |
|---|---|---|---|---|---|---|---|---|---|
| | Location[a], km | Work performed, kN · km | Relative wear, % | Location, km | Work performed, kN · km | Relative wear, % | Location, km | Work performed, kN · km | Relative wear, % |
| 1 | 7.5 | 60000 | 1.1 | 8.75 | 168000 | 2.8 | 10.00 | 180500 | 3.2 |
| 2 | 7.25 | 87000 | 1.2 | 8.50 | 193800 | 3.2 | 9.75 | 213040 | 3.8 |
| 3 | 7.0 | 98000 | 1.2 | 8.25 | 217800 | 3.4 | 9.50 | 252700 | 4.6 |
| 4 | 6.75 | 168750 | 2.6 | 8.00 | 384000 | 4.3 | 9.25 | 360290 | 5.8 |
| 5 | 6.5 | 195000 | 3.1 | 7.75 | 492900 | 4.8 | 9.00 | 444600 | 6.5 |
| 6 | 6.25 | 231000 | 3.4 | 7.50 | 576000 | 5.9 | 8.75 | 548620 | 8.2 |
| 7 | 6.00 | 252000 | 3.8 | 7.25 | 582900 | 7.3 | 8.50 | 557170 | 9.1 |
| 8 | 5.75 | 258750 | 4.6 | 7.00 | 596400 | 10.6 | 8.25 | 595650 | 9.6 |
| 9 | 5.5 | 269000 | 5.3 | 6.75 | 639900 | 12.2 | 8.00 | 615600 | 9.8 |
| 10 | 5.25 | 299250 | 6.0 | 6.50 | 702000 | 13.5 | 7.75 | 677350 | 10.1 |
| 11 | 5.00 | 290000 | 6.2 | 6.25 | 705000 | 13.0 | 7.50 | 662620 | 10.3 |
| 12 | 4.75 | 289750 | 5.4 | 6.00 | 698400 | 12.6 | 7.25 | 654310 | 9.7 |
| 13 | 4.50 | 288000 | 5.4 | 5.75 | 690000 | 12.0 | 7.00 | 651700 | 9.5 |
| 14 | 4.25 | 276000 | 5.2 | 5.50 | 679800 | 11.4 | 6.75 | 641250 | 8.6 |
| 15 | 4.00 | 264000 | 5.0 | 5.25 | 667800 | 11.0 | 6.50 | 636020 | 8.2 |
| 16 | 3.75 | 247500 | 4.3 | 5.00 | 660000 | 11.2 | 6.25 | 623430 | 8.0 |
| 17 | 3.50 | 238000 | 4.1 | 4.75 | 644100 | 10.8 | 6.00 | 621300 | 7.4 |
| 18 | 3.25 | 237250 | 3.6 | 4.50 | 626400 | 9.5 | 5.75 | 617260 | 7.1 |
| 19 | 3.00 | 222000 | 2.4 | 4.25 | 601800 | 8.7 | 5.50 | 606100 | 6.7 |
| 20 | 2.75 | 209000 | 2.2 | 4.00 | 585600 | 8.2 | 5.25 | 588520 | 6.2 |
| 21 | 2.50 | 192500 | 2.0 | 3.75 | 571500 | 6.3 | 5.00 | 584250 | 5.6 |
| 22 | 2.25 | 180000 | 1.6 | 3.50 | 554400 | 5.6 | 4.75 | 577600 | 5.2 |
| 23 | 2.00 | 162000 | 1.2 | 3.25 | 530400 | 4.8 | 4.50 | 566020 | 5.0 |
| 24 | | | | 3.00 | 511200 | 3.9 | 4.25 | 549100 | 4.6 |
| 25 | | | | 2.75 | 488400 | 2.8 | 4.00 | 539600 | 4.1 |
| 26 | | | | 2.50 | 453000 | 2.6 | 3.75 | 527250 | 3.9 |
| 27 | | | | 2.25 | 415800 | 2.1 | 3.50 | 508720 | 3.2 |
| 28 | | | | | | | 3.25 | 481650 | 2.8 |
| 29 | | | | | | | 3.00 | 447450 | 2.4 |
| 30 | | | | | | | 2.75 | 415390 | 2.2 |
| 31 | | | | | | | 2.5 | 384750 | 2.0 |
| 32 | | | | | | | 2.25 | 350550 | 1.9 |
| 33 | | | | | | | 2.00 | 311600 | 1.8 |

[a] Location is a distance between the middle portion of a pipe set and the well head.

**Table 3.22.** Data on drill pipe consumption in the first stage of SG-3 drilling down to 7263 m

| Parameter | LBT-147 × 11 | LBT-147 × 13 | LBT-147 × 15 | SBT |
|---|---|---|---|---|
| Total of pipes used | | | | |
| m | 19025 | 4314 | 1525 | 7659 |
| t | 310 | 77 | 25 | 288 |
| Pipes discarded or downgraded for the following reasons: | | | | |
| worn-out pipe thread | | | | |
| m | 4277 | 4314 | – | – |
| t | 68 | 77 | – | – |
| worn-out tool-joint thread | | | | |
| m | 14748 | – | – | 7959 |
| t | 242 | – | – | 288 |
| worn-out body or upset portion, | | | | |
| m | – | – | 1525 | – |
| Average life of pipes, runs | 365 | 150 | 108 | – |
| Consumption rate of metal per 1 m penetration[a], kg | 42.3 | 32.0 | 52.4 | 40 |

[a] Consumption rate per 1 m penetration does not include 4277 m of LBT-147 × 11 mm pipes and 2260 m of LBT-147 × 13 mm pipes due to their short life (damaged pipe thread).

**Table 3.23.** Data on drill pipe consumption in the second stage of SG-3 drilling from 7263 to 11500 m

| Parameter | LBTVK-147 (D16T) | LBTVK-147 (O1953) | LBTVK-147 (AK4-1) | SBT |
|---|---|---|---|---|
| Total of pipes used | | | | |
| m | 19200 | 15100 | 5900 | 7700 |
| t | 306 | 256 | 89 | 252 |
| Pipes discarded or downgraded for the following reasons: | | | | |
| worn-out tool-joint thread | | | | |
| m | 4922 | 15100 | – | 1550 |
| t | 77 | 256 | – | 50 |
| worn-out body or tool joint | | | | |
| m | 9644 | – | 5900 | 6150 |
| t | 153 | – | 89 | 202 |
| wash-out, m | 4434 | – | – | – |
| Average life of pipes, run | 291 | 367 | 250 | 470 |
| Consumption rate of pipes per 1 m penetration, kg | 72 | 60 | 20.9 | 59.3 |

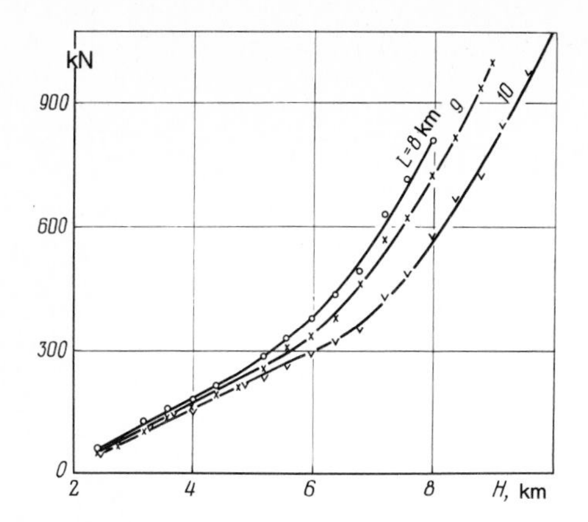


**Fig. 3.67.** Graph showing resistance forces versus hole depth

Drag forces measured in pulling a drill string out of the SG-3 hole at a 10,000 m depth are given in Table 3.24, and drill string torque at a 10,250 m depth is given in Table 3.25.

To determine distribution of drag forces in the hole, the use was made of a stage-wise method in pulling a drill string out of the bottom hole. If $F_{L1}$, $F_{L2}$, $F_{L3} \ldots$, $F_{Ln}$ are drag forces as determined by hook load at different drill string lengths (in pulling out the first, second, third, etc. drill pipe stand) drag forces at depth $L$ are

$$F_L = \frac{G_{L\mathrm{h}} - G_{\mathrm{bt}}}{\eta_{\mathrm{bth}}} - P_{L\mathrm{d}} \; ,$$

where $G_{L\mathrm{h}}$ is the hook load as read by the instrument; $G_{\mathrm{bt}}$ is the weight of the block and tackle system; $\eta_{\mathrm{bth}}$ is the efficiency of the block and tackle system in hoisting; $P_{L\mathrm{d}}$ is the design weight of a drill string to be pulled out.

Drag force increment in the depth range between the first and second measurements is as follows:

$$\Delta F_1 = F_{L1} - F_{L2} = \left(\frac{G_{L1\mathrm{h}} - G_{\mathrm{bt}}}{\eta_{\mathrm{bth}}} - P_{L1\mathrm{d}}\right) - \left(\frac{G_{L2\mathrm{h}} - G_{\mathrm{bt}}}{\eta_{\mathrm{bth}}} - P_{L2\mathrm{d}}\right) \; ;$$

$$\Delta F_1 = \frac{G_{L1\mathrm{h}} - G_{L2\mathrm{h}}}{\eta_{\mathrm{bth}}} - (P_{L1\mathrm{d}} - P_{L2\mathrm{d}}) \; .$$

In other depth ranges

$$\Delta F_2 = \frac{G_{L2\mathrm{h}} - G_{L3\mathrm{h}}}{\eta_{\mathrm{bth}}} - (P_{L2\mathrm{d}} - P_{L3\mathrm{d}}) \; .$$

$$\Delta F_n = \frac{G_{Ln\mathrm{h}} - G_{L(n+1)\mathrm{h}}}{\eta_{\mathrm{bth}}} - (P_{Ln\mathrm{d}} - P_{L(n+1)\mathrm{d}}) \; .$$

The smaller the interval between two successive measurements, the higher is the accuracy of determination of resistance forces in the well.

**Table 3.24.** Drag force versus depth in the Kola well

| Drill pipe stand no. | Drill string length, m | Hook load, kN | | Design weight of drill string, kN | Hoisting speed, m $s^{-1}$ | Resistance force, kN |
|---|---|---|---|---|---|---|
| | | Per instrument | Accounting for efficiency of block and tackle system | | | |
| 268 | 9990 | 2600 | 2670 | 1550 | 0.3 | 1120 |
| 265 | 9866 | 2680 | 2760 | 1510 | 1.0 | 1250 |
| 260 | 9681 | 2600 | 2670 | 1450 | 1.0 | 1220 |
| 255 | 9495 | 2500 | 2560 | 1390 | 1.0 | 1170 |
| 250 | 9308 | 2360 | 2410 | 1330 | 1.0 | 1080 |
| 245 | 9121 | 2220 | 2250 | 1270 | 1.0 | 980 |
| 240 | 8934 | 2120 | 2140 | 1200 | 1.0 | 940 |
| 235 | 8749 | 2040 | 2060 | 1160 | 1.0 | 900 |
| 230 | 8564 | 2000 | 2010 | 1130 | 1.0 | 880 |
| 225 | 8378 | 1900 | 1900 | 1100 | 1.0 | 800 |
| 220 | 8193 | 1830 | 1830 | 1070 | 1.0 | 760 |
| 215 | 8006 | 1750 | 1740 | 1040 | 1.0 | 700 |
| 210 | 7818 | 1680 | 1660 | 1020 | 1.0 | 640 |
| 205 | 7633 | 1620 | 1600 | 990 | 1.0 | 610 |
| 200 | 7445 | 1560 | 1530 | 960 | 1.0 | 570 |
| 195 | 7258 | 1500 | 1470 | 930 | 1.0 | 540 |
| 190 | 7074 | 1460 | 1420 | 910 | 1.0 | 510 |
| 185 | 6889 | 1400 | 1360 | 880 | 1.0 | 480 |
| 180 | 6705 | 1340 | 1290 | 860 | 1.0 | 430 |
| 175 | 6514 | 1300 | 1250 | 830 | 1.0 | 420 |
| 170 | 6333 | 1260 | 1200 | 810 | 1.0 | 390 |
| 165 | 6145 | 1240 | 1180 | 780 | 1.0 | 400 |
| 160 | 5960 | 1160 | 1090 | 760 | 1.0 | 330 |
| 155 | 5772 | 1120 | 1050 | 740 | 1.0 | 310 |
| 150 | 5587 | 1080 | 1010 | 720 | 1.0 | 300 |
| 145 | 5400 | 1050 | 970 | 690 | 1.0 | 290 |
| 140 | 5212 | 1030 | 950 | 670 | 1.0 | 280 |
| 135 | 5027 | 990 | 910 | 650 | 1.0 | 260 |
| 130 | 4840 | 950 | 860 | 620 | 1.0 | 240 |
| 125 | 4654 | 910 | 820 | 600 | 1.0 | 220 |
| 120 | 4463 | 880 | 790 | 580 | 1.0 | 210 |
| 115 | 4277 | 850 | 750 | 560 | 1.0 | 190 |
| 110 | 4090 | 820 | 720 | 530 | 1.0 | 190 |
| 105 | 3902 | 790 | 690 | 510 | 1.0 | 180 |
| 100 | 3715 | 760 | 660 | 490 | 1.0 | 170 |
| 95 | 3528 | 710 | 600 | 470 | 1.0 | 130 |
| 90 | 3343 | 680 | 590 | 440 | 1.0 | 130 |
| 85 | 3163 | 660 | 550 | 420 | 1.0 | 130 |
| 80 | 2980 | 630 | 510 | 390 | 1.0 | 120 |
| 75 | 2798 | 600 | 480 | 370 | 1.0 | 110 |
| 70 | 2616 | 580 | 460 | 340 | 1.0 | 120 |
| 65 | 2434 | 550 | 430 | 310 | 1.0 | 120 |
| 60 | 2250 | 510 | 380 | 290 | 1.0 | 90 |
| 55 | 2069 | 480 | 350 | 270 | 1.0 | 80 |

**Table 3.25.** Drill string torque versus depth in the Kola well

| Drill pipe stand no. | Drill string length, m | Rotor torque at n = 2 rpm, N · m |
|---|---|---|
| 80 | 3011 | 500 |
| 108 | 4060 | 3100 |
| 134 | 5010 | 6000 |
| 161 | 6026 | 10500 |
| 188 | 7030 | 13000 |
| 215 | 8031 | 17000 |
| 239 | 8920 | 21300 |
| 269 | 10035 | 27300 |
| 274 | 10213 | 27400 |

**Table 3.26.** Experimental data and drag force in the Kola well

| Sensor installation depth, m | Length of drill string below sensor, m | Design weight of drill string below sensor, kN | Recorded drill stand elongation, mm | Elongation factor in calibration | Actual load on reference drill stand, kN | Resistance forces, kN | |
|---|---|---|---|---|---|---|---|
| | | | | | | For reference drill stand | For hook |
| 1028 | 9590 | 1286 | 127.0 | 1.75 | 2230 | 940 | 1020 |
| 2295 | 8323 | 1094 | 128.0 | 1.46 | 1870 | 780 | 820 |
| 4304 | 6314 | 814 | 115.0 | 1.26 | 1450 | 640 | 560 |
| 6584 | 4034 | 535 | 69.0 | 1.52 | 1050 | 520 | 470 |

The same method is used for measuring the drill string torque increment

$\Delta M_n = M_n - M_{n+1}$ .

The accuracy of the stage-wise method for determining drag forces within the length of a drill string has been checked experimentally (see Table 3.26) using self-contained downhole instruments positioned in different cross-sections of a drill string and recording its elongation within the length of a drill pipe stand depending on load applied.

The experimental tests have confirmed high accuracy (error up to 10%) in determination of drag forces as a function of the length of a drill string. The stage-wise method has been found of practical use in determining actual tensile loads for a uniform-strength drill string.

When analyzing depth-dependent variation of drag forces in SG-3 drilling, one may draw the following conclusions:

– the depth-dependent variation of drag forces is of parabolic configuration, the dependence in the upper sections of the hole being close to linear with a gradual increase of drag forces according to depth (maximum drag forces are observed in the bottom hole region – about 1000 m from the bottom hole);

– as drilling progresses and the overall length of the tool-side wall contact increases, the drag force increment tends to decrease, which is due to the smoothing of the side walls and, consequently, by the lower influence of the cross-section configuration;

**Table 3.27.** Drilling line performance in the Kola well

| Line length, m | Depth range, m | Number of runs made | Average life of drilling line, kN · km | Drilling line consumption, m | |
|---|---|---|---|---|---|
| | | | | Per run | Per 1 m penetration |
| 8850 | 7284 – 8010 | 200 | 495620 | 44.3 | 12.2 |
| 13084 | 8010 – 9025 | 219 | 425920 | 59.7 | 13.0 |
| 10086 | 9025 – 9993 | 157 | 487800 | 68.8 | 11.2 |
| 13406 | 9993 – 11059 | 180 | 490100 | 74.5 | 12.6 |
| 3192 | 11059 – 11514 | 36 | 516880 | 88.7 | 7.0 |
| 49338 | 7284 – 11514 | 792 | 489870 | 62.3 | 11.7 |

– friction of the drill string against the sidewalls is of a boundary type, which has been confirmed in studying the distribution of the friction coefficient in the hole, as well as in analyzing the influence exerted by contact pressure on the friction coefficient under laboratory conditions;

- drag forces are largely influenced by sludging-up of the hole and, especially, of the bottom hole region;
- shape of the hole cross-section and radial dimensions of drill string components exert great influence on drag forces;
- as the hoisting speed of tools in the elliptical section hole increases, drag forces tend to grow.

## Drilling Line Performance

The performance of a drilling line 38 mm in diameter is recorded by a special counter of the Kola expedition design. The counter provides for actual data on work performed by a drilling line in each run and helps to ascertain the relationship between this work and line wear. The drilling line performance enables us not only to predict its consumption at large depths, but also to assess round trip operations in general. This helps optimize drilling line performance using a counter and to predict its consumption depending on actual drilling conditions.

Data characterizing drilling line performance at depths ranging from 7284 to 11,500 m are given in Table 3.27.

## Coring

A total of 9235.2 m have been core drilled in the Kola superdeep hole (depth range 0 to 11,500 m); 3700.1 m of cores have been recovered, which makes a total of 40%. Table 3.28 presents data on core recovery.

| Depth range, m | 0–500 | 500–1000 | 1000–1500 | 1500–2000 | 2000–2500 | 2500–3000 | 3000–3500 | 3500–4000 | 4000–4500 | 6179–6202 |
|---|---|---|---|---|---|---|---|---|---|---|
| Average length of core sample, cm | 13.4 | 11 | 17.4 | 13.1 | 17.1 | 13.9 | 23.9 | 19.7 | 17.8 | 8.9 |
| Depth range, m | 6202–6300 | 6300–6403 | 6403–6411 | 7263–7299 | 7324–7410 | 7410–7508 | 7508–7558 | 7637–7647 | 7706–7809 | 7813–7902 |
| Average length of core sample, cm | 6.7 | 6.6 | 7.8 | 5.9 | 3.9 | 4.5 | 3.6 | 4.9 | 3.8 | 4.3 |
| Depth range, m | 7902–8006 | 8007–8101 | 8101–8190 | 8202–8302 | 8302–8400 | 8400–8494 | 8522–8589 | 8618–8701 | 8710–8803 | 8803–8901 |
| Average length of core sample, cm | 8 | 6.5 | 2.8 | 5.2 | 7.3 | 6 | 5 | 4.5 | 8.3 | 8.9 |
| Depth range, m | 8901–9003 | 9003–9103 | 9103–9202 | 9202–9299 | 9303–9404 | 9404–9510 | 9510–9610 | 9610–9682 | 9902–10000 | |
| Average length of core sample, cm | 3.7 | 3.9 | 4.3 | 2 | 1.8 | 2.7 | 2.2 | 2.5 | 2.2 | |
| Depth range, m | 10001–10105 | 10105–10202 | 10202–10301 | 10301–10360 | 10424–10502 | 10502–10608 | 10608–10703 | 10703–10760 | | |
| Average length of core sample, cm | 2.5 | 1.9 | 2 | 1.8 | 1.7 | 1.9 | 0.7 | 0.6 | | |

As is seen from Table 3.28, the hole can be divided into four sections.

Section 1 (depth range 0 to 4673) provides core samples with linear dimensions more than 10 cm fractured in conformity with rock bedding. Sample length varies from 11 to 23.9 cm. Penetration per run decreases with depth from 8.7 to 5.9 m. Core disking and self-disintegration are not observed. With an average penetration of 7 m per run, the linear core recovery is relatively high, reaching 3.5 to 3.6 m or 50 to 55%. Poor recovery is mostly due to core sticking in the inner tube.

Section 2 (depth range 4673 to 7263 m) features noticeable hole enlargement and high core crushability from about 5000 m downwards with the drilling techniques unchanged. Average penetration per run has grown by 30 to 70%, and the linear core recovery decreased by 1.5 to 2.5 times. High rock crushability added to core self-sticking in the inner tube. The average length of a core sample ranges from 6.6 to 9 cm. To prevent core self-sticking, experimental operations were

**Table 3.28.** Core recovery data

| Depth range, m | Section No | Number of runs | Penetration, m | Core recovery | | Average per run, m | |
|---|---|---|---|---|---|---|---|
| | | | | m | % | Penetration | Core recovery |
| 0–1059 | 1 | 108 | 942 | 386 | 41.0 | 8.7 | 3.6 |
| 1059–2805 | | 239 | 1674 | 913 | 54.5 | 7.0 | 3.8 |
| 2805–4673 | | 265 | 1570 | 940 | 59.7 | 5.9 | 3.5 |
| Total: | | 612 | 4186 | 2239 | 53.4 | 6.8 | 3.7 |
| 4673–5624 | 2 | 80 | 644 | 103 | 16.0 | 8.1 | 1.3 |
| 5642–6823 | | 125 | 865 | 260 | 30.1 | 6.9 | 2.1 |
| 6823–7263 | | 35 | 334.7 | 46.8 | 14.0 | 9.6 | 1.3 |
| Total: | | 240 | 1843.7 | 409.8 | 22 | 7.7 | 1.7 |
| 7263–7943 | 3 | 39 | 291.3 | 105.8 | 36.3 | 7.5 | 2.7 |
| 8043–9008.4 | | 105 | 742.5 | 308.1 | 41.4 | 7.1 | 2.9 |
| Total: | | 144 | 1033.8 | 413.9 | 40 | 7.2 | 2.9 |
| 9008.4–10028 | 4 | 104 | 835.8 | 248.3 | 30 | 8.0 | 2.4 |
| 10028–10772 | | 56 | 612.2 | 163.9 | 26.8 | 10.9 | 2.9 |
| 10772–11500 | | 61 | 723.7 | 225.2 | 31.1 | 11.9 | 3.7 |
| Total: | | 221 | 2171.7 | 637.4 | 29 | 9.8 | 2.9 |
| Grand total: | | 1217 | 9235.2 | 3700.1 | 40.1 | – | – |

Depth-dependent core crushability is specified below.

extensively performed in the section (see Table 3.29). While analyzing their results, one can draw the following conclusions. Core recovery is accompanied by its self-disintegration and -sticking in the inner tube. The lowest percentage (13%) of core recovery was, therefore, observed in the case of conventional core barrels. The best results were obtained with reverse circulation (53 and 40%), oil-filled core barrel (40.6%) and core barrels with ground plates featuring a low friction coefficient (35.5%), i.e. with core barrels designed to ensure forced core movement or with a lower friction coefficient at a core-inner tube contact. The highest linear core recovery per run was achieved using intensive reverse circulation (5 m) and a core barrel with ground plates (3.34 m). The use of tapered core barrels to reduce outward thrust did not essentially increase core recovery (18.8%).

Notwithstanding some positive results, the use of oil-filled core barrels is not rational, since this eliminates reverse flushing to force cores along the tube. Reverse flushing in combination with a core receiving portion lined with some antifriction material with a low friction coefficient (metallographite, vitreous enamel) is found more effective.

Section 3 (depth range 7263 to 9008.4 m) is also characterized by mining and geological conditions resulting in high core crushability. The average length of core samples decreases with depth to between 2.8 and 8.9 cm, core samples (disks) of 4 to 5 cm in length being most common (Fig. 3.68). Poor core recovery results mostly from its sticking in the inner tube. MAG-195-214/60 core

**Table 3.29.** Data on experimental coring in the Kola well

| Parameter | KDM-195-214/60 | | | | | KTD4-195-2M/60 – 80 | |
|---|---|---|---|---|---|---|---|
| | Direct flushing through core barrel | | | Reverse flushing | | | |
| | Conventional core barrel | Tapered core barrel | Core barrel with ground plates | Body 195 mm in dia, conventional core barrel, circulation rate $= 1 - 2\ l\ s^{-1}$ | Body 203 mm in dia, conventional core barrel, circulation rate $= 6 - 8\ l\ s^{-1}$ | Without flushing oil-filled core barrel | Fixed core barrel with check valve |
| Number of runs | 11 | 9 | 5 | 29 | 8 | 6 | 4 |
| Total penetration, m | 75.4 | 64.8 | 47 | 238.5 | 75.3 | 39.1 | 28.3 |
| Total core recovery | | | | | | | |
| m | 9.84 | 12.2 | 16.7 | 95.25 | 39.85 | 15.9 | 5.65 |
| % | 11 | 18.8 | 35.5 | 40 | 53 | 40.6 | 20 |
| Penetration per run, m | 6.83 | 7.2 | 9.4 | 8.2 | 9.4 | 6.5 | 7.1 |
| Core recovery per run, m | 0.89 | 1.35 | 3.34 | 3.3 | 5 | 2.65 | 1.4 |

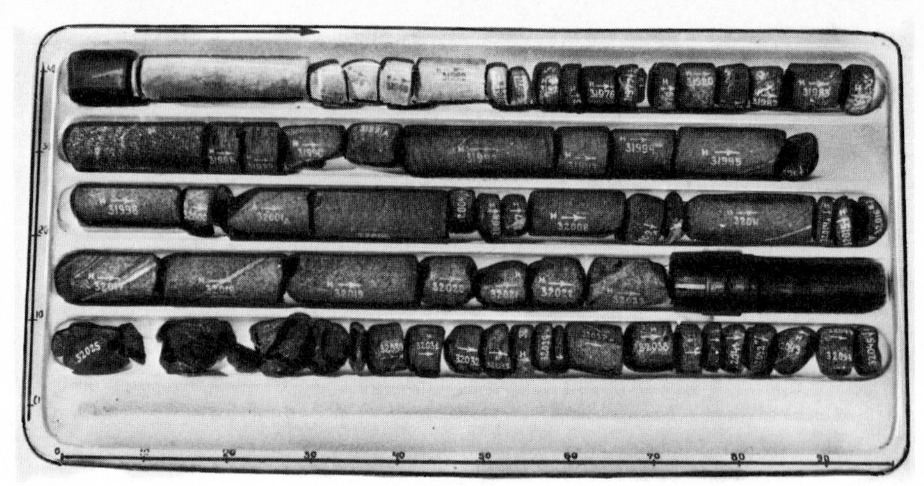

**Fig. 3.68.** Core samples recovered from a depth of 8958 m

samplers with hydraulic transportation have been tested for the first time in this section.

Extensive use was made of core samplers with a low friction coefficient at a core-inner tube contact. Fourteen runs have been made using KN-195 (ShUK-195) core sampler with an oil-filled core barrel used as an accessory for the shaft of A7S, A7GTSh or TRM-95 turbodrills for full-hole drilling and 52 runs have been made using KDM-195-214/60 core sampler with an oil-filled core barrel and A7Sh, A7GTSh, D2-172M and TRM-95 downhole motors.

Data on core sampler performance are given in Table 3.30.

As is seen from Table 3.30 the best results in drilling at depths exceeding 7000 m under core disking conditions have been achieved using core samplers with cores forced into the sampler chamber, i.e. MAG-type core samplers with hydraulic transportation. A total of 78 runs have been made, 593.9 m drilled with 45% of cores recovered using these samplers. The performance of MAG core samplers in terms of penetration per run and core recovery exceeds that of other similar units by 1.2 to 1.5 times.

Yet the performance of different units with respect to core recovery varies, depending not only upon their design features, but also upon their rational utilization under specific mining and geological conditions of the Kola well; 1149.1 m have been drilled with MAG core samplers within 139 runs in a depth range of 7469.4 to 9651 m.

As is seen from Table 3.31, 98 runs produced no cores in the sampler chamber as these were stuck in the core guiding tube. The core sampler featured partial reverse flushing with no hydraulic transportation. As a result, the percentage of core recovery was the same (21%) as the one ensured by conventional core samplers and the linear core recovery did not exceed 2 m (1.7 m) with a penetration per run over 8 m. The remaining runs resulted in almost all the core samples brought into the chamber with hydraulic transportation in use and no core sticking in the core guiding tube. In this case, core recovery exceeded 70% (6.1 m) at an average penetration of 8.5 m per run.

**Table 3.30.** Core sampler performance data

| Depth range, m | KN-195 | | | | | | KDM-195-214/60, SKU-195/80 | | | | | | MAG-195-214/60 | | | | | |
|---|---|---|---|---|---|---|---|---|---|---|---|---|---|---|---|---|---|---|
| | In the depth range | | | Per run | | | In the depth range | | | Per run | | | In the depth range | | | Per run | | |
| | N[a] | H, m | $L_c$, m | h, m | l, m | core recovery, % | N | H, m | $L_c$, m | h, m | l, m | core recovery, % | N | H, m | $L_c$, m | h, m | l, m | core recovery, % |
| 7469 – 7943 | 1 | 5.1 | 3.2 | 5.1 | 3.2 | 61 | 14 | 99.3 | 29.5 | 7.1 | 2.1 | 30 | 25 | 192 | 76.3 | 7.7 | 3.1 | 40 |
| 8043 – 8465.8 | 2 | 12.6 | 2.2 | 6.3 | 1.1 | 17 | 13 | 72.6 | 25.8 | 5.6 | 2.0 | 35 | 33 | 226.4 | 90.7 | 6.9 | 2.8 | 40 |
| 8506.5 – 9008.4 | 11 | 83.9 | 27.6 | 7.6 | 2.5 | 33 | 25 | 170.2 | 59.8 | 6.8 | 2.4 | 35 | 20 | 175.5 | 101.9 | 8.8 | 5.1 | 58 |
| 9008.4 – 9344.4 | 1 | 9.7 | 2.65 | 9.7 | 2.65 | 27 | 4 | 33.7 | 2.1 | 8.4 | 0.5 | 6 | 37 | 275.2 | 91.5 | 7.5 | 2.5 | 33 |
| 9344.4 – 9651 | | | | | | | | | | | | | 24 | 280 | 58 | 11.7 | 2.4 | 21 |
| 9738 – 9910 | 1 | 7.3 | 0.4 | 7.3 | 0.4 | 5 | 6 | 47.3 | 2.2 | 7.9 | 0.4 | 5 | 8 | 66.9 | 33.5 | 8.4 | 4.2 | 50 |
| 9918 – 10028 | | | | | | | | | | | | | 22 | 104.4 | 55.8 | 4.8 | 2.5 | 54 |
| 10028 – 10136 | | | | | | | | | | | | | 12 | 106.3 | 26.3 | 8.9 | 2.2 | 25 |
| 10136 – 10243 | | | | | | | | | | | | | 10 | 100.9 | 41.6 | 10.1 | 4.2 | 41 |
| 10248 – 10360 | | | | | | | | | | | | | 8 | 102.7 | 31.6 | 12.8 | 4.0 | 31 |
| 10427 – 10622 | | | | | | | 1 | 12.7 | 0 | 12.7 | 0 | 0 | 14 | 174.3 | 39.7 | 12.3 | 3 | 24 |
| 10656 – 10772 | | | | | | | 4 | 31.3 | 0.6 | 7.8 | 0.15 | 2 | 7 | 82.1 | 26.7 | 11.7 | 3.8 | 33 |
| 10633 – 10792 | | | | | | | | | | | | | 12 | 132.2 | 34.3 | 11 | 2.9 | 26 |
| 10800 – 11003 | | | | | | | | | | | | | 12 | 147.2 | 37.1 | 12.3 | 3.1 | 25 |
| 11003 – 11254 | | | | | | | | | | | | | 14 | 180.3 | 49.2 | 12.9 | 3.5 | 27 |
| 11254 – 11499 | | | | | | | | | | | | | 16 | 185.4 | 73 | 11.6 | 4.6 | 39 |

[a] N is number of runs; H is total penetration; $L_c$ is core recovery; h is average penetration per run; l is average core recovery per run.

**Table 3.31.** MAG-type core sampler performance data

| Depth range, m | Without cores in the sampler chamber | | | | | | With cores in the sampler chamber | | | | | |
|---|---|---|---|---|---|---|---|---|---|---|---|---|
| | In the depth range | | | Per run | | | In the depth range | | | Per run | | |
| | N[a] | H, m | $L_c$, m | h, m | l, m | core recovery, % | N | H, m | $L_c$, m | h, m | l, m | core recovery, % |
| 7469.4 – 7943.0 | 19 | 142.4 | 45.6 | 7.5 | 2.4 | 32 | 6 | 49.6 | 30.8 | 8.3 | 5.1 | 62 |
| 8043.0 – 8465.8 | 23 | 154.2 | 33.4 | 6.7 | 1.5 | 22 | 10 | 72.2 | 57.3 | 7.2 | 5.7 | 79 |
| 8506.5 – 9008.4 | 10 | 71.6 | 22.4 | 7.2 | 2.2 | 31 | 10 | 103.9 | 79.5 | 10.4 | 8.0 | 77 |
| 9008.4 – 9344.4 | 26 | 197.3 | 34.3 | 7.6 | 1.3 | 17 | 11 | 77.9 | 57.2 | 7.1 | 5.2 | 73 |
| 9344.4 – 9651.0 | 20 | 234.4 | 34.8 | 11.7 | 1.7 | 15 | 4 | 45.6 | 23.6 | 11.4 | 5.8 | 51 |
| Total | 98 | 799.9 | 170.5 | 8.2 | 1.7 | 21 | 41 | 349.2 | 248.4 | 8.5 | 6.1 | 71 |

[a] For explanations see Table 3.30.

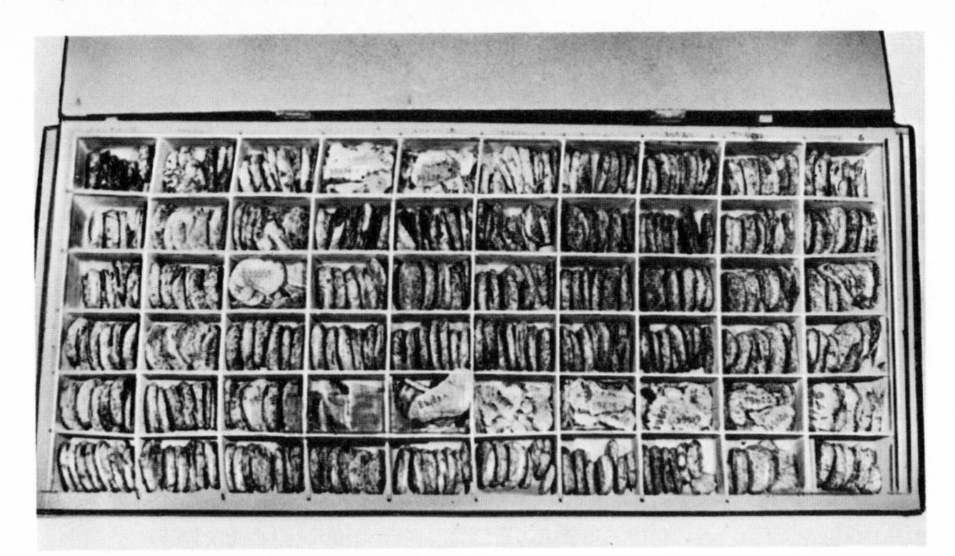

**Fig. 3.69.** Core samples recovered from a depth of 10,730 m

Thus, core recovery may be increased (notwithstanding disking) up to more than 70% using MAG-type samplers with an adequate circulation rate in reverse flushing. This is hampered by intensive caving, which reduces drilling mud returns so that hydraulic transportation of cores is stopped. The effectiveness of core sampling can also be enhanced through a lower rotation speed of the sampler, higher axial load and circulation rate.

Section 4 (depth range 9008.4 to 11,500 m) features extremely high core crushability (Fig. 3.69), the length of core samples ranging from a few centimeters up to the size of drill cuttings. MAG-type core samplers with hydraulic transportation were mostly used. The test results have confirmed their high performance under the conditions of composite-stressed bottom hole and core disking. MAG-type core samplers account for 43% of core recovery in the section, whereas KN, KDM and SKU units with conventional drainage and an oil-filled core barrel are responsible for only 2 to 6% of recovered cores (see Table 3.30). The highest (over 53) percentage of core recovery was ensured by a MAG-195-214/60 core sampler and a 2V-K214/60TKZ core drilling bit with high level of core entry which corresponds structurally to drilling with reverse bottom-hole flushing.

The following conclusion can, therefore, be drawn from studying core recovery techniques:

Notwithstanding up-to-date devices used in core sampling, its effectiveness decreased with depth by 30 to 50%, which is accounted for mostly by depth-dependent core disking resulting in poor recovery because of sticking and disintegration.

# Conclusion

The Kola superdeep well, situated in the north-eastern part of the Baltic Shield has for the first time exposed the Earth's ancient continental crust of the Archean-Proterozoic and has provided unique information on its structure and inherent geological processes. Comprehensive geological, petrographical, geochemical, mineralogical, structural, radiometric, seismic acoustic, nuclear physical, magnetic, electrical and thermal studies of core samples and surrounding rock mass have been performed throughout the entire SG-3 well down to a depth of 11,600 m. The borehole condition has been continuously checked.

Major conclusions are given below.

## 1. Geology

The SG-3 section is composed of Proterozoic (0 to 6842 m) and Archean (6842 to 11,662 m) rock complexes covering a long period of the geological history of the Earth ranging from 3 to 1.6 Ga. The study of the Proterozoic complex has led to a better understanding of the geology of the Pechenga ore district. The Pechenga volcano-sedimentary complex consists of two series. The overlying Nickel series is represented by picrite-basalt and greywacke formations, and the underlying Luostari series by trachyandesite-basalt and quartzite-carbonate formations. The major stratigraphic units of the Pechenga complex have been traced down to great depths. The Archean complex consists of seven rhythmically alternating sequences, each rhythm comprising biotite-plagioclase gneiss with high-alumina minerals and subjacent biotite-plagioclase gneiss and amphibolite. The gneiss with high-alumina minerals is sedimentary in origin and belongs to the sand-clay formation. The biotite-plagioclase gneiss and amphibolite are attributed to the basalt-andesite formation.

Drilling of the SG-3 well has provided voluminous information on a metamorphic zonality in a range from the prehnite-pumpellyite to amphibolite facies. The zonality is affected by rock lithology, depth and temperature. The temperature shows an increase with depth from 300° to 650 °C. Hence, during metamorphism, the geothermal gradient was five to seven times higher. Not only the temperature of prograde metamorphism, but also the degree of equilibrium of mineral assemblages is found to grow with depth. It has also been established that the prograde metamorphism of the Pechenga complex was contemporaneous with imbricate thrusting of tectonic blocks. In large fault zones, these processes have resulted in a sharp anisotropy of metamorphic rocks in terms of their inner structure and elastic properties, and have involved granite gneiss of the

Kola Series and almost entirely obliterated the earlier mineral paragenesis of the granulite facies. The polymetamorphic Archean complex has been subjected to regional granitization in the amphibolite facies and to a wide development of muscovite in metamorphic rocks.

The first attempt has been made to construct a reliable vertical geochemical cross-section of the Precambrian crust. Reflecting varying contents of petrogenic, rare and trace elements, the section shows that the crustal composition changes with depth, from femic to essentially sialic. Regular variations of rock acidity and alkalinity, depending on original lithology and superimposed processes, have been established. Metamorphism has been found to be of isochemical nature for petrogenic elements, including potassium and sodium. The high aqueous content (up to 6.8%) in rocks of the prehnite-pumpellyite facies has been shown to decrease to 2.5% in the greenschist facies and to 1.5% at the contact between the greenschist and epidote-amphibolite facies.

Mineralized subsurface fissure water has for the first time been found to occur in deep-seated zones of the continental crust. The vertical hydrogeological zonality of the water reveals that essentially chloride-calcium water changes with depth to hydrocarbonate-sodium water. Gas composition is also depth-dependent, featuring a higher percentage of hydrogen, helium and a lower percentage of hydrocarbon gas, which is associated mostly with sedimentary units of the Proterozoic complex.

Drilling of the Kola superdeep well has shown that the continental crust is rich in minerals and that the distribution of different types of mineralization is governed by a successive change of sedimentary, magmatic, metamorphic and hydrothermal processes. The well has intersected the so far unknown layer of ultrabasic rocks with sulphide copper-nickel ores in a depth interval from 1540 to 1800 m. Studies of the core samples have confirmed the polygenic origin of the ore and revealed the persistence of its mineral composition, structure, texture and geological position to 2.5 km up-dip. The Kola series contains ferrugenous quartzite, metamorphosed magmatic iron-titanium ores and sulphide copper-nickel mineralization. A low-temperature hydrothermal pyrite-chalcopyrite-pyrrhotite mineralization of the mantle origin has been observed in zones of retrograde fault metamorphism at depths from 600 to 1100 m. The quantitative data have proved that mineralized fissured structures actually occur at depths three to four times greater than those anticipated theoretically. All this makes it possible to suggest that tectonic blocks, and different faults and folds holding promise of ore deposition may occur even at great depths, due to vertical and horizontal displacements.

The SG-3 cross-section does not cover the whole Precambrian period. With the drilling data available, however, it is possible to verify the reliability of the alternative hypotheses on the structure of deep-seated zones of the continental crust using a three-dimensional geological model of the Pechenga ore district, and to consider it as a reference standard for interpretation of data obtained from geological and geophysical studies of ancient metamorphic units occurring at great depths. As is shown by this model, the geological history of the area includes two major cycles – Archean and Proterozoic. The Archean cycle in its turn has two stages, namely: andesite-basalt volcanism and sedimentation;

folding, metamorphism of the granulite facies and ultrametamorphism. The Proterozoic cycle consists of four stages: initiation of the intracontinental rifting mobile belt, andesite-basalt volcanism, picrite-basalt volcanism, folding and faulting and zonal metamorphism ranging from the prehnite-pumpellyite to amphibolite facies.

The emplacement of nickel-bearing intrusives is connected with generally epicratonic magmatism rather than with the final stage of the formation of the Proterozoic geosyncline. The fact that the flatlying nickel-bearing intrusives have been broken into fragments as a result of imbricate thrusting of tectonic blocks adds to the identification of copper-nickel ores in the Pechenga structure and changes their criteria for prospecting and appraisal. Other types of mineralization intersected by the Kola well, include hydrothermal sulphide mineralization which is associated with zones of retrograde metamorphism along dislocations.

As for the deep seismic boundaries, it becomes evident from geological and geophysical studies that most of the dipping seismic boundaries found in the upper portion of the section coincide with contacts between geological bodies and large faults. The horizontal seismic boundaries in the lower portion of the section cannot be explained by transition from acid to basic rocks, intensive metamorphism or basification, or by subhorizontal tectonic zones. They are of isochemical nature and result from changes in the physical properties of rocks composing deep crustal layers.

## 2. Geophysics

Geophysical and petrophysical studies of the Kola well section have made it possible to:

– obtain direct evidence of the physical state, properties and composition of rocks in situ to a depth of 11,600 m and on the basis of recovered cores;
– ascertain the geological nature of geophysical (including seismic) boundaries in the Precambrian continental crust;
– determine temperature conditions of the Earth's interior;
– study the stressed state of rock mass intersected by the hole;
– check the technical condition of the borehole.

One of the major results of these studies is the identification of a low-velocity zone in the Proterozoic Pechenga complex at a depth of 4500 m. The low velocities are due to low-temperature metamorphism superimposed on basaltic rocks. No vertical gradient of propagation velocity of longitudinal $V_p$ and transverse $V_s$ waves has been observed down to 4500 m. The thick Kola complex, with properties typical of a granite-gneiss layer, has been exposed below the Pechenga complex at a depth of 6840 m. This contradicts the assumption based on deep seismic sounding made prior to drilling. The discovery of this complex has been verified on the elasticity-density model based on core measurements. The model clearly exhibits three depth intervals composed of rocks with different elastic and density properties, namely: 0 to 4500 m (Proterozoic basic volcanics), 4500 to 6840 m (thick zones of folidation) and below 6840 m (Archean rocks).

Igneous rocks occurring in the upper part of the section exhibit maximum density $\sigma$ and $V_p$ and $V_s$ velocities. There is a stepwise decrease in $\sigma$, $V_p$, $V_s$ and increase of porosity, permeability and anisotropy parameters at a depth of 4500 m. A similar variation of these properties is observed at the boundary between the Proterozoic and Archean complexes.

Analysis of the elastic properties throughout the section has shown that their variations are due to abrupt changes in rock composition and to an increase in degree of foliation in large fault zones with depth.

Abrupt interfaces deduced from the density, elastic and other properties of rocks are also evident by electromagnetic characteristics. A distribution of magnetic minerals throughout the section is controlled by the following regularity: at depths from 0 to 4586 m (sulphide mineralization zone), dominating magnetic minerals are pyrrhotites, except ultrabasic rocks containing large amounts of magnitude; in an interval of 4586 – 5642 m, pyrrhotite is practically absent, giving way only to magnetite and hematite; at deeper levels, negligible concentrations of magnetite and pyrrhotite are observed.

Detailed temperature measurements down to 11,600 m have been carried out, and the heat conductivity of samples of characteristic rocks has been determined. These data have shown that both the geothermal gradient and density of the heat flow increase with depth. The geothermal gradient averages 1.6 °C/100 m in the Proterozoic and 2 °C/100 m in the Archean complexes.

All the layers, beds, sequences and formations have been sampled for radioactive elements: U, Th and K. Radiogenic heat has been found to account for more than 50% of the overall heat flow balance.

The theoretical and experimental methods have revealed an irregular distribution of rock pressure throughout the hole section. There are low- and high-stress zones as compared with lithostatic pressure. The boundaries of these zones are traced by seismic methods.

The effectiveness of the geophysical studies has been ensured by new logging techniques, unique thermobaroresistant geophysical equipment manufactured in the USSR, 12,000-m-long logging cable, as well as by the comprehensive analysis of the information obtained.

## 3. Drilling

Effective penetration deep into the Earth's interior has been made possible by the Soviet-made high-performance drilling equipment and tools and advanced technologies. Surface and downhole equipment and tools feature light weight and simple design. They can be successfully employed wherever there is insufficient information on drilling conditions in crystalline rocks. The breakthrough has largely been due to new technical approaches, such as open-hole advance drilling, low-speed turbodrilling, the use of light-alloy aluminium drill pipes and instruments for monitoring the turbodrill operation at the bottomhole with data transmitted to the surface, etc.

The drilling technology has proved highly effective – a depth of 12 km was reached using a drilling rig with a load capacity of only 4 MN, the open pilot hole had an unprecendented length of more than 9.5 km, the weight of the run-in drill string was within 160 – 170 t and the full cycle of round trip operations did not exceed 18 h.

Previously existing purely theoretical concepts of rules of variation of the drilling process have been verified. For instance, rock drillability at great depths was not affected at all. The maintaining of the stability of the crystalline rock mass in the zone surrounding the hole has proved to be a more complex problem than expected. The procedures for drill string design had to be essentially modified to allow for resistance forces and loss of strength of light-alloy pipes at high temperatures. A technique of core recovery from the bottomhole in a state of composite stress has been improved. New features have been established to control hole path and other drilling parameters. With the sample information available, one can effectively plan both the drilling process and development of high-performance equipment with prescribed characteristics on a scientific basis.

Up-to-date drilling equipment, advanced technologies and the experience gained by the USSR in superdeep drilling add to the effectiveness of deep drilling for oil, gas and minerals.

With further drilling of the Kola well, the investigations will be continued, and data and conclusions will be specified more exactly.

The Inter-Departmental Scientific Council of the State Committee for Science and Technology, together with the USSR Ministry of Geology and the Academy of Sciences, have worked out a long-term comprehensive programme aimed at exploring the mineral resources of this country on the basis of results obtained from drilling of the Kola well.

The programme envisages a breakthrough in studying both the structure and evolution of different types of the Earth's crust in the USSR, as well as in predicting mineral deposits. We believe that it will also greatly contribute to the development of the Earth sciences in general, in view of the diversified geology of this country.

# References

**works published in the USSR are in Russian.**

## Section 1


Akhmedov AM, Ozhegova GR (1974) Ore elements in pyrites from metasedimentary strata of the Pechenga complex. In: Proc on mineralogy of the Kola peninsula, issue 10. Leningrad Nedra, pp 62–65

Archean geology (1981) Geol Soc Aust Spec Publ No 7:515 p

Atlas of structures and textures of sulphide copper-nickel ores of the Kola peninsula (1973) Nauka, Moscow

Balashov YA (1976) Geochemistry of rare earth elements. Nauka, Moscow

Barker C, Torkelson BE (1975) Gas adsorption on crushed quartz and basalt. Geochim Cosmochim Acta, No 2, 39:212–2218

Bayduk EI, Volorovich MP, Lovitova FM (1982) Elastic anisotropy of rocks at high pressures. Nauka, Moscow

Belkov IV, Yelina NA (1979) Rare earth elements in the most ancient granitoids of the Kola Peninsula. In: The most ancient granitoids of the Baltic Shield. USSR Acad Sci Kolsky, pp 116–123

Belousov VV (1966) Basic tasks in study of the Earth's interior. In: History and prospects of superdeep drilling. Moscow, Nauka, pp 7–27

Belousov VV (1982) Programme of investigation of the deep interior of the USSR territory. Vestn Akad Nauk No 3:100–114

Belyaevsky NA, Fedynsky VV (1961) Study of the Earth's deep interior and tasks of superdeep drilling. Sov Geol

Belyayev OA, Petrov VP (1980) Cyclicity and facies of regional metamorphism in Precambrian time on the Kola Peninsula. In: Methodology of mapping of metamorphic complexes. Novosibirsk Nauka, pp 86–91

Belyayev OA, Zagorodny VG, Petrov VP, Voloshina ZM (1977) Facies of regional metamorphism of the Kola Peninsula. Nauka, Leningrad

Bezmen NI, Tikhomirova VI, Kosogova VP (1975) Pyrite-pyrrhotite geothermometer: distribution of nickel and cobalt. Geokhimiya No 5:700–714

Bezrodnov VD (1979) Chloride subsurface waters of the Kola peninsula. Sov Geol No 8:67–71

Bezrodnov VD, Borevsky LV (1975) Peculiar approach to the study and mapping of hydrothermal conditions of zones of the exogenic fissuring on ancient shields. In: Proc of the scientific-technical meeting on geothermal methods of investigations in hydrogeology. VSEGINGEO, Moscow, pp 141–142. No 11:55–70

Bibikova EV, Tugarinov AI, Gracheva GV (1973) About age of granulites of the Kola Peninsula. Geokhimiya No 5:664–675

Bondarenko LP, Dagelaisky VB (1968) Geology and metamorphism of the Archean rocks of the Kola Peninsula. Nauka, Leningrad

Borodin LS (1981) Geochemistry of the main series of igneous rocks. Nedra, Moscow

Bralia A, Salatini, Troja RA (1979) A re-evaluation of the Co/Ni ratio in pyrite as geochemical tool in ore genesis problems. Evidence from southern Tuscan pyrite deposits. Miner Deposita No 3, 14:353–374

Burkov YK (1973) Associations of chemical elements as indices of the nature of metamorphosed sedimentary rocks. In: Proc of the 5th All-Union meeting on lithology. Moscow, Nauka, pp 197–199

Coombs DS (1961) Some recent work on the lower grades of metamorphism. Aust J Sci No 5, 24:203–207

Correlation of the Precambrian rocks (1977) Vol 1, 2. Nauka, Moscow

Dobretsov NL (1981) Global petrological processes. Nedra, Moscow

Dobretsov NL, Sobolev VS, Khlestov VV (1972) Facies of regional metamorphism of moderate pressures. Nedra, Moscow

Dobrzhinetskaya LF (1978) Structural-metamorphic evolution of the Kola series (Baltic Shield). Nauka, Moscow

Duk GG (1977) Structural-metamorphic evolution of rocks of the Pechenga complex. Nauka, Leningrad

Eastern part of the Baltic Shield, geology and deep structure (1975) Nauka, Leningrad

Early history of the Earth (1980) Mir, Moscow

Firky F, Gibson A, Gibson EK (1979) Release and analysis of gases from geological samples. Am Mineral 64:543 – 563

Fyfe WS, Price NJ, Thompson AB (1981) Fluids in the Earth's Crust. Mir, Moscow

Galdin NE (1984) Velosity seismograms of the Earth's Crust in the eastern part of the Baltic shield. Geol Razvedka No 5:3 – 9

Genkin AD, Loginov VP, Organova NI (1965) Interrelation of hexagonal and monoclinic pyrrhotite in ores. Geol. Rudnykh Mestorozhdenii No 3:3 – 24

Gerling EK, Matveeva II (1964) Age of basic rocks determined with the help of the K-Ar technique. In: Absolute age of geological formations. Problem 3. Nauka, Moscow, pp 328 – 341

Gerling EK, Koltsova TV, Duk GG (1983) Anomalous content of radiogenic isotopes and helium in minerals of the Pechenga complex rocks (Kola Peninsula). In: Methodological problems of nuclear geology. Nauka, Leningrad, pp 5 – 21

Glagolev AA, Rusinov VL, Plusnina LP, Troneva NV (1983) Mineral associations and metamorphism of basites from the Pechenga series (northwestern part of the Kola Peninsula). Izv Akad Nauk SSSR Ser Geol No 1:29 – 47

Glebovitsky VA (1977) Thermo- and barometry studies of metamorphic rocks. Nedra, Moscow

Gorbunov GI (1968) Geology and genesis of sulphide copper-nickel deposits of the Pechenga region. Nedra, Moscow

Gorbunov GI, Makievsky SI, Nikolaeva KA (1978) Metallogenic zonality associated with the tectonomagmatic activization of the Baltic Shield. Sov Geol No 4:15 – 26

Gorokhov IM, Krylov IN, Baikova VS, Lobach-Zhuchenko SB (1976) Geochronologic study of the polymetamorphic complex of the Kola series rocks. In: Development and application of methods of the nuclear geochronology. Nauka, Leningrad, pp 177 – 192

Gorokhov IM, Dagelaisky VB, MOrozova ZM, Varshavskaya ES (1981) Age position of the Olenegorsk ferrum-ore deposit (Kola Peninsula) according to data of Pb-Sr and K-Ar techniques. Geol Rudnykh Mestorozhdenii No 3: 67 – 79

Gorokhov IM, Varshavskaya ES, Kutyavin EP, Melnikov NN (1982) Low-grade metamorphism impact on Rb-Sr systems in sedimentary and volcanogenic rocks. Litol Poleznye Iskopaemye No 5:81 – 91

Henley RW, Sheppar DS (1975) Hydrothermal activity and hydrothermal chemistry in the metamorphic environment. In: Ellis AI (ed) Geochemistry, 400 p

Hey MN (1954) A new review of the chlorites. Miner Mag No 30:27

Kajiawava I, Krause HR (1971) Sulfur isotope partitioning in metallic sulfide systems. Can J Earth Sci No 11, 8:1397 – 1408

Kalyuzhny VA, Svoren IM (1979) Principles of rational application of methods for analysing gas components of fluid inclusions (a problem of hydrogene determination). Miner Sb, issue 1, No 33:35 – 41

Kazansky VI (1980) Protoactivation and ore formation. In: Metallogeny and mineral deposits. Nauka, Moscow, pp 30 – 40

Kazansky VI (1982) Evolution of the Precambrian ore-bearing structures: Archean cratons and areas of protoactivation. In: Precambrian ore-bearing structures. Nauka, Moscow, pp 7 – 66

Kazansky VI, Smirnov YP, Kuznetsov YI, Kuznetsov AV (1980) Fissuring vein mineralization and rock anisotropy of the Pechenga complex. Geol Rudnykh Mestorozhdenii, No 4:21 – 31

Khitarov NI (1961) On a problem of superdeep drilling on the USSR territory. Sov Geol No 6:134 – 138

Kozlovsky YeA (1981) Record depth reached at the USSR Kola well. Drilliste No 3:28 – 35

Kozlovsky YeA (1981) Ten thousand meters of discoveries. Nauka, Moscow, pp 6 – 11

Kozlovsky YeA (1982) Kola superdeep: internal results and prospects. Episodes, Vol 4:8 – 11

Kozlovsky YeA (1982a) Joint programme of deep study of the Earth's interior. Sov Geol No 9:3 – 12
Kozlovsky YeA (1982b) Geology of the Soviet Union and tasks of further investigations. Sov Geol No 12:3 – 18
Kratsko (1978) Earth's crust of the eastern part of the Baltic Shield. Nauka, Leningrad
Mineral deposits of the Kola Peninsula (1981) Nauka, Leningrad
Murmansk district, Geological description (1958) Geology of the USSR, Vol. 27. Gosgeotekhizdat, Moscow, p 1
Muslimov RK (1977) Results of geological exploration for oil and gas and task of the 10th five-year period for the Tatar ASSR. Geol Nefti Gaza No 5:12 – 20
Myasnikov VS (1977) Some peculiar features of the titanium-magnetite ore deposits of the Southern Urals and signs of metamorphism. Geol Rudnykh Mestorozhdenii No 2:49 – 62
Nalivkina EB (1977) The Early Precambrian ophiolite associations. Nedra, Moscow, Nauka
Nalivkina EB, Vinogradova NP (1980) Regularities in alteration of rock-forming minerals in the Precambrian deep section. Zapiski vsesoyuznogo mineralogicheskogo obshchestva, p 109, pp 531 – 544
Nalivkina EB, Vinogradova NP (1981) Rock-forming minerals in the deep section of the Precambrian Earth's Crust. In: Proc of the 11th Congr of IMA, Moscow, pp 217 – 226
Nikolaevsky VN (1982) Review: Earth's crust, dilatancy and earthquakes. In: Mechanics, No 28, Moscow, Nauka, pp 133 – 215
Nockolds SR (1954) Average chemical compositions of some igneous rocks. Bull Geol Soc Am No 10, 65:1007 – 1032
Norris RI, Henley RW (1976) Dewatering of a metamorphic pile. Geol J No 6, 4:333 – 336
Ohmoto H, Rye RO (1979) Isotopes of sulfur carbon. In: Geochemistry of hydrothermal ore deposits, 3rd Edn, Hofts Rinehart and Winston, pp 509 – 567
Perchuk LL (1970) Equilibrium of rock-forming minerals. Nauka, Moscow
Perchuk LL, Ryabchikov ID (1976) Phase conformity in mineral systems. Nedra, Moscow
Predovsky AA, Fedotov ZA, Akhmedov AM (1974) Geochemistry of the Pechenga complex. Nauka, Leningrad
Price NI (1975) Rates of deformation I. Geol Soc Lond 131:553 – 575
Pushkarev YD, Kravchenko EV, Shestakov GI (1978) Precambrian geochronological markers of the Kola Peninsula. Nauka, Leningrad
Rozanov YA (1962) Experimental studies of rock deformations at high pressures and at a temperature up to 200 °C. Proc IGEM USSR Acad Sci, issue 66, pp 1 – 83
Seismic models of the lithosphere for main geotectonical structures of the USSR territory (1980) Nauka, Moscow
Seismic models of the lithosphere for main geotectonical structures of the USSR territory (1980) Nauka, Moscow
Sherman SI (1978) Physical regularities in development of the Earth's Crust. Nauka, Novosibirsk
Sidorenko AV, Bilibina TV (1980) Metallogeny of the eastern part of the Baltic Shield. Nedra, Leningrad
Simmons ES, Nanson GN, Lumbers SB (1980) Geochemistry of the Shawmere Anorthosite Complex, Kapys-Kasing, structural zone. Ontario, Precamber Res No 1/2, 2:43 – 71
Sitdikov BS, Glagolev AA, Kazansky VI, Troneva NV (1980) The Archean polymetamorphic complex in a section of the Minnibayev borehole No. 20000. In: The Early Precambrian metamorphism. USSR Acad Sci Kolsky, Apatity, pp 130 – 139
Sobotovich EV, Grishchenko SM, Alexandruk VM, Shats MM (1963) Dating of the most ancient rocks with the help of leadisochron and isotope-spectral strontium techniques. Izv Akad Nauk SSSR Ser Geol No 10:3 – 14
Structures of copper-nickel ore fields and deposits of the Kola Peninsula (1978) Nauka, Leningrad
Suslova SN (1976) Komatiites from the Lower Precambrian metamorphosed volcanogenic strata of the Kola Peninsula. Dokl Akad Nauk SSSR, No 3, 288:697 – 700
Sviridenko LP (1980) Granite formation and problems of the Precambrian Earth's Crust formation. Nauka, Leningrad
Taylor L (1970) A low-temperature phase relations in the Fe-S system. Carnegie Inst Wash Year Book 68:259 – 270
Tectonic layering of the lithosphere (1980) Nauka, Moscow

Teilor KP (1977) Application of oxygen and hydrogen isotopic studies for problems of hydrothermal alteration of enclosing rocks and ore formation. In: Stable isotopes as applied to problems of ore deposits (Translated from English). Mir, Moscow, pp 464 – 509
The most ancient granitoids of the Baltic Shield (1979) USSR Acad Sci Kolsky, Apatity
The most ancient granitoids of the USSR (1981) Complex of grey gneisses. Nauka, Leningrad
The Earth's tectonosphere (1978) Nauka, Moscow
Timofeev BV (1979) Microfossils of the Pechenga series. In: Precambrian and Early Precambrian paleontology. Nauka, Leningrad
Tugarinov AI, Bibikova YV (1980) Geochronology of the Baltic Shield according to the zirconometry. Nauka, Moscow
USSR Acad Sci Kolsky, Apatity (1979) The most ancient granitoids of the Baltic Shield
Väyrynen H (1938) Petrologie der Nickelerdfelder Kaulatunturi-Kamikivitunturi in Petsamo. Bull Commiss Geol Finnl No 116
Vetrin VP, Vinogradov AN, Vinogradova GV (1975) Petrology and facial-formation analysis of the Litsk-Araguba diorite-granite complex. In: Intrusive charnockites and porphyry granites of the Kola Peninsula. USSR Acad Sci Kolsky Apatity, pp 149 – 316
Vinkler G (1979) Genesis of metamorphic rocks. Nedra, Moscow
Vinogradov AN, Vinogradova GV (1979) Enderbites of the Kaneitjavri massif and a problem of a geochemical type of the primary-crust granites. In: The most ancient granitoids of the Baltic Shield. Apatity, Moscow, Nedra, pp 91 – 111
Vinogradova AP, Tarasov LS, Zykov SZ (1959) Isotope composition of ore leads of the Baltic Shield. Geokhimiya No 7:571 – 607
Yakovlev YN, Yakovleva AK (1974) Mineralogy and geochemistry of the metamorphosed copper-nickel ores (exemplified by the Allarechinsk region). Nauka, Leningrad
Yeliseev NA, Gorbunov GI (1961) Ultrabasic and basic intrusions of the Pechenga region. Publ House USSR Acad Sci, Moscow, Leningrad
Zagorodny VG (1980) On peridotization and tectonic regimes of the north-eastern part of the Baltic Shield in Archean time. In: Geology and development of the Precambrian structural zones of the Kola Peninsula. USSR Acad Sci Kolsky, Apatity, pp 36 – 46
Zagorodny VG, Mirskaya DD, Suslova SN (1964) Geological structure of the Pechenga sedimentary-volcanogenic series. Nauka, Moscow
Zaraisky GP, Balashov VN (1978) On disaggregation of rocks in heating. Dokl Akad Nauk SSSR No 4, 240:926 – 929
Zhdanov VV (1966) Metamorphism and deep structure of the norite-diorite (granulitic) series of the Russian Laplandia. Nauka, Moscow

**Section 2**

Bat M (1980) Spectral analysis in geophysics. Nedra, Moscow
Bott M (1974) Earth inner structure. Mir, Moscow
Brodskaya SY, Vetoshkin ID, Zherdenko ON (1973) Relation between magnetic and other properties of natural pyrrhotites. Izv Akad Nauk SSSR Ser Fizika Zemli No 3:112 – 128
Earth crust of the Baltic shield eastern part (1978) Nauka, Leningrad
Eastern part of the Baltic Shield (1975) Geology and deep structure. Nauka, Moscow
Geophysical methods of investigating the wells (1983) Manual for geophysicists. Nedra, Moscow
Karus EW, Shkerina LV (1974) To construction of a geoacoustic model of a medium. Izv Vuzov Ser Geol Razvedka No 10:131 – 141
Karus EW, Rudenko GE, Khudzinski LL (1977) Three-component four-point seismic downhole tool and some results of its testing. Izv Vuzov Ser Geol Razvedka No 12:147 – 153
Karus EW, Kurznetsov OL, Kuznetsov YI et al (1981) On the operational procedure of superdeep well logging. Izv Vuzov Ser Geol Razvedka No 5:81 – 89
Kazanski VI, Smirnov YP, Kuznetsov YI et al (1980) Fracturing, mineral streaks and anisotropy of the Pechenga sequence rocks. Geol Rudnikh Mestorozhdeniy No 4:21 – 31
Khramov AN, Gontcharov GI, Komissarova RA et al. (1982) Paleomagnetogeology. Nedra, Leningrad
Khudzinski LL, Rudenko GE (1975) Complex program for digital processing the results of well reflection wave seimics. Ivz Vuzov Ser Geol Razvedka No 6:119 – 123

Kobranova VN, Izvekov BI, Ptsevich SA, Shvartsman MD (1977) Deriving the petrophysical parameters from samples. Nedra, Moscow
Kuznetsov YI (1971) To determination of elastic wave velocity on samples of fractured rocks. Tr VNIGRI 230:50–63
Lebedev TS, Poznanskaya NF (1970) Magnetic properties variation for ferromagnetic minerals at elevated pressure and temperature. Geofiz Sb No 33:49–54
Leshchuk VV (1977) Geoacoustic investigations in the borehole vicinity. Naukova Dumka, Kyiv
Litvinenko IV (1971) Method of studying complex structures in the upper part of consolidated continental crust by seismic. In: Problems of geophysical exploration. Gorny Institute Leningrad, pp 21–36 (Zap LGI, 61:2)
Lyubimova EA, Karus EW, Firsov FV et al. (1972) Earth heat flow in Pre-Cambrian shields of the USSR. Sov Geol No 8:10–22
Magnitski VA (1978) Earth inner structure and physics. Nauka, Moscow
Markov GA (1977) Tectonic stresses and overburden pressure in the Khibin Massif mines. Nauka, Leningrad
Nikolaev PN (1977) Method of statistical analysis of fractures and reconstruction of tectonic stress fields. Izv Vuzov Ser Geol Razvedka No 12:103–115
Panasyan LL (1981) On possible changes in structure of stress fields caused by a combined action of gravity and tectonics. In: Problems of non-linear geophysics. Izd VNIIYaGG, Moscow, pp 147–155
Parkhomenko EI, Bondarenko AT (1972) Electrical conductivity of rocks at high pressure and temperature. Nedra, Moscow
Physical properties of rocks and minerals (1976) Petrophysics. Manual of geophysicist. Nedra, Moscow
Recommendations on procedure of digital processing of well seismic data (1983) Reflection wave. Izd VNIIYaGG, Moscow
Smirnova EV (1979) On the use of the method of comparison in determining the rock heat conductivity. In: Experimental and theoretic studies of heat flows. Nauka, Moscow, pp 113–122
Smislov AA, Moiseenko GI, Tchadovitch ZP (1979) Heat balance and radioactivity of the Earth. Nedra, Leningrad
Turtchaninov IA, Medvedev RV (1973) Combined study of rock properties. Nauka, Leningrad

**Section 3**

Aluminium alloys (1972) Reference book. Metallurgiya, Moscow
Alexandrov MM (1982) Interaction between strings and hole walls. Nedra, Moscow
Anderson BA (1980) Polymer drilling muds in foreign countries. Ser Bureniye VNIIOENG, Moscow
Baklashov IV, Kartoziya BA (1976) Rock mechanics. Nedra, Moscow
Basovich VS (1979) The effect of the block and tackle system on tool weight accuracy determined by the pull of a dead line of the drilling rope. Mashiny i Neftyanoye Oborudovaniye No 9
Basovich VS, Vorozhbitov MI, Guberman DM (1979a) Method of drill pipe assessment. Trudy VNIITneft', issue 11
Basovich VS, Vorozhbitov MI, Guberman DM (1979b) The effect of radial dimensions of drill string components on resistance forces in round trips. Trudy VNIITneft', issue 11
Bergshteyn OY, Vugin RB (1974) Methods of prediction of drilling parameters with coring of hard abrasive rocks. Neftyanoye Khozyaistvo No 3)
Bergshteyn OY, Vugin RB, Velikosel'sky MA (1977) Updating of core sampling techniques and equipment in deep drilling. Nedra, Moscow
Gorbatsevich FF (1979) Non-linearity of rock deformation in static loading. Inzhenernaya Geologiya, No 4
Katsaurov IN (1972) Rock pressure, issue 2, Rock mechanics. Izd MGI Moscow
Kister EG (1972) Chemical treatment of drilling muds. Nedra, Moscow
Krupennikov GA, Filatov PA, Musin BZ, Barkovsky VM (1972) Stress distribution in rock masses. Nedra, Moscow
Kuznetsov YI (1977) Methods and techniques of petrophysical studies. Apatity Izd Kola branch of the Akad Nauk SSSR

Markov GA (1977) Tectonic stresses and rock pressure in mines of the Khibiny massif. Nauka, Leningrad

Medvedev RV, Kuznetsov YI (1982) Rock mass properties as determined by deep drilling. Nauka, Novosibirsk

Neymark AS (1975) On the problem of determination of anisotropic parameters of orthotropic materials. Izv Vuzov Mashinostroyeniye No 6

Neymark AS, Danelyants SM, Fayn GM (1979) Drill string assembly for super-deep holes. Mashiny i Neftyanoye Oborudovaniye No 6

Paus KF (1973) Drilling muds. Nedra, Moscow

Shcherbyuk ND, Yakubovsky NV (1974) Thread joints of oil drill pipes and downhole motors. Nedra, Moscow

Sinyavsky VS (1977) Corrosion and protection of aluminium alloys. Metallurgy, Moscow

Timofeyev NS, Vugin RB, Yaremeychuk RS (1971 a) Hole wall fatigue strength. Nedra, Moscow

Timofeyev NS, Vorozhbitov MI, Simonyants LY (1971 b) Problems, methods and possible solutions in drilling and operation of super-deep holes. Vnezhtorgizdat, Moscow

Tumanov AG, Kvasov FI, Fridlyander IN (eds) (1972) Aluminium. Physical metallurgy, treatment and application of aluminium alloys (translated from English). Metallurgiya, Moscow

Turchaninov IA, Markov GA, Kasparyan EN (1977) Fundamentals of rock mechanics. Nauka, Moscow

Turchaninov IA, Markov GA, Ivanov VI et al (1978) Tectonic stresses in the earth crust and stability of mine workings. Nauka, Leningrad

Vorozhbitov MI, Golubev GR (1976) Determination of rational distribution of deviation intensity in deep well intervals. Neftyanoye Khozyaistvo No 5

Vorozhbitov MI, Guberman DM, Ivannikov VI (1975) On the problem of assessment of major parameters of drilling super-deep holes within a specified period of time. Trudy VNIIBT, issue 34

Springer